PRINCIPLES OF PLANT NUTRITION

Principles of Plant Nutrition

5th Edition

by

Konrad Mengel

Justus Liebig University, Giessen, Germany

and

Ernest A. Kirkby

School of Biology, University Leeds, U.K.

with the support of

Harald Kosegarten

Justus Liebig University, Giessen, Germany

and

Thomas Appel

University of Applied Sciences, Bingen, Germany

KLUWER ACADEMIC PUBLISHERS
DORDRECHT / BOSTON / LONDON

A C.I.P. Catalogue record for this book is available from the Library of Congress.

ISBN 0-7923-7150-X

Published by Kluwer Academic Publishers,
P.O. Box 17, 3300 AA Dordrecht, The Netherlands.

Sold and distributed in North, Central and South America
by Kluwer Academic Publishers,
101 Philip Drive, Norwell, MA 02061, U.S.A.

In all other countries, sold and distributed
by Kluwer Academic Publishers,
P.O. Box 322, 3300 AH Dordrecht, The Netherlands.

Printed on acid-free paper

Printed in the Netherlands.

TABLE OF CONTENTS

Chapter 6
Fertilizer Application

PREFACE

The main object of this book is to explain basic processes and relationships of relevance to the scientific understanding of Plant Nutrition. These include diffusion, mass flow and interception of plant nutrients in soils, nutrient buffer power, cation exchange, anion adsorption in relation to soil minerals, water potential and redox potentials in soils, assimilation of nutrients by plants and assimilation and mineralization of organic matter by soil microorgansisms, ion pumps and transporters and related electrochemical potentials of plant cells, osmosis and plant water potentials. All are involved in Plant Nutrition and the complex process of growth, and hence also in crop production and in crop quality.

More than 10 years have passed since the last edition of *Principles of Plant Nutrition* was published. During this time the rate at which relevant literature has appeared has increased exponentially and to keep pace with this ever increasing flow has required continuous and tenacious study. The reader will therefore appreciate that although we have tried to incorporate all significant developments into the book we may have omitted to quote some important works in our reference list for which we duly apologize. We must say though that we have tried to avoid a mere compilation of published work since the predominant aim of the book is to provide a working guide and overview of Plant Nutrition.

Plant Nutrition is an essential discipline of Crop Science which is the science of crop production. In energetic terms crop production is the conversion of solar energy into a storable energy form of chemical energy. Most living organisms on earth are directly or indirectly dependent on this fundamentally important process of energy conversion. This also holds for humans who in modern times consume more energy than they harvest, taking into account all the detrimental effects they bring about, such as the exploitation of limited raw material sources and environmental pollution. It is for this reason that in the near future the world-wide importance of crop production will greatly increase bearing in mind too that every year millions of hectares of fertile land are ruined by desertification, acidification, erosion and salinization. Energy conversion by cultivation of efficient crop species and the maintenance of soil fertility are therefore paramount tasks of agriculture and present a major challenge for agricultural scientists. To meet this challenge requires the application of both relevant knowledge and its further elaboration. It is the prominent intention of this book to contribute to this goal.

The book is essentially a text-book for students of agriculture, horticulture, forestry and environmental sciences. In addition it should also serve as a guide for all those who are interested in plant science and crop production. In the preparation of this new edition we have included in the team of authors two scientists of the younger generation, Dr. habil. Harald Kosegarten from the Justus Liebig University at Giessen and Prof. Dr. Thomas Appel from the University of Applied Sciences, Bingen.

We gratefully acknowledge and thank those who provided us with various photgraphs and micrographs; Prof. Dr. I. Cakmak, Sabanci University, Istanbul, Turkey, Dr. M. Canny, Canberra, The Australian Nat. University, Dr. H. Kosegarten and co-workers, Justus Liebig University Giessen, Dr. H.W. Koyro, Justus Liebig University Giessen and Dr. W.H. Schröder, Forschungszentrum Jülich. We would also like to thank various colleagues throughout the world who supported us by their encouragement and by the stimulating discussions they held with us. The drawings were made by Frau Edith Hutzfeld. Her precise work is gratefully acknowledged especially by the senior author. Last but not least we are very grateful to Frau Helga Mengel who did a tremendous job in writing the text, correcting it and checking numerous pages of the literature. Finally we should like to thank Kluwer Academic Publishers for enjoyable and efficient co-operation in preparation of the manuscript for publication. The senior author apologizes that his optimistic publication deadlines did not turn out to be realistic. Nevertheless at no phase in the preparation of the new edition did he feel pressed by the Publishers for which he is most grateful. We are looking forward to further successful co-operation.

Giessen, Leeds, Spring 2001

K. Mengel, E.A. Kirkby, H. Kosegarten, Th. Appel

CHAPTER 1

PLANT NUTRIENTS

1.1   Definition and Classification

An outstanding feature of life is the capability of living cells to take up substances from the environment and use these materials for the synthesis of their own cellular components or as an energy source. The supply and absorption of chemical compounds needed for growth and metabolism may be defined as nutrition and the chemical compounds required by an organism termed nutrients. The mechanisms by which nutrients are converted to cellular constituents or used for energetic purposes are metabolic processes. The term *metabolism* encompasses the various reactions occurring in a living cell in order to maintain life and growth. Nutrition and metabolism are thus very closely interrelated.

The essential nutrients required by higher plants are exclusively inorganic, a feature distinguishing these organisms from man, animals and many species of microorganisms which additionally need organic foodstuffs to provide energy. By contrast, plants absorb light energy from solar radiation and convert it to chemical energy in the form of organic compounds, whilst at the same time taking up mineral nutrients to provide the chemical elements also essential for growth. For an element to be described as essential three criteria must be met. These are:

1. A deficiency of the element makes it impossible for the plant to complete its life cycle.
2. The deficiency is specific for the element in question.
3. The element is directly involved in the nutrition of the plant, as for example as a constituent of an essential metabolite or required for the action of an enzyme system.

Based on these criteria as proposed by Arnon and Stout (1939), the following chemical elements are now known to be essential for higher plants:

Carbon C	Potassium K	Zinc Zn
Hydrogen H	Calcium Ca	Molybdenum Mo
Oxygen O	Magnesium Mg	Boron B
Nitrogen N	Iron Fe	Chlorine Cl
Phosphorus P	Manganese Mn	(Nickel) Ni
Sulphur S	Copper Cu	(Sodium) Na
(Cobalt) Co	(Silicon) Si	

Cobalt, Nickel, Sodium, and Silicon have been established as essential elements for some but not all higher plant species. These elements are therefore shown above in brackets.

Nickel is the most recent addition to the list of elements and is indispensable for the activity of the enzyme urease which is required when plants are fed with urea (Gerendas *et al.* 1999). Chloride is a most unusual nutrient in that it is often present in plants in high concentrations yet in essential requirement is very much lower. Indeed, in many plant species it can be difficult to induce Cl deficiency because of contamination. From experiments with isolated chloroplasts it well established that Cl is essential for the water splitting system of photosystem II. In intact leaves, however, Cl deficiency does not inhibit photosynthetic activity and according to Terry (1977) the effect of Cl deficiency in decreasing leaf surface area is primarily an effect in reducing the rates of cell devision and extension (Flowers 1988). Chloride like Na^+ and K^+ plays an important role as osmoticum. It can also affect growth indirectly by stomatal regulation as a mobile counterion to K^+. The high demands of oil and coconut palms for Cl^- (Daniel and Ochs 1975, von Uexküll 1985) presumably relate to these effects.

Sodium in relatively low concentrations is essential for some C_4 plant species (Brownell and Crossland 1972). The basis of essentiality is that it appears to protect the mesophyll chloroplasts and in particular regulate the import of pyruvate by these organelles (Ohnishi *et al.* 1990, Brownell and Bielig 1996) thus controlling the regeneration of phosphoenolpyruvate (PEP), the primary CO_2 acceptor in C_4 photosynthesis with the malic - NADH carboxylation (see page 155). Sodium can act as an important osmoticum in both C_3 and C_4 species especially in halophytes (Greenway and Munns 1980) in which it occurs in high concentrations. In this role it can partially substitute for K^+ and can be more effective than K^+ in generating turgor and cell expansion, since Na^+ accumulates preferentially in cell vacuoles. This behaviour largely explains specific effects of Na^+ in increasing plant growth as for example in the C_3 crop plant, sugar beet (Marschner *et al.* 1981), even though there is no evidence of an essential role of Na^+ in C_3 plants. According to Epstein (1994, 1999) Si is essential for species of the *Equisetaceae* family and for wetland grasses including rice and sugar cane.The list of essential elements shown above may well not be complete and other elements, in very low concentrations, may yet be shown to be essential for higher plants. The involvement of unusual elements has been discussed by Asher (1991).

The plant nutrients may be divided into macronutrients and micronutrients. Macronutrients are found and needed in plants in relatively higher amounts than micronutrients. The plant tissue concentration of the macronutrient N, for example, is over a thousand times greater than the concentration of the micronutrient Zn. Using this classification based on the element concentration in plant material, the following elements may be defined as macronutrients: C, H, O, N, P, S, K, Ca, Mg (Na, Si). The micronutrients are: Fe, Mn, Cu, Zn, Mo, B, Cl, Ni. This division of the plant nutrients into macro- and micronutrients is somewhat arbitrary and in many cases differences between the concentration of macronutrients and micronutrients are considerably less well defined than the example cited above. The Fe or Mn concentration of plant tissues for example is sometimes nearly as high as the concentration of S or Mg. The concentration of the micronutrients is also often far in excess of physiological requirements. This is true for Mn *e.g.* and demonstrates that the nutrient concentration of plant organs (leaves, stems, fruits, roots) provides little indication as to nutrient demand for physiological and biochemical processes. Plants may even contain high concentrations of non essential elements some of which may be toxic (Al, Ni, Se, and F).

From a physiological viewpoint it is difficult to justify the classification of plant nutrients into macronutrients and micronutrients depending on element concentration in plant tissues. Classification of plant nutrients according to biochemical behaviour and physiological function seems more appropriate. Adopting such a physiological approach one possible classification of plant nutrients is shown in Table 1.1.

Table 1.1   Classification of plant nutrients

Nutrient element	Uptake	Biochemical functions
1st group C, H, O, N, S	in the form of CO_2, HCO_3^-, H_2O, O_2, NO_3^-, NH_4^+, N_2, SO_4^{2-}, SO_2 The ions from the soil solution, the gases from the atmosphere.	Major constituents of organic material. Essential elements of atomic groups which are involved in enzymic processes. Assimilation by oxidation-reduction reactions.
2nd group P, B, Si	in the form of phosphates, boric acid or borate, silic acid from the soil solution.	Esterification with native alcohol groups in plants. The phosphate esters are involved in energy transfer reactions.
3rd group; K, Na, Ca, Mg, Mn, Cl	in the form of ions from the soil solution.	Non-specific functions establishing osmotic potentials More specific reactions in which the ion brings about optimum conformation of an enzyme protein (enzyme activation). Bridging of the reaction partners. Balancing anions. Controlling membrane permeability and electrochemical potentials.
4th group Fe, Cu, Zn, Mo	in the form of ions or chelates from the soil solution.	Present predominantly in a chelated form incorporated in prosthetic groups. Enable electron transport by valency change.

The first group includes the major constituents of the organic plant material: C, H, O, N, and S. Carbon is taken up in the form of CO_2 from the atmosphere and possibly in the form of HCO_3^- from the soil solution. Both forms are assimilated by carboxylation, with the formation of carboxylic groups. This incorporation of C is also accompanied by the simultaneous assimilation of O, for not only C but CO_2 or HCO_3^- are assimilated. Molecular oxygen may directly serve as substrate for hydroxylation of organic compounds in which a hydroxyl group is formed. Hydrogen is taken up in the form of water from the soil solution or under humid conditions from the atmosphere. In the course of photosynthesis H_2O is reduced to H (photolysis) which is transferred *via* a series of steps to an organic compound resulting in the reduction of nicotinamide adenine dinucleotide (NADP$^+$) which in turn is reduced to (NADPH). This is a very important coenzyme of universal significance in oxidation-reduction processes as the H from NADPH can be transferred to a large number of different compounds. Plants take up nitrogen in the nitrate or ammonium form from the soil solution or as gaseous NH_3 from the atmosphere. Dinitrogen N (N_2) is taken up from the atmosphere by particular microorganisms living in symbiosis with plants. The assimilation of N_2 by these microorganisms is termed fixation of molecular N_2. It is dependent on the presence of specific microorganisms some of which (*Rhizobium*, *Actinomyces*) are symbiotically associated with higher plants. The N of NO_3^- is assimilatd in the process of reduction and subsequent amination. Ammonium-N assimilation also involves an amination process. The incorporation of N from molecular N_2 depends on an initial reduction of N_2 to NH_3, which is again metabolized by the amination process. The assimilation of sulphate-S is analogous to NO_3-N incorporation, *i.e.* a reduction of SO_4^{2-} to the SH-group. Sulphur is not only taken up from the soil solution in the form of SO_4^{2-}, but can also be absorbed as SO_2 from the atmosphere. The reactions which result in the incorporation of C, H, O, N, and S into organic molecules are fundamental physiological processes of plant metabolism. These will be described in more detail later. In this context it need only be mentioned that these main constituents of the organic plant material are assimilated by well defined physiological reactions, and in this respect they differ considerably from the other plant nutrients.

Phosphorus, B, and Si constitute another group of elements which show similarity in biochemical behaviour. All are absorbed as organic anions or acids, and occur as such in plant cells or are bound largely by hydroxyl groups of sugars forming phosphate-, borate, -and silicate esters.

The third group of plant nutrients is made up of K, Na, Ca, Mg, Mn, and Cl. These elements are taken up from the soil solution in the form of their ions. In the plant cell they are present in the free ionic state or are adsorbed to indiffusible organic anions, as for example Ca^{2+} by the carboxylic groups of the pectins. Magnesium may also occur strongly bound in the chlorophyll molecule. Here the Mg^{2+} is chelated being bound by covalent and coordinate bonds (the term chelate is discussed in more detail on page 6). In this respect Mg more closely resembles the elements of the 4th group: Fe, Cu, Zn, and Mo. These elements are predominantly present as chelates in the plants. The division beween the 3rd and 4th group is not very clear-cut, for Mg, Mn, and Ca may also be chelated.

1.2   General Functions

As described above, C, H, O, N, and S are constituents of organic material. As well as this, however, they also are involved in enzymic processes: C and O mainly as components of the carboxylic group, H and O in oxidation-reduction processes, N in the form of NH_2-, NH=, and even -N^+- and S in the form of the SH group. They are therefore reactants in fundamental biochemical processes. Some general examples of the reaction involved are shown below.

Carbon is assimilated by plants as CO_2. This process is called carboxylation. Decarboxylation is the reverse process and provides the basic mechanism by which CO_2 is released as shown in Figure 1.1A for the decarboxylation of oxaloacetic acid. Considering the reactions shown in Figure 1.1 one has to bear in mind that carboxylic groups may be protonated or deprotonated the same is true for N containing groups including amino groups. Protonation and deprotonation depends on the pH of the environment. The pK of terminal carboxylic groups is in the range of 3 which means that at a pH of 3 half of the solute is protonated and the other half not. Since cytosolic pH is about 7, organic acids present are mainly deprotonated and it is more precise to talk of oxaloacetate than of oxaloacetic acid and of pyruvate instead of pyruvic acid. In this book we generally use the anion designation, *e.g.* pyruvate but the reader should always bear in mind that protonation and deprotonation is dependent on the prevailing pH and that for organic acids in most cases both forms are present.

In Figure 1.1B a nicotinamide derivate is shown, namely nicotinamide-adenine dinucleotide (= NAD$^+$, Figure 1.1B) which is a very important coenzyme. The nicotinamide moiety has one N atom integrated in the ring which in the oxidized form lacks one electron and therefore the N has a positive charge and is bound by four ligands. In the course of the reduction of NAD$^+$ *e.g.* by malate as H donor, one H atom is bound to the pyridine ring of the nicotinamide and the other H atom is split into one H$^+$ and one e$^-$ the latter neutralizing the posively charged N in the ring, so that in the reduced form (NADH) the N atom in the ring is bound by three ligands.

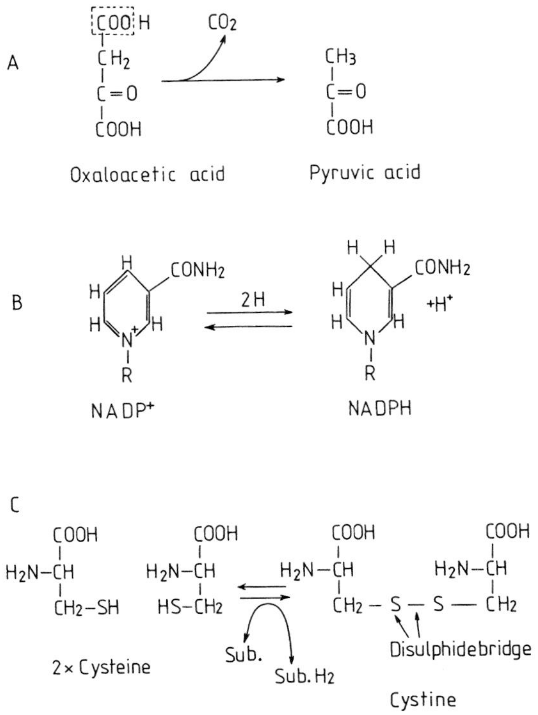

Figure 1.1 Important biochemical reactions — A: Decarboxylation of oxaloacetate to pyruvate — B: Reduction of NADP$^+$ to NADPH. — C: Oxidation of 2 cysteine molecules to cystine forming a S-S bridge.

The sulphhydryl group (SH-group) is the essential group of coenzyme A. It can also be involved in oxidation/reduction processes as shown in Figure 1.1C. The reaction of the two SH-groups of two molecules of cysteine results in the synthesis of one molecule of cystine, a compound characterized by the presence of an S-S-bridge. In the reaction the cysteine molecules are oxidized as two H atoms are removed. The S-S-group is very common in proteins, serving as a link between polypeptide strands (see page 440).

The third and fourth group of plant nutrients (Table 1.1) have nonspecific ionic cellular functions such as establishing osmotic potentials in cell organelles or maintaining ionic balance. In addition these nutrients may carry out specific functions. In an excellent review paper Clarkson and Hanson (1980) have adopted a system in which this third group of nutrients is divided into four categories. These are:

1. Trigger and control mechanisms (Na^+, K^+, Mg^{2+}, Ca^{2+}, Cl^-) by controlling osmotic potentials, membrane permeability, electropotentials and conductance.
2. Structural influences (K^+, Ca^{2+}, Mg^{2+}, Mn^{2+}) by binding to organic molecules particularly enzymes and thus altering their conformation.
3. Formation of a coordinative bond by binding to an electron pair of another ligand (Mg^{2+}, Ca^{2+}, Mn^{2+}, Fe^{2+}, Cu^{2+}, Zn^{2+}).
4. Redox reactions (Cu^{2+}, Fe^{2+}, Co^{2+}, Mn^{2+}). These ions are essential components of prosthetic groups which mediate an electron transfer.

Calcium and Mg^{2+} have a high affinity for carboxylic and phosphate groups, whereas the transition metals (Fe, Mn, Cu, Zn) are attracted more specifically to the lone electron pair occurring on N and S atoms. Since Ca^{2+} possesses a thiner hydratation shell than Mg^{2+}, it can bind more strongly to ligands. Calcium is particularly bound to carboxylic and phophate groups of cell walls and membranes. Magnesium forms a typically strong orientation with pyrophosphate groups which is of particular importance for the binding of Mg^{2+} to adenosine triphosphate (ATP). As shown in Figure 1.2a Mg^{2+} is ionically bound to the phosphate groups and coordinatively bound to the lone electron pair of a N atom of an enzyme protein. This is the means by which the universal coenzyme ATP is bound to an enzyme.

The heavy metals constitute the last group of plant nutrients (Table 1.1) and they occur in plants predominatly in chelated form. A chelate is an organic molecule which contains two or more atoms which are able to bind to the same metal atom thus forming a ring structure. An example is shown in Figure 1.2 for ethylene diamine tetraacetate, the chelator which binds with Ca^{2+}. The Ca^{2+} is bound within the molecule by four bonds, two covalent bonds with two acetate groups and two coordinative bonds with the N atoms (see arrows). An analogous chelate is formed between Fe^{III} and a chelator called mugineic acid. Mugineic acids belongs to the group of siderophores which are chelates binding to Fe (*sideros*, Gk. = iron). As shown in Figure 1.2d the Fe^{III} is bound by three covalent bonds to the carboxylic groups of the molecule and coordinatively bound to two N atoms and one O atom. Siderophores play an important role in plant nutrition (see page 555) Metal chelates formed in this way are highly soluble in water and stable over a broad pH range. Compounds capable of chelating heavy metals, mainly produced by microorgansims, are present in the soil and

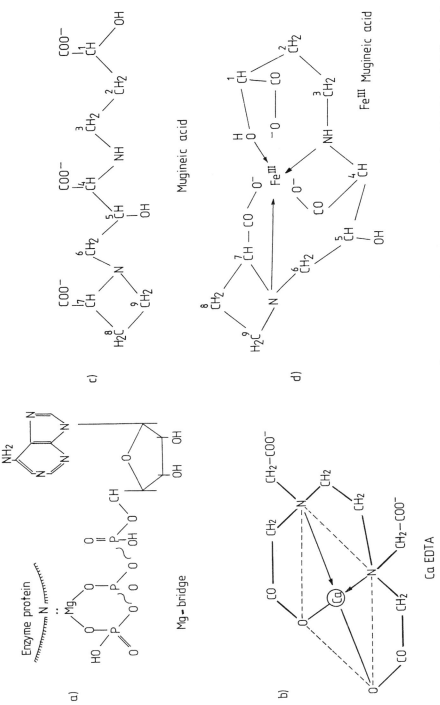

Figure 1.2   a) Magnesium bridging an enzme with ATP — b) Chelation of Ca²⁺ by ethylene diamine tetraacetate — c) Mugineic acid (MA) — d) Fe^{III} mugineic acid complex

extremely important in the acquisition of heavy metals by plant roots. Figure 1.3 shows the haem structure in which Fe is chelated and the structure of the chlorophyll in which Mg is chelated. In chlorophyll b the -CH$_3$ group is substituted by the -CHO group. The four pyrrole rings are indicated by I, II, II, IV. The Mg in the centre of the molecule is bound by two covalent and two coordinate bonds to the N atoms of the pyrrole rings.

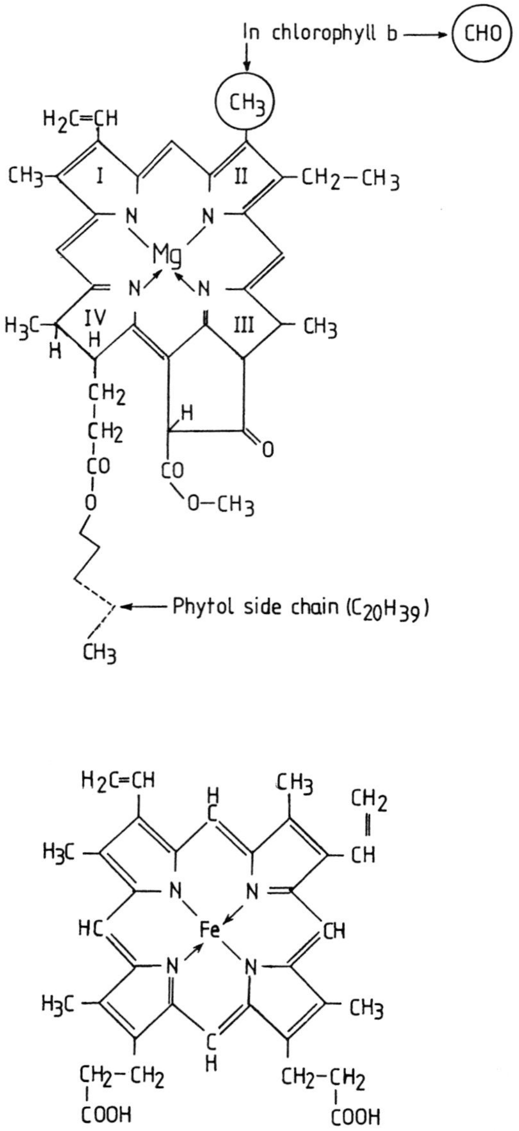

Figure 1.3 Structure of the chlorophyll (above), structure of the haem complex (below).

As described in more detail on page 137 chlorophyll has a unique function in trapping and converting light energy. The haem group forms the prosthetic group of a number of enzymes (catalase, peroxidase, cytochromes, cytochrome oxidase). The Fe present in the haem moiety can change its valency from Fe^{II} to Fe^{III}

$$Fe^{II} \Leftrightarrow Fe^{III} + e^-$$

This enables the transfer of electrons, the principle function of this prosthetic group. In the reduced state (Fe^{II}) the group is called haem and in the oxidized state (Fe^{III}) haemin. Other metal atoms such as Cu, Co and Mo also function in enzyme systems in an analogous way to that described for Fe.

1.3   Mineral Concentrations in Plant Material

The material of living plants consists of organic matter, water and minerals. The relative amounts of these three components may vary, but for green plant material, water is always present in the highest proportion and the minerals in the lowest. The percentage distribution of these three components is in the following order of magnitude:
-   water 700 mg/g fresh matter
-   organic material 270 mg/g fresh matter
-   minerals 30 mg/g fresh matter

More detailed data showing the content of water in plant material is presented in Table 1.2. The minerals make up only a comparatively small proportion of the dry matter. They are nevertheless of extreme importance because they enable the plant to build up organic material (photosynthesis).

The terms "content" and "concentration" are frequently used synonymously and in the strict sense one should distinguish between these terms. The SI system (= systeme international) is based on metric units. If a quantity of weight or volume is related to a definite unit, *e.g.* gram (g) cubic metre (m^3) this is referred as a concentration such as mg kg^{-1} or mL m^{-3}. From this it is clear that a given weight is related to a unit of weight and a given volume to a unit of volume. This need not to be the case in expressions of % or ppm (parts per million). Both these terms can cause confusion and are outdated. A concentration of 50 ppm Fe in plant dry matter is 50 mg Fe kg^{-1} dry matter or 50 mg Fe/kg dry matter and a soil with 18% clay is a concentration of 180 g kg^{-1} soil. The term content may be used if a quantity is not based on a defined unit, *e.g.* the quantity of a nutrient in a leaf. From a physiological aspect it is more appropriate to express concentrations in moles, *e.g.* 0.5 mol NO_3^- m^{-3} in the soil solution (= 0.5 mmol L^{-1}) which in the older terminology is 0.5 mmolar = 0.05 mM. The new SI system no longer uses the term *molar* but quotes the quantity in weight or in moles per unit volume (Incoll *et al.* 1977). Hence a 5 mM solution is (= 5 mmol per liter) is 5 mol m^{-3}. The denominator has a negative sign which means "divided by" and can be substituted by a slash (/). This avoids the small superscript and for the sake of convenience this is frequently used as is the case in this book. In a strict sense instead of litre (L) the term dcm^3 should be used. Since litre is a metric unit and much used in measuring solutions we shall use it as well as the smaller units of L (mL, (L). Instead of tonne (t) the SI terminology uses Mega gramme (Mg) which is 10^6 g. The Mega gramme (Mg) should not be confused with the chemical symbol for magnesium (= Mg).

The SI system (Système International d'Unités) was adapted in its present form in 1960 and is based on the use of metric units which eminated from "decimal measures" as advocated for international use by the great French Scholar Antoine Laurent de Lavoisier (1743—1794) in the preface of his "Treatise on Chemistry" in 1789. In S I nomenclature the basic unit of length is the metre (m), the unit of area the square metre, and the unit of volume mass the kilogran (kg). The advantages of SI units in publications in Plant Science have been considered by Incoll *et al.* (1977) but the use of SI units is still by no means universal.

The mineral composition of plants is generally expressed on a dry weight basis where fresh plant material has been dried at 105°C until all water has been removed. In the older terminology nutrient concentrations in the dry matter were expressed in a variety of ways *e.g.* 4,3% K, 3mg P g^{-1} or 130 ppm Mn. The term ppm means parts per million *e.g.* 130 parts of Mn (weight) per 1, 100, 000 parts (weight) dry matter. As already mentioned confusion can arise using "%" and "ppm" and it is preferential to describe concentrations in terms of unit weight or unit volume. For example 4.3% K is better expressed as 43 mg K g^{-1} dry wt (see Table 1.3). The mineral concentration of plants and plant organs is of physiological and practical significance.

Table 1.2   Water content of various plant tissues and materials in % of the fresh weight

Young green material	90—95	Tomato fruits	92—93
Young roots	92—93	Oranges	86—90
Old leaves	75—85	Apples	7—81
Mature cereal straw	15—20	Banana fruits	73—78
Hay	15	Potato tubers	75—80
Cereal grains	10—16	Sugar beet roots	75—80
Rape seed	7—10		

The main factor controlling the mineral concentration of plant material is the specific, genetically fixed nutrient uptake potential for the various nutrients. This accounts for the fact that N and K concentrations of green plant matter is about 10 times higher than that of P and Mg which in turn is about 10 to 1000 times higher than that of the micronutrients. This general pattern occurs in all species of higher plants. Within plant species, however, considerable differences in mineral concentrations do occur, which are also genetically determined. This question was studied by Collander (1941), who grew 20 different plant species in the same nutrient solution and determined the mineral composition of the resulting plants. It was found that the concentration of K did not differ greatly beween species but marked differences in concentrations in Ca, Mg, and, Si occurred. The greatest interspecies differences were found for Na and Mn.

Plant genera with a high uptake potential for these minerals such as *Atriplex* and *Vicia* for Na$^+$, *Lactuca* and *Pisum* for Mn^{2+}, in extreme cases contained 60 times more Na or Mn than those genera with a low uptake potential (*Fagopyrum* and *Zea* for Na, *Salicornia* and *Nicotiana* for Mn).

Plant nutrient concentrations in tissues and organs depend to a large extent on the relative proportions of cellular material present ie cell wall, cytoplasm and vacuole. Cell walls are rich in Ca^{2+}, and also contain appreciable concentrations of Mg^{2+} (Williamson and Ashley 1982), but they are low in N, P, and K$^+$. Hence older plant tissues with thick, frequently lignified cell walls are rich in Ca^{2+}, which is mainly bound to carboxylic groups but the tissues are low in N, P, and K. The cytoplasm is rich in N, P, and K. The cytosol is the solum of the cytoplasm, *i.e.* the cytoplasm not including the organelles such as nucleus, mitochondria, and plastids. The cytosol contains high concentrations of N and in K$^+$ the latter being present in concentrations of about 150 mol/m^3 (Leigh and Wyn Jones 1986); but is extremely low in Ca^{2+} with concentrations of 10^{-3} to 10^{-5} mol/m^3 (Hanson 1984). The vacuole provides a location for the storage of numerous solutes including plant nutrients such as phosphates, NO$_3^-$, K$^+$, Mg^{2+}, Ca^{2+}, and osmotica such as Na$^+$, and Cl$^-$. The concentration of a plant nutrient in the vacuole usually reflects the nutritional status of the plant *e.g.* if the K$^+$ concentration in the vacuole is high, the plant is well supplied with K$^+$. The same is true for phosphate and nitrate. Low concentrations of plant nutrients in the vacuole indicate poor nutritional status (Leigh and Wyn Jones 1986). This physiological fact is the basis for plant and tissue analysis as a means of assessing nutritional status of plants and plant organs (see page 103). Therefore the concentration of nutrients in the plant depends also on the availability of these plant nutrients in soils.

Plant analysis expression of nutrient concentration is generally based on plant dry matter, because dry plant material can be stored more easily than fresh material. Tissues made up with a relatively large proportion of water, such as young leaves and young roots are characterized by cells with large vacuoles a high proportion of cytoplasm and a low cell cell wall fraction. These tissues show high concentrations of N, P, and K expressed on dry matter basis, since water loss occurring during drying means an actual loss of tissue weight. This is the main reason why young plant tissues and organs show high concentrations of N, P and K$^+$. Older leaves, roots and stems with a higher proportion of cell material are relatively rich in Ca, Mn, Fe, and B (Smith 1962) since these nutrients are bound to the cell walls in relatively high amounts. The relationship between plant age and nutrient concentration in oats is shown in Figure 1.4 from the work of Scharrer and Mengel (1960) for the nutrients N, P, K. From plant emergence until tillering nutrient uptake was higher than the increase in dry matter and concentrations in N, P, and K were raised. After tillering, and the rapid onset of plant growth a steep decrease in N, P, and K concentrations occurred. In the later stages, rates of dry matter production of oats were higher than rates of nutrient uptake.

Mineral nutrient concentrations in oats and rape are shown in Table 1.3. Concentrations of N, P, Mg and K are fairly constant for different plant species provided the same physiological stage is considered. Ca concentrations are generally much higher in dicots than in monocots because of the difference in cell wall structure between both groups

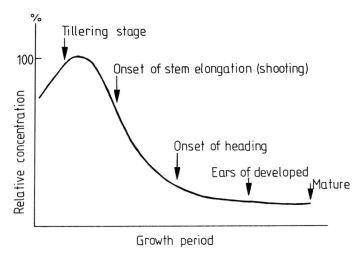

Figure 1.4 Relative concentrations of N, P and K in shoots of oats during the course of the growing period (after Scharrer and Mengel 1960).

of plants. Sodium concentration may differ widely and is generally much higher in halophytes than in glycophytes. Iron and Mn may be bound in relatively large amounts in the cell wall and hence the concentrations of Fe and Mn may also differ much depending on species and environmental conditions. In Table 1.3 nutrient concentrations are expressed in mg g^{-1} DM (= dry matter) for nutrients present in higher concentrations and in μg g^{-1} DM for nutrients occurring in plants in lower concentrations. It must be emphasized that these are total concentrations thus comprising the different forms of the nutrient element present, *e.g.* N in the form of nitrate, ammonium, amino N and other organic N forms. In the case of young plant material most of the N is amino N, originating from enzyme proteins mainly present in the cytoplasm. In the case of grains, *e.g.* oat grains, most of the N is present in the form of amino N of storage proteins. Sulphur similarly may be present in plant tissues in organic and inorganic form. Phosphate occurs as phosphate esters or inorganic phosphate, $H_2PO_4^-$ or HPO_4^{2-} or both depending on the ambient pH. The concentration of phosphate in plant material is simply referred as mg P g^{-1} (= mg/g) plant material. Potassium occurs almost exclusively in ionic form and is therefore designated in this book as K^+. The same is true for Na^+ and Cl^-. Calcium, Mg, and Fe may be present in a complexed form as shown above and therefore the concentrations of these elements are referred to as mg Ca, Mg, and Fe per unit plant material. Potassium is mainly present in the tissue water so if K^+ concentration is expressed in terms of dry material organs low in water content such as seeds and grains are also low in K^+ and organs with high water contents such as fruits are rich in K^+.

For practical purposes such as the calculation of total nutrient uptake of crops or the use of plant analysis as a tool for diagnozing the nutrient avalability in soils, mineral concentration based on dry matter is most appropriate. For physiological considerations, however, it is often more convenient to express nutrient concentrations in the plant on a fresh matter basis in the form of milli mol *i.e.* 5 mmol Ca kg⁻¹ fresh material. This can give a more realistic impression of the actual mineral concentration in plant cells. It is also particularly useful when expressing the concentrations of organic molecules such as free amino acids, organic acids and sugars. In addition, by basing concentrations on the fresh material and expressing the values in mol m⁻³ = (mol/m³), it is often easier to recognize physiological relationships. One such example is the effect of age on the mineral concentration of plant tissues. Generally the water content of plant material is high in younger tissues. Young tissues of plants are thus not so rich in N, P and K as is often believed from dry matter analysis. Jungk (1970) showed in the case of *Sinapis alba* that the concentration of K^+ and NO_3^- based on fresh weight, remains fairly constant throughout the growing season, provided that the plants are adequately supplied with these two nutrients. Similar results showing this constancy of K^+ concentration when expressed on a fresh weight basis have also been reported by Leigh and Johnston (1983). In experiments with barley plants they observed that whereas the K^+ in the dry matter generally declined during growth, the K^+ concentration in the tissue water remained reasonably constant at about 200 mmol kg⁻¹ tissue water. Plant nutrients and their concentrations in a number of plant species are described by Jones *et al.* (1991) in a valuable monograph.

Table 1.3   Mineral concentration of different plant materials (Mengel 1979)

Element	Oat tops at the tillering stage	Oat grains	Oat straw	Rape at the vegetative stage
	mg/g DM			
N	39	17	4.5	56
P	4.4	4.3	1.2	4.9
S	3.2	2.8	3.3	9.3
Cl	15	2.7	14	12
K	43	6.4	14	46
Na	5.3	0.2	3	1.3
Ca	9.4	2.2	9.0	29
Mg	2.1	1.2	1.0	2.0
Si	3.5	1.8	3.3	3.4
	μg/g DM			
Fe	74	53	85	550
Mn	130	80	50	250
Cu	7	3	2.3	7
B	6	1.1	7	35
Mo	2	1.6	1.0	—

Chapter 2

THE SOIL AS A PLANT NUTRIENT MEDIUM

2.1   Important Physico-Chemical Properties

2.1.1   General

Soil is a heterogeneous material consisting of three major components: a solid phase, a liquid phase and a gaseous phase. All three phases specifically influence the supply of nutrients to plant roots. The solid phase is the main nutrient reservoir. The inorganic particles of this phase contain cationic nutrients such as K, Na, Ca, Mg, Fe, Mn, Zn, and Cu, whilst the organic particles form the main reserve of N and to a lesser extent also of P and S. The liquid phase of the soil, the soil solution, facilitates nutrient transport in the soil, *e.g.* for the transport of nutrients from various parts of the bulk soil to plant roots. Nutrients transported in the liquid phase are mainly present in ionic form, but O_2 and CO_2 are also dissolved in the soil solution. The gaseous phase of the soil mediates the gaseous exchange which occurs between the numerous living organisms of the soil (plant roots, bacteria, fungi, animals) and the atmosphere. This process results in the supply of living soil organisms with O_2 and the removal of CO_2 produced by respiration from the soil atmosphere. For plant species living symbiotically with N_2 fixing bacteria, N_2 supply to the root nodules is also mediated by the soil atmosphere. Mineral nutrient behaviour in the soil is closely dependent on nutrient interactions between solid liquid and gaseous phases. These interactions and their influence on the accessibility of mineral nutrients to plant roots are considered in more detail in this chapter.

2.1.2   Cation adsorption and exchange

Adsorption and exchange processes regulate nutrient interactions between the solid and liquid phases of the soil. Colloidal soil particles both clays and humus play an important role since because of their minute size they expose a very large surface area in the soil which mostly bears negatively charged sites. The negative charge on clay mineral surfaces arises largely because of isomorphous replacement of cations in the crystalline lattices where trivalent cations are substituted by divalent cations. Negative charges can also result from the dissociation of H^+ from weak acids. This is particularly important in producing negatively charged sites on organic soil particles.

The negatively charged surfaces of these various soil particles attract cations such as Ca^{2+}, Mg^{2+}, K^+, Na^+ as well as Al cation species (Al^{3+}, $AlOH^{2+}$, $Al(OH)_2^+$) and Mn^{2+}. Cations electrostatically adsorbed to the negatively charged surface of colloidal particles dispersed in an electrolyte solution are subjected to both interionic (coulombic) and kinetic forces. Interionic forces tend to bind the cations tightly to the surface of the particle. On the other hand, kinetic forces in the form of random thermal motion tend to dissociate cations from the surface. As a result of these forces an electrical potential gradient is set up near the clay surface. On equilibrium a characteristic ion distribution pattern results between the clay surface, exchangeable cations and free solution (Figure 2.1). In the immediate vicinity of the negatively charged surface the so called Stern layer, there is a high cation concentration, and the anion concentration is approximately zero. With increasing distance from the colloid surface the cation concentration decreases at first rapidly and then asymptotically. This decrease in cation concentration is accompanied by a reciprocal increase in anion concentration to the free solution where cation and anion concentrations become equal, as shown in the lower part of the figure. The double layer as described above which contains an excess of cations extends from the negatively charged surface to the free solution and is known as the Gouy-Chapman layer as it was at first described by these workers (Gouy 1957, Chapman 1957). It is also referred to as the diffuse double layer. Its thickness from the clay surface to the free solution is usually about 5 to 10 nm.

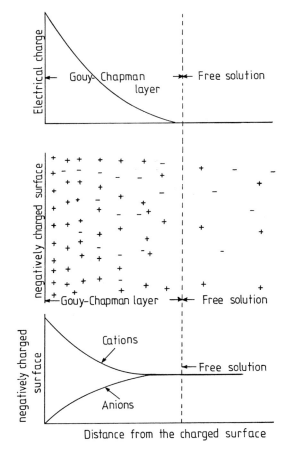

Figure 2.1 Decline of the electrical field with increasing distance from a charged surface and the resulting cation anion distribution pattern (Gouy-Chapman-model).

The equilibrium between the ions of the diffuse double layer and the free solution is a dynamic one. The ions of the free solution are in rapid exchange equilibrium with ions in the Gouy-Chapman layer. When the concentration of the outer solution is diluted some ions from the diffuse double layer diffuse into the free solution thus causing a new equilibrium to be set up. By progressive dilution the outer solution virtually becomes free of ions and the total cations adsorbed to the surface are equivalent to the negative charge. This process occurs when a cation exchanger loaded with a salt solution is leached with water. Surplus anions and cations are washed off and finally only those cations are retained that are equivalent to the negative charge of the exchange capacity of the ion exchanger.

For cation adsorption as described above one cation species replaces another. This reversible process which usually occurs between the liquid (soil solution) and the solid phases (colloidal fraction of the soil) is called cation exchange. The principle of this exchange process, which is stoichiometric, is shown in Figure 2.2 where one Ca^{2+} is replaced by two K^+. Generally all cation species are interchangeable but the degree to which one cation species may replace another cationic species depends on the strength of retention of the adsorbed cation. According to Coulomb's law, the interionic bond is stronger the closer the ionic partners are located, and varies inversely as the square of the distance between the charges. The binding is also stronger the higher the charge of ions. This means that trivalent cations are more strongly bound than divalent cations which in turn are more tightly held than monovalent cations. This preference increases the more dilute the solution, and the higher the charge density of the adsorbing surface. In addition the degree to which an ion is hydrated also influences the bonding strength since a hydrated cation can not be attracted so closely to the negatively charged surface because of the presence of its hydration shell. Smaller cations have thicker hydration shells because they have a higher charge density and they are thus not so tightly bound to charged surfaces. Table 2.1 shows the diameters of several cations in a hydrated and nonhydrated form.

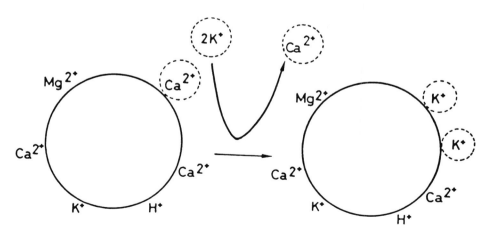

Figure 2.2 Principle of cation exchange. Ca^{2+} is replaced by 2 K^+.

Table 2.1   Diameter (nm) of hydrated and non-hydrated cations

	hydrated	non-hydrated
Rb^+	0.51	0.30
K^+	0.53	0.27
NH_4^+	0.54	0.29
Na^+	0.76	0.20
Li^+	1.00	0.15
Mg^{2+}	0.64	0.16
Ca^{2+}	0.56	0.21

Hydrogen ions (protons) in aqueous solution are also hydrated as water is a proton acceptor. As shown below the hydroxonium ion is formed when water is protonated:

$$H_2O + H^+ = H_3O^+ \text{ (Hydroxonium ion)}$$

Although it is more precise to describe the hydrogen ion as hydroxonium ion, in practice hydrogen ion or proton is used with the understanding that hydroxonium ion is meant in most cases. A pecularity of mineral soils is that their H^+ concentration in the soil solution and also the quantity of adsorbed H^+ to soil particles is low because H^+ is rapidly neutralized by Al- and Fe oxy/hydroxy complexes (see page 54 Table 2.8, Schwertmann *et al.* 1987) as shown in the example below for gibbsite $[Al(OH)_3]$:

$$Al(OH)_3 + H^+ \Leftrightarrow Al(OH)^+_2 + H_2O$$

$$Al(OH^+)_2 + H^+ \Leftrightarrow Al(OH)^{2+} + H_2O$$

$$Al(OH)^{2+} + H^+ \Leftrightarrow Al^{3+} + H_2O$$

Besides the Al complexes shown above, hydroxy-Al-polymers play a major role in neutralizing H^+ in mineral soils. In this respect one of the most important hydroxy-Al-polymer is $(Al_6(OH)_{15})^{3+}$ which occurs in solution as well as being adorbed to negatively charged surfaces. Its H^+ neutralizing reaction is shown below:

$$2[Al_6(OH)_{15}]^{3+} + 30\ H^+ \Rightarrow 12\ Al^{3+} + 30\ H_2O$$

When cationic Al species are titrated by a base (OH^-) they consume OH^- and therefore behave like H^+ as shown below for Al^{3+}:

$$Al^{3+} + 3\ OH^- \Rightarrow Al(OH)_3$$

Therefore titratable acidity in mineral soils can be attributed mainly to cationic Al- and Fe species and not so much to H^+.

In organic colloids the carboxylic- and OH groups represent the negatively charged surface sites. The anionic group can form a covalent bond with H^+ a reaction is favoured by a high H^+ concentration as is obvious from the equation below

$$R\text{-}COO^- + H^+ \Leftrightarrow R\text{-}COOH$$

In organic soils the number of cation exchange sites is therefore highly dependent on the prevailing pH, decreasing as the pH falls. Protonation and deprotonation is of particular importance for humic acids. Their negative charge mainly originates from carboxylic groups (humates the anionic designation) and at high pH levels also from deprotonated phenolic hydroxylic groups. Since protonation and deprotonation depends on the prevailing pH, the negative charge of humates is variable, high at high pH and decreases as the pH falls. These relationships are of particular importance in organic soils in which the cation exchange capacity mainly originates from humates.

Cation species which are only weakly adsorbed can readily be exchanged and *vice versa*. Thus the relative replacing power of a particular cation species depends on its strength of binding. The following sequence of relative replacing power was established by Hofmeister and is called Hofmeister's cation sequence.

$Li^+ < Na^+ < K^+ < Rb^+ < Cs^+$
$Mg^{2+} < Ca^{2+} < Sr^{2+} < Ba^{2+}$
Increase in relative replacing power
Decrease in the degree of hydration

⎯⎯⎯⎯⎯⎯⎯⎯⎯⎯⎯⟶

This rule is not universally applicable as clay mineral structure also affects the binding power. This is particularly true for K^+ adsorption. Schachtschabel (1940) investigating the replacement of NH_4^+ adsorbed to various adsorption complexes by several cation species found the following sequence of replacing power:

Kaolinite	$Na^+ < H^+ < K^+ < Mg^{2+} < Ca^{2+}$
Smectite	$Na^+ < K^+ < H^+ < Mg^{2+} < Ca^{2+}$
Mica	$Na^+ < Mg^{2+} < Ca^{2+} < K^+ < H^+$
Humate	$Na^+ < K^+ < Mg^{2+} < Ca^{2+} < H^+$

⎯⎯⎯⎯⎯⎯⎯⎯⎯⎯⎯⟶

Increase in the relative ease of replacement of NH_4^+

This example demonstrates that K^+ is more strongly adsorbed by mica than can be predicted by its valency and degree of hydration. In an analogous way other 2:1 clay minerals (illite, vermiculite) can adsorb K^+ and NH_4^+ rather specifically. This is considered in more detail on page 32.

The relative replacing power of one cation species by another one depends not only on the nature of the two cation species in question but also on their concentrations or more precisely on the activities of the ions present. For the sake of simplicity, ion concentrations are usually considered. It should be remembered, however, that ion activities rather than concentrations apply to exchange reactions and equilibrium conditions between free and adsorbed ions. If ion concentrations are low they approximate to their activities. At higher concentrations, however, deviations from the predicted behaviour of ideal solutions occur. When Ca^{2+} concentrations are high deviations between Ca^{2+} activity and Ca^{2+} concentration may be considerable. Such deviations result from interionic forces and the formation of associated ions in the solution under high concentration conditions. The concentration is therefore corrected by a factor (activity coefficient) which is always <1. The relationship is expressed by the equation:

$$a = f \times c$$

where a = activity, c = concentration, f = activity coefficient

The activity coefficient thus decreases as the ionic strength of the solution increases. The influence of increasing activity or concentration on the replacement of one cation species by another may be considered by reference to a simple system where one cation species completely saturating a colloidal particle is being replaced by an increasing concentration of another cation species in the bathing solution. If K^+ is the ion originally saturating the colloid and Ca^{2+} the ion replacing K^+, the relationship between Ca^{2+} adsorption, K^+ desorption and Ca^{2+} activity may be represented as shown in Figure 2.3. It is clear that as the concentration or activity of free Ca^{2+} increases, the adsorption of Ca^{2+} also increases in the form of a saturation or exchange curve. This can be described in another way by saying that the lower the concentration of the replacing ion, the greater is its replacing power in relation to its concentration. This relationship between activity or concentration and adsorption as reflected in the asymptotic curve shown is applicable to all cation adsorption processes.

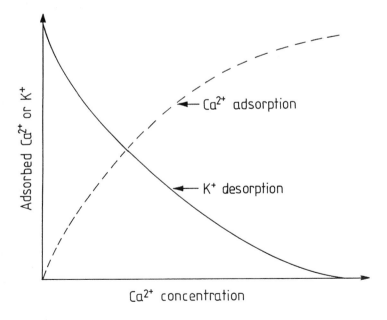

Figure 2.3 Relationship between increasing Ca^{2+} concentration, Ca^{2+} adsorption and K^+ desorption.

In the example cited above a monovalent cation (K^+) was replaced by a divalent cation (Ca^{2+}) or in other words a strongly bound cation (Ca^{2+}) replaced a less strongly bound cation (K^+). Exchange processes also occur, however, where a more strongly adsorbed cation species is replaced by a cation species which is on average less strongly adsorbed. Such exchanges occur due to the fact that kinetic as well as electrostatic forces act upon the ions. As already discussed, provided there are no specific adsorption sites on the charged surface, the binding strength of a cation species depends on its valency and on its degree of hydration. This is true, where the mean behaviour of a large number of ions is being considered as is usually the case. The relative replacing power of a single ion, however, depends on its kinetic energy. This is not the same for all ions of a given species. When a large number of ions are present, energy distribution follows the Maxwell energy distribution curve. This means that for every cation species a small proportion of so called "high energy" cations occur which can take place in exchange reactions not open to the bulk of other ions. In the cation exchange process the energy distribution of cations is thus of particular significance as it enables the replacement of cations more strongly bound by those which are on average more weakly bound. Thus Ca^{2+} can be replaced from a clay mineral or ion exchanger by a highly hydrated monovalent cation such as Na^+, as some Na^+ is always present at high enough an energy level. Replacement of Ca^{2+} by Na^+ may also occur on biological membranes under saline conditions. The disturbing effect on membrane stability is detrimental to cellular activity (Lynch *et al.* 1987). In order to achieve complete replacement of a more strongly bound cation species, however, large amounts of the more weakly bound cation species are required. This is the principle which underlies the regeneration of a cation exchange column where the column is treated with a large excess of a replacing cation species, so that "high energy" cations are also present in excess. In this context it should be remembered that cations are not bound tightly to the surface, but are present as a diffuse layer which facilitates exchange processes based on the individual energy levels of the cations.

Cation exchange capacity (CEC) is expressed as a measure of the moles of positive charge per unit mass. This may be expressed in terms of centimoles of positive charge per kg of soil (cmol/kg). If a soil has a cation exchange capacity of 20 cmol/kg this means that it can adsorb or exchange 20 cmol of H^+ for any other univalent cation *e.g.* K^+ or Na^+. For a divalent cation such as Ca^{2+} or Mg^{2+} the value is half that figure (*i.e.* 10 cmol) as adsorption and exchange take place on the basis of chemical equivalents. For a trivalent ion the value is one third. Thus

$$1 \text{ mol } H^+ (K^+, Na^+) = \tfrac{1}{2} \text{ mol } Ca^{2+}(Mg^{2+}) = 1/3 \text{ mol } Al^{3+}$$

The cation exchange capacity can be measured without estimating the individual elements and the determination is made frequently in soil analysis.

Sometimes the term percentage base saturation is used. This represents the total exchangeable cationic species without H^+ and without the Al cations because they behave like H^+ as shown above (see page 18). The base saturation is expressed as a percentage the sum of the adsorbed cation species without the H^+ and Al cations of the total cation exchange capacity.

As the proportion of clay in a soil increases the soil system becomes more dispersed and hence the total surface area of soil particles is also increased. This means that soils rich in clay minerals are able to adsorb more water and cations than soils low in clay. Clay rich soils have thus a higher cation exchange capacity and a higher water holding capacity than soils low in clay. Cation exchange capacity values can vary considerably. Brady and Weil (1999) give values from 2.0 in a sand to 57.5 centimol/kg in a clay soil. Values between 10 and 20 cmol/kg soil are common.

2.1.3 The Gapon equation

One of the most well known equations which quantitatively describes the relationship between adsorbed monovalent and divalent cations and their concentrations in the equilibrated solution is the Gapon equation Gapon (1933). This may be described as follows:

$$C_{ads}{}^+/C_{ads}{}^{2+} = k\, a_c{}^+/(a_c{}^{2+})^{\frac{1}{2}}; \quad a_c{}^+/(a_c{}^{2+})^{\frac{1}{2}} = AR$$

where

$C_{ads}{}^+$, $Ca_{ads}{}^{2+}$ = adsorbed monovalent and divalent cations respectively

$a_c{}^+$, $a_c{}^{2+}$ = activity of the monovalent and divalent cations respectively mol/l

AR = Activity ratio = ratio of the activities of the two cation species in the equilibrium solution. For a monovalent and divalent cation species, this is the ratio of the activity of the monovalent cation over the square root of the activity of the divalent cation.

k = Gapon coefficient or selectivity coefficient.

The value of k is a measure with which a monovalent cation (C^+) is adsorbed in comparison with a divalent cation (C^{2+}). This could for example be K^+/Ca^{2+} or K^+/Mg^{2+} (activity ratio = AR). For a given exchange system k is constant within limits, and various authors have used the equation to describe quantitative relationships between cations in adsorbed and equilibrium solution forms (Bolt 1955, Lagerwerff and Bolt 1959, Ehlers et al. 1968). The use of the Gapon equation by the US Salinity Laboratory for the expression of cation exchange studies in saline and alkaline soils has also meant that the equation is fairly well accepted for the prediction of adsorbed ions and solution composition of these soils (Fried and Broeshart 1969).

It was originally held that k should be constant. This value, however, is dependent on a number of factors including the degree of cation saturation of the clay mineral and its charge density (Schwertmann 1962). It is also considerably influenced by clay mineral structure and the presence of specific binding sites. Such specific sites particularly relate to the binding of K^+ and $NH_4{}^+$ to the 2:1 clay minerals. Where for example, K^+ is specifically adsorbed, the Gapon "constant" for K^+/Ca^{2+} and K^+/Mg^{2+} is higher than normal because of the higher quantity of adsorbed K^+(see p 485).

2.1.4   Anion adsorption

The anion adsorption capacity of most agricultural soils is generally small as compared with the cation exchange capacity (CEC). However, a number of soil minerals and also amorphous soil colloids are capable of adsorbing anions very strongly. These anion adsorbers include hydrous Fe and Al oxides (haematite, goethite, gibbsite, amorphous hydroxides), 1:1 clay minerals, 2:1 clay minerals, Fe and Al organo complexes and Ca carbonates. Surface AlOH and FeOH groups are particularly important sites for anion adsorption occurring on both inorganic and organic complexes. *e.g.* humates and fulvic acids.

Two kinds of adsorption may be distinguished, ligand exchange with OH groups and adsorption to protonated groups (Parfitt 1978). Ligand exchange may be described in a simple way by the equation:

$$Me\text{-}OH + An^- \Leftrightarrow Me\text{-}An + OH^-$$

Phosphate adsorption by Fe- and Al-oxides as well as by clay minerals may occur in two steps, a mononuclear adsorption followed by a binuclear one. The principle of this kind of adsorption is shown in Figure 2.4.

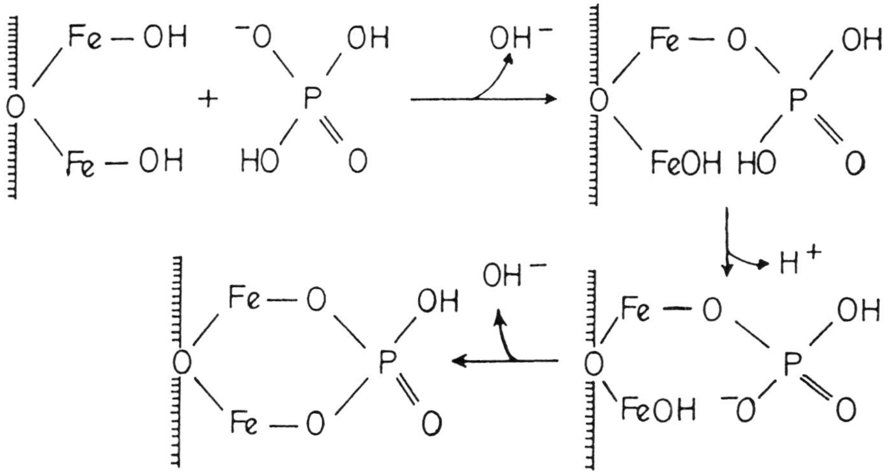

Figure 2.4   Phosphate adsorption at the surface of an Fe oxide. In the 1st. step mononuclear adsorption occurs, in the 2nd step adsorption becomes binuclear.

In the first step $H_2PO_4^-$ is adsorbed by exchanging one OH^- from the adsorbing surface (Hingston *et al.* 1974, Parfitt and Smart 1978). The phosphate so adsorbed may then dissociate a H^+ and then exchange for a further OH^- from the surface with the production of a molecule of water. This binuclear surface bridging of the phosphate means that it is very strongly adsorbed and scarcely available to plants (Barekzai and Mengel 1985).

From the reactions shown in Figure 2.4 it can be seen that the binuclear adsorption of one $H_2PO_4^-$ results in a net release of one OH^-. This means that the reaction is pH dependent, being favoured under low pH conditions and *vice versa*. Infrared spectroscopy studies (Nanzyo and Watanabe 1982) have revealed that phosphate adsorption to goethite is binuclear over the wide pH range from 3.3 to 11.9. The quantity of phosphate adsorbed also depends on the specific surface of the minerals as has been shown by Lin *et al.* (1983) for gibbsite, goethite and kaolinite. Since adsorption and desorption of phosphate are pH dependent, a pH increase, *e.g.* by liming, may result in an increase of phosphate availability (Haynes 1984).

Sulphate may also be adsorbed to soil particles (Nodvin *et al.* 1986). According to Turner and Kramer (1991) goethite (FeOOH) and hematite (Fe_2O_3) are strong sulphate adsorbents. Similarly to phosphate, sulphate may be adsorbed in a mononuclear and binuclear configuration. Adsorption begins by protonation of a hydroxyl of the goethite. Protonation is favoured by low pH. The protonated hydroxyl attracts a sulphate followed by a ligand exchange by which a water molecule is released:

$$\equiv Fe\text{-}OHH^+ + {}^-OSO_2OH \Rightarrow \equiv Fe\text{-}O\text{-}SO_2OH + H_2O$$

The hydroxyl of the adsorbed sulphate may exchange with another neighbouring hydroxyl so that a binuclear bond is formed. This configuration is particularly stable. The mononuclear binding dominates at low pH because under these conditions the deprotonation of the HSO_3^- is depressed. This pH dependence of sulphate adsorption is in accord with the observation of Nodvin *et al.* (1986) that sulphate adsorption increases in the pH range from 3 to 4 followed by a linear decrease with a further rise in pH. Over this higher pH range protonation of the Fe[III] hydroxyl complex is supposedly the limiting factor.

Protonation of hydroxyls located on mineral surfaces results in the formation of positively charged points at which anions may be adsorbed. This second kind of adsorption mechanism occurs under low pH conditions. Here OH groups may become protonated thus allowing anion adsorption by electrostatic interaction.

$$MeOH + H^+ \rightarrow MeOHH^+ \text{ (protonation)}$$
$$MeOHH^+ + An^- \rightarrow MeOHH\text{-}An \text{ (neutralization)}$$

Both processes of protonation and neutralization are extremely pH dependent being enhanced when the H^+ concentration is raised. Highest anion exchange capacities are thus found in acid soils rich in hydrous Fe and Al oxides or in clay minerals or in both.

The two mechanisms described above for anion adsorption differ in anion specificity. Anion adsorption to protonated groups largely involves an electrostatic interaction and as such is almost totally nonspecific. Ligand exchange on the other hand is associated with a chemical interaction in which an anion becomes coordinated to a metal ion, and is for this reason much more anion specific. The pronounced specificity of anions in ligand exchange reactions is responsible for the marked differences in adsorption capacity of soils for specific anions. Phosphate which is largely adsorbed by ligand exchange is a very strongly adsorbed anion. On the other hand for nitrate and chloride where ligand exchange plays little if any role, these anions are only weakly adsorbed. Parfitt (1978) gives the following selectivity order of anion adsorption by soil soils:

phosphate > arsenate > selenite = molybdate > sulphate = fluoride > chloride > nitrate.

Other anions than those mentioned above may participate in exchange reactions. Under high pH conditions boric and silicic acids may form anions.

$$H_3BO_3 + H_2O \Rightarrow B(OH)_4^- + H^+; \ H_4SiO_4 \Rightarrow H_3SiO_4^- + H^+$$

In addition some organic anions may also compete for adsorption sites. In this respect citrate plays an important role in desorbing phosphate and releasing it into the soil solution which can then be taken up by plants (Gerke 1994).

The Langmuir equation is frequently used to describe quantitative relationships in anion adsorption. This is expressed mathematically as:

$$A = A_{max} \times kc/(1 + kc)$$

where A = amount of anion adsorbed; A_{max} = maximum of anion adsorption; c = anion concentration in solution, k = a constant related to adsorption energy and is higher the stronger the adsorption.

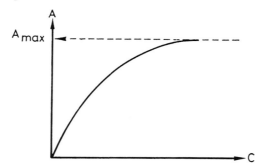

Figure 2.5 Relationship between solution concentration (c) and adsorption (A) according to the Langmuir equation.

The relationship is shown in Figure 2.5. The equation is not completely obeyed for anion adsorption by the soil matrix since the Lamgmuir model was derived for monolayer gas adsorption to solid surfaces. In anion adsorption both a charged particle and a charged surface are involved. Additionally adsorption is associated with chemical reactions. Also adsorption is dependent on pH and salt conditions as both these factors affect A_{max} and k (Bowden et al. 1977). As an approximate measure, however, the Langmuir equation is valuable in quantifying anion adsorption processes and has been used by a number of workers including Holford and Mattingly (1975) who investigated phosphate adsorption on calcite surfaces.

2.1.5   Water adsorption

Dipolar molecules are also bound to surfaces. The forces which bring about adsorption, however, are van der Waal's forces. This kind of adsorption therefore differs fundamentally from cation or anion adsorption to charged surfaces, as the adsorbed molecules are not so strongly bound and there is no strong equivalency between the surface charge and the quantity of adsorbed molecules. This indicates an absence of quantitative exchange between adsorbed and free molecules. An example of this type has already been mentioned in the adsorption of water molecules by charged surfaces in ion hydration (see page 17).

The most important example of adsorption of this kind is the adsorption of water to various particles such as clay minerals or organic matter in the soil or to protein complexes in cells as well as cell wall material. The asymmetric water molecule is a dipolar molecule, and has a negative zone or side associated with the O-atom and a positive zone or side associated with the H-atoms. A negatively charged surface such as provided by a clay mineral attracts the positive sides of the water molecules and binds them strongly to its surface as shown in Figure 2.6. The first adsorbed monomolecular layer of water molecules thus exposes another negatively charged surface. This again attracts more water molecules thus forming a series of layers of water molecules over the adsorbing surface. The water layer in direct contact with the surface is most strongly bound. Adsorption strength decreases with increasing distance of the water layer from the adsorbing surface. Films of water built up in the way described above are bound to soil particles and to particles in living cells. The attraction by which water molecules are bound to surfaces by means of H bonding can be expressed in terms of water potential (see page 182) and measured in units of pressure (*i.e.* the Pascal, Pa). The water potential is basically an expression of the thermal movement of water molecules. At normal air pressure and room temperature the water potential of pure water is zero.The movemant of water molecules is restricted by adsorption and therefore the water potential of adsorbed water is below zero. The pressure has therefore a negative sign and the water adsorption is greater the more negative the value. The first monolayer of water held to the soil surface is in the range of -10^5 to -10^6 kPa (*i.e.* kilo Pascal). The water status of plant tissues is generally expressed in terms of MPa (*i.e.* mega Pascal, 1 Mpa = 10^6 Pa or 10^3 kPa).

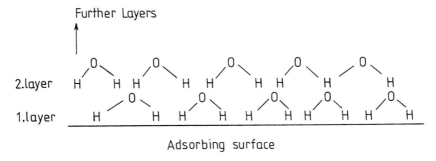

Figure 2.6 Layers of water molecules adsorbed to a matrix (surface).

Molecular adsorption of water is temperature dependent. The higher the temperature, the more thermal energy can be transferred to the adsorbed molecules. By increasing temperature a point is thus reached when the energy becomes high enough to dissociate molecules from the adsorbing surface into the vapour phase. This is one reason why soils dry out more rapidly under high temperature conditions. In this respect molecular adsorption differs basically from ion adsorption, which under soil conditions is largely independent of changes in temperature.

2.1.6   Colloidal systems

All the processes described above, involve reactions between diffusible particles (ions, molecules) and surfaces. The significance of these surface reactions is greater the larger the exposed surface area. The ratio between the surface area and the bulk of a material depends on the degree of its dispersion. Smaller particles clearly expose a relatively larger surface area. In systems made up of extremely small particles, surface forces play a dominant role. Such systems are called colloidal systems, the particles of which have a diameter in the order of $0.1-10$ μm and therefore belong to the clay fraction (<2 μ) or fine silt fraction ($2-63$ μm). The main feature of colloidal systems is not so much the chemical composition of the particles but rather the degree to which the particles are dispersed. A colloidal system consists of a disperse phase, made up of the small particles, and dispersion medium, which can be a gas or a liquid. In soils, colloidal systems are mainly made up of clay minerals dispersed in water. Water too is the dispersion medium in plant tissues but here proteins and polysaccharides represent the disperse phase. When the particles of the disperse phase of a colloidal system are discrete and homogeneously dispersed throughout the dispersion medium the system is called a sol. If the dispersed phase is in a coagulated state it is called a gel. In many cases colloidal systems are reversible. This means that they can be converted from sol to gel and *vice versa*.

Colloidal particles in the sol form are all either negatively or positively charged, so that individual particles electrostatically repel each other. The disperse phase can thus remain in suspension and the system does not coagulate. In a system with water as the dispersion medium, the particles are surrounded by a hydration shell because of their electrical charge, and this also prevents coagulation. As a rule negatively charged particles can be neutralized by the addition of cations and particularly by H^+. Positively charged colloidal particles can be neutralized by the addition of anions, and especially by OH^-. As soon as the charge of the particles is neutralized they lose their water shell, aggregate together and coagulation occurs. This type of reaction means that the stability of a sol depends on the pH of the surrounding medium. Other ion species are also capable of coagulating colloidal systems. The extent of coagulation depends on the valency and the degree of hydration of the ion species concerned. As described above, in water systems, ions are adsorbed in hydrated form. Highly hydrated ions as for example Na^+ cannot therefore be closely bound to the surface of colloidal particles because the water shell of the ion and the water layer of the surface prevent the close approach of the opposing charges. The neutralizing effect between the ion and the oppositely charged colloidal surface is thus weak. The colloidal particle, therefore, retains a relatively high amount of its charge and for this reason repels other colloidal particles, and coagulation does not occur.

In contrast, ions with a higher valency and a thinner water shell, *e.g.* Ca^{2+}, are adsorbed closely to the charged surface. They thus tend to neutralize the colloidal particles and lose their water shell; coagulation occurs and a gel system is formed. In soils, this is called flocculation and it has a very important bearing on soil structure (see page 38). Colloids in gel structure bind larger particles and thus form soil aggregates with pores of different size. These aggregates are the basis for a good soil structure (page 39) and pores formed are important for water storage as well as solute and air movement. In the formation of soil aggregates organic matter is also involved and Ca^{2+} and other polyvalent cations frequently bridge a carboxylic group of the organic matter with the negative charge of a clay mineral surface (Figure 2.7). Besides Ca^{2+}, Fe and Al cations also stabilize soil aggregates and soil organic matter (Schulten and Leinweber 1995). When Ca^{2+} is exchanged by monovalent cations, mainly H^+ or Na^+, the bonds are broken, the aggregate disperses and the gel shifts over to the sol status. This transition from a gel into a sol condition is called peptization (derived from pepsin, which brings coagulated proteins into solution in the stomach). As already indicated an increase in valency enhances coagulation. According to Schulze Hardy the relative coagulating capability of Na^+: Ca^{2+} :Al^{3+} is in the ratios of 1:20:350. Coagulation also increases at higher ion concentrations. This means that even ions which do not readily coagulate can induce coagulation if present in high enough concentrations. This occurs because a high ionic concentration in the vicinity of the colloidal particle surface disturbs the surface water shell and can induce neutralization. It is for this reason that proteins can be coagulated by ammonium sulphate.

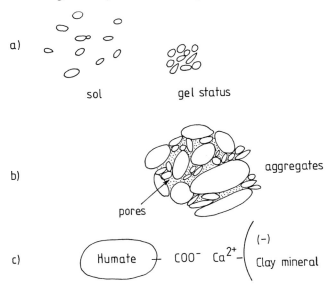

Figure 2.7 a) Sol and gel status — b) Aggregate formation — c) Ca^{2+} bridging a humate with a negatively charged clay particle.

2.2   Important Soil Characteristics

2.2.1   Soil texture and clay minerals

The solid phase of the soil is made up of inorganic and organic components. The inorganic fraction consists of particles of different sizes which range from clay (<2, μm) to silt (2 μm to 63 μm) to sand (63 μm to 2 mm) to gravel (2 mm to 2 cm) and to rocks. The relative proportions of these particles determines the texture of a given soil. This soil property is of extreme importance in determining the physical behaviour of the soil. In addition soil texture is also of major importance in determining nutrient storage in the rooting profile. Figure 2.8 shows the distribution of the most important soil minerals in the three main particle size groups, sand, silt, and clay (Schroeder 1984). It is evident that quartz is dominant in the sand fraction. This sterile mineral contains virtually no plant nutrients with exception of Si and has neither the capacity to adsorb nor to store plant nutrients. As the particle size decreases from sand to silt and to clay the quartz proportion becomes smaller. The silt and sand fractions generally contain appreciable amounts of primary silicates, especially micas and feldspars. Both are K^+ bearing minerals and release K^+ during weathering. In some soils in which the silt fraction is to a large extent made up of mica, K^+ release can contribute substantially to K^+ supply of crops (Mengel *et al.* 1998). Of the micas biotite (Mg mica) in particular weathers easily and thus releases K^+ and also Mg^{2+} under acid soil conditions. Muscovite (K mica) is more stable and weathers at a much lower rate (Feigenbaum *et al.* 1981). The clay fraction predominantly contains secondary minerals that is to say minerals formed by weathering of primary minerals. The most important secondary clay minerals are the 2:1 phyllosilicate silicate clay minerals (Figure 2.10) which adsorb cations and therefore play a major role in storing plant nutrients such as K^+, Ca^{2+}, Mg^{2+} and NH_4^+. The sand, silt, and clay fractions also contain a group of oxides and hydroxides predominantly of Fe, Al, and Mn. These are end products of weathering processes and therefore dominate in highly weathered soils. A well known example of such soils are the Oxisols mainly developed under tropical conditions and characterized by desilification *i.e.* a loss of Si (van Wambeke 1991). They are therefore low in secondary phyllosilicate clay minerals but rich in Fe and Al oxides/hydroxides. These strongly adsorb phosphates but are incapable of adsorbing cations. Phosphorus deficiency can thus frequently occur on these soils. Under acid soil conditions the Al oxides/hydroxides may form Al cation species which are toxic to plant roots (page 55). On the other hand Al- and Fe oxides/hydroxides are less sticky, plastic and cohesive than some phyllosilicate clay minerals which can be beneficial for soil structure (Brady and Weil 1999).

The most important clay minerals in soils are the layer silicates (= phyllosilicate minerals). The basic molecular building blocks of these minerals are the tetrahedron and octahedron (see Figure 2.9). Tetrahedra linked together in the same plane form a tetrahedral sheet and in an analogous manner octahedra form an octahedral sheet. The tetrahedron consists of four closely packed equally spaced oxygen atoms surrounding one centrally situated atom which is usually Si. A pyramidal structure is thus formed. The octahedral structure (8 faced structure) is made up of 6 OH⁻ groups coordinated around a central cation. Generally the centre of the octahedron is occupied by Al^{3+} although it can be replaced by Mg^{2+} and Fe^{2+}. Sometimes the central atom may be absent.

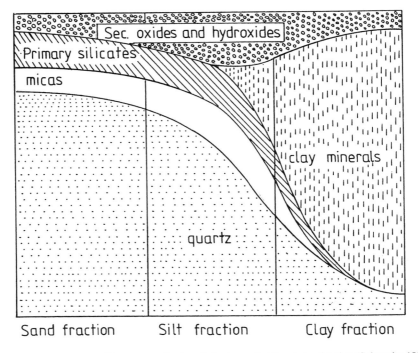

Sand fraction   Silt fraction   Clay fraction

Figure 2.8 Soil texture and important soil minerals in the sand, silt,and clay fraction (Schroeder 1984).

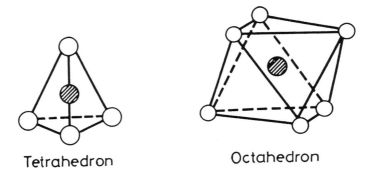

Tetrahedron   Octahedron

Figure 2.9 Tetrahedron and octahedron as structural elements of phyllosilicates.

The layer silicates carry a negative charge which enable them to adsorb cations. Both a variable and a permanent negative charge are present. The variable charge originates from H^+ dissociation of OH^- groups and is therefore pH dependent, being high at high pH and *vice versa*. This is mainly true for kaolinite and for organic colloids. The permanent negative charge of layer silicates arises by isomorphic substitution, *e.g.* Si^{4+} by Al^{3+}, Mg^{2+} or Fe^{2+}. This substitution means a lack of positive charges, hence the mineral carries a negative charge which is pH independent and therefore permanent. Permanent negative charge is an essential characteristics of 2:1 clay minerals.

Three major types of silicates are represented in Figure 2.10 by kaolinite, micas, and smectites. Kaolinite is made up of one tetrahedral layer alternating with one octahedral layer and is hence termed a 1:1 type crystal lattice (see Figure 2.10). Both layers form the *unit layer* with a basal spacing between the unit layers of 0.72 nm under standard conditions.The tetrahedal and octahedral layers are strongly bound to each other by mutually shared O atoms, each unit being firmly attached to the next by oxygen hydroxyl linkages (H bonds).There is therefore little possibility of expansion and water and cations cannot move between the units.

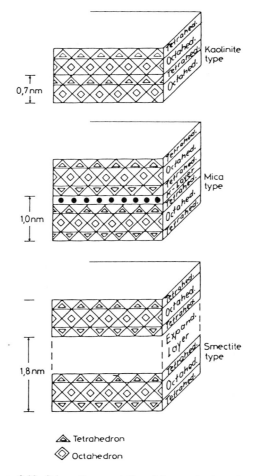

△ Tetrahedron
◇ Octahedron

Figure 2.10 Schematic presentation of three main clay minerals.

The micas differ from kaolinite in that the unit layer is made up of two tetrahedral layers with an octahedral layer held between them (see Figure 2.10) by mutually shared O atoms. This structure is typical of 2:1 phyllosilicates. A high negative charge occurs in the tetrahedral layer which is satisfied by K^+. This K^+ which is present in a non-hydrated form binds the 2:1 unit layer strongly together and thus represents an integral component of micas. The basal spacing between the unit layers under standard conditions is 1.0 nm. The two main types of mica, biotite and muscovite are different in structure. Biotite is a trioctahedral mica, and muscovite a dioctahedral mica. The term trioctahedral indicates that each centre of the octahedron is occupied by a cation (Mg, Ca, Al, Fe) whereas in the dioctahedral structure (muscovite) only two out of three octahedral cation sites are occupied (Fanning *et al.* 1989). Because of this difference in structure, both mica types differ much in stability, biotite weathering much more rapidly than muscovite. For this reason older soils contain little if any biotite.

Illite is also a 2:1 phyllosilicate which is similar in structure to mica with a basal spacing between the unit layers of 1.0 nm. Basically illite is a weathered mica but contains, however, less K^+ and more water than mica. Since its particle size is $<2\mu m$ it is also called clay mica (Allen and Hajek 1989). At the edges and also sometimes within this mineral the layers can be expanded forming "wedge zones" (see Figure 10.2) to which other cations than K^+ are adsorbed (Schroeder 1978). Because of its small particle size and the wedge zones at the edges of the mineral, it plays a substantial role in cation adsorption and particularly in the specific adsorption of K^+ and NH_4^+. Illite is mainly formed by weathering of muscovite and therefore is dioctahedral in structure. It is the most important clay mineral in European soils derived from loess and plays an important role in K^+ turnover in these soils in relation to specific K^+ adsorption and K^+ release by weathering.

In contrast to mica and illite, vermiculite is an expandable 2:1 phyllosilicate mineral so that in common with smectites (see Figure 2.10) the interlayer space may take up or release H_2O and therefore expand or shrink depending on soil moisture because the interlayer lacks the K^+ to contract and stabilize the unit layers. Hence the basal distance between unit layers, which is an important criterion of minerals, is not constant but may change depending on soil water status. Under standard conditions the basal distance between the unit layes of vermiculite is 1.4 nm. However, because of this wide interlayer zone, vermiculite possesses an inner surface and is therefore characterized by a high CEC. Additions of K^+ or NH_4^+ lead to a shrinkage of the mineral to a basal spacing distance of 1.0 nm similar to that of mica. By this process K^+ and NH_4^+ may be fixed (page 484) so that these nutrients become less available to plant roots. The phenomenon is of major importance for the fixation and release of K^+ and particularly NH_4^+ as shown recently by Steffens and Sparks (1999). Vermiculites are found in many soil orders and are present in the clay, silt and sand fractions (Douglas 1989). There are two types of vermiculites, one formed from biotite by weathering and possessing a trioctahedral structure and another with a dioctahedral structure presumably derived from muscovite.

Smectites are a further group of expandable 2:1 phyllosilicate minerals (Figure 2.10) which occur almost exclusively in the clay fraction (Borchardt 1989). Like vermiculite the mineral may shrink and expand depending on soil moisture status. As a result of the high specific surface of smectites (small particle size) and particularly the presence of inner surfaces, smectites have a high CEC (Table 2.2). The basal distance between unit layers is about 1.8 nm (Figure 2.10) and because of this larger distance, smectites in contrast to vermiculite do not fix K^+ or NH_4^+. The three principal types of smectites are montmorillonite with Mg, beidellite with Al, and nontronite with Fe as main cation species at the centre of octahedrons. Smectites occur in various soil types and are frequently found in high concentrations in alluvial soils. The loess derived soils of the Midwest of the USA have smectites as the most important phyllosilicate mineral. In most soils with a high cation exchange capacity (=CEC) smectites are the dominant clay minerals. Important properties of smectites, illite and kaolinite are shown in Table 2.2.

Table 2.2   Comparative properties of three major types of silicate clay (after Brady 1974)

Property	Type of clay		
	Smectite	Illite	Kaolinite
Size (μm)	0.01—1.0	0.1—2.0	0.1—5.0
Shape	regular flakes	irregular flakes	hexagonal crystals
Specific surface (m²/g)	700—800	100—200	5—20
External surface	high	medium	low
Internal surface	very high	medium	none
Cohesion, plasticity	high	medium	low
Swelling capacity	high	medium	low
CEC (cmol/kg)	80—100	15—0	—

Chlorites are secondary clay minerals with a vermiculite structure in which the negative charge of the interlayer surfaces is neutralized by cationic hydroxy species of Al, *e.g.* $Al(OH)_2^+$, Mg^{2+} and/or Fe^{2+} and Fe^{3+} (Barnhisel and Bertsch 1989). This neutralization stabilizes the structure of the mineral so that its swelling potential is much reduced or even completely blocked. Because of this charge neutralization, the CEC of chlorites is much lower than that of expandable 2:1 minerals such as smectites or vermiculite. The basal distance of unit layers is 1.4 nm. The cationic hydroxy interlayers may adsorb anions at their edges which may particularly affect phosphate availability. Chlorites develop from vermiculite and smectites under acid soil conditions with prevailing Al, Fe, and Mg hydroxy cationic species. Chlorites are less stable than other secondary clay minerals and as they weather in an acid environment release appreciable amounts of Mg^{2+} which may be utilized by plants.

The *allophanes* are generally regarded as amorphous, *i.e.* non crystalline showing a featureless X ray diffraction pattern. Recent research has shown, however, that they are hydrous aluminosilicates consisting of irregulary spherical particles with a diameter of 3 to 5 nm and an Si/Al ratio of 1:2 to 1:1 (Wada 1989). They occur mainly in volcanic soils and may be associated with imogolite which is a paracrystalline mineral.

34

Both allophanes and imogolite have similar chemical properties and carry variable positive and negative charges depending on pH. At pH 7 the CEC of allophanic clays is in the range of 10 to 40 cmol/kg and at pH 4 the anion exchange capacity (AEC) is from 5 to 30 cmol/kg. These values demostrate the enormous impact of pH on the adsorption potential of allophanes and imogolite for anions and cations. The physical properties of soils rich in allophanes and imogolites, the so called Ando soils, are unique. These soils have a low bulk density with a porous system which can store enormous quantities of water and their mineral-organic particles are of high stability. Such soils therefore make up some the most fertile soils in the world.

2.2.2   Soil organic matter

Mineral soil components are rather stable and changes occur mainly over geological time spans. This is not true for the organic soil component, however, which is characterized by a steady turnover and by an input and decomposition of organic matter. Figure 2.11 shows a flow diagram of organic C in soils elaborated by Bradbury *et al.* (1993). The input of organic matter basically originates from phototrophic organisms, mainly higher plants.

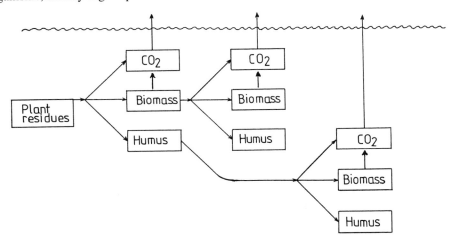

Figure 2.11   Flow diagram of organic carbon (after Bradbury *et al.* 1993).

This organic C serves as food for microbes, fungi and bacteria, and under aerobic soil conditions decomposition is indicated by the release of CO_2. A part of organic matter is easily decomposable, such as amino acids and organic anions (Yan *et al.* 1996b). Additionally non structural carbohydrates such as sugars and polysaccharides as well as proteins can be rapidly decomposed by miroorganisms. Lignified cellular material, however, is resistant to decomposition and its rate of breakdown is low. Microbial attack may thus be slowed down by thick and lignified cell walls encapsulating the more readily decomposable proteins and carbohydrates. Lignins and cellulose are broken down only by a few specialized fungi with white rot fungi being one of the most important (*Phanerochaete chrysosporium*, Carlile and Watkinson 1994, p. 272—73). Polyphenols also retard the decay of organic carbon by inhibiting enzyme activation (Handayanto *et al*. 1994).

The resistant decomposable organic C is the precursor for the formation of humic substances. The process of decomposition is not only brought about by enzymes but also by purely chemical reactions. In the formation of humic substances lignin is believed to play an important role. Humic substances are divided into humic acids (humates) and fulvic acids. Humates have a molecular weight of several thousand Da whereas as fulvic acids are much smaller with a molecular weight of several hundreds. Fulvic acids occur predominantly in acid soils. Recent investigations have shown that principal building blocks of humates are aromatic rings crosslinked by aliphatic chains carrying phenolic OH groups and carboxyl groups (Schulten and Schnitzer 1995). These groups give the humates their acidic character (Schnitzer and Skinner 1965). Because of their low pK, carboxylic acids more readily dissociate a proton at a lower pH than do phenolic OH-groups. This difference means that humates carry a pH dependent negative charge.

Humates are associated with carbohydrates and peptides which may serve as source of nutrition for microbes. The binding mechanisms of peptides and carbohydrates to humate are still not completely understood (Schulten and Schnitzer 1993). In a recent investigation Schulten and Schnitzer (1995) presented a three dimensional model of a standard humate obtained by means of numerous computer runs using recent analytical data of humates. This standard humate has the following composition:

$$C_{342}H_{388}O_{124}N_{12}$$

According to weight proportions this is 618 mg C, 59 mg H, 298 mg O and 25 mg N/g humate and with a molecular weight of approximately 7000 Da. In the model humate, nitrogen is mainly present as heterocyclic N and nitriles. A void is present in the model which traps either a hexapeptide with the sequence Asp-Gly-Arg-Glu-Ala-Lys or a trisaccharide. This organo- mineral complex is capable of binding the hexapeptide by means of 14 hydrogen bonds. The finding is of particular interest for plant nutrition because the release of the hexapeptide may quickly lead to its microbial decomposition and mineralization of the organic nitrogen. The standard humate may also bind clay minerals to form organo mineral complexes by means of bridges with Al, Fe, Ca, and Mg.

The turnover of organic C in soils depends much on environmental conditions such as O_2 supply, temperature and soil water and for this reason the concentration of organic C, mainly accumulated in the upper soil layer, differs considerably for various soil types as shown in Table 2.3. Lack of oxygen delays the decomposition and therefore hydromorphic soils and flooded soils are usually high in organic carbon. Under continental climatic conditions soils are frozen during winter so that microbial actvity is inhibited and soils rich in organic matter develop. A well known example of this kind are the Chernozems (black earths) belonging to the Soil Order Mollisols. Because of their high content of humate these soils have a dark appearence and are known as the world's most productive soils. Under tropical conditions with warm soil temperature all the year around and with adequate amounts of water for most of the time, organic matter is quickly decomposed. This is particularly the case after clearing of tropical natural forests. According to research findings of Cadish and Giller (1996), the initial level of 100% soil organic matter at the time of clearing a tropical forest dropped to a level of almost 60% under a grass-legume vegetation after cultivation for six years. The fast decomposition of organic C leads to a loss of water holding capacity and the aggregates become less stable and the soils more prone to erosion, by wind and water with ensuring loss of soil fertility (Picard 1994) which is extremely difficult to restore (Dayly 1995).

Table 2.3   Concentration of organic matter in various soil types

Soil Types	g/kg dry matter soil
Mineral soils	<20
Humic soils	20−150
Anmoorpeat	150−300
Peat soils	>300

Microbial decomposition of organic matter also depends on soil texture. As already mentioned above humates as well as other organic matter constituents may be bound to clay minerals and therefore become less susceptible to enzymic attack. Long-term field trials in Canada showed that during the 90 years of cultivation of a prairie soil the initial organic C concentration decreased by about 50% with highest losses occurring during the first 12 years. Stabilization of the organic C was brought about by clay minerals (Schulten *et al.* 1995). The close correlation between clay mineral concentration and organic matter in soils may be accounted for by the binding of the organic C to clay minerals (Loll and Bollag (1983). Soil mangement and crop species in rotation are of utmost importance for maintaining organic C levels in agricultural land. Perennial grassland leads to an increase of organic matter, whereas a decrease occurs by intensive tillage with annual crops like sugar beet, potatoes and vegetables. Frequent incorporation of straw into soils, particularly when combined with mineral nitrogen fertilizer results in an increase of soil organic matter (Amberger and Schweiger 1971). The long-term application of farm yard manure (FYM) increases soil organic C considerably (Figure 2.12) as was found in the classical experiments over 150 years at Rothamsted in the UK (Jenkinson *et al.* 1994). Microbial biomass is the biomas of living and dead biomass of soil bacteria and fungi was also doubled by FYM application in the Rothamsted field trials (Jenkinson and Rayner 1977). Most of the organic matter input in the form of FYM is oxidized. Michael and Djurabi (1964) found in a 25 year old field experiment that of the total of 52 t C/ha applied in the form of FYM only 6 t/ha were recovered in the form of soil organic matter and not even 1% of the total C input was found as humate. Similar results were reported by Schmalfuss and Kolbe (1963) from the 80 year old field trials of the "Ewiger Roggenbau" (continuous rye cultivation) in Germany.

The long-term field trials in Illinois (USA) clearly showed that crop rotation also influenced the content of soil organic matter (Odell *et al.* 1984). The level of soil organic C and N was highest in the rotation maize — oats — clover, and lowest in the permanent maize rotation. Also in this US field trial (Morrow Plots) the treatment with FYM + lime produced by far the highest level of soil organic matter. This observation is in line with recent investigations of Yang and Janssen (1997) in northern China who emphasize that straw and FYM are important organic inputs which are neccessary for maintaining or even increasing the content of soil organic matter. Their model describing the rate of decomposition of organic matter incorporated into the soil *vs* time shows that the rate decreases the longer the organic matter is in the soil.

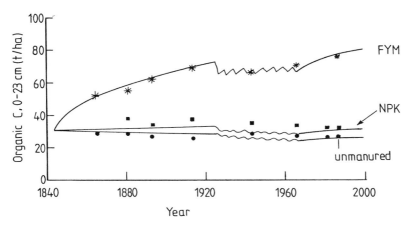

Figure 2.12 Effect of long-term farm yard manure application and mineral fertilizer on the amount organic C in the top 23 cm of soil in Mg/ha (after Jenkinson *et al.* 1994).

Organic matter in soils has two principal functions; a nutritional one resulting from mineralization of organic nitrogen, sulphur and phosphorus and a physical one relating to the improvement of the physical properties of soils, such as water holding capacity, water infiltration rate, stabilization of soils and thus a protection against wind and water erosion. The physical functions are mainly attributable to the humates and particularly to the aggregates they form with clay minerals. Nutrient supply as a consequence of mineralization is mainly related to the decomposition of microbial biomass (see Figure 2.11) but, as discussed above, humate complexes may also release organic nitrogen susceptible to mineralization. In addition, organic matter constituents carry a negative charge and are hence able to adsorb cations. Cation adsorption is non specific and follows the Hofmeister cation sequence (see page 19). Divalent ions are therefore preferentially adsorbed to monovalent cations. In this respect the behaviour of H^+ is exceptional because the binding of H^+ to these groups represents the formation of a chemical bond. The cation exchange capacity (CEC) of humates is in the range of 200 to 400 cmol/kg. This seems very high compared with that of the clay minerals. This CEC value, however, is based on weight and as the density of organic matter is considerably lower than that of the clay minerals, organic soils *in situ* often do not have such a high CEC as soils rich in clay. In humic mineral soils frequently about 50% of the CEC is made up of organic matter in the surface layer. As shown in Table 2.3, however, soil organic matter concentration can differ considerably depending on soil type. The contribution of organic matter to the CEC of soils can therefore vary substantially between soils.

2.2.3   Soil structure

Soil structure may be defined as the arrangement of soil particles into groups or aggregates (Brady and Weil 1999). The capability of the soil to form aggregates, the size and shape of aggregates, and the stability of the aggregates produced is closely related to the colloid content of the soil. Light textured soils low in soil colloids are structureless as the coarse sand material does not form aggregates.

The higher the clay concentration the more important is soil structure. Clay minerals adsorb water which causes swelling of the soil. Swelling and shrinkage depend on the water available. Thus clay soils swell under wet conditions and shrink under dry conditions producing cracks and fissures in the profile. This behaviour is typical of soils rich in smectites as this clay mineral readily adsorbs water. An example of soils in which clay swelling and drying occurs are the Tropical Black Earths. These soils have a high clay concentration consisting almost exclusively of smectites. In the wet season they swell and become sticky whereas in the dry season they dry out to such an extent that they turn rock hard and crack. Despite the high nutrient content of these soils, the poor structure therefore limits their agricultural potential because they are so difficult to work (Plate 2.1).

Shape and particle size depend to a large extent on the type of clay minerals present in the soil. Smectite clays tend to produce prismatic angular structures whereas kaolinite and hydrous oxides are associated with more granular aggregates.

Plate 2.1   Shrinking and cracking of a black cotton soil in India (Photo: Mengel).

Soil structure develops from the flocculation of soil colloids (see page 28) and the formation of aggregates in which the flocculated colloids are bound to mineral particles and organic matter. Binding mineral components to organic matter is mainly by cation bridges of Al, Fe, Mg, and Ca (Schulten *et al.* 1995). A porous system is thus built up which is of utmost importance for water storage and for the supply of plant roots with air and water, in which the mineral plant nutrients are dissolved. Both in the formation and stability of soil aggregates, organic matter plays a decisive role as can be seen in Figure 2.13 which depicts the relationship between concentration of organic C in soils and maximum bulk density for a number of soils (Thomas *et al.* 1996).

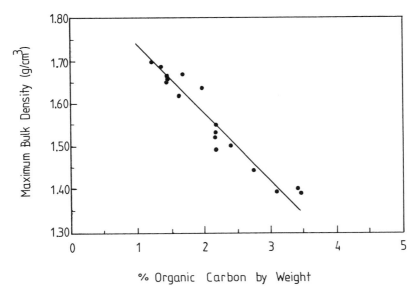

Figure 2.13   Relationship between soil organic C and the maximum bulk density of soils (after Thomas *et al.* 1996).

The bulk density of a soil may be definied as the weight of an undisturbed soil volume per unit volume, *i.e.* g soil/cm^3, a low bulk density indicating the presence of a high proportion of pore space and *vice versa*. This is at least true for mineral soils with a relatively small proportion of organic matter which by nature has a low specific weight. Figure 2.13 shows the "maximum bulk density" values of soils with a different tillage history. The density values were obtained after the soils had been exposed to a standard compaction procedure. The data therefore also provide information about aggregate stability; the lower the value the more stable the aggregates. As shown in Figure 2.13 there is a close negative linear correlation between the maximum bulk density and the concentration of organic C in the soils. Thus soils were more resistant to compaction the higher their concentration in organic matter. Thomas *et al.* (1996) emphasize that this relationship was not only due to the lower specific weight of the organic matter but also the influence of the organic matter in stabilizing the soil structure. A well known example of soils with a very good soil structure are the Andisols (v.Wambeke 1991). In these soils, relatively frequent in Japan, aggregates are formed by the flocculation of allophanes and Fe/Al oxides/hydroxides which bind to humates and protect the organic matter from decomposition. The soils are characterized by a crumb and porous structure which allows a high storage of water and which can be easily penetrated by plant roots.

Flocculation of soil colloids is a prerequisite for a good soil structure. Sandy soils have no real structure because they lack mineral soil colloids, mainly clay. In these soils, structure is not a major problem because they are pervious and can easily be penetrated by plant roots. For soils with higher clay concentrations, particularly in the form of 2:1 clay minerals, the cation exchange sites need to be saturated with Ca^{2+} of 60 to 80% of the CEC in order to maintain a satisfactory structure. For the more acid tropical soils (Oxisols) with a high concentration of Fe/Al oxides/hydroxides, a Ca saturation of the CEC of 20% is sufficient due to the flocculating effect of Al and Fe cationic species (Broyer and Stout 1959).

Table 2.4 Percentage proportions of various cation species saturating soil colloids of different soil orders (modified after Hoagland 1948)

Soil order	Climate	Na^+	K^+	Mg^{2+}	Ca^{2+}	H^+ or cationic Al
Aridisols (alkali soils)	arid	30	15	20	35	0
Mollisols (cherozem)	semi-arid	2	7	14	73	4
Spodosolo (podsol)	humid	trace	3	10	20	67

Cation saturation of soil colloids depends much on soil type and prevailing climatic conditions. This is shown in Table 2.4, in which cation saturation is compared in three very different soil types. In the alkali soil the soil colloids are saturated by a high proportion of alkali cations and by Na^+ in particular. Such soils obviously have a poor structure (see page 28). The cation saturation of the chernozem soil represents an ideal situation where Ca^{2+} is the dominant cation species. The particularly good granular crumb structure of these soils is well known. In the podzol H^+ and Al cationic ion species are present in excess of other cation specices. The prevalence of Al cationic species has already been discussed on page 18. It may amount to 20 to 50% of the CEC as found by Bromfield et al. (1983) in acid mineral soils.

Soil structure also depends on the vegetative cover. Under permanent grassland in particular, very good soil structures occur. This results from the effect of the high organic matter content and soil fauna. In this respect the earthworm contributes considerably to the formation of stable aggregates. Arable soils are often lower in organic matter and soil fauna and for this reason often are poorer in soil structure.

2.2.4   Soil water

The availability of water to plants depends generally on two major soil factors. These are the total water content in the rooting zone and the extent to which the water present is bound to soil particles. The degree to which water is bound to soil particles is expressed as soil water potential, which in older terminology was called water tension. Water tension was generally measured in terms of pF. In this older concept the suction of water to soil surfaces is expressed in terms of height (cm) of a water column the pressure of which is equal to the suction. Thus a high column is associated with a strong suction. The pF value is equal to the common logarithm of the height of the water column measured in cm. A water column of 10 cm is thus equal to pF 1. pF values are positive in contrast to water potential values which are mostly negative.

The concept of the water potential was introduced as a basic means of describing water status and water movement. The concept is particularly useful in allowing a uniform treatment of soil and plant water relations. It is considered in more detail on page 181 in connection with the water potential in plant tissues. As already outlined when discussing water adsorption on page 26 water potential is an expression of the thermal energy of liquid water (Brown's molecular movement). The water potential of pure water at room temperature and atmospheric pressure is defined as zero. The most important process restricting the movement of water molecules is their adsorption to surfaces as their potential to move is restricted and hence the potential is below zero. Adsorbing surfaces are soil particles, aggregates, colloids, walls of soil pores as well as solutes such as molecules and ions dissolved in soil solution. Water adsorbed to solid soil particles is called matrix water since it is adsorbed to the soil matrix and water adsorbed to the surfaces of solutes is osmotically bound by which water movment is also resticted. Hence soil water potential is influenced by soil matrix forces and by osmotic forces. In the old water terminology (pF scale) soil water is regarded as being pressed to soil surfaces which is not correct; the soil water is virtually sucked to soil surfaces a phenomenon which may be regarded as a negative pressure. Therefore water influenced by adsorption processes and not exposed to a hydrostatic pressure other than that of the atmospheric has a negative water potential. Water flow is always from the higher to a lower water potential since the potential difference is the driving force for water flow.

In modern terminology the Pascal (Pa) is the unit of pressure. 1 bar = 10^5 Pa = 10^2 kPa = 0.1 Mpa. The unit is named after the French philosopher Blaise Pascal who was the first to measure atmospheric pressure. Water potentials in soils are usually expressed in terms of kPa, whereas for plants MPa is used.

Water bound at a tension of pF 3 — a suction pressure of 1 bar — is equivalent to a water potential of -100 kPa. The relationship between pF and kPa is given by the equation:

$$10^{pF-1} = kPa$$

To a large extent water supply to plants is regulated by water retention and water movement in soil. The water holding capacity rises as the concentrations of inorganic and organic colloids increase (page 44). In addition water is retained by small and medium sized pores. Soil structure therefore directly affects the soil water holding capacity (WHC) which is an important characteristics of soils. The water holding capacity under field conditions is known as *field capacity*. Veihmeyer and Hendrickson (1931) defined the field capacity as "the amount of water held in the soil after the excess gravitational water has drained away and after the rate of downward movement of the water has materially decreased". The water potential at field capacity ranges between -10 to -30 kPa. The technique of measurement of the field capacity is to saturate a soil completely with water so that all pores are water filled and then to allow the soil to drain off for a period of about two to three days under conditions where no evaporation occurs. The amount of water remaining in the soil represents the field capacity. This is expressed as unit weight of water in the unit dry weight of the soil. If for example the soil contains 200 g water per kg dry soil the field capacity is 200 g/kg. For organic soils measurement by volume is preferable because of the high water holding capacity and the low density of the dry soil. Field capacity gives an indication of the storage potential of a soil for water. This corresponds to the capability of a soil to supply plants with water during dry seasons.

The supply of plants with water does not only depend on the field capacity but also on the strength by which water is adsorbed to soil particles. As already pointed out on page 26, water layers directly adjacent to the adsorbing surfaces may be bound by forces as high as -100 kPa and higher.

Water present at field capacity includes adsorption water and water held in the capillary soil pores. This latter fraction is termed capillary water and the intensity by which it is held depends on the diameter of the pores. As the diameter decreases the water binding or suction tension becomes stronger (Table 2.5). Pores with a diameter in excess of 0.05 mm are too large to retain water after the soil has been brought to field capacity and they are termed non capillary pores. The forces binding capillary water are related to the adsorption of water molecules to the walls of the pores and the cohesive forces by which the water molecules are attracted to each other. As a result of adsorption and cohesive forces water in a soil can rise in the same way as water in a capillary tube. The height of the capillary rise increases as the diameter of the capillary becomes smaller. This capillary rise of water is of importance in the water supply of plants particularly when the water has to be transported from the deeper soil layers.

Table 2.5   Relationship between the diameter of soil pores and the water potential of the water held in the pores (De Boodt and De Leenheer 1955)

Size of pores	Diameter, μm	Water potential, kPa
Coarse	>50	> -6.3
Medium	50 to 10	-6.3 to -31
Fine	10 to 0.2	-31 to -159
Very fine	>0.2	-159 >

Plants growing in soil first absorb weakly bound water. This is the fraction with a relatively high but already negative water potential and which is most available. As water is taken up, the remaining soil water becomes progressively less available as it is more strongly held. Thus as a soil dries out the water availability declines finally reaching a point at which the water is so strongly held by adsorption that plant roots are unable to utilize it, and plants growing in the soil begin to wilt. The water potential at which wilting occurs is called the wilting point. One may distinguish between a temporary wilting point and a permanent *wilting point*. In the former case plants are able to recover when water is supplied to the soil, whereas when the permanent wilting point has been reached wilting is irreversible and the plant dies. The wilting point can not be precisely defined in terms of water potential or water content of the soil. Generally, however, for many plant species the permanent wilting point is reached when the water potential is in the order of about -1000 to -1500 kPa. The exact value depends on plant species and environmental conditions.

Water bound by forces corresponding to water potentials lower than -1000 to -1500 kPa is thus not available to plants. The maximum amount of available water which a soil can contain is therefore the difference between the water at field capacity and the water fraction held by forces higher than -1000 kPa. It is clear that soils containing high amounts of clay and thus having a high surface area capable of adsorption, also contain appreciable amounts of water which is unavailable to plants.

The relationship between water amount (g H_2O/kg soil) and water potential (kPa) is shown in Figure 2.14 for a clay soil and a silt loam soil. In the clay soil the fall in water potential from -20 to -80 kPa is associated with only a small decrease in soil water content. Even at the very low water potential of -1500 kPa (wilting point), 260 g water/kg soil are unavailable to the plant. For the silt loam soil the relationship between soil water content and soil water potential is quite different. The drop in water potential is accompanied by a substantial decrease in soil water content and at a water potential of -1500 kPa, the soil water content amounts to only 80 g H_2O/kg soil. This shows that the water in the silt loam can be depleted by plant uptake to a much higher degree than the water in the clay soil (Heatherly and Russell 1979).

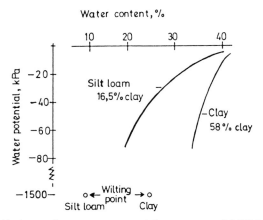

Figure 2.14 Relationship beteen the water content and the water potential (kPa) of a clay soil and a silt loam soil (after Heatherly and Russell 1979).

The amount of available water that can be stored by soils is that water held between the permanent wilting point (-1000 to -1500 kPa) and the field capacity (-10 kPa). From Figure 2.14 it can be calculated that the storage capacity of available water is much higher in the silt loam soil than in the clay soil. In the latter, total water storage capacity is high, but only a small proportion of the storage water is available for plants. In sandy soils the reverse is true. Water storage capacity is low, but most of the stored water is available.

Water storage capacity of a soil not only depends on the difference of stored water between field capacity and wilting point but also on the rooting depth of a soil profile *i.e.* the soil depth penetrated by plant roots. On shallow soils the proportion of a unit soil volume capable of storing available water may be high, yet the total water stored is low. Assuming a homogeneous soil depth of 1m penetrated by roots and a storage volume for available water of 100 L/m^3 soil, the soil can store 100 L of water per m^2 surface which is exactly the quantity which falls with 100 mm of rainfall per m^2 soil surface. Hence the storage capacity for available soil water may also be expressed in mm rainfall which can be stored by a soil in an available form. This measure for available water has the advantage that the rooting depth and the properties of soil horizons are also included because top soil and subsoil layers may differ in their potentials for storing plant-available water. Spodosols usually have a poor capacity for storing water because their field capacity is low and also their rooting depth is restricted. Andosols in contrast may have rooting depths of 2 m associated with a high storage capacity for available water per soil volume. Table 2.6 shows water storage capacities of plant-available water for different soil orders.

Table 2.6 Plant-available water capacity of some soil orders (after Scheffer and Schachtschabel 1982 and v.Wambeke 1991)

Soil order	Available water, kg H$_2$O/m^2 soil surface = mm rainfall
Spodosols (sandy soils, mainly acid	50—120
Alfisols (silty soils, mainly derived from loess)	170—220
Vertisols (soils rich in clay)	50—150
Andosols (rich in organo-mineral aggregates)	120—260

Storage capacity of soils for plant-available water is a very important soil criterion which in many cases plays a decisive role in the productivity of soils. In numerous regions worldwide, crop yield level is primarily dependent on the available soil water. This is particularly true for rain-fed agriculture not only under arid and semi arid climate conditions but even under humid climatic conditions (Gales 1983).

In addition water movement in soils also plays an important role in the water supply of plants. Downward movement caused by gravity occurs only when the upper soil layers have higher water contents than field capacity. This downward movement brings about the leaching of nutrients into the deeper soil layers and out of the profile. This excess water is usually of no great importance in plant nutrition. The upward movement of water in the soil, results from capillary forces. Generally in coarse textured soils a substantial fraction of the pores are of large diameter, whereas in fine textured soils there is a higher proportion of smaller pores. The rise in water brought about by larger pores is generally more rapid but the height to which the water rises is lower than in soils containing smaller pores. This relationship was observed by Wollny (1885) in the last century.

This finding is of practical importance. Medium textured soils are able to transport water from the ground water table to the upper layers of the soil more readily than sandy soils. On the other hand, in fine textured soils (clay soils) the capillary rise from the ground water is often so slow that it is unable to meet the water demands of plants. Because of the higher capillary rise of water in medium textured soils, relatively deep ground water may form part of the water supply of crops growing in these soils.

Water supply to plants only becomes critical when the water status of the soil is below field capacity. The amount of water which becomes accessible to plants under these conditions by interception as roots force their way through the soil is relatively low in relation to the total demand. Water therefore must be transported to the plant roots. Capillary water rise and the lateral movement of water due to capillary action are thus of importance.

2.2.5   Soil atmosphere

The composition of soil air differs from that of the atmosphere. The CO_2 level of the atmosphere is about 300 mL/m^3 air whereas in the soil it is higher and in the order of 2 to 10 L/m^3 air in the surface layers. Soil air also contains a correspondingly lower O_2 content of about 200 L/m^3 air as compared with that of 210 L/m^3 air in the atmosphere. Higher levels of CO_2 result from the respiration of living organisms such as plants and particularly plant roots and aerobic living fungi and bacteria which consume O_2 and release CO_2. The main function of respiration is the generation of ATP and therefore a shortage of O_2 severely interferes with the energy metabolism of aerobic living organisms and may be lethal.

Oxygen and water supply to terrestial plants are interrelated. Both are at optimum at field capacity of soils *i.e.* at a water potential of -10 to -30 kPa (see page 42). At field capacity coarser soil pores are filled by air and the smaller ones by water so that both air and water may easily reach plant roots. Soil water concentrations higher than field capacity depress the O_2 supply to roots and may finally lead to anaerobiosis *i.e.* a complete lack of ogygen. Such waterlogging conditions frequently occur transiently after heavy rainfall, in a humid temperate climate in spring. Plants affected by waterlogging are much retarded in growth, show a premature leaf senescence, the chlorophyll is destroyed and the leaves have a greyish-brown appearance. Oxygen shortage may also occur after excessive irrigation.

Under anaerobic conditions, the pyruvate produced in the glycolytic pathway of roots is converted to ethanol which in some plant species is stored in the roots, and by others excreted into the outer medium, the rhizosphere. Anaerobiosis is detrimental to the energy status of plants (Drew 1988). Their energy charge, usually in the order of 0.80 to 0.95, may fall too as low as 0.2. Energy charge is the ratio

ATP+½ADP/ATP+ADP+AMP. Under such conditions the metabolic processes in root tips are especially much impaired. As the consequence of a lack of ATP, synthesis of amino acids and proteins is almost blocked and the polysomes dissociate. Anaerobiosis leads to a decline of cytosolic pH and an increase of the vacuolar pH because of a back diffusion of H^+ from the vacuole into the cytosol as a consequence of a shortage of ATP and pyrophosphate required to drive the tonoplast proton pump (page 119). The plasmalemma ATPase is also retarded by a lack of ATP and the uptake rates of cations and anions are therefore severely depressed. This is the main reason why plants suffering from anaerobiosis have low concentrations of N, P, and K^+. Not all plant species are sensitive to anaerobiosis. The mechanism for this resistance to low oxygen supply has not yet been completely elucidated. Such species are characterized by high carbohydrate reserves and it is supposed that these species are capable of providing sufficient ATP by the glycolytic pathway. Rice (*Oryza sativa*) is not one of these resistant species. The tolerance of rice to anaerobic soils lies in its ability to allow the diffusion of air from the atmosphere through the aerenchymatous tissues of leaves and roots to the rhizosphere (see page 48). A useful review article on the question of oxygen deficiency in plants and plant nutrition was published by Drew (1988).

Under anaerobic conditions aerobic microorganisms are much depressed and anaerobs dominate with metabolic end products frequently toxic to plants such as ethylene, methane, hydrogen sulphide, cyanide, butyric acid and a number of other fatty acids. Plants affected by these toxins are impaired in growth and often show wilting symptoms. The detrimental effect of waterlogging on plant growth is frequently more severe than can be accounted for by a simple lack of O_2. Useful reviews dealing with plant growth and nutrient relationships in waterlogged and submerged soils have been published by Marschner (1972) and Ponnamperuma (1972).

A measure of reducing conditions in submerged soils can be assessed by the redox potential. This is regulated by the concentrations of reduced and oxidized substances according to the following equation:

$$E = E_0 + RT/nF \cdot \ln (Ox)/(Red)$$

where

(Ox) = Concentration of oxidized substances, R = Gas constant

(Red) = Concentration of reduced substances, T = Absolute temperature

Eo = Standard redox potential, F = Faraday constant

E_0 is equal to E, if the ratio (Ox)/(Red) is 1. n = Valency

The term (Ox) comprises all ions molecules and other substances which are capable of accepting electrons (e^-) and the reverse is true for the term (Red) which comprises all redox partners which donate e^-. The redox partners may be of pure chemical but also of biological nature. Both types are present in soils. Fe^{III} or Mn^{IV} may function as e^- acceptors, Fe^{2+} or Mn^{2+} may serve as e^- donors. The index in Roman numbers (Fe^{III}, Mn^{IV}) means that the Fe or Mn is in a solid state and not dissolved in the soil solution; the index with Arabic numbers indicates that the redox partner is dissolved.

Fe^{III}/Fe^{2+} and Mn^{IV}/Mn^{2+} are basically pure physico-chemical redox systems but frequently microbes are involved in the transfer of e^- from the donor to the acceptor (Munch and Ottow 1983). In aerobic soils with an active microbial life the most important e^- donor is cytochrome c oxidase which mediates the e^- transfer in the respiration chain to O_2 a very strong e^- scavenger. It is for this reason that aerobic soils do not have a surplus of e^-.

The redox potential in soils is generally measured using a platinum electrode against a reference electrode and is expressed in terms of voltage. Often the potential thus obtained is denoted as 'Eh' rather than 'E'. In modern terminology the term *pe* is used for the redox potential of flooded soils (Savant and De Dattta 1982). "e" means the concentration or more accurately the activity of reducing electrons. Since this activity covers a broad range it is expressed as negative decadic logarithm and hence it is analogous to the pH scale. Redox potentials expressed in E (voltage) can be converted into pe units using the equation:

$$E \text{ (Volt)}/0.0591 = pe \text{ or } E \text{ (Volt)} \times 16.9 = pe$$

The redox potential may be positive or negative. The lower the value, the higher is the reducing power. Therefore regardless whether the redox potenial is expressed as E (Volt) or pe, a negative potential means a strong reducing power. A pe profile of a paddy soil is shown in Figure 2.15.

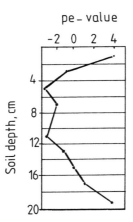

Figure 2.15   Redox potentials expressed as pe in a paddy soil profile (modified after Savant and De Datta 1982).

In the thin upper layer of several cm, the pe of 2 to 4 represents a low e^- density of 1/100 to 1/10000 mol e^-/L and therefore aerobic conditions prevail. In the profile depth of 5 to 11 cm, the pe is about -2 which is an e^- density of 100 mol/L indicative of highly reducing conditions. Below a depth of about 11 cm the redox potential increases in an almost linear way indicating that the reducing power of the soil is declining.

During rice cultivation paddy soils are kept under submergence to a depth of several cm. As soon as the soil is flooded anaerobic conditions set in and a specific sequence of reaction steps can be observed which can be divided into two stages. These two stages and the individual steps are shown in Table 2.7. Anaerobiosis begins with the disappearance of O_2 and the microbial reduction of NO_3^- to N_2 and N_2O (denitrification, see page 349). In the denitrification process the O of the NO_3^- is used as e^- acceptor. Some aerobic bacteria may adapt to anerobiosis by expressing an additional enzyme which mediates the e^- transfer from the respiration chain to nitrate. In the next steps Mn^{IV} and Fe^{III} are reduced to Mn^{2+} and Fe^{2+}. This reduction is associated with a pH increase because it consumes H^+ as shown for Fe in the following example:

$$Fe^{III}(OH)_3 + e^- + 3H^+ \Rightarrow 3H_2O + Fe^{2+}$$

$$M^{IV}O_2 + 2\ e^- + 4\ H^+ \Rightarrow Mn^{2+} + 2H_2O$$

The dissolved Mn^{2+} and Fe^{2+} may be translocated by percolating water into deeper soil layers with a higher redox potential and here become oxidized and thus precipitate and gradually form a less permeable layer (Gong 1986). Manganese and Fe^{III} reduction are to a large extent biological processes brought about by microorganisms which use Fe^{III} and Mn^{IV} as e^- acceptors for respiration (Munch and Ottow 1983). This first stage of reduction in paddy soils is not detrimental to rice plants provided that Fe^{2+} and Mn^{2+} concentrations do not reach toxic levels (van Bremen and Moormann 1978). The second step of reduction is associated with a substantial drop of redox potential indicating that reducing substances have been produced. At this stage sulphate is reduced to sulphide (see page 437) and at an even lower redox potential molecular H_2 and methane are formed. Accumulation of additional toxic substances such as formic acid, acetic acid, propionic acid and isopropionic acid and butyric acid occur at this low redox potential. At low pH these fatty acids are protonated and may penetrate the plamalemma of root cells thus depressing the cytosolic pH which is greatly detrimental to cell metabolism. The toxicity of these acids increases with their chain length (Tadano and Yoshida 1978). Plants suffering from the toxic substances are much restricted in growth, show a premature senescence and chlorosis as well as necrosis occurs beginning at the tips and margins of leaves. This kind of toxicity in rice is known as suffocation disease.

This second stage should be avoided in rice cultivation. Soils to which organic matter has been applied or which are naturally high in organic matter are prone to low redox potentials due to the fact that the organic matter favours the growth and metabolism of anaerobic microorganisms (Pomnamperuma 1965). Low redox potentials may lead to very high Fe^{2+} concentrations which are harmful to rice plants and induce a bronzing of the plants (see page 564). Plants suffering from Fe toxicity are characterized by extremely high Fe concentrations $(300-1000\ \mu g/g$ dry weight) as well as high Mn concentrations $(>1000\ \mu g/g$ dry weight). Soils on which this kind of iron toxicity occurs often have a low cation exchange capacity and are poor in Ca and K (Ottow *et al.* 1991). The appearance of Fe toxicity depends on the oxidizing *power* of the rice root which results from the excretion of O_2 by roots and induces a rise in the redoxpotential in the shizospheres and hence Fe^{2+} is oxidized. The Fe^{3+} is partially precipitated as Fe oxide at the root surface giving the roots a red/brown colour (Trolldenier 1988). This colour is indicative of healthy roots. Under anaerobic conditions FeS is precipitated at the root surface and the roots are black as shown in Plate 2.2. The oxidizing power of rice roots is due to the aerochyma, a porous root tissue with large intercellular spaces. Air is translocated through the spaces from the leaves into roots and can easily diffuse and even be excreted into the rhizosphere thus rendering the the soil layer adjacent to the root surface aerobic. This aerochymatous tissue is a particular characteristics of rice roots which enables rice to grow under flooded conditions. When rice is inadequately supplied with K the oxidizing power of roots is poor and Fe toxicity is likely to occur (Trolldenier 1973).

Table 2.7   Steps of microbial metabolism in waterlogged soils (Takai *et al.* 1957)

Step	Main reaction	Initial redox, potential (V)
	First stage	
1st	O_2 disappearance	+0.6 to +0.5
2nd	Nitrate reduction	+0.6 to +0.5
3rd	Mn^{2+} formation	+0.6 to +0.5
4th	Fe^{2+} formation	+0.5 to +0.3
	Second stage	
5th	sulphide formation by sulphate reduction	0 to -0.19
6th	H_2 formation	-0.15 to -0.22
7th	CH_4 formation	-0.15 to -0.19

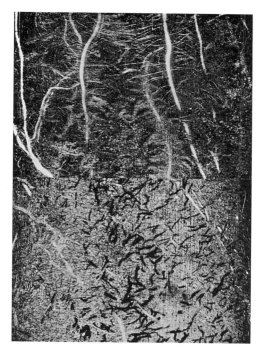

Plate 2.2 Upper part, normal rice roots growing under an optimum redox potential; lower part, rice roots affected by too low a redox potential, lateral roots are covered with a black coating of FeS (Photo: Trolldenier).

In paddy soils a characteristic profile may be observed as shown diagramatically in Figure 2.16. At the uppermost surface of the flooded soil there is a thin zone with a maximum depth of 1 cm. This is known as the oxidized layer as aerobic conditions prevail due to oxygen from the water. The layer is a reddish colour, Fe being present mainly in the Fe^{III} form. This tendency to oxidation is also indicated by a rather high redox potential (E > +0.4 V, pe > +6.8) and N is present as NO_3^-. Underlying this layer is a broader zone which is exclusively anaerobic and is blue grey in colour due to the presence of Fe^{2+}. It is here that NO_3^- originating from the oxidized layer is reduced to N_2 or N_2O. As these compounds are volatile they may be lost from the system. It is for this reason that NO_3^- fertilizers are not recommended for paddy soils (Matsubayashi *et al.* 1963).

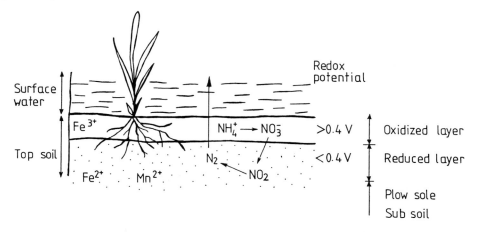

Figure 2.16 Soil profile of a submerged paddy field soil (after Matsubayashi *et al.* 1963).

Ammonium N regardless of whether applied as fertilizer or produced during the decomposition of organic N compounds, can also be lost *via* denitrification. Patrick and Reddy (1977) have demonstrated that NH_4^+ may diffuse from the lower layer to the oxidized thin surface soil layer where it can be oxidized to NO_3^-. If this NO_3^- is then transported back into the reducing deeper soil zone it can be denitrified and thus lost. In this process NH_4^+ diffusion in the soil is the limiting step (Reddy and Rao 1983).

Submergence also influences the availability of plant nutrients other than N (Ponnamperuma 1978). Generally phosphate availability increases. This is due partially to the release of occluded phosphates after the reduction of Fe^{III} to Fe^{2+} on the Fe^{III} oxide skin (see page 457) and also partially to hydrolysis of $Fe(OH)_3$. In addition the dephosphorylation of inositol hexaphosphate is promoted by submergence (Dalal 1977). The concentration of soluble cations rises as a result of cation exchange induced by soluble Fe^{2+} and Mn^{2+}. Hydrogen sulphide produced under reducing conditions forms precipitates (sulphides) with Fe, Cu, Zn, and Mn and thus depresses the availability of these nutrients. By this formation of FeS, plants are protected from toxic levels of Fe^{2+}. The formation of sulphides has no major influence on S availability, as sulphides can be oxidized in the rhizosphere by bacteria (Ponamperuma 1972). As mentioned above, flooding results in an increase of pH because the reduction of Fe^{III} and Mn^{IV} as well as the reduction of nitrate (see page 54 and 350) require H^+. Calcareous soils and sodic soils (Aridisols), however, decrease in pH when flooded. These soil types are characterized by a high pH in which the dissolution of CO_2 in water forms H^+ and HCO_3^- (bicarbonate = monohydrogen carbonate) according to the following reaction:

$$H_2O + CO_2 \Rightarrow H^+ + HCO_3^-$$

At high pH the equilibrium of the reaction shifts to the right side of the equation and it is for this reason that only at elevated soil solution pH does the dissolution of CO_2 produce H^+. The source of CO_2 is microbial and plant root respiration. Respiration rates in flooded soils, however, are generally lower than in aerobic soils because of the limited O_2 supply under submergence. According to Mikkelsen et al. (1978) the pH of the flooding water follows a diurnal rhythm with a steep increase during the day when pH levels as high as 10 may be attained and a pH decline beginning at sunset. This typical diurnal pH pattern is only observed if photosynthetically active algal growth prevails in the flooding water. It is assumed that these organisms split bicarbonate into CO_2 and OH^- by means of a carbonic anhydrase (Badger and Price 1994), (see the equation below), CO_2 is used for photosynthesis and the remaining OH^- increases the pH.

$$HCO_3^- \Leftrightarrow CO_2 + OH^-$$

The pH increase favours the formation of ammonia which is volatile and therefore escapes into the atmosphere. As can be seen from the following equation the equilibrium between ammonia and ammonium is the shifted to the side of ammonia the higher the pH.

$$NH_4^+ \Leftrightarrow NH_3 + H^+$$

Considerable amounts of NH_4-N can then be lost from the soil by NH_3 volatilization particularly after ammonium or urea application on the flooding water of a young rice crop (Schnier et al. 1988) but also from calcareous and altaline soils. Useful monographs on nitrogen dynamics in flooded rice soils have been published by Savant and De Datta (1982) and by De Datta and Patrick (1986).

2.2.6 Soil pH

The H^+ concentrations in soil solutions as well as in biological liquids are absolutely low (<1 mol per L) and comprise a broad range and are therefore measured on a logarithmic scale. The H^+ concentration is expressed in mol H^+/L. The pH is defined as the negative decadic logarithm of the H^+ concentration since using this definition gives the pH as a positive value.

$$pH = \log (1/H^+)$$

Soils with a pH around 7 have a neutral pH and soils with a >7 are alkaline. Acid soils have a pH around 5 to 6 and strongly acid soils a pH lower than 5. In soils on a world scale the H^+ concentration may vary between 10^{-3} and 10^{-10} which means pH 3 to pH 10. Cultivated soils generally have a pH range from 4 to 8. A soil pH scale is shown in Figure 2.17 (Brady 1984)

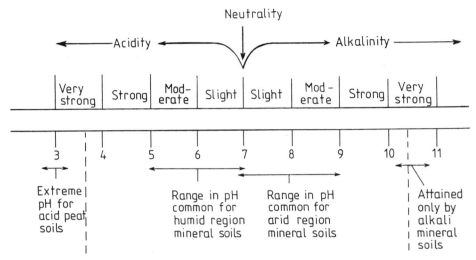

Figure 2.17 Scale of soil pH (modified after Brady 1984).

The H^+ concentration of the pH scale relates to the dissolved H^+ which is mainly the H^+ in the soil solution and not the H^+ adsorbed to soil colloids. Both, however, are related to each other and form a H^+ buffer system as will be considered below. Generally the pH in soils is measured in soil suspension obtained with water or KCl or $CaCl_2$ extracts. Extraction with water better reflects the H^+ concentration of the soil solution than extracts with the KCl or $CaCl_2$ solution which give somewhat lower pH values because of the exchange of some of the adsorbed H^+ by K^+ or Ca^{2+}. The water extract may give slightly lower H^+ concentration than those in the soil solution because the water extract dilutes the soil solution. This difference, however, is of no practical importance.

In analogy to the intensity/quantity concept (I/Q ratio) for plant nutrients (see page 74) the H^+ concentration in the soil solution may be considered as an intensity component and the adsorbed H^+ as the quantity component. The H^+ concentration in the soil solution is represented by the actual acidity which has a direct impact on physico-chemical as well as biological processes. The adsorbed H^+ together with adsorbed Al cation species represent the potential acidity which is the quantity component. This also includes Al cation species since these consume H^+ when titrated with OH^- as shown on page 53. The potential acidity in cultivated soils is much higher than the actual acidity the latter generally is <1 mmol/kg soil, the former being in a range of 10 to 100 mmol/kg soil.

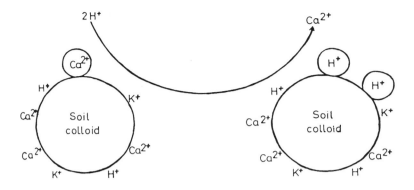

Figure 2.18 Principle of H⁺ buffering, 2H⁺ exchange for another cation species (Ca²⁺).

The principle of this buffering process is shown in Figure 2.18. Hydrogen ions exchange for other cation species adsorbed to soil colloids. From this it is clear that soils rich in organic and inorganic colloids have a high buffer capacity for H⁺, provided that a high proportion of the exchange capacity is saturated with cation species other than H⁺ and/or Al cation species. The H⁺ buffer power of a soil may be defined as the quantity of H⁺ required to depress the soil pH by one unit or the quantity of OH⁻ required to increase the soil pH by one unit. This relationship is shown for a medium-textured soil in Figure 2.19 (Yan *et al.* 1996a). In this particular case the curve is linear because the ordinate is on a logarithmic scale.

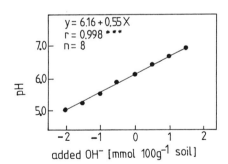

Figure 2.19 Relationship between soil pH and the addition of bases (= OH⁻), modified after Yan *et al.* (1996a).

Soils may contain different inorganic and organic substances capable of neutralizing and therefore of buffering H⁺. These various H⁺ buffer systems are shown in Table 2.8 (Schwertmann *et al.* 1987).

Table 2.8   Soil H^+ buffer system (modefied after Schwertmann *et al.* 1987)

H^+ buffer system	Reaction	pH range
Carbonate buffer		
Carbonate	$MeCO_3 + H^+ \rightarrow Me^{2+} + HCO^-_3$	6.8—8.0
Bicarbonate	$HCO_3^- + H^+ \rightarrow H_2O + CO_2$	4.5—7.0
Cation exchange buffer		
Clay minerals	Exchanger-$OMe^+ \rightarrow$ Exchanger-$OH + Me^+$	<5—8
Humates	$R\text{-}COOMe + H^+ \rightarrow RCOOH + Me^+$	<3—6
Amino groups	$R\text{-}NH_2 + H^+ \rightarrow R\text{-}NH_3^+$	4—7
Silicate buffer		
Primary minerals	$-(SiO)Me + H^+ \rightarrow -(SiO)H + Me^+$	<7
Clay minerals	$-(SiO_3)Al + 3H^+ \rightarrow -(SiO_3)H_3 + Al^{3+}$	<7
Oxide/hydroxide buffer		
Al hydroxides	$Al(OH)_3 + 3H^+ \rightarrow Al^{3+} + 3H_2O$	3—4.8
Fe oxides/hydroxides without reduction	$FeOOH + 3H^+ \rightarrow Fe^{3+} + 2H_2O$	<3
Fe oxides/hydroxides with reduction	$FeOOH + 3H^+ + e^- \rightarrow Fe^{2+} + 2H_2O$	<7
Mn oxides/hydroxides	$MnO_2 + 4H^+ + 2e^- \rightarrow Mn^{2+} + 2H_2O$	<8

Me = metal cation

The carbonate buffer plays a major role in calcareous soils, characterized by high concentrations of Ca and Mg carbonates. Such soils also contain high bicarbonate concentrations in their soil solution which are involved in inducing of Fe chlorosis in plants grown on calcareous soils (see page 568). Cation exchange is the principal buffer system in most arable soils which do not contain carbonates. The silicate buffer comes into play as soon as the buffer power of the carbonates and the cation exchange buffer power are exhausted. At this stage a severe dissolution of important minerals begins which in the long term is detrimental to soil fertility. The dissolution is particularly intense under low pH conditions (Feigenbaum *et al.* 1981, Tributh *et al.* 1987). This soil degeneration is accelerated by the entry of strong acids from the atmosphere as reported for forest soils in Sweden by Hallbäcken(1992). Low soil pH also favours the dissolution Ca phosphates and of heavy metals some of which may be plant nutrients such as Mn, Zn, and Fe. The same in true of toxic heavy metals and their uptake by plants, especially the uptake of Cd and Ni (Sauerbeck and Lübben 1991). The availability of Mo decreases as soil pH falls while that of B is raised. The solubility of salts including carbonates, phosphates, sulphates is greater in the lower pH range.

The release of Al from various minerals is highly pH dependent. This pH effect on the release of soluble Al has already been shown for gibbsite on page 18 (Bache 1985). The relationship between soil pH and soluble Al is depicted in Figure 2.20 from the results of Lathwell and Peech (1964). In many cases it is not so much the high H^+ concentration as the soluble Al cation species such as Al^{3+}, $Al(OH)^{2+}$, $Al(OH)_2^+$ which are toxic (Haynes 1984, Kinraide 1991).These Al containing cations are particularly detrimental to root growth (see page 660). Roots remain short and are thickened and lateral roots are hardly developed (Mühling *et al.* 1989). The amounts of these cationic Al species also depends on the soil type and are low in organic soils and it is for this reason that on such soils even under conditions of low soil pH (pH 4.4 to 5) plant growth is little affected. Silty soils and clay clay soils are generally rich in Al compounds and should therefore be cultivated at a pH range 6.5 to 7.5 where the solubility of Al is much depressed (Figure 2.20).

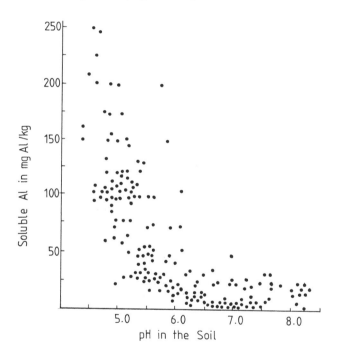

Figure 2.20   Relationship between soil pH and soluble Al (after Lathwell and Peech 1964).

The various physiological disorders related to toxic Al are considered on page 660. Plant species and even cultivars of the same species may differ considerably in their sensitivity to Al. Some species are capable of Al exclusion or Al sequestration (Foy *et al.* 1978), whilst others may adsorb Al ions in the mucilage of root tips (Horst *et al.* 1982). According to investigations of Grimme (1983) cationic Al depressed Mg^{2+} uptake of oats very specifically whereas the uptake of Ca^{2+} and K^+ was scarcely affected. There is evidence that Al impairs membrane channels (Ding *et al.* 1993) an observation which may explain the influence of Al on ion uptake. Besides Al toxicity Mn toxicity may also occur on acid soils (page 580)

In organic soils the pH should not be too high for these soils are by nature poor in a number of plant nutrients, and their availability is suppressed by high pH conditions Lucas (Lucas and Davis 1961). This is especially the case for phosphate, borate, Mn, Cu and Zn.

All living organisms respond in some way to their environmental pH but there are marked differences in response between the various groups and species. Generally bacteria are more sensitive to low pH than are fungi (Trolldenier 1971a). This means that in acid soils the bacterial breakdown of organic matter is affected as well as ammonification, nitrate and nitrite formation and the fixation of dinitrogen (N_2) by symbiotic and free-living N_2 fixing bacteria. Also bacterial reduction of nitrate (denitrification) is depressed under low soil pH conditions. It is hence obvious that soil pH may have a substantial impact on the nitrogen turnover in soils. This is particularly true at pH levels < 5. The reason why the low pH restricts the activity of various bacterial species is not yet completely understood. Schubert *et al.* (1990) reported that at low soil pH the multiplication of *Rhizobium* was retarded. Plant growth, and particularly root growth is also impaired by low soil pH. According to Yan *et al.* (1992) a high H^+ concentrations in the nutrient solution leads to a damage of root tips and decreases the concentration of organic anions in roots as shown in Table 2.9. It was found that at low pH in the nutrient solution, the net proton pumping by the plasmlemma ATPase is depressed and that the cytosol has to cope with a surplus of H^+ which presumably detrimentally affects cytosolic enzymes.

Table 2.9   The effect of the pH in the nutrient solution on the concentration of organic anions in the roots of maize and *faba* beans (Yan *et al.* 1992)

Nutrient solution	pH	Malate	Pyruvate	Lactate
			mmol/g dry matter	
Maize	7.0	128	8	15
Maize	3.5	33	8	13
Beans	7.0	415	5	42
Beans	4.1	81	5	62

The data in Table 2.9 show that at low pH the malate concentration in particular was greatly depressed in the roots. Malate is required for the stabilization of the cytosolic pH and is decarboxylated in order to neutralize H^+ (see page 130). Low malate concentration as found in the treatments with low pH in the nutrient solution are indicative that the capacity for neutralizing a surplus of H^+ in the cytosol is exhausted. The results also show that *Vicia faba* was more sensitive than *Zea mays* to low pH.

It has still to be shown whether such a surplus of H^+ in the cytosol is a general phenomenon for living organisms when subjected to low pH conditions. At such low pH levels the uptake of cations is restricted presumably by partial depolarisation of the plasmamembrane (see page 119).

Plant species are able to cope to a varying degree with differences in H^+ in the soils and the accompanying effects induced by pH. The optimum pH ranges for maximum growth of individual crops therefore differ. The pH limits presented in Table 2.10 only serve as a guideline. There is considerable variation, due to the effects of differences in crop species and cultivars and the influence of climate and soil conditions. The optimum pH ranges shown in Table 2.10 are from data obtained in temperate climate conditions (Klapp 1951). The pH limits appear rather wide, but it must be remembered that often it is not the soil pH itself that is the growth limiting factor but rather one or more secondary factors which are pH dependent, especially the solubility of cationic Al species.

Table 2.10 Optimum pH ranges (pH in KCl extract) of various crop plants, according to Klapp (1951)

Crop	pH range
Lucerne	6.5—7.4
Barley	5.3—7.4
Sugar beet	6.4—7.4
Clover (*Trifolium pratense*)	5.3—7.4
Wheat	4.1—7.4
Peas (*Pisum sativum*)	5.3—7.4
Oats	4.0—7.0
Potatoes	4.1—7.4
Rye (*Secale cereale*)	4.1—7.4
Lupins	4.1—5.5

Mahler and McDole (1987) carrying out field experiments in Idaho for five years on 39 diffferent sites found that *Lens culinaris* was the most sensitive crop to low soil pH followed by *Pisum sativum*. Cereals were more resistant to low soil pH and showed marked differences in this respect between cultivars. The sensitivity was not only due to toxic Al since even at pH < 5 only very low concentrations of soluble Al were found. The pronounced effect of soil pH on the grain yield of winter wheat is shown in Figure 2.21.

The pH of the soil can be corrected by the addition of chemicals which bring about a decrease in H^+ concentration. The most common treatment is liming. This is dealt with in detail on page 533.

Soil pH can be altered by physicochemical and biological processes. The most important physicochemical process in this respect is the vertical movement of H^+

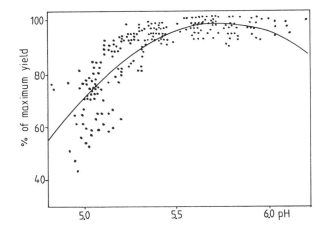

Figure 2.21 Effect of soil pH on the grain yield of winter wheat (after Mahler and Mc Dole 1987).

and bases, the latter mainly OH^-, HCO_3^-, CO_3^{2-} and organic anions by leaching. The movement of CO_3^{2-} and HCO_3^- plays only a role in soils with a pH > 7 due to the equilibrium shown below:

$$CO_2 + H_2O \Leftrightarrow H^+ + HCO_3^- \Rightarrow 2H^+ + CO_3^{2-}$$

Therefore leaching of HCO_3^- and CO_2^{2-} occurs only on soils of high pH such as calcareous and saline soils. Organic anions are potential bases because they consume H^+ when decarboxylated (Barekzai and Mengel 1993) as shown in the following equation:

$$R\text{-}CO\text{-}COO^- + H^+ \Rightarrow RCHO + CO_2$$

In soils with pH < 7 it is mainly the leaching of organic anions produced by microbial breakdown of organic matter that achieves a loss of bases thus lowering soil pH. Counter cation species of these organic anions are mainly those cationic species which dominate in the soil solution, and these are predominantly Ca^{2+}, Mg^{2+} and K^+. Hence leaching losses of these cation species are frequently associated with an acidification of soils and therefore these cation species are also denoted as bases. This is incorrect in a strict sense because if the counter anions leached with these so called bases are nitrate or chloride this will have no impact on soil pH. It is the counter anion which is of importance for the change in pH. It thus appears that leaching of organic anions is the main reason why under humid climate conditions all soils tend to acidification. The higher annual rainfall rates and the more pervious the soils the greater is the impact of organic anion leaching on soil pH. Under arid climate condition there is a tendency for soils to become more alkaline since on average organic anions are not leached and in some cases even bicarbonate containing water is brought to the soil surface by evaporation from deeper aquifers. Such conditions may lead to the salinization and even alklinization of the upper soil layer (see page 535).

Microbial activity has a strong impact on soil pH. Under aerobic conditions the decomposition of organic matter is frequently associated with the production of protons (H^+). The CO_2 produced plays no major role in this respect because as shown above the dissolution of CO_2 in water produces only H^+ if the soil pH is <5 (Oenema 1990). The decomposition of organic nitrogen finally leading to the formation of nitrate is associated with the production of a strong acid, namely nitric acid as shown below:

$$Organic\text{-}N \Rightarrow NH_3 + 2O_2 \Rightarrow H^+ + NO_3^- + H_2O$$

For microbial details of this reaction see page 409. Incorporation of organic matter into soils has therefore an impact on soil pH through microbial decomposition. Organic matter rich in organic nitrogen such as green manure depresses the soil pH because of nitrate formation. Material poor in organic N but rich in organic anions mainly counterbalanced by Ca^{2+} and/or Mg^{2+} increase soil pH as shown by Kretzschmar et al. (1991) after the incorporation of pearl millet straw into a sandy soil. According to Yan et al. (1996b) decarboxylation of organic anions occurs rapidly after the incorporation of leaf material into soils immediately raising soil pH followed by a pH decline due to the formation of nitrate.

In an analogous way to organic N, organic sulphur bound to carbon, but not sulphate esters, produces protons when oxidized by microbes:

$$Organic\text{-}S \Rightarrow H_2S + 2O_2 \Rightarrow SO_4^{2-} + 2H^+$$

In this case too, a strong acid is produced, namely sulphuric acid. Since biological organic matter contains much less organic S than organic N organic, however, sulphur oxidation is generally of no major importance for soil pH. In soils, however, subjected to frequent changes of flooding and drying large amounts of organic sulphur may be oxidized and thus lead to strong acidification as was reported by Ahmad and Wilson (1992) for sulphate acid soils in the costal area of Surinam. The upper layer had a pH of 4, in the deeper layer the pH was as low as 2.3. These layers are extremely rich in soluble Al and therefore such soils provide enormous difficulties for crop growing. For their amelioration huge quantities of lime, in the range of 20 to 30 t/ha $CaCO_3$ and phosphate, about 1000 kg P/ha are required (Ahmad and Wilson 1992).

While microbial oxidation proceeding under aerobic conditions may produce H^+ as in nitrate and sulphate formation, reduction of nitrate (denitrification) and of sulphate (desulphurication) produce OH^- and therefore have an alkalinizing effect as shown in the following equations:

$$2NO_3^- + 10e^- + 12\,HOH \Rightarrow N_2 + 6\,H_2O + 12\,OH^-$$

$$SO_4^{2-} + 8e^- + 10\,HOH \Rightarrow H_2S + 4\,H_2O + 10\,OH^-$$

The e^- required for the reduction mainly originate from respiration. The same is true for the reduction of Fe^{III} and Mn^{IV} as already shown above (see page 48). It thus appears that under anaerobic conditions soil pH increases.

Soil acidity is also produced by plants and it has been known for many years that roots can acidify the soil. It is relatively recently, however, the main source of H^+ excretion by roots has been identified as the plasmamembrane proton pump (see page 118). This pumps H^+ out of the cytosol into the outer medium, into the immediate vicinity of the root surface, which for soil-grown plants is the rhizosphere (Mengel and Schubert 1985, Serrano 1989). This plasmamembrane located proton pump is an ATPase which splits water into H^+ and OH^-. The H^+ is pumped out of the cell, the OH^- remains in the cytosol where it favours the synthesis of organic anions. Therefore the amount of H^+ released into the soil equals the amount of bases (OH^- + organic anions) in the plant (Mengel and Steffens 1982, Yan et al. 1996b; see page 408). Counter ions of these organic anions are mainly Ca^+, K^+, and Mg^{2+}. As shown above these organic anions are potential bases. Thus in theory for a given plant growing on the soil, by returning all the plant material containing organic anions to the soil, potential bases are also returned to the soil which means that the organic anions when broken down by microorganisms consume H^+ (see page 58). If, however, the organic material is removed as is the case when crops are harvested, organic anions are also removed so that a loss of bases occurs which means that the potential of the soil to neutralize H^+ is lowered. The same is true when organic anions are leached from the crop or plant residues into deeper soil layers.

The net release of protons is not only a question of the plasmalemma ATPase activity but depends also on the uptake of anions by root cells since these are taken up by proton cotransport (see page 126). This means that each anion traversing the plasmamembrane is accompanied at least by two or three H^+. By this anion/proton cotransport substantial amounts of H^+ excreted by the ATPase into the rhizosphere are recycled back into the cytosol. Since in growing plants about 80% of the total of anion uptake may be in form of nitrate the recycling of H^+ into the cytosol is mainly a question whether plants are fed with nitrate, ammonium or by symbiotically fixed N_2. With nitrate nutrition the recycling of H^+ into the cytosol is generally higher than the rate of proton excretion by the plamamembrane ATPase. Hence nitrate nutrition leads to a pH increase in the outer medium (soil, nutrient solution) while with ammonium nutrition pH declines as has been reported by various authors (Kirkby and Mengel 1967, Mengel et al. 1983, Marschner et al. 1991). This effect of nitrate or ammonium nutrition on soil pH is of practical importance as was found by Smiley (1974) in field experiments in which nitrate application raised and ammonium application decreased rhizosphere soil pH. A most prominent example of this kind is the "Park Grass" long-term experiment of the Rothamsted Experiment Station which, as depicted in Table 2.11, showed that ammonium application decreased soil pH to a very low level while nitrate application had the reverse effect (Johnston et al. 1986).

Symbiotically grown leguminous species also have a soil acidifying effect as found by several authors (Israel and Jackson 1978, Knight et al. 1982). This finding is explained by the absence nitrate uptake and hence the limited recycling of H^+ into the cytosol (Hauter and Steffens 1985). Therefore cultivation of leguminous species leads to soil acidification which is particularly severe if the aerial plant parts are removed from the field as is the case with forage legumes. If only the seeds are removed the potential bases (organic anions) in the vegetative plant parts remain in the soil/plant system and soil acidification is less severe. This is of practical importance as reported by Bromfield et al. (1983). These authors found a dramatic decrease in soil pH associated with long term cultivation of subterranean clover as shown in Figure 2.22.

Table 2.11 Effect of long term ammonium and nitrate fertilizer application and effect of no N fertilizer application on the development of soil pH in a permanent grassland soil (after Johnston *et al.* 1986)

Treatment	1876	1923	1959	1976	1984
			Soil pH		
No N fertilizer	5.3	5.7	5.2	5.2	5.0
Ammonium	5.3	4.8	4.0	4.1	3.7
Nitrate	5.4	6.3	5.7	5.9	5.7

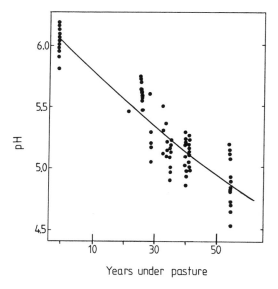

Figure 2.22 Relationship between pH in the upper soil layer and the duration of subterranean clover cultivation in pastures on granitic soil (after Bromfield *et al.* 1983).

The pH decrease in these Australian pasture soils was associated with an increase of soluble and exchangeable Mn and Al which in some cases reached toxic levels (Williams 1980, Bromfield *et al.* 1983). This example shows that permanent growth of leguminous species without liming may ruin soil fertility.

Finally it should be emphasized that a substantial proportion of acid entry into soils is of anthropogenic origin. It is the SO_2 and No_x (NO_x = NO or NO_2) produced by burning fuel and petrol which are potential acidifyers. Once in the atmosphere these oxides may react with ozone and H_2O and thus producing acids as shown in the following equations:

$$SO_2 + O_3 \Rightarrow SO_3 + O_2, \quad SO_3 + H_2O \Rightarrow H_2SO_4 \text{ (strong acid)}$$

$$2NO_2 + H_2O \Rightarrow HNO_3 + HNO_2; \quad HNO_3 \text{ being a strong acid}$$

$$HNO_2 + O_3 \Rightarrow HNO_3 + O_2$$

These reactions occur in small atmospheric droplets and the oxidations may involve ozone as shown in the equations above. These strong acids formed in the form of rain are deposited on soils and vegetation (wet proton deposition) and may contribute considerably to soil acidification (Asche 1988). According to Hallbäken (1992) about 50% of the acidification in forest soils in Sweden occurring over the past thirty years was of anthropogenic origin. Johnston et al. (1986) reported that in England wet proton deposition makes up about 70% of total soil acidification. The emission of NH_3 mainly originating from excessive lifestock density (Roelofs et al. 1985, Isermann (1990) contributes to the neutralization of strong acids in atmospheric droplets by forming ammonium sulphate and ammonium nitrate (Zobrist 1987). The addition of these ammonium salts from the atmosphere does not directly cause soil acidification but when high amounts are deposited they may induce acidification via nitrification (Roelofs et al. 1988).

2.2.7 Salt affected Soils

Soil salinity is a world-wide problem. Salt affected soils are characterized by an excess of inorganic salts and mainly occur in arid and semi arid regions. As a result of their high salt concentration these soils are charaterized by a high electrical conductivity. Under arid conditions salts accumulate in the upper soil layer. This accumulation usually results from evapotranspiration causing a rise in ground water which contains salts. The process may be promoted by irrigation particularly if the irrigation water is rich in salts. The effect is particularly marked where the ground water is near the surface as occurs in depressions or low lying sites and when irrigation water comes into contact with the ground water. Salt accumulation in soils results in poor crop growth and both yield and quality of crops are low. This problem and salt tolerance of plants has been reviewed by Flowers et al. (1977).

Two major soil types of salt affected or halomorphic soils may be distinguished; the saline soils (solonchak) and the black alkali soils (solonetz). Saline soils contain an excess of neutral salts such as the chloride and sulphates of Na^+, K^+, Ca^{2+} and Mg^{2+}. The quality of soils depends much on the proportions of these cation species and if the Ca^{2+} is relatively high and the Na^+ proportion low, soil structure is usually satifactory. If the proportion of Mg^{2+} is high, crops may suffer from Mg toxicity. The exchangeable sodium percentage ESP is a measure of the % proportion of Na^+ of the cation exchange capacity (CEC) i.e. exchangeable Na/CEC × 100%.

$$ESP = \text{exchangeable Na/CEC} \times 100$$

If the ESP is below 15% the soils are called saline soils or "white alkali soils" because in dry periods these soils often show a white efflorescence of salts on the surface.

The main anion species present are chloride and sulphate, and in some cases nitrate (Charley and McGarity 1964). The pH of saline soils is in the range of 7 to 8.5. Nitrate accumulation occurs when organic soils are drained, the access of oxygen bringing about a vigorous microbiological mineralization of organic N (Giskin and Majdan 1969). Substantial amounts of NO_3^- can thus accumulate in soils in arid regions giving an excessive supply of nitrate to crops and also providing a hazard to the environment because of leaching during periods of high rainfall. On particular sites the borate concentration may reach toxic levels and crops may suffer from boron toxicity. The high accumulation of neutral salts typical of saline soils means that the clay is highly flocculated so that generally a good soil structure results which is uniform down the profile.

If the ESP is >15%, the soil structure becomes poorer and a high proportion of exchangeable Na^+ may exchange with H^+ from water molecules as shown below:

$$Exch.-Na^+ + HOH \Rightarrow Exch.-H^+ + Na^+ + OH^-$$

The soil solution thus becomes alkaline, the process being favoured by low concentrations of divalent cations. Such soils with a ESP > 15 are called "saline-sodic soils" and represent the precursor of sodic soils or black alkali soils. As shown above the Na^+ desorption results in a pH increase. A further pH increase is brought about by the reduction of nitrate and sulphate or both during periods of transient waterlogging. The reduction processes of denitrification and desulfurication as discussed on page 59 consume H^+. The elevated resulting pH favours the formation of bicarbonate and carbonate as shown below:

$$CO_2 + H_2O \Rightarrow H^+ + HCO_3^- \Rightarrow 2H^+ + CO_3^{2-}$$

The equilibrium is shifted more to the side of carbonates the higher the pH. For this reason in sodic soils basic anion species such as bicarbonate and carbonate dominate and therefore the pH is in a range of 8.5 to 10. The alkalinity (ALK) of a soil may be described by the following equation which shows the components of alkalinity (Sposito 1989).

$$ALK = (HCO_3^-) + 2(CO_3^{2-}) + (OH^-) - (H^+)$$

The ionic components in brackets refer to the activities of the ion species.

High pH conditions together with the high Na^+ concentrations favour the dispersion of soil aggregates as a consequence of the dissolution of humates, since Na humates unlike Ca-humates are water soluble. During dry periods these Na-humates may effloresce on the soil surface giving a blackish brown appearance. For this reason the sodic soils are also called "black alkali soils". In the Russian literature they are known as Solonetz and the white alkali soils as Solonchak.

The negative effect of the high Na^+ concentration may be moderated by Ca^{2+} and Mg^{2+} and therefore the degree of soil alkalinization is measured by the ratio of Na^+ to $Ca^{2+} + Mg^{2+}$ according to the following equation (Sodium Adsorption Ratio = SAR):

$$SAR = Na^+/(Ca^{2+} + Mg^{2+})^{\frac{1}{2}}$$

The concentration of the cation species in the equation are expressed in (mol m^{-3}) in a soil extract or in the irrigation water (Sposito 1989).

Sodic soils are difficult to cultivate. The high bicarbonate concentration may induce iron chlorosis (Sahu *et al.* 1987). An even greater problem of sodic soils, however, is their structure. The dissolution of soil aggregates is associated with a movement of clay down the profile to form a pan in the B horizon and this horizon often becomes compacted impeding plant root penetration (Szabolcs 1971). The black alkali soils are very sticky when wet and form hard unworkable compact clods when they dry out. This, together with the high pH conditions are the primary effects bringing about poor crop growth. Sodium and carbonates are the most abundant ions in the profile (Raikov 1971).

The total ion concentration of the soil solution of saline and alkali soils can reach levels which can bring about plasmolysis of plant root cells. The radicles of germinating seeds are particularly sensitive to high ion concentrations in the soil solution. A further drawback of the high salt concentration in the soil solution is its osmotic effect. A high concentration of solutes means a low water potential and therefore water uptake by plant roots is restricted.

The degree of salinity is usually measured in the water extract of a soil paste as electrical conductivity expressed in dS/m(deci Siemens) and the conductivity increases with the electrolyte concentration. Salt affected soils usually show conductivities in the range of 4—20 dS/m. In the USA a value of 4 dS/m is considered critical for crops. In Table 4.12 on page 240 crop species are listed according to their sensitivity to saline conditions.

Salinity problems arise especially where irrigation is applied to impermeable soils. All irrigation waters contain salts and these can remain in the upper soil layer and accumulate. In this respect light textured soils are easier to handle because of their high permeability. For heavy textured soils it is often necessary to provide artificial drainage, along with irrigation. Useful discussions on halomorphic soils and the associated salinity problems have been presented by Szabolcs (1971) and Poljakoff-Mayber and Gale (1975).

2.3   Factors Relating to Nutrient Availability

2.3.1   General

The term "nutrient availability" is often used in plant nutrition but is frequently not precisely defined. As a first approach, the available nutrients of a soil may be considered as that nutrient fraction in the soil which is accessible to plant roots. The term nutrient availability thus encompasses both the chemical and physical status of the nutrients in the soil as well as a plant root relationship which involves plant metabolism. It is for this reason that in a strict sense the amounts of soil nutrients cannot be precisely measured and expressed in quantitative terms. Indeed a nutrient which is accessible to one plant species may not be accessible to another because of differences in root morphology and metabolism. Nevertheless it is important to know the factors and their causal relationships which contribute to "nutrient availability", and this is discussed below.

2.3.2 Interception and contact exchange

An important question in nutrient availability is whether nutrients need to be transported to the plant roots or whether they come in contact with roots by interception as the roots push their way through the soil. One theory which held favour for many years in support of root interception, for the uptake of cations at least, was the contact exchange theory as proposed by Jenny and Overstreet (1938). It was envisaged that a close contact between root surfaces and soil colloids enabled direct exchange of H^+ released from the plant root with cations from soil colloids. It was argued that by releasing H^+ produced in plant metabolism, plant roots should be able to strip off and mobilize cationic nutrients absorbed to clay minerals. This contact exchange process was envisaged as the first step of cation uptake.

Today it is well known that the plasmalemma H^+ pump (Serrano 1989) pumps H^+ out of the cell into the root medium. This H^+ released by the pump, however, does not come in direct contact with the surface of clay minerals. Figure 2.23 shows the sites and dimensions of the cell membrane (plasmalemma) and the surface of a clay mineral. It is evident that, if at all, only cations at the very outer surface of the cell wall can exchange for cations adsorbed to the clay mineral. By an exchange of H^+ from the cell wall, K^+ or other cation species can be mobilized from a clay mineral as indicated in Figure 2.23. Even if this does occur, however, the exchanged K^+ is still at the outer surface of the cell wall of an epidermal cell and far from the real sites of uptake systems located in the outer cell membrane (plasmalemma). The thickness of a cell wall is in the range of 500 to 1000 nm. There is no evidence that K^+ is able to move across the cell wall by further exchange processes but rather moves by diffusion from the soil solution adjacent to the cell wall surface along pores and intercellular spaces to the plasma membrane. The carboxyl and phosphate groups of the cell wall structures represent a strong buffer to the H^+ released by the plamsmalemma H^+ pump (Grignon and Sentenac 1991, Kosegarten et al. 1999). The pH prevailing in the liquid phase of cell walls present in pores and intercellular spaces has an impact on nutrient uptake (see page 557) especially on the uptake of Fe (Toulon et al. 1992, Kosegarten et al. 1998). In comparison with the size of inorganic ions the cell wall is immense. The fully developed cell wall is a rigid relatively thick structure (Carpita et al. 1996) which normally contains channels filled with air and soil solution. These channels almost certainly allow ions an accessible route to the plasmamembrane. It seems likely therefore that K^+ and other cations act as counter ions in anion transport with exchange movement being relatively unimportant.

Generally the amount of nutrients which directly contacts plant roots (interception) is small as compared with the total nutrient demand. This is particular true for nutrients required in high quantities (Barber et al. 1963). For this reason mass flow and diffusion rather than interception are by far the most important processes by which plant nutrients are transported from the soil to the root surfaces. This view is supported by the data of Drew and Nye (1969) who found that only 6% of the total K^+ demand of *Lolium perenne* was supplied by the soil volume in the immediate vicinity of the root hair cylinder. Ninety-four % of the K^+ taken up had therefore originated from beyond the limit of root hairs, and thus must have been transported to the roots.

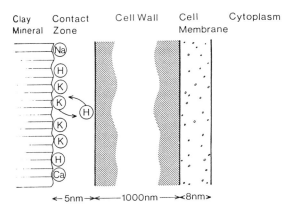

Figure 2.23 Contact exchange between a clay mineral and an epidermal root cell, showing the dimensions involved.

2.3.3   Mass-flow and diffusion

Nutrients in the soil can be transported by two different mechanisms: by mass-flow and by diffusion. Mass-flow occurs when solutes are transported with the convective flow of water from the soil to plant roots. The amount of nutrients reaching the root is thus dependent on the rate of water flow or the water consumption of the plant and the average nutrient concentration in the water. The level of a particular nutrient around the root may be increased, decreased or remain the same depending on the balance between the rate of its supply to the root by mass flow and the rate of uptake by the root.

Diffusion occurs when an ion is transported from a higher to a lower concentration by random thermal motion. Diffusion comes into operation when the concentration at the root surface is either higher or lower than that of the surrounding solution. It is directed towards the root when the concentration at the root surface is decreased, and away from the roots when it is increased. Diffusion follows Fick's first law:

$F = -D \cdot dc/dx$

F = Diffusion rate (quantity diffused per unit cross section and per unit time),

dc/dx = concentration gradient

c = concentration

D = diffusion coefficient

x = distance

If the concentration (c) is given in mole per m^3 and the distance (x) in m, the diffusion coefficient is expressed in terms of m^2 sec^{-1}.

Plant roots absorbing nutrients from the adjacent soil can create a sink to which nutrients diffuse (Drew *et al.* 1969, Kuchenbuch and Jungk 1984). The nutrient depletion depends on the balance between the supply from the soil and the demand by the plant. A high plant requirement or a high root absorbing power as it has been termed by Nye (1968) gives rise to a strong sink and the low nutrient concentration at the root surface may promote the dissolution and desorption of plant nutrients (Kong and Steffens 1989, Dou and Steffens 1993). This indicates that the root itself and its metabolism also influence nutrient availability.

If diffusion is the main process by which a plant nutrient is transported to the root surface, the quantity of the nutrient absorbed by the root can be described approximately by the following equation (Drew *et al.* 1969):

$$Q = 2\pi \, a \, \alpha \, c \, t$$

where

Q = Quantity of nutrient absorbed per cm root length

a = Root radius in cm

α = Nutrient absorbing power of the root in cm root length

c = Average nutrient concentration at the root surface

t = Time of nutrient absorption.

The nutrient concentration (c) at the root surface may change. At the beginning of the absorption period (t) it may be relatively high and then gradually decline, the degree at which it falls depending on the capacity of the soil to replenish the soil solution with the nutrient. This capacity for nutrient replenishment is referred to as the nutrient buffer capacity and is discussed in more detail on page 76. In this context it need only be understood that the term c (= average nutrient concentration at the root surface) not only depends on the bulk nutrient concentration of the soil but also on the nutrient buffer capacity. A soil with a high nutrient buffer capacity is more capable of maintaining a high nutrient concentration at the root surface than is a soil with a low nutrient buffer capacity.

The term (α = root absorbing power) in the above equation represents the proportion of a nutrient absorbed of the total nutrient flux to the root surface. The root absorbing power is not a constant but is much dependent on root metabolism and the nutrient status of the plant (Barber 1979, Glass 1983).

Nutrients taken up rapidly by plant roots and which are generally present in the soil solution in low concentrations such as NH_4^+, K^+, and phosphate are mainly transported to plant roots by diffusion. The contribution of mass flow to the transport of these nutrients can be calculated as the product of solution concentration and transpiration rate. Values thus obtained are far too low to meet the needs of the plant in any of these elements (Barber *et al.* 1963). Diffusion also dominates when transpiration is low.

Mass flow plays an important role for nutrients present in soil solution in high concentration and when transpiration is high. Under such conditions considerable quantities of water are moved to the roots carrying various solutes. Occasionally ion accumulations can even occur around roots as is sometimes the case with Ca^{2+} (Barber 1974). In the case of NO_3^-, transport can take place either by mass flow or diffusion. Investigations of Strebel *et al.* (1983) with sugar beet have shown that under field conditions at the beginning of the growth period mass flow is the major process in transporting NO_3^- towards plant roots. In the later stages of growth, however, when the NO_3^- concentration in the soil solution is low, diffusion becomes the more important process.

If the rate of nutrient uptake is higher than the rate of nutrient transport towards an plant root, nutrient depletion occurs around the root (Barber 1974). This is typical for K^+, NH_4^+ and phosphate (Lewis and Quirk 1967, Bhat and Nye 1974). A depletion pattern resulting from diffusion is demonstrated in Plate 2.3. In this experiment from the work of Barber (1968) a study was made on the uptake of the radioisotope ^{86}Rb from the soil by maize roots. Rubidium and K^+ have closely related chemical properties so that the plate can also be taken to illustrate the behaviour of K^+. A photograph of root growth is shown on the right-hand side of the plate with an corresponding autoradiograph on the left-hand side. The light areas of the autoradiograph show the depletion of the labelled Rb which follow the root growth pattern. The dark lines indicate ^{86}Rb accumulation in the roots.

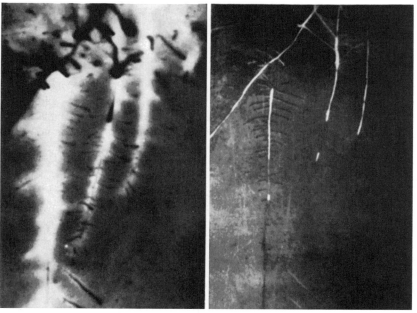

Plate 2.3   A photograph (right) and an autoradiograph (left) showing the effect of maize roots on the distribution ^{86}Rb in the soil. Light areas show Rb depletion around the maize root (Photo: Barber)

The relative depletion of plant nutrients declines with increasing distance from the root surface (Lewis and Quirk 1967). This can be seen in Figure 2.24 which shows the relationship between P concentration and distance from the root surface in soil supplied with different P fertilizers, the horizontal lines indicate the phosphate concentration in the bulk soil appropriate to each fertilizer In this experiment bulk soil concentration of P was highest in the treatment with basic slag (see page 473) application and the lowest in the treatment with no phosphate fertilizer (Steffens 1987). It is clear, that the treatments with the higher phosphate concentration in the bulk soil had the steeper concentration gradient and therefore the rate of diffusion to the plant roots was greater. The higher nutrient level in the bulk soil also gives a higher concentration at the root surface which results in a more rapid uptake and the larger gradient in turn allows this to be maintained.

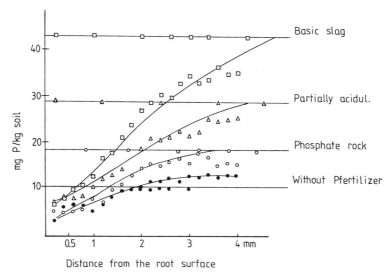

Figure 2.24 Phosphate depletion in the rhizosphere of rape fertilized with different P fertilizers (after Steffens 1987).

The nutrient concentration at the root surface directly controls nutrient uptake as has been shown for the K^+ uptake of young maize plants by Claassen and Barber (1976). Hendriks *et al.* (1981) studying the depletion of phosphate around maize roots, found that maize roots had a depletion zone which extended from about 1.6 mm from the root surface and was thus twice the value of the length of the root hairs (0.7 mm). The phosphate concentration at the root surface was about 1 mmol/m³, a value about 40 times lower than the phosphate concentration in the bulk soil solution. For K^+ the depletion zone extends further from the root surface than that for phosphate, Claassen *et al.* (1981) measuring a K^+ depletion extension of 3 to 5 mm around maize roots, with a K^+ concentration at the root surface of about 2 mmol/m³ (Claassen and Jungk 1982). Ammonium may also be depleted around roots as has been shown by Liu and Sheng (1981) for rice roots. Depletion is particularly evident in flooded soil in which NH_4^+ is the major N source. A valuable review article on nutrient diffusion in soil-root system and nutrient depletion at root surfaces was published by Jungk and Claassen (1997).

The mobility of ions is defined in terms of the diffusion coefficient. Measurements made in water are not directly applicable to movement in soils as other factors come into play. Nye and Tinker (1977) describe the diffusion of ions in soil in terms of the effective diffusion coefficient D which takes into account three factors and is described by the following equation:

$$D = D_i \, \phi \, f \times dc_s/dC$$

where

D = diffusion coefficient of the ion in the soil medium

D_i = Diffusion coefficient of the ion in the water phase

(ϕ = Fraction of the soil volume occupied by solution giving the cross section for diffusion

f = Impedance factor

c_s = Concentration of ion in the soil solution

C = Total concentration of ion in the soil

dc_s/dC = the reciprocal of the buffer power for the ion

From the equation it follows that the fraction of soil volume filled by solution is of particular importance in determing the diffusion coefficient. As soils dry out a drastic reduction occurs in the cross section for diffusion (ϕ) and hence also in the rate of diffusion. The impedance factor represents the tortuosity of the pathways through water filled pores along which diffusing solutes have to follow to reach the plant root. Drying out the system makes the diffusion pathways more tortuous, so that the value of "f" also falls. This approaches zero when diffusion occurs only in a monomolecular water film in the soil system. The importance of soil moisture in determining the rate of diffusion of an ion may be appreciated from the observation that a fall in water potential from -10 to -1000 kPa may be associated with a hundred fold decrease in the product of ϕ f. c_s is the ion concentration in soil solution and C the total concentration of an ion species directly or indirectly involved in ion transport. C thus represents all soil fractions of the particular ion which can equilibrate with the ion in solution. For K^+, this is the exchangeable K^+ and a proportion of the nonexchangeable K^+, for phosphate, it is the adsorbed fraction. The term "dC_s/dC", is the reciprocal of the buffer power. From this it follows that the diffusion coefficient of an ion decreases as its buffer power increases. Table 2.12 shows some diffusion coefficients of ions in different systems.

Table 2.12 Diffusion coefficients (m^2/sec) for some ion species in different systems (after Nye 1979)

Ions in water	$0.5{-}2.0 \times 10^{-9}$
Cl^- and NO_3^- in moist soils	10^{-10}
$H_2PO_4^-$ in moist soils	10^{-11} to 10^{-12}
K^+, release from micas	10^{-19}
Cation exchange between interlayer position and the external solution	10^{-13}

In dry soils diffusion coefficients can be lower than those in moist soils by a factor of 10 to 100. Phosphate, since it can be adsorbed by soil particles, is relatively immobile and its diffusion coefficient is therefore considerably lower than that of NO_3^- or Cl^-. The same is true for K^+ where the diffusion coefficient falls the higher the cation exchange capacity of the soil. The release of K^+ from micas or illites as well as the release of cations from interlayers of 2:1 clay minerals is dependent on exchange and diffusion. The rates of ion diffusion in interlayers are much lower than in the other soil systems (see Table 2.12). Since the diffusion distances are short, however, appreciable amounts of cations may be released from interlayer sites. This release is also dependent on soil moisture as has been shown by Scherer and Mengel (1981) in the case of interlayer NH_4^+ in alluvial soils. The diffusion coefficients of ions in the apoplast (free space) of plant roots are in the order of 10^{-10} m²/sec (Pitman 1977) and thus similar to the diffusion coefficients for NO_3^- in moist soils.

Ion diffusion in the soil is even considerably restricted at water potentials at which the water availability is still adequate for normal plant growth. It is for this reason that in dry periods poor nutrient mobility rather than the direct effect of water may often be the growth limiting factor. Useful discussions on ion movement in soil have been presented by Nye (1979), Nye and Tinker (1977) and Barber (1995).

2.3.4   Soil solution

Plants only take up mineral nutrients dissolved in the soil soution. Nutrient mobility in the soil and nutrient supply to plant roots depend considerably on the nutrient concentration in the aqueous phase, nutrient transport being generally faster the higher the concentration. Only a very small fraction of the soil nutrients are present in the soil solution and the soil solution concentration appears low in terms of the needs of the plants. However, the solution is in equilibrium with the soil phase including inorganic and organic complexes so that removal of nutrients by uptake can be replenished by release of nutrients into the soil solution from desorption, dissolution and mineraliza- tion. These reactions are shown in Figure 2.25. The main processes decreasing the nutrient concentration in the soil solution are uptake by plants, leaching, fixation by clay minerals and precipitation as well as immobilization *i.e.* assimilation by soil microorganisms. Important processes increasing the nutrient concentration of the soil solution are fertilizer application, weathering and dissolution of minerals and microbial decomposition of organic matter. The soil solution is not a homogenous solution but characterized by marked spatial and temporal variability. It differs in concentration and composition and is much influenced by physical, chemical, and biological processes *e.g.* by nutrient uptake of plant roots as shown above. The water filled spaces and pores of soils are not completely interconnected so that solutes cannot easily diffuse from higher to lower concentrations. Such solutes may comprise ions, charged complexes and non charged molecules. Ions such as K^+ dissolved in water form a solvation complex attracting water molecules because of the electrical charge and the dipol properties of the H_2O molecule. Chelates are a further group of easily water soluble complexes which are generally characterized by an organic ligand which is bound to a central metal with more than one bond. Such chelates are of particular importance for the dissolution of heavy metals, and particularly iron as shown more in detail on page 556. The formation of soluble complexes is dependent on soil pH as can be seen from Table 2.13 (Sposito 1989).

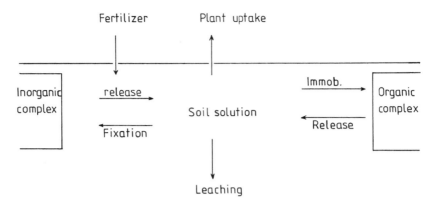

Figure 2.25 Factors affecting nutrient concentration in the soil solution.

Table 2.13 Representative chemical species in soil solution (modified after Sposito 1989)

Cation	Acid soils	Alkaline soils
Na^+	Na^+	Na^+, $NaHCO_3$, $NaSO_4^-$
Mg^{2+}	Mg^{2+}, $MgSO_4$, Org.	Mg^{2+}, $MgSO_4$, $MgCO_3$
Al^{3+}	$AlOH^{2+}$, AlF^{2+}, Org.	$Al(OH)_4^-$, Org.
Si^{4+}	$Si(OH)_4$	$Si(OH)_4$
K^+	K^+	K^+, KSO_4^-
Ca^{2+}	Ca^{2+}, $CaSO_4$, Org.	Ca^{2+}, $CaSO_4$, $CaHCO_3^-$,
Mn^{2+}	Mn^{2+}, $MnSO_4$, Org.	Mn^{2+}, $MnSO_4$, $MnCO_3$, $MnHCO_3$, $MnB(OH)_4^+$
Fe^{2+}	Fe^{2+}, $FeSO_4$, $FeH_2PO_4^+$	Fe^{2+}, $FeCO_3$, $FeHCO_3^+$, $FeSO_4$
Fe^{3+}	$FeOH^{2+}$, $Fe(OH)_3$. Org.	$Fe(OH)_3$, Org.
Cu^{2+}	Cu^{2+}, Org.	$CuCO_3$, $CuB(OH)_4^+$, Org.
Zn^{2+}	Zn^{2+}, $ZnSO_4$, Org.	$ZnHCO_3^+$, $ZnCO_3$, Zn^{2+}, $ZnSO_4$, $ZnB(OH)_4^+$, Org.
Mo^{6+}	H_2MoO_4, $HMoO_4^-$	$HMoO_4^-$, MoO_4^{2-}

Org = organic complex of the ion species in question

From the data in Table 2.13 it is clear that plant nutrients may be dissolved in the soil solution in the form of different complexes. This is particular true for solutes at alkaline soil pH such as carbonate and even borate complexes. Not all of the soil solution solutes are electrically charged, *e.g.*$Si(OH)_4$ and some carbonates and sulphates. Metallo-organic complexes play an important role in the dissolution and transport of heavy metals and also Al which as an organic complex is not phytotoxic in contrast to the inorganic Al cation species. The concentrations of various plant nutrients found in soil solution are shown in Table 2.14 (Recalculated data of Peters, 1990 from Marschner 1995).

Table 2.14 Annual average concentrations of mineral nutrients in the soil solution (topsoil 0—20 cm) of an arable soil (Luvisol, pH 7.7). Recalculated data of Peters (1990) from Marschner (1995)

	Concentrations in mmol/m^3		
Nutrient	Concentration	Nutrient	Concentration
K	510	SO_4-S	590
Ca	1650	PO_4-P	1.5
Mg	490	Zn	0.48
NH_4-N	48	Mn	0.002
NO_3-N	3100		

Concentrations vary widely between nutrients. A most striking feature is the low concentration of phosphate. This is indicative of the strong fixation by soil particles and hence the extremely low mobility of phosphate in the soil either by leaching in the soil profile or transport to roots *via* mass flow (see page 460). The same applies to a lesser extent to NH_4^+. Because of the promoting effect of organic molecules and microbial activity in mobilizing phosphate (see page 87) a substantial proportion — as much as 20 to 70% of the soil solution P — can be organically bound (Ron Vaz *et al.* 1993).

Of the cations Ca^{2+} and Mg^{2+} there is often a high correlation between soil solution and exchangeable values as observed by Nemeth *et al.* (1970) in investigating seventy arable soils in Germany. In the case of K^+, however, this was not the case. The type of soil clay played an important role since 2:1 clay minerals present in many soils in temperate climate can specifically adsorb K^+. Of all the ions NO_3^- is frequently present in highest concentrations. As NO_3^- is unbuffered, marked and rapid fluctuations occur (Barraclough 1986).

The concentrations of micronutrients Fe, Mn, Zn, and Cu depend mainly on soil pH, redox potential and soil organic matter content. Concentrations increase as the pH falls. Micronutrients (heavy metals) are frequently present in the soil solution as organic complexes. Of the total amount in the soil solution this can amount to 50—55% for Mn, 75 to 85% for Zn and 80 to 90% for Cu (McGrath *et al.* 1988). These metallo-organic complexes play an important role on the transport of heavy metals in soils. The same is true of organically bound Al.

Composition and concentration of the soil solution depend considerably on soil moisture. In a wet soil (field capacity) the soil solution is diluted and as the soil dries out the solution becomes more concentrated. Some of its ion species may even reach concentrations higher than their solubility products and precipitation of these solutes may occur. Calcium, sulphate and phosphates are particularly susceptible to precipitation. In order to compare soil solutions of different soils they must be based on equivalent soil moisture levels. This is usually taken at field capacity (Adams 1974). According to Hantschel *et al.* (1988) marked differences in composition occur between equilibrated soil solution samples and percolation extracts.

Isolating representative samples of soil solution still presents difficulties (Sposito 1989). For routine analysis the extration of a water saturated soil sample by vaccuum extrationn is a convenient technique. Although there are drawbacks the method provides reasonable reliable data of concentration ranges of plant nutrients prevailing in the soil solution. Nutrient concentrations obtained with the water saturated extraction technique may differ widely depending on soil properties.

2.3.5 Intensity, quantity and buffer power

Plants must be supplied adequately with nutrients during their entire growth period. For this reason the concentration of plant nutrients in the soil solution must be maintained at a satisfactory level for plant growth. Nutrient availability depends therefore not only on the nutrient concentration of the soil solution at any given time but also on the ability of the soil to maintain the nutrient concentration. This capability of a soil to 'buffer' the nutrient concentration of the soil solution is a further important factor in nutrient availability.

Generally those nutrients required by plants in high amounts, are present in the soil solution in relatively small concentrations. This is particularly the case for phosphate and K^+. Calculated on an area basis the soil solution contains in the order of only about $0.5-1.0$ kg P/ha and $10-30$ kg K/ha, whereas the total demand for these nutrients is considerably higher. A cereal crop for example requires about 20 kg P/ha and 100 kg K/ha. As a cereal crop growing under the soil conditions described does not necessarily become deficient in P or K, this shows that the removal of these nutrients from the soil solution by the crop must be accompanied by a substantial replenishment of the soil solution from the solid phase of the soil. One may thus distinguish between two nutrient fractions in the soil: the quantity or capacity factor (Q) which represents the amount of a potentially available nutrient and the intensity factor (I) which is directly available and represented by the concentration of the soil solution.

The concept of nutrient intensity and nutrient quantity was first proposed by Schofield (1955). He compared the availability of phosphate with the availability of soil water. Soil water availability depends not on the total amount of water present in the soil but rather on the strength by which the water is bound to the soil matrix (see page 26). The same holds true for phosphate and also for some other plant nutrients. Therefore nutrient intensity and quantity factors are interrelated. The relevant relationships are illustrated in Figure 2.26 showing that various fractions and parameters are involved in nutrient quantity as described for phosphate by Williams (1970) in a useful paper.

Plant roots are dependent on the nutrient intensity or soil solution concentration. This is usually regulated by a much larger labile pool of easily exchangeable or desorbable nutrients. Generally the labile pool represents the main component of the quantity of available nutrients. For cations this is mainly the exchangeable fraction. However, this is not always the case as nutrient release from more slowly available sources can sometimes contribute considerably to nutrient supply of plants. This is particularly true for nutrients which become available by microbial activity such as the mineralization of organic nitrogen or sulphur or the dephosphorylation of organic phosphates by phosphatases. For these enzymic processes soil conditions such as pH, temperature, aeration and moisture are pertinent. The quantity factor is therefore very much dependent on weather and soil conditions as well as on the prevailing climate. In addition it also depends on the volume of soil penetrated by roots which may differ considerably between plant species. This means of course that all factors influencing the distribution of roots in the profile contribute to the quantity of nutrients accessible to roots (Figure 2.26).

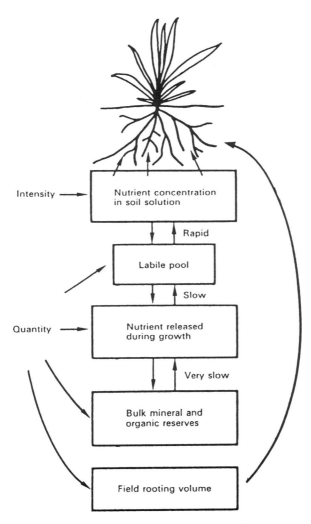

Figure 2.26 Intensity, quantity, and nutrient sources (after Williams 1970).

Another important factor in nutrient availabiliy is the ability of the soil to maintain the nutrient intensity, *i.e.* the nutrient concentration in the soil solution. This capability is called the buffer capacity or buffer power of soils for plant nutrients. It indicates how the nutrient intensity varies with quantity or the amount of available nutrients per unit nutrient concentration in the soil solution. In Figure 2.27 the quantity of K^+ is plotted against intensity for two soils differing in K^+ adsorption capacity, soil A and soil B. For both soils increasing intensity (K^+ concentration of the soil solution) is accompanied by an increase in quantity. Soil A, however, shows a steeper rise in the slope than soil B. This means that for one unit of intensity (ΔJ) soil A provides a higher K^+ quantity than soil B. Therefore the decrease of the K^+ concentration (intensity) is slower in soil A than in soil B. Hence soil A has a higher K^+ buffer power.

In quantitative terms the buffer capacity is expressed as the ratio $\Delta Q/\Delta I$.

$$B_K = \Delta Q/\Delta I$$

where B_K = buffer capacity for K^+. The higher the ratio of $\Delta Q/\Delta I$, the more the soil is buffered for K^+. The buffer power is represented by the steepness of the curve. Since the curves are not linear but rather saturation types of curves the buffer power is the higher the lower the nutrient concentration in the soil solution. Figure 2.27 also shows the so called Q/I relationship in which the Y-axis represents Q and the X-axis I. The slope of the curve represents the buffer power which is higher the steeper the slope. From this it follows that the buffer power *per se* gives no information about the actual availability of a given nutrient because the same curve is related to a wide range of intensity (I) and quantity (Q) values.

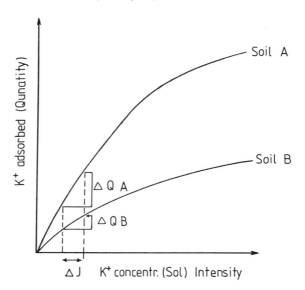

Figure 2.27 Relationship between the K^+ intensity, K^+ quantity, and K^+ buffer power for two soils with different cation exchange capacity; soil A a high and soil B a low one. The steepness of the curves represents the K^+ buffer power.

Generally the rate of K^+ uptake by roots is higher than the diffusive flux of K^+ towards the roots (see page 483). The K^+ concentration at the root surface may thus decline during the uptake period. This fall in K^+ concentration is dependent on the K^+ buffer capacity of the soil. If this is high, the decline is likely to be small because of the efficient K^+ replenishment of the soil solution. On the other hand for soils with a poor K^+ buffer capacity, the K^+ concentration at the root surface may fall considerably over the growth period. For optimum plant growth, nutrient concentration in soil solution should be maintained above a certain level. This concentration may be termed the critical level as concentrations below this value result in yield depression. It has been shown by Mengel and Busch (1982) that the critical K^+ level in the bulk soil solution is related to the K^+ buffer capacity (= buffer power), critical concentration is higher, the lower the K^+ buffer capacity.

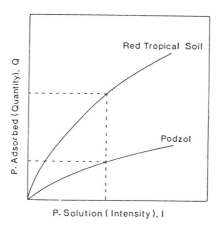

Figure 2.28 Phosphate adsorption curves for two soils differing greatly in their ability to adsorb phosphate and therefore also in the P buffer power.

The quantity/intensity concept (Q/I relationship) is also useful to characterize phosphate availability. Soils may differ much in their ability to adsorb phosphate (see page 471). Red tropical soils (Oxisols) generally adsorb huge quantities of phosphate whereas the phosphate adsorption potential of podzols (Spodosols) is low. The phosphate adsorption curves for two such soils are shown in Figure 2.28. It can be seen that the red tropical soil requires a much higher amount of adsorbed phosphate than the podzol in order to maintain the same phosphate concentration in the equilibrated solution. From this it follows that soils with a high phosphate adsorption capacity should contain more adsorbed phosphate than soils with a low phosphate adsorption capacity, generally sandy and organic soils, in order to provide a satisfactory phosphate intensity. The ratio "adsorbed phosphate/phosphate concentration in solution" at least partially represents the phosphate buffer capacity since in neutral to alkali soils phosphate in the soil solution is mainly buffered by Ca phosphates of varying solubility whereas in acid soils adsorbed phosphate plays the same role. The quantity of phosphate which can be taken up by plants is related to the phosphate buffer power of soils as has been shown by Olsen and Watanabe (1970), Holford (1976) and Nair and Mengel (1984).

The quantity/intensity concept may be also applied to NH_4^+-N particularly in flooded rice soils (Schön *et al.* 1985). The Q/I concept is not relevant to NO_3^- supply since by far the greatest amount of soil N is present in the organic form. This is biologically converted to NO_3^- which is only weakly adsorbed to the soil matrix and the nitrate concentration of the soil solution is thus not buffered.

2.3.6   Root growth and morphology

Roots have three main functions, a) anchoring the plant in the soil, b) absorption and translocation of water and nutrients, c) synthesis of phytohormones and other organic compounds. In the context considered here water and nutrient uptake are of special interest. The capability of plants to exploit the soil for nutrients and water depends much on root morphology. This term comprises root depth, root branching, the number of root hairs, the root tips, and root diameter. Root morphology is genetically controlled but is also influenced by a number of environmental factors. Monocotyledons and dicotyledons differ fundamentally in root growth and morphology. In the dicotyledons a tap root is formed at an early stage which extends deeply into the soil. Later lateral roots are developed. In monocots, especially in grasses lateral roots develop from the seminal roots a few days after germination and generally form a dense root system with numerous slender roots. Mengel and Steffens(1985) comparing *Lolium perenne* with *Trifolium repens* grown under field conditions found that most root parameters including root length and density were greater for *Lolium* than for *Trifolium*. This was particularly true for root length, the roots of *Lolium* being 4 to 6 times longer than those of *Trifolium*.

Root depth may also differ considerably between species. Thus perennial plants species generally root deeper than annuals. For agricultural crops a rooting depth of 50 to 100 cm is common but some species may have root depths of 2 m and more. Root growth, root morphology and root depth are influenced by external factors, especially by the soil atmosphere, by mechanical impedance, and by plant nutrient status. This question has been considered by Drew and Goss (1973). These authors stress that O_2 supply is essential to root growth and metabolism, but that low O_2 partial pressures such as 5 kPa generally suffice for normal root development. Some of the O_2 required can be taken up from the atmosphere by the leaves and then transported into the roots (Greenwood 1971). Anaerobic soil conditions, however, may not only affect O_2 supply to plant roots, but may also result in the formation of toxic substances which inhibit root growth and may lead to severe root damage. Such toxic substances include ethylene and volatile fatty acids (see page 48).

Mechanical impedance may restrict root growth considerably (Keita and Steffens 1989). Generally as roots grow through the soil they follow soil pores and fissures. In this process the roots have to enlarge some soil pores which are initially smaller than themselves. The root tip has therefore to displace soil particles. This only occurs if the mechanical impedance of the soil is not high, otherwise roots are not able to penetrate the soil (Bennie 1996).

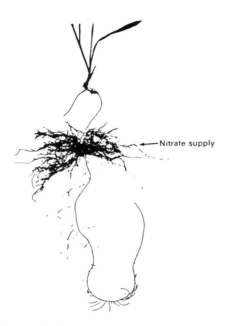

Nitrate supply

Figure 2.29 Effect of localized supply of nitrate on the growth of lateral roots of barley (Photo M.C. Drew).

Root growth is unlikely to be restricted by a lack of water, provided that enough water can be supplied to the root tip by other plant parts. In general root growth is less restricted in dry than in wet soils. Root proliferation depends much on plant nutrient distribution in the soil. Enhanced root growth in zones surrounding fertilizer granules has thus often been observed. An example of such an effect is shown in Figure 2.29 from the paper of Drew and Goss (1973). In this culture solution experiment all the root system received a complete nutrient solution but only the middle zone received 1.0 mol/m^3 nitrate. In the upper and lower zones the nitrate concentration was 0.1 mol/m^3. This higher nitrate supply resulted in marked root proliferation but this was restricted to the middle zone where the higher nitrate level had been provided. Too high salt concentrations in the soil medium, however, may restrict or even prevent root growth. In particular roots are sensitive to high concentrations of NH_3 and cationic Al species (Bennett and Adams 1970).

In annual crops the highest proportion of roots is generally present in the upper soil layer (0 to 20 cm). Rooting density decreases with soil depth as is shown in Figure 2.30 for maize grown at two different sites. From Figure 2.30 it is also evident that the total root mass was much greater in the Luvisol (silty loam) as compared with the Ultisol (loamy clay). This difference in rooting density coincides with the air pore volume of the soils which was about 13% in the upper layer of the Luvisol and therefore more than twice as high as the air pore volume in the upper layer of the Ultisol. Mohr (1978) suggested that especially in gleys and pseudogleys the air pore volume is a limiting factor in root growth, whereas in well aerated soils the pore volume does not restrict root proliferation.

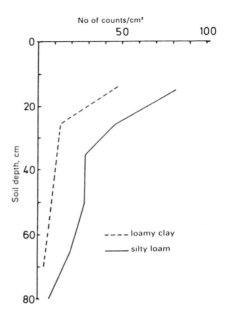

Figure 2.30 Root density of maize on a Luvisol (silty loam) with a high pore volume and on an Ultisol (loamy clay) with a low air pore volume. Density was measured by isotopes and is referred to as number of counts (after Mohr 1978).

The parameters by which root systems may be measured such as root mass, root length, root surface and rooting density are of varying importance. Root weight generally is a poor parameter as old and thick roots may contribute much to the weight but play only a minor role in nutrient and water uptake. Rooting density is a more important parameter. This is the length of root (cm) per unit volume of the soil (cm³) where Lv = rooting density and m is the arithmetic mean of the number of root axes intercepted per unit area of the three principle planes.

$$Lv = 2m$$

If for example the number of root axes intercepted is 6, 8 and 4 per cm² the rooting density is 12. Rooting density has a considerable influence on the extent to which a soil is depleted of water and nutrients. In dense rooting systems depletion zones resulting from nutrient uptake may overlap. This implies competition between neighbouring roots for available nutrients. As the depletion zones for K^+ and NH_4^+ are about 10 times higher than those for phosphate, K^+ and NH_4^+ uptake is more affected by the competition of neighbouring roots than is phosphate uptake. Thus Newman and Andrews(1973) found that a dense rooting system provided by a limited soil volume decreased the K^+ uptake of young wheat plants to a higher degree than phosphate uptake.

A further important root parameter is root length (L_A). This is defined as the total root length per unit soil surface.

$$L_A = \text{Total root length (km)/Soil surface (m}^2\text{)}$$

From this value the root surface area can be calculated, if the average root diameter is known. Soil type may considerably affect crop growth and root morphology. This is shown in Table 2.15 in which several root parameters are compared in field grown wheat from three soil types.

Table 2.15 Root parameters of winter wheat grown in the field (after Rex 1984)

Soil Type	Root length km/m² soil surfrace	Root density cm length/cm³	Rooting depth soil cm
Inceptisol (brown earth)	8.7	2.8	45
Ultisol (pseudogley)	12.4	4.0	45
Luvisol (grey brown podsolic	21.3	4.8	100

Besides root length and root density the number of root tips is also of importance, since some plant nutrients such as Ca^{2+}, Mg^{2+} and iron are mainly taken up by young root tissues in which the cell walls of the endodermis are unsuberized (Clarkson and Sanderson(1978), Clarkson and Hanson (1980). Behind the root tip, the calyptra following along the root axis are the division and elongation zones then comes the root hair zone which begins about 1 to 2 cm from the tip. The division and the elongation zones are free of root hairs as can be seen from Plate 2.4 from investigations of Kosegarten *et al.* (1999a).

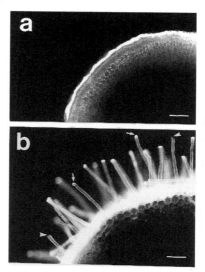

Plate 2.4 Fluorescence micrographs of a cross section of young maize roots showing boronate coupled fluorescein fluorescence in the apoplast of peripheral root cells. A: root elongation zone — B: Root hair zone (Photo: Kosegarten *et al.* 1999a)

The calyptra and the root hair zones have a relatively low microorganisms surface cover and for this reason there are no major interactions between bacteria and roots in these section of the root axis. The zone of root hairs is the most important segment for nutrient uptake (Claassen and Jungk 1984, Uren and Reisenauer 1988). Root hairs also allow the close contact between the soil and the root forming a water continuum between the soil and the root tissue. Root hairs are capable of penetrating moderately resistant clays and may thus contribute to nutrient exploitation of less accessible soil particles. They also play a special role in the acquisition of nutrients of limited mobility in the soil as they substantially increase the root surface area. Itoh and Barber (1983) reported that the contribution of root hairs to phosphate uptake depended on their length. In species such as wheat, lettuce, and carrots with relatively short root hairs they were of little influence. In species with long root hairs, however, including tomato, and Russian thistle (*Salsola kali*) root hairs play an important role in the uptake of phosphate. The life span of root hairs and their density may differ considerably between species. Generally root hairs collapse after a few days, but they may also persist for a longer period especially in grasses. Differences in root hair density may occur between the cultivars of a particular species (Bole 1973). This is shown in Plate 2.5. The wheat cultivar shown in section C of the Plate has a root hair density which is about four times greater than that of the cultivar shown in section A.

The nutrient uptake potential of a root system (roots per number of plants or roots per unit soil surface) generally by far exceeds the nutrient requirement of the plant. A small portion of the total root system is thus capable of absorbing the water and nutrients required for the whole plant provided that it is abundantly supplied with nutrients and water. This has been shown by Maertens(1971) for maize and the same is probably also true for other crop species (Drew and Goss 1973). The relatively high nutrient uptake potential of root systems enables the plant to absorb adequate amounts of nutrients even under conditions, where the level of accessible nutrients in the soil is low.

It must be borne in mind that under field conditions the total surface of a root system is unlikely to be in direct contact with the soil solution. The root surface/soil solution contact is mainly restricted to areas where soil pores extend to the root surface. This implies that at any given time only a part of the entire root system absorbs water and nutrients.

Considering the nutrient demand of plants per unit root system *e.g.* per unit root length, it is obvious that it is the young plant which has the highest nutrient requirement. This question has been studied by Mengel and Barber (1974) with maize under field conditions. The most important results of this investigation are shown in Table 2.16. It can be seen from this Table that the nutrient requirement per m root length and per day was especially high in the early stage of growth and declined rapidly as the crop developed. Similar results were reported by Adepetu and Akapa (1977) who found that the uptake rates for P and K (uptake per m root length) were four to five times higher in 5 days old cowpea plants (*Vigna unguiculata*) as compared with 30 days old plants. Vincent *et al.* (1979) also found that the K^+ uptake rates of root tips of soya beans were much higher in the vegetative stage of plant growth as compared with the reproductive stage. From these observations it is clear that it is the young plant in particular which needs to take up a high nutrient quantity per unit root length and it is for this reason that young plants especially require a relatively high level of available nutrients in the soil.

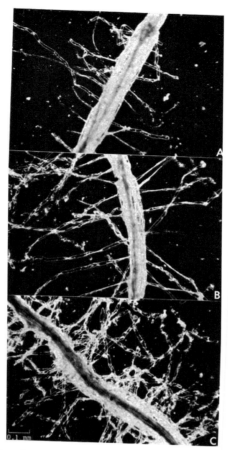

Plate 2.5   Root hair density of three different wheat cultivars. A: Chinese Spring (20 root hairs/mm)
— B: Chromosome substitution line (35 root hairs/mm) — C: S - 615 (80 root hairs/mm) (Photo: Bole)

Table 2.16   Nutrient requirement of maize per unit root length in relation to the age of plants (Data from
Mengel and Barber 1974)

Age of plants, days	μmol/m root length				
	N	P	K	Ca	Mg
20	227	11.3	53	14.4	13.8
30	32	0.9	12.4	5.2	1.6
40	19	0.86	8.0	0.56	0.90
50	11	0.66	4.8	0.37	0.78
60	5.7	0.37	1.6	0.08	0.29

Root studies are not easy since the measurement of root systems and root parameters is complicated. Böhm (1978) described and commented on the various techniques by which root systems are measured under field conditions. A very interesting method for measuring root growth and root systsems was developed by Sanders and Brown (1978) in which a highly refined fiber optic duodenoscope is used for observing and photographing root development patterns within a soil profile.

2.3.7 The rhizosphere and root exudation

The immediate vicinity of plant roots is of particular importance for plant nutrient turnover and availablility. This part of the soil which is directly influenced by roots, is called the rhizosphere and extends about 1 to 3 mm from the root surface into the bulk soil. The effect of the root on the adjacent soil medium is mainly brought about by the release of organic and inorganic material into the soil. Organic matter arises from the sloughage of root material and also from direct root exudation (Bowen and Rovira 1991). As plant roots push their way through the soil some of the outer tissues are sloughed off and decomposed by autolysis or by microorganisms. The total amount of organic carbon thus released into the soil medium is considerable. Several workers have shown that as much as 10 to 30% of the total C photosynthesized by plants is released from the roots into the soil (Whipps and Lynch 1986, Dinkelacker et al. 1989). A gelatinous material called mucigel is formed on the surface of the roots. This is made up of original and modified plant mucilages, bacterial cells and their metabolic products as well as colloidal mineral and organic matter from the soil (Bowen and Rovira 1991). This large amount of organic material is readily decomposed by rhizosphere microorganisms. Root sloughage is considered to be the main source of carbon released by roots. Besides this, production of mucilage contributes much to the transfer of organic carbon from roots to the soil. Mucilage is a layer of granular and fibrillar material covering the surface of roots and root hairs. This slimy material consists mainly of polysaccharides with galactose, fucose and uronic acids as the most important monomeric building blocks (Paul et al. 1975). It is now well established from the investigations of Paul and Jones (1975, 1976) with maize root cap slime, that the polysaccharides are produced in the root cap cells. It is believed that the dictyosomes are the site of polysaccharide synthesis and that these microbodies are also involved in polysaccharide transport and secretion.

Bacteria may feed on this material on the root surface. Bacteria embedded in mucigel are shown in Plate 2.6 from the work of Guckert et al. (1975). Analysis of epidermal cell walls by lysates, hydrolytic enzymes, also allow the bacteria to invade the epidermal and corrical cells of the root. The mucigel is capable of adsorbing clay minerals (Breisch et al. 1975), toxic Al cation species (Horst et al. 1982) and heavy metals such as Cu^{2+}, Cd^{2+}, and Pb^{2+} (Morel et al. 1986). It is thicker at the root tip and may thus protect the root meristematic tissue from such toxicities. The gels produced by roots and microorganisms promote close contact between the root and the soil, with the gel filling spaces between roots and soil particles. The close contact between roots and soil is of importance for nutrient and water supply particularly in preventing the formation of a gap between roots and soil when roots shrink under conditions of high transpiration.When a plant is transplanted this close contact formed by mucigel between root hairs and soil particles is destroyed. On replanting, this has to be reestablished by the production of mucigel and the development of new root hairs. The period after transplanting is thus a critical one as water supply is disturbed which in turn affects metabolic processes. This phenomenon is known as transplanting shock and plays a particular role in transplanting rice seedlings (Schnier *et al.* 1990).

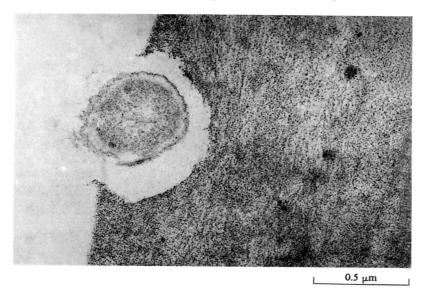

0.5 μm

Plate 2.6 Bacterium embedded in the slime layer of a root. The slime (mucigel) has been dissolved around the bacterium. (Electronmicrograph: Guckert *et al.* 1975).

Besides polysaccharides, roots exude a number of low molecular weight organic compounds of which amino acids, sugars and organic acids are the most frequent in occurrence. The release of these compounds by plant roots provides a source of food for microorganisms in the rhizosphere. Microbial activity in the rhizosphere is thus to a large extent dependent on plant metabolism since this regulates exudate release. Plants growing under favourable conditions for example translocate considerable quantities of photosynthates to the roots so that root metabolic activity is high and the rhizosphere microorganisms are well supplied with organic carbon amd nitrogen compounds (Merbach *et al.* 1999). This promotes multiplication of microorganisms increasing the microbial biomass (Joergensen 1994) associated with the assimilation of inorganic nitrogen (Janzen 1990, Mary *et al.* 1993, Jensen and Sorensen 1994).

The enrichment of the rhizosphere with organic compounds (sugars, amino acids), utilized by microbes, results in a considerably higher microbial density in the rhizosphere than in the bulk soil. The ratio between the microbial densities of the rhizosphere and soil (rhizosphere = R, bulk soil = S) may be as high as 100 (Katznelson 1946). Table 2.17 shows this R/S ratio and the bacterial density in the root free soil and in the rhizosphere for a number of crops (Rovira and Davey 1974).

Table 2.17   Bacterial colony counts in the rhizosphere of crop plants and
in the bulk soil (after Rovira and Davey 1974)

Crop	Colony count, 10^6/g soil		
	Bulk soil	Rhizosphere	R/S ratio
Red clover	134	3255	24
Oats	184	1090	6
Flax	184	1015	6
Wheat	120	710	6
Maize	184	614	3
Barley	140	505	3

The N_2 fixers live in close association with the plant root. Under favourable conditions they may fix considerable amounts of nitrogen (Neyra and Döbereiner 1977) as considered in more detail on page 399. The ammonifying bacteria produce NH_3 from the exuded amino acids and the proteins present in the debris of root cells. Ammonium thus released may be reabsorbed by the plant, incorporated into microbial bodies or even fixed by clay minerals. The denitrifying bacteria produce volatile N_2 and N_2O under anaerobic conditions (see page 349). Species of *Nitrosomonas* and *Nitrobacter* which produce nitrite or nitrate respectively are less frequent in the root vicinity (Rovira and Davey 1974). Nitrification can be inhibited in the rhizosphere of some plant species, especially grasses and trees. This process is probably of particular importance in the supply of N to these species (Rovira *et al.* 1983).

The excretion of inorganic and organic compounds by plant roots may have a considerable effect on the ability of plant roots to exploit the soil for nutrients. One of the most important mechanisms in this respect is the ATPase proton pump which is located in the plasmalemma of root cells and is the driving force for nutrient uptake (Serrano 1989, for details see page 126). Protons pumped out of the cell into the rhizosphere render it acid and the pH attained depends largely on the H^+ buffer power of the soil. If this is high the pH in the rhizosphere is buffered and *vice versa* (Hauter and Mengel 1988).

The pH in the rhizosphere is greatly influenced by the uptake of anions mediated by H^+/anion cotransport (see page 127) by which protons are transported back into the cytosol. In this removal of H^+ from the rhizosphere nitrate plays a major role since it represents about 80% of all the anions taken up by the plant provided nitrate is the major N source. It is for this reason that the rhizosphere is usually alkaline in plants supplied with NO_3-N (Smiley 1974, Marschner *et al.* 1991). By contrast NH_4 nutrition acidifies the rhizosphere. The same is true for plants supplied exclusively with N_2 fixed symbiotically *e.g.* legumes (Hauter and Steffens 1985). The pH in the rhizosphere influences microbiological activity; a low pH depressing bacterial growth, but favouring the proliferation of fungi including mycorrhizal fungi (Paul and Clark 1996). An example of the benefit of low rhizospheric pH induced by NH_4-nutrition is the control that it offers for "take all" the bacterial root disease of wheat. Low pH particularly retards nitrification and denitrification. The activities of some root enzymes can be greatly increased. This is the case for phosphatases which cleave phosphate esters to provide a source of inorganic phosphate for plants (Tarafdar and Jungk 1987, Tarafdar and Claassen 1988). Low pH in the rhizosphere may favour the dissolution of less soluble compounds such as phosphates or Fe oxides/hydroxides and even induce the decomposition of minerals so that cations such as Mg^{2+} and K^+ are released (Feigenbaum et al. 1981). Even toxic heavy metals may be dissolved (Sauerbeck and Lübben 1991). On the other hand low pH restricts the desorption of phosphates and sulphate.

As already discussed on page 69 nutrient concentrations may be very low near the root surface as a result of nutrient uptake. Such low concentrations promote the dissolution of less soluble compounds and also the desorption of nutrients including the release of interlayer K^+ (Kuchenbuch and Jungk 1984, Hinsinger and Jaillard 1993) and interlayer NH_4^+ (Scherer and Ahrens 1996). In Ca-rich soils Ca compounds may accumulate at the root surface because much more Ca^{2+} is transported to the root surface via mass flow with the water than can be taken up by roots (Barber 1974). Such Ca accumulations may lead to root calcification as reported by Jaillard et al. (1991).

The excretion of organic anions or acids by plant roots merits particular interest since it can be related to the nutrient status of plants. Hoffland et al. (1989) reported that rape plants suffering form insufficient phosphate supply excreted malic and citric acids and thus decreased the pH in the rhizosphere which promoted the dissolution of rock phosphate. Liu et al. (1990) also found that rice plants excreted citrate when suffering from poor phosphate nutrition. Similar findings have been made for leguminous plants as reported by Gerke et al. 1994 for red clover Trifolium pratense. Since citrate is also very potent in the desorption of phosphate from soil particles citrate release from plant roots can mobilize adsorbed phosphate (Gerke 1994). According to Ae et al. (1990) pigeon peas have a particular potential to exploit phosphates from Fe compounds by excreting piscidic acid which is a derivative of tartrate and which is capable of chelating iron and hence breaking down Fe-phosphate compounds.

The most prominent species capable of exploiting soil phosphate by citrate excretion is the white lupin (Lupinus albus) which has proteoid roots. These are characterized by a cluster of highly branched short lateral roots and which are capable of rapid excretion of relatively high amounts of citrate into the rhizosphere (Gardner and Parbery 1982). Citrate may mobilize soil phosphate not only by desorption of adsorbed phosphate and by chelating of Fe^{III} of mineral Fe compounds (Gardner et al. 1983) but also in calcareous soils by chelating Ca^{2+} and thus dissolving Ca phosphates as shown by Dinkelacker et al. (1989). These authors found that even in a soil with 20% $CaCO_3$ the pH in the immediate vicinity of the proteoid roots was 4.8 while in the bulk soil the pH was 7.5. The formation of proteoid roots is dependent on the P status of the plants, developing as a response to a lack of phosphate. Proteoid roots are mainly found in species of the Proteaceae which grow on very poor acid soils and the growth of proteoid roots and accompanying release of citrate provides a means of adaption to an adverse environment. A valuable review on the physiology of proteid roots has been published recently by Watt and Evans (1999). According to Uren and Reisenauer (1988) citrate and malate may also be involved in reduction processes since both anions may provide reducing H atoms which may reduce Fe and Mn compounds. The reduction process, presumably requiring microbial enzymes, is shown below for malate and $Mn^{IV}O_2$.

$$HOOC\text{-}CH_2\text{-}HCOH\text{-}COO^- + Mn^{IV}O_2 + 2H^+ \Rightarrow HOOC\text{-}CH_2\text{-}CO\text{-}COO^- + Mn^{2+} + 2H_2O$$
$$\text{malate} \qquad\qquad\qquad\qquad \text{oxaloacetate}$$

An analogous example of adaption to an insufficient supply of nutrients is the response of plant species to an insufficient supply of Fe. Under Fe stress dicotyledons release increased rates of protons and grasses excrete phytosiderophores (Römheld and Marschner 1986). These organic compounds, such as avenic acid and mugineic acid, dissolve Fe^{III} out of mineral Fe compounds by chelation. The water soluble Fe^{III} siderophore may then diffuse to plant roots to be taken up (further details see page 566). The mechanism of anion excretion by plant roots is still not clear; the problem is considered in the chapter of ion uptake (see page 126).

The mechanism of the excretion of uncharged molecules such as sugars by roots is also not yet understood. It is feasible that this is a simple leakage. A substantial proportion of sugars leaking out of the root tissue into the apoplast may be retrieved (Mühling et al. 1993) by proton cotransport. Where there is a shortage of protons in the apoplast, considerable amounts of sugars may diffuse into the rhizosphere.

The acquisition of plant nutrients from the rhizosphere has been considered by Marschner and Römheld (1996) and Hinsinger (1998) in useful review articles.

2.3.8   Mycorrhiza

The term mycorrhiza was at first used by the German scientist A.B. Frank in 1885 to describe the fungal hyphae closely associated with plant roots. He called them "Mykorrhizen" which literally means "fungus roots". It was later found that the fungi live symbiotically with higher plants. The higher plant provides organic carbon to the fungus, mainly as carbohydrates and the fungus exploits the soil providing the root system with additional nutrients and water. The fungal hyphae are in close contact with the plant root and extend into the soil. By this the contact surface of the mycorrhizal root with the soil is enlarged considerably so that the plant is able to exploit a larger soil volume for nutrients and water. Plants with a poor root system generally develop a large mycorrhiza. This symbiosis between the fungi and the plant is some hundred million years old and it is believed that the fungi enabled aquatic plants to adapt to terrestrial conditions (Barea and Azcon-Aguilar 1983).

Mycorrhizal fungi are currently grouped into three major groups: *ericoid mycorrhizae*, *ectomycorrhizae* (EM) and *arbuscular mycorrhizae* (Paul and Clark 1996). The ericoid mycorrhizae are mainly associated with plant species living on poor acid humic soils, especially heather plants such as *Calluna* and *Vaccinium* species. Accordingly the ericoid mycorrhizae are adapted to a soil pH as low as 3.5 to 4.2. Present in the hyphae are proteases and polyphenol oxidases, enzymes which degrade organic soil matter with a high C/N ratio (>100).

The ectotrophic mycorrhizae cover roots and rootlets with a thick mantel of hyphae. This is shown in Plate 2.7 for roots of *Pinus silvestris* (Trolldenier 1971a). The fungal sheath spreads between the cortical cell of the roots thus enabling the fungus to maintain a close contact with the plant.

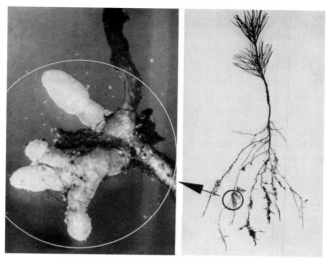

Plate 2.7 Young *Pinus silvestris* with mycorrhiza (right), hyphae mantel around root tip (left) (Trolldenier 1971a).

The fungi depend on carbohydrates which are supplied by the root. It has been shown that ^{14}C labelled photosynthates are rapidly translocated not only to the roots but also into the sheath and hyphae of mycorrhizal fungi (Harley 1971). Carbohydrates, mainly in the form of st ~ose, obtained from the host root are rapidly converted into typical fungal sugars such as trehalose, mannitol or even glycogen. In this way the organic carbon is trapped because these carbohydrates are only poorly reassimilated by the plant root. It has been shown that the transport of assimilates from the aerial plant parts towards the roots is higher in mycorrhizal infected plants than in non infected plants. The mycorrhiza is thus a sink for photosynthates.

The thick sheath of hyphae covering the roots favours the uptake of water and inorganic nutrients, especially phosphate since it effectively enlarges the surface area of the root in direct contact with the soil. The thin mycorrhizal hyphae (diameter 2 to 4 μm) are able to penetrate soil pores inaccessible to the root hairs which have a diameter about 5 times greater than that of the hyphae. According to Gerdemann (1974) infected roots live longer than non infected ones. The finest lateral rootlets, which have a very short life in uninfected conditions and remain unbranched, respond to mycorrhizal infection by growing for a longer period of time and by branching. In this way the mycorrhizal fungi assist the root in exploiting the soil for water and nutrients in particular phosphate. The fungus is also capable of accumulating plant nutrients and supplying these nutrients to the plant in periods of low soil nutrient availability. Thus host plant supplies the mycorrhiza with photosynthates and the fungus in turn supplies the plant with inorganic nutrients.

Ectomycorrhizae are mainly found on roots of trees and shrubs and are of economic importance for the growth of forest trees. They make up only a few percent of all the mycorrhizae. Numerous investigations have shown that the ectomycorrhizae promote the growth of trees when grown on soils low in available phosphate. There are cases where the mycorrhiza is essential for growth. It has thus been reported that seedlings of

Pinus and *Picea* planted on newly drained organic soils only grew when the roots were infected with the appropriate mycorrhizal fungus (Schlechte 1976).

The physiological relationships between the host plant and the endomycorrhiza are analogous to those described above for the ectomycorrhiza and according to Paul and Clark (1996) the same fungus producing an ectomycorrhiza (EM) on one host plant may form an endomycorrhiza on a different host plant or under different conditions. Endo-mycorrhizas have some characteristics of both EM and arbuscular mycorrhiza (AM). The latter belong to the endomycorrhiza and mainly comprise species of the Glomales. The hyphae of these fungi, in contrast to those of the ectomycorrhizae, penetrate the cells of the root cortex forming an internal hyphal network. Some hyphae also extend into the soil (see Figure 2.31).

For many plant species including most agricultural crops the predominant type of fungal infection is the arbuscular mycorrhiza (AM) previously called vesicular arbuscular mycorriza (VAM). This name derives from the occurrence of two types of structures characteristic of the infection, vesicles and arbuscules (Harley 1971). The vesicles are used for storage particularly for lipids produced from sucrose provided by the host plant (Barea and Azcon-Aguilar 1983). The arbuscles persist in the individual plant cells for only several days. It is assumed that they mediate the nutrient transfer between their branched hyphae and the highly invaginated plasmamembrane. A diagrammatic representation of vesicular arbuscular mycorrhiza is shown in Figure 2.31.

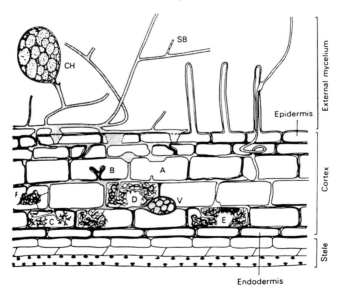

Figure 2.31 Schematic diagram of the association of vesicular arbuscular mycorrhizal fungi and a plant root. The external mycelium bears large chlamydospores (CH) and occasional septate side branches (SB). Infection of the plant can occur through root hairs or between epidermal cells. Arbusculae at progressive stages in development and senescence are shown (A—F) as is also a vesicle (V). To avoid confusion cell walls of the root are not indicated when they underlay fungal hyphae (from a drawing by F.E. Sanders, p.129, Plant Root Systems by R. Scott Russell 1977 by courtesy of The McGraw Hill Book Company).

The fungus may be considered as a two-phase system with a mycelium in the cortex connected to an external mycelium in the rhizosphere and soil. The figure shows hyphae penetrating the cortex by producing a series of branches both outside and within the cortex cells. It also shows the presence of structures at different stages in development with a shrub-like appearance called arbuscules (*arbuscula*, Latin = shrub). These structures are similar to haustoria but are produced by dichotomous branching of hyphae. The second type of structure the vesicle (*vesicula*, Latin = small bladder) may be formed by swelling of the hyphae and may occur within or between cells. External vesicles also develop on the external hyphae.

The symbiosis between the host and fungus is well co-ordinated. Young feeder roots show no sign of damage when infected. Indeed the living root is essential for the culture of these fungi (Hayman 1980). This symbiotic relationship between host and fungus in principle is the same as that between the ectomycorrhiza and host plant, the host plant supplying the fungus with organic carbon compounds and the fungus assisting the roots in exploiting the soil for water and inorganic nutrients. The relationship is of particular importance for phosphorus nutrition since phosphate depletion zones can readily occur around plant roots (see page 69). The network of hyphae extending from the root into the soil, as with the ectomycorrhizae, enlarges the contact area between the soil and the fungus-host root association and hence facilitates a greater uptake rate of phosphate. This has been demonstrated by Sanders and Tinker (1973) for onion.

Mycorrhizal fungi may take up phosphate from less soluble phosphate forms (Tinker 1984). Plants with root systems of low specific surface areas with fleshy roots and few root hairs profit most from phosphate uptake from arbuscular mycorrhiza. These plants species include onions, citrus and grapevine. Arbuscular mycorrhizae appear to be of particular importance for leguminous species since their presence can enhance N_2 fixation in a tripartite symbiosis. Mosse *et al.* (1976) found that clover did not nodulate in a grossly P deficient soil unless it was mycorrhizal. As well as increasing P uptake mycorrhizas may possibly also increase the uptake of other elements including Zn, Mn, Cu (Marschner and Römheld 1996) as well as Mo, which is involved in N_2 fixation. This may explain results of El-Hassani and Lynd (1985) who found that infection of *Vicia faba* with the mycorrhizal fungus *Glomus fasciculatum* significantly increased N_2 fixation, nodulation and nitrogenase activity. Nodules themseves are not colonized by mycorrhizae (Barea-Azco-Aguilar 1983).

Most crop species, except those of the Brassica family and sugar beet have arbuscular mycorrhizae. Infection is widespread in the families of the legumes and grasses and hence found in pasture and forage legumes, in maize, wheat barley and many vegetable species. Plantation crops such as coffee, tea, cocoa, oil palm, papaya, and rubber are known for their mycorrhizal infection (Hayman 1980). Cassava is believed to be obligately associated with an arbuscular mycorrhiza and for this reason is able to grow reasonable well even on phosphate deficient soils.

The interactions between the host plant species and the mycorrhizal fungus are very specific. Best effects in nutrient exploitation of soils are only obtained if the host plant root is infected by the appropriate fungus species and even ecotype. Lioi and Giovannetti (1987) inoculated *Medicago sativa* with various species of *Glomus* with varying success. Using *Glomus occultum* the lucerne yields were increased by a factor of 17 as compared with the non inoculated crop but with *Glomus caledonium* the fungus had no effect. Analogous results were obtained by Diederichs (1991) who inoculated *Zea mays* with AM species and who also found growth responses depended markedly on the mycorrhizal species. This finding highlights the severe problem of utilizating mycorrhiza for the exploitation of soil nutrients. Appropriate fungal species are required and these may differ for various crops, causing great difficulties when crops are grown in rotation. Hall (1987) emphasizes, in a valuable review article, the importance of mycorrhizae for pasture species, especially for legumes. Since pastures cover about 2/3 of the total global agricultural land this aspect merits attention. Hall (1987) reported that the response to inoculating pastures with mycorrhiza was greater the poorer the soil as can be seen in Table 2.18 which shows that the yield increment obtained by inoculation was dependent on the fertility status of soils.

Table 2.18 Effect of soil inoculation with mycorrhizae on yield increase of forage as related to the fertility status of soils (Hall 1987)

	Yield increment, kg DM/ha
Very poor fertility	1000
Poor fertility	750
Medium fertility	825
High fertility	640

Inoculation using mycorrhizal, fungi in the field still is difficult, not least because high quantities of inoculum are required in the range of 0.8 to 100 t/ha (Hall 1987). This technical problem still remains to be solved.

2.4   Determination of Nutrient Availablity

2.4.1   General

Ever since man has grown crops it has been well known that soils differ widely in their fertility. Understanding the factors which underlie the phenomenon of soil "fertility", or the capability of soils to produce good crop growth has therefore been of interest for a very long time. It is only relatively recently, however, that it has become increasingly evident that soil fertility is dependent both on physical and chemical properties of the soil. The discovery made in the 19th century, that plants receive most of their chemical constituents from the soil, revealed that one of the components of soil fertility is the content of plant nutrients present within a soil. As already discussed in detail, however, the total amount of plant nutrients in the soil is not of primary importance in this respect, but rather the content of soluble and easily accessible nutrients. The determination of this available nutrient fraction can be carried out by various techniques. These differ basically in principle and three different approaches are provided by the methods of soil analysis, plant analysis, and plant experiments such as pot and field trials.

2.4.2 Soil sampling and interpretation of soil tests

All soil analytical methods depend very much on careful soil sampling, as the nutrient concentration of a soil can differ markedly not only within the same profile but also in the same horizon. For a given soil, sufficient sub-samples must therefore be collected in order to obtain a representative sample. This is particularly true for the determination of available N, as the N concentration of soils can differ widely from one site to another. For phosphate and K^+, 25 sub-samples per ha are generally regarded as sufficient in obtaining a representative sample (Hanotiaux 1966).

In interpreting soil analytical data the density of the soil should also be considered. The most important factor influencing this is the soil organic matter content. Generally the higher the organic matter content of a soil, the lower is its density. For an organic soil for example, 1kg of soil has a considerably higher soil volume than that of 1 kg of mineral soil. The quantity of available nutrients as determined by extraction methods on a weight basis therefore refers to a larger soil volume. Caution must therefore be exercised when comparing available nutrients from organic and mineral soils. In order to overcome this difficulty calculations of available nutrients are often expressed on a unit area basis, *e.g.* "available kg P or K per ha". For the necessary calculation generally only the top soil layer (20 to 30 cm) is taken into account.

There are other difficulties associated with the interpretation of soil analytical data. This holds true in estimating a critical level above which no further yield increase results from additional fertilizer application. Generally values lower than 80 mg of exchangeable K^+/kg soil are indicative of soils capable of significant yield responses following application of a K fertilizer. Fertilizer responses may well be obtained, however, in soils with higher concentrations of available K^+ than for the critical value given above. So much depends on other factors such as rooting depth, soil type and texture and also on climatic conditions. As a general rule the critical level for an available nutrient, its threshold value, is that level above which there is no yield response to fertilizer application. The higher this is the more favourable are other growth conditions including soil and climatic factors. The higher the potential growth rate of a crop the greater is the critical level for the nutrient availability because high growth rates also require high rates of nutrients supplied to plant roots.

For nutrient availability two components must be considered, nutrient intensity and nutrient quantity (see page 74). Unfortunately most soil analytical methods give results which are dependent on both. In extreme cases two soils may have the same amount of available phosphate as determined by one particular method, but both their phosphate intensity and quantity measurements may differ considerably. For this reason application of phosphate fertilizer may give rise to quite different responses on the two soils concerned. The same is also true for K^+. Account must therefore be taken of both quantity and intensity measurements.

Another important aspect which must be considered in relation to nutrient availability is the difference in nutrient requirement between crops. High yielding crops and intensive cropping systems place a greater demand on available nutrients in the soil and particularly on phosphate and K. This is especially the case where high N levels are applied.

Nutrient availability is related to physico-chemical factors such as nutrient concentration in the soil solution, buffer power, and soil moisture and also to biological factors such as root parameters and microbial activity in soils. Silberbush and Barber (1983a, b) in evaluating the importance of the various factors in a mechanistic-mathematical model found that for phosphate and K^+ uptake, biological factors such as root length and root radius were as important as the physico-chemical factors. For phosphate uptake the following order of importance was found: Root length > phosphate concentration of the bulk soil solution > root radius > phosphate buffer power > diffusion coefficient. Both the estimation and prediction of biological factors are difficult and it must always be remembered that the determination of the physico-chemical factors can only give an indication of nutrient availability in the soil-plant-root system.

2.4.3 Estimation of cations

In soil analysis most methods of available nutrient estimation involve treatment of the soil with a suitable extractant to remove an accessible nutrient fraction. This is the case for the cationic nutrients. For these nutrients the easily accessible ions in the soil are made up by cations either dissolved in the soil solution or adsorbed on inorganic and organic exchange complexes. The greater proportion are in exchangeable form. It is for this reason that the determination of available cations is based on the analysis of exchangeable cations. The extractants used contain an excess of a cation species which is able to exchange with the adsorbed cation nutrient. This principle is applied in the determination of exchangeable soil K^+ and Mg^{2+}. Using this type of extraction a high proportion of the total exchangeable and nearly all the dissolved cations of the soil solution are removed from the soil and then determined in the extract. This estimation gives an indication of the quantity factor of the nutrient concerned.

A number of different extractants are used in various methods but the principle is the same. Generally 1 $kmol/m^3$ NH_4^+ acetate or 1 $kmol/m^3$ NH_4Cl (kmol = kilo mol) are employed in the determination of exchangeable K^+ and Mg^{2+} (Hesse 1971). "Exchangeable K", however, is often not a satisfactory parameter for measuring K^+ availability since in addition to the "exchangeable K^+" the "non exchangeable K^+" (K^+ in the interlayers layer silicates, page 484) is of substantial importance of supplying a plant with K^+. Some of this non exchangeable K^+ can be extracted by dilute acids. To assess this portion of non exchangeable K^+ that is plant available. Schachtschabel (1961) proposed that the soil be extracted with 10 mol/m^3 (= 10 mM) HCl, a technique similar to that using 10 mol/m^3 HNO_3 according to Pratt (1965) and McLean and Watson (1985). In the Netherlands available K^+ is determined by extracting soil samples with 1 mol/m^3 HCl. These acid extractants remove the exchangeable K^+ and also extract some of the non exchangeable K^+. Non exchangeable K^+ may also be extracted by the method of electro-ultrafiltration (EUF) as described on page 99 (Mengel and Uhlenbecker 1993, Mengel et al. 1999). The method, predominantly used in Germany for the determination of available K^+ is based on the extraction of the soil with an acid Ca lactate solution. This extractant is of a Ca lactate solution + HCl with a lactate concentration of 0.25 mol/m^3 and a pH of 3.6 (Hoffmann 1991).

Available Mg^{2+} is usually determined by extraction with 10 mol/m³ NH_4 acetate or NH_4Cl or by the extraction with 0.125 mol/m³ $CaCl_2$ (Schachtschabel 1954). Details of the various methods and extractants used for exchangeable cations can be found in the standard texts on soil chemical analysis (Hesse 1971, Sparks et al. 1996).

2.4.4   Estimation of phosphates

The most important soil phosphates are the Ca phosphates, phosphates adsorbed to soil colloids, and organic phosphates. The choice of a suitable method for the determination of available phosphate in a given soil depends on which phosphate forms are dominant in the soil. In acid and neutral soils a wide range of extractants are used. Anions present in an extractant mainly exchange the adsorbed phosphate and H^+ dissolves phosphate precipitates, predominantly Ca phosphates. Hence solutions with a low pH are suitable for soils with a higher proportion of Ca phosphates and solutions with organic anions or fluoride are suitable to exchange the adsorbed phosphate. Solutions with a low pH may also favour the hydrolysis of organic phosphates. The extractant proposed by Bray and Kurtz (1945) consists of 0.03 mol/m³ NH_4F + 0.025 mol/m³ HCl. Hence this extractant desorbs adsorbed phosphate and also dissolves Ca phosphates. Truog (1930) used a 1 mol/m³ H_2SO_4 + a small amount of ammonium sulphate (pH 3.0) an extractant which mainly dissolves Ca phosphates. Olsen's extractant consists of an 500 mol/m³ $NaHCO_3$ solution (pH 8.5). The high pH and presence of HCO_3^- in this extractant both favour phosphate desorption (Olsen et al. 1954). In addition some of the protons dissociated from the HCO_3^- may contribute to the dissolution of Ca phosphates. The Olsen method is well suited for the determination of available phosphates in calcareous soils. In these soils acid extractants are inappropriate since protons are quickly neutralized by the surplus of the carbonates present.

In Germany the Ca lactate + HCl extractant (pH 3.6) mentioned above for the extraction of available K^+ is also used for the extraction of phosphate (Hoffmann 1991). This solution extracts too high amounts of phosphate in soils fertilized with rock phosphates (Werner 1969). For such soils the CAL method is recommended (Schüller 1969). This extractant consists of a solution of Ca lactate + Ca acetate + acetic acid (pH 4.1) and the pH is well buffered.

According to investigations of van der Paauw (1962, 1969) the extraction of soil phosphate by water has proved a useful tool in determining available soil phosphate. Phosphate extraction with water is hardly affected by the soil properties (concentration of humus, clay, lime, pH value). This method is characterized by a high water/soil ratio and for this reason the suspension is much diluted and therefore promotes the dissolution of Ca phosphates and the desorption of adsorbed phosphates. The absolute quantities of phosphate extracted by water are low and reflect well the phosphate concentration in the soil solution i.e. the phosphate intensity (see page 74) According to Willliams and Knight (1963) and to more recent experiments of Steffens (1994) phosphate intensity is a better indicator of phosphate availability than phosphate quantity. Williams and Knight (1963) who analysed a selection of different soils using various extractants, found that mild extractants of intermediate pH with fairly short extractant periods, and hence more indicative of intensity than quantity measurements, gave the highest correlations with yield. Intensity measurements are particularly useful in phosphate enriched agricultural soils where the concentration of phosphate in the soil solution can be sustained by a high phosphate capacity. The intensity factor is also important in relation to early growth and in highly responsive rapidly growing crops such as potatoes.

Another approach to assess plant available soil phosphate status is to extract phosphates from the soil using synthetic anion exchange resins. Results obtained by this technique correlated well with plant P uptake (Sibbesen 1978). The reason why the method gives a closer correlation with P uptake than that given by the extraction methods is that the exchange resin may more satisfactorily simulate the absorbing root which is a sink for mobile soil phosphate. The use of various extractants and their merits in available phosphate estimation has been discussed in detail by Hesse (1971) and Page *et al.* (1982).

2.4.5 Tracer techniques

About fifty years ago a new approach was made to determining available soil nutrients with the aid of radioactive elements. The technique is particularly applicable to nutrients elements with suitable radioactive isotopes such as ^{32}P and ^{45}Ca with half lives and radioactive emissions such to enable their fate to be followed in the soil. The principle of the methods used is based on the concept that a radioactive isotope which is added to the soil equilibrates with that fraction of its stable element in the soil which is accessible to the plant. This fraction is termed the *labile pool* and includes nutrients which are in solution or can readily pass into it. When for example radioactive phosphate is added to a soil suspension it will mix with soil phosphate and exchange with soil colloids and with the solid phosphates such as Ca and adsorbed phosphates. Eventually isotopic equilibrium will be attained. The study of nutrient availability and particularly that of phosphate has largely been aided by observations of dilution and exchange of isotopes under such equilibrium conditions.

If a solution of a radioactive substance is added to a solution of the same substance in unlabelled form, after the attainment of equilibrium, the ratio of labelled to unlabelled atoms will be constant throughout the system. This may be expressed as an equation:

$$\frac{\text{Total unlabelled substance}}{\text{Total labelled substance}} = \frac{\text{Unlabelled substance in any sample}}{\text{Labelled substance in any sample}}$$

The process is called isotopic dilution and it offers a very simple means of measuring the total quantity of an unlabelled substance. It is only necessary to add a known amount of a labelled form and withdraw a sample for analysis after equilibration.

Where soil is being considered the presence of both a liquid and a solid phase means that both dilution and exchange processes are involved. For a surface from which ions are in constant exchange with chemically identical ions in solution, the addition of an isotope to the same solution will result in the following reaction.

$$E \text{ exch.} + E^* \text{ sol.} \Leftrightarrow E^* \text{ exch.} + E \text{ sol.}$$

where E is an ion and E* is the isotope.

At equilibrium the following equation holds:

$$\text{E* exch./E* sol.} = \text{E exch./E sol.}$$

E exch.= exchangeable ion,
E sol.= ion in solution

This principle was developed for the determination of phosphorus availability in soils by McAuliffe et al. (1947). If P is substituted into the above equation the following holds

$$^{32}\text{P exch./}^{32}\text{P sol.} = {}^{31}\text{P exch./}^{31}\text{P sol.}$$

Using this equation exchangeable ^{31}P may be calculated as the other terms may be determined experimentally. ^{31}P (solution) and ^{32}P (solution) are the concentrations of stable phosphate and radioactive phosphate respectively in the soil solution after equilibration between the isotopic and stable phosphate has been attained. The term "^{32}P(exchangeable)" is equal to the difference between the amount of ^{32}P added to the soil suspension and the amount in the soil solution at equilibrium.

McAuliffe et al. (1947) used this technique to estimate what they described as 'surface phosphate' as they held the view that the more rapid stage of isotopic dilution in a soil suspension only involved phosphates on the surface of solid particles. The work was extended by Russell et al. (1954) who developed a rapid laboratory method for estimating available phosphates. These workers made no assumptions concerning the origin of the labile phosphorus. It was realised that isotope equilibration is never fully attained under laboratory conditions but recognised that after some time the rate decreases considerably. They therefore chose an arbitrary time of 48 hours for shaking suspensions of soil with radioactive phosphates and estimated total exchangeable phosphate which they termed the E or exchangeable value.

The estimation of exchangeable nutrients by this isotopic dilution procedure is not limited to phosphate. It is equally applicable to other nutrients and efforts have been made to determine available K^+ with the use of isotopic K^+ (Graham and Kampbell 1968), although the short half life of the K^+ isotope limits the method.

A further approach to determine available phosphate with the help of ^{32}P was made by Larsen (1952). This was to follow the specific activity of phosphorus taken up by a test crop grown in a soil labelled with ^{32}P. In his method the soil is thoroughly mixed with the carrier free ^{32}P present as soluble phosphate. Ryegrass is usually taken as the test crop so that several cuts can be taken, the growth period of the plant allowing for isotopic dilution to occur. Using the equation for isotopic dilution an L (labile value) can be calculated:

$$L = (C_o/C - 1)X$$

where C_o and C are the specific activities of the applied phosphorus and plant phosphorus respectively, and X the amount of phosphorus added. In practice it was found that the L value was independent of the amount of P added and became independent of time suggesting that isotopic equilibrium had been attained during the growth period. The L value itself is a measure of the total quantity of plant available soil phosphorus. It has been defined as "the amount of P in the soil and soil solution on the attainment of isotopic equilibrium that is exchangeable with orthophosphate ions added to the soil as measured by the plant growing in the system" (Larsen 1967a). In principle the concept of the E and L values is the same. Both are quantity measures of labile phosphorus but whilst the E value is calculated from chemical estimations on the soil solution, the plant is used for the L value. This accounts for the high correlation between both estimations. As has been pointed out by Larsen (1967a), however, the estimations are not identical, as isotopic exchange occurs under different environmental conditions for E and L values determination. The E value refers to a soil where no phosphorus removal takes place. For the L value the soil is at a moisture level below field capacity, and some removal of phosphorus is taking place which might well cause more extensive dilution than by isotopic exchange alone. Conceptual differences between E and L value measurements have been discussed very concisely by Fried (1964) and Larsen (1967b).The L value gives the best measurement of the quantity factor. In the field this is dependent on the volume of soil effectively utilized by plant roots, soil depth, physical conditions in the profile and the quantity and distribution of moisture (Williams (1970). The tracer technique for the estimation of nutrient fractions in soils is an important scientific tool. The expectations of researchers over forty years ago that the isotopic dilution technique would gain importance for practical soil tests, however, have not been fulfilled. Non isotopic methods mostly provide similar information, and are easier and cheaper to handle.

2.4.6   Electro-ultrafiltration technique (EUF technique)

This technique was mainly developed by Nemeth (1979) and his co-workers following earlier investigations by Köttgen (1933). In this method ions and molecules are extracted from soils by means of electro-ultrafiltration (EUF), the principle of which is shown in Figure 2.32. The soil suspension is exposed to an electrical field so that the cations migrate to the cathode and the anions to the anode where they are collected and thus removed from the soil suspension. The electrodes consist of a platinum sieve, covered by a filter, which dips into a vessel which is filled with a soil suspension. This is generally a suspension containing dry soil dissolved in distilled water. The electrical voltage between the electrodes and the temperature can be altered. In the standard programme the suspension is extracted at 20°C and 200 Volt the first 30 min followed by an extraction of 5 min at a temperature of 80°C and 400 Volt. This change in temperature and voltage increases the intensity of extraction as shown in Figure 2.33 from the work of Mengel and Uhlenbecker (1993). At 20°C and 200 V only low rates of K^+ are extracted and the curves level off after an extraction period of 30 min. The treatment following at 80°C and 400 V results in higher K^+ extraction rates and the cumulative K^+ release thus obtained is reflected by the Elovich function shown below (Mengel and Uhlenbecker (1993).

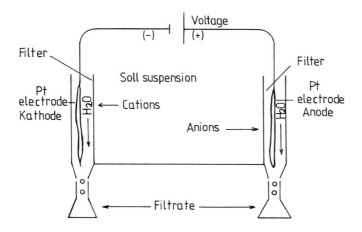

Figure 2.32 Scheme of the electro-ultrafiltration (EUF) apparatus.

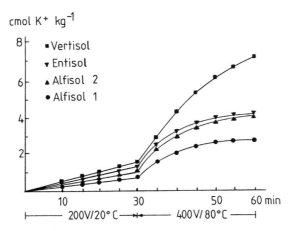

Figure 2.33 Cumulative extraction of K^+ from several soils by electro-ultrafiltration (EUF), after Mengel and Uhlenbecker (1993).

$$Y = a + b \, \ln x \text{ (Elovich equation)}$$

where Y is the cumulative quantity of extracted K^+, "x" the extraction time, and "a" and "b" constants. "a" is the intercept on the y axis and "b" represents the steepness of the slope and therefore an indication of available interlayer K^+. According to Nemeth (1971) the extraction at 20°C and 400 V mainly recovers the planar K^+. If soils are low in planar K^+, however, small amounts of interlayer K^+ also may be extracted (Mengel and Uhlenbecker 1993). More important is that the K^+ extracted at 80°C and 400 V almost completely originates from interlayer K^+ and thererfore the cumulative curves and the b values of the corresponding Elovich equation give a reliable indication of the release of interlayer K^+. In the standard technique of Nemeth et al. (1987) applied for routine soil tests extraction at 80°C and 400 V lasts only 5 min. Nevertheless, the value obtained from this 5 min extraction of K^+ correlated significantly with K^+ uptake values of ryegrass grown in soil in pot experiments (Mengel and Uhlenbecker 1993). In numerous field experiments with sugar beet Nemeth et al. (1987) showed that the routine EUF soil test for available soil K^+ precisely indicated the threshold value of available soil K^+ above which no further sugar yield was obtained. Simard et al. (1991) also reported that the uptake of K^+ by alfalfa grown in pot experiments was significantly correlated with the K^+ extracted by EUF.

The mechanism of plant nutrient extraction by EUF may be likened to soil exploitation of plant nutrients by roots. The extraction of cations and anions from the soil suspension in the EUF apparatus brings about a dilution of the nutrient concentration which gives rise to the desorption of adsorbed nutrients as well as to the dissolution of precipitated nutrients. Plant roots in a similar way dilute the nutrient concentration at the root soil interface (see page 68) as a result of uptake, which in turn promotes the desorption and dissolution of nutrients. This aspect is of particular importance for phosphate because the most important phosphate fractions accessible to plant roots are adsorbed and precipitated phosphates as considered on page 69. Presumably this is the reason why regardless of the solubility of phosphate fertilizers applied to soils (superphosphate, rock phosphate etc.) phosphate extracted by EUF reflects the phosphate availability of the soil (Obigbesan and Mengel 1981, Judel et al. 1985, Steffens 1994).

Mühling et al. (1988) used the EUF method for the determination of phytotoxic aluminium and found it a better indicator than the standard method in which the Al is extracted by a 1 mol/m^3 KCl solution (Hesse 1971). The reason for this superiority is that using EUF only the cationic species of Al are recovered at the cathode and not the organic Al compounds which are not toxic and are included in the standard method because of the extraction technique i.e. the extraction by a KCl solution cannot differentiate between the cationic and the organically bound Al.

The EUF technique also has advantages in the determination of available nitrogen. It extracts not only NH_4^+ and NO_3^- but also an organic nitrogen fraction which is closely related to N mineralisation in soils (Mengel et al. 1999).This is described in the following section.

2.4.7 Estimation of available nitrogen

A fertile mineral soil contains about 7000 kg N/ha in the rooting profile. Of this about 1500 kg may be present in the form of NH_4^+ mainly in the interlayers of phyllosilicates, the rest is more or less organic N. Only 10 to 20% of the organic fraction is hydrolysable N (Campbell 1978) which can be mineralized. Catroux and Schnitzer (1987) found that about half of the total soil nitrogen in an arable soil was in an unidentifiable form. About 40% of the N consisted of amino N, 4% of amino sugars, and 14% as ammonium N, the latter being present mainly in interlayers of phyllosilicates. The bulk of organic

N is in the humic and fulvic acid fraction and is therefore relatively stable. Of the total hydrolysable organic N only a small percentage is mineralized over a growth period. This percentage of N net mineralization, however, can vary considerably between soils. Incubation methods for measuring the available N in the soil, such as the anaerobic incubation technique of Waring and Bremner (1964), the aerobic incubation method of Stanford and Smith (1972) and also Ellenberg's (1964) incubation test, which is carried out in the field, give N mineralization rates which are often much higher than those occurring in undisturbed agricultural soils. These incubation methods are also not practicable for large numbers of routine soil tests for available nitrogen. It appears that at the present time the EUF-method as well as the extraction method of Houba *et al.* (1986) with $CaCl_2$ provide the most realistic techniques for estimating the net N mineralization capability of soils (Appel and Mengel 1990, 1992, 1993, 1998, Groot and Houba 1995, Mengel *et al.* 1999). It has been shown by various researchers that the organic N extracted by EUF, or by a $CaCl_2$ solution according to Houba *et al.* (1986), correlates with the net mineralization of organic soil nitrogen (Kohl and Werner 1987, Appel and Steffens 1988, Appel and Mengel (1990, 1992). According to recent results of Mengel *et al.* (1999) it is predominantly the amino nitrogen (peptides and amino acids) and to a lesser extent the amino sugars both extracted by EUF or by a $CaCl_2$ solution which serve as a substrate for net mineralization. Net mineralization results from the difference between the total mineralization rate and the rate of inorganic N reassimilation by soil microorganisms over a given period, *e.g.* the growth period of a crop (Mengel 1996).

Soil tests for available nitrogen should provide a measure of the quantity of fertilizer N which needs to be applied to a given crop under specific soil and climatic conditions. Ziegler *et al.* (1992) have shown in field trials with winter wheat that this target can be achieved by the EUF method. Barekzai *et al.* (1992) elaborated a N fertilizer recommendation concept, known as the Giessen model, in which the N fertiizer rates given to winter cereals in a split application are calculated based on inorganic and organic N extracted by EUF. This model was tested for two years in field experiments yielding satifactory results. The same applies for field trials with sugar beet, maize, winter wheat and barley carried out over three years in which the N fertilizer rates were given as recommended by the EUF method (Fürstenfeld *et al.* 1994).

The determination of the nitrate content in the rooting zone of the soil profile has been used successfully by several authors as a measure of available nitrogen (Soper and Huang 1962, Borst and Mulder 1971, Wehrmann and Scharpf 1979). In this technique (N_{min} method) soil samples are taken in early spring (February, March) from rooting depth, which may be as deep as 1 m, and the fresh samples (about 150 g non dried soil) are extracted with 600 ml NaCl + $CaCl_2$ (1 mol/m^3 + 0.1 mol/m^3) for a period of one hour. This procedure extracts nitrate and non specifically adsorbed ammonium. Both these N fractions are determined. Generally nitrate is by far the greater fraction and only where organic manure, particularly slurry, has been applied, are high amounts of NH_4^+ found (Barekzai *et al.* 1993). The quantities of NO_3^- + NH_4^+ extracted are calculated on a hectare basis allowing for the water content of the soil sample, the compactness of the soil and the rooting depth. The quantity of available N thus obtained is expressed as kg N/ha (Scharpf and Wehrmann 1975) and this is used as a basis on which to make N fertilizer recommendations. Numerous field experiments carried out by Wehrmann and Scharpf (1979) have shown for winter cereals that the quantity of available N thus found in early spring must be made up by a mineral N application so that the sum of both, available N + applied N, is in the order of 120 to 140 kg N/ha. This amount guarantees optimum N supply during the vegetative growth stage of winter wheat under the growing conditions of Germany. In addition to this early N application a further treatment of 40 to 60 kg N/ha is also recommended at a later stage in growth. A useful review article on the prediction of mineralizable nitrogen in soils on the basis of an analysis of extractable organic N has been published by Appel and Mengel (1998).

This technique of estimating mineral N in the soil profile in early spring is known as the *Nmin-method*. It has proved a useful tool in estimating available N. A drawback of the Nmin-method is that the soils samples must be taken in spring from relatively deep soil layers over a relatively short time, in contrast to the EUF method where soil samples are generally taken from the upper soil layer in autumn. In paddy soils NH_4^+ is the most important N form for the rice crop. In these soils exchangeable NH_4^+ is a reliable indicator of N availability (Schön et al. 1985).

Nitrogen nutritional status of plants is related to the chlorophyll concentration in leaves. Leaves well supplied with nitrogen are deep green in colour and leaves suffering from nitrogen deficiency have a yellow appearance. Based on this phenomenon various authors have tried to use the leaf greenness as an indicator of the nitrogen nutritional status of crops (Walburg et al. 1982, Piekilek and Fox 1992). In a recent investigation Ma et al. (1996) measured the reflectance of light with a wavelength of 600 nm and of 800 nm in a maize crop stand. Light of 600 nm is of yellow/orange colour and photosynthetically active. Therefore the reflectance of this wavelength decreases as the chlorophyll concentration increases due to the fact that it is absorbed by the chorophyll. Light with a wavelength of 800 nm is infrared light and not absorbed by leaf pigments. The reflectance of light of 800 nm was therefore positively correlated with the chlorophyll concentration and in addition also with the leaf area (Ma et al. 1996) because with a larger leaf area more light of 800 nm can be reflected. "Leaf area" × "chlorophyll concentration" gives the field greenness. Based on these relationships Ma et al. (1996) established the *normalized difference vegetation index* (NDVI):

$$NDVI = (800 - 600)/(800 + 600)$$

800 and 600 denote the intensities of light reflectance of the 800 and 600 nm wavelenths. The NDVI is a quotient in which the numerator increases with the field greenness because the reflectance of 800 nm light increases and the reflectance of 600 nm light decreases with the field greenness. The denominator, however, decreases due to the fact that the reflectance intensity of 600 nm light decreases as the photosynthetic light absorption increases. It is for this reason that the NDVI is a sensitive indicator of field greeness and of the nitrogen nutritional status of crops as has been shown by Ma et al. (1996) in three-years field experiments with maize.

For nitrogen fertilizer application it is pertinent that measurements can be made early enough in the vegetation so that nitrogen fertilizer application is not too late. N rates required must be calibrated with the NDVI data.

2.4.8   Leaf analysis, plant analysis and tissue analysis

The analysis of plant material presents another type of approach in determining the nutrient availability of soils. This technique is based on the concept that the content of a particular nutrient in the plant is greater the higher its availability in the soil. The method was elaborated by Lundegardh (1945) more than five decades ago. In principle the concept is sound, because plant nutrients present in the plant must originally have been available in the soil. Unfortunately, however, the technique also has its drawbacks, as the mineral concentration in the plant not only depends on nutrient availability in the soil, but is also affected by various other factors, which are discussed in more detail below.

There is a basic relationship between the concentration of a plant nutrient in the plant matter and the growth or yield of the plant as shown by Smith (1962) and depicted in Figure 2.34 in the slightly amended version of Jones et al. (1991). When the nutrient concentration in the plant tissue is very low, the rate of growth is also low. As the growth rate increases by increased nutrient supply to the plant a decrease in nutrient concentration in the plant can at first sometimes but not always be observed as a result of a dilution brought about by the increase in production of plant material (section A on the curve). At the next stage (section B) the growth rate is improved without any marked change in the nutrient concentration. As the nutrient supply is increased further the growth rate and nutrient concentration also increase until the so called 'Critical Level' is attained. In excess of this concentration nutrient supply to the plant does not have any significant effect on the growth rate whereas the nutrient concentration is enhanced (section D, adequate luxury range). For practical purposes the point of importance is the critical level at which yield declines from an increase in nutrient concentration, the so called Toxic Range (E) concentration. Extremely high levels of nutrient supply impair growth and high nutrient concentrations are observed as shown in section E of the curve. The different stages in nutrient concentrations correspond to severe deficiency (A), deficiency (B), mild deficiency (C), adequate to luxury range (D) and

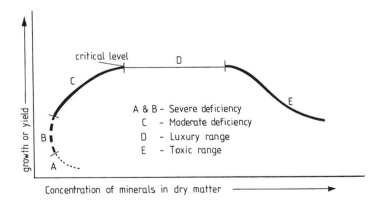

Figure 2.34   Relationship between growth or yield and the nutrient concentration in plant tissue (after Jones et al. 1991).

toxic range (E). Deficiency symptoms are observable below the Critical Level and toxicity symptoms in the section (E).

The nutrient concentrations in plant tissues and organs not only reflect soil availability. They are also affected by other factors, such as the kind of plant organ or tissue, the age of the plant, the supply of the plant with other plant nutrients, and the mobility of the particular plant nutrient within the plant. A concentration of 2 mg P/g dry matter of cereal straw *e.g.* may be regarded as a high P concentration, but the same concentration in the young plant would be too low to ensure optimum plant growth. A K^+ concentration of 6 mg K/g cereal grain is considered as high, but the same concentration in the vegetative plant tissues would also be too low for adequate growth. These examples demonstrate that for plant or tissue analysis the age of the plant or the plant organ in question must also be considered. Without this it is impossible to make comparisons between samples from different sites in relation to fertilizer recommendations.

As already discussed on page 11, plant nutrient concentrations if expressed on a dry weight basis are dependent on the water content of the tissue or organ. Young tissues are rich in water and therefore generally rich in nutrients which are dissolved in water, mainly in the vacuole and in the cytosol. These nutrients are predominantly N, P and K. Their concentrations decrease with the age of the plant or plant organ, whereas the concentrations of Ca, Mg, Mn and B often increase with ageing. Young leaves therefore show relatively high concentrations of N, P and K, whilst in older leaves an accumulation of Ca can often be observed. For this reason leaf samples for tissue analysis should be of the same physiological age or originate from the same point of insertion in the stem.

In contrast to soil analysis, leaf or tissue analysis reflects nutrient uptake conditions of the soil. For example as the absorption of various plant nutrients depends on root respiration, low nutrient uptake can also result from poor soil aeration. On the other hand optimum soil moisture conditions favour the nutrient supply of the roots and hence also nutrient uptake. The resulting high nutrient concentrations in the plant found under such conditions may therefore result primarily from the optimum uptake conditions and only to a lesser degree from the high nutrient status of the soil (Fries-Nielsen 1966). A high concentration of a certain nutrient in the plant may also result from an inadequate supply of another plant nutrient. Where N is in short supply for example, growth depressions may result in the accumulation in other plant nutrients, as N deficiency usually has a greater effect on growth than on the uptake of nutrients. In interpreting plant analytical data antagonistic and synergistic relationships between plant nutrients must also be taken into account. An antagonistic effect is one in which the uptake of one plant nutrient is restricted by another plant nutrient. A synergistic relationship is the reverse effect where the uptake of one plant nutrient is enhanced by another. The nature of antagonism and synergism is discussed more in detail on page 134.

In most cases leaf or tissue analytical data correlate fairly well with soil tests (Hipp and Thomas 1968), and it is therefore often held that leaf analysis can replace soil analysis and *vice versa*. As already discussed, however, leaf or tissue analysis also reflects conditions of uptake. A further difference between both techniques is the fact that the relationship between nutrient concentration in the plant and nutrient availability in the soil generally follows an asymptotic curve, as shown in Figure 2.35. This means that above a critical nutrient level in the plant only small changes in plant nutrient concentration may occur despite marked increases in nutrient availability in the soil. From this it follows that leaf or tissue analyses are particularly useful in the range of low nutrient availability. In the higher range of availability, however, leaf analysis is not sensitive enough. Here soil analysis is more appropriate since soil analysis also gives an indication of soil nutrient reserves.

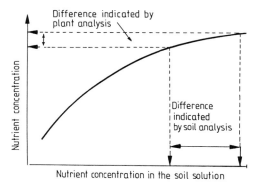

Figure 2.35   Relationship between nutrient concentration in the soil solution and the corresponding nutrient concentration in the plant tissue.

The relationship between nutrient availability in the soil and nutrient concentration in the plant is shown in Table 2.19 from the work of Kovacevic and Vukadinovic (1992).

Table 2.19   Effect of K⁺ fertilizer rates on K⁺ concentratins in leaves and grain yield of maize, obtained from a K⁺ fixing soil (Kavacevic and Vukadinovic 1992)

K⁺ fertilizer rates kg K/ha	K⁺ concentration in leaves mg K/g DM	Grain yield Mg/ha
125	6.4	1.75
275	7.8	2.57
460	8.6	4.66
650	10.3	6.95
835	14.3	7.76
1580	17.1	8.98
2200	18.6	8.88

In this particular case when K⁺ was a growth limiting nutrient in the soil, the availability of K⁺ was increased markedly by K⁺ fertilizer application. It is evident that with the increase in fertilizer rates both K⁺ concentration in the leaves as measured in the ear leaf and grain yields were raised but these increases were not linear and levelled off at the higher fertilizer rates. Potassium fertilizer rates were very high owing to very strong fixation of K^+ in the soil (see page 484) The authors emphasize that at this growth stage a K^+ concentration range of 15—18 mg K^+/g in the dry matter of the ear leaf is sufficient for an optimum yield.

Table 2.20 from the work of Gollmick *et al.* (1970) shows the concentrations of the major nutrients in maize leaves. Optimum K^+ concentration in the ear leaf are in the same range as those shown in Table 2.19 as reported by Kovacecic and Vukadinovic (1992). It is obvious that even for the same plant species, however, that adequate or critical concentrations may cover a wide range. Such results may originate from the broad spectrum of environmental conditions under which these critical levels have been established. Generally it appears that with greater intensity of cropping there is a tendency for higher critical levels to be required.

Table 2.20 Appraisal of the nutrient status of the ear leaf of maize at the flowering stage (Christensen, cited by Gollmick *et al.* 1970)

Element	mg/g dry matter				
	Deficient	Low	Adequate	High	Excess
N	<20	20—25	25—35	>35	
P	<1	1—2	2—5	5—8	>8
K	<10	10—15	15—30	30—35	>55
Ca	<1	1—2	2—10	>10	
Mg	<1	1—2	2—10	>10	

	μg/g dry matter				
	Deficient	Low	Adequate	High	Excess
Mn	<10	10—20	20—200	200—250	350
Fe	<10	10	10—300	300—550	
B	<2	3—5	6—40	40—55	>55
Cu	<2	3—5	6—50	50—70	>70
Zn	<5	15—20	20—70	70—150	>150

Leaf or tissue analysis provides a particularly useful means of assessing the nutritional status of perennial plants such as fruit trees, vines, tea, forest trees (Baule and Fricker (1970) and various plantation crops (Turner and Barkus 1974). As these plants are grown for years or even decades on the same site and under the same climatic conditions, critical levels are more easily established than for annual crops. For apple trees such critical values are well known and Table 2.21 gives a range of the mineral concentrations of apple leaves (Neubert *et al.* 1970). It must be emphasized, however, that even varieties of the same species may show considerable differences in critical levels (Chapman 1966, Gollmick *et al.* 1970, Bergmann 1992).

Table 2.21 Nutrient appraisal of apple leaves, sampled at the base of new shoots (Neubert *et al.* 1970)

Element	Low	Adequate	High
		mg/g dry matter	
N	<18	18—24	>24
P	<1.5	1.5—3.0	>3.0
K	<12	12—18	>18
Ca	<10	10—15	>15
Mg	<2.5	2.5—4.0	>4.0
		μg/g dry matter	
B	<25	25—50	>50
Cu	<5	5—12	>12
Mn	<35	35—105	>150
Zn	<25	25—50	>50
Fe	<50	5—150	>150

Ulrich and co-workers in California have carried out numerous investigations to test whether the nitrate concentraion of the petioles of sugar beet can serve as a guideline to evaluate the N nutrition of the crop (Ulrich *et al.* 1967). The determination of critical values for nitrate concentration is difficult, for this like soil NO_3^- can vary considerably even over short periods. The nitrate concentration in the petioles of sugar beet also falls with increasing age. According to Ulrich and Hills (1973) recently matured sugar beet leaves should contain about 1.0 mg nitrate- N/g dry matter in the petioles for satisfactory growth. Schnug (1991) holds the view that soil tests for available sulphate are questionable since crops may take up sulphate from soil water present in deeper soil layers. Leaf analysis is therefore recommended for assessing the available sulphate. The assessment of an adequate S supply is of particular importance for rape which has a relatively high sulphur demand. According to Schnug (1989, 1991) the critical S concentration in recently fully developed leaves of rape is 6.5 mg S/g dry matter.

As shown in Table 2.20 and 2.21 the range of optimum nutrient concentrations in plant tissues can be broad and depends on various factors affecting growth. In order to stabilise a more precise optimum a large number of field trials must be carried out in which all growth factors except the nutrient in question must be approximately optimum, which is not easy to achieve. An interesting approach to overcome this difficultiy was proposed by Walworth *et al.* (1986) is using a *boundary line approach*. The rationale of this method is the supposition that a maximum yield can only be attained if the nutrient concentration in the plant is optimum. If a higher concentration is present some other factor must be growth limiting. Lower concentrations than optimum are indicative of limited growth caused by a shortage of the nutrient. Such a boundary

108

line approach was reported by Walworth *et al.* (1986) using over eight thousand ear leaf samples of maize of worldwide origin for which mineral composition and corresponding grain yield data were known. Using this information material scatter diagrams were established by plotting the yield against the ear leaf concentrations of specific nutrients. One such plot is shown in Figure 2.36 for N.

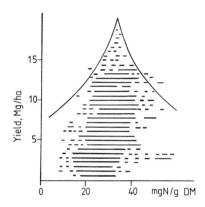

Figure 2.36   Boundary line approach for the N concentration in the leaf ear of maize. Samples shown as points came from various maize growning countries (after Walworth *et al.* 1986)

The scatter of points is delinitated by two lines where the peak, the point of intersection, represents the maximum yield and the corresponding concentration, the optimum concentration. In this case for N it is 34 mg N/g dry matter. Optimum concentrations obtained for K and P in an analogous way give values of 23 mg K and 3.2 mg P per g dry matter. This method has a universal significance since the yield data and plant nutrient analysis relate to maize crops grown worldwide under greatly different climates and soil conditions. The authors hold the view that this concept of the boundary lines can also used for other data, *e.g.* for the evaluation of soil tests, although a large number of samples are required.

Generally the amount of nutrient present in a plant organ is expressed as a concentration and therefore related to fresh or dry weight. In an analogous way one nutrient may be related to another one, *e.g.* the P concentration plotted against the N concentration. Using this technique the effect of age of the plant matter can be eliminated (Walworth and Sumner 1988). The *Diagnosis and Recommendation Integrated System* (DRIS) uses such nutrient ratios (Walworth and Sumner 1988). As shown by Walworth *et al.* (1986) nutrient ratio data can also be treated using the boundary line concept in order to obtain the optimum nutrient ratios.

A very promising approach to measure the P nutritional status of plants was made by Bollons and Barraclough (1997). The concept of the method to establish critical inorganic phosphate (Pi) concentration for maximum shoot growth is based on the finding that inorganic phosphate (P_i) can be stored in the vacuole (see page 129) at high concentrations indicative of an excess for immediate growth requirement and low concentrations at which growth is restricted. Bollons and Barraclough (1997) carried out solution culture experiments with young wheat plants. Dry matter yields and P_i concentration in the shoots were measured in relation to the P concentrations in the external solution. P_i concentrations in the shoots were made on microwave dried plant material as hydrolysis of P esters was found to occur in conventional oven dried plant material thus overestimating P_i. A clear relationship was obtained between the P_i concentration in the plant shoots and their dry matter production. Maximum dry matter production was obtained at a P_i concentration of 3 mol/L tissue water. With increasing phosphate supply the level of dry matter production remained constant until the highest Pi concentration in the shoots was attained which was 40 mol/L tissue water. From this finding it is evident that the optimum of P_i concentration for plant growth is very broad. Figure 2.37 shows the relationship between the P_i concentration in the nutrient solution and the shoot dry weight and the P_i concentration in the shoot.

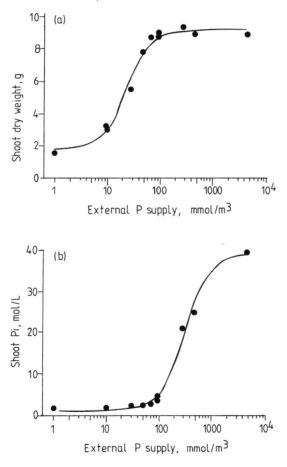

Figure 2.37 Relationship between the P concentration in the nutrient solution and a) the shoot dry weight of young wheat plants — b) the P concentration in the shoots calculated from the content of shoot water. (Bollons and Barraclough 1997).

Maximum shoot growth was already obtained at a concentration as low as about 100 mmol/m^3 P in the nutrient solution (upper part of Figure 2.40) with a corresponding concentration of about 3 mol P$_i$/L tissue water in the shoot (lower part of Figure 2.40). From these findings of Bollons and Barraclough (1997) the so called *storage pool approach* seems to be a sensitive method for measuring the nutritional status of plant tissues and could be developed into a useful routine plant test.

In some techniques the total amount of nutrients taken up by plants is used as an indicator of the soil fertility status. In the Neubauer method rye is grown for a period of two weeks in a mixture of quartz sand and of the soil under investigation. The quantities of K and P taken up from the soil by young plants over this period serve as a measure for the nutrient status of the soil (Neubauer and Schneider (1923). Quantities greater than 20 mg K and 3 mg P per 100 g soil are regarded as satisfactory levels.

The Mitscherlich-method also uses a mixture of quartz sand and soil, but the plants are grown in larger pots and for a longer period as compared with the Neubauer-method. The nutrient availability of the soil is calculated (Mitscherlich 1954), according to the Mitscherlich equation from the yield increments brought about by the nutrient concerned (see page 306). Useful monographs on the problem of nutrient concentrations in plant tissues and the nutritional status of numerous crop species have been published by Bergmann (1983) and Jones *et al.* 1991).

2.4.9 Microbiological methods

Microbiological methods can also be used in the assessment of nutrient availability in soils. Here a fungus is used rather than a higher plant. This technique has been used particularly in the determination of the availability of micronutrients such as Fe, Cu, Zn and Mo. A nutrient solution deficient in the particular nutrient in question is added to the soil under investigation. The soil suspension is then infected with a microorganism, usually *Aspergillus niger* and incubated for several days at constant temperature. The growth of the fungus as estimated by the weight of the mycelium produced is used as a measure of nutrient availability (Stapp and Wetter 1953, Nicholas 1960). Today these microbiological techniques are hardly ever used as micronutrient availability in soils (Cu, Zn, Fe, Mn) can be assessed much more rapidly by soil extraction with chelates. This is discussed in the respective chapters dealing with these nutrients.

CHAPTER 3

NUTRIENT UPTAKE AND ASSIMILATION

3.1   Ion Uptake and Ionic Status of Plants

3.1.1   General

It is now over 50 years ago since Hoagland and co-workers reported their fundamental findings on ion uptake by plants (Hoagland 1948). In experimenting with fresh water alga *Nitella* and the sea water alga *Valonia* they found that the ion concentrations in the vacuoles of these two algae did not correspond to the concentrations in the respective algal nutrient environments (Figure 3.1).

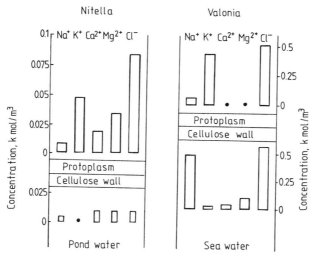

Figure 3.1   Ion concentrations in the vacuole of *Nitella* and *Valonia* in relation to the ion concentrations in outer medium (after Hoagland 1948).

In the vacuole of *Nitella* several ion species and particularly K⁺ and Cl⁻ were concentrated to a considerably high degree. The same was also true for *Valonia* with the exception of Na⁺ where the concentration was higher in the sea water than in the vacuole. From these findings important conclusions could be drawn which have since been confirmed by numerous authors.

1. The plant takes up ions selectively. Thus K⁺ which is lowest in concentration of all the cations in the pond water is the cation which is accumulated to the greatest extent by far in the vacuole of *Nitella*. On the other hand the concentrations of Na⁺ in the vacuole of *Valonia* is kept to a relatively low level even though the concentration of Na⁺ in the sea water is high.

2. Ion accumulation can take place so that the concentration of mineral nutrients in the plant such as N, P, K, can be very much higher than in the external solution, *e.g.* soil solution.

Additionally it is now known that there are distinct differences between plant species in patterns of ion uptake. For example the uptake of K^+ as compared with Ca^{2+} per unit dry weight plant is much greater in the graminaceous monocots than in dicots. Even within the same plant species marked differences can occur in the mineral composition of cultivars grown under the same nutritional conditions. Similarly some cultivars of plant species accumulate particular mineral elements as for example Se (see page 666) or Zn (see page 593). A further important consideration is that energy in the form of ATP is required for the uptake process.

From these general observations a number of questions may be raised. How are ions transported into cells and how is ion uptake and distribution regulated within cells and plants? What is the role of ATP? Mineral nutrients are taken up by roots but utilized for growth predominantly by the shoot. Do shoot/root interactions occur to control uptake? These and other intriguing questions are considered.

3.1.2 The plant cell

Before describing the various processes of nutrient uptake and assimilation it seems appropriate to give a simplified picture of the plant cell, for the two previous chapters have dealt primarily with inorganic materials. Ion uptake may be considered as a boundary process in which the inorganic domain impinges upon the living world. The smallest viable unit of living matter is the cell. Figure 3.2 shows a much simplified diagram representing a mesophyll cell. The most important organelles are depicted. The cell wall structure is made up of pectic substances, cellulose and hemicellulose. Cellulose tends to aggregate to form chain like structures known as microfibrils. Intermicrofibrillar spaces allow the entry of water, air and solute particles into the cell wall. The plasma membrane or plasmalemma is the membrane boundary between the cytoplasm and the cell wall; the tonoplast is the membrane which separates the cytoplasm from the vacuole. Membranes and their structure are considered in more detail later. Located within the cytoplasm are the most important organelles within the cell. These include the nucleus, chloroplasts and mitochondria. Chloroplasts are the organelles in which light energy conversion and CO_2 assimilation take place. In the mitochondria, enzymes are present which control the various steps of the tricarboxylic acid cycle, respiration and fatty acid metabolism. The ribosomes are supramolecular assemblies composed of ribosomal nucleic acid and proteins which enable the synthesis of polypeptides from free amino acids. Many of the ribosomes are attached to the endoplasmic reticulum (ER). This is a folded sheet like structure which gives rise to a series of membranous channels permeating the cytoplasm and often leading from one cell to another. It is regarded as a major route in symplastic transport. The ER with attached ribosomes is called the rough ER in contrast to the smooth ER which has no ribosomes.

The main function of the ER is to produce materials used in the synthesis of membranes, *e.g.* the plasma membrane. Cells are connected together by the plasmodesmata and the continuous plasmatic connection which occurs in the cells of a tissue is called the symplasm. It is surrounded by the plasmalemma and comprises the cytosol and the various organelles embedded in the cytosol. The cytoplasm without the organelles is called the cytosol. Also the vacuole is an organelle which makes up a very large proportion of the cell volume. It contains an aqueous solution consisting mainly of inorganic ions as well as some low molecular weight organic substances such as organic anions, amino acids sugars, and molecules generated in secondary plant metabolism as well as enzymes, especially hydrolases are also present (Kreis and Hölz 1992). The vacuole has an important bearing on the water economy of the cell as well as providing a site for the segregation of water and the end products of metabolism. It also serves as a storage pool, *e.g.* for sugars, nitrate, and phosphate and K^+.

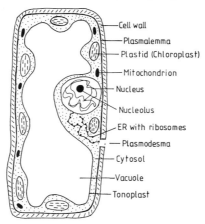

Figure 3.2 Simplified representation of a mesophyll cell (not to scale).

The size of living cells varies considerably for different tissues and plant species. Mesophyll cells and cells of the root cortex tissue are about 20—100 micron (μm) in length. The diameter of chloroplasts and plastids is in the range of 8 μm, whilst that of the mitochondria is about 1 μm and that of the ribosomes 23 nm (1μm = 10^3 nm). Compared with these organelles the size of the low molecular weight substances appear rather small. The diameter of a sucrose molecule is 1nm, a glucose molecule 0,6 nm, and the various inorganic ions in their hydrated form have diameters in the order of 0.5—1.0 nm. These values are given simply to indicate the minute size of the inorganic ions in comparison with the various cell organelles.

The cell organelles such as the nucleus, chloroplasts, plastids, and mitochondria and also the vacuole are surrounded by membranes, which are very effective barriers for water soluble substances and thus compartmentalize the cell. Some organelles, such as the nucleus (Newport and Forbes 1987), plastids and mitochondria, are surrounded by a double membrane between which is located an inter membrane space which is of importance for pH control and uptake processes. This compartmentalization is essential for the normal functioning of individual organelles as well as the whole cell since specific biochemical processes occur within the different organelles. Understanding the transport mechanisms by which organic and inorganic substances are transferred between cellular compartments is therefore of outstanding importance. It is in this context too that the transport of inorganic plant nutrients from the outer medium, the soil solution, across the plasmalemma into the cytosol, the cell must be considered. This membrane tranport represents the essential step of nutrient uptake. It is the plasmalemma, also referred to as the plasma membrane and not the cell wall, which forms the boundary of the cell to the outer medium and which presents the effective barrier to the uptake of all ions and molecules dissolved in the aqueous outer medium.

3.1.3 Membranes

Biological membranes consist of protein and lipid molecules in approximately equal proportions and are about 7 to 10 nm thick. For decades the lipid protein sandwich structure proposed by Danielli and Davson (1935) was considered as the universal structure model of a biological membrane. It is now generally accepted that membrane structure is more intricate than that mentioned above. The basic molecules making up biological membranes are amphiphilic, that is to say they possess both hydrophilic (OH groups, NH_2 groups, phosphate groups, carboxylic groups) and hydrophobic regions (hydrocarbon chains, mainly consisting of $-CH_3$ and $=CH_2$ groups) and these characteristics are incorporated into the membrane. An example of an amphiphilic molecule is shown in Figure 3.3 for lecithin.

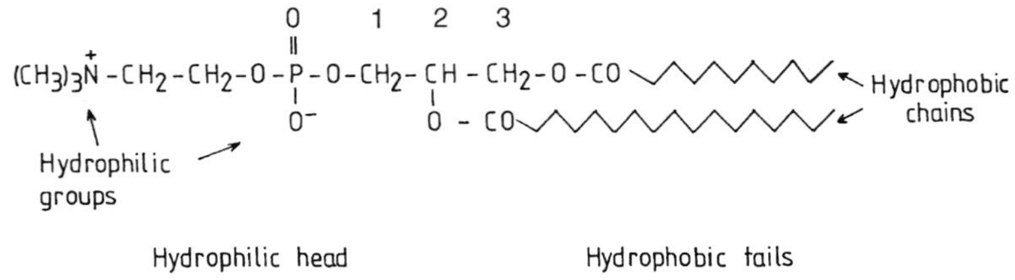

Figure 3.3 Molecular structure of lecithin with a hydrophilic head and hydrophobic tails.

The most important lipids present in biological membranes are phospholipids, glycolipids, and steroids. Glycolipids dominate in the outer chloroplast membrane (Heber and Heldt 1981). Sphingolipids are important components of animal membranes, but are probably of no major significance in plant membranes (Morré 1975). Membranes are composed of two lipidic layers (double layer) which consist of amphiphilic molecules. The molecule shown in Figure 3.3 has two lipidic tails (hydrocarbon chains) and one hydrophilic head, the phosphate amino complex. This hydrophilic part of the molecule also carries a positive and a negative charge under physiological pH conditions. The axis of the glycerol moiety lies vertical to the membrane plane and the fatty acid chain at the C_2 position is located at the surface of the lipidic phase of the membrane. The chain from the C_3 position projects from the membrane into the aqueous phase (Figure 3.3). The negatively charged phosphate group may bind cations, thereby influencing both lipid conformation and membrane stability. Calcium is of particular importance in the maintanance of membrane stability as it may bridge two negatively charged phosphate groups as shown in Figure 3.4. The lipidic monolayer consists of amphiphilic layers orientated in such a way that the heads form a plane. In the double layer or bilayer, the tails are orientated towards each other (see Figure 3.4) with each lipidic monolayer representing a two dimensional liquid. The bilayer is not symmetrical so that different types of lipids occur in the two layers. The head groups protrude from both sides of the membrane into the outer and inner media. In the case of the plasmalemma the cell wall or apoplast makes up the outer medium and the cytosol the inner medium. Stabilization of bilayers is brought about by the binding of the head groups to each other mainly by ionic bonds as shown in Figure 3.4 where the phosphate anion is bound to the positively charged amino group of phosphatidyl ethanol amine. Calcium ions also stabilize the membrane by binding two anions (Figure 3.4). The fatty acid chains of the various amphiphilic membrane lipids are bound to each other by hydrophobic bonds which also contribute to membrane stability.

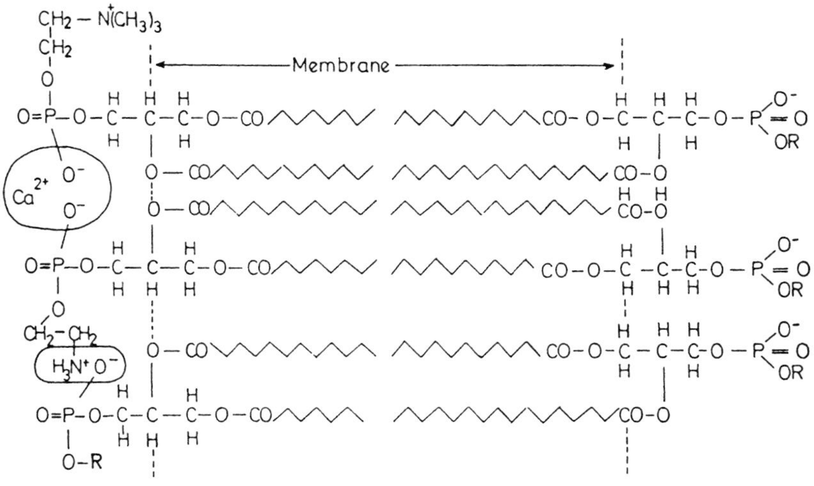

Figure 3.4 Schematic representation of a membrane bilayer.

From these types of stabilization it follows that membrane stability can be weakened or even broken by non polar liquids which sever the hydrophobic bonds or by extreme pH conditions, such as low pH < 4 when H^+ replaces Ca^{2+} thus breaking the Ca^{2+} bridge and neutralizing anionic groups. High pH levels may lead to the deprotonation of amino groups thus also severing ionic bonds of protonated groups. Such destabilization makes the membranes leaky and solutes lost from the cell or compartment are detrimental to metabolism.

Biological membranes have a high turnover rate. The half life of phosphatidyl choline (lecithin) is about 14 hours and the half life of membrane proteins is in the range of 2 to 400 hours. The fatty acid composition of membrane lipids differs between plant species and is also influenced by environmental factors particularly temperature.

The composition controls the fluidity of membranes; fluidity being increased by short fatty acid chains and non saturated fatty acids and decreased by saturated long fatty acid chains. The resistance of plants to low temperature *i.e.* cold resistance is dependent on membrane fluidity and is raised by increasing fluidity of the membranes. Indeed plants may adjust to low temperature by shifts in membrane fatty acid composition which maintain membrane fluidity as has been shown by Ashworth *et al.* (1981). Similarly species with a high proportion of unsaturated fatty acids such as spinach or *Arabidopsis thaliana* grow well at low temperatures as compared with cucumbers which are characterized by a low proportion of unsaturated fatty acids in the chloroplast membrane. Genes responsible for cold resistance have been transferred from *A. thaliana* to tobacco thus greatly improving the cold resistance of this crop plant (Murata *et al.* 1994).

The lipid double layer is the matrix in which proteins are embedded. Most of these proteins are transport proteins, channel proteins, and redox proteins and most are transmembrane proteins that is to say, with the polypeptide strand tranversing the bilayer from one side to the other. Generally the strand transverses the bilayer several times forming loops on both sides of the membrane as shown in Figure 3.5.

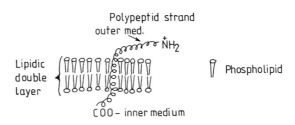

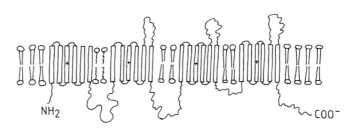

Figure 3.5 Upper part: Polypeptide strand transversing the lipidic double layer of the membrane. — Lower part: Putative cation channel formed by a polypeptide strand with 4 repeating units. Each repeat folds into 6 transmembrane helices. At some sites the polypeptide strand protrudes from the membrane into the outer or inner medium forming large loops (modified from Stryer 1995, p. 301).

The upper part of Figure 3.5 shows the carboxylic end of the strand located in the cytosol and the amino end in the outer medium. This is always the case if the polypeptide has an odd number of transmembrane sections or domains when it may transmit signals from the outer medium to the cytosol and *vice versa* (Baringa 1995).

The transmembrane section of the polypeptide strand consists of hydrophobic amino acids which form hydrophobic bonds with the lipid double layer. In this way the membrane proteins are tightly incorporated into the lipid double layer. The transmembrane section comprises about 20 amino acids (Tanner and Caspari 1996). A cation channel is shown schematically in the lower part of Figure 3.5. It comprises a single polypetide strand which transverses four identical units, each unit having six domains (transmembrane sections), the amino end and the carboxylic end of the polypeptide strand both being located in the cytosol. At some sites on the lipid double layer surfaces large loops of the strand protrude into the outer and cytosolic medium. These loops supposedly recognize substrates suitable to pass through the channel or they may catalyze transport across the channel (Tanner and Caspari 1996). In Figure 3.5 the channel is represented two dimensionally for as yet the sterical structure of such channels is not exactly known. However, if the two outer domains, one carrying the amino and one carrrying the carboxylic group (lower part of Figure 3.4) were brought together by bending the membrane a barrel-like structure would be formed in which the transmembrane sections (domains) would represent staves of the barrel. Such a structure has been described for a maltoporin channel by Schirmer et al. (1995). This channel allows the passage of maltose and linear polysccharides. Its diameter of 0.5 to 0.6 nm is also similar in to that of cation channels. In a model described by Schirmer et al. (1995) the transmembrane sections of the polypeptide have a β-structure and run antiparallel. Berendes et al. (1993) have described an ion channel in which the hydrophilic pore was formed by four helix bundles. If the glutamate located at the centre of the pore was substituted by serine the conductance of Ca^{2+} through the channel was much decreased and the conductance for K^+ and Na^+ greatly increased. This example shows that minor alterations in the polypeptide chain may change the selectivity of the channel considerably. Potassium channels play an important role in K^+ transport of plant metabolism. These channels are also highly important in animal tissues and particularly in neurobiology where they transfer electrical messages. Potassium channels are structurally the simplest in the superfamily of ion channels (Hille 1996) and are characterized by a relatively short chain of the polypeptide strand. In the schematic representation of a cation channel shown above, one domain of each unit carries a positive spot resulting from protonated arginine and lysine residues and it is supposed that these positive charges are involved in voltage gating (see page 125).

The aquaporins (Chrispeels and Maurel 1994) form another group of important membrane channels. These have been found in plant and animal membranes and consist of a single polypeptide strand with six putative transmembrane domains. Aquaporins located in the plama membrane enhance the flow rate of water by a factor of 10 and over and thus favour symplastic water transport. These channels can be regulated by phosphorylation (Chrispeels et al. 1999) and allow water flow but exclude the passage of ions and metabolites. Transcellular water conductance may be improved during water stress by the incorporation of newly synthesized aquaporins being integrated into the plasma membrane.

Other membrane proteins include various transporters, ion pumps, carriers and redox systems as occur in the inner membranes of mitochondria and chloroplasts (see page 113 and page 145) and redox enzymes in the plasmamembrane where in particular they are responsible for the reduction of Fe^{III} chelates (Rubinstein and Luster 1993).

3.1.4 Ion pumps, electrochemical potentials, and ion channels

3.1.4.1 *Ion pumps*

Ion pumps are membrane proteins which catalyze the transport of an ion from one side of the membrane to the other with the consumption of energy. Generally this transport across the membrane is an uphill process, that is to say transport against the prevailing electrochemical gradient (page 120). Sodium-ions, Ca^{2+}, K^+, and H^+ can all be transported by such pumps. The most prominent proton pump in plant metabolism however, is, the plasmalemma H^+- ATPase (EC 3.61.35) which due to its universal importance has been called the master enzyme (Serrano 1989). Its activity is about 100 to 1000 times higher than the activity of plasmalemma Ca-ATPase (Ca^{2+} pump). The term ATPase signifies that the pumping mechanism is associated with the hydrolysis of ATP. The plasma membrane H^+ -ATPase is a highly regulated enzyme with multiple physiological functions as described by Michelet and Boutry (1995) in an interesting review article. Plasmalemma H^+- ATPase is shown schematically in Figure 3.6 (modified after Serrano 1989).

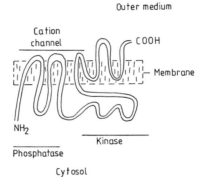

Figure 3.6 Plasmalemma ATPase, shown schematically with the proton pump located in the plasmalemma (modified after Serrano 1989).

The core of the pump is a transmembrane polypeptide which traverses the membrane and is made up of nine transmembrane sections comprising a proton channel, a kinase, and a phosphatase. The reaction starts with the transfer of a phosphoryl group from the ATP to the polypeptide (kinase reaction) associated with the release of ADP. This phosphorylation supposedly induces a change in conformation of the polypeptide section. Phosphorylation is followed by a phosphatase reaction splitting off of inorganic phosphate (P_i) from the polypeptide strand. This phosphatase reaction is not a simple hydrolytic reaction but rather phosphorylation inducing the conformation followed by dephosphorylation bringing about the expulsion of protons from the cytosol into the outer medium, the cell apoplast. Although the exact mechanism of this proton transport (proton pumping) is not yet understood the main characteristics are well documented. The proton pump is inhibited by vanadate which competes with ATP at the binding site of the polypeptide section and is stimulated by fusicoccin and auxins. Fusicoccin is a diterpene glycoside produced by the fungus *Fusicoccum amygdali* (Marre 1979). Both fusicoccin and auxins are believed to lower the cytosolic pH and hence decrease the energy required for the proton pumping process (Senn and Goldsmith 1988). According to Berczi *et al.* (1989) plasmalemma ATPase is also inhibited by Ca^{2+}.

This proton pumping function in all plant tissues is a characteristic of all living plant cells. Plasmalemma ATPase is responsible for cytosolic pH regulation, cell elongation and division, stomatal movement, and the uptake of various metabolites including inorganic nutrients. According to Bouche-Pillon *et al.* (1994) H^+ ATPase is not evenly distributed in the plasma membranes but is particularly frequent in transfer cells which are characterized by cell wall ingrowths thus increasing the interface between cell wall and plasmalemma. In the overall reaction

$$ATP \Rightarrow ADP + P_i + H^+ \text{ transport } (P_i = \text{inorganic phosphate})$$

it is not yet clear how many H^+ ions are pumped per ATP hydrolyzed. This may depend on the energy required for the pumping process, a steep electrochemical gradient between the cytosol and the outer medium requiring much more energy than a slight gradient. High H^+ concentrations in the outer medium may even induce a proton reflux from the outer medium into the cytosol (Yan *et al.* 1992).

The vacuolar membrane contains two types of H^+ pumps, H^+ATPase (V-ATPase, EC 3.6.1.3.) and pyrophosphatase (V-PP ase, EC 3.6.1.1.) the latter which is ubiquitous in plants but otherwise known only in a few phototrophic bacteria (Rea and Poole 1993). The vacuolar H^+ pump (V-ATPase) is different from the plamalemma H^+ pump and resembles the F-ATPase in mitochondria (Barkla and Pantoja 1996). This pump also uses ATP as an energy source for the pumping process but is not inhibited by vanadate but by halides and N-ethylmaleimide (NEM). Pyrophosphatase (= V-PPase) is supplied by pyrophosphate as an energy source. The pump has a strict requirement for Mg^{2+} which probably functions as an allosteric modulator. It not only pumps H^+ but also K^+ which may substitute for H^+. By this mechanism K^+ may be translocated from the cytosol across the tonoplast into the vacuole. Although the stoichiometric relationships have not yet been completely elucidated there is evidence that per molecule of pyrophosphate, three monovalent cations can be translocated and that the enzyme has three negative binding sites for which H^+ and K^+ may compete. Since K^+ is an important osmoticum V-PPase may also play a role in osmoregulation.

All three pumps, the plasmalemma H^+ pump and the two vacuolar H^+ pumps, pump protons out of the cytosol and since the pumps are electrogenic *i.e.* cation transport is not associated with concomitant anion transport, the H^+ pumps render the cytosol negatively charged relative to the outer medium or the vacuolar liquid as shown in Figure 3.7. The electropotential differences thus obtained are in the range of 100 to 200 mV for the cytosol/outer medium and 30 to 50 mV for the cytosol/vacuole. This negatively charged cytosol may attract cations and since there are cation channels in the membranes, cations may traverse the membrane and enter the cytosol. Such cation transport depresses the electropotential difference, a process which is called depolarization of the membrane.

The potential thus built up by the proton pumps consists of a chemical and an electrical component. The free energy available in the proton electrochemical potential difference ($\Delta\mu H^+$) is a function of both the proton concentration difference (ΔpH) and the transmembrane electrical potential (Bush 1993):

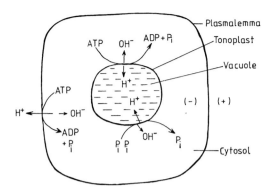

Figure 3.7 Plasmalemma and vacuolar H⁺ pumps shown schematically rendering the cytosol negatively charged relative to the outer medium.

$$\Delta\mu H^+ = -2.3\ RT\ \Delta\ pH + F\ \Delta\ \varphi$$

where
R = gas constant
T = absolute temperature in Kelvin
F = Faraday constant.

$\Delta\ \mu$ is an energy term and the force correponding this energy is the proton motive force (pmf), which can be described by the following equation (Poole 1978).

$$pmf = \Delta\ pH + \Delta\ \varphi$$

where pmf = proton motive force or electrochemical gradient of H⁺ ions across the membrane
Δ pH= difference in H⁺ concentration across the membrane
Δ φ = electrical potential difference across the membrane

The proton motive force (= pmf) is the force generated by proton separation by which other cation species are transported across the plasma membrane into the cytosol. This kind of ion transport across a membrane is called *facilitated diffusion* because it follows the electrochemical gradient and is facilitated by channels which allow a selective transport (Bush 1993). Such a transport along the electrochemical gradient is also called *downhill* transport which is passive transport.

3.1.4.2 *Electrochemical potentials*
Ions in solution are subjected to two main physical forces. One arises from a chemical potential gradient when ions move from a higher to lower concentration, and the other an electrical potential gradient when cations are attracted to a negative electropotential and anions to a positive electropotential (Dainty 1962). These forces may be expressed quantitatively as follows:

$d\mu/dx$ = chemical driving force; $z\ F\ d\varphi/dx$ = electrical driving force where

μ = chemical potential
z = valency of the ion

φ = electrical potential
F = Faraday constant, 92 J/mV/mol

The chemical potential depends on the concentration or more precisely the activity of the specific ion:

$$f \times c = a$$

where

a = activity   c = concentration   f = the activity coefficient

The activity coefficient is less than 1 and approaches unity as the concentration of the solution falls. In plant tissues ion concentrations are low so that concentration rather than activity is used in calculating chemical potential.
The chemical potential of an ion is given by the formula:

$$\mu = \mu^\circ + R\,T\,\ln a$$

where

μ° = the standard chemical potential of the ion or molecule species
R = gas constant, 7.95 J/°C/mol
T = absolute temperature

The electrochemical potential of the ion can be defined by taking into account the electrical potential:

$$\mu = \mu^\circ + R\,T\,\ln a + z\,F\,\varphi$$

When a cell is in equilibrium so that the electrochemical potential of an ion is the same on either side of the plasma membrane *i.e.* the cytosol and the outer solution, the following equation holds:

$$R\,T\,\ln a_o + z\,F\,\varphi_o = R\,T\,\ln a_i + z\,F\,\varphi_i$$

Where

φ_i = electrical potential of the cytosol
φ_o = electrical potential of the outer solution

a_i = ion activity in the cytosol
a_o = ion activity of outer solution

The electrical potential difference across the cell membrane $\varphi_i - \varphi_o$ can be measured in milli volts (mV) and is referred to as E.

$$E = \varphi_i - \varphi_o = \frac{RT}{zF} \ln \frac{a_o}{a_i} \qquad \text{Nernst equation}$$

The Nernst potential (E) is a measure of the electrical potential difference needed to maintain an asymmetric distribution of an ion across a membrane at equilibrium.
The Nernst equation can be further simplified by converting natural logs to logarithm base 10 and considering the gas constant and the absolute temperature. This gives:

$$E = \frac{58}{z} \log \frac{\text{(concentration outside)}}{\text{(concentration outside)}} \quad \text{milli volts (mV)}$$

As discussed above, living cells are negatively charged as compared with the outer medium. For this reason the passage of ions through the plasmalemma or tonoplast must also be considered in relation to the prevailing electrical potential gradient as well as to the concentration gradient between the "outer solution" (medium) and "inner solution" (cytosol). Facilitated cation diffusion discussed above is a typical example of where cations are moved down an electrochemical gradient. This net inwardly directed movement of cations terminates as soon as the equilibrium between the electrical and kinetic driving forces is attained. As described by the *Nernst equation*.

A simple case may be considered in which an aqueous solution of KCl is separated by a membrane which is permeable to both ions, K^+ and Cl^-. Assuming that the electrical potential across the membrane is E, equilibrium for K^+ and Cl^- is attained as soon as the concentrations of these ions at either side of the membrane satisfy the Nernst equation which is shown above (page 121). From this equation it follows that when $E < 0$ (the cell is negatively charged) the term $[K_o^+] / [K_i^+]$ must be <1. This means that under equilibrium conditions an accumulation of K^+ occurs in the inner solution (cytosol). It further follows that the term $[Cl_i^-] / [Cl_o^-]$ must be <1. This implies that under equilibrium conditions the Cl^- concentration of the outer solution is higher than that of the inner solution.

Cation concentration in the cytosol may thus be several times higher than that of the outer solution without requiring an "uphill transport" of cations, *i.e.* a transport against an electrochemical gradient. If for example the K^+ concentration of the inner solution (cytosol) is 100 times higher than that of the outer solution the term log $[K_o^+] / [K_i^+] = -2$. The corresponding electrical potential difference for equilibrium conditions is then -116 mV, since the value of the term RT/zF at 20°C with z (valency) 1, is equal to 58 (Dainty 1962)".

This electrical potential difference of -116 mV lies in the normal range of measured values and the example shows that under equilibrium conditions the K^+ concentration in the cytosol may be 100 times higher than in the outer medium. As shown on page 491 a K^+ concentration in the cytsol of 100 mol/m^3 is normal and not very high and a K^+ concentration of $1 mol/m^3$ in the soil solution is realistic, although relatively high (see page 73). The example shows that K^+ can be accumulated in the cytosol by the factor of 100 relative to the nutrient solution by passive (down hill) uptake. Only, if the concentration in the cytosol is higher than that indicated by the equilibrium condition (Nernst equation), must an "uphill transport" have occurred, *i.e.* a transport against an electrochemical gradient. In a strict thermodynamical sense the transport against an electrochemical gradient is called *active transport*, whereas the transport down or along an electrochemical gradient is a *passive transport*. Active transport needs additional energy and cannot be brought about merely by kinetic and electrical forces (Etherton and Higinbotham 1961, Etherton (1963).

As outlined above, for chloride equilibrium is already attained at a relatively very low Cl^- concentration in the cytosol, relative to the outer medium. If *e.g.* the Cl^- concentration in the cytosol is 1 mol/m^3 and in the outer solution is 100 mol/m^3 the equations read as follows:

$$E = \frac{RT}{(-z)\,F} \ln a_a/a_i$$

in this equation z has a negative sign because it is an anion. When calculated as decadic logarithm, since the term $RT/F = 58$ and $\log 100/1 = 2$

$$E = 58/\text{-}1 \quad 2 = \text{-}116 \, \text{mV}$$

This example makes it evident that the accumulation of Cl^- as well as other anion species in the cytosol relative to the outer medium requires active transport. As will be shown below the anion transport across the plasmalemma is brought about by a particular transport mechanism called cotransport.

In order to test, whether an ion species has been moved actively or passively into the cell, the concentrations of the particular ion species in the outer medium and in the cytosol must be measured as well as the electropotential (E_m) between the cytosol and the outer medium. This can be achieved using a microelectrode. By substituting the measured ion concentrations into the Nernst equation an electrical potential difference (E_{ca}) can be calculated. Where E_m designates the measured potential, the difference between E_m and E_{ca} indicates whether a passive or an active transport has occurred.

$$E_m - E_{cal} = E_d$$

E_d is the driving force. For cations a negative value of E_d indicates a passive uptake and a positive value an active uptake. For anions the reverse is true. A negative value is indicative of active transport, and a positive value of passive transport. It must be remembered that the test of whether an ion species has been transported actively or passively is only valid, if equilibrium conditions have been maintained in the system. This is often difficult to achieve in whole plant studies as plant tops provide a very strong sink for ions taken up by roots.

Spanswick and Williams (1964) measured the electropotential differences and the ion concentrations of the outer medium and in the cell of *Nitella*. The data of this experiment are shown in Table 3.1. For Na^+ at equilibrium a potential difference of -67 mV would have been sufficient for the prevailing Na^+ concentrations. As the electrical potential difference was higher (-138 mV) Na^+ uptake was passive. For K^+ an electrical potential difference of -179 mV was required for equilibrium conditions. As the measured potential difference was lower (-138 mV) K^+ uptake was active. For Cl^- a potential difference of +99 mV (positively charged cell) would have been needed to maintain the equilibrium at the measured Cl^- concentrations. As the measured electrical potential was considerably lower, Cl^- must have been taken up actively.

Table 3.1 Measured (E_m) and calculated electropotential differences (E_{cal}) and the resulting driving forces (E_d). The data refer to experiments with *Nitella translucens* (Spanswick and Williams 1964)

Ion species	E_m	E_{cal}	E_d	type of uptake
Na^+	-138	-67	-71	passive
K^+	-138	-179	+41	active
Cl^-	-138	+99	-237	active

Figure 3.8 shows the results from mung bean root tips (*Phaseolus aureus*) in which measured Cl⁻ uptake values are compared with the calculated ones derived from the *Nernst equilibrium* (Gerson and Poole 1972). The measured values were several times higher than the calculated ones and therefore Cl⁻ uptake must have been active. The fact that living cells are always negatively charged implies that anion uptake is more subjected to active transport than cations and the reverse is true for the ion efflux out of the cell since the negatively charged cytosol my easily expel anions across the plasmalemma provided there are channels which allow their passage. For cations the situation is quite different. Due to the negatively charged cytosol, cations may accumulate in the cytosol merely by physical, non metabolic forces. Higinbotham (1973) cites several experiments indicating that the concentrations of Na^+, Ca^{2+} and Mg^{2+} in plant cells do not exceed the physical equilibrium level and are thus obviously absorbed passively. Only in the case of K^+ have experimental data been obtained which are indicative of active uptake. The results shown in Table 3.1 provide evidence of this kind. A further demonstration of active K^+ transport has been provided by Davis and Higinbotham(1976) who showed that K^+ was transported against an electrochemical gradient from the parenchyma cells into the xylem vessels of maize roots. The type of active K^+ tranport will be dicussed on page 489.

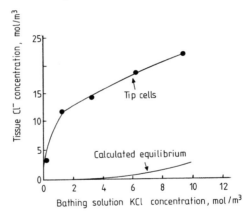

Figure 3.8 Chloride concentrations in mung bean root tips compared with the maximum concentration (calculated equilibrium) which could be due to passive uptake (after Gerson and Poole 1972).

3.1.4.3 *Ion channels*

Passive movement of ions through a membrane depends on the prevailing electrochemical gradient across the membrane and may take place in either direction. Membranes thus allow the passive influx and efflux of ions (Bush 1993). The rate of these fluxes depends on membrane permeability. This may be defined as the quantity of ions which may be transported per unit membrane surface, per unit time. This permeability depends mainly on ion channels. Most of these channels have a high selectivity and are regulated so that they may open and close. During the past decade the use of the *patch-clamp-technique* has revealed the presence of channels in the membranes of animals, plants, fungi and bacteria (Hedrich and Schroeder 1989). The basis of the technique is that a microelectrode is brought in contact with a cell membrane surrounding an isolated protoplast *i.e.* a cell with the wall removed, to investigate ion transport across a restricted area of the membrane. The small region in which the tip of the electrode is in contact with the membrane is called the patch and this is attached to the tip of the electrode which is bathed in a solution of known composition. A small electrode (about 1 μm) may contain a single ion channel. Using an electrical circuit, the potential difference across the membrane can be "clamped" or held at a required value, and the flow of current through the membrane patch measured.

The channels of the plasmalemma of plant tissues are fairly similar to those in the plasma membrane of animals and fungi while those of the tonoplasts are more related to channels in the lysosomes. Opening and closing of channels (channel gating) depends much on the prevailing electrical potential difference across the membrane, the Ca^{2+} concentration in the cytosol and also pressure gradients due to turgor differences (Tyerman 1992). In addition Ca $^{2+}$ proteins and protein kinases and phosphatases may be involved in the channel gating. Opening of anion channels in the plasmalemma is induced by a relatively high Ca^{2+} concentration in the cytosol. A voltage dependent anion channel in guard cells requires Ca^{2+} and nucleotides and opens in the range of electrical potential difference between -80 and -20 mV. Channel transport rates are generally 100 to 1000 higher than carrier transport (see page 129; Hedrich and Schroeder 1989). Chloride channels in particular contribute to the electrical potential difference across the plasma membrane since an open channel induces a strong Cl^- efflux and hence reduces the electrical charge which in turn reduces the retention capacity for cations, especially K^+. Therefore the opening of the plasmalemma chloride channel induces an efflux of Cl^- and K^+ resulting in a loss of turgor of the cell (Tyerman 1992). This example shows how the activation of ion channels may quickly alter the physiological conditions of a cell. The turgor regulation of guard cells is brought about by ion fluxes across the plasmalemma with K^+ being the most important cation and malate the most important anion in glycophytes. Potassium ion channels are structurally the simplest of cation channels. There is experimental evidence from various plant tisssues that two principal types of rectifying K^+channels exist, one outwardly and one inwardly directed (Schroeder and Hagiwara 1989, Fairly-Grenot and Assmann 1992. The term rectifying means that the K^+ flux occurs along the electrochemical gradient (passive transport). The outwardly directed channel is blocked by Ca^{2+} or Ba^{2+} present in the outer solution (Bouteau *et al.* 1996). This is an interesting finding which shows that the K^+ efflux out of the cell may be blocked by a Ca^{2+} concentration > 5 mM. As shown by Bouteau *et al.* (1996) with protoplasts of lactifers isolated from *Hevea brasiliensis*, K^+ efflux may occur under physiological stress inducing a decline of the electrical potential difference across the plasma membrane. This could lead to considerable K^+ losses from plants growing in acid soils where the Ca^{2+} concentration in the soil solution is low. This Ca^{2+} effect accords well with earlier results of Mengel and Helal (1967) who reported that Ca^{2+} did not affect K^+ uptake by roots but blocked K^+ efflux. This phenomenon was known as the *Viets effect* in the older literature (Viets 1944). Abscisic acid (ABA) together with extracellular Ca^{2+} has also been shown to block

K⁺ channels of the plasmalemma. This finding is of particular significance since it demonstrates how a phytohormone may influence plasma membrane permeability.

Several ion channels including Ca^{2+}- and Na^+- activated K^+ channels and K^+ channels open in the plasma membrane when it becomes hyperpolarized *i.e.* when the electrical potential difference between the cytosol and the outer medium is very high. The tonoplast is able to maintain high concentration levels between the cytosol and the vacuole as is particularly true for Ca^{2+}, H^+, and malate (Hedrich and Schroeder 1989). Anion accumulation in the vacuole may be brought about by anion rectifying channels since the electrical charge of the vacuole is higher relative to the cytosol. Nitrate accumulation in the vacuole is of particular interest since it may serve as a resevoir for available nitrogen. Most of the ion channels of the tonoplast are also volt-gated and hence depend on the activity of tonoplast proton pumps. The tonoplast comprises slow and fast activating channels which allow ion transport in both directions (Barkla and Pantoja 1996). The role of the slow acting channels is not yet clear. Useful review articles on ion channels have been published by Hedrich and Schroeder (1989), Tyerman (1992), and Barkla and Pantoja (1996).

3.1.5   Mechanisms of membrane transport

Channels do occur which simply allow transport of solutes across the membrane along the electrochemical gradient (Barkla and Pantoja 1996). Such channels include the aquaporins which are membrane pores allowing a facilitated diffusion of water molecules (Chrispeels and Maurel 1994) through the membrane. This kind of transport mechanism is called facilitated diffusion. Solutes may also be transported through channels together with a cation. The complex thus formed between the cation and the solute is positively charged and therefore also follows the electrochemical gradient when transported across the membrane. This mechanism is called *symport* or *cotransport*. It is a secondary active transport because the transport of the solute alone is against its electrochemical gradient and only by combining the solute with a cation is transport along the electrochemical gradient achieved (Bush 1993). Proton cotransport plays a major role in plant metabolism. In this case one or several protons are combined with the solute which may be an ion, a sugar, or an amino acid. Nitrate, phosphate and sulphate are also mainly transported across membranes by proton cotransport (Ullrich 1992, Clarkson *et al.* 1993, Ullrich-Eberius 1981).Since these are anions and carry negative charges they require an excess of protons in order to form a positively charged complex. The principle of proton cotransport is shown in Figure 3.9, which also illustrates various other types of nutrient transport across a membrane. From this it can be seen that the uptake of anions, amino acids, and sugars also require H^+ which is provided by the proton pump in the plasmalemma and the proton pumps in the tonoplast. Plasmalemma ATPase activity is also of importance for the retrieval of anions and metabolites. As a consequence of the negatively charged cytosol, anions may diffuse through channels out of the cytosol along the electrochemical gradient. These anions may be reabsorbed, by proton cotransport the so called *retrieval mechanism*. Blocking the ATPase by vanadate results in a major release of anions since there is a lack of H^+ in the outer solution for the retrieval process (Matzke and Mengel 1993).

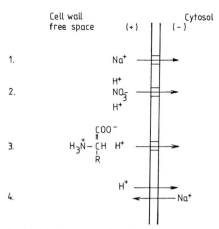

1. Faciltated diffusion of Na⁺
2. H⁺/nitrate cotransport
3. H⁺/amino acid cotransport
4. H⁺/Na⁺ antiport

Figure 3.9 Types of nutrient transport across a membrane.

In an analogous way the retrieval of sugars has been shown to be inhibited if the plasmalemma proton pump is blocked (Mühling *et al.* 1993).

Ullrich and Novacky (1981) reported that the uptake of nitrate is associated with a depolarization of the plasma membrane or in other words the protons transported together with the nitrate into the cytosol neutralize the negative charge of the cytosol. Nitrate uptake increases the pH of the outer solution (Kirkby and Mengel 1967, Smiley 1974). This phenomenon is particularly evident with nitrate since it is taken up at high rates as compared with other anion species. Experiments of Schachtman and Schroeder (1994) showed for the first time that K^+ may also be taken up by proton cotransport. This mechanism is of particular importance if the nutrient solution has a low K^+ concentration so that K^+ cannot be taken up by facilitated diffusion. Such uptake systems have a high affinity to K^+ and therefore are called high affinity-transporters in contrast to uptake by facilitated diffusion which requires a relatively high K^+ concentration with low affinity-transporters (Maathuis and Sanders 1996). According to Fox and Guerinot (1998), who have written a useful review article on the molecular biology of cation transport, high affinity K^+ transporters function in a concentration rage of 1 to 200 mmol/m³ K^+. The finding of a secondary active K^+ uptake mechanism agrees well with earlier findings that K^+ uptake may exceed the Nernst equilibrium (Table 3.1 see page 123). The K^+ channel found for the proton/K^+ transport can also be utilized by NH_4^+ although it is more selective for K^+. Ammonium transport across the plasma membrane is of interest since ammonium is an important plant nutrient some plant species such as rice living almost exclusively from NH_4^+ as a mineral nitrogen source. Ammonium and K^+ are chemically related to each other but in a physiological sense they differ considerably. Potassium ions accumulate in the cytosol to high levels in the range 100 to 150 mol/m^3 whereas ammonium nitrogen entering the cytosol is quickly assimilated (Kosegarten et al. 1997). This means that the ammonium concentration in the cytosol is very low and the electrochemical gradient therefore favours facilitated diffusion of ammonium across the plasmalemma. Gazzarini et al. (1999) found three constitutive NH$_4^+$ transporters in roots of *Arabidopsis thaliana* inducible by N starvation. In root hairs and leaves of tomatoes different genes coding for ammonium transporters have been identified von Wiren et al. (2000). These examples show that plants are equipped with efficient NH$_4^+$ uptake systems which may quickly adapt to nitrogen supply. There is evidence also that Na$^+$ may function instead of H$^+$ as co-cation in the uptake of K$^+$ (Rubio et al. 1995). The potassium ion may also function as co-cation in membrane transport of sugars and amino acids (van Bel and van Erven 1979). According to Saftner and Wise (1980) sucrose is transported across the tonoplast by K$^+$/sucrose cotransport coupled with a K$^+$/H$^+$ antiport. This antiport cycles K$^+$ back from the vacuole into the cytosol where it again may function in the cotransport of sucrose.

Figure 3.9 shows the *antiport mechanism* for Na$^+$/H$^+$. By this mechanism the electrical potential is not altered in contrast to cation-cotransport in which the membrane is depolarized. The Na$^+$/K$^+$ antiport favours the maintainance of a high cytosolic K$^+$ concentration and may also protect the cytosol from too high a Na$^+$ concentration in that K$^+$ is taken up and Na$^+$ removed from the cytosol. The 2H$^+$/Ca^{2+} antiport is responsible for the very low Ca^{2+} concentration in the cytosol and the Ca^{2+} accumulation in the vacuole (Barkla and Pantoja 1996).

Besides these three mechanisms of membrane transport, facilitated diffusion, cation cotransport, and antiport, *carrier transport* may also occur in biological membranes. Carrier transport may be described as the transport of an ion from one side of the membrane to the other by means of a carrier molecule which is able to diffuse through the membrane. Until about a decade ago this type of transport was believed to play a major role in plant metabolism. More recent results and particularly those obtained using the patch-clamp-technique, however, have shown that channels are of much greater importance. Hedrich and Schroeder (1998) reported that carriers transport 10^4 to 10^5 ions per carrier/s whereas channel transport allows a passage of more than 10^6 ion/s.

3.1.6   Compartmentation of ions in plant cells

Figure 3.10 schematically shows various transport systems occurring in the plasmalemma and the tonoplast. Systems in the plasmalemma are ATPase, H$^+$/sucrose, H$^+$ glucose and H$^+$ amino acid cotransport systems. Those shown in the tonoplast are V-ATPase and V-PPase which provide protons for a H$^+$/glucose, a H$^+$/amino acid, and H$^+$/sucrose antiport system transporting amino acids, glucose and sucrose into the vacuole. All these mechanisms are required to bring about compartmentation of ions and metabolites in the cell which is important for normal metabolic status. The Ca^{2+} concentration in the cytosol is kept extremely low and generally <1 mmol/m^3 (Glass ans Siddiqi 1984) otherwise Ca phosphates would precipitate at the prevailing high pH. Excess Ca^{2+} moving into the cytosol is pumped back into the apoplast or sequestered in the vacuole by proton antiport (Leigh and Wyn Jones 1986). The cytosolic

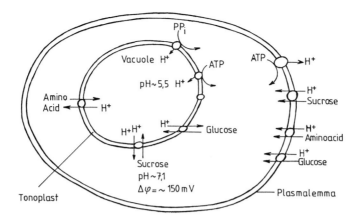

Figure 3.10 Membrane transport systems in the plasmalemma and tonoplast (modified after Bush 1993).

K$^+$ concentration is high and in the range of 100 to 150 mol/m^3. This concentration is required for optimum protein synthesis (Wyn Jones 1981). The vacuolar K$^+$ concentration depends much on the rate of K$^+$ supply as can be seen from the data in Table 3.2 (Fernando *et al.* 1992).

Table 3.2   K$^+$ concentrations in the cytosol and vacuole of barley root cells as related to the K$^+$ concentration in the outer solution (after Fernando *et al.* 1992)

Outer solution	Vacuole mol/m^3	Cytosol
1.2	85	144
0.1	61	140
0.01	21	131

From the data in Table 3.2 it is evident that at high K$^+$ supply, K$^+$ is stored in the vacuole. The same is true for phosphate and nitrate. The vacuole is thus a storage site from which the main plant nutrients may be mobilized if required for metabolic purposes. The vacuole may also contain considerable Na$^+$ concentrations which together with nitrate are of importance for the cell turgor (McIntyre 1997). Accumulation of Na$^+$ in the vacuole is particularly important under saline conditions (Barkla and Pantoja 1996).

A further important characteristic is the pH in plant cells. The apoplast generally has a slightly acid pH which is in the range of 5.0 to 5.5 (Hoffmann and Kosegarten 1995, Grignon and Sentenac 1991). The apoplast has a relatively strong H$^+$ buffer capacity due to the presence of carboxylic and phosphoryl groups (Grignon and Sentenac 1991). Proton buffer systems in the cytosol represent phosphates, carbonates, amino groups of proteins and carboxylic groups(Smith and Raven 1979). Besides these, the H$^+$ pumps in the plasmalemma and tonoplast contribute to the maintenance of a high pH level in the range of 7.3—7.6 in the cytosol. Additionally stabilization of this high cytosolic pH is achieved by production and consumption of protons by the formation and removal of carboxylic groups (Smith and Raven 1979). This *pH stat system* as it has been called by Davies (1994) is based on the carboxylation of phosphoenlpyruvate (PEP) and the decarboxylation of oxaloacetate. Both reactions are shown in Figure 3.11.

The carboxylation reaction is favoured under conditions of OH^- production. Monohydrogencarbonate is produced by the binding of CO_2 to OH^-. The reaction is catalyzed by a carbonic anhydrase (Badger and Price 1994). This reaction consumes one OH^-

$$CO_2 + OH^- \Leftrightarrow HCO_3^-$$

and therefore prevents a rise in cytosolic pH. Conversely the decarboxylation of the oxaloacetate consumes one H^+ and therefore prevents a fall in cytosolic pH. A low pH in the cytosol may occur when the pH in the outer medium (nutrient solution) is also low. In this case the energy required by the proton pump is high due to a steep pH gradient and the coupling ratio of the H^+ pump *i.e.* the number of H^+ pumped out of the cytosol per ATP hydrolized by the H^+ pump may be low (Barkla and Pantoja 1996). In this case oxaloacetate may be decarboylated at a higher rate in order to maintain the optimum cytosolic pH. Eventually, however, malate the precursor of oxalacetate is exhausted. At this point root growth in particular is severely depressed (Yan *et al.* 1992).

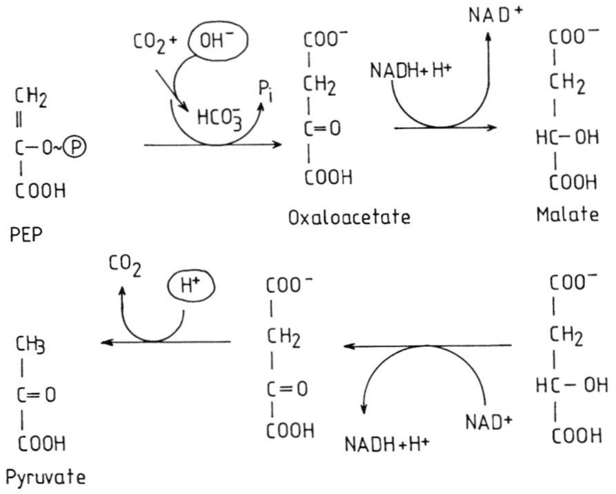

Figure 3.11 PEP/Malate H^+ buffer system, after Smith and Raven (1979). The upper reaction sequence shows the carboxylation of PEP consuming one OH^- and the lower reaction sequence shows the decarboxylation of oxaloacetate consuming one H^+.

3.1.7   Cation/anion balance and ion antagonism

The observation that plant cells are negatively charged as compared with the outer medium implies that they contain a surplus of negative charges. This surplus, however,

is extremely small so that it cannot be measured by chemical analytical methods
even when electropotential differences are high, but only by measuring cellular
electropotentials. The total sum of anion equivalents in a cell or tissue is therefore
virtually equal to the total sum of cations. In this balance organic anions represent an
considerable proportion of the cation and anion equivalents. The production of organic
anions is related to the activity of plasmalemma and and tonoplast proton pumps.
As shown in Figure 3.12 proton pumps produce 1 OH^- per H^+ pumped into the outer
medium or vacuole. Continuous pumping would thus result in an immense rise of
the cytosolic pH if the OH^- were not be used for the synthesis of malate according to
the PEP/malate buffer system (Figure 3.11). Hence the negative charge of the OH^- is
shifted to oxaloacetate and from there to malate and other organic anions.

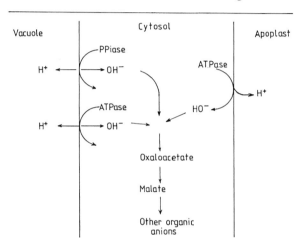

Figure 3.12   Plasmalemma and tonoplast proton pumps producing OH^- in the cytosol which is used for
the synthesis of organic anions.

Earlier work of Coic *et al.* (1962) and Kirkby (1968) showed that when plants were
fed with nitrate-N, the concentrations of cations and organic anions in the plants were
higher than when they were fed with ammonium-N. This phenomenon were explained
in terms of cation/anion balance. Not only ion concentrations were affected, however,
but also plant growth which was poorer in the ammonium treatment. According to more
recent investigations of Yan *et al.* (1992), the depression in growth occurring at low pH
associated with ammonium nutrition is a first indication of NH_4^+ toxicity. Net release
of H^+ from the roots under conditions of low pH is severely depressed and hence pH
regulation in the cytosol is disturbed (Michelet and Boutry 1995). As a consequence
of the decrease in cytosolic pH, expected under such conditions, more malate must
be decarboxylated according to the pH-stat (page 130). This depletes the malate pool
as found by Yan *et al.* (1992) and reported earlier by Kirkby (1968). The depression
in release of H^+ by roots induced by the low pH resulting from ammonium nutrition
also means that the electropotential difference across the plasma membrane falls and
so too therefore the retention capacity of the cell for cations, especially for K^+.

Cation concentrations in the plants are therefore lower as found by Kirkby (1968). When plant growth is not depressed by ammonium nutrition as was the case in experiments of Dijkshoorn and Ismunadji (1992) with rice the concentrations of organic anions in the plants did not differ much between the two forms of nitrogen nutrition and net cation uptake was only somewhat higher in the nitrate fed plants. Further evidence of the importance of low pH in depressing growth of ammonium fed plants can be inferred from the work of Mengel *et al.* 1983, who showed that the growth of maize was unaffected by the form of nitrogen supply, nitrate or ammonium, provided that an appropriately high pH was maintained in the nutrient solution in which the plants were grown.

Exclusive ammonium nutrition leads to a pH decrease in the nutrient medium because the H^+ pumped out into the nutrient medium by the plasmalemma H^+ pumps is not recycled back into the cytosol by H^+/nitrate cotransport (see page 127). This pH decrease caused by ammonium nutrition is particularly evident in solution culture since, in contrast to soils, solutions have virtually no H^+ buffering capacity. This effect of nitrate and ammonium nutrition on pH in the nutrient solution is clearly shown by research work of Breteler and Smit (1974) some data of which are presented Table 3.3.

Table 3.3   Effect of the form of nitrogen nutrition on the pH in the nutrient solution. Young wheat plants after an uptake period of 16h (Breteler and Smit 1974)

N in the nutrient solution	pH in the nutrient solution
4 mol/m^3 nitrate	7.18
4 mol/m^3 nitrate + 1 mol/m^3 ammonium	3.72
4 mol/m^3 nitrate + 2 mol/m^3 ammonium	3.38
4 mol/m^3 nitrate + 3 mol/m^3 ammonium	3.07
4 mol/m^3 nitrate + 4 mol/m^3 ammonium	3.17
pH in the nutrient sol. at the beginning of the experiment	5.31

These data show how effectively the pH of the nutrient solution was depressed by ammonium nutrition even in the presence of nitrate. The addition of up to four times as much nitrate as ammonium in the nutrient medium could not prevent the drop in pH. This is because ammonium considerably depressed the uptake of nitrate whereas nitrate had hardly any effect on ammonium uptake (Mengel and Viro 1978). Nitrate when supplied without ammonium resulted in a marked pH increase. Plants may adapt to low pH in the nutrient medium as reported recently reported by Yan *et al.* (1998). Maize plants were shown to adapt gradually to low pH by a change in the characterisics of the plasma membrane ATPase in that the K_m and V_{max} values (see page 135) were raised. The adapted plants were thus still able to depress the pH in the nutrient solution at pH of 3.4 whereas the non adapted plants were not. As shown in Figure 3.13 it is evident that the non adapted plants were unable to excrete H^+ into the outer medium with the presumably detrimental consequence of depressing cytosolic pH. The pH of nutrient solution or soils is markedly depressed by N fixing leguminous species. This results primarily from a lack of nitrate uptake which recycles H^+ back into the cytosol. Figure 3.14

shows the development of soil pH under red clover living symbiotically with *Rhizobium* and ryegrass fertilized with ammonium nitrate (Mengel and Steffens 1982). The plants were grown in pots on a medium textured soil with a medium H^+ buffer power and under the same environmental conditions. With red clover, soil pH decreased markedly over the seven cuts taken, whereas with ryegrass the soil pH remained at the same level. Since the plants were grown on non sterile soils it may supposed that a substantial proportion of ammonium applied as ammonium nitrate was nitrified and the nitrate thus produced and taken up was also able to recycle the H^+ into the cytosol.

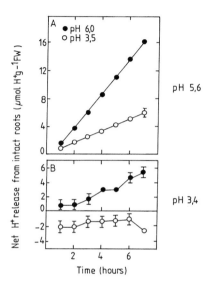

Figure 3.13   Net release of H^+ from roots of intact maize plants adapted to low pH (filled circles) and plants non adapted to low pH (open circles) — Upper part of the *Figure* H^+ release in an outer medium with pH 5.6 — Lower part in a medium with pH 3.4 (after Schubert *et al.* 1998).

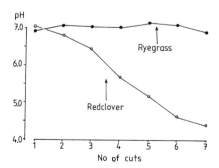

Figure 3.14   Development of soil pH under rye grass and clover. The rye grass was dressed with ammonium nitrate and the clover was supplied with symbiontically fixed nitrogen (after Mengel and Steffens 1982).

The older literature provides data that the total sum of cations in a plant or plant tissue is little changed despite variations in the levels of the individual cations in the nutrient medium. Increasing the supply of one cation species in the nutrient medium can thus depress the levels of other cation species in the plant. This phemomenon that one ion species present in excess in the nutrient medium may depress the uptake of other ion species is called *ion antagonism*. Table 3.4 shows a typical example of this kind. Increasing the level of Mg^{2+} application to sunflower plants resulted in a decrease of Na^+ and Ca^{2+} concentrations and a corresponding increase in the Mg^{2+} concentrations thus maintaining a fairly constant total sum of cations in the plant of all treatments. Interestingly K^+concentrations were not depressed by Mg^{2+} applications a finding presumbly due to the very efficient uptake systems for K^+, *e.g.* a H^+/K^+ symport (page 489).

The same pattern occurs when the supply of any major cationic nutrient species is increased. It is the general rule therefore that increasing the supply of one cation species results in lowering the concentrations of other cation species.

Table 3.4 The effect of an increasing Mg^{2+} application on the concentration of various cation species in sunflower plants (Scharrer and Jung 1955)

	K^+	Na^+	Ca^{2+}	Mg^{2+}	Sum
		me/kg dry matter			
Mg_1	490	40	420	490	1440
Mg_2	570	30	310	610	1520
Mg_3	570	20	230	680	1500

This antagonistic behaviour of cation species is mainly associated with the uptake of cations into the vacuole whereas cytosolic cation concentrations are not much affected so that the predominately high K^+ and low Ca^{2+} concentrations in the cytosol are maintained. This interpretation is in line with the finding that the tonoplast cation channels are not very selective (Barkla and Pantoja 1996).

Potassium ions are strong competitors with other cation species because of their efficient uptake systems. In the absence of K^+ in the nutrient solution the uptake of other cation species is thus much enhanced because competition for uptake is less severe. This relationship is shown in Table 3.5 from an experiment of Forster and Mengel (1969) in which barley plants were grown in a complete nutrient solution. In one treatment K^+ was withheld from the nutrient solution for 8 days during the growth period. After this time the cation concentrations of roots and shoots were determined in samples of both treatments (control and interrupted supply). As shown in Table 3.5 this interruption in K^+ supply resulted in a drop in the K^+ levels in the roots and shoots, whereas the concentrations of Ca^{2+}, Mg^{2+}, and Na^+ increased considerably. The total amount of the four cation species was not significantly affected by the K^+ interruption. This indicates that the deficient equivalents of K^+ were more or less made up by the other cation species. These cations species, however, were unable to substitute for the physiological functions of K^+. Presumably the K^+ concentration in the cytosol was not optimal so that growth and particularly protein synthesis were affected. The yield data of Table 3.5 show that interruption in K^+ supply, during the tillering stage, resulted in a highly significant grain yield depression.

In anion uptake, antagonistic effects are less common although Cl^-, SO_4^{2-} and $H_2PO_4^-$ uptake can be stimulated when NO_3^- uptake is strongly depressed (Kirkby and Knight 1977). The most common anion antagonism is between NO_3^- and Cl^-. High chloride supply in the nutrient medium lowers the nitrate uptake and *vice versa*. The effects are particularly marked in plants species which accumulate nitrate and chloride in their vacuole such as the *Chenopodiaceae*. Wehrmann and Hähndel (1984) reported that nitrate uptake of spinach was much depressed by chloride.

Table 3.5   The effect of an interruption in the K^+ supply on the cation concentrations of young barley plants; interruption period 8 days (Forster and Mengel 1969)

	Roots		Shoots	
	Control	Interruption	Control	Interruption
	me/kg dry matter			
K	1570	280	1700	1520
Ca	90	120	240	660
Mg	360	740	540	210
Na	30	780	trace	120
Total	2050	1920	2480	2510

Grain yield (g/pot): Control 108; Interruption 86[***]

3.1.8   Relationship between uptake rate and the ion concentration in the nutrient solution

Basically the rate of nutrient uptake is controlled by the growth rate of plants and related to the physiological requirement. Cereals *e.g.* require nitrogen in the grain filling stage and therefore take up nitrogen at this late stage in growth in contrast to K^+ which is mainly taken up in the vegetative phase and not in the reproduction phase. High growth rates are associated with high nutrient uptake rates and *vice versa*. Plants which have been starved of a particular nutrient may take up the nutrient at a high rate and if such plants are exposed to nutrient solutions with increasing concentrations of the lacking nutrient, the plot of uptake rate *versus* nutrient concentration in the uptake solution may be reflected by a Michaelis-Menten curve (Glass and Siddiqi 1984). From this observation ion uptake may be seen as a process analogous to that of an enzyme process. This may be acceptable if uptake relates to one particular uptake system. Recent research work on ion channels has shown, however, that numerous systems may be responsible for the uptake of a certain ion species and that the ion concentration in plant cells is more or less independent of the corresponding ion concentration in the soil or nutrient solution. Only if the concentration in the cell is lowered below a threshold level is the ion taken up at higher rates. For K^+, Na^+, $H_2PO_4^-$ and Cl^- the rate of uptake is lower, the higher the concentration of these ion species in the cytosol (Glass and Siddiqi 1984). These findings provide evidence that nutrient uptake rates are regulated depending on the concentration of the particular nutrient in the plant tissue. The mechanism of this regulation is as yet not understood. There are, however, indications that where demand is high, genes coding for membrane transport systems are expressed (Glass und Siddiqi 1984, Schachtman and Schroeder 1994). Glass and Siddiqui reported (1984) that the apparent Michaelis-Menten constant (K_m) was higher the higher the K^+ concentration in the nutrient solution. According to Claassen and Barber (1976) the V_{max} for the Michaelis-Menten curve describing K^+ uptake was not constant but lower the higher the K^+ status of plants. From this it follows that K^+ uptake is finely tuned according to the K^+ nutritional status of the plant and that various transporters are involved in K^+ uptake. This same fine tuning of uptake determined by demand is also true for other plant nutrients (Marschner et al. 1997). Recent research has shown that in plants requiring nitrogen, genes are activated which code for nitrate uptake rates are quickly upregulated followed by downregulation as soon as the nitrogen demand is met (Glass 2000). The signal responsible for this regulation is glutamine which regulates nitrate uptake as well as the uptake of NH_4^+.

Finally it should be emphasized that nutrient uptake requires energy, mainly in the form of ATP for driving the proton pumps. If oxygen supply is limiting or the carbohydrate reserves of plant tissues are low so too is nutrient uptake. Such an example is shown for the ammonium uptake by rice plants which in some treatments were exposed to darkness before uptake was measured (Table 3.6).

Table 3.6   Uptake of ^{15}N labelled NH_4^+ by rice plants exposed to darkness before the uptake test (Mengel and Viro 1978)

	Shoots	Roots
	μg ^{15}N/g fresh weight	
Control, no darkness	42.4	108.9
24h darkness	17.0	45.3
48h darkness	8.4	34.2

3.2   Photosynthesis and CO_2 Assimilation

3.2.1   General

Nutrition has already been defined as the supply of an organism with its essential foodstuffs. For all animals and most microorganisms these foodstuffs not only contain essential chemical elements but are also a source of chemical energy by which the energy demands of the organism are satisfied. For green plants the situation is quite different because the nutrient sources, CO_2, H_2O and inorganic ions, are of low energy status and therefore not able to meet the plant's energy requirement. The assimilation of these inorganic nutrients does, in fact, need energy. In green plants this requirement is satisfied primarily by the absorption of light. This unique ability of green plant cells to absorb light energy and convert it into chemical energy is one of the most important biological processes. All other organisms with the exception of a few microorganisms are dependent on this energy conversion.

Intense research activity over the past thirty to forty years has resulted in a comprehensive understanding of photosynthesis. It is beyond the scope of this book, however, to give a detailed account of photosynthesis. Here it is our aim simply to present the principles by which radiation energy is transformed into chemical energy. This conversion is the most important function of green plants and it is by this unique capability that crops are grown. The conversion of light energy into chemical energy is closely related to the conversion of CO_2 into organic compounds. For decades both reactions — energy conversion and CO_2 fixation — were regarded as one reaction complex described by the equation:

$$6 \ CO_2 + 6 \ H_2O \overset{\text{light energy}}{\Rightarrow} C_6H_{12}O_6 + 6 \ CO_2$$

A clear distinction is now drawn between energy conversion and CO_2 assimilation. In modern terminology, the term photosynthesis has been applied to the process whereby a pigment system absorbs electromagnetic radiation and converts this into chemical forms of energy which are available for growth in a particular environment.

3.2.2 Light absorption and electron flow

In higher plants light absorption is brought about by chlorophylls and carotenoids (Figure 3.15). These are located in thylakoid membranes of the chloroplasts (Plate 13.4). The unique feature of these pigments is the ability to absorb light and convert it to chemical energy. Other coloured particles are also capable of light absorption but the energy is dissipated either as heat or as light in the form of fluorescence or phosphorescence. Chlorophyll a and chlorophyll b are the major chlorophyll molecules in higher plants. Both molecules consist of a porphyrin ring comprising four pyrrole rings (Figure 3.15), in the centre of the which is a Mg atom. This is bound covalently to the two N atoms of the pyrroles and coordinatively to the other two N atoms of the pyrrole rings. In chlorophyll b one methyl group of the porphyrin ring of chlorophyll a is substituted by a formyl group as indicated by an arrow in Figure 3.15. There is a longer side chain in which the alcohol phytol is esterified to the molecule. In some bacteriochlorophylls this phytol rest can be substituted by the alcohol farnesol. These long chains are lipidic in character and it is supposed that they anchor the molecule to the membrane (Marshall 1993). Carotenes and carotenoids are build up by eight isoprene rests. Two of the isoprenes may form a ionone ring. Carotenes are build up only by C and H atoms while carotenoids additiononally contain O as shown for β-carotene and lutein in Figure 3.15. Carotenes and carotenoids have a yellow to orange and red colour which

138

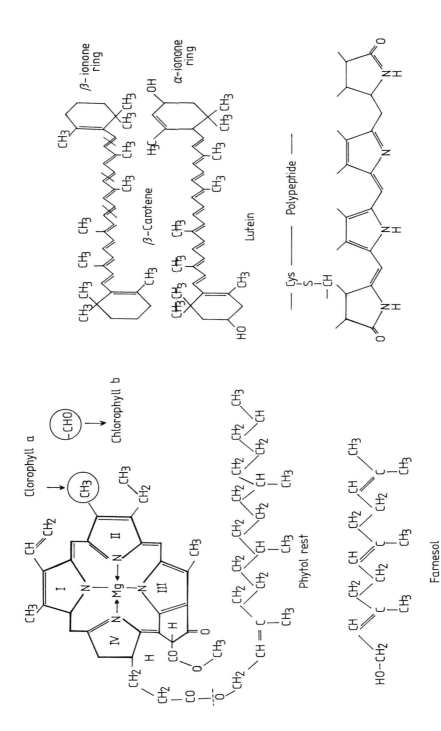

Figure 3.15 Molecular structure of photosynthetic pigments. Left: chlorophylls, Right: carotenoids, phycobilins.

is generally masked by the green of the chlorophylls. In autumn, however, when the chlorophylls are degraded these beautiful colours become visible. Phycocyanobilin and phycoerytrobilin are pigments of the *Cyanophyceae* and of some eukaryotic algae. Phycocyanobilin is blue (cyanic, gr. = dark blue) and phycoerythrobilin reddish (erythros, gr. = red).

The photosynthetic pigments are characterized by their absorption spectra. That of chlorphyll a is shown in Figure 3.16. It has two absorption maxima; one in the range of blue light (400 to 430 nm) and one in the range of red light (640 to 700 nm). Green light is not absorbed and it is for this reason that photosynthetic organs appear green to the human eye. The absorption spectra of carotenes and carotenoids are more in the range of lower wavelengths (430 to 600 nm) and since the radiation energy is reciprocal to the wavelength, carotenes and carotenoids absorb a more energy-rich radiation than chlorophylls. The most frequently occurring carotenoids and carotenes in higher plants are lutein and β-carotene. Also the phycobiliproteins (phycocyanobilin and phycoerythrobilin) have maximum light absorption in the range of 500 to 600 nm which means that they also mainly absorb energy-rich radiation.

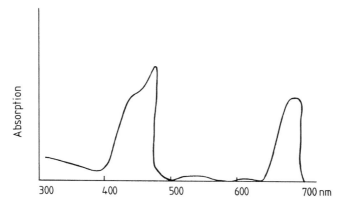

Figure 3.16 Absorption spectrum of chlorophyll.

These pigments are integrated into membrane proteins which transverse the thylakoid membrane similar to the transport membrane proteins shown in Figure 3.5. The arrangement of the pigments is such that the energy-rich light absorbed is funnelled to pigments which absorb light of lower energy. Such pigment arrangements consisting of 10 to 20 chlorphyll molecules (chlorophyll a and b) and several carotene and carotenoid molecules are called light harvesting centres (LHC). The function of these pigments consists of absorbing light energy and transferring it by inductive resonance to a special *chlorophyll a* molecule, which has a unique function in the photosystem.

In higher plants two photosystems are responsible for the energy conversion; photosystem I and II. Both these photosynthetic units consist of about 400 chlorophyll molecules and some hundred molecules of carotenes and carotenoids. Both systems contain several light harvesting centres. The energy absorbed by them is focused on to a particular *chlorphyll a* molecule which in photosystem I is known as pigment 700

(= P-700) as its absorption maximum is at 700 nm. P-700 differs from the bulk of chlorophyll a molecules in that it has an absorption maximum at a somewhat longer wavelength. The light energy absorbed by the carotenes, carotenoids, chlorophyll b and by ‚normal‘ chlorophyll a molecules in photosystem I is transferred to P-700 so inducing the emission of an electron. It is assumed that one photon hitting the system induces the release of one e^-,

$$Energy + P\text{-}700 \Rightarrow (P\text{-}700)^+ + e^-$$

This is the basic process that initiates electron flow. The pigments in the photosynthetic unit thus act as a funnel directing absorbed energy to one P-700 molecule. This brings about the oxidation of P-700 and the release of an electron. as shown above. The electron acceptor in this particular process is supposedly bound ferredoxin. Its standard redoxpotential is low (-0.44 V) in comparison with that of P-700 (+0.46 V). The electron ejected from P-700 has thus to be moved against an electrical gradient.

$$(P\text{-}700)^+ \quad \overset{e^-}{\Rightarrow} \quad (bound\ ferredoxin)$$
Redoxpotential +0.46 V   Redoxpotential -0.44 V

This 'uphill transport' requires energy which is ultimately derived from the light energy absorbed by the pigments of photosystem I.

Photosystem II functions in an analogous way to photosystem I. In photosystem II the electron emittor is also a *chlorophyll a* molecule which has the absorption maximum at 682 nm. It is hence known as P-682. The electron acceptor of P 682 is usually designated as Q (= quencher) due to the fact that its presence ‚quenches‘ the fluorescence. The primary e^- acceptor of P-682 is a quinone (= Q). As the standard redoxpotential of P-682 is +0.8 V, electron transfer from P-682 to Q is also an uphill transport requiring energy, from the light absorbed by the pigments of photosystem Il.

$$(P\text{-}682)^+ \quad \overset{e^-}{\Rightarrow} \quad (Q)$$
Redoxpotential, +0.8V   Redoxpotential, -0.1 V

In higher plants these two photosystems function in series and are components of an electron transport pathway transferring electrons from water to $NADP^+$. This means that ultimately water is the electron donor and $NADP^+$ the electon acceptor in the overall process as shown below.

light energy
$$H_2O + NADP^+ \Rightarrow NADPH + H^+ + \tfrac{1}{2} O_2$$

Figure 3.17 shows this electron transport chain and its redoxsystems. Photosystem II is closely associated with the splitting of water which serves as the electron donor. P-682 in its oxidized form $(P\text{-}682)^+$ is a very strong oxidant and thus capable of oxidising H_2O. In this reaction H_2O is split in O_2, H^+ and electrons.

$$2H_2O \Rightarrow 4H^+ + 4e^- + O_2$$

This process is called photolysis and this is the origin of O_2 produced in photosynthesis. P-682 does not react directly with water. The process is mediated by further redox systems in which Mn is involved. A bound form of Mn in the chloroplasts appears to undergo photooxidation from Mn^{2+} to Mn^{3+}.

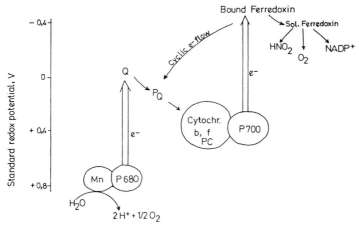

Figure 3.17 Photosynthetic e⁻ transport chain. Redox systems arranged according to their standard redox potentials.

The electrons supplied by H_2O are transferred to P-682 and from there *via* various redox systems to pheophytin, a chlorophyll molecule lacking Mg^{2+} and then passed on to a quinone Q which is a bound redox system of photosystem II. From Q (quinone) the electrons move energetically downhill following the increasing redoxpotentials of plastoquinone, cytochrome b/cytochrome f complex, plastocyanin and P-700 as shown in Figure 3.17.

Plastoquinone and its reduced form plastoquinol are low molecular weight compounds which are strongly lipophilic in character due particularly to the isoprenic tail (Figure 3.18). For this reason these molecules are very mobile in the thylakoid membrane and able to shuttle H atoms from the bound quinone of photosystem II to the cytochrome b, f complex (Figure 3.17).

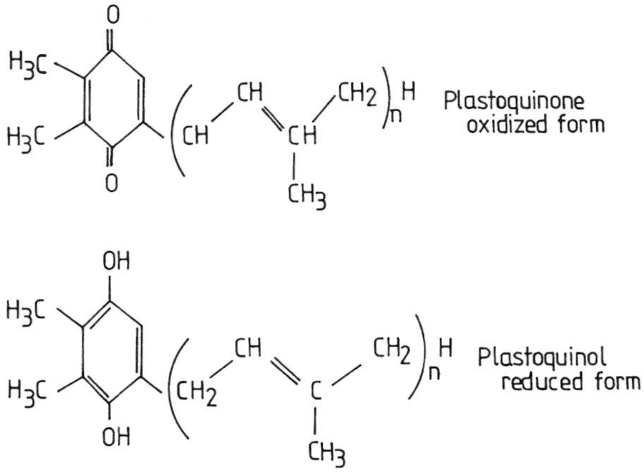

Figure 3.18 Plastoquinone, oxidized form and plastoquinol, reduced form.

Plastocyanin is an acidic protein containing 2 Cu atoms per molecule. It is called the *blue protein*. From P-700 a further uphill transport occurs as already indicated raising the electron from a redox potential of +0.46 V to a redoxpotential of -0.44 V. In the scheme shown in Figure 3.16 both these uphill transports from P-700 and P-682 are indicated by vertical arrows pointing from a more positive to a more negative redox potential. Because of the zigzag pattern of the scheme it is sometimes known as the Z scheme.

The electron acceptor of P-700 is supposedly bound ferredoxin. Electrons are then transferred to soluble ferredoxin. Ferredoxins are stable 2Fe-2S proteins. As can be seen in Figure 3.19, Fe is bridged by S atoms (covalent and coordinative bonds) and covalently bound to the S atoms of cysteine residues. This particularly structure enables the molecule to release and to take up one e^-. There is both a soluble form of ferredoxin and a form integrated into the thylakoid membrane, which is also known as the 2Fe-2S complex. The soluble form has a molecular weight of about 12000 Da and contains 2 atoms of Fe and 2 atoms of inorganic S per molecule as shown in Figure 3.19.

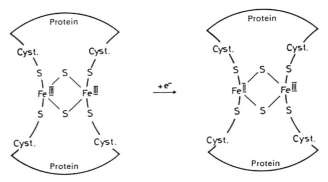

Figure 3.19 The 2Fe-2S complex (ferredoxin) bound to protein by the S of cysteine rests. Left oxidized form, right reduced form.

In the oxidized form both of the Fe atoms are present as Fe^{III} whereas in the reduced form one Fe atom occurs as Fe^{II}. Soluble ferredoxin thus acts as a one e^- carrier. Electrons bound by ferredoxin can be recycled to plastoquinone (Figure 3.16) thus inducing a cyclic e^- flow which gives rise to *cyclic photophosphorylation*. This process is considered in the next section.

Ferredoxin is the first stable redox compound of the photosynthetic electron transport chain. Its high negative redox potential (-0.43 V) means that it has a strong reducing power and can reduce various substances such as $NADP^+$, nitrite, O_2, sulphate and haem proteins (see Figure 3.20). The soluble ferrexdoxin receives its e^- directly from photosystem I. Hence the reduction of nitrite and sulphate, each representing a first step in nitrogen and sulphur metabolism, respectively, are directly linked with photosynthesis. NADPH produced by the reduction of $NADP^+$ is required for the assimilation of CO_2 and is also required in lipid metabolism. It is thus evident that most important metabolic processes of plants depend directly on photosynthetic activity.

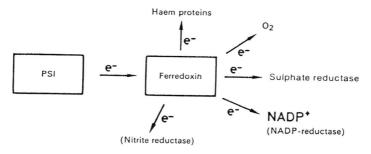

Figure 3.20 Ferredoxin providing e⁻ for various important reduction processes such as NADP⁺ reduction, nitrite and sulphate reduction and the reduction of haem proteins.

In Figure 3.20, O_2 is also shown as a potential e⁻ acceptor provided by ferredoxin. The reduction of O_2 may produce superoxide which is an aggressive radical (Cadenas 1989). Molecular oxygen is known to be a strong e⁻ scavenger and the *superoxide radical* may be formed

$$O_2 + e^- \Rightarrow O_2^{\cdot -}$$

The dot on the O_2 denotes that the O_2 is a radical and not molecule at it has an additional unpaired e⁻ The negative sign denotes that the radical is negatively charged. The superoxide radical as shown above may be the precursor of further oxyen radicals as descibed by Cadenas (1989) in a valuable review article. For example

$$O^{\cdot}_2 + H^+ \Rightarrow HOO\cdot \text{ hydroperoxyl radical}$$

In this case the uptake of H^+ by the $O^{\cdot}_2$ neutralizes the negative charge of the $O^{\cdot}_2$ and its protonation results in the formation the *hydroperoxyl radical* (HOO·). The superoxide radical is quickly detoxified by the enzyme *superoxide dismutase* according to the following reaction:

$$2O^{\cdot}_2 + 2H^+ \Rightarrow H_2O_2 + O_2$$

By this reaction hydrogen peroxide (H_2O_2) is formed which may be reduced by Fe^{II} complexes as shown below:

$$H_2O_2 = HO{:}OH + Fe^{II} \Rightarrow Fe^{III} + OH^- + OH\cdot$$

In this reaction the *hydroxyl radical* OH· is formed. From the equations shown above it is clear that in photosynthetic e⁻ transfer from ferredoxin to O_2 three major toxic ogygen radicals may be produced which are very aggressive and may destroy pigments and biological membranes and especially the thylakoid membranes. There is evidence that the chlorophyll of the photosynthetic reaction centre is especially susceptible to this photooxidation (Boardman 1977).

The risk of the production of oxygen radicals increases with light intensity because under such conditions the rate of e⁻ flow in the photosynthetic e⁻ transport chain is high so that the chance of the formation of $O_2^{\cdot -}$ is great. The damage, thus produced, is visible by the destruction of pigments, apparent as pigment lesions, and is known as photooxidation. It may lead to the so called *sunscald* in vegetables and fruits with a harmful effect on the quality of these crops (Bowler *et al.* 1992). The problem is particularly severe in arid areas with high light intensity. Oxygen radicals can be detoxifyed by carotenes and carotenoids. These pigments present in the thylakoid membrane are therefore not only of importance for light absorption but also for the detoxification of oxygen radicals and plant species well supplied with carotenes and carotenoids are more resistant to photooxidation than are poorly supplied species. Plant species living in shaded habitats, so called shade plants, are particularly prone to photoxidation and tolerate light intensities which are only 20 to 30% of the light intensity tolerated by non shade plant species (Boardman 1977).

A substantial amount of ferredoxin is oxidized enzymatically by the ferredoxin-NADP$^+$ reductase which reduces NADP$^+$ to NADPH. As the e$^-$ reducing NADP$^+$ to NADPH originate from H$_2$O the overall process of this reduction is described by the equation:

$$NADP^+ + H_2O \Rightarrow NADPH + H^+ + \frac{1}{2} O_2$$

This is an endergonic process requiring approximately 220 kJ per Mol NADPH produced. The energy is provided by the light energy trapped by photosystems I and II. It is supposed that the ejection of one e$^-$ (one photochemical event) from P-700 as well as one from P-682 each requires one photon. As the reduction of NADP$^+$ requires 2 e$^-$ a total of 4 photons is needed. One mol of photons (= 1 Einstein = photon mol) of red light at a wavelength of 680 nm has an energy content of about 176 kJ. Hence the total energy of 4 photon mols absorbed amounts to 704 kJ. This energy quantity exceeds the total chemical energy produced by the photosynthetic process. This total chemical energy produced is the sum of the energy required for the production of NADPH (220 kJ) and the synthesis ATP (32 kJ). Hence the Z scheme is energetically feasible, allowing a reasonable loss in the conversion of light energy to chemical energy.

The photosynthetic e- transport chain as given in Figure 3.17 shows only an outline of the reactions taking place. The details are more complicated. It is now generally accepted that the thylakoid membrane is traversed severalfold by long polypeptide strands as occurs similarly in the plasmamembrane (Figure 3.5). There are three supramolecular membrane-spanning complexes; the photosystem II complex, the cytochrome b, f complex, and the photosystem I complex. Electron transport between photosystem Il and the cytochrome b, f complex is linked by plastoquinone and plastoquinol which are present in abundant quantities in the membrane.

A real problem in understanding the e$^-$ transport chain of photosynthesis is the fact that photosystem I and II are not so closely connected as shown in Figure 3.17. They function in series, although the spatial arrangement of both systems does not favour such a process. The inner chloroplast membrane (thylakoid membrane) has sections where the membranes are stacked thus forming grana. Other sections occur when only a single membrane is present (stroma section). There is now much evidence that photosystem II is located in the stacked thylakoids (grana) whereas photosystem I is in the stroma thylakoid. Electron transport from the grana thylakoid to the stroma thylakoid may occur at contact sites, the transport being mediated by the plastoquinone/plastoquinol shuttle which basically transports H. However, since H can produce H$^+$ + e$^-$, the plastoquinnone/plastoquinol system also transports e$^-$.

3.2.3   Photophosphorylation

Before describing the process of photophosphorylation some major features of the chloroplast anatomy should be mentioned. The chloropast is surrounded by a double membrane,
the inner membrane of which may produce vesicles by invagination as shown in Figure 3.21.

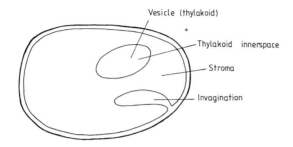

Figure 3.21   Simplified scheme of the invagination of the inner chloroplast membrane and the formation of
the vesicle (=thylakoid) with the stroma and the thylakoid inner space.

Generally chloroplasts contain numerous thylakoids but for the sake of simplicity reason
only one vesicle (thylakoid; *thylakos, gr.* bag like structure) is shown. Photosynthetic
redox systems including pigments are exclusively located in the thylakoid membrane
which separates the stroma from the thylakoid inner space. This strict separation plays
a decisive role in photophosphorylation.

   According to the hypothesis of Mitchell (1961, 1978) the flow of e⁻ through the e⁻
transport chain of photosynthesis drives protons (H^+) across the thylakoid membrane.
This results in an electrochemical proton gradient across the membrane. The gradient
consists of two components, a H^+ concentration and an electrical potential, analogous of
that produced by proton pumps (see page 118). The electrochemical potential difference
(proton motive force = pmf) of the gradient may thus be described as

$$pmf = \Delta H^+ + \Delta \varphi$$

where
   ΔH^+ = proton gradient across the membrane
   $\Delta \varphi$ = electrical gradient across the membrane
This proton motive force provides the energy for ATP synthesis which is the reverse
of an ATPase reaction (proton pump) in which ATP is hydrolzed and H^+ ions are
separated (see page 120).

$$ADP + P_i \Rightarrow ATP + H_2O$$

The mechanism by which electron transport is able to induce a proton gradient is
dependent on the nature of the electron transport components and their location in
the membrane. Figure 3.22 shows this arrangement. Proton separation comes about
if a H carrier (plastoquinol) reacts with an electron carrier. This is the case when
photosystem II transfers the e⁻ to plastoquinone. Reduction of plastoquinone to
plastoquinol requires e⁻ + H^+:

$$Q + 2e^- + 2H^+ \Rightarrow QH_2$$

146

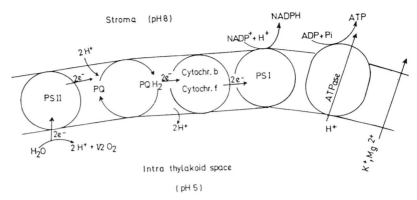

Figure 3.22 Coupling of electron flow, proton separation and ATP synthesis in the thylakoid membrane.

Thus H⁺ in this reaction is extracted from the stroma. The reduction of the cytochrome b, f complex requires only e⁻. Thus when it is reduced by plastoquinol (QH_2), H⁺ is released and secreted into the thylakoid inner space.

$$\text{Cytochrome b, f} + QH_2 \Rightarrow \text{Cytochr. b, f} + 2e^- + Q + 2H^+$$

These processes render the stroma alkaline and the intrathylakoid space acid. Additional acidification of the intrathylakoid space is brought about by the photolysis of H_2O.

$$H_2O \Rightarrow 2H^+ + 2e^- + \tfrac{1}{2} O_2$$

By these reactions a proton gradient is generated across the thylakoid membrane and the proton motive force thus established drives ATP synthesis. Along the electropotential difference across the thylkaoid membrane protons flow down-hill through a proton channel located in the ATPase complex from the thylakoid space into the stroma and this H⁺ flow brings about the synthesis of ATP fom ADP + inorganic phosphate (P_i). The mechanism of this ATP synthesis catalyzed by an ATP synthase is not yet completely understood. The synthesis occurs at the stromal site of the enzyme (Figure 3.22).

The proton gradient across the thylakoid membrane represents a storage of free energy with a chemical component (ΔH^+) and an electrical component ($\Delta \varphi$). The number of protons required for the synthesis of 1 ATP depends on the steepness of this gradient. For a steep gradient 1 H⁺ may suffice for the synthesis of 1 ATP molecule but generally 1 to 3 H⁺ are needed.

The electrical potential gradient also depends on the counter flux of cations particularly K⁺. There is evidence by research work of Berkowitz and Peters (1993) that the inner chloroplast membrane contains an ATPase which drives the import of K⁺ and the export of H⁺ out of the stroma and thus contributes to the high stromal pH and thus promotes ATP synthesis. Generally the stroma pH is in a range of 7.5 to 8 and the pH in the inner space of thylakoids in a range of 5.0 to 5.5.

As shown above in Figure 3.17 the proton electrochemical potential across the thylakoid membrane is brought about by electron transport from water to NADP⁺. Besides this *linear* electron flow, however, *cyclic* electron flow can also occur. In this case bound ferredoxin reduces plastoquinone (Q), which again supplies electrons into the photosynthetic electron transport chain (see Figure 3.17). Electron transport from plastoquinone to cytochrome f results in a proton separation and thus also in the build up of an electropotential across the membrane. This cyclic flow of electrons drives the so-called *cyclic photophosphorylation*. During cyclic electron flow no NADPH and no O_2 is produced. The system can thus adjust to physiological requirements (Arnon1977). Raghavendra and Das (1978) reported that the ratio *cyclic/non cyclic* phosphorylation increases with increasing light intensity.

Oxidative phosphorylation in mitochondria is brought about by an analogous mechanism. In respiration as in photosynthesis, electron flow occurs along an electropotential gradient made up of a number of redoxsystems. As electron flow in respiration is coupled with the synthesis of ATP and with the consumption of oxygen it is called *oxidative phosphorylation*. Both phosphorylating processes can be uncoupled by a number of chemicals. The most well known uncoupler of oxidative phosphorylation is 2.4 dinitrophenol (2.4 DNP), other uncouplers are: arsenate, ouabain and long chain fatty acids. Photophosphorylation is uncoupled by a number of reagents including NH_3 and the aliphatic amines. These chemicals diffuse from the stromal side across the thylakoid membrane into the thylakoid inner space. Here they are protonated due to the prevailing low pH and by this the proton motive force required for ATP synthesis is lowered.

3.2.4 CO$_2$ assimilation and the Calvin cycle

Photosynthetic CO_2 assimilation is the primary process by which inorganic C is converted into organic form simultaneously trapping energy provided by the light reaction of photosynthesis. Carbon dioxide assimilation is thus of paramount importance for the production of organic material as well as for the storage of energy in a chemical form. Carbon dioxide is assimilated by the carboxylation of ribulose bisphosphate (RuBP). The actual process is more complicated than that presented in the scheme below (Figure 3.23).

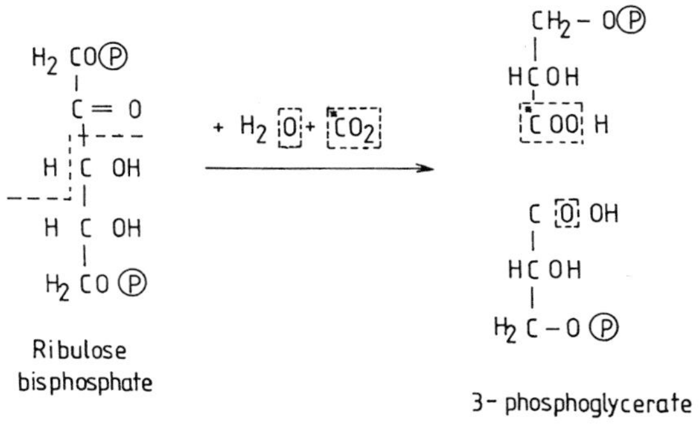

Figure 3.23 Assimilation of CO_2 and H_2O by ribulose bisphosphate forming 2 molecules of 3-phosphoglycerate.

The reaction is catalyzed by *ribulose bisphosphate carboxylase* (= Rubisco). The overall reaction shows that in addition to a CO_2 molecule, one water molecule is accepted by RuBP and required for the synthesis of 2 molecules of 3-phosphoglycerate. The enzyme requires Mg^{2+} and an alkaline pH.. As shown in Figure 3.22 the photosynthetic light reaction triggers the import of Mg^{2+} into the stroma and the e^- flow in the photosynthetic e^- transport chain renders the stroma pH alkaline, thus providing optimum conditions for RuBP carboxylase.

The carboxylation reaction is a highly exergonic reaction with a standard free energy of about -50 kJ/mol and therefore does not require additional energy. This is also evident from the fact that in the reaction a sugar (RuBP) is converted to an acid (3-phosphoglycerate). Rubisco catalyzes the first step of a cyclic sequence of reactions which was elucidated by Calvin and his co-workers (Calvin1956, Bassham and Calvin 1957). All enzymes catalyzing the various steps of this cycle are located in the stroma of the chloroplast.

The two steps following the carboxylation reaction require energy in the forms of ATP and NADPH and it is at this stage of the CO_2 assimilation process at which the products formed during the light reaction (see Figure 3.22) are required. The phosphoglycerate (PGA) is reduced to glyceraldehyde 3-phosphate. As this reduction is an energy consuming reaction, the phosphoglycerate must be loaded with energy before the reaction can proceed. This *priming reaction* is brought about by ATP. The 1,3 bisphosphoglycerate thus synthesized is then reduced by NADPH to glyceraldehyde 3-phosphate. This is the first sugar (triose phosphate), synthesized in the CO_2 assimilation reaction sequence (Figure 3.24). Glyceraldehyde 3-phosphate is readily converted to its isomer dihydroxyacetone phosphate, and in the presence of aldolase both molecules react to form fructose-1,6-bisphosphate a phosphorylated hexose. Fructose-1,6-bisphosphate is the precursor of all other hexoses including glucose and its polymers. A direct series of reactions can be thus traced leading from CO_2 assimilation *via* the triose phosphates (glyceraldehyde 3-phosphate and dihydroxyacetone phosphate) to fructose-1,6-bisphosphate and all other carbohydrates. Some of these compounds such as starch, sucrose, and inulin are energy storage compounds, whereas others like cellulose, hemicellulose and pectins play a significant role as structural cellular constituents. Triose phosphates can be used either for the synthesis of carbohydrates and lipids required in metabolism or in the regeneration of the CO_2 acceptor ribulose bisphosphate. The sequence of reaction by which ribulose bisphosphate is regenerated is known as the *Calvin cycle*. This is shown in Figure 3.25.

The regeneration is a sequence of reactions in which sugar phosphates with different numbers of C atoms are involved. Following the reaction series as indicated by step 1 in Figure 3.25 it can be seen that two triosephosphate molecules are condensed to form one molecule of fructose bisphosphate which by splitting off a phosphate group is transformed to fructose monophosphate. In the reaction indicated by step 2 one triose phosphate molecule reacts with fructose monophosphate to form a tetrose phosphate and a pentose phosphate (xylosephosphate). The tetrosephosphate (erythrose phosphate) and a further molecule of triose phosphate are condensed to form a heptose bisphosphate (sedoheptulose bisphosphate) as indicated by step 3. A phosphate group is split off from heptose bisphosphate thus forming a heptose monophosphate. This reacts with a further

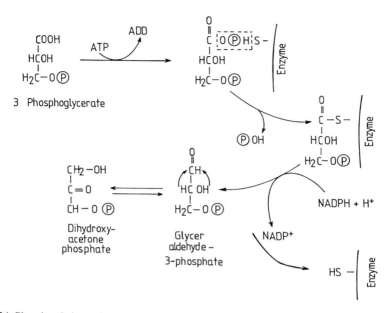

Figure 3.24 Phosphorylation and reduction of 3-phosphoglycerate to glyceraldehyde 3-phosphate (upper part), the isomeration of glyceraldehyde 3-phosphate to dihydroxyacetone phosphate (lower part).

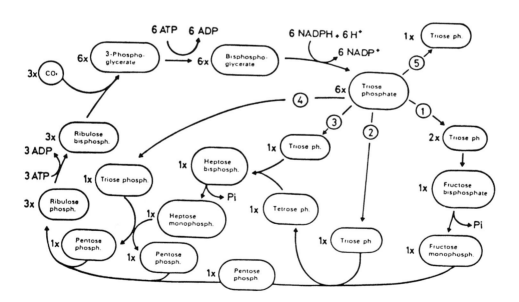

Figure 3.25 Regeneration of ribulose bisphosphate, the Calvin cycle.

triose phosphate molecule as shown in reaction step 4 to give two pentose phosphates (xylose phosphate and ribose phosphate). All these reaction steps ultimately result in the formation of pentose phosphates which are further converted to the isomer ribulose monophosphate. Thus in the whole reaction cycle 5 triose phosphates are converted to 3 pentosephosphates. These are phosphorylated by ATP thus forming the primary CO_2 acceptor ribulose bisphosphate.

This assimilation of 3 molecules of CO_2 requires 9 molecules of ATP and 6 molecules of NADPH. Thus for the assimilation of 1 molecule of CO_2, 3 molecules of ATP and 2 of NADPH are needed. This shows that the process requires more ATP than NADPH. In order to supply this ratio both cyclic and non-cyclic photophosphorylation must therefore occur as non-cyclic photophosphorylation only supplies ATP/NADPH in a ratio of 1:1 (Figure 3.22). Cyclic photophosphorylation adjusts to the required ATP/NADPH ratio in the C-reduction cycle.

The levels of NADPH and ATP regulate the Calvin cycle. In the dark the concentrations of NADPH and ATP in the chloroplasts drop so that the reduction of 3-phosphoglycerate is not only stopped, but even a reverse reaction takes place, namely the oxidation of the triosephosphate to 3-phosphoglycerate. Under dark conditions fructose bisphosphate phosphatase and sedoheptulose bisphosphate phosphatase are also inactived. The Calvin cycle is thus blocked. If the light is turned off an immediate appearance of 6-phosphogluconate is observed. This is an important intermediate of the so called oxidative pentose phosphate cycle(OPP), in which hexose phosphates are degraded and eventually intermediates of the glycollate pathway (see below) are formed. Thus dark conditions initiate respiration and accumulated chloroplast starch is oxidized, whereas under light conditions the triosephosphates produced are converted to starch and stored in the chloroplast. Conversely to the oxidative pentose phosphate cycle (OPP cycle), the Calvin cycle is characterized by a reduction step and is therefore called the reductive pentose phosphate cycle (RPP cycle).

3.2.5 Photorespiration and the glycollate pathway

Not all the energy and reducing power produced in the photosynthetic pathway described above can be fully utilized by plants. It has been shown that although there is a net uptake of CO_2 by green plants, a release of CO_2 occurs which is greater in the light than in the dark. The efflux of CO_2 is accompanied by enhanced O_2 uptake. This additional respiration induced by light is called *photorespiration*. The glycollate pathway (C_2 pathway) is the major process in photorespiration. It is shown in Figure 3.27. The individual steps of this reaction sequence have been reviewed by Tolbert (1979). The reaction sequence is started by ribulose bisphosphate carboxylase. Interestingly this enzme not only reacts with CO_2 but also with O_2. Therefore it calalyzes not only a carboxylation but also an oxygenation and it is for this reason that the enzyme is also called *Rubisco* in which the c stands for carboxylation and the o for oygenation reaction. Obviously CO_2 and O_2 compete for the ribulose bisphosophate and for the same site at the enzyme as shown in Figure 3.26. Rubisco may therefore oxidize RuBP especially under conditions where O_2 is present in abundant and CO_2 only in low concentrations.

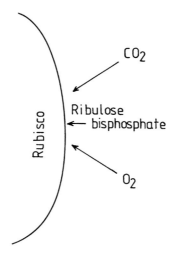

Figure 3.26 CO_2 and O_2 competing for the same substrate and reaction site of the Rubisco.

As shown in Figure 3.27 the first step (reaction 1) is the cleavage of RuBP into phosphoglycollate and phosphoglycerate. This step is an oxygenase reaction and consumes one molecule of O_2 per RuBP molecule. The resulting phosphoglycerate is a metabolite of the Calvin cycle and may thus be reduced to glyceraldehyde 3-phosphate. The phosphoglycollate is converted to glycollate by a phosphatase which splits off inorganic phosphate (reaction 2). The glycollate so formed may be released by the chloroplast and enter a peroxisome. Peroxisomes are microbodies frequently associated with chloroplasts and separated from the cytoplasm by a membrane. Peroxisomes contain several enzymes including oxidases and catalase. In the peroxisome, glycollate is oxidized to glyoxylate by a glycollate oxidase (reaction 3). The reaction consumes 1 molecule of O_2 and produces 1 molecule of H_2O_2 per glycollate molecule oxidized. The H_2O_2 formed is split into ½ O_2 and H_2O by a catalase (reaction 4). The glyoxylate is then aminated by an aminotransferase and is thus converted to glycine (reaction 5). Glycine thus formed may be transferred to a mitochondrion where it is subjected to oxidative decarboxylation. This is a complex reaction which can be subdivided into several steps. In the first step (reaction 6a) glycine is oxidized (dehydrogenated) by a glycine dehydrogenase forming the corresponding imino acid. In a second step (reaction 6b) the imino acid is deaminated. Ammonia is thus released and a glyoxylate molecule formed. This C_2 compound is then broken down into two C_1 components, namely CO_2 and = CH-OH by decarboxylation (reaction 6c). The radical = CH-OH (hydroxymethylene) is transferred to tetrahydrofolic acid (THFA), which is a coenzyme achieving the transfer of C-1 groups (reaction 6d). THFA transfers this C-1 group (hydroxymethyl) to a further glycine molecule and serine is thus formed (reaction 7). Serine may then be converted by a series of steps to phosphoglycerate which may be taken up by the Calvin cycle.

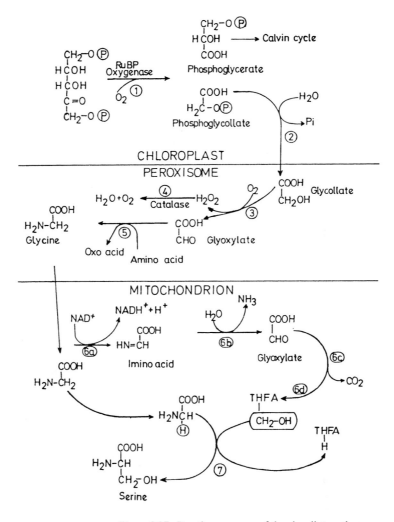

Figure 3.27 Reaction sequence of the glycollate pathway.

From the reaction sequence of the glycollate pathway as shown in Figure 3.27 it is clear that O_2 is absorbed and CO_2, H_2O and NH_3 are released. This ultimately results in a loss in organic C and N. Photorespiration thus appears to be a wasteful process and may drain off nearly 50% of the C assimilated. At normal atmospheric concentrations of CO_2 and O_2 at 25°C the ratio of carboxylation/oxygenation is about 4:1 (Ogren 1984). The physiological significance of this process is not yet clear. Suggestions that it is important in amino acid synthesis, the disposal of excess reducing power or protection of the plant from high O_2 toxicity have been considered (Tolbert 1979, Zelitch 1979). Woolhouse (1978) has discussed the impact of photorespiration on crop production.

It should be emphasized that in the final part of the glycollate pathway in the mitochondria, NH_3 is released by the degradation of glycine (reaction 6b). It is believed that this NH_3 is reassimilated more or less completely (Keys *et al.* 1978). As the turnover in the glycollate pathway can be very high under optimum conditions, it is feasible, however, that some of the NH_3 produced may be released into the atmosphere. This suggestion is consistant with results of Stutte *et al.* (1979), who found that the loss of gaseous NH_3 of soya beans increased linearly with an increase in temperature.

Photorespiration is influenced to a marked extent by external factors. Generally all factors which favour the light reaction of photosynthesis such as high light intensity and high temperature lead to conditions which are favourable for the glycollate pathway. High light intensity combined with high temperature results in an abundant level of RuBP. This is associated with a high level of O_2 produced by photolysis, and a low concentration of CO_2 in the leaf tissue because of a high CO_2 assimilation rates. These conditions are exactly those which promote the reactions of the glycollate pathway. Schäfer *et al.* (1984) reported that wheat and sugar beet showed maximum CO_2 assimilation before the highest light intensity was attained at noon. This observation could be well becaused by an increasing rate of photorespiration with the advance of the day. In addition when photosynthetic activity is high this is associated with an increase in Mg^{2+} concentration and a raised pH in the stroma of the chloroplast (see page 146), conditions which are essential for the activity of phosphoglycollate phosphatase (Figure 3.23 reaction 2). These stromal conditions provide a regulatory mechanism switching on the enzyme and thus the glycollate pathway under light conditions and blocking it in the dark.

Depressing photorespiration, *e.g.* by the inhibitor glycidate (= 2,3-epoxypropionate), results in an increase in net CO_2 assimilation. Whether glycidate is an actual inhibitor of photorespiration has been questioned by Poskuta and Kochanska (1978). These authors found that glycidate favours both net CO_2 assimilation and photorespiration. According to Zelitch (1979) intermediates, such as glutamate, aspartate and glyoxylate are natural inhibitors of photorespiration.

The most important ‚natural‘ inhibitor of photorespiration is CO_2. In C_4 plants, a category of plant species which is considered in more detail in the following section, photorespiration is practically absent. In these species high CO_2 concentrations prevail in those chloroplasts in which the Calvin cycle reactions take place. These high CO_2 concentrations result from a particular CO_2 trapping mechanism.

3.2.6 C_4 pathway

Ribulose bisphosphate is not the only CO_2 acceptor in photosynthesis. In a number of plant species, mainly tropical grasses, a particular reaction pathway (C_4 pathway) precedes the Calvin cycle. In this pathway molecules with four C atoms play a major role and for this reason the reaction sequence is called the C_4 pathway and plants possessing this sequence are called *C_4 plants*. This distinguishes them from the more usual *C_3 plants* in which phosphoglycerate is the primary fixation produce and RuBP the primary CO_2 acceptor. Prominent C_4 crop plants are maize, sugar cane, and sorghum. Also some dicots mainly of the families of *Amaranthaceae*, *Chenopodiaceae* and *Portulaceae*

154

have this particular CO_2 assimilation mechanism of which phosphoenolpyruvate (PEP) is the primary CO_2 acceptor. These species are characterized by a particular form of carboxylation (CO_2 assimilation) and decarboxylation (release of CO_2). Carboxylation and decarboxylation occur at different sites in the leaf tissue, carboxylation in the mesophyll cells and decarboxylation in the bundle sheath cells. In most plant species in which the C_4 pathway is operative these two cell types are arranged in the so called *Kranz type leaf anatomy*. This anatomy is shown in Plate 3.1. The vascular leaf tissue, comprising phloem and xylem strands is surrounded by large bundle sheath cells, virtually forming a ‚Kranz‘ or wreath. The bundle sheath cells are surrounded by layers of mesophyll cells.

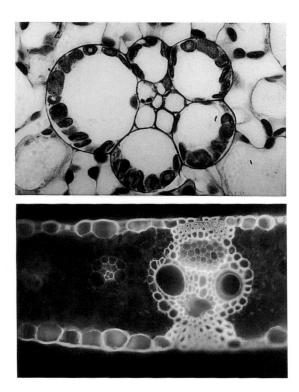

Plate 3.1 Transverse section of maize leaf showing veins with the characteristic "Kranz" antomy. The two veins are from the same leaf. In the upper part a small vein is shown. The five large cells surrounding it are the bundle sheath cells. In the lower part a large vein is shown which one can see as parallel lines on the surface of a fresh leaf. Between these large veins there are many small veins such as shown in the upper part of the photo. These run parallel to the larger veins. — The two large lumens in the larger vein are big metaxylem vessels. The small cells above the metaxylem is the phloem. (Courtesy of Dr. Martin Canny).

Monohydrogen carbonate (HCO_3^-) reacts with phosphoenolpyruvate (= PEP) thus forming a primary C_4 intermediate, the oxaloacetate (OAA) (Figure 3.28).

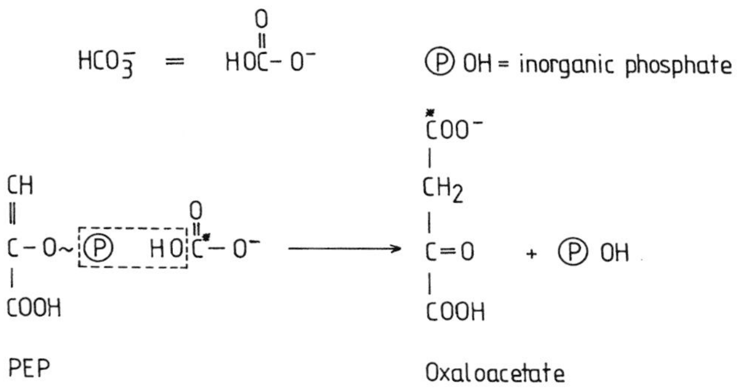

$$HCO_3^- \quad = \quad HO\overset{\overset{\displaystyle O}{\|}}{C}-O^- \qquad \text{ⓟ } OH = \text{inorganic phosphate}$$

Figure 3.28 The reaction of the PEP carboxylase catalyzing the formation of oxaloacetate from PEP and HCO_3^-.

The enzyme catalyzing the reaction is *phosphoenol carboxylase* (PEP carboxylase). The HCO_3^- is quickly provided by carbonic anhydrase, a Zn-containing metalloenzyme which catalyzes the reaction (Badger and Price 1994):

$$CO_2 + H_2O \Rightarrow HCO_3^- + H^+$$

Carbon dioxide fixation by PEP is strongly exergonic as the high energy phosphate bond (60 kJ/mol) is split off from PEP and HCO_3^- is assimilated and oxaloacetate formed. This is converted to a further C_4 acid by reduction to form malate. Oxaloacetate may also be aminated by means of an amino acid as amino donor. By this transamination aspartate is produced (step 4 in Figure 3.29).

The C_4 pathway was mainly elucidated by Hatch and his colleagues. Its main steps are shown in Figure 3.29 (Slack and Hatch1967, Andrews and Hatch1971):

1. PEP is formed by the phosphorylation of pyruvate with ATP and Pi. The reaction comprises two phosphorylations. Therefore the enzyme is called pyruvate phosphate dikinase with pyruvate as the substrate. This enzyme occurs in all C_4 plants and during the reaction the ATP is split into adenosine monophosphate (AMP) and pyrophosphate.

2. PEP reacts with HCO_3^- and H^+ to yield oxaloacetate as already described. The assimilated C is marked with an asterix.

3 Oxaloacetate is reduced to malate by the NADPH- specific malate dehydrogenase

4. Oxaloacetate may be aminated to aspartate by the enzyme aspartate aminotransferase. In this case the CO_2 assimilation is directly linked with the amino acid metabolism.

5. Malate produced in reaction 3 is decarboxylated by the $NADP^+$ — specific malic enzyme to produce CO_2 pyruvate and NADPH. The CO_2 thus produced is used for carboxylation in the Calvin cycle and the NADPH for the reduction of phosphoglycerate to glyceraldehyde 3-phosphate (Figure 3.25). Pyruvate is then recycled and used as substrate by the PEP carboxlylate as shown in step 1 of Figure 3.29. The CO_2 produced by the decarboxylation of malate (step 5) is fixed by RuBP into phosphoglycerate and metabolized in the Calvin cycle as already described above (Figure 3.25). Details of the pathway and the enzyme systems involved have been discussed by Kelly *et al.* (1976), Ray and Black (1979) and Coombs (1979).

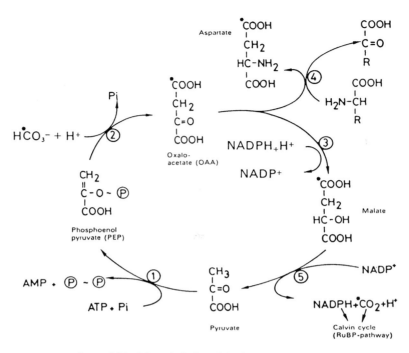

Figure 3.29 CO$_2$ assimilation of C$_4$ plants, the C$_4$ pathway.

The reaction sequence shown in Figure 3.29 occurs in the ‚malate' type of C$_4$ plants. For this C$_4$ plant type, the NADPH malic enzyme is characteristic and malate is the major form in which the trapped C is translocated from the location of CO$_2$ assimilation (mesophyll cells) to the location of decarboxylation (bundle sheath cells, see Plate 3.1 and Figure 3.30). In some C$_4$ plant species aspartate rather than malate is the main transport form for organic C from the mesophyll cells to the bundle sheath cells. Decarboxylation of aspartate is brought about by an NADH malic enzyme. Oxaloacetate produced in this reaction sequence may be decarboxylated by a PEP carboxykinase as shown in Figure 3.30. Coombs (1979) suggests that aspartate is converted to oxaloacetate in the mitochondria of the bundle sheath cells. Oxaloacetate thus formed may be reduced to malate which is eventually decarboxylated by NADH specific malic enzyme in the chloroplasts of bundle sheath cells. The pyruvate produced is recycled to the mesophyll cells (see Figure 3.30). Pyruvate may also be aminated by an alanine aminotransferase to alanine, which then may be recycled to the mesophyll cells (Figure 3.30).

The metabolic cycle (C$_4$ pathway) as shown in Figure 3.29 does not make much sense if it is not considered in relation to the particular anatomy of C$_4$ species. In the C$_4$ cycle (Figure 3.29) HCO$_3^-$ is assimilated and later CO$_2$ is produced by decarboxylation, the whole cycle requiring energy in the form of ATP. Hence there is no net assimilation

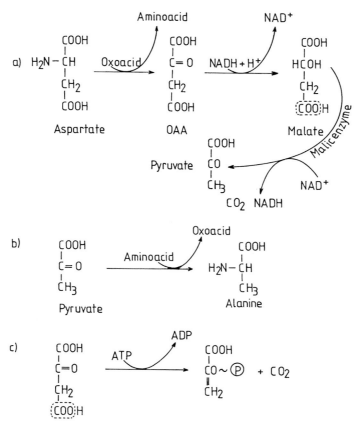

Figure 3.30 Reactions of C$_4$ metabolites (dicarboxylic acids) in bundle sheath cells: a) Aspartate providing indirectly CO$_2$ and NADH for the Calvin cycle after having been deaminated to malate. — b) Formation of alanine by transamination of pyruvate. — c) Formation of PEP by decarboxylation of oxaloacetate catalyzed by the PEP carboxykinase. The CO$_2$ thus produced is assimilated by Rubisco.

of inorganic C although energy is consumed. The important characteristic of the system is that there is a spatial separation of carboxylation and decarboxylation and by this arrangement CO$_2$ assimilation becomes very efficient. Phosphoenolpyruvate (PEP) carboxylation catalyzed by the PEP carboxylase occurs in the cytosol of mesophyll cells as shown in Figure 3.31. The resulting oxaloacetate is imported into the mesophyll chloroplast where it is reduced by photosynthetically produced NADPH. The malate thus formed is exported from the chloroplast *via* an antiporter with malate leaving and oxaloacetate entering the mesophyll chloroplast. Malate is then transported into the chloroplast of the bundle sheath cells mainly following the plasmodesmatal pathway.

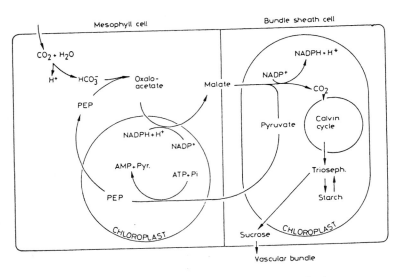

Figure 3.31 Carbon pathway and compartmentation in C_4 plants.

The oxidative decarboxylation of malate in the chloroplasts of the bundle sheath cells provides NADPH and CO_2 for the Calvin cycle and hence C and reducing equivalents (NADPH) for CO_2 assimilation. Pyruvate resulting from the oxidative decarboxylation of malate is recycled into the mesophyll cell and imported into the chloroplast. According to Brownell and Bielig (1996) the uptake of pyruvate by the mesophyll chloroplasts requires Na^+ supposedly to maintain the chloroplast integrity. In the chloroplast the photosynthetically produced ATP provides the energy for the production of PEP *via* pyruvate phosphate dikinase producing PEP. This enzyme is termed dikinase because it catalyzes two phosphorylation steps according to Coombs (1979) as shown below:

Enzyme + ATP + P_i $\Rightarrow$ Enzyme-P + AMP + Pyrophosphate
Enzyme-P + Pyruvate $\Rightarrow$ PEP + Enzyme

Net reaction : ATP + P_i + Pyruvate $\Rightarrow$ PEP + AMP + Pyrophosphate

The PEP thus produced is exported into the cytosol of the mesophyll cell *via* a phosphate/PEP antiporter (Brownell and Bielig 1996). Here it functions as a substrate for PEP carboxylase and the oxaloacetate produced is imported into the mesophyll chloroplast where it is reduced to malate by means of NADPH. From this it is evident that the light harvesting potential of the mesophyll chloroplasts is used for the reduction of oxaloacetate and the phosphorylation of pyruvate. The energy trapped by these processes is eventually translocated in the form of malate to the bundle sheath cells. Here the rate of malate decarboxylation exceeds the rate of RuBP carboxylation and the CO_2 concentration is thus maintained at a level at which Rubisco carboxylation activity is high and the oxygenase activity low (Ogren1984). It is for this reason and also because of the low O_2 partial pressure that the competion between CO_2 and O_2 for Rubisco is mainly in favour of CO_2 and that carboxylation and not oxygenation prevails (see page 151). Hence CO_2 is used at high rates for assimilation and the assimilation is very efficient.

Chloroplasts of the bundle sheath cells differ from those of the mesophyll cells. The former are larger, can accumulate starch and generally do not contain grana. Grana which are present in the smaller chloroplasts of mesophyll cells are rich in photsystem II constituents and involved in photolysis and thus in O_2 formation. Mesophyll cells are thus supposedly relatively rich in O_2 in contrast to the bundle sheath cells. Chloroplasts of the mesophyll cells do not produce starch.

The C_4 pathway is light dependent because the NADPH malic enzyme and pyruvate phosphate dikinase are light activated. The latter is also sensitive to cold temperatures and this may be the reason, why C_4 species do not thrive as well as C_3 plants in temperate climate if the temperature is low. This may frequently be observed in spring when maize seedlings exposed to a cold period may be much depressed in growth and show symptoms of leaves turning yellow whereas C_3 species grow well. As soon as the weather warms up the maize plants grow vigorously and at much higher growth rates than the C_3 species.

Bundle sheath cells are surrounded by mesophyll cells rich in PEP carboxylase. Thus CO_2 released by mitochondrial respiration or even photorespiration is rapidly refixed by PEP carboxylase. Carbon dioxide diffusing from the apoplast into the cytosol of the mesophyll cells has to traverse only one membrane, the plasmalemma, and then has a high chance of being trapped by PEP carboxylase. In C_3 plants the CO_2 additionally has to cross the double membrane of the chloroplast in order to come in contact with Rubisco. This and the low oxygenase activity of RuBP carboxylase in C_4 species explains, why these species have a low CO_2 compensation point i.e.the CO_2 concentration at which CO_2 consumption and CO_2 production are in equilibrium. C_4 plants can therefore use relatively low atmospheric CO_2 concentrations. The CO_2 compensation point of C_4 species is in a range of 5 to 10 cm^3 CO_2/m^3 whereas for C_3 plants it is between 30 to 50 cm^3 CO_2/m^3. For this reason C_4 plants can depress the CO_2 concentration in the leaf tissue to much lower level than the C_3 plants and in contrast to C_3 plants still show net CO_2 assimilation.This situation is particularly true under favourable photosynthetic conditions such as high light intensity and warm weather when the conditions for photorespiration are also favourable (see page 303).

A further reason to account for the evolution of plants with a highly efficient CO_2 fixing system may relate to their water economy. Many C_4 plant species occur naturally in arid, semi-arid and tropical conditions, where the closure of stomata to prevent water loss is essential for growth and even survival and therefore CO_2 entry must thus also be restricted. Under such environmental conditions species may well have evolved which are very efficient utilizers of water and CO_2. This view is consistent with the findings of Downes (1969) who observed that the ratio of weights of CO_2 assimilated to water transpired (the water use efficiency, see page 230) of C_4 plants was often twice that of C_3 plants.

3.2.7 Crassulacean acid metabolism

Today about 300 plant species are known which are characterized by *Crassulacean Acid Metabolism (CAM)* by which plants are adapted to arid conditions with low soil water availability (Osmond 1978). Most of the species showing CAM belong to the families *Crassulaceae, Agavaceae, Bromeliaceae, Cactaceae, Euphorbiaceae, Liliaceae* and *Orchidaceae*. Both obligate and facultative CAM species, occur the latter shifting over to CAM under arid conditions characterized by low soil water potentials.

As this form of CO_2 assimilation was first discovered in *Bryophyllum calycinum* a member of the *Crassulaceae* it has become known as Crassulacean Acid Metabolism (CAM). The most important feature of CAM is that by maintaining stomata closed during the day major water losses due to transpiration are prevented. Stomata are open at night and CO_2 diffuses into the leaf tissue and is assimilated by PEP carboxylase located in the cytosol of leaf cells (Figure 3.32, step 1).

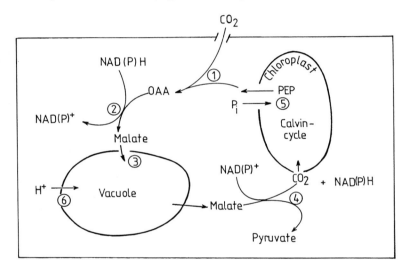

Figure 3.32 Scheme of the Crassulacean Acid Metabolism.

The oxaloacetate produced by this reaction is reduced to malate with NAD(P)H as H donor Figure 3.32, step 2) and the malate is stored in the vacuole (Figure 3.32, step 3). Carbon dioxide assimilation by PEP carboxylase and the formation of malate are analogous to that in the C_4 plants (Figure 3.29) with the difference that the reduction of oxaloactate occurs in the cytosol and not in the chloroplast as is the case for C_4 plants. Carbon dioxide fixation *via* PEP carboxylase and oxaloacetate reduction are not light dependent and therefore the processes including malate storage in the vacuole also occur in the dark. During the day when the stomata are closed malate is transported into the cytosol where it is decarboxylated (Figure 3.32, step 4) and the resulting CO_2 is assimilated by Rubisco and the photosynthetic light reaction supplies the Calvin cycle with ATP and NADPH. Decarboxylation of malate may be catalyzed by the NAD^+ or $NADP^+$ malate enzyme by which pyruvate, NAD(P)H, and CO_2 are produced (Figure 3.32, step 4). The latter is fed into the Calvin cycle. Oxaloacetate may also be decarboxylated by PEP carboxykinase with the production of PEP and CO_2 for the Calvin cycle (Figure 3.30).

During net CO_2 assimilation at night, the CO_2 acceptor (PEP) must be generated in high quantities. It is now generally accepted that PEP is produced in the chloroplasts by the glycolytic breakdown of starch and exported into the cytosol *via* a PEP/phosphate antiporter (Figure 3.32, step 5). For this reason the starch level in the chloroplasts drops considerably during the night.

During the night malate is stored in the vacuole and may attain concentrations as high as 200 mol/m³. The transport of malate across the tonoplast supposedly occurs *via* facilitated diffusion brought about by tonoplast proton pumps (Figure 3.32, step 6) which render the vacuole positive relative to the cytosol (Hedrich and Schroeder 1989) and also bring about a low pH in the vacuolar liquid. The plant water potential is the main factor controlling CAM (Osmond 1978). Under water stress conditions plant species with a facultative CAM may shift over to this type of CO_2 assmilation with stomata closed during the day and open at night. Under such conditions CAM plants are capable of maintaining a water potential (see page 184) of -0.5 to -1.0 MPa even if the soil water potential is as low as -2.2 MPa, a water potential at which for most plant species reach the permanent wilting point (see page 43). Since most CAM plant species are characterized by a low stomatal frequency and by a thick cuticle, water loss by transpiration is very low and plants can withstand long dry periods during which other plant species would die because of lack of water. During such periods net CO_2 assimilation rates of CAM plants are low and the growth rates may be limited by the capacity of storing malate in the vacuole which serves as CO_2 source for photosynthesis at night. In addition some of the CO_2 produced by decarboxylation at night escapes into the atmosphere, particularly at high temperature (Kluge 1979). The growth rate of CAM plants is therefore generally low. A valuable review article on CAM was published by Osmond (1978).

3.3 Nitrogen and Sulphur Assimilation

3.3.1 General

The acquisition of CO_2 is not the only process of assimilation by which plants are able to synthesize large amounts of organic compounds from an inorganic source. The same is also true for the assimilation of both N and S which are essential elements for all organisms. Nitrogen is present in all amino acids, nucleic acids, proteins and coenzymes; sulphur occurs in some of them. The processes by which plants convert inorganic N (NO_3^-, NH_4^+, N_2) and inorganic S (SO_4^{2-}) to organic forms are important in biology, for animals are dependent on a dietary source of organic N and S originating from plants and microorganisms.

3.3.2 Nitrate and nitrite reduction

Nitrate is often the major source of N available to plants. Before this can be metabolized, however, it must be reduced to NH_3. One distinguishes between assimilatory nitrate reduction and a dissimilatory nitrate reduction. The former is responsible for the assimilation of nitrate and is described below. In dissimilatory nitrate reduction nitrate serves as an e^- acceptor in bacterial respiration under anaerobic conditions (see page 349). The assimilation of nitrate consists of two steps, the reduction of NO_3^- to NO_2^-, and the further reduction of NO_2^- to NH_3. The generally accepted basic mechanism of NO_3^--assimilation in green plant tissues in the light is shown in Figure 3.33. The two enzymes involved in the process are nitrate reductase and nitrite reductase (Hewitt 1975).

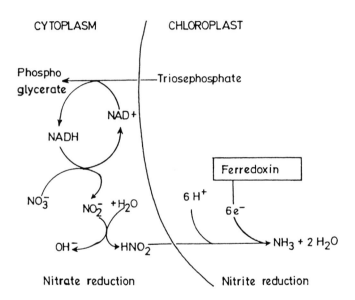

Figure 3.33 Scheme of nitrate and nitrite (nitrous acid) reduction.

Nitrate reductase catalyses the first step from NO_3^- to NO_2^-, which takes place in the cytosol. The further reduction of NO_2^- to NH_3 occurs in the chloroplasts and is brought about by the enzyme nitrite reductase (Shingles et al. 1996). Both enzymes, nitrate reductase and nitrite reductase, function in series so that no appreciable nitrite accumulation occurs.

Nitrate reductase (NR) activity has been detected in a number of plant organisms including bacteria, blue green algae, green algae, fungi and higher plants (Guerrero et al. 1981). In recent years the most important details of the enzyme structure have been clarified (Campbell 1996). Nitrate reductase (EC 1.6.6 1-3) is a homodimeric complex (two identical monomers) with each subunit containing a 100 kDa-polypeptide, a polypeptide with Mo (Mo pterin), a polypeptide with Fe (haem), and a polypeptide with FAD. The reaction sequence of these redox systems (prosthetic groups) as proposed earlier and shown in Figure 3.34 has been confirmed in more recent research (Campbell 1996).

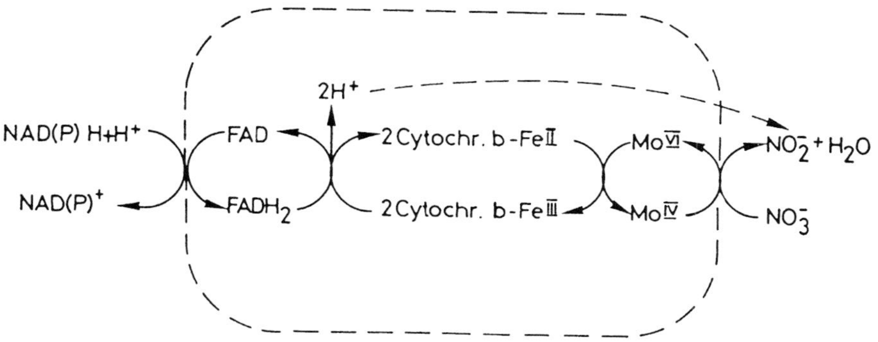

Figure 3.34 Prosthetic groups of the nitrate reductase and the sequence of reactions (according to Guererro *et al.* 1981). NAD(P)H denotes that both coenzymes NADH and NADPH may function as H donors.

Nitrate reductase has two active sites: The FAD complex where NAD(P)H is bound and the Mo pterin complex in which nitrate is bound. The enzyme has a two site steady state kinetics where the enzyme "pings" and "pongs" between the oxidized and the reduced form. This means that the NAD(P)H pings on the FAD site to which the NAD(P)H is bound. After the transfer of H atoms from the NAD(P)H to the FAD it "pongs" from the FAD which means the $NAD(P)^*$ (oxidized form) is released. In an analogous way nitrate "pings" on the site of the Mo-pterin and nitrite "pongs" from it.

The reducing power required is NAD(P)H, denoting that either NADH and NADPH may serve as H donors. Various types of NR, however, are highly specific for NADH or NADPH, with only a few NR types which are bispecific, *i.e.* using either NADH or NADPH as H donors. The reason for this specifity is not yet completely understood. NADH may originate from glycolysis and may also indirectly be supplied by triose phoshates (3-phosphoglyceraldehyde) exported by the chloroplast. The oxidation of 3-phosphoglyceraldehyde to 3-phosphoglycerate yields NADH (Figure 3.33). In the mesophyll cells of C_4 plants the chloroplast may provide malate *via* the oxaloacetate-malate antiport (Figure 3.31). the reduction of malate to oxaloacetate in the cytsol yields NAD(P)H which may also serve as donor for nitrate reduction. During the transfer of reducing equivalents from $FADH_2$ to the cytochrome, H atoms are split into H^+ and e^- and the latter are transferred to the cytochrome and from there to the Mo-pterin complex which directly reduces nitrate to nitrite. In Figure 3.34 for simplicity's sake the Mo redox reaction is shown as an e^- transfer between Mo^{IV} and Mo^{VI}. As yet, however, this has not been completely clarified and it is supposed that both Mo^{VI}/Mo^V and Mo^V/Mo^{IV} are involved (Solomonson and Barber 1990).

The overall reaction of the nitrate reduction according to the sequence shown in Figure 3.32 is described by the equation:

$$NO_3^- + NADH + H^+ \Rightarrow NO_2^- + H_2O + NAD^+$$

The main characteristics of NR do not differ greatly between various eukaryots, *i.e.* higher plants, fungi, green algae, in contrast to these of the NR of cyanobacteria. As can be seen from Table 3.7 the e$^-$ donor of the cyanobacteria *Anacystis nidulans* is ferredoxin and not NAD(P)H and the optimum pH and the K_M of the cyanobacteria are much higher than for green plants.

Table 3.7 Main characteristics of nitrate reductase in various organisms (modified after Guerrero *et al.* 1981)

Characteristic	*Anacystis nidulans*	*Clorella vulgaris*	*Spinacea oleracea*	*Neurospora crassa*
MW, kD	75	280	197	228
e- donor	Ferredoxin	NADH	NADH	NAD(P)H
pH optimum	10.5	7.6	7.5	7.5
K_M, mmol/m^3	690	84	180	200

Control of NR activity can be achieved by the synthesis and degradation of the enzyme and also by altering its activity. This is brought about by an inhibitor protein which is bound to the NR by means of a kinase reaction (Kaiser and Heber 1984). Crawford *et al.* (1986) showed that nitrate induced the synthesis of mRNA coding for NR. Hence nitrate nutrition has an impact on the level of NR in cells as shown in Table 3.8 from the work of Mengel *et al.* (1983).

Table 3.8 Nitrate reductase activity in maize roots after an exposure of young plants to nutrient solutions containing NO_3^-, NH_4^+, or NH_4NO_3 (after Mengel *et al.* 1983)

Treatment	NR activity, μmol NO_2^-/h $\times$ g fresh weight
NH_4^+	0.10 a
NH_4NO_3	0.95 b
NO_3^-	3.71 c

Different letters (a, b, c) indicate a significant difference at $p < 0.01$

The turnover of NR is rapid (Oaks *et al.* 1972) and its half-life is about 4h (Schrader *et al.* 1968).

Light plays an important role in nitrate assimilation. When green plants are transferred from light to dark conditions, the activity of NR is depressed even when NO_3^- is present in adequate amounts. There is now convincing evidence that accumulation of NO_3^- in plant tissue is not indicative of a lack of enzyme proteins but depends rather on the lack of reducing power. Even barley mutants characterized by a very low activity of nitrate reductase are able to reduce enough NO_3^- to grow normally (Warner and Kleinhofs 1981). Aslam and Huffaker (1984) found that exposing excised barley leaves to a source of permanent light reduced nearly all the NO_3^- taken up. At low light intensity however, only 25% of the NO_3^- absorbed was reduced. This example indicates that NO_3^- reduction is much more sensitive to low light intensity than is nitrate uptake. The source of reducing power for nitrate reduction is NAD(P)H. This is produced either by the reduction of 3-phosphoglyceraldehyde or malate

$$3\text{-phosphoglyceraldehyde} + NAD^+ \Rightarrow \text{Phosphoglycerate} + NADH + H^+$$

$$\text{Malate} + NAD(P)^+ \Rightarrow \text{Oxaloacetate} + NADH + H^+$$

In C_4 plants malate is the main source of reducing power whereas in C_3 plants it is 3-phosphoglyceraldehyde. In C_3 plants the chloroplast mainly exports dihydroxyacetone phosphate in the light and phosphoglycerate in the dark (Figure 3.35). There is thus a lack of reducing power for NO_3^- reduction in the dark and the rate of NO_3^- reduction is therefore low and NO_3^- may accumulate. This is demonstrated in Table 3.9 which shows the effect of the time of day on the nitrate concentration of spinach (Steingröver *et al.* 1982).

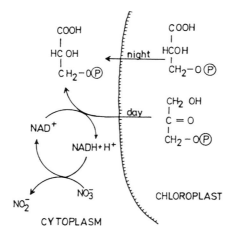

Figure 3.35 Relationship between assimilate export out of the chloroplast and NO_3^- reduction in the cytoplasm.

Table 3.9 Effect of time of day on the nitrate concentration of spinach (after Steingröver *et al.* 1982)

Time of day	Stems	Leaf	Petiole
	mg NO_3^--N/kg fresh matter		
8.30	372	228	830
13.30	207	101	546
17.30	189	91	504

The assimilation of nitrate by plants is influenced by mineral nutrition and in particular by Mo. When Mo is deficient, nitrate accumulation takes place and the concentrations of soluble amino N compounds are depressed because of the lack of NH_3 available for amino acid synthesis (see page 174).

Nitrate reductase occurs predominantly in the cytosol of meristematic cells. Young leaves and root tips are thus rich in the enzyme. Van Egmond and Breteler(1972) reported that the nitrate reductase activity in fully developed young sugar beet leaves was about 10 times higher than in older leaves. According to Hewitt (1970) nitrate reductase tends to rise to a maximum in moderately young leaves. In roots maximum activities occur in the younger tissues and decline markedly in the older root regions (Hewitt1970). Nitrate reductase activity also depends on the supply of the tissue with organic carbon as was shown by Pace et $al.$ (1990). This makes physiological sense because only if organic carbon is available can the assimilated organic N be utilized for growth. This adjustment of the NR activity to physiological demand also protects the plant tissue, for the accumulation of nitrite, the product of the NR could be toxic at high concentration. As reported by Kaiser et $al.$ (1999) in a useful review article the activity of the NR is rapidly modulated by environmental concditions, such as light intensity, CO_2 concentration and oxygen supply.

The site of NO_3^--reduction — roots or above-ground plant parts — differs between plant species. In tomato plants between 80—90% of the N in the xylem sap is present in the form of nitrate (Lorenz1976) so that NO_3^- reduction must take place primarily in green plant parts. Similar findings have been reported by Wallace and Pate (1967) in cocklebur (*Xanthium pennsylvaticum*). In this species nitrate reductase activity was found to be absent in the roots. Most plant species however, are able to reduce NO_3^- both in the roots and the upper plant parts. According to Pate (1971) who investigated NO_3^- reduction in a number of crop plant species, the proportions of NO_3^- reduced in the roots decreased in the following sequence:

Oats > Maize > Sunflower > Barley > Oil Radish

The leaves of trees and shrubs do not contain nitrate, and according to Sanderson and Cocking (1964) nitrate reduction takes place exclusively in the roots. Klepper and Hagemann (1969), however, were able to detect NR in the leaves of apple trees which had been subjected to high levels of nitrate fertilizer. As these leaf tissues were able to bring about nitrate reduction it seems probable that the leaves of other shrub and tree species are also potentially capable of inducing nitrate reductase activity.

The activity of nitrate reductase is potentially capable of influencing yield production, as to some extent it controls the rate of assimilation of nitrate. Croy and Hagemann (1970) investigating the nitrate reductase activity in the leaves of 32 different wheat cultivars found no clear relationship between the enzyme activity and the grain protein concentration. These authors suggest that besides the nitrate reductase activity in the flag leaf, the translocation rate of amino compounds to the grains is a major factor influencing grain protein concentration. More recent investigations of Traore and Maranville (1999) with diverse sorghum genotypes differing in nitrate reductase activity (NRA) showed that NRA was not correlated with shoot biomass or grain yield but with the N concentrations in sorghum grains. From this finding one may conclude that during the grain filling stage NRA had a direct or indirect influence on the translocation of amino acids to the developing grains.

The NO_2^- produced by NR is an anion of a weak acid and is partially protonated according to the equation:

$$NO_2^- + H^+ \Leftrightarrow HNO_2 \quad pK = 5.22$$

There is now evidence that HNO_2 rather than NO_2^- penetrates the inner envelope of plastids including chloroplasts (Heber and Purceld 1977, Shingles $et\ al.$ 1996). The acid space between the outer and inner envelope of chloroplasts is supposedly generated by a proton pump located in the inner chloroplast envelope (Shingles $et\ al.$ 1996). Hence the equilibrium between NO_2^- and HNO_2 is shifted to the side of the non dissociated form which is able to penetrate the chloroplast membrane. On entering the chloroplast or plastid HNO_2 is reduced by the nitrite reductase which is probably located at the outer side of the thylakoid membrane and thus may directly accept the e^- for reduction from ferredoxin (Figure 3.20) which in turn receives e^- from photosystem I (Figure 3.36). There is thus a direct relationship between photosynthetic activity and nitrite reduction. At low light intensity O_2, $NADP^+$ and HNO_2 may compete for the e^- from photosystem I (Mir $et\ al.$ 1998). Nitrite reductase of higher plants consists of a single polypeptide strand to which an Fe_4-S_4-cluster and a sirohaem (Fe-S-haem) is attached (Figure 3.36).

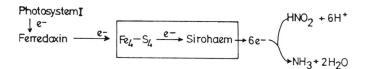

$Figure\ 3.36$ Nitrite reductase - e^- flow in nitrite reduction.

The reduction as shown in Figure 3.36 is reflected by the equation:

$$HNO_2 + 6H^+ + 6e^- \Rightarrow NH_3 + 2H_2O$$

The reaction is analogous to the assimilatory nitrite reduction of bacteria (Beevers and Hageman 1983) shown in Figure 3.37. Here a putative means of nitrite reduction is also shown in which only two e^- are required for the reduction of nitrous acid to NH_3 and hence less energy is needed. In this reduction, as proposed by Heber and Purczeld (1977) O_2 is produced. This reaction is in line with more recent results of Mir $et\ al.$ (1995) who found that nitrite assimilation increased net O_2 evolution and hence supports the concept of Heber and Purczeld (1977).

Nitrite is also reduced in plastids of roots where the reduction process can not be directly supplied from the photosystem with e^-. Suzuki $et\ al.$ (1984) isolated a ferredoxin-like e^- carrier enzyme from maize roots that can transfer e^- from NAD(P)H to ferredoxin. The enzyme is located in the plastids (Oaks and Hirel 1985) and hence nitrite reduction in roots is analogous to the reduction in green tissues with the exception that in the roots the e^- donor is NAD(P)H. The synthesis of nitrite reductase is induced by nitrite as well as by nitrate. Both anions are also supposedly taken up by the same membrane transport systems, the synthesis of which can be induced by either anion (Aslam *et al.* 1993). Nitrite reductase activity is also dependent on the supply of photosynthates as reported by Suzuki *et al.* (1995). Various forms of nitrite (nitrous acid) reduction to NH_3 are shown in Figure 3.37.

1. $ON-OH + NADH + H^+ \longrightarrow \left[HO-\overset{H}{N}-OH \right] + NAD^+$
    Nitrous acid

    $\left[HO-\overset{H}{N}-OH \right] + NADH + H^+ \longrightarrow HO-NH_2 + H_2O + NAD^+$
    Hydroxylamine

    $HO-NH_2 + NADH + H^+ \longrightarrow NH_3 + H_2O + NAD^+$
    Ammonia

2. $ON-OH + 2e^- + 2H^+ \longrightarrow \left[HO-\overset{H}{N}-OH \right]$

    $\left[HO-\overset{H}{N}-OH \right] + 2e^- + 2H^+ \longrightarrow HO-NH_2 + H_2O$

    $HO-NH_2 + 2e^- + 2H^+ \longrightarrow NH_3 + H_2O$

3. $ON-OH + 2e^- + 2H^+ \longrightarrow NH_3 + O_2$

Figure 3.37 Putative ways of nitrous acid reduction to ammonia. 1) Bacterial reduction. — 2) Reduction requiring 6e⁻. — 3) Reduction requiring 2e⁻ associated with the release of O_2.

3.3.3 Nitrogen fixation

The atmosphere provides a vast reservoir of molecular nitrogen (N_2). This is not directly available for use by plants, however, and before it can be assimilated it must first be converted to a so-called fixed form either by oxidation to NO_3^- or by reduction to NH_4^+. As N_2 is highly inert, these conversions are not readily brought about, and require a considerable amount of energy. Only prokaryotes are capable of reducing N_2 to NH_3 thereby directly using atmospheric N_2 as a N source. Of the 47 known families of bacteria 11 are able to reduce N_2 to NH_3 and of the 8 families of cyano-bacteria six can carry out this reduction (Werner 1980).

These microorganisms play a unique role in the whole N-cycle of nature for by the conversion of molecular N_2 into an organic form atmospheric N is rendered available to other organisms. This process is called N_2-fixation. The quantity of N_2 reduced in this way on a world scale is immense. According to Delwiche (1983) total world biological fixation amounts to about 118×10^6 Mg/year while the total technical production is 73×10^6 Mg/year (Brändlein 1987). These figures show that to date technical N_2 fixation is about 60% of the global biological N_2 fixation. Microorganisms capable of N_2 fixation may be divided into those which are free living and those which live symbiotically with higher plants. These microorganisms are considered in Chapter 7.

The main features of the biochemistry of N_2 fixation are fairly well established and the basic concepts elucidated more than 20 years ago by Evans and Barber (1977) have been confirmed. Fixation occurs in the bacteroid and the metabolic processes and reactions in these organelles are shown in Figure 3.38. The term *Bacteriod* describes a *Rhizobium* bacteria which has undergone several changes following the infection of the host root cell (see page 401). The bacteroid is embedded in a vacuole, the so called infection vacuole which is formed by invagination of the plasmalemma of the host cell. The bacteroid contains ATPases and transport systems, and is mainly supplied by organic anions such as malate and succinate which serve as H donors for enzymes of the tricarboxylic cycle (Figure 3.38) located in the bacteroid. These enzymes, malate dehydrogenase and succinate dehydrogenase, reduce NAD^+ to NADH. The latter is H donor for ferredoxin and for the respiration chain located in the bacteroid membrane. Respiration provides the ATP required for the N_2 reduction process. Leghaemoglobin located in the infection vacuole guaranties a controlled supply of O_2 to respiritory chain.

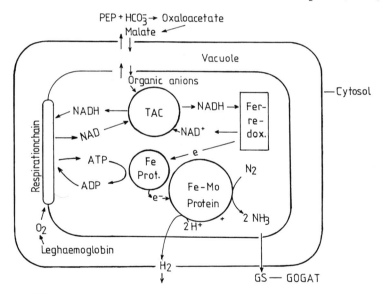

Figure 3.38 Metabolism of the N_2 fixing bactereoid (TAC = Tricarboxylic Acid Cycle).

This O_2 carrier leghaemoglobin is analogous to haemoglobin in the blood of vertebrates. A key feature of the bacteroid is the presence of an enzyme nitrogenase which is unique to N_2 fixing microorganisms and capable of catalizing the reduction of several substrates as well as molecular N_2. Nitrogenase consists of 2 O_2 sensitive proteins, a small Fe protein and a larger FeMo protein. The Fe protein has a molecular mass of about 60 kDa and consists of 2 subunits bridged by a single Fe_4S_2 cluster. This cluster receives e⁻ from ferredoxin and undergoes a one e⁻ redox cycle between the Fe^{II} and Fe^{III} state (Howard and Rees 1994). Electrons are then transferred from the Fe protein to the Fe-Mo protein. This is an up-hill process and the energy required for the reduction of the N_2 is provided in the form of ATP produced by respiration (Figure 3.38). ATP is bound to the Fe protein by a Mg^{2+} bridge and then hydrolyzed during which process a conformational change occurs. Each of two subunits of the Fe protein binds one ATP before dephosphorylation of the bound phosphate brings about the transfer of one e⁻ from the Fe protein to the Fe-Mo protein. Since the reduction of one molecule of N_2 to 2 molecules of NH_3 — as shown below — requires 8 e⁻, there is a requirement of 16 molecules of ATP for the reduction of one molecule of N_2. This high energy demand is necessary to break the N ≡ N bond which contains about 900 kJ/mol. The putative reaction centre of the Fe-Mo protein is an aggregation consisting of a Fe_4-S_4 and an Fe_3Mo-S_4 cluster in which, one atom of Fe is replaced by one atom of Mo (Chan et al. 1993). The clusters are linked together by two S atoms and thus form a cavity at the centre of which the N_2 is bound (Figure 3.39).

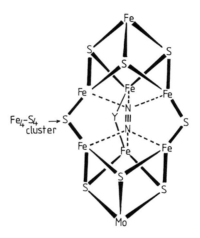

Figure 3.39 Cage consisting of an Fe_4-S_4 cluster linked by 2 S atoms to an Fe_3Mo-S_4 cluster. In the center of the cavity, N_2 is bound. Y denotes an as yet unknown ligand (after Chan et al. 1993).

Dinitrogen (N_2) receives e⁻ for reduction from the Fe and Mo of this structure. Before N_2 can be bound in the centre, H_2 must first also be bound for reasons as yet unknown. The putative reduction cycle is shown in Figure 3.40 and follows a sequence of steps:
1. Two e⁻ and two H⁺ (= H_2) are bound at the centre
2. H_2 thus bound is replaced by N_2
3. N_2 is reduced to an imide by accepting one e⁻ and one H⁺
4. The imide is then reduced to the hydrazine by the uptake of one e⁻ and one H⁺
5. Uptake of one further e⁻ and one H⁺ reduces the hydrazine to NH_3 which is released
6. Uptake of three e⁻ and three H⁺ reduces the bound N to NH_3 which is released

The centre of the cavity is then free and a new cycle of reduction may begin. The most critical step in this sequence is the reduction of N_2 to the imide (Howard and Rees 1994). The balance of this reaction sequence is reflected by the follwing equation:

$$8 \, H^+ + 8 \, e^- + N_2 + 16 \, ATP \Rightarrow H_2 + 2 \, NH_3 + 16 \, ADP + 16 \, P_i$$

From this equation it is evident that the reduction of one dinitrogen (N_2) to NH_3 requires 16 ATP and that the prodution of two NH_3 is associated with the release of one H_2.

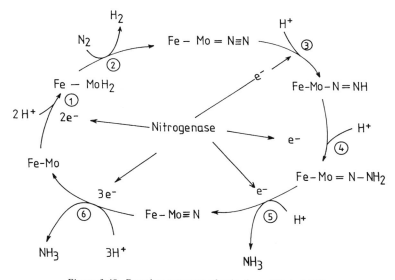

Figure 3.40   Reaction sequence of reduction of N_2 to 2 NH_3.

The Fe protein and the Fe-Mo protein are in structure and function highly conserved among different organisms (Howard and Rees 1994). A surprising feature of the nitrogenase reaction is its slowness in activity in comparison with other enzymatic processes. According to Postgate (1985) this is attributable to the fact that the oxidized Fe protein dissociates from the Mo-Fe protein every time one e^- is transferred from the Fe protein to the Mo-Fe protein. Dinitrogen fixers have therefore to compensate for this by containing substantial amounts of the enzyme. In some cases up to 20% of the cell protein of a diazotroph (N_2 fixing organism) can be accounted for by nitrogenase.

Under non optimal conditions for fixation such as low reductant supply, low ATP supply, low temperature or deficiency of N_2, the conversion of protons to H_2 is increased at the expense of N_2 reduction. When N_2 is absent then H_2 alone is formed from the protons. Therefore the ratio of N_2/H^+ reduction is not a constant one but depends on the conditions for N_2 reduction. Moloney *et al.* (1994) found that a decrease of this ratio (= decrease of N_2 efficiency) was associated with a decrease of N_2 partial pressure, an increase of H_2 partial pressure and a decrease in nodule permeability to O_2 diffusion. This result also shows that an optimum O_2 supply is required for an efficient N_2 reduction. This is in accord with the data of Layzell *et al.* (1990) who found that a low O_2 supply to *Bradyrhizobium* infected root nodule cells led to an enormous decrease in nitrogenase activity because of a lack of O_2 required for respiration. The utilization of ATP to make H_2 as a by product of N_2 fixation is an apparently wasteful process. However, some of the most efficient diazotrophs are able to split the H_2 by hydrogenase.

$$H_2 \Rightarrow H^+ + e^-$$

The e^- thus produced is then cycled back to ferredoxin so that the electrons may be recycled to nitrogenase to be used in the reduction of N_2. According to Schubert *et al.* (1978) this recycling of e^- may to some extent determine the N_2 fixing efficiency of *Rhizobium* bacteria.

The intensity of N_2 reduction depends very much on continuous supply of photosynthates *via* the phloem to the root nodules. Presumably the C source to the bacteroid is mainly organic anions such as malate and succinate formed from sucrose. There is evidence that increased PEP carboxylase activity in the cytosol of host cells may provide malate for the bacteroid (Vance and Heichel 1991). Drought stress may severely depress nitrogenase activity (Castillo *et al.* 1994) but according to Plies-Balzer *et al.* (1995) the effect of mild water stress in depressing nitrogenase activity was not the limiting factor for plant growth. The overriding factor was the impact of water stress on the growth rate of the host plant. Schubert *et al.* (1995) reported that nitrogenase activity was well adjusted to water supply of alfalfa. The authors suppose that the reduced turgor due to water stress was the reason for the retarded growth and that the increased concentration of amino acids in the nodules resulting from lower shoot demand depressed the nitrogenase activity by a feed back mechanism. Similarly the well known depression in nitrogenase activity in legumes supplied with mineral N to the roots may also result from an increase in reduced N cycled from leaves to root and attached nodules.

Nitrogenase is extremely sensitive to O_2 because it breaks the disulphide bridge in the cavity in the Fe-Mo protein (Figure 3.39) and may thus irreversibly inactivate the enzyme. Oxygen (O_2) concentrations are extremely low in the nodule interior and O_2 supply is strictly controlled by leghaemoglobin. These low O_2 concentrations are achieved by the presence of an O_2 diffusion barrier in the nodule cortex (Becana and Rodriguez-Barrueco 1989) and the rapid respiration rates in the nodules. Nitrogenase can reduce a number of substrates other than N_2 including acetylene (HC = CH) to yield ethylene. This reaction is used to measure biological N_2 fixation.

The NH_3 produced by nitrogenase is exported from the bacteroid into the cytosol of host cells, where it is used for the synthesis of amino acids and amides. In some legumes, particularly the tropical legumes ureides are formed (Oaks and Hirel 1985). The enzyme responsible for NH_3 assimilation are glutamine synthetase and glutamate synthase. The reactions are considered in more detail on page 174. Glutamate and glutamine do not occur in high levels in the bacteroid cell, and this has a physiological implication because high levels of both metabolites depress the so called ‚nitrogen fixation genes‘ (= nif genes). These genes are located in the genom of the bacteria and they code the synthesis of the nitrogenase proteins. The bacteroid may provide the host plant with more than 90% of the NH_3 produced. This is one reason, why symbiotic living N_2 fixing bacteria are so efficient in N_2 fixation in comparison with free living bacteria in which nitrogenase activity is strictly controlled by the endogenous level of glutamate. It is supposed that the formation of nitrogenase also depends on the level of NH_3 in the bacteroid, a high level inhibiting nitrogenase synthesis. Exogenous NH_3 may also enter the bacteroid and thus depress the intensity of N_2 fixation (Latimore *et al.* 1977).

It is well established that *Rhizobium* bacteria require Co. From the reports of Evans and Russell (1971), Co is essential in the propionate pathway which probably controls the synthesis of leghaemoglobin (see page 563). Copper may also to be essential for N_2 fixation. Additionally, however, Cu may also play a role in leghaemoglobin synthesis and hence has a controlling influence on carbohydrate supply to the nodule. Similar indirect effects on N_2 fixation can be caused by a lack of other plant nutrients (page 406).

3.3.4   Ammonia assimilation

Both the assimilation of nitrate and the fixation of molecular N_2 give rise to ammonia. For ammonia assimilation three enzymes are of importance:

Glutamate dehydrogenase (GDH)
Glutamine synthetase(GS)
Glutamate synthase(GOGAT)

Glutamate dehydrogenase (GDH; EC 1.41.3) catalyzes the reaction between NH_3 and 2-oxoglutarate. The enzyme has been shown to be present in many higher plants. As shown in Figure 3.41 oxoglutarate reacts with NH_3 to form imino-glutarate which is then reduced by NAD(P)H to glutamate. Both these steps are reversible as shown in the reaction sequence. The process is an amination of the 2-oxoglutarate coupled with a reduction brought about by NAD(P)H. For this reason it is termed reductive amination. The reaction, however, may be reversed when NH_3 concentrations are low and glutamate is degraded. This is generally the case when amino acids are metabolized in mitochondria and NH_3 released (Figure 3.41). If, however, a high rate of NH_3 is produced by photorespiration (Figure 3.27). the reaction runs in the direction of NH_3 assimilation. There is now evidence that GDH located in the mitochondria plays a major role in recycling NH_3 released by photorespiration (Oaks 1994a).

For NH_3 assimilation in higher plants *glutamine synthetase* (GS, EC6.3.1.2) is of principal importance. It is an octameric protein and two major isoenzymes of GS are known, GS_1 in the cytosol and GS_2 in the chloroplast (Lea *et al.* 1992). Glutamine synthetase brings about a reaction in which glutamate functions as an NH_3 acceptor to produce glutamine (Figure 3.41). The reaction is an endergonic process needing ATP and also Mg^{2+}. For GS_2 (chloroplast) ATP is supplied by photosynthetic phosphorylation, whereas for GS_1 respiration is the main source for ATP. In the presence of a reducing source the amino group from glutamine is transferred to 2-oxoglutarate. The enzyme catalyzing this reaction is *Glutamate 2-Oxoglutarate Amino Transferase* (GOGAT,) also called glutamate synthase (Figure 3.42). Two forms of GOGAT exist, one receiving reducing equivalents from ferredoxin, and the other from NAD(P)H, the former dominating in green tissues and the latter in non photosynthetic tissues

(Lea *et al.* 1992). Glutamine synthetase functions in series with GOGAT in that the 2-oxoglutarate is the amino acceptor for the amino group donated from glutamine.

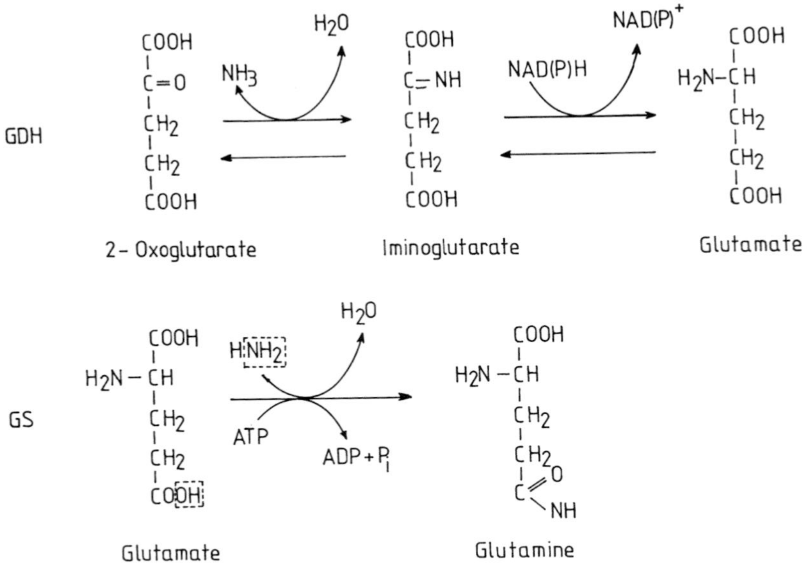

Figure 3.41 Reactions of glutamate dehydrogenase (GDH), and of glutamine synthetase (GS).

The two reactions may be written:

Glutamate + NH_3 + ATP $\Rightarrow$ Glutamine + ADP + P_i

Glutamine + 2-oxoglutarate + 2 e^- + $2H^+$ $\Rightarrow$ 2 Glutamate

---

NH_3 + 2-oxoglutarate + ATP + $2e^-$ + $2H^+$ $\Rightarrow$ Glutamate + ADP + Pi

One glutamate molecule is thus produced from one molecule of 2-oxoglutarate and NH_3. The system requires a reduction step brought about by ferredoxin or NAD(P)H as shown in Figure 3.42. The major difference between NH_3 assimilation by GDH and by GS + GOGAT is in the affinity for NH_3. This is much higher for glutamine synthetase which is capable of incorporating NH_3 present in very low concentrations into 2-amino N (Miflin 1975). The presence of glutamine synthetase in the chloroplasts ensures that the concentration of NH_3 formed by nitrite reductase can be incorporated without building up levels which would uncouple photophosphorylation (see page 147). The reaction sequence of NH_3 assimilation by GS followed by the GOGAT reaction (Figure 3.42) represents the most important route of NH_3 assimilation in the leaves and roots of green plants. The net product shown in this reaction sequence is glutamate which functions as amino donor in numerous transamination reactions and most of the N present in plants originates from the amino N of the glutamate produced in the GS + GOGAT reactions.

NADH-glutamate synthase is a key enzyme in leguminous root nodules in which the *Rhizobium* infected cells receive much NH_3 from the bacteroid. The NH_3 translocated into the cytosol of the infected cell is then assimilated *via* the GOGAT system (Trepp *et al.* 1999). NADH- GOGAT protein is found throughout the infected cell region and there is evidence that the presence of the bacteroid induces the expression of the NADH- glutamate synthase.

GOGAT

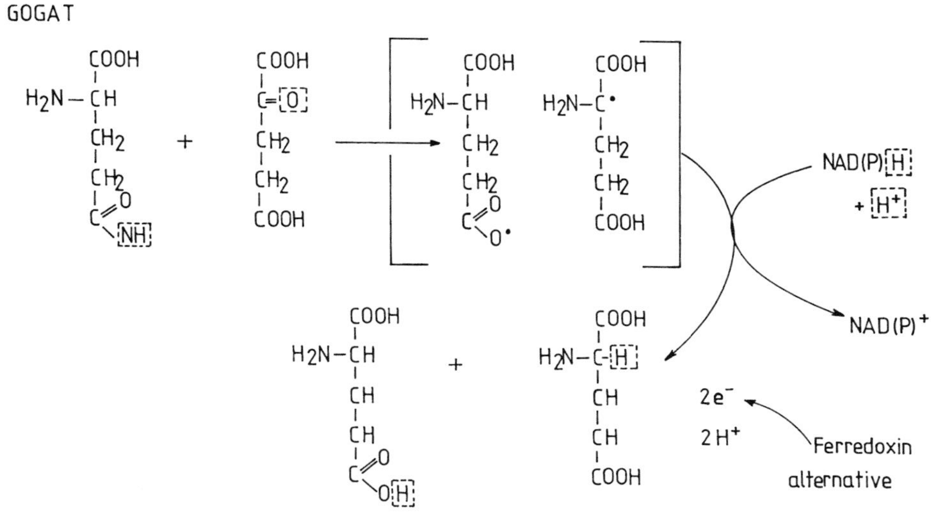

Figure 3.42   Reaction sequence of Glutamine 2-Oxo Glutarate Amino Transferase (GOGAT). In brackets the radicals are shown which are intermediary products of the process of amino transfer and glutamate as net product.

The efficiency of GS may differ between plant species. Assimilation rates of NH_3 must be particularly rapid high if concentrations of NH_3 penetrate the plasmalemma. This should be the case in the root cells of plants growing in a nutrient medium supplied at elevated levels of NH_4-N at high pH. Kosegarten *et al.* (1997) have shown that in rice NH_4^+ assimilation may be controlled by the highly efficient GS in the roots whereas the maize in contrast this is not so because of inhibition of GS by the high concentrations of NH_3. This example shows that rice is adapted to high pH levels in flood water associated with an abundant supply of ammonium N.

3.3.5   Amino acids and amides

The amino N in glutamate can be transferred to other oxo acids (keto acids) by the process of transamination. The enzymes catalyzing this reaction are amino transferases. An example is shown below on the left hand side of Figure 3.43 where the NH_2 group from glutamate is transferred to the oxo acid pyruvate to form the amino acid alanine. In the reaction a new amino acid (alanine) is synthesized. The resulting 2-oxoglutarate may again be used as amino group acceptor in the GOGAT reaction. Transamination provides a means for the synthesis of a number of amino acids and the amino group of many amino acids is originally derived from glutamate. The most important NH_2 acceptors (oxo acids) of the transamination process and their corresponding amino acids are shown in Table 3.10.

Table 3.10   Oxo acids and their amino acid analogues

Oxo acid	Amino acid
2-oxoglutarate	glutamate
oxaloacetate	aspartate
glyoxylate	glycine
pyruvate	alanine
hydroxy pyruvate	serine
glutamate semialdehyde	ornithine
succinate semialdehyde	gamma-amino-butyrate
2-keto 3-hydroxy-butyrate	threonine

When excess NH_3 is available, glutamate and aspartate can serve as NH_3 acceptors. Asparagine *synthetase* (AS, EC6.3.5.4) may use NH_4^+ for amination as well as glutamine to serve as amino donor (Oaks 1994b). This is a transamination reaction as shown on the right hand side of Figure 3.43.

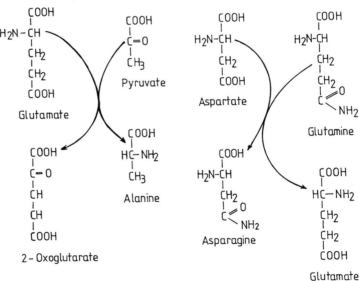

Figure 3.43   Transamination. Transfer of the amino group from glutamate to pyruvate resulting in the formation of alanine and 2-oxoglutarate and the formation of asparagine by the transfer of the amino group from glutamine to aspartate.

Glutamine and aparagine are important N carriers both in the long-distance transport in phloem and xylem, respectively. In cereals glutamine is the principal N transport form. It is synthezised by GS in the vascular bundles (Kamaguchi *et al.* 1992). Valuable review articles on ammonium and N_2 assimilation have been published by Vance and Heichel (1991), Lea *et al.* (1992), and Oaks (1994).

3.3.6 Sulphur assimilation

The most important source of S for higher plants is sulphate. In several respects its assimilation resembles that of nitrate although the detailed mechanism is not so well understood. Sulphate taken up by plant cells must be reduced because in the major S containing organic molecules, S is present in a reduced form. These organic compounds include cysteine, cystine and methionine as well as the proteins containing these amino acids (see page 440). Sulphur assimilation starts with the formation of adenosine phosphosulphate in a reaction catalyzed by *ATP sulphurylase* (ATP: sulphate adenyltransferase, EC 2.7.7.4). The enzyme has been found in many higher plant species and is thought to be the predominat sulphotransferase in higher plants (Brunhold 1990). This is an exchange reaction in which the sulphuryl group is transferred from sulphate to replace the pyrophosphate group from ATP resulting in the formation of *adenosine phosphosulphate* (APS) and pyrophosphate as shown in Figure 3.44.

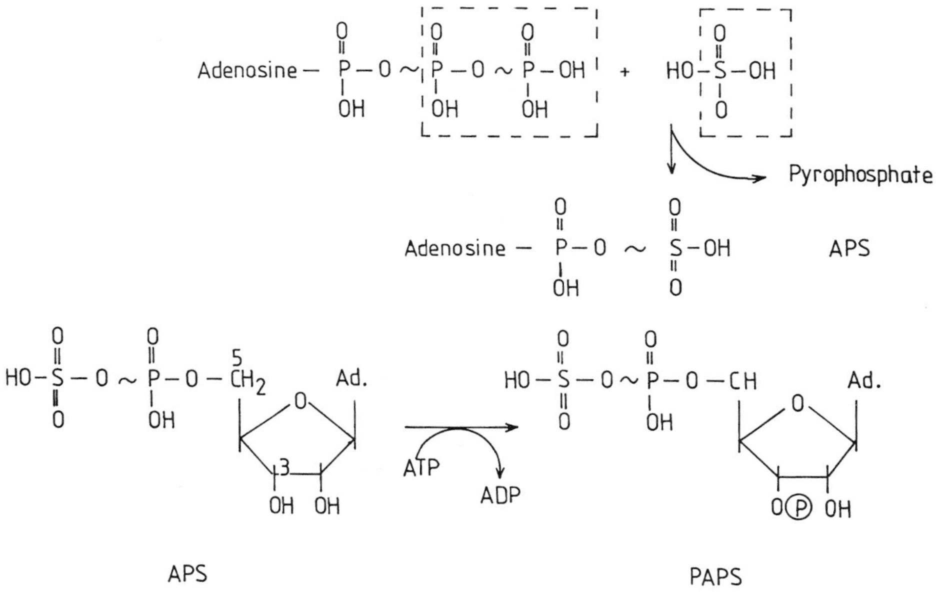

Figure 3.44 Formation of adenosine phosphosulphate (APS) and adenosine 3'-phospho 5'phosphosulphate (PAPS).

Adenosine phosphosulphate may be phosphorylated by ATP and thus adenosine 3'-phospho 5'-phosphosulphate (PAPS) is formed which is the the universal donor of sulphuryl groups for the formation of sulphate esters (Figure 3.44) (Schiff *et al.* 1993). A useful review on sulphate assimilation and transport was recently published by Leustek and Saito (1999). PAPS sulphotransferase has been found in bacteria, cyanobacteria as well as in spinach. The assimilation (reduction) of the sulphuryl group is a more complicated process and has yet to be completely elucidated. The most important reaction steps are described below and shown in Figure 3.45 according to Schiff *et al.* (1993). In this reaction sequence APS serves as suphuryl donor. Plants starved of sulphur showed a five fold increase of the ATP sulphurylase activity and also an increased plasma membrane S transporter activity (Hawkesford *et al.* 1995) observations indicating that uptake and reduction of sulphate are regulated.

Assimilatory sulphate reduction appears to be present in the chloroplasts of all photosynthetic plant organs. The reducing equivalents required are directly supplied by ferredoxin (Figure 3.20) so that the reduction is dependent on photosythetic activity.

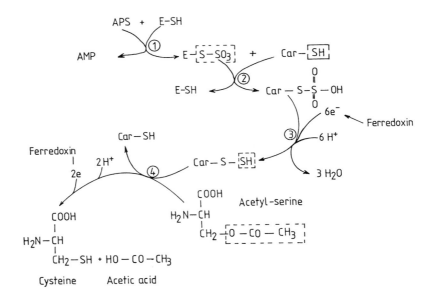

Figure 3.45 Putative reaction sequence of the sulphate reduction (modified after Schiff *et al.* 1993). — Step 1: The sulphuryl group of APS is transferred first to an enzyme (E-SH) forming the sulphurylated enzyme. — Step 2: The sulphuryl group is then transferred to an unknown carrier containing a SH group the H of which is substituted by the sulphuryl group thus forming a disulphide bridge. — Step 3: The sulphuryl group of the carrier is then reduced by e⁻ originating from ferredoxin and an SH group bound to a S atom on the carrier complex is thus produced. — Step 4: This SH (thiol) group is transferred to acetylserine which splits to produce cysteine and acetic acid with the regeneration of the carrier SH. The splitting reaction requires two e⁻ which are provided by ferredoxin. The enzyme catalyzing this step, an O-acetyl serine sulphurhydrolase (EC 4.299.8) is well known.

Sulphate is mainly reduced during the light period, due to the fact that the sulphate reducing enzymes are located in the chloroplast membrane. Whether other organelles are also capable of reducing sulphate is not yet known. Since isolated roots can also grow on sulphate as a sole sulphur source, it has been suggested that sulphate reduction also occurs in nongreen tissue, probably in proplastids (Schmidt 1979).

Sulphate reduction of prokaryotes may differ from that of higher plants in so far that adenosine 3'-phosphate 5'-phosphosulphate (PAPS) is an indispensable intermediate in the reduction process (Schiff 1983) which transfers the sulphuryl group to a dithiol carrier. In the reduction sequence sulphite is produced as shown in Figure 3.46.

Sulphate reduction is carried out by a number of organisms including higher plants, algae, fungi, blue-green algae and bacteria. Animal metabolism, however, is dependent on the intake of a source of reduced S and hence depends on the S-assimilation of the organisms mentioned above. Cysteine is the first stable product in which S is present in a reduced organically bound form. Cysteine is a precursor of methionine, another important S-containing amino acid.

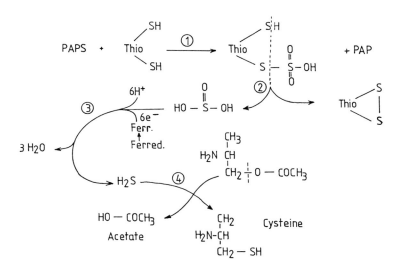

Figure 3.46 Reduction of sulphuryl to sulphite and the formation of cysteine (after Brunhold 1990). — 1. The sulphuryl group is transfered to a dithiol complex (thioredoxin). — 2. The sulphurylated thioredoxin is oxidized in which process H_2SO_3 and the oxidized form of thioredoxin are produced. — 3. Sulphite (H_2SO_3) is reduced to hydrogen sulphide (H_2S) by e- provided by ferredoxin. — 4. H_2S reacts with O-acetylserine and cysteine and acetate are formed.

CHAPTER 4

PLANT WATER RELATIONSHIPS

4.1   Basic Processes in Plant Water Relationships

4.1.1   General

Life is intimately associated with water, and particularly with water in its liquid phase. Water is the form in which the H atom, an essential element of all organic molecules, is absorbed and then assimilated in the course of photosynthesis (page 147). About 500 g of water is absorbed by roots to produce 1g organic material and thus may be considered as a plant nutrient, in the same way as CO_2 or NO_3^- are also plant nutrients. The quantity of water required for the photosynthetic process, however, is small and amounts to only about 0.01% of the total quantity of water used by the plant. Most functions in which plant water is involved, are of a physical nature. Water is a solvent for many substances such as inorganic salts, sugars and organic anions. It is also the medium in which all biochemical reactions take place. Water molecules are adsorbed at the surfaces of particles forming hydration shells, which influence physical and chemical reactions. Water in liquid form allows the diffusion and mass flow of solutes, and for this reason is essential for the translocation and distribution of nutrients and metabolites throughout the entire plant. Water is also important in the vacuoles of plant cells as it generally exerts an intracellular pressure on the protoplasm and cell wall (so called turgor pressure) thus maintaining the rigidity of leaves, roots and other plant organs. These few examples indicate the overall importance of water in plant physiology.

4.1.2   Water potential

Growing plant tissue consists typically of 80 to 95% water. The major part, about 90% of the water is present in the vacuoles and the remainder is distributed between cytoplasm and apoplast. For a more basic understanding of water movement at the cell and also at the whole plant level it is important to appreciate the concept of water potential (Slatyer 1967). The term water potential (chemical potential of water) describes the free energy status of water and was introduced into the literature to provide a unified terminology for studies in soil-plant-water relationships. It may be defined as the difference in chemical potential of water (J/Mol) per unit volume (J/m^3) between a given water sample and pure free water at the same temperature. This may be expressed in an equation as follows:

$$\Psi = \frac{\mu_w - \mu^o_w}{v_w}$$

where
Ψ = water potential (greek letter psi)
μ_w = chemical potential of the water
$\mu°w$ = chemical potential of pure free water at the same temperature
V_w = partial molar volume of water in the system

The units of chemical potential are Joules per mole, whereas the partial molar volume of water is expressed in terms of volume (m^3) per mole. Substituting these units into the above equation it becomes:

$$\frac{J \ Mol^{-1}}{m^3 \ Mol^{-1}} = \frac{J}{m^3}$$

Per definition: Energy (J) = Force (N) × distance (m)
where
N = The Newton (the unit of force)

Substituting this expression for J in the above equation the water potential is given by the following term:

$$\frac{N \times m}{m^3} = \frac{N}{m^2}$$

This represents force per unit area and as such is the definition of pressure. The international standard unit of which is the Pascal (Pa).

$$Pa = N \times m^{-2}$$

As the Pascal is a very small unit, kPa or MPa is the most frequently used unit in water potential measurements.

$$10^6 \ Pa = 10^3 \ kPa = 1 \ MPa; \ 1 \ bar = 10^5 \ Pa$$

The chemical potential of water is an energy form based on the Brownian's movement of water molecules and is represented by the equation:

$$\mu = R \ T \ ln \ a$$

μ = chemical potential, R = gas constant, T = absolute temperature, a = activity of water molecules (activity see page 121).

Water potential is thus greatly dependent on temperature. For liquid water, which we consider here, the span of temperature is theoretically between the ice and water vapour; for biological systems, however, it is much narrower. Since a given water potential is related to a standard water potential which is exposed to the same ambient conditions, namely temperature and atmospheric pressure, these two parameters generally are not considered and for practical measurements the standard water potential is pure water with no solutes dissolved in it at room temperature and normal atmospheric pressure. This standard water potential is defined zero.

Water in soils and plant tissues is present in pores and intercellular spaces and cellular compartments where it is exposed to adsorption forces (see page 26); hence, the Brownian's movement of water molecules is restricted and therefore their chemical energy reduced. Such water adsorption may occur at macroscopic surfaces, such as the surface of soil pores or the surface of cell walls. Adsorption of water molecules also occurs at microsurfaces, such as the surface of solutes *i.e.* organic molecules and inorganic ions etc. dissolved in the water. The higher this degree of water adsorption the lower the water potential and since the standard water potential (pure water) is zero the water potential in plants and soils is < 0 because in such systems a portion of the water is adsorbed to surfaces.

The hydrostatic pressure within the cell (pressure potential ψ_p) has a tendency to press water away and has therefore a positive sign in the water potential equation shown below, in contrast to the adsorbed water (solute and matrix potential) which is retained from movement and which has therefore a negative sign.

$$\Psi = \psi_p + \psi_s + \psi_m$$

Ψ = water potential, ψ_p = pressure potential (positive sign), ψ_s = solute potential (negative sign), ψ_m = matrix potential (negative sign)

The solute potential (ψ_s) represents the water adsorbed to dissolved solutes. Dissolution of water soluble compounds leads to a reduction of the activity of free water which means a reduction in the water potential. For example, the water potential of a 100 mol/m³ sucrose solution is 0.244 MPa lower than pure water at the same temperature and pressure.

The matrix potential (ψ_m) results from the water adsorbed to macromolecular structures (matrix), *e.g.* onto thin surfaces of membranes, cell walls, soil particles etc. Water can be bound to large molecules such as proteins and polysaccharides by H bonding. In this way water can be held to cell walls, membranes, cell organelles and other macromolecular structures. Dry plant material also tends to adsorb water molecules from its surroundings. This is a process of biological significance. Dry seeds as for example seeds embedded in soil, rapidly adsorb water from the surrounding medium. Water molecules penetrate the intercellular spaces of the cell wall and also other cellular compartments due to adsorption forces. These forces cause the swelling of the seed which, if confined, can exert negative pressures in the order of about -100 MPa (Sutcliffe 1979). This kind of water uptake provides seeds with enough water for the biochemical reactions which initiate the germination process. The swelling of gelatine brought about by water is an analogous process. Uptake of water resulting from adsorption forces is called imbibition.

The pressure potential (ψ_p) of a solution is numerically equal to the hydrostatic pressure. The pressure potential results from a phenomenon called *osmosis* which is described in more detail on page 185. As a result of osmosis plant cells can establish fairly high pressures which press the cytoplasm to the cell walls. This pressure is called *turgor*. Increases of the pressure potential within cells (after water uptake) lead to increases of the water potential and for this reason the term ψ_p has a positive sign. In well watered plants it is between $+ 0.1$ to $+ 1$ MPa and depends largely upon the solute potential. A positive turgor is important for plant growth (page 219) in order to stretch the cell walls and supports the mechanical rigidity of plant tissues. Slight water loss from fully turgid tissues for example which may have little effect on cell volume and the tissue water content can cause a large decrease in turgor pressure leading to a considerable decrease in water potential (Figure 4.2). Such changes of turgor pressure are usually accompanied by small changes in ψ_s. When the turgor pressure falls to approach zero wilting occurs. The hydrostatic pressure becomes negative in the apoplastic solution surrounding the cells, in cell walls and in the xylem, where a water tension (negative hydrostatic pressure) is present due to *transpiration* (page 200).

Water potentials in leaves of well-watered plants are usually in the range of -0.2 to -0.4 MPa. In extreme cases, in arid climates values between -2 to -5.0 MPa have been recorded which shows that these values are much dependent on the plant environment. The solute potentials in crop plants may also vary considerably and are between -1 to -2 MPa under conditions of optimal water supply (Slatyer 1963). Plant organs that store high concentrations of sugars (*e.g.* sugar beet or grape berries) may show ψ_s values up to -2.5 MPa (Wyse *et al.* 1986, Lang *et al.* 1986). The solute potential of halophytes may reach values between -2 to -10 MPa and even lower than -10 MPa (Slatyer 1963, Flowers *et al.* 1977) in order to absorb water from the saline environment. Generally, the apoplastic water solution and the xylem water contains only low concentrations of solutes and hence the ψ_s in these compartments is much higher, between -0.1 to 0 MPa. Here, low water potentials are caused by a negative hydrostatic pressure which may attain values of -1 MPa and lower in the xylem of rapidly transpiring plants. In particular in cell walls the water potential of the apoplastic liquid is also determined by a low matrix potential. In fresh plant material the proportion of matrix bound water is usually very low (Wiebe and Al-Saadi 1976). In many plant species therefore the matrix potential (ψ_m) only plays an important role where tissues have been depleted of more than 50% of their normal water content (Hsiao 1973). In plants the matrix potential can thus often be neglected with the exception of seeds and cell walls. In the soil system, however, the matrix potential is an important component of the soil water potential (Gardner 1965).

One of the most important aspects of the use of water potential is that it represents the driving force for water transfer. Water movement in cells, tissues and whole plants takes place from a higher to a lower water potential. Different plant tissues tend to have characteristic water potentials. The water potential of leaves is usually lower (more negative) than that of roots. According to the catenary hypothesis of van den Honert (1948) water movement between any two points depends on the difference in water potential and on the hydraulic resistance to flow. Such resistances in plant systems are related to water retention by cell walls, to the hydraulic conductivity of membranes and the xylem vessels, to the presence of cuticles etc. Thus, the rate of water flow may be described by the equation:

$$F = \frac{\Psi_1 - \Psi_2}{R}$$

where

Ψ_1 and Ψ_2 = the difference in water potential between two points
F = flow rate
R = resistance

Water movement in the plant occurs predominantly by diffusion across membranes (osmosis) along a water potential gradient or by bulk flow along a pressure difference, *e.g.* in the apoplast and xylem from the soil through the plant to the atmosphere. These mechanisms are described in the following sections.

4.1.3   Osmosis

Osmosis is the diffusion of water across membranes. It occurs when two solutions with different water potentials are separated by a semi-permeable membrane, which allows the penetration of water molecules but not the penetration of the solutes dissolved in the water. This is demonstrated in Figure 4.1. In the past the means by which water moves through semipermeable membranes was not clear and it was supposed that diffusion through the bilayer may occur. Recently, however, integral membrane proteins have been characterized which form water-selective pores (so-called aquaporins, Figure 4.1 A) and which may allow major water movement across the membrane (Chrispeels and Maurel 1994, Weig *et al.* 1997, Chrispeels *et al.* 1999). For the sake of simplicity in the example presented in Figure 4.1 B and C one part of the system consists of pure water (left hand side of the Figure 4.1 B and C) and the other of a sucrose solution (right hand side). Provided that both the pure water and the sucrose solution are exposed to the same temperature and atmospheric pressure, the water potential of the sucrose solution is lower than that of the pure water, because the movement of water molecules is restricted due to the adsorption of water molecules by the sucrose molecules. From this follows that per unit time more water molecules impinge on the semiperme-able membrane from the left hand side in Figure 4.1 (high water potential) than from the right hand side (low water potential). A net water flow from the pure water to the sucrose solution thus results. This net flow of water leads to a gradual increase in the number of water molecules in the sucrose solution. There is also a simultaneous increase in the rate at which water molecules impinge on the sucrose side of the membrane. Finally a state is obtained, where the rate of water molecules impinging on both sides of the membrane is equal and the rate of water flow in both directions is the same so that no net transport of water occurs and equilibrium is attained. At this stage water pressure in the sucrose solution is much higher than in the pure water because in the sucrose solution the water molecules are more tightly "squeezed together". This hydrostatic pressure built up by osmosis is called turgor presure and is a unique and essential phenomenon of living cells.

Such a turgor can only be established if two solutions are separated by a semi-permeable membrane which allows the passage of water molecules, but not the passage of solutes. The turgor which is built up depends on the concentration difference of solutes at either side of the membrane. The higher the solute concentration at one side of the osmotic system the more water is taken up and the higher the resulting turgor. Plasmalemma and tonoplast of intact plant cells are not semipermeable membranes in a strict sense, since they also bring about the uptake of solutes, such as inorganic nutrients and assimilates and thus provide the conditions for osmotic water flow. Solute movement usually depends on active metabolism, is independent of water movement and much slower. In this sense, solutes normally are present inside a cell and the membranes can be considered to be semipermeable. This means that the permeability for the passive movement of most solutes is negligible and the hydraulic conductivity of membranes is almost entirely related to water movement (Kramer and Boyer 1995). It is a simplification, but holds in many cases.

The net uptake of water by the plant cell increases the water potential (ψ) in two ways. It dilutes the cell solute concentration so the solute potential (ψ_s) in the cell increases (it becomes less negative). As a consequence of the rigidity of the cell wall it also raises the turgor pressure (ψ_p) of the cell (it becomes more positive). In turgid cells, water molecules exert an outward pressure on cell walls. Cell walls are rigid, so that a change in the water potential of the cell is associated with a larger change in cell turgor than in solute potential. Figure 4.2 shows the changes which take place in turgor pressure, solute potential and water potential when cells imbibe water (Höfler diagram). Because cell walls are to some extent elastic, however, and can enlarge the cell volume by as much as 20 to 30%, the extent of wall rigidity can have a major impact on changes in cell water potential (Steudle 1989). Thus in cells with highly elastic walls, the cell water potential changes very little as water is removed from the cell, whereas in cells with more rigid walls, a very small loss in water results in a large fall in cell water potential. This difference in behaviour is of ecological and agricultural significance for plant species with rigid cell walls, because they can take up water by increasing the water potential gradient from soil to leaf with only little dehydration. Thus under drought conditions plant species with lignified leaf cell walls such as olives, heathers, and rhododendron are able to absorb water from drying soil without excessive leaf dehydration (Kramer and Boyer 1995).

The concentration of solutes in plant cells is usually in the range of 0.2 to 0.8 kmol/m^3; thus where concentrations in the outer solution are higher (hypertonic solution) than that of the cell (lower water potential in the ambient medium) net water movement proceeds from the cell into the outer medium. This water loss results initially in a shrinkage of the cell volume, which at a later stage is followed by a contraction of the cytoplasm and the plasmalemma pulls away from the cell wall. This phenomenon is called *plasmolysis*. If the cell has not been damaged, the process can be reversed and such plasmolyzed cells are able to attain their full turgor when exposed to pure water or to hypotonic solutions.

Both inorganic (in particular K$^+$ and NO$_3^-$; Na$^+$ and Cl$^-$ in halophytes) and organic solutes (*e.g.* sugars) may contribute to changes of the solute potential in plant cells, and both play a major role in osmoregulation (Morgan 1984, Hsiao and

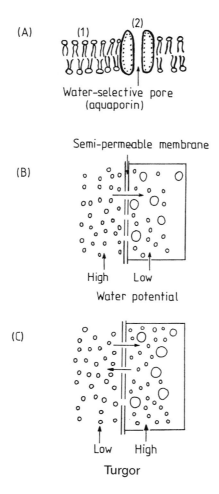

(A) (1) (2)

Water-selective pore
(aquaporin)

Semi-permeable membrane

(B)

High   Low
Water potential

(C)

Low   High
Turgor

Figure 4.1   Osmotic system: (A) Water movement occurs through the membrane bilayer (1) and through water-selective pores formed by aquaporins (2). (B) Net water movement from a higher to a lower water potential. The lower water potential of the sucrose solution on the right hand side is caused by a low solute potential. (C) Water uptake results in a turgor increase (right hand side) reaching an equilibrium where net water movement ceases. At equilibrium the water potential at both sides is equal. The solution with the higher solute concentration has the lower solute potential (ψ_s) which is balanced by the higher pressure potential (ψ_p), both vectors having an opposite direction. In this example the volume of the sucrose solution remains constant (is not expanded).

Läuchli 1986, Munns 1988, McIntyre 1997) to changing ambient water conditions. Under drought stress, plants are able to maintain water adsorption by increasing the cellular solute concentration, a process which is called *osmotic adjustment*.

188

The accumulation of inorganic ions, in particular under saline conditions occurs predominantly in the vacuole, since in the cytoplasm they may inhibit enzymes (Hajibagheri and Flowers 1989). Due to this compartmentation specific solutes, such as glycine betaine etc. (page 226) accumulate in the cytoplasm to maintain the intracellular equilibrium of the water potential. These organic solute are so called *compatible solutes* (page 226) as they maintain enzymatic functions.

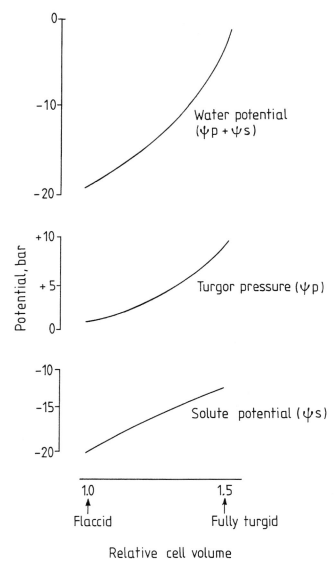

Figure 4.2 Relationship between cell volume, turgor pressure, solute potential and water potential (Höfler diagram).

4.1.4 Bulk flow of water

In contrast to diffusion, *bulk flow* or *mass flow* is independent of solute concentration gradients. It is driven by a pressure gradient and is the predominant mechanism for water transport in the plant *via* the xylem (page 190). It is also of major importance for water movement in the soil and in the free space of plant cell walls. Bulk flow for a lesser amount of the water occurs *via* plasmodesmata, in the root (page 194), *via* the phloem and in all other symplastically connecting plant organs (page 210).

4.2   Water Balance

4.2.1   General

Water movement from the soil through the plant to the atmosphere occurs across membranes, through cells and cell walls and along the xylem which is a part of the apoplast and represents a long and dead conduit system inside the plant. Thus water movement mainly includes osmosis and bulk flow. Finally, water is lost from the leaves as vapour by diffusion. Along this pathway water transport is passive because it moves from higher to a lower water potential of the atmosphere (corresponding to a decrease of free energy of water). This difference in water potential is the driving force which brings about the translocation of water from the soil solution through the plant to the atmosphere. Different components are responsible for the differences in water potential along the water pathway. Generally, the leaf water potential is not very much lower than that in the soil. However, a large difference in water vapour concentration occurs across the boundary layer around the leaf surface with their stomatal cavities and the atmosphere. Water in the leaf apoplast escapes as a vapour into apoplastic air spaces to the stomatal cavity and diffuses to the atmosphere driven by a gradient in water vapour concentration (transpiration). The rate of water transfer across this leaf - atmosphere interface is proportional to the difference in water vapour concentration between both sides of the boundary. Water transport in the xylem system is driven by a hydrostatic pressure gradient. Water transport throughout the root tissue is complex and responds to water potential gradients across the tissue. The *soil — plant — atmosphere continuum* is of utmost importance in the water supply of all plant organs and tissues. On its path from the soil to the tips of stems and leaves, water has to overcome a number of resistances. These and the forces involved in the water translocation process will be discussed subsequently.

Three main steps in water translocation may be distinguished; the centripetal transport from the soil solution through the cortex tissue of the roots towards the xylem vessels of the central cylinder, the vertical xylem transport from the roots towards the leaves, and the release of water as gaseous molecules at the plant atmosphere interfaces. These three principal steps are shown in Figure 4.3.

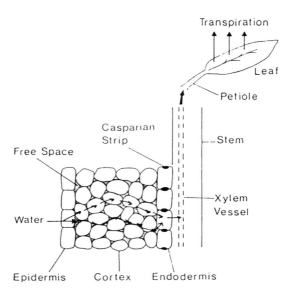

Figure 4.3 Water pathways in the higher plant.

4.2.2   Soil-plant-water relationships

The water potential of the soil is mainly determined by the matrix potential and is most important for dry soils. The solute potential of soil water is typically not very low, *e.g.* at -0.02 MPa. Only in saline soils as a result of high salt concentrations may ψ_s reach values of around -0.2 MPa and lower. Sandy soils contain large pores so that rain or irrigation water tends to drain and thus have limited storage capacity for water. In clay soils the pores are finer so that water is held more tightly and is less available to the plant roots (Table 4.1). Water potential and water movement in soils are thus largely dependant on the soil type.

Table 4.1   Sizes and function of soil pores and the corresponding soil water potentials (modified after Kuntze *et al.* 1994).

pore size (μm)	function		soil water potential (MPa)
>50	large pores	fast draining	> -0.006
10—50	large pores	slow draining	-0.03 to -0.006
0.2—10	middle pores	available water	-1.5 to -0.03
<0.2	fine pores	dead water	< -1.5

In a loam soil consisting of sand, silt and clay in equal proportions, all pore types are present and hence soil water movement and availability for the plant is optimal. The importance of soil texture on the available soil water and important soil water parameter such as field capacity and wilting point of plants were already considered on p. 42 and 43. Under drought conditions the amount of free water in the soil decreases and thus also the soil water potential. Plant growth generally can be maintained at potentials above about -0.7 MPa (about 20% water per volume) (Kaufmann 1968). Plant water driven by the water potential gradient moves mainly via the medium sized pores by bulk flow towards the root. As the soil water content declines the large soil pores and spaces become gradually filled with air so that water movement is restricted to fewer and finer pores. Thus, the so-called hydraulic conductivity in a drying soil decreases rapidly. At water potentials (< -0.8 MPa crop plants often lose turgor in the middle of the day during highest radiation and thus growth mainly occurs during the night. In very dry soils, the water potential falls below the so-called permanent wilting point. At this point the soil water potential is so low that plants are unable to take up water, and remain wilted, even at night when transpiration stops. The permanent wilting point is not determined only by the soil (Slatyer 1957). It depends also on the solute potential in the plant and hence on plant adaptation mechanisms to drought and salinity. For most crop species it is around -1.5 MPa (Richards and Wadleigh 1952) and is considerably lower for xerophytes and halophytes.

The soil water potential is clearly reflected by the water potential of the plant growing in the soil as is shown in Table 4.2 for the leaf water potential of soya beans (Adjei-Twum and Splittstoesser 1976). A low root water potential is the prerequisite to absorb the water from the soil. The differences in water potential between root and leaf tissue are usually small. A large difference in the water potential exists between leaf air and outside air (Table 4.3). Water evaporation from the cell walls of the leaf mesophyll generates large negative pressures (tensions) in the apoplastic water (see further data in Nobel 1991) which are transmitted to the xylem (page 202).

Table 4.2   Relationship between soil water potential and the water potential of
             soya bean leaves (data from Adjei-Twurm and Splittstoesser 1976)

Soil water potential (MPa)	Leaf water potential (MPa)
0 to -0.01	-0.2
0 to -0.02	-0.4
0 to -0.04	-1.2
0 to -0.1	-1.9

4.2.3   Water and solute uptake by roots and their centripetal movement

Water moves passively through the root in a centripetal direction along a water potential gradient and is thus transported very differently from ions. Water movement through the root cortex tissue towards the xylem vessels into the central cylinder (centripetal water transport) is complex which is in keeping with the complex root anatomy. Figure 4.4 shows schematically the tissues of a young root including a living epidermis, the cortex tissue, the endodermis, and the central cylinder with the vascular tissues of the xylem and phloem.

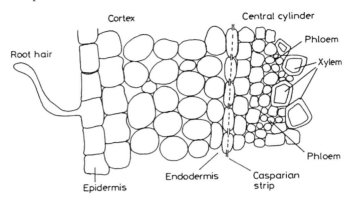

Figure 4.4 Transverse section of a young root.

Water movement in the root is dependant on the extent of suberization of the endodermis (Figure 4.5). This layer of cells which separates the cortex from the central cylinder is generally characterized by thickened cell walls and in the mature part of a root, some distance behind the root tip, by a suberized band of the radial cell walls (Casparian strip). Suberin is a hydrophobic molecule which represents a considerable resistance to water transport. The endodermis in the apical part of a root is unsuberized and thus at the root tip provides a primary site for water uptake (Figure 4.5). In the more mature root regions a similar layer of cells, the exodermis, with thickened walls and impregnated with hydrophobic material (Peterson 1988), also sometimes develops below the epidermis and represents an additional barrier to water movement. The development and function of the endo- and exodermis are discussed in detail by Clarkson (1993) and by Steudle and Peterson (1998).

Water movement throughout the root cortex to the xylem occurs *via* three different pathways: the apoplast, the transmembrane pathway, and the symplast (Steudle and Peterson 1998, Figure 4.6). The apoplast (apo, Greek = away, away from the plasma) is the continuous system of water and air filled spaces of the cell wall and is in close contact with the soil medium, the contact often being enhanced by numerous root hairs. In the apoplastic pathway (Figure 4.6) soil water extends into the free space of the cortex tissue by moving through the water filled pores and spaces of the cell wall without crossing any membranes. The pores and intercellular spaces of the cell wall allow the free movement of water (water free space) which has to be distinguished from the Donnan free space for solutes (Laties 1959). Cations interact with the fixed anions in the cell wall (pectins and proteins) and hence their motion is much restricted. This part of the apoplast is also called apparent free space as this space appears to be free for diffusion. Apoplastic water movement is blocked by the Casparian strip of the endodermis (Figure 4.6); and thus seems to play a major role in young unsuberized root parts which also seems of particular importance for Ca^{2+} transport (see page 522). However, as shown in Figure 4.5 some water enters roots even where the endodermis is suberized. Besides suberin lignin is another major component of the Casparian strip (Schreiber 1996) and since this macromolecule is rather hydrophilic, some water may also traverse the Casparian band.

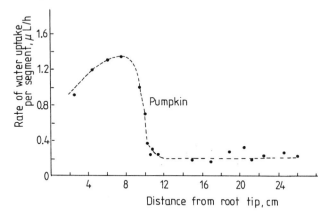

Figure 4.5   The rate of water uptake at various distances behind the root apex of pumpkin (modified after Kramer and Boyer 1995).

Water may also travel through the root cortex *via* the cellular pathway including membrane transport (transmembrane pathway) or through plasmodesmata (symplastic pathway). Water uptake across the plasmalemma mainly results from osmosis. As discussed on page 185 biological membranes function as semi-permeable membranes and allow osmotic water movement. The solute potential in the cytoplasm depends on metabolism. Processes such as the active uptake of ions and the synthesis of organic acids and sugars decrease the solute potential in the cell and thus result in an increased net uptake of water. This demonstrates that water uptake is linked to metabolism and therefore all factors influencing root metabolism may have an indirect impact on water uptake. Low temperature, lack of oxygen and toxic substances depress water uptake, because of their detrimental effect on metabolism (Kramer 1955). Thus Ehrler (1962) found a reduction of about 70% in water uptake when lucerne plants were subjected to a low temperature of about 5°C. A significant positive correlation between water uptake and O_2 uptake by the roots of bean plants (*Phaseolus vulgaris*) has been reported by Holder and Brown (1980). Also treatments inhibiting root respiration restrict water transport (Mengel and Pflüger 1969) and blocked root respiration may also be the mechanism inducing plant wilting in waterlogged soils. It should be borne in mind that the effect of metabolism on water uptake and retention is an indirect one. Metabolism generates the solute potential. Solutes arising from metabolism decrease the symplastic water potential. Thus the water potential difference between the symplast and the apoplast is raised and a higher water uptake results. Water flow between the cytosol and the vacuole is also controlled by the water potential difference between these compartments. As the tonoplast functions as a semipermeable membrane, osmotic forces are also mainly responsible for the net movement of water between the cytoplasm and the vacuole.

The cytoplasm of one plant cell is generally connected to the cytoplasm of neighbouring cells by numerous plasmodesmata. These form a cytoplasmatic continuum which is called the symplast (Arisz 1956), and provides another transport pathway for water. In the symplastic pathway, water moves between cells through plasmodesmata without traversing membranes. The conductivity of water through plasmodesmata is still not well understood (Kramer and Boyer 1995). Plasmodesmata in the endodermis as well as pits (regions lacking the secondary wall and with thin porous primary wall, Figure 4.7) are of low resistance for water movement and thus allow cellular water movement from the root cortex to the central cylinder when apoplastic water movement is restricted at the endodermis.

The upper part of Figure 4.6 A shows water movement through the root cortex at the tip where the endodermis is still unsuberized (no Casparian strip). This lack of a Casparian strip allows water movement to take place from the soil solution into the vascular tissues (xylem vessels) either in the apoplast or from cell to cell. Figure 4.6 B shows the pathways of water transport for upper root parts, where the endodermis has a well developed Casparian strip. As mentioned, the suberized zone of the endodermis cell walls represents a strong barrier to water movement and therefore continuous water flow from the soil through the apoplast to the central cylinder is prevented. Water flow across the endodermis to the central cylinder has thus mainly to follow the cellular pathway as shown in Figure 4.6 B.

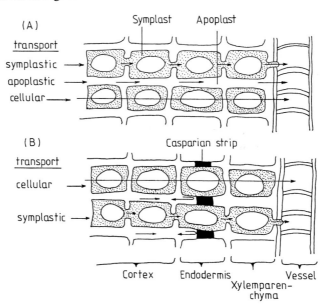

Figure 4.6 Centripetal transport of water through the root towards the xylem vessel. — (A) Unsuberized root tip allowing apoplastic and cellular transport (transmembrane and symplastic pathway). — (B) Suberized upper root parts with Casparian strip allow only cellular water transport.

The apoplastic water path is presumably important when high rates of transpiration allow a rapid water movement. Under such conditions water flow is a bulk flow driven by a hydrostatic pressure gradient between root xylem and soil solution. The cellular pathway which is mainly driven by gradients in solute potential dominates not only when apoplastic water transport is blocked by Casparian strips, but also at low rates of transpiration (at night and under drought and salinity). Under these conditions roots may be protected from water loss because of a high water retention (low solute potential) (Weimberg *et al.* 1982) and high hydraulic resistance (Steudle 1994). Roots in drying soils show increased lignification and suberization (North and Nobel 1991) as well as decreased maturation of metaxylem and reduction in size and number of xylem vessels (Cruz *et al.* 1992).

In particular at low transpiration rates a considerable resistance to water transport is evenly spread over the living root tissue which comprises: epidermis, (exodermis), cortex parenchyma, endodermis and parenchyma cells of the central cylinder (Peterson and Steudle 1993). Water movement from the soil medium towards the xylem vessels is limited by several resistances. The formation of apoplastic barriers as Casparian bands (in the endodermis and sometimes in the exodermis) may result in a high resistance of apoplastic water flow. Capillary forces arise due to the narrow pores and channels of the cell wall and thus some of the free space water is bound very strongly and has a very low water potential (about -10 MPa) largely as a result of matrix forces. This low potential means that the water can be very strongly withheld. This implies that the free space of the root tissue presents a considerable resistance to water flow (Newman 1974). Because of the narrow pores of the cell wall material (most pores have a diameter < 10 nm, the matrix potential of the root may be considerably lower (more negative) than that of the surrounding soil. Therefore, at the same water potential the water content of the cell wall material is higher than that of a medium-textured soil. For example at a water potential of -0.1 MPa, cell walls contain about 2 g H_2O g dry matter, whereas the water content of a medium-textured soil is only 0.1 to 0.2 g H_2O g dry matter. This difference is particularly evident for plants with a high water consumption growing in dry soils (*e.g.* -0.3 MPa soil water potential). Under such conditions the root water potential may become considerably lower because of transpiration than the soil water potential. Generally, the resistance to water movement from soil to the root is smaller than the resistance to centripetal water movement in roots (Taylor and Klepper 1975, Reicosky and Ritchie 1976, Blizzard and Boyer 1980). As the soil dries, the resistance to water movement through the soil increases; thus water absorption decreases and soil shrinkage reduces the contact at the root-soil interface (Faiz and Weatherley 1978).

The high resistance of roots to centripetal water transport is also the reason for diurnal shrinking and swelling of roots. Under high transpiration conditions at noon, the rate of water transport out of the roots into the upper plant parts exceeds that of water uptake from the soil, and root shrinking occurs. The reverse process takes place as transpiration declines during the late afternoon and evening when the water deficit of the roots is made up and the roots swell (Taylor and Klepper 1978). In contrast to the relatively high resistance to centripetal water transport there is little resistance to water flow in the xylem (Frensch and Steudle 1989, Frensch and Hsiao 1993, Steudle and Peterson 1998). Taylor and Klepper (1978) emphasize the fact that the resistance to water transport is higher the deeper the soil layer from which the roots absorb the water. For this reason water depletion from the upper soil layers occurs earlier than that of deeper layers. Plant species which are able to take up water from deeper soil layers are believed to have lower axial root resistances to water.

Solute flow is coupled to water flow. Ions move to the root surfaces by bulk flow and diffusion (Barber 1962, Jungk and Claasen 1997). Many of the factors influencing ion movement from soil to root are thoroughly discussed by Jungk and Claasen (1997) which make clear that the soil as a multiphasic system impedes ion fluxes as does the root which does not simply act as a sink for nutrients. At low transpiration rates ions diffuse along a water potential gradient. In the cell to cell movement passive and active processes at the plasma membrane and nutrient transport across plasmodesmata are involved. By these processes solute potentials are established which promote water uptake by root cells (osmosis). Because of their permeability to ions the root cells do not act as a perfect osmometer, although for simplicity it is frequently regarded as one. The permeability of the endodermis depends on its degree of suberization and lignification (Figure 4.5). It is lower for ions than for water (Peterson *et al.* 1993) and therefore ions predominantly move through the symplast before they can enter the central cylinder. Here, after cell to cell movement the nutrients diffuse into the xylem. Thus, as suggested by Steudle and Peterson (1998) a relatively high hydraulic conductivity, *e.g.* under high transpiration may be coupled with low solute permeability. The Casparian strip also functions as a barrier to back-diffusion of ions (Peterson *et al.* 1993) that have been released into the apoplast of the central cylinder so that ion transport to the shoot may be efficient and root pressure (page 199) can be established at low transpiration. For this latter reason the nutrient concentration in the xylem is usually higher than in the soil water surrounding the roots.

4.2.4   Water release into the xylem vessels and xylem loading of nutrients

Water moving through the root cortex in a central direction must cross the vessel wall before reaching the xylem lumen. Most of the water passes through pits (Figure 4.7). Plate 4.1 shows a metaxylem cell and bordering xylem parenchyma cells. In the thickened cell wall of the metaxylem a pit is visible and in this region the metaxylem is separated from the parenchyma cell only by the plasmalemma and the primary wall.

The mechanism by which water is released into the xylem is not yet completely understood. It is generally accepted, that this is controlled by osmosis and therefore closely linked to ion transport. Of all the ions taken up by roots K^+ transport has been most widely examined. Läuchli *et al.* (1974) using electron probe analysis have shown that K^+ accumulates in the parenchyma cells and suggest that the xylem parenchyma cells play a crucial role in secreting K^+ into the vessel lumen. K^+ is released straight into the xylem vessel through such pits (Figure 4.7). Several K^+ transporters (see for explanation page 489) and channels (one K^+ inward rectifying channel and two outward rectifying channels) at the symplast/xylem interface have been discovered and their physiological roles in respect to uptake and to H^+ATPase activity energising the plasmalemma of xylem parenchyma cells are under discussion (De Boer 1999). The outward rectifying K^+ channels may be responsible for xylem loading (Wegner and Raschke 1994) and are regulated by the membrane potential and by cytosolic Ca^{2+} (Roberts and Tester 1997a; De Boer and Wegner 1997). This implies that the flux of ions from the xylem parenchyma into the xylem is under tight metabolic control *via* plasma membrane H^+-ATPase and ion efflux channels and thus does not represent a simple leakage. Also three anion conductances have been found in the xylem parenchyma cells which are highly permeable to NO_3^- (Köhler and Raschke 2000). These studies were conducted with barley which is known to be a salt includer (Wolf and Jeschke 1986) and thus comparable permeability for Cl^- and NO_3^- was measured. In the study of Köhler and Raschke (2000) ion transport into the xylem was found to occur along an electrochemical gradient with the membrane potential in the range in which anion and cation channels are open; thus concluding that xylem loading with K^+, NO_3^- and Cl^- is passive.

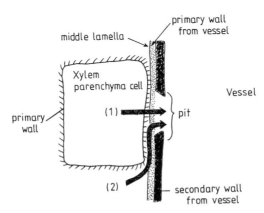

Figure 4.7   Pathways for K^+ entry into a xylem vessel through a pit (after De Boer 1999). — (1) Transmembrane path for K^+, *via* transporters and channels in the plasma membrane of xylem parenchyma cells directly facing the pit in the vessel wall. — (2) Apoplastic path, through the apoplast of primary walls of xylem parenchyma cells and neighbouring xylem vessels.

The secretion of ions into the vessel causes a fall in the water potential in the vessel and hence induces a net water flow into the xylem. Once in the xylem water and solutes are rapidly moved in an axial direction towards the leaves. In maize for example both early and and late metaxylem vessels are present to conduct the flow. Early metaxylem vessels which mature at about 25 mm above the root tip have a diameter of about 23 μm. Late vessels have a diameter of about 100 μm and are present at about 250 mm from the root tip. The late vessels are particularly important for the transport of water in older root systems where water from many lateral roots is collected in the main root. Because of their larger diameter there is less resistance to water flow in the late as compared with the early vessels (Poiseuille's law). There is, however, much less restriction to water movement in the early metaxylem than across the cortex from the soil solution to the xylem vessel. This comparative efficiency in water transport is attributed to a lack of membranes and the presence of open cross- walls in the early metaxylem (Steudle and Peterson 1998). Different driving forces bring about water and solute transport in the xylem, namely *root pressure* and hydrostatic gradients set up by transpiration and which are considered in the following discussions.

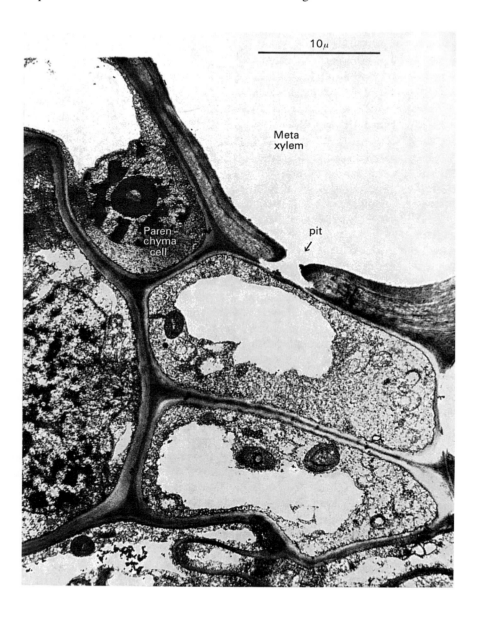

Plate 4.1 Parenchyma cells bordering a metaxylem vessel with a pit in the cell wall. (Photo: Kramer).

4.2.5   Root pressure

Excreting solutes into the root xylem leads to a decrease in the xylem solute potential (ψ_s) and thus to a lowering in the water potential. The Casparian strip may impair back diffusion of ions (Peterson *et al.* 1993) and thus at the end of the centripetal path solutes accumulate in the xylem so that roots function as a nearly perfect osmometer (Steudle and Frensch 1989; Steudle and Peterson 1998). Thus water is absorbed, which has a positive hydrostatic pressure which is called root pressure. In contrast to normal cells, where water uptake is limited by cell volume, the xylem vessels show no major restriction as absorbed water can move in an upward direction. For this reason when water uptake is increased, the hydrostatic pressure (ψ_p) in the vessels does not rise to such an extent as is the case in living cells. Water is thus absorbed comparatively easily by the xylem vessels as a consequence of ion uptake.

Root pressure can be demonstrated by observing the exudation sap accumulating on the stumps of decapitated plants (Bollard 1960). The rate of exudation depends considerably on prevailing metabolic conditions. It is decreased by the effects of inhibitors or anaerobiosis which depress metabolically mediated ion uptake (Kramer 1955, Vaadia *et al.* 1961). The rate of exudation is also influenced by the presence of specific ion species and their concentration in the nutrient solutions. Mengel and Pflüger (1969) found highest exudation rates when KCl was present in the external solutions due to the fact that both K^+ and chloride are taken up rapidly. Lowest exudation rates were observed when the external solution was of pure water. The promoting effect of K^+ on water uptake and transport has also been demonstrated by Baker and Weatherley (1969) in exuding root systems of *Ricinus communis*. These observations agree well with the results of Wegner and Raschke (1994) and De Boer (1999), discussed above.

Root pressure is typically at +0.1 MPa (Kramer and Boyer 1995) and undoubtedly contributes to the upward translocation of soluble organic and inorganic material, particularly in young plants under conditions of sufficient water supply and high humidity where transpiration is low (Locher and Brouwer 1964). In seedlings root pressure can often cause guttation, the exudation of liquid water from leaves. Water is pumped through the entire plant and released as droplets at the leaf tips. Guttation is indicative of intense root metabolism and a high root pressure. It is often observed in the early morning partly as a result of the low water deficit of the atmosphere during the night which restricts evapotranspiration. The droplets contain solutes and Oertli (1962) has reported that young plants exude boron by this means. Generally, however, with the exception of a number of tropical rain forest trees which guttate continuously, guttation is of little significance in older plants including trees, in particular when transpiration rates are high during the day and under dry conditions. At high transpiration, water is lost so rapidly *via* the leaf surface to the atmosphere that a positive pressure never develops in the xylem.

4.2.6   Transpiration and water movement in the xylem

Apart from young plants root pressure and capillary forces are too weak to play a major role in the upward transport of water in the xylem. In tall trees, for example, water may be raised to a height of almost 100 m. The water at the top of the tree develops a large negative hydrostatic pressure (a tension) which pulls water up along the whole xylem duct. To understand this movement several aspects of plant water must be considered. In the first place a continuous water phase, the so called soil-plant-atmosphere continuum extends from the soil solution through the entire plant (see Figure 4.3). The water molecules in this continuous phase are bound by cohesive forces. At the leaf atmosphere boundary, water is present in the fine pores and the intercellular spaces of cell walls in the stomatal cavities (see Figure 4.8). As transpiration into the atmosphere takes place, capillary action and the cohesive properties of the water support the large tensions in the xylem water columns. It has been calculated that water in small capillaries can resist tensions more than -20 MPa which is much larger than about -3 MPa (-2 MPa pressure difference from the ground to the top of the tree which is needed to overcome the viscous drag of water in a vessel plus -1 MPa pressure difference due to gravity) the negative tension being necessary to pull water up the tallest tree (Canny 1998). Thus it is ensured that water flows through the plant to replace that water which is being lost. This normally means that as water is transpired at the leaf surface, water moves into the root from the soil. This concept of water movement is called the cohesion-tension theory and proposed almost a century ago is still accepted on the basis of recent results (Holbrook *et al.* 1995, Pockman *et al.* 1995, Wei *et al.* 1999).

The rate at which water molecules evaporate from the plant atmosphere boundary increases with temperature and is higher when the water potential of the atmosphere is low (higher atmospheric water deficit). It exerts the tension in the water columns of the xylem. This tension is higher under conditions of rapid transpiration or low water availability in the root medium. The large tensions lead to transport velocities of 1 to 60 m/h (Kuntz and Riker 1955, Bloodworth *et al.* 1956) which largely depends on the diameter of the conducting vessel. The fall in the hydrostatic pressure in the xylem induces a small, but detectable shrinkage of the stems. In tree trunks there is often a typical diurnal variation in circumference, a maximum being observed in the early morning, when the water deficit is at a minimum (Ahti 1973). The circumference also depends on the water availability in the root medium. The lower the water potential of the soil, the more the circumference of the trunk is reduced (Ahti 1973).

However, the large tensions may create problems. Weak cell walls could collapse under the influence of large tensions. However, the xylem tissue is well equipped to withstand the negative hydrostatic pressure of the water column. The wooden elements of the xylem cell walls are rigid enough to prevent any major compression of the xylem vessels by adjacent cells. If this were not so, too large a negative hydrostatic pressure of the xylem water could result in a break in the water file and a blockage in the transpiration stream by air bubbles. There is still another mechanism by which water flow in the xylem may be blocked. Under increased tension air may be pulled through air filled spaces in the xylem cell wall into the xylem lumen which results in cavitation (embolism). Cavitation may block water movement in the xylem and thus breaks water movement to the leaves (Tyree and Sperry 1989, Canny 1998). Generally cavitation occurs only in a few vessels. Water can move around the blocked vessels through neighbouring, connected xylem vessels and thus in total the axial water flow in the xylem system is not completely blocked. Cavitated vessels may also be refilled with water (Canny 1997, Tyree *et al*. 1999); the mechanism of cavitation repair is still unclear. One mechanism under discussion is that similar to the establishment of root pressure water is released into the gas-filled vessels from adjacent living cells. In addition, under less negative pressure or even positive pressure (root pressure) in the xylem, gas bubbles may dissolve back in the xylem water. Also, new xylem vessels formed in each year may repair the cavitated ones.

As mentioned, the resistance to water flow along the xylem vessels is relatively low. For this reason the main water flow is along the xylem vessels and follows the major and minor vein systems of the xylem. Xylem cell walls, however, are permeable to water molecules. Some water is imbibed into the cell walls of the xylem and the free space of adjacent tissues. Water uptake by neighbouring tissues can then occur by osmosis. The resistance to water flow in the walls of the xylem vessels is considerably higher than in the xylem vessels themselves (Newman 1974). Lateral movement of water in plant tissues thus proceeds at a much lower rate than the upward translocation.

Higher plants expose a considerable leaf surface area to the atmosphere. This is necessary for the capture and assimilation of CO_2. On the other hand it means that the rate of water loss by transpiration is high. Many higher plants, therefore consume large quantities of water. Actively transpiring leaves may exchange all of its water within 1 hour. A mature maize plant, for example, contains about 3 litres of water although during its growing period it may have transpired more than a hundred fold that amount.

Water is transported to the leaves *via* the xylem of the leaf vascular bundle which branches from the midrib vein into a fine network of veins so that all leaf cells are more or less adjacent to minor veins in a distance of about 500 μm (Kramer and Boyer 1995). In small-diameter vessels (8 μm) most of the transpiring water passes out through the pits supplying the leaf cells along the apoplast (Canny 1995). Thus, analogous to the soil, a fine capillary water film covers the cell wall of mesophyll cells which are in direct contact with the atmosphere through the intercellular spaces in the leaf. As water transpires, the residual water is drawn into the thin spaces of the cell wall and thus a tension, a negative pressure, the driving force for xylem transport is generated at this leaf atmosphere boundary.

By far the largest amount of water transpired by crop plants is released through the stomatal pores. Figure 4.8 shows a section of a mesophyll leaf with open stomata. Stomata are mainly located on the undersides of leaves and enable gaseous exchange between leaf and atmosphere. It can be seen that intercellular air spaces are interspersed between the mesophyll cells. These pores represent the main pathway for the transpiration of water (Slatyer 1967, Sheriff 1984), because the upper and lower epidermis of a leaf are generally covered by a waxy layer, called the cuticle. This consists of a lipidic material and presents a considerable barrier to the transpiration of water molecules. Thus, for most crop plants, most of the transpiration water passes from the intercellular spaces through the stomata (Meidner 1976, Maier-Maercker 1983).

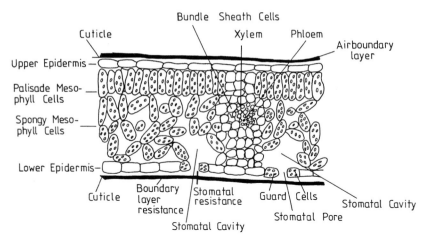

Figure 4.8 Schematic cross section of a leaf showing guard cells, stomatal pores, stomatal cavity, bundle-sheath cells, xylem and phloem. The dark lines above and below the leaf surface show the air-boundary layer.

Table 4.3   Relative humidity and the corresponding water potentials of air inside and outside the leaf at 25°C (data from Nobel 1991).

	relative humidity	air water potential (MPa) (inside and outside the leaf)
inner air spaces	0.99	-1.38
inner air (stomatal pore)	0.95	-7.04
outer air (stomatal pore)	0.47	-103.70

Water loss from the leaf depends on the difference in water vapour concentration between leaf air and external air and on factors that may limit diffusion: the stomatal resistance which is more important than the boundary layer resistance (Holmgren *et al.* 1965). When water has escaped into the intercellular spaces of a leaf it then diffuses into the stomatal cavity. The driving force for leaf water loss to the outside atmosphere is driven by the concentration gradient of water vapour which is high inside and low outside (Table 4.3). The difference in water vapour concentration depends on leaf temperature because as the temperature rises the leaf air holds more water at the same relative humidity. Thus with rising leaf temperature the concentration difference driving water loss from the leaf is increased. Relative humidity is the ratio between actual water vapour concentration and water vapour concentration at saturation and is only important for calculating the water potential in the gas phase. The air of intact leaves normally shows a relatively high humidity, at around 100% with water potentials in the physiological range. The relative humidity of external air is typically much lower, *e.g.* at 50% and shows dramatically low water potentials (Table 4.3).

Stomatal opening and closing is therefore an important process, not only for supplying sufficient CO_2 for assimilation, but also for moderating water consumption of the plant. Therefore, at night stomata are closed, to prevent unnecessary water loss. In the morning around 3 hours before maximal solar irridation when typical crop species of humid climates show maximal net photosynthetic rates (Schäfer *et al.* 1984) stomatal pores are wide open to take in CO_2 for photosynthesis. Simultaneously, water consumption is large and is only restricted by stomatal closure, which in particular occurs at noon in the summer during high solar radiation accompanied by maximal transpiration rates (van Bavel *et al.* 1963). Therefore, typical C_3 plants (page 153) lose large quantities of water. Around 500 molecules of water are lost for every CO_2 molecule incorporated by photosynthesis (corresponding to a very low water use efficiency see page 230). This dilemma is explained by the much larger concentration gradient driving water loss than that driving CO_2 uptake (around 50 times smaller which is explained by the high water vapour concentration of leaf air and the small CO_2 concentration of air outside the leaf). In addition, CO_2 diffusion in air is more restricted and the path is longer than that of water (Holmgren *et al.* 1965, Nobel 1991).

Once water has escaped the stomatal pore, the last step of movement occurs across the boundary layer on the leaf surface (Figure 4.8) which is a film of air. When the air is calm the boundary layer is large (so called unstirred layer) and thus transpiration may be limited dramatically even when stomatal pores are open. This is not the case in moving air when the layer is thin (Bange 1953). Typically the air surrounding the leaf is not calm and thus control of stomatal apertures is of pivotal importance in the regulation of leaf transpiration.

Holder and Brown (1980) have demonstrated that water loss by transpiration is much dependent on root metabolism and to a lesser extent on leaf area. Using entire bean plants (*Phaseolus vulgaris*) these authors obtained a good correlation ($r = 0.89$) between oxygen uptake by the roots and water uptake. Measuring transpiration rates at successive stages of defoliation it was also observed that as the leaves were removed the transpiration rate per unit leaf area of the remaining leaves increased. This indicates that for plants well supplied with oxygen, root uptake and not leaf surface area controls the water flux.

Long distance flow of water in plant tissues is not only brought about by transpiration but also by the growth of meristematic tissues (Boyer 1988). Westgate and Boyer (1984) in studying transpiration and growth of maize plants found that the water potential of meristematic tissues (leaf, culm, root) was about 0.3 MPa lower than that of the corresponding mature tissues. This difference results from the lower solute potential (higher concentration of osmotica) and from the higher elasticity of cell walls of meristematic and elongating cells. This water potential difference between mature and meristematic cells represents a driving force for water and solutes towards the growing points.

4.2.7   Stomatal opening and closure

The idea that the opening-closing process of stomata is dependent on changes in turgor of the guard cells is old and was firstly proposed by von Mohl in 1856 (see for details in Meidner 1987). High turgor results in opening and low turgor in closure. Stomatal opening is largely controlled by the alignment of the microfibrils around the circumference of guard cell walls which restrict expansion of the circumference as the guard cells swell. Pairs of guard cells are joined at both ends and with increase in turgor the cells increase in lenght and bend outward to open the stomata. An increase in turgor in the guard cells results from a fall in solute potential with a consequent flow of water into the cells. As turgor pressure rises cell volume may double due to the elasicity of the cell wall. Under drought, guard cells lose their turgor and stomata close.

Besides water, several environmental factors such as light, temperature and intracellular CO_2 concentrations trigger stomatal movement. Light is the most important environmental factor when plants are well-watered. Srivastava and Zeiger (1995) have shown that stomatal aperture of leaves of *Vicia faba* closely follows the intensity of photosynthetically active radiation applied to the leaves (Figure 4.9A).

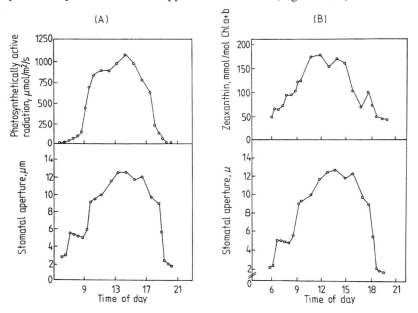

Figure 4.9 (A) Stomatal apertures (lower part) in the leaf surface of *Vicia faba* as related to light radiation (400 to 700 nm). In the upper part of the graph the photosynthetically active radiation (400 to 700 nm) is shown. (B) Concentration of zeaxanthin in guard cells and stomatal apertures in the same leaf of *Vicia faba*. The incident radiation, the zeaxanthin concentration and the stomatal apertures are closely related which supports the view that zeaxanthin tracks blue light radiation (modified after Srivastava and Zeiger 1995).

Research in recent years has shown that stomatal movement is driven by red and blue light. Red and blue light application leads to stomatal opening. Red light increases the production of sucrose by photosynthesis in the guard cells and thus lowers the solute potential. Stimulated photosynthesis probably also leads to increased levels of the carotenoid zeaxanthin which specifically senses blue light. Figure 4.9B shows the close relationship between the content of zeaxanthin of guard cells and stomatal aperture. Blue light stimulates the plasmalemma H^+-ATPase of the guard cells (Assmann et al. 1985; Shimazaki et al. 1986) and the resulting electrochemical gradient provides the driving force for ion uptake. Blue light also stimulates the breakdown of starch and favours malate synthesis. By these reactions the solute potential is decreased and water is taken up with the resulting increase in turgor pressure leading to stomatal opening.

The early experimental data of Fischer (1968) and Fischer and Hsiao (1968) have shown that the turgor of guard cells is related to metabolically dependent ion uptake and particularly to the uptake of K^+ (Humble and Hsiao 1969). Fischer (1968) reported that with an increasing concentration of labelled K^+ in the guard cells the stomatal openings of Vicia faba became wider. This finding was supported by electron probe analysis of Humble and Raschke (1971) who demonstrated that K^+ is accumulated in the guard cells of open stomata, whereas in closed stomata no K^+ accumulation was observed (see Figure 10.4). Potassium ion concentrations can increase markedly from 100 mol/m³ in the closed state to up to 800 mol/m³ when stomata are open (Outlaw 1983). The K^+ ion is balanced mainly by Cl^- which is also taken up into the guard cells during opening and excluded during closure. Malate, the other counter anion, is synthesized on stomatal opening and then metabolized or extruded during stomatal closing. K^+ is taken up passively along the electrical gradient of about -50 mV via potassium channels (Schroeder et al. 1994). Chloride also moves along anion channels and is presumably taken up by H^+ cotransport. A H^+/Cl^- transporter has been shown for the first time in root hair cells of Sinapis alba (Felle 1994). As K^+ plays a dominant role in stomatal opening, the K nutritional status of the plant influences water loss by transpiration. Zech et al. (1971) observed enhanced rates of transpiration in Pinus silvestris suffering from K^+ deficiency. The beneficial effect of K^+ preventing water loss has been reported by Brag (1972) for Triticum aestivum and Pisum sativum.

Recent studies of Talbott and Zeiger (1998) have provided interesting results on osmoregulation in guard cells and changes in stomatal aperture during day time (Figure 4.10). An increase of the K^+ concentration is, as suggested by different authors, paralleled by stomatal opening in the morning. It then decreases after noon under conditions when the stomata are still open. During this phase of K^+ efflux the sucrose concentration in the guard cells increases markedly becoming the major osmotically active solute. Sucrose may be produced by starch hydrolysis as was suggested about a hundred years ago or by guard cell photosynthesis or by uptake from neighbouring mesophyll cells (Talbott and Zeiger 1998). These data suggest that stomatal opening is induced by K^+ uptake at sunrise before photosynthesis starts, but later in the main period of CO_2 assimilation, sucrose is the major osmoticum which regulates stomatal opening.

High temperatures are often associated with high water consumption and thus may also cause stomatal closure. Stomatal closure under these conditions may result from increased levels of CO_2 in the stomatal cavities due to enhanced respiration. It has been shown by Talbott et al. (1996) that stomata can be very sensitive to ambient CO_2 concentrations. Stomatal closure thus protects the plant against excessive water loss. This mechanism of opening and closing provides a very efficient means of regulating the water balance of the entire plant. Under drought conditions abscisic acid plays a central role in stomatal closure which is considered on page 273.

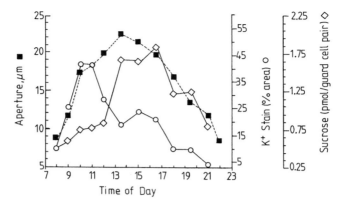

Figure 4.10 Changes in stomatal aperture, and in potassium and sucrose content in guard cells of intact *Vicia faba* leaves (after Talbott and Zeiger 1998).

4.3   Long Distance Transport of Mineral Nutrients

4.3.1   General

In higher plants an adequate transport of minerals between sites of uptake and production and sites of consumption is essential. In this transport process water plays a dominant role. The most important pathways for long distance transport are the vascular tissues of the xylem and phloem. The main materials which are transported are water, inorganic ions and organic compounds. Water and minerals are taken up from the root medium and translocated towards the upper plant parts by the xylem. The phloem is the tissue that mainly translocates organic compounds, the products of photosynthesis and amino acids from mature leaves to areas of growth and storage driven by a turgor pressure gradient (see page 248). Phloem-mobile mineral nutrients are also transported in the phloem and are translocated both in an upward and downward direction (page 253). This section deals with nutrient cycling *i.e.* the retranslocation of nutrients from the upper plant part to the root and with nutrient recycling (translocation of the cycled nutrients back in the xylem to the shoot). It also deals with xylem transport of nutrients and the mobilization and phloem translocation of nutrients (redistribution) from source leaves. The importance of nutrient remobilization is also considred as occurs under nutrient deficiency, during generative growth or during leaf senescence or in perennial plants in spring when nutrients are mobilized from the stem or root.

Although water movement in the plant and particularly in the xylem can considerably affect the transport of plant nutrients, it should be remembered that water uptake and ion uptake are two separate processes. A high rate of water uptake is not necessarily associated with a high rate of ion uptake. Hanson and Bonner (1954) thus demonstrated with artichoke tubers, that lowering the rate of water uptake by a low water potential in the outer solution did not affect the uptake of Rb^+. Under conditions, where water traverses biological membranes, ion uptake precedes water uptake thus enabling osmotic water transport (see page 197). However, where water movement is not across a membrane, as in the transpiration stream, plant nutrients may be carried passively along with the water by mass flow. It is generally accepted (*e.g.* Smith 1991) that transpiration is the main driving force not only for water transport, but also for nutrient translocation in the xylem.

4.3.2 Xylem transport

The relatively high rate of water flow along the xylem vessels in an upward direction may cause a rapid translocation of solutes dissolved in the xylem sap. Inorganic ions once transported into the xylem vessels (see page 197) are thus quickly translocated to the upper plant parts. This effect was clearly demonstrated by Rinne and Langston (1960) in an experiment in which one part of the root system of a peppermint plant was fed with labelled phosphate (^{32}P). Labelled phosphate was detected in the upper plant after some hours, but only in those plant parts directly supplied by xylem vessels from that portion of the root system which had received labelled phosphate. This is shown schematically in Figure 4.11. The shaded parts of the leaves show the accumulation of labelled P. The distribution pattern of ^{32}P observed in the leaves is typical for this kind of labelling experiment. It also demonstrates that inorganic ions are mainly distributed along the vascular xylem system towards the leaf and that lateral movement proceeds from the respective leaf bundle side into adjacent cells.

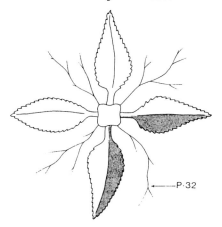

Figure 4.11 Distribution of labelled P (shaded area) in the leaves of a peppermint plant. ^{32}P was suplied to one part of the root system (after Rinne and Langston 1960).

Along the xylem path cations and particularly divalent cations may be adsorbed to cell wall surfaces and exchanged for other cations (Biddulph *et al.* 1961, Isermann 1970). Thus, the rate of upward translocation depends mainly on the transpiration intensity and water uptake by roots. Linser and Herwig (1963) showed with young maize plants that reducing the rate of transpiration resulted in a decrease in phosphate translocation from the roots to the upper plant parts. Michael et al. (1969) growing tobacco in a growth chamber at extremely low transpiration rates induced B deficiency in the youngest leaves, because of the low rate of B translocation in the transpiration stream. Not all plant nutrients depend on transpiration when translocated in an upward direction. Some of the major plant nutrients are also transported in the phloem sap. This kind of transport is treated in more detail on page 247.

The xylem sap is a rather dilute solution, which is made up largely of inorganic ions and also of amino acids. Different techniques have been used in the collection of xylem sap which have resulted in variations in the ion concentrations that have been obtained. Xylem sap has often been sampled by exudation, a process which is driven by root pressure the samples being collected from root stumps (Bollard 1960). Because the water flow rate in actively transpiring plants is much higher than water flow of detopped plants water flow determinations from root pressure exudates generally overestimate ion concentration in the transpiration stream and significant changes in their relative concentrations occur (Table 4.4), in particular when sampling occurs over a long period. The root pressure chamber technique is currently the only method available to collect xylem sap representative of an intact, transpiring plant (Schurr 1999) which is a prerequisite for studying the dynamics of nutrient transport in the xylem under transpiring conditions. Table 4.4 shows a comparison of the inorganic ion composition of xylem sap in *Ricinus communis* obtained using the root pressure chamber and by exudation. In both cases K^+, NO_3^- and Ca^2 are present in highest concentrations. The concentrations of the individual inorganic ions are highly variable and also depend on plant species. Ion concentration and the relative composition depend on the rates of uptake, their concentration in the nutrient medium, the water flow rate in the xylem and thus also on the intensity of transpiration.

Large diurnal variations of xylem sap compounds including the xylem sap pH have been found in different crop plants (Schurr and Schulze 1995, Delhon *et al.* 1995). Generally, the availability of individual nutrients at the root has a major impact on the amplitude of the diurnal pattern (Schurr 1999). Nutrient concentrations are commonly high during the night associated with a low flow rate; during the day low concentrations coincide with high water flux rates showing a peak at midday. In spite of constant transpiration, xylem nutrient concentrations and flow rate also vary during the day which indicates that additional processes are involved in nutrient translocation. It is unclear which processes contribute to the temporal variation of xylem sap composition. Rapid transpiration causes a low solute concentration in the xylem because the incoming water dilutes the xylem solution. At low transpiration rates, ion uptake into the xylem continues and hence the ion concentration in the xylem solution may increase. In this case the concentrations change proportionally to transpiration, although other processes can influence the dilution effect (*e.g.* nutrient uptake, metabolism of the root, xylem loading).

Lateral exchange of cations and particularly divalent cations with xylem cell wall surfaces (Biddulph *et al.* 1961, Isermann 1970) as well as uptake processes into neighbouring cells may have an essential impact on diurnal variations of mass flow driven by transpiration. Some inorganic nutrients can be taken up rapidly from the cells adjacent to the xylem vessels. They thus can decrease in concentration as xylem water is transported along the xylem vessels (*e.g.* $H_2PO_4^-$ and K^+). On the other hand, other nutrients (*e.g.* NO_3^-) are absorbed relatively slowly from the xylem sap by adjacent cells (Schurr, unpubl. data).

Table 4.4 Composition of xylem sap in *Ricinus communis* after applying the root exudation and the root pressure chamber technique. Mean concentrations over 9 h light period. The concentrations in the exudates from the root stump were significantly higher than the concentrations of xylem sap obtained from intact plants (root pressure chamber) (factorial increase, right column). Notably the increase in concentration is not constant between solutes and can thus not be explained by a simple dilution effect. Exudation rates averaged to 20 μL/min in the root exudation technique and transpiration rate was at 450 μL/min in average (adopted from Schurr and Schulze 1995).

Nutrients	Intact plant (mol/m³)	Root exudation (mol/m³)	Root exudation/ Intact plant
K^+	6.6	17	2.6
Ca^{2+}	1.8	5	2.8
Mg^{2+}	0.7	4	5.5
NO_3^-	4.7	20	4.3
SO_4^{2-}	0.2	1.8	7.5
PO_4^{3-}	0.2	4	27.7
Cl^-	0.09	0.4	4.6
Gln	0.57	5	8.8
Arg	0.004	0.2	16.5
Glu	0.009	0.06	23.3
H^+	0.00031 (pH 6.5)	0.0036 (pH 5.4)	11.6

Nutrient uptake has a strong impact on xylem sap concentrations. Increasing K^+ levels in the nutrient medium considerably increase the exudation rate and thus induces water flow and enhance further uptake of ions into the xylem (Kirkby et al. 1981). The rate of ion uptake by the root is usually independent of the rate of water uptake and varies diurnally. Schulze and Bloom (1984) found that NO_3^- uptake is independent of transpiration when transpiration is rapid. Only at low transpiration, e.g during the night may NO_3^- uptake be affected because of the high nitrate levels in the xylem. In general, NO_3^- uptake is more or less independent of transpiration and largely depends on shoot demand (Imsande and Touraine 1994) as is also the case for potassium uptake (Mengel 1999). Interestingly the uptake of nitrate and potassium are closely interrelated (Macduff *et al.* 1997). They both show the same diurnal variation and a similar amplitude during the day (Schurr and Schulze 1995). Since a significant amount of K^+ is cycled back in the phloem (page 212) less K^+ is taken up as compared with nitrate. Since shoot demand and nutrient assimilation vary during the day, it is tempting to suggest that the diurnal variations in nutrient uptake are coupled with plant demand. Nutrient uptake is also regulated by phloem cycling to the root and the nature of signals as well as the mechanism of signalling is still not clear (page 214).

Nitrate is typically the main inorganic N form in the xylem (Pate 1973), although with different forms of nitrogen supply xylem composition changes (Van Beusichem *et al.* 1988). Increasing NH_4^+ concentrations in the nutrient solution or under exclusive ammonium nutrition, *e.g.* at 1 mM NH_4^+, an upper concentration which occurs in the soil solution of agricultural soils (Wolt 1994), the concentrations of NH_4^+ and amino acids in the xylem sap increase and thus may also contribute to leaf nitrogen nutrition (Kosegarten *et al.* 1999a). Husted and Schjoerring (1995) found NH_4^+ concentrations up to 0.8 mM in the leaf apoplast of *Brassica napus* grown on sandy soil and reported high uptake rates for NH_4^+ from the apoplast into the leaf cell. Under exlusive ammonium nutrition glutamine may even become the dominating N compound in the xylem sap (Kosegarten *et al.* 1999a). Thus nitrogen supply to a young leaf may also be provided in the form of amino acids *via* the xylem and *via* the phloem. In the phloem, levels in nitrate and ammonium are low (Pate 1980, Jeschke *et al.* 1991, Peuke *et al.* 1996, see Table 4.5). The proportion of organic N in the xylem depends much on the capacity for nitrate reduction in the roots and is generally high in woody plant species and lower in herbaceous species. The main types of amino acids found in the xylem sap are amides (glutamine and asparagine) and their precursors glutamate and aspartate (Pate 1980). In legumes fixing molecular N_2 the organic N forms of amides or ureides (allantoin and allantoic acid) are most important compounds in the xylem sap transported from the nodules to the shoots (Pate 1973, Pate 1980). Basic amino acids, *e.g.* arginine, are quickly absorbed by the tissue surrounding the xylem; asparagine and glutamine are absorbed by stems at a moderate rate, while aspartate and glutamate are hardly absorbed at all and are thus transported to the leaves. These amides are used for the synthesis of proteins in young leaves or transported with the excess nitrogen continuously arriving *via* the xylem to mature leaves for redistribution *via* the phloem to fruits or younger leaves. Amino acids absorbed from the stem tissue are mainly used for storage. The level of amino acids in the xylem sap also depends on the physiological age of the plant. Concentrations are often high in spring, when storage proteins of roots and stems are mobilized for the formation of young leaves (Sauter 1976, Pate 1980).

4.3.3 Cycling and recycling of mineral nutrients between phloem and xylem transport

The phloem tissue contains the pathways which allow the translocation of photosynthates and so called phloem mobile nutrients. The phloem often runs parallel with the xylem strand (vascular tissue) and is generally located on the outer side. Phloem tissue contains sieve elements, companion cells and parenchyma cells. The sieve elements are most important; they are highly specialized for long distance translocation and in contrast to the dead xylem vessels they are living. The anatomy of the phloem tissue and the mechanism of phloem transport is described in more detail in Chapter 5 on page 250. Basically the transport in the phloem is driven by a turgor pressure gradient from the source area of the plant (typically the mature leaf) with a low solute potential (corresponding to a high turgor) to several sink areas with a high solute potential (*e.g.* developing leaves, roots and storage sinks). In contrast to translocation in the xylem, which occurs only in one direction, phloem transport is bidirectional.

Generally photosynthates synthesized in the older source leaves are translocated in a downward direction mainly to the roots, whereas younger source leaves mainly supply photosynthates to the apex, youngest sink leaves and fruits. This pattern of distribution has been shown by Major and Charnetski (1976) in rape plants. The direction of transport also depends on physiological conditions. An example of this kind is shown in a schematic way in Figure 4.12.

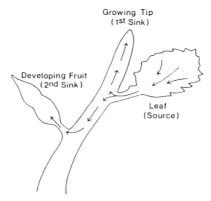

Figure 4.12 Distribution of photosynthates in relation to the strength of the physiological sink.

A mature source leaf is shown first to provide assimilates for a young leaf (I^st sink). This means that assimilates from the old leaf are transported in an upward direction. As fruit setting begins a new physiological sink develops and the assimilates are diverted in a downward direction to the young developing fruit (2nd sink). Such source-sink relationships are of major importance in crop production as discussed in Chapter 5. The mature leaves providing the photosynthates are regarded as the *physiological source*, whereas tissues in which the assimilates are stored or consumed for growth or by respiration are termed *physiological sinks*.

Table 4.5 shows the composition of the phloem sap of the castor oil plant (Hall and Baker 1972). Phloem saps obtained from other plant species differ only in detail from the results presented here. The main pattern is always the same: sucrose is the dominant component by far with concentrations in the range of 200 to 900 mol/m^3 and is the form in which carbon is translocated in the plant. There are some plant species, however, in which other nonreducing sugars as raffinose, stachyose, verbascose (sucrose bound to one, two, three galactose molecules, respectively) or the sugar alcohols such as mannitol and sorbitol are translocated instead of sucrose. The concentrations of amino acids can vary considerably depending on physiological conditions (Mengel and Haeder 1977). The main forms are glutamate and aspartate and their respective amides, glutamine and asparagine. Of the inorganic ions, K^+ in particular and to a lesser extent Mg^{2+}, phosphate and chloride occur in relatively high concentrations. In contrast, nitrate, ammonium, sulphate, Ca^{2+} and $Fe^{2+/3+}$ are present only in minor concentrations or almost completely excluded. The phytohormones such as auxins, gibberellins, cytokinins and abscisic acid also occur in the phloem sap. The pH of the phloem sap is relatively high (about 8), probably due to the presence of HCO_3^-.

Table 4.5   Composition of phloem sap of *Ricinus communis* (from Hall and
Baker 1972). Keto acids were calculated as divalent.

Dry matter	100—125 mg/g
Sucrose	234—304 mol/m^3
Reducing sugars	—
Amino acids	35.2 mol/m^3
Keto acids	15—24 mol/m^3
Phosphate	2.5—3.8 mol/m^3
Sulphate	0.3—0.5 mol/m^3
Chloride	10—19 mol/m^3
Nitrate	—
Bicarbonate	1.7 mol/m^3
Potassium	60—112 mol/m^3
Sodium	2—12 mol/m^3
Calcium	0.5—2.3 mol/m^3
Magnesium	4.5—5.0 mol/m^3
Ammonium	1.6 mol/m^3
Auxin	0.60×10^{-1} mmol/m^3
Gibberellin	0.67×10^{-2} mmol/m^3
Cytokinin	0.52×10^{-1} mmol/m^3
ATP	0.4—0.6 mol/m^3
pH	8—8.2
solute potential	-1.4—-1.5 MPa
conductance	13.2 mS
viscosity	1.34 cP at 20°C

Unlike higher animals, which have a closed system of circulation, the main
transport pathways in higher plants, the phloem and the xylem, are not directly linked
to one another. Thus in the translocation between these two pathways water and
solutes must pass through the connecting tissues. The phloem absorbs water from the
surrounding tissues which in turn obtain water from the xylem. Thus on average about
5% of the water transported in an upward direction in the xylem is retranslocated
via the phloem to the lower plant parts (Zimmermann 1969). The water potential
in these connecting tissues may exert an influence on the flow rate in phloem and
xylem. A high rate of transpiration lowers the water potential of tissues surrounding
the xylem and thus also depresses water transport into the phloem and the turgor
of sieve tubes.

Research in the last decade has shown that continuous nutrient cycling *i.e.* the
retranslocation of nutrients in the phloem from the shoot to the root and the recycling *i.e.*
the translocation of cycled nutrients back in the xylem to the shoot is of great importance
for nutrients which show a high phloem mobility, such as for nitrogen and potassium.
The same is also true to a lesser extent for phosphorus, sulphur and magnesium (Marschner

et al. 1996, 1997). In contrast, recycling of nutrients from sink leaves to source leaves does not take place because xylem transport is upwardly directed.

One major function of this cycling is to cover nutrient root demand for growth. This is clearly of particular importance in plants supplied exclusively with nitrate and where most of this nitrate is reduced in the shoot. However, even when most nitrate is reduced in the root large amounts of nitrogen are cycled in the form of amino acids from the shoot to the root far in excess of root demand so that the excess recycles back to the shoot (Jeschke and Pate 1991a). In barley grown with nitrate around 50% of the cycled amino acids were found to be recycled back in the xylem (Simpson et al. 1982). When plants are fed with ammonium N, assimilation occurs in the roots to cover nitrogen demand and thus less of the amino N in the xylem derives from phloem cycling (Peuke and Jeschke 1993). Sulfate is mainly reduced in the leaves and this nutrient can also be cycled in its reduced form as glutathione via phloem in order to cover root demand (Herschbach and Rennenberg 1994). This cycling-recycling process not only applies to nutrients which are largely assimilated in the leaves, but is also very important for K^+. Potassium, as can be seen from Table 4.5 is present in the phloem sap in a high concentration and for this reason it can be translocated to various plant parts very quickly. Table 4.6 shows the very high proportions of the total uptake of K^+ that are both imported via the xylem and exported via the phloem in the leaves of both white lupin and castor bean. Potassium thus contributes significantly as a driving force to solute flow in both xylem (see page 197) and phloem (see page 249). Seventy eight % of total K^+ taken up was recycled via the xylem to the shoot of castor bean which underlines its important role in promoting xylem transport, particularly at low transpiration. In this experiment white lupin was exposed to NaCl stress and a substantial Na^+ cycling in the phloem to the roots was observed. According to Jeschke et al. (1995) this process may counteract the accumulation of Na^+ in source leaves.

Table 4.6 K^+ cycling and recycling in castor bean and NaCl-stressed white lupin (denoted from Jeschke and Pate 1991b and Jeschke et al. 1987). In NaCl-stressed white lupin relatively large amounts of Na^+ are cycled back in the phloem to the roots.

| | Proportion of total uptake | | | |
| Path | white lupin | | castor bean | |
	K	Na	K	Na
Xylem import to the leaf	96	45	138	11
Xylem export from the leaf	72	33	93	9
Phloem transport to the root	59	33	85	9
Cycling through the root	39	—	78	—
Total uptake (mmol per plant)	1.07	1.23	2.88	0.48

Another major function is the maintainance of cation-anion balance. In this context the K^+ is of central importance. In plants where nitrate is mainly reduced in the leaves malate and other organic acid anions are produced. Potassium ions partially cycled in the phloem to the root maintains the charge balance of the organic anions (Ben-Zioni *et al.* 1971, Kirkby and Knight 1977). Cycling of mineral nutrients as well as of metabolites to the root may act as a signal to control root nutrient uptake. Amino acids as well as organic acids may play this role in the case of nitrate. It has been suggested that at high shoot demand a lower proportion of nutrients cycles back to the root acting as a feedback signal for enhanced uptake. On the other hand when shoot demand is low the increase in cycling acts to repress nutrient uptake and xylem transport. This is still a hypothesis, but there is already some evidence of feedback control for K^+ (Engels and Marschner 1992), for nitrogen (Touraine *et al.* 1992), for phosphorus (Drew and Saker 1984) and for sulphur (Herschbach and Rennenberg 1994).

4.3.4   Redistribution of mineral nutrients in the phloem

The retranslocation of plant nutrients *via* the phloem under nutrient deficiency depends much on the capability of sieve tubes to take up nutrients rapidly. Retranslocation of K^+ from older to younger leaves in barley plants has been shown by Greenway and Pitman (1965). Magnesium is also translocated in the phloem and is thus rather mobile throughout the whole plant (Steucek and Koontz 1970). Schimanski (1973) reported that the translocation pattern of Mg^{2+} in the phloem is similar to that of the photosynthates. Because of the high concentrations of K^+ and Mg^{2+} in the phloem sap, fruits and storage tissues, which are mainly supplied by the phloem sap, are relatively rich in K^+ and Mg^{2+}. Thus tomato fruits (Viro 1973) and potato tubers (Addiscott 1974) are comparatively rich in Mg^{2+} as compared with Ca^{2+}. Like the transport of photosynthates (see page 544), Mg^{2+} translocation in the phloem seems to be enhanced by K^+ (Addiscott 1974). Calcium occurs only in minute concentrations in the phloem sap (Table 4.5) and in contrast to K^+ and Mg^{2+} is not phloem mobile. Thus Ca^{2+} translocated by the transpiration stream to the upper plant parts, is scarcely moved in a downward direction in the phloem (Loneragan and Snowball 1969). The difference in phloem mobility between K^+ and Mg^{2+} on the one hand and Ca^{2+} on the other is also reflected in the appearance of deficiency symptoms. Calcium deficiency first becomes visible in the youngest leaves because Ca^{2+} cannot be retranslocated from mature leaves. This also means that it cannot be transported to the root against the transpiration stream and thus root growth is much dependent on external Ca^{2+} supply. For K^+ and Mg^{2+} the reverse holds true. Both ion species can be removed from older plant parts *via* the phloem pathway and thus translocated to the younger tissues. When external K^+ or Mg^{2+} supply is inadequate, the K^+ and Mg^{2+} in the older plant parts are mobilized and translocated *via* phloem into the younger growing tissues. This is the reason why K- and Mg -deficiency symptoms appear first in the older leaves.

As the Ca supply to a plant organ mainly depends on transpiration intensity, the transpiration rate of a given plant organ is of particular importance in determining its Ca^{2+} concentration. Where transpiration is low, Ca supply may be inadequate and Ca deficiency may thus result. Fruits and storage organs generally have a lower transpiration rate than leaves. This gives rise to blossom end rot in tomatoes, bitter pit in apples and blackheart in celery (see page 531). The younger internal and low transpiring leaves of lettuce are also prone to Ca deficiency disorders (Collier and Tibbits 1982). All these physiological disorders are caused by an undersupply of Ca to these organs.

Phosphate may be translocated in the phloem in inorganic and organic forms. Quite a substantial amount of P in the phloem sap is present in the form of ATP (Gardner and Peel 1969). Redistribution of nitrogenous compounds is an essential part of N metabolism in most plant species. Many species are capable of storing N in the form of proteins at early stages in development, which at a later stage are hydrolyzed and translocated *via* the phloem from mostly mature leaves to fruits and storage organs (Pate *et al.* 1979). C_3 plants in particular store high amounts of leaf protein as RuBP carboxylase which may also serve as a storage protein, as well as a CO_2 fixing enzyme. Brown (1978) argues that C_4 plants have a greater N use efficiency (biomass production per unit of N in the plant) than C_3 plants because of the smaller investment of N in the photosynthetic carboxylation enzymes. Fruits are mainly nourished by phloem sap. In an excellent review paper Pate (1980) notes that phloem sap provides 98% of the C, 89% of the N and 40% of the water of the fruits of *Lupinus albus*. Relatively high contents of amino acids in the phloem are often observed at the end of the growing period, as proteolysis occurs during leaf senescence. The resulting amino acids are transported from the leaves to storage tissues.

4.4   Physiological Aspects of Drought Stress

4.4.1   General

Drought stress in plants is one of the major environmental limitations that affect crop production throughout the world. In arid and semiarid regions water availability is the predominant factor limiting plant growth and in these regions of hunger crop yield is greatly depressed. However, even in humid climatic zones a lack of water can restrict growth (Table 4.7) as a consequence of irregular distribution of rainfall (Boyer 1982, Gales 1983). This can often be observed in the field during summer when crop plants lose turgor or show wilting symptoms.

Drought stress is often related in literature to salinity stress. However, the environmental conditions and plant reactions differ considerably between drought and salinity stress. Under salt stress, which is considered in detail later, the soil may contain abundant water and toxic ion levels. Under drought stress the environment is completely different and is mainly characterized by a low water content. Plant species show a great variety of mechanisms for the adjustment to drought stress. The physiological reactions of plants to drought stress are complex (Figure 4.13) depending on moderate (*e.g.* decrease of leaf water content, loss of turgor, growth reduction) and severe dehydration effects (*e.g.* reduction of leaf surface, CAM metabolism, succulence, seed formation and survival). The major and most sensitive process affected is cell growth. A reduction of growth can occur either by a decrease of the tissue water content and/or by a decrease of the organic mass. The decrease of the tissue water content depends on water availability in the soil. The formation of organic mass depends additionally on the supply with CO_2 for photosynthesis and on the degree of water use efficiency of the plant (page 230).

Table 4.7 Crop yield of maize and soybean in U.S.A. between 1979 to 1988, expressed as percentage of the average yield throughout this period (from U.S. Department of Agriculture 1989). In 1980, 1983 and 1988 water deficit was severe and thus crop yield in these years was reduced by about 10 to 20%.

Relative crop yield (10-year average = 100%)		
Year	maize	soy bean
1979	104	106
1980	87	88
1981	104	100
1982	108	104
1983	77	87
1984	101	93
1985	112	113
1986	113	110
1987	114	111
1988	80	89

A moderate decrease of the soil water potential can easily be balanced by the plant through a decrease of the solute potential (osmotic adjustment) and matrix potential. For example, osmotic compounds are synthesized (Weimberg *et al.* 1982) leading to a net increase in concentration of cellular solutes (compatible solutes), usually associated with a decrease of the solute potential. Under these conditions the turgor is still undisturbed. The higher the drought stress, the more energy is consumed in the synthesis of these cellular solutes, which is diverted from the needs of growth and biomass production so that both are greatly impaired (Wyn Jones 1981, Yeo 1983).

Severe drought stress can lead to water loss and a reduction in turgor pressure (ψ_p) within cells. The most well known response to drought stress is stomatal closure which reduces water release as well as CO_2 uptake from the atmosphere. The decrease of turgor can be delayed by a controlled cellular water release, so called *cell shrinkage* (Santakumari and Berkowitz 1989). Recently, it has been observed that the turgor pressure in the epidermis of young leaves in *Beta maritima* decreases under drought stress, but due to cell shrinkage a certain turgor pressure is maintained and plasmolysis is avoided (Plate 4.2). Thus by controlled cellular water release the wall and turgor pressure is maintained to a certain level and cell shrinkage is associated with spatial arrangements of membranes.

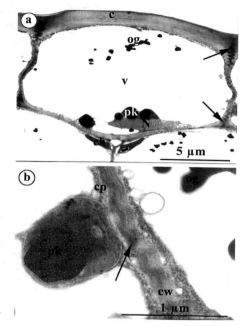

Plate 4.2   a: Shrinkage of a typical epidermal cell of juvenile leaves of *Beta vulgaris* after severe drought stress. The lateral cell walls are contracted, see arrows. c cuticle, N nucleus, os osmophilic globules, pk protein crystal, v vacuole b: Protein crystal in the chloroplast adjacent to the cell wall which shows contracted cell wall structures typical for drought stress. cp cytoplasm, cw cell wall, pk protein crystal (Photo: Koyro).

The mechanisms mentioned above may not be sufficient to retain the water within the plant. As soil dries out the water potential decreases markedly and this is accompanied by a decrease of soil hydraulic conductivity. Under such severe conditions even water may flow back from plant roots to the soil. This may be avoided by considerable reduction of water conductance in the plant. The closure of plasmodesmata and aquaporins may increase the resistance to symplastic water movement (Oparka *et al.* 1992, Steudle 1994, Tyerman *et al.* 1999) and a rise of apoplastic solute concentrations (Zhang *et al.* 1996) may also impede the water flow in the apoplast. Another important mechanism is xylem cavitation in perennial plants which breaks the continuity of the water column in the xylem and thus stops water transport (Tyree 1997, Tyree and Sperry 1989). In annual plants the diameter of the xylem vessels, in particular in young leaves can be reduced (Allaway and Ashford 1996, Zotz *et al.* 1997) and this is one reason for the higher drought resistance of these tissues compared with fully expanded leaves.

Besides growth reduction and changes of the water relations in the plant further processes are influenced by drought: growth reduction is followed very closely by a decrease in cell wall and protein synthesis in tissues with high growth potential (Figure 4.13). With a further decrease in water potential, cell division may decline and the levels of some enzymes decrease. Abscisic acid starts to accumulate even with moderate drought stress and is one of the most well-known responses. Stomata may then close with a consequent reduction in transpiration and photosynthesis (Boyer 1970, 1995). Very low plant water potentials may also reduce respiration (Bell et al. 1971).

4.4.2   Interdependence of the parameter of leaf water potential

The low leaf water potential under conditions of drought stress is not a reliable indicator of the physiological water status of the plant and plant tissue. A fall in water potential may occur as a consequence of water release and thus turgor loss, but may also result from the decrease of the matrix potential and the production of osmotica. In the former case plants show wilting symptoms, but in the latter turgor can be maintained under drought stress. Plants under these conditions often show high undisturbed turgor, a relatively low leaf water potential due to a very low solute potential and leaf growth is already depressed (see Figure 4.14). Under stress, walls are less extensible ("wall hardening"); as far as is known complex metabolically regulated changes in the cell wall are influenced mainly by osmotic factors (Passioura and Fry 1992).

In plant species which do not show osmotic adjustment stomata may close very quickly, thus restricting water loss and maintaining a water potential even higher than that in species capable of osmotic adjustment. Adjusted species can maintain turgor at lower water potentials than nonadjusted ones and therefore growth of adjusted species may be favoured (page 226).

4.4.3   Physiological processes and parameters affected by drought stress in plants

As mentioned, drought stress can bring about very different physiological effects. In an excellent review on plant response to drought stress Hsiao and Acevedo (1974) has suggested the following very tentative scheme of the development of these effects in tissues with drought stress and indicates the generalized sensitivity of these processes (Figure 4.13).

Plant growth and cell wall properties
The first change suggested is the reduction in shoot and leaf growth brought about by a reduced water potential and occurs even when only small changes in plant water status are involved. In mesophytic crop plants, for example, a water loss of only about 10-15% which corresponds to a lowering of the leaf water potential (Ψ) by around 0.6 MPa can markedly influence different metabolic processes (Hsiao et al. 1976). Acevedo *et al.* (1971) reported that the elongation of young maize leaves was depressed when the water availability of the root medium was slightly depressed from -0.01 MPa to -0.02 MPa accompanied with a large decrease in the corresponding leaf water potentials from -0.28 MPa to -0.7 MPa, respectively (see page 188 and table 4.2).

Growth comprises cell production and expansion. In particular cell expansion which is a turgor-driven process is extremely sensitive to drought stress. Cell expansion is dependent on water uptake and cell wall properties and is described by the following equation (Lokhardt equation according to Boyer 1987).

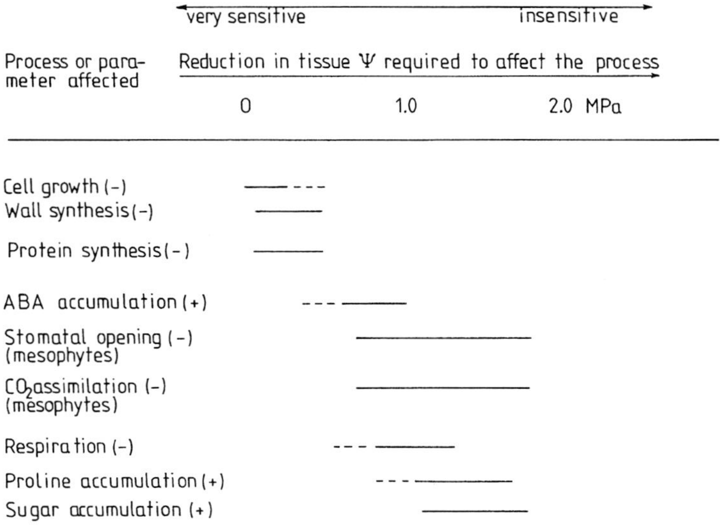

Figure 4.13 General sensitivity of different metabolic processes and parameters to drought stress; (-) reduced, (+) increased (after Hsiao and Acevedo 1974).

$$GR = m \, (\psi_p - Y)$$

GR = growth rate
ψ_p = turgor pressure (typically between 0.3 to 1.0 MPa).
Y = represents the threshold pressure at which wall stretchening ceases.
m = wall extensibility which represents the responsiveness of the cell wall to pressure. It depends on the chemical structure of the wall and the physiological conditions of the apoplastic liquid.

The equation shows that a fall in turgor decreases growth rate. The turgor need not fall to zero to stop leaf expansion; when the turgor has reached the threshold pressure Y growth is already totally impaired. Y is frequently only 0.1 to 0.2 MPa lower than turgor pressure. For this reason a small water loss can be associated with a very rapid growth reduction.

Cell expansion comprises a complex sequence of reactions. Turgor pressure increases due to water uptake and stretches the cell wall acting as a counterforce or tension (= negative pressure) of less than -10 MPa (Cosgrove 1997).

Apart from turgor reduction, as a result of drought stress, the wall extensibility (m) decreases and the threshold pressure (Y) increases. The wall extensibility in leaves is maximal when apoplastic pH is low (at around pH 5, Bogoslavsky and Neumann 1998). It has been suggested that IAA (page 267) is involved in this process (Taiz 1984, Kutschera 1999). When water availability to expanding leaf cells is low H$^+$ pumping ceases and thus apoplastic pH under drought stress is increased (Hartung et al. 1988). Changes of cell wall extensibility can be explained by the acid growth hypothesis.

It has been shown that at acidic apoplastic pH cell expansion is faster than at neutral pH (Rayle and Cleland 1992) and this phenomenon appears to be related to all higher plants (Cosgrove 1996). The low apoplastic pH results from H^+ extrusion by plasmalemma H^+ pumping. Proteins, so-called expansins (McQueen-Mason *et al.* 1992) are probably involved in wall extension at low apoplastic pH catalyzing the "loosening" of H-bonding between the macromolecules of the cell wall (Cosgrove 1997).

With cell wall "loosening" the turgor pressure and, simultaneously the wall tension are reduced with almost no change in wall dimensions. This process is called *stress relaxation* (Cosgrove 1997). Consequently, due to low turgor, the cell water potential becomes more negative, water thus flows into the cell and subsequently there is an expansion of the "loosened" cell wall and an increase of the cell volume. Continuous growth therefore needs simultaneous wall stress relaxation and a reduction of turgor pressure. These processes are followed by water uptake which in turn increases turgor pressure again.

Cell wall synthesis as measured by the incorporation of labelled glucose into wall material is substantially depressed by drought stress by a drop of only a few tenths of an MPa (Cleland 1967). Together with complex structural changes of the wall and unfavourable environmental conditions in the apoplast, the mechanical extensibility of the wall may be reduced ("wall hardening") under drought stress. In contrast to the fast turgor relaxations, cell wall changes during drought ("wall hardening") are relatively slow to be reversed and for this reason these processes are of utmost importance for plant growth. Consequently, the threshold pressure increases and a higher turgor is needed to maintain the same growth rate under drought stress (Figure 4.14). A high turgor in plant tissues is sometimes only attained at night. This results in an enhanced growth rate at night as compared with during the day (Boyer 1968). However, because of changes in wall extensibility and threshold pressure under drought during the day, the growth rate is still lower than that of unstressed plants at the same turgor (Figure 4.14).

There is a correlation between plant hydraulic conductivity and water availability in the soil. If soil water potential is high, the plant hydraulic conductivity is also high and water can move easily into the plant tissue. In case of severe drought stress water loss can be hindered by a low plant hydraulic conductivity. As shown recently hydraulic limitations on drought stress may also limit water movement into the shoot (Lu and Neumann 1999) and this is suggested as a limiting factor for leaf growth. Therefore, parameters such as hydraulic conductivity and those ascribing wall properties are useful indicators of drought stress, in particular when it is moderate and plant growth is inhibited because of a lack of water.

Plant growth is also dependent on the developmental stage. Apart from the meristematic growth of developing leaves, it has been shown that during flowering in the phase of cell expansion and production, drought stress is most sensitive in depressing grain yield, *e.g.* in spring wheat (Christen *et al.* 1995) and in faba bean (Plies-Balzer *et al.* 1995). Drought stress during grain filling had a smaller effect and it is suggested by Caley *et al.* (1990) that the low water potential may inhibit starch synthesizing enzymes. A low water potential during grain growth may induce the synthesis of ABA (Zeevaart and Creelman 1988; Tuberosa *et al.* 1992) and thus premature termination of grain development. Cereal grains remain small under these conditions and starch production is much restricted, but the synthesis of storage proteins is little affected (Mengel *et al.* 1985, Renelt 1993).

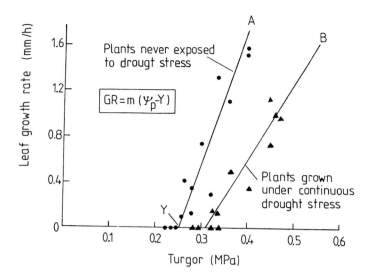

Figure 4.14 Leaf growth rate (GR) as related to leaf turgor (ψ_p) and changing threshold pressure (Y) of well watered plants (treatment A) and plants under continuous drought stress (treatment B). After these treatments plants were rewatered and then plants were continously stressed by water withholding, thus turgor and leaf growth rate decreased. However, at the same turgor leaf growth in plants which had not been pre-stressed (treatment A) was higher than in treatment B (after Matthews *et al.* 1984).

Protein synthesis and enzyme activity

Drought stress and in particular salinity (page 236) have a large impact on protein synthesis. Apart from salt effects, dehydration effects under drought (Frota and Tucker 1978, Teermaat and Munns 1986) may affect protein synthesis as well as the native protein structure, *i.e.* on removal of the hydration shell the protein may dissociate into inactive subunits during drying (Crowe and Crowe 1992). It is known that the incorporation of amino acids into leaf proteins is inhibited even at moderate drought stress. Results of Dhindsa and Cleland (1975) using a double labelling radio technique (^{14}C and ^{3}H) with *Avena* coleoptiles showed that moderate drought stress induces both a qualitative change in the types of protein produced as well as a quantitative reduction in the rate of incorporation of leucine into proteins. Impaired protein synthesis may therefore largely affect leaf growth under drought stress.

Drought stress also may directly influence enzymatic reactions. Enzyme activities associated with photosynthesis and respiration as well as ATPase activity are frequently increased by moderate drought stress in order to maintain increased energy demand (Gale 1975, Ayala *et al.* 1996, Koyro 1997). Increasing stress results in most cases in a decrease in enzyme activity. For example, nitrogenase activity in faba beans is considerably depressed at a soil water potential of around -0.85 MPa (Plies-Balzer *et al.* 1995).

It is tempting to speculate that the protein synthesis is affected. A direct influence on enzyme conformation, *i.e.* the removal of the hydration water only proceeds during extreme dehydratation. Since particular amounts of proline are synthesized in faba bean during drought stress (Plies-Balzer *et al.* 1995) water loss to the point of desiccation should not occur.

An intriguing question is whether enzymes located in the plasma membrane are affected by the turgor (ψ_p) which presses the membrane to the cell wall. Zimmermann (1978) discussing the question in a useful review paper suggests that the pressure results in a change of membrane thickness which in turn could influence the membrane located enzymes. It is thus feasible that the activity of the plasma membrane H^+-ATPase is controlled to some extent by the turgor which also implies that membrane potentials are dependent on turgor. Sucrose uptake by H^+-cotransport (phloem loading) decreases with increasing turgor of sieve tubes (see page 248). Wyse *et al.* (1986) have shown that in sugar beet taproots at low turgor, plasma membrane H^+-ATPase is stimulated and accordingly so too H^+-sucrose cotransport across the plasma membrane. Also sucrose uptake in other storage organs, such as potato tubers or legume seeds has been shown to be sensitive to cell turgor (Oparka and Wright 1988, Patrick and Offler 1995). A similar relationship was found between the turgor and K^+ uptake in *Valonia utricularis* (Zimmermann 1978). K^+ uptake increased as the cell turgor decreased and *vice versa*. It thus appears that turgor pressure has a dual function in growth processes. It is required to stretch the walls and to facilitate the breaking of chemical bonds and in a following phase it controls the uptake of solutes required for growth.

Abscisic acid accumulation

The relationships between drought stress and phytohormones are complex. Some general points are, however, fairly clear. It is well established (Figure 4.13) even under moderate drought stress there is a rapid and dramatic accumulation of abscisic acid (ABA, formula see page 274) (Daie *et al.* 1984, Schulte-Altedorneburg 1990). Accumulation of ABA induces stomatal closure (see next paragraph) and thus inhibits transpiration (Herppich und Peckmann 1997). Abscisic acid is synthesized *via* the terpenoid pathway. The pathway starts within the chloroplast and is probably completed in the cytosol (Sindhu and Walton 1987). The levels of ABA in the leaf can change dramatically over some hours with changing plant water status (Harrison and Walton 1975) and is the result not only of biosynthesis, but also of compartmentation and transport. As is described in the next paragraph, cytosolic ABA increases during drought stress as a result of synthesis in the leaf, distribution in the mesophyll cell and import from the root. On rewatering, ABA is degraded and then exported out of the leaf. Apart from saving water by regulating the stomatal water release, with drought stress ABA increases the water flow in the plant by increasing the hydraulic conductivity. Abscisic acid seems to modify membrane properties (Van Steveninck and Van Steveninck 1983) by promoting ion uptake and osmotical water flow. Abscisic acid is also clearly involved in leaf senescence and according to Zacarias and Reid (1990) ABA induces the senescence process which at a later stage is also influenced by ethylene (Gepstein and Thimann 1981).

Stomatal aperture and photosynthetic activity

It is well known that drought stress inhibits stomatal opening which in contrast to growth reduction is one of the short-term effects of drought stress. Moderate drought stress appears to have little effect on stomatal closure (Hsiao 1973, Gruters *et al.* 1995, Niinemets *et al.* 1999). Soya beans for example, showed no reduction in gaseous exchange, indicative of substantial closure of stomata, until the water potential of the leaves had fallen to as low as -1.0 MPa. The corresponding values for sunflowers and maize were -0.7 MPa (Boyer 1970). These so-called *threshold values* indicate, that due to turgor loss at relatively low water potentials gaseous exchange is inhibited and thus also the diffusion of CO_2 from the atmosphere through the stomata into the leaf tissue (Boyer 1995, Herppich *et al.* 1996, Lorenzini *et al.* 1997).

The mechanisms leading to stomatal closure are different. Direct water loss of the guard cells to the atmosphere (low humidity) results in turgor loss and closure (*hydropassive closure*). Where the whole leaf is affected by drought stress the stomata close by a mechanism called *hydroactive closure*. The mechanism behind this is a reversal of stomatal opening described on page 204, which means that solute loss of the guard cells leads to water release and turgor decrease and thus to stomatal closure. This process is probably triggered by abscisic acid. ABA is synthesized in mesophyll cells and accumulates mainly in the chloroplast. On dehydration of the cell ABA is released from the chloroplasts into the leaf apoplast. Under drought stress the stromal pH of mesophyll cells is depressed and hence ABA becomes protonated and as such is highly membrane permeable. Chloroplastic ABA thus moves into the cytosol and leaf apoplast. In contrast to the stroma, the pH of the leaf apoplast is raised on dehydration perhaps by a reduction in H^+-ATPase activity (Hartung *et al.* 1988) and ABA is present here in a deprotonated state. Less ABA is thus taken up by neighbouring mesophyll cells on its path to the guard cells (Cornish and Zeevart 1985) as in the case of acidic apoplast of the cells of unstressed leaves. ABA synthesis is increased after closure has begun.

Several cellular reactions in guard cells are involved leading to hydroactive closure by ABA. Cytosolic Ca^{2+} is of pivotal importance in signalling these reactions (McAinsh *et al.* 1997) and represents an impressive example for its cellular secondary messenger function (page 529). Abscisic acid probably binds to an uncharacterized receptor which is located in the plasmamembrane of the guard cell (as shown in Figure 4.15) or which is intracellularly located. The binding of ABA (step 1) initiates several reactions. The first response is a transient membrane depolarization due to activation (step 2) of calcium channels (Schroeder and Hagiwara 1990) which leads to an increase of cytosolic Ca^{2+}. Abscisic acid has also been shown to open Cl^- channels (step 3) which further depolarize the membrane (Pei *et al.* 1997). Cytosolic calcium increases also derive from intracellular stores; ABA increases the level of inositol 1,4,5-triphosphate (IP_3), a secondary messenger which releases Ca^{2+} from *e.g.* vacuoles (step 4). Both effects, the transient depolarization and the increase in cytosolic Ca^{2+} then trigger long-term depolarization of the membrane (Ward *et al.* 1995). Elevated cytosolic Ca^{2+} has a marked effect on several ion channels; it activates the opening of further Cl^- channels (step 5) and in addition restricts inward K^+ channels (step 6; Schroeder and Hagiwara 1989) leading to a permanent net efflux of negative charges accompanied by a large and long-term depolarization. Apart from cytosolic Ca^{2+} increase, cytosolic pH rises on ABA application (step 7; Irving *et al.* 1992) and both factors inhibit the plasmamembrane H⁺-ATPase (step 8; Shimazaki et al. 1986) which may contribute to depolarization. The increase in cytosolic pH opens outwardly directed K⁺ channels (step 9; Blatt and Armstrong 1993) and K⁺ diffuses out of the guard cell due to large and permanent membrane depolarization which is primarily caused by Cl⁻ efflux. Because of massive K⁺ loss the turgor of the guard cell falls and is finally too low to keep stomata open, probably even when sucrose is present in large amounts (Figure 4.10).

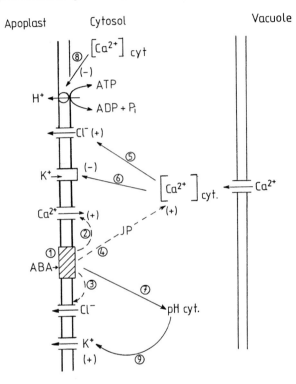

Figure 4.15 Model for cellular responses upon ABA in stomatal guard cells (modified according to Blatt and Thiel 1993). The single steps leading to stomatal closure are described.

It is suggested that on drought stress ABA is synthesized in the leaf and that some ABA may also be produced in the roots. Formation of ABA in the root and probably ABA signalling *via* the transpiration stream in the xylem may occur before the low soil water potential causes any change in the leaf water potential. Interestingly, stomatal reponses are often more closely related to soil water potential than are other leaf responses. Abscisic acid may therefore be regarded as an 'early signal' indicating that the soil is becoming dry and inducing stomatal closure (Davies and Zhang 1991). The xylem sap concentration of ABA in sunflower is between 1 to 15 μmol/m³ and when droughted ABA levels of about 3 mmol/m³ are found (Schurr *et al.* 1992) which are sufficient for stomata closure. A rise in leaf apoplast pH and xylem sap pH means that ABA is in the deprotonated state and hence less ABA enters leaf mesophyll cells and more reaches the guard cell to promote early closure. In this respect the pH increase may also be regarded as an 'early signal' (Wilkinson and Davies 1997).

Photosynthesis is also usually not impaired by moderate drought stress. The results of Boyer (1970) for sunflower plants provide evidence that a fall in the leaf water potential due to turgor loss is more sensitively registered by changes in leaf enlargement than by net photosynthesis (Figure 4.16).

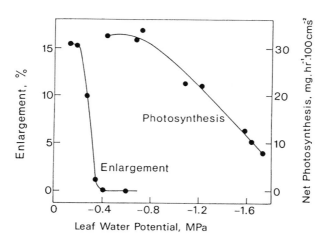

Figure 4.16   Effect of leaf water potential on leaf enlargment and net photosynthesis in sunflowers (after Boyer 1970).

As can be seen from Figure 4.16 leaf enlargement which, as already discussed, is very sensitive to drought stress, dropped rapidly when the water potential was depressed below -0.3 MPa. Net photosynthesis, however, was not lowered appreciably until values under -0.8 MPa were reached. In experiments with *Phaseolus vulgaris* O'Toole *et al.* (1977) found highest CO_2 uptake and transpiration rates at a leaf water potential of -0.3 MPa. At a water potential as low as -0.9 MPa, CO_2 absorption and transpiration were almost inhibited. Brevedan and Hodges (1973) found that maize grown under field conditions with drought stress resulting in a water potential of -1.7 to -2.2 MPa showed an inhibited CO_2 assimilation rate. Different mechanism depressing photosynthesis may play a role. In the first place there may be a simple lack of the substrates: CO_2 and at more severe stress also a lack of water may occur. As a further consequence, in particular at high radiation toxic photoproducts *e.g.* oxygen radicals are produced (see page 143), which may especially damage photosystem II (Asada 1996). In addition, during cell shrinkage induced by drought stress, Mg^{2+} concentrations may increase in the chloroplast stroma as has been shown by Rao *et al.* (1987). Photosynthesis is restricted much earlier with decreasing leaf water potential in the presence of high Mg^{2+} concentrations and it has been suggested that high levels of Mg^{2+} may disturb the electrical potential at the thylakoid membrane and thus also the driving force for ATP production in chloroplasts.

In order to cope with drought and salt stress (page 219) different metabolic adaptations have been observed. When plants subjected to drought or to saline conditions they may switch from C_3 to CAM photosynthesis in succulent species (Schäffer *et al.* 1995). Thus, the water consumption is much lowered (see page 161). This switch is a remarkable adaptation to water deficit and salinity, because substantial metabolic changes are involved including the *de novo* synthesis of enzymes (PEP carboxylase etc.), transport of malate through the tonoplast and changes in stomatal response during the day/night cycle.

Osmotic adjustment

Species capable of osmotic adjustment can maintain turgor at lower leaf water potentials than in nonadjusted species. Thus, osmotic adjustment clearly promotes drought tolerance, because such plants show improved growth when supplied with limited water than do non-adjusted species. Turgor maintenance under drought may sustain cell expansion and stomatal conductance.

The synthesis of solutes starts rapidly when stomatal closure is not sufficient to retain plant water under drought stress (Figure 4.13) and is accompanied by tissue salt accumulation, in particular under salinity (page 237). The concentrations of osmotica may change during the course of the day with changing water supply and transpiration. The very high losses of water from plant tissues which occur in the middle of the day may to some extent be prevented by the production of osmotica since they increase water retention. For different crop species (*e.g.* sunflower, maize, sorghum and soybean) clear diurnal changes of the solute potential have been observed in the range of -0.3 to -0.7 MPa (Morgan 1984). Osmoregulation is of particular importance in younger leaves which need water in order to expand. The low solute potential at midday protects the leaf from excessive water loss by transpiration. Plant species differ considerably in their solute potentials. For mesophytic shade plants solute potentials are in the order of -0.5 MPa. The corresponding values for most crop species are in the range of -1.0 to -2.0 MPa (Slatyer 1963), values which reflect the ability of the various crops to maintain turgor pressure (Hanson and Hitz 1982). Soya bean is not very efficient in maintaining leaf turgor pressure under drought stress. Cowpea is a typical nonadjusting species which under conditions of drought stress conserves water by stomatal closure. Under extreme environmental conditions of drought xerophytes often attain solute potentials between -3.0 to -4.0 MPa and under saline conditions some halophytes may show values less than -10 MPa (Slatyer 1963).

The most important osmotica are K^+, NO_3^-, Na^+ and Cl^- and organic acids, sugars and specific solutes (so-called compatible solutes), which are synthesized under drought stress or salinity in the cytoplasm to adjust the equilibrium water potential in this compartment. It at least means that water is retained in the cytoplasm when ions are partitioned inside the vacuole or in the case of saline conditions also in the apoplast. These more specific solutes include quaternary NH_4^+ compounds such as the betaines (McNeil *et al.* 1999, see Figure 4.17), polyols (Stoop *et al.* 1996) and amino acids such as proline (Delauney and Verma 1993). Apart from in the cytosol large amounts of glycine betaine have also been found in chloroplasts (Mc Neil *et al.* 1999).

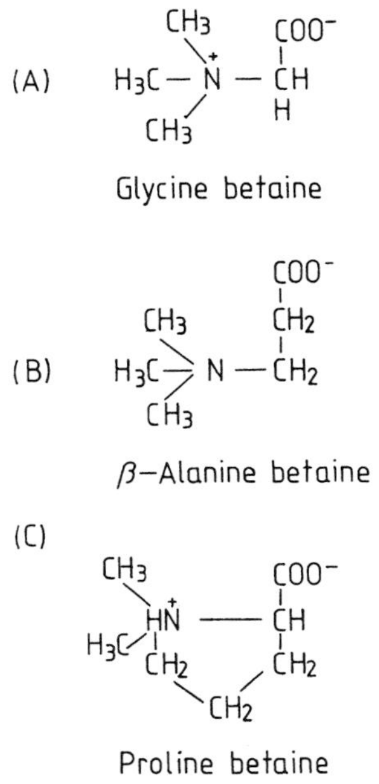

Figure 4.17 Chemical structures of betaines; (A) glycine betaine, (B) ß-alanine betaine, (C) proline betaine. Glycine betaine is widespread, whereas the other betaines are found only in some families.

Crop species such as rice, soybeans and potato lack significant amounts of betaines or related osmotica and therefore their adaptation to drought stress is poor. To improve their tolerance to drought and salinity, it has long been a target of plant breeders to introduce genes for the synthesis of such osmotica into these species. It is possible to do this now, although the betaine levels obtained to date in transgenic plants are very small (Nuccio *et al.* 1998). However, non-adjusted plant species are not always much inferior in productivity to species with osmotic adjustment and fully opened stomata (McCree and Richardson 1987). Stomatal closure to conserve water in non-adjusted plants is often not 100% and since a partial closure of stomata reduces transpiration more than CO_2 uptake (Slatyer and Bierhausen 1964, Gale and Hagan 1966) non-adjusted plant species may still have a fairly high CO_2 assimilation rate at relatively low water loss (McCree and Richardson 1987). As synthesis of osmotic solutes requires much energy (Wyn Jones 1981, Yeo 1983), in adjusted species a substantial proportion of photosynthates and energy is not available for growth and biomass production.

Apart from turgor maintenance, the compatible solutes in the cytoplasm protect cellular structures such as membranes, protein complexes and enzymes from dehydration and high salt concentrations. Dehydration of these structures as well as the presence of small ions, such as Na^+ and Cl^- (salinity, page 233) which interact directly with the protein surfaces can lead to unfolding and denaturation. The compatible solutes are strong water structure formers and are preferentially solubilized in the cell bulk water and they may also attract the smaller inorganic ions still being in the cytoplasm. Thus by these mechanisms the detrimental effect of dehydration and salts on the native, folded protein structures is avoided. During extreme dehydration solutes such as trehalose directly interact with the dry protein in order to stabilize enzyme activity (Carpenter and Crowe 1989). Thus, compatible solutes enable metabolic functions or retain the native structure of cellular macromolecules at dehydration and high cytoplasmic concentrations of inorganic ions (see for details in Somero *et al.* 1992).

It is interesting that plants with a high drought tolerance cannot necessarily cope with high salt concentrations (page 237). Figure 4.18 very clearly shows the effect of NaCl salinity on the concentration of different compatible solutes in leaves of *Sorghum bicolor* (Cavalieri 1983) which is a typical drought resistant species and in leaves of *Spartina townsendii* Grov. (Storey and Wyn Jones 1978) and of *Spartina alterniflora* Loisel (Cavalieri 1983) which both are salt tolerant. In both salt tolerant species glycine betaine and proline are synthesized to different relative proportions in order to cope with the salinity stress. Interestingly under severe stress (600 to 800 mol/m³ NaCl) a total level of around 160 mmol/kg FW was attained in both species. Increasing salt stress above this level which is higher than sea water concentrations (page 237) restrict further adaptation. Too high energy costs and too high N demands for the synthesis of glycine betaine and proline are probably the most important reasons. In contrast, *Sorghum bicolor* is salt sensitive, but will survive at soil water potentials in the range of -2 MPa under drought (Jorden and Sullivan 1982). In the experiment with *Sorghum* application increasing NaCl concentrations up to 180 mol/m³ brought about by a soil

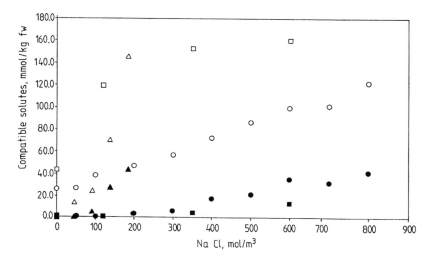

Figure 4.18 Effect of increasing NaCl salinity on the concentration of compatible solutes in the leaves of the drought tolerant, but salt sensitive species *Sorghum bicolor* and of the salt tolerant species *Spartina townsendii* Grov. and *Spartina alterniflora* Loisel. Compatible solutes in *Sorghum*: sucrose (△) and proline (▲); *S. townsendii:* glycine betaine (○) and proline (●); *S. alterniflora:* glycine betaine (□) and proline (■) (Data from Storey and Wyn Jones 1978 and from Cavalieri 1983).

water potential of -0.8 MPa which was associated with a rapid increase of proline and of sucrose in the leaf reaching fairly high levels of 40 and 140 mmol/kg FW, respectively. Thus in *Sorghum* at a relatively low salt stress even higher total levels of compatible solutes have been attained than in the *Spartina* species. Drought tolerant plants are not capable of excluding high amounts of NaCl or sequestering high amounts of salts in the vacuole. Increasing salt stress above 180 mol/m^3 NaCl in *Sorghum* would lead to toxic effects of Na$^+$ and Cl$^-$ (page 228) and the plant would die, although the soil water potential would be above -2 MPa.

Osmotic adaptation to drought stress requires additinal energy as a result of the synthesis of organic solutes, *e.g.* betaines, proline and sucrose as under salt-stress (page 227). Apart from the synthesis of compatible solutes, many salt-tolerant species are able to adjust to salt stress by increased uptake of inorganic ions, particularly Na$^+$ and Cl$^-$ (Koyro and Stelzer 1988). This process of adaptation requires much less energy than the synthesis of organic molecules. According to Wyn Jones (1981) the energy costs for the synthesis of 300 mol/m^3 sugar corresponding to about 20% plant dry matter is around 100 times higher than for the uptake of equivalent NaCl concentrations.

4.5   Practical Aspects of Drought Stress

4.5.1   Water potential values and irrigation

When the water availability in the soil is poor and transpiration is high, a negative water balance results, *i.e.* the loss of water by the plant is greater than its uptake (ψ falls; it becomes more negative). Drought stress inhibits growth even before reduction in turgor and if water loss becomes excessive the plants wilt.

Obviously the most usual way of balancing a water deficit and avoiding drought stress is by irrigation. According to Padurariu et al. (1969) the leaf water potential for maize during the main growing period should not be lower than about -0.6 to -0.7 MPa. The respective value for sugar beet is -0.5 MPa, showing that the latter crop responds more sensitively to drought stress. For most crop species optimum soil moisture conditions are in a range of -0.02 to -0.05 MPa (pF 2.3 to 2.7). Insufficient water supply has a tremendous impact on the growth of most crop species (see Figures 4.14 and 4.16).

The water requirement of crops differs between growth stages. Maize, for example, is particularly sensitive to drought stress at the tasseling stage. Irrigation at this growth stage has a substantial effect on grain yield under conditions where soil water supply is poor. An example of this kind is shown in Table 4.8. Irrigation of 150 mm water in July and August raised grain yields considerably (Buchner and Sturm 1971). Increasing the rate of nitrogen application had a similar effect provided that it was accompanied by irrigation. In the treatment without irrigation, nitrogen application even depressed grain yield.

Sionit *et al.* (1980) reported that wheat was especially sensitive to drought stress during anthesis. At this period of development the plants were less capable of adjusting to low water availability in the soil than at an earlier growth stage. Drought stress at anthesis leads to a reduction in the number of ears and number of grains per ear and is thus detrimental to grain yield.

Table 4.8 Interactions between N application and irrigation and the effect on the grain yield of maize (Buchner and Sturm 1971).

Rate of N appl.	grain yield (Mg/ha)			
kg N/ha	1969		1970	
	without	with irrigation	without	with irrigation
60	5.1	8.8	3.4	6.1
120	3.8	9.7	3.7	8.7
120+60	3.3	9.6	2.3	10.0

4.5.2   Transpiration and crop yield

Crop yield is closely related to utilization of soil water and is dependent on both transpiration and *evaporation*. Transpiration represents the water flow through the plant and released into the atmosphere and evaporation the water directly released from the soil to the atmosphere. Both together are known as *evopotranspiration*. The *transpiration coefficient* is defined as the amount of water in weight or volume unit which is required for the production of a weight unit of plant dry matter *i.e.* g water per g dry matter. For the calculation of water consumption generally the evapotranspiration and not the transpiration is used (Ehlers 1997). Transpiration coefficients quoted in the older literature are in the range of 300 to 900 g water per g plant dry matter. The more recent findings of Roth *et al.* (1988a), however, report much lower values for the European climate conditions. These authors found a transpiration coefficient for winter wheat of 300 and for sugar beet of 200. The transpiration coefficient is lower the better the growth condition and nutrient supply including water. Thus Roth *et al.* (1988b) found that the transpiration coefficient was related to the water holding capacity of soils, decreasing with an increase of the soil capacity to store plant available water. Ehlers (1997) discussing the transpiration coefficient in an interesting review article emphasizes that climatic conditions have a strong impact on the transpiration coefficient. For a wheat crop with a total above-ground biomass production of 15 Mg/ha the transpiration coefficient found in Schleswig-Holstein (humid area of the Northern part of Germany) amounted to 193 and for the arid conditions in Akron (Colorado) to 571. The transpiration coefficient is generally lower for C_4 than for C_3 crop species.

In modern terminology *Water Use Efficiency* (WUE) rather than transpiration coefficient is used. The water use efficiency is the reciprocal of the transpiration coefficient; it is the weight unit of plant dry matter produced per weight unit water consumed. Water use efficiencies are in the range of 1.5 to 6.0 mg dry matter/g water (Tanner 1981). The range of dry matter per g water consumption for C_4 plants and for C_3 plants is about 3-5 mg and around 2-3 mg, respectively. Pineapple, the most important

CAM crop plant (other than agave which is used for tequilla production) is even a higher water saver than C_4 plants and shows a water use efficiency of about 20 mg dry matter/g water (Kramer and Boyer 1995). For potatoes grown with irrigation on a loamy sand under the climatic conditions of Wisconsin Tanner (1981) found WUE values between 3.6 to 5.5 for the whole crop. Interestingly the WUE for the production of potato tubers (tuber dry matter yield per unit water consumed) were also similar, namely 4.3 to 6.3 mg/g water. Schapendonk et al. (1997) related the weight units of CO_2 assimilated per weight unit of H_2O transpired. WUE values thus found for a permanent sward of Lolium perenne differed considerably and ranged between 5 and 30 (mg CO_2 assimilated/g H_2O transpired). These WUE values are higher than those related to the production of dry matter since a substantial amount of the assimilated CO_2 can be respired during crop production. Schapendonk et al. (1997) found a significant increase of the WUE when the CO_2 concentration was doubled and therefore twice as high as in the atmosphere. Interestingly, the transpiration of both treatments, normal CO_2 concentration and doubled CO_2 concentration, did not differ much. All measurements favouring growth generally increase the WUE in contrast to conditions where plant growth is retarded and evapotranspiration is less affected.

Kallsen et al. (1984) in studying grain yield of spring barley in relation to transpiration under the arid conditions of New Mexico found the following relationship:

$$y = -222 + 16.21 \text{ T}$$

where y = grain yield, kg ha^{-1}, T = transpiration

In a crop stand, water losses not only result from transpiration but also from evaporation (direct water loss from the soil by volatilization of water). The relative importance of evaporation may be particularly high if a crop stand is thin, as may be the case when nitrogen is deficient. The relationship between N fertilizer application, grain yield, and water loss is shown in Table 4.9 from the work of Kallsen et al. (1984).

Table 4.9   Relationship between N fertilizer rate, grain yield of barley, evaporation, and transpiration (Kallsen et al. 1984).

N fertilizer rate kg N/ha	Grain yield Mg (t)/ha	Transpiration mm	Evaporation mm
30	1.02	85	235
125	1.65	121	278
225	2.69	217	212

At the highest N fertilizer rate, the unproductive water loss (evaporation) was lowest. In this treatment the production of 1 Mg grain consumed 159 mm water whereas in the treatment with 30 kg N ha^{-1}, 313 mm water were required in the production of

1 Mg grain. Thus N fertilization which is a prerequisite for optimal grain yield helps to save water. This is true although with yield increases, transpiration rises due to increased foliar plant material. This aspect is of utmost interest where water is a growth limiting factor and where irrigation is necessary.

The linear relationship which exists between transpiration and yield is shown in the above equation by Kallsen *et al.* 1984 and is also true for the equation proposed by De Witt (1958):

$$y = M \, T/E$$

where y = yield, T = transpiration, E = evaporation, M = crop factor.

This second equation shows also that an increase in evaporation is associated with a yield decrease (see Table 4.9). The M value is an indicator of efficiency for crop water use. As would be expected C_4 plants have higher M values than C_3 plants since they are more efficient in water utilization. A valuable review paper on the question of crop yield and water consumption has been published by Hanks and Rasmussen (1982) and a further valuable review article was published by stanhill (1986).

4.6 Salinity

4.6.1 General

Soil salinity is an increasingly worldwide problem in crop production which is predominantly induced by irrigation in arid and semiarid regions (irrigation agriculture). About one-third of the irrigated land on earth is affected by salinity. Evapotranspiration removes water from the soil and thus solutes gradually start to concentrate in the soil surface, in particular when salts cannot be flushed out by a drainage system. Saline soils are characterized by high salt levels in the soil profile (see page 62). Different ion species (Na^+, Cl^-, HCO_3^-, PO_4^{3-}, Ca^{2+}, Mg^{2+}, SO_4^{2-} and borate) with changing composition are present often in concentrations affecting crop growth.

To meet the food demands of increasing global population is a severe problem as freshwater resources become limited and irrigation steadily increases in order to produce biomass. Plants can be divided broadly into two groups which differ in growth response to high salt levels. Halophytes grow naturally in saline enviroments (around 300 kmol/m^3 to 600 kmol/m^3 NaCl) and show several adaptation mechanisms (salt tolerance). On the other hand in non-halophytes (glycophytes) which include most crop species salt tolerance is relatively low. Different strategies have therefore to be adopted to combat the increasing problems associated with soil salinity. One strategy is selection and breeding of crop species for salt tolerance (Marschner *et al.* 1981a). Another important issue is the use of halophytes for food production. Apart from their productivity on saline soils, cultivation of halophytes includes the use of saline waste water, natural brackish water or even sea water. Thus, greenification of desert land in coastal areas by use of the unlimited resource of seawater is one provoking practical application (Lieth *et al.* 1997) and, the conversion of salt tolerant plants into crops an important challenge for the future.

Plants growing on saline habitats are often stunted with small and dull bluish green leaves and the shoot growth is usually more reduced than root growth (Greenway and Munns 1980). In general the presence of soluble salts in the nutrient medium can affect plant growth in several ways. In the first place plants may suffer from water stress. In contrast to a drying soil, a saline soil usually contains abundant amounts of water, but of low availability water. Thus in this context the term *water stress* is used. Secondly, high concentrations of specific ions can be toxic and induce physiological disorders (Na^+, chloride borate) and thirdly intracellular ionic imbalances (K^+, Ca^{2+} and Mg^{2+}) can be caused by high salt concentration.

Salt tolerance has been defined in different ways by various authors; in respect to crop production according to Kinzel (1982) plants are salt tolerant until a growth reduction of 50% occurs. Glycophytes are typically not adapted and salt tolerance in most crop species is low. The mechanisms of salt tolerance are different. In terrestrial halophytes salt tolerance is mainly based on compartmentation of Na^+ and Cl^- in the vacuole and synthesis and accumulation of compatible solutes in the cytoplasm (see Figure 4.18) thus maintaining turgor pressure and protecting the cytoplasmic macromolecular functions. Typical examples are the *Gramineae* and the *Chenopodiaceae* with many halophytic tolerant species. Simultaneously, the concentrations of Na^+ and Cl^- are kept at low levels in the cytoplasm. Where high cytoplasmic Na^+ concentrations occur the metabolic functions of K^+ may be replaced (Koyro *et al.* 1997). However, a permanent inclusion as for example in *Beta vulgaris ssp. maritima* throughout the vegetation period may lead to toxic Na^+ and Cl^- effects and to K^+ deficiency (Koyro, unpubl. data). Therefore, typical halophytic adaptations include leaf succulence in order to dilute toxic ion concentrations, ion accumulation in mature leaves prior to young leaves and the development of salt glands (*e.g.* in *Spartina townsendii* Grov.). In grasses, exclusion of toxic ion concentrations may also be important for high salt tolerance (*e.g.* in *Puccinellia peisionis*). Under these conditions internal water deficit has to be avoided by osmotic adjustment.

Wilting symptoms are seldom observed (Bernstein and Hayward 1958) in salt stressed plants; however, the low water potential in saline soils is one important factor for the reduced growth of glycophytes. Most studies have shown that the leaf turgor of salt-sensitive plant genotypes is usually higher than that of salt-tolerant ones (*e.g.* Munns 1993). Salt sensitivity is often accompanied by poor salt exclusion by roots and thus salt concentrations are high in leaves of sensitive genotypes. The growing tissue responds to saline conditions by rapid osmotic adjustment (page 228) and thus maintains the turgor pressure of the plant cells (Matsuda and Riazi 1981).

All these results imply that the turgor does not limit leaf expansion when sensitive species are exposed to saline conditions (Cramer and Bowman 1991). In spite of adequate turgor of plants growing under saline conditions the detrimental effect of soluble salts on plant growth is complex and results from water stress as well as from salt induced physiological disorders. This has been confirmed experimentally in solution culture studies. When the nonpenetrating PEG (polyethylene glycol MW approx. 20 000) is added to a culture solution the water potential is depressed. The same effect is observed on the addition of inorganic salts. It is thus possible to compare plant growth from solutions of the same low water potential with and without the effects of salts. Such experiments have shown very clearly that plant growth is poorer in the presence of high salt concentrations than with an isoosmotic solution of PEG (Chazen and Neumann 1994).

4.6.2   Water and salt stress effects of salinity in crop plants

Since NaCl is often the dominant salt in saline soils, almost all research on salinity has been conducted by use of NaCl. NaCl salinity affects most sensitive plant growth and different metabolic processes, such as CO_2 assimilation, protein synthesis, respiration and often promotes the synthesis of compatible solutes. In the following sections these processes are considered.

Plant Growth
Growth is affected before photosynthesis (Yeo *et al.* 1991). Processes which restrict growth have been subdivided into water and salt stress effects (Munns 1993, Figure 4.19). The water stress effects are rapidly developed and most experiments have shown that they are presumably related to changes in the mechanical properties of growing cell walls (Kramer and Boyer 1995). Under stress walls are less extensible ("wall hardening") (see page 220). It is suggested that the water stress effects are linked to the soil water potential and signals from the root involving ABA may trigger early events in the leaf associated with reduction in leaf growth (Saab *et al.* 1995). However, the short-term response of growth reduction is far from being understood. When salinity is applied, leaf elongation in maize is immediately blocked, but then it returns to a new, but lower steady-state growth rate (Cramer 1992). It is quite possible that there is a rapid turgor reduction which within hours returned to the initial turgor level. Yeo *et al.* 1991 have shown that after only one minute of salt exposure to the root, leaf elongation is inhibited. They speculate that the early growth inhibition may result from hydraulic limitations to water uptake from the root to the shoot which is not related to inhibitory effects on the activity of water channels in the root (Lu and Neumann 1999).

Ion imbalance and toxicity affects plant growth and contributes mainly to the long-term effects. As mentioned, glycophytes which suffer from salinity show a K^+/Na^+ imbalance of ions in the plant tissue, often accompanied with a large excess of Na^+. The capacity of plants to maintain K^+ homeostasis and low Na^+ concentrations in the cytoplasm appears to be one important determinant of plant salt tolerance (Yeo 1998, Matthius and Amtmann 1999, Läuchli 1999). Helal and Mengel (1979) have shown that NaCl salinity impaired leaf growth and protein synthesis (Table 4.10) which interestingly was improved by K^+ supply to the nutrient solution. It is therefore hypothesized that homeostatic K^+ levels (a high K^+/Na^+ ratio) in the cytoplasm favouring protein synthesis is of utmost importance in the process of leaf expansion.

According to Munns (1993) growth of rapidly expanding leaves is first depressed by water stress and genotypes differing in salt tolerance do not show growth differences in this first phase. In the second phase, plant growth is affected by ion imbalances and toxicity and probably leads to the long-term growth differences between the salt-tolerant and salt-sensitive crop species (Figure 4.19). For example in maize, Schubert and Läuchli (1986) did not find a positive correlation between salt tolerance and Na^+ exclusion in a short-term experiment of 17 days. However, experiments of Cramer

(1992) comparing a Na includer maize variety with an excluder are not in accord with these observations and salt-specific effects may have overlapped during this first phase.

Table 4.10   Effect of NaCl salinity and K$^+$ supply on leaf and root dry matter and ^{15}N labelling in leaf proteins of young barley plants (Helal and Mengel 1979). Shoot growth was clearly reduced, while root growth was only marginally affected. Salinity was applied for 20 days and then followed for a 24 hour period ^{15}N labelling.

Treatment	roots	leaves	leaf proteins
	mg DM/plant		(^{15}N, % of total labelled N)
Control	126	371	43.9
80 mol/m³ NaCl	122	286	28.7
80 mol/m³ NaCl + 5 mol/m³ KCl	127	324	39.9
80 mol/m³ NaCl + 10 mol/m³ KCl	124	323	49.0

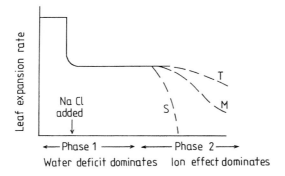

Figure 4.19   Response of leaf expansion to water deficit (phase 1) and to ion toxicity (phase 2). The response of three genotypes for a given species that differ in salt tolerance is shown (S, sensitive; M, moderately tolerant; T, tolerant). Adopted from Munns (1993).

In contrast to the water stress effects, occurring in the meristematic region of younger leaves, salt toxicity predominantly occurs in mature leaves. Under severe salinity stress, the cytoplasm as well as the apoplast can be overloaded with Na$^+$, as is the case when the vacuole is filled up and leaves continue to transpire. Sodium ion accumulation in the leaf apoplast may at least lead to tissue dehydratation. Excess of Na$^+$ over K$^+$ may result in K$^+$ deficiency. High Na$^+$ levels may affect enzymes and organelles present in the cytoplasm as well as enzymes in the apoplast. Some enzymes are more sensitive to salt than others, but when toxic cytoplasmic levels are built up over a range of about 24 hours all metabolic functions are impaired by Na$^+$ (Munns 1993). Sodium ions may interfere with cationic sites involved in binding of K$^+$, Ca^{2+} and Mg^{2+}. In pariticular, Na$^+$ toxicity might be related to the large interaction with weak Mg^{2+} binding sites inhibiting enzme activity which are counteracted by K^+ supply. This has been shown for nucleotidases and ribonucleases. Chloride is also toxic (see Figure 4.20), but only little is known about the physiological targets; it is suggested that it interferes with anionic binding sites, *e.g.* of RNA (Serrano *et al.* 1999).

Salinity stress and Ca^{2+} functioning are related and probably several phenomena are involved. For example, Cramer *et al.* (1985) have emphasized that under saline conditions Na^+ can displace Ca^{2+} from the plasmamembrane associated with a change of membrane permeability and a leakage of K^+. It has long been known that application of Ca^{2+} alleviates salinity stress symptoms and is the prerequisite for a high K^+/Na^+ selectivity (Hajibagheri *et al.* 1987). It has also long been known that with increased external Ca^{2+} the efflux of K^+ (so-called Viets-effect) is inhibited (Bouteau *et al.* 1996) and recently clear physiological basis has been provided which shows that Na^+ flow through non-selective channels is strongly impaired by application of external Ca^{2+} (Roberts and Tester 1997b, Davenport *et al.* 1997) and thus a high cytoplasmic K^+/Na^+ ratio is favoured. Apart from such direct effects, the intracellular Ca^{2+} homeostasis is of pivotal importance in signalling salt stress (Lynch and Läuchli 1988). As a secondary messenger, cytosolic Ca^{2+} may trigger (page 528) a large number of biochemical reactions including transport activities of K^+ and Na^+ fluxes. Cytosolic Ca^{2+} in plants is increased in response to salt stress and by tuning a phosphorylation process the activity of the K^+ and Na^+ transport systems may be influenced (Knight *et al.* 1997). The regulation might occur *via* transcription factors and/or directly by altering the transport activities. Several genes have recently been isolated and shown to be responsible for a higher salt tolerance in respect to K^+/Na^+ homeostasis (*e.g.* Dubcovsky *et al.* 1996).

Protein synthesis, photosynthesis and respiration
Protein synthesis is largely impaired by drought (page 221) and in particular by salinity. Dehydration effects by salts, imbalances of the cytoplasmic K^+/Na^+ ratio and Cl^- toxicity usually leads to a larger reduction of protein concentration than under drought (Frota and Tucker 1978, Thiyagarajah *et al.* 1996, Sanchez-Blanco *et al.* 1998). For example, an unfavourable K^+/Na^+ ratio probably affects the binding of tRNA to ribosomes (Wyn Jones *et al.* 1979, Gibson *et al.* 1984, Wyn Jones 1999).

As already mentioned, plants suffering from salinity often show a high ATPase activity and high respiration rates to keep the increased energy demand (Ayala *et al.* 1996, Glenn *et al.* 1999). Under severe conditions when photosynthesis is affected, probably because leaves are small, storage carbohydrates may be used to a greater extent than in plants grown under non saline conditions. Thus, plants suffering from salinity are poor in their energy status. This relationship between energy supply and salinity has been demonstrated by Chimiklis and Karlander (1973) for *Chlorella* and by Helal and Mengel (1981) for *Vicia faba*. In both cases it was shown that the toxic effect of NaCl salinity was less severe when the plants were grown under high light intensity as compared to low light intensity. Under high light intensity the plants were able to maintain balanced cation concentrations in the plant organs in contrast to low light intensity, where an excess of Na^+ and low K^+ concentrations were found. This imbalanced ionic status was associated with impaired protein synthesis (Table 4.10), but also a lack of energy may have affected this process (Helal and Mengel 1981).

Marked differences in salt tolerance occur between plant species. Figure 4.20 from a useful review paper of Greenway and Munns (1980) shows the differential growth response to increasing Cl⁻ concentrations denoted on the x-axis. Line 1 in the figure shows the growth response of the halophyte *Suaeda maritima* in relation to the Cl⁻ concentration. It is obvious that maximum growth was obtained at concentrations as high as 150 to 300 kmol/m³ Cl⁻, clearly higher than the lethal level for sensitive species (line 4). Line 2 shows the growth response of sugar beet, a species which is related to the halophytes and thus highly tolerant, but growth is retarded. Line 3 represents moderately salt tolerant species, like cotton and barley which do not grow at higher salt concentrations (around 250 kmol/m³). Line 4 shows salt sensitive plants such as maize, beans, soybeans, lettuce, many fruit trees etc. (see Table 4.13) which are totally inhibited in growth by low salt concentrations (around 80 kmol/m³).

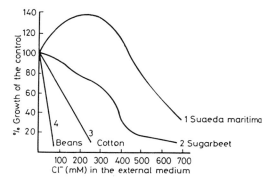

Figure 4.20   Growth response of species with different salt tolerance to increasing Cl⁻ concentrations in the external mediun (modified after Greenway and Munns 1980).

As mentioned, halophytes are able to cope with coastal seawater salinity concentrations, due to their capability of taking up large amounts of ions and sequestering them in vacuoles to act as omotica. Halophytes even need an excess of salts for maximum growth and for attaining solute potentials around -2.0 to -5.0 MPa (Flowers *et al.* 1977). In these plant species, so-called salt includers, the ions taken up are effective in osmotic adjustment and concentrations of up to about 250 mol/m³ NaCl stimulate growth (Greenway and Munns 1980). Other halophytes, salt excluders, adapt to saline conditions by ion exclusion so that osmotically active solutes have to be synthesized within the plant to meet turgor pressure demands.

Non halophytic, but salt-tolerant plants species possess the same mechanisms but developed to a lesser extent. The growth of sugar beet is enhanced by NaCl salinity and the concentrations of Na⁺ and Cl⁻ increase in the shoot with increasing external supply (Lessani and Marschner 1978). For soyabean, salt tolerance appears to depend on the restriction of Cl⁻ transport from root to shoot. Growing soyabean cultivars on a saline soil, the Cl⁻ concentration of the leaf dry matter of the salt-sensitive *Jackson* was about 9 mg/g whereas it was only about 5 mg/g in the salt-tolerant *Lee* (Läuchli and Wieneke 1979).

For barley, effective compartmentation of Cl⁻ in the leaf appears to be important in relation to salt tolerance (Huang and van Steveninck 1989). These workers found less Cl⁻ in the mesophyll than the leaf epidermis which they interpreted as a protective mechanism for photosynthesis.

Salt-tolerant-species of the *Chenopodiaceae* are well known to accumulate large amounts of Na^+ in the shoot associated with a low K^+/Na^+ ratio which is around 2.2 (Haneklaus *et al.* 1998). In salt-tolerant sugar beet and barley increased Na^+/H^+ antiporter activity has been found in the tonoplast (Blumwald and Poole 1987, Garbarino and Dupont 1989) which is increased under NaCl stress and thus in these species Na^+ deposition in the vacuole driven by the tonoplast H^+ gradient is of major importance. Therefore, it is rather the ability to keep cytoplasmic Na^+ at a low level and cytoplasmic K^+ at a high level than the overall salt concentration of leaves which plays a role in salt tolerance. This is also evident from the work of Lessani and Marschner (1978) who reported similar high concentrations of Na^+ and K^+ in the salt sensitive pepper plant and the salt tolerant sugar beet.

High salt tolerance in crop species is often related to the exclusion of toxic ions (Greenway and Munns 1980). A striking example of salt tolerance of closely related plant species was reported by Rush and Epstein (1976) for tomato. The salt tolerant ecotype *Lycopersicon cheesmanii* was able to survive in a full strength sea water nutrient solution whereas *Lycopersicon esculentum* was not able to withstand more than 50% of sea water. Many legume species are sensitive to high accumulations of Na^+. Tolerance to high salt concentrations in a number of species is related to Na^+ exclusion from the shoot. Difference in ability to restrict Na^+ from the shoot occurs both between species and between cultivars of the same species. This difference is pronounced in maize cultivars as shown in Table 4.11 which gives data comparing the effect of NaCl salinity on the cultivar *Pioneer*, a strong Na^+ excluder, and the cultivar *Across* in which Na^+ transport to the shoot is much less restricted than in *Pioneer*. The same effect was observed in plants treated with Na_2SO_4 indicating that Na^+ is the most toxic ion to maize. These findings are supported by a study made by Hajibagheri *et al.* (1987) which showed that for 26 maize cultivars subjected to salt stress, the shoot Na^+ concentration and degree of toxicity were positively correlated. It is supposed that exclusion of excess Na^+ from the developing leaf protects the cytoplasm from too high Na^+ concentrations. Interestingly the K^+/Na^+ ratio in the shoots of grasses is generally much higher than in dicots (Kinzel 1982).

Sodium can be excluded from the shoot by processes occurring in the rhizodermal cells of the root endodermis and in the xylem parenchyma cells. Several controlling mechanisms may thus be involved. As yet there are no indications of a Na^+ efflux pump in the rhizodermal cells. A Na^+/H^+ antiporter activity in the plasmamembrane has been found in the halophyte *Atriplex* (Hassidim *et al.* 1990) and in the salt-tolerant ecotype *Lycopersicon cheesmannii* (Wilson and Shannon 1995). Also in other crop plants, *e.g.* in carrot (Reuveni *et al.* 1987), in wheat (Allen *et al.* 1995) and in maize (Fortmeyer 2000) a Na^+/H^+-antiport has been found (Table 4.11). Effective Na^+ exclusion in maize may also be related to low passive Na^+ permeability of the root cell membranes (Gorham *et al.* 1985). In contrast to highly selective transport systems for K^+ (see page 489), Na^+ appears to move through less selective channels. In particular, non-selective channels may allow massive Na^+ influx at high external Na^+ concentrations in saline environments. For further detailed information about the channel and transporter systems involved in Na^+/K^+ influx the reader is directed to the valuable review paper of Maathuis and Amtmann (1999).

Table 4.11 Long-term effect of NaCl salinity on shoot concentrations of K^+, Na^+ and Cl^- of the two maize cultivars Pioneer 3906 and Across 8023, 52 d after the beginning of the treatment. Significant differences between cultivars. (Data from Fortmeier and Schubert 1995).

Treatment	Na^+	K^+	Cl^-
Pioneer 3906			
Control	—	1.21	0.36
100 mol/m³ NaCl	0.44	1.00	0.91
Across 8023			
Control	—	1.90**	0.56**
100 mol/m³ NaCl	1.73*	0.73	1.53

In principle, the Casparian strip of the endodermis is an effective barrier against passive solute movement into the central cylinder. A high K^+/Na^+ ratio in the leaf is achieved by low xylem transport of both ion species and high K^+ transport through the phloem of mature leaves (Wolf *et al.* 1991). In addition, a high rate of Na^+ retranslocation in the phloem from the leaf to the root decreases sodium import into the shoot (page 213). Moreover, effective back transport of Na^+ from the xylem which may be driven by K^+ import into the xylem may play a significant role for the control of Na^+ translocation to the leaves (Läuchli 1975, Lacan and Durand 1996). Interesting structural adaptations in the halophyte of *Aster tripolium* may keep Na^+ shoot concentrations at a low level. Plate 4.3 shows transfer cells along the phloem which with their invaginations increase the adjacent cell surface and thus also the lateral movement of Na^+ into the adjacent phloem parenchyma cells (dark cells in Plate 4.3).

In such Na^+ and Cl^- excluding species, however, a lack of solutes may result in adverse effects on water balance, so that water deficiency rather than salt toxicity may be the growth limiting factor (Greenway and Munns 1980). To achieve a low root water potential which is the prerequisite for water uptake in saline environments with a relatively low soil water potential, the root solute potential in these species is decreased by the synthesis of organic solutes and by highly selective K^+, Mg^{2+} and Ca^{2+} uptake. Thus, on one hand the synthesis of osmotically active solutes is necessary for the water potential equilibrium and cell growth, on the other hand, however, the synthesis of organic solutes is energy demanding and the formation of these solutes decreases the energy status of the plant (page 174). Thus, for plant survival, growth depression is a necessary compromise in NaCl excluding crop species because of increased energy consumption.

In summary, a low cytoplasmic Na^+ concentration in salt tolerant species may be attained in some of these species by secreting excess of Na^+ into the vacuole *via* H^+/Na^+ antiport. In the majority of glycophytes exclusion of Na^+ from the cytoplasm results from a low Na^+ permeability of the plasma membrane and/or by H^+/Na^+ antiport.

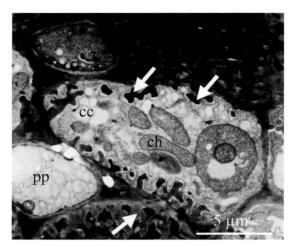

Plate 4.3 Transfer cells along the phloem strand in *Aster tripolium* L. (Photo Koyro). The arrows point to the characteristic invaginations of companion cells (cc) but are also visible in the adjacent parenchyma cells (dark cells). cc = companion cell, ch = chloroplast, pp = protophloem.

Overexpression of a Na^+/H^+ antiporter in yeast (Jia *et al.* 1992) and in *Arabidopsis* (Aspe *et al.* 1999) has resulted in a greater Na^+ tolerance; thus it is feasible to argue that transgenic plants with overexpression of Na^+/H^+ antiporters in the root epidermis may be better able to exclude Na^+. Also, lowering plasmamembrane permeability by lowering the number of non-selective channels may be crucial in salt tolerance.

4.6.4   Crop production

In crop production studies soil salinity is measured as the electrical conductivity (EC) of a water saturated soil extract. The conductivity, expressed in deci Siemens / m, is higher the more concentrated the ionic composition of the extract. Table 4.12 gives a survey showing the degree of salinity in terms of electrical conductivity (EC) in relation to crop species (Bernstein 1970). Considerable differences occur between crop species and cultivars in relation to salt tolerance.

Table 4.12   Crop response to salinity (from Bernstein 1970)

Salinity (EC dS/m at 25°C)	Crop responses
0 to 2	Salinity effects mostly negligible
2 to 4	Yields of very sensitive crops may be restricted
4 to 8	Yields of many crops restricted
8 to 16	Only tolerant crops yield satisfactorily
above 16	Only a few very tolerant crops yield satisfactorily

Generally, most fruit crops are more sensitive to salinity than are field, forage or vegetable crops. Table 4.13 shows the response of various field crops to salinity decreasing from the top of the table downwards (Bernstein 1970). The detrimental effects of salinity are also often dependent on the stage of plant growth. For many species germination is very sensitive to salinity (Okusanya and Ungar 1983, Chartzoulakis and Loupassaki 1997). Growth and development of young leaves are often favoured over mature leaves which can accumulate high concentrations of toxic ions leading to premature death in order to guarantee the development of young leaves (Chartzoulakis and Loupassaki 1997, Fung et al. 1998). In most cereal crops, grain yields are less affected than straw yields. In rice, tillering and especially ear development and grain production are affected by salinity (Lehman et al. 1984). Salinity may also affect crop quality. In sugar beet, for example, high levels of sugar may result in the storage tissue (Saftner and Wyse 1980, Willenbrink 1987, Barkla et al. 1990). In tomato the concentration of sugars and titratable acids in the fruits is increased (Lapushner et al. 1986, Auerswald et al. 1999); however, plant and fruit growth is decreased.

Table 4.13   Salt tolerance of various field crops indicated as conductivity at which the yield is reduced by 25% (data of Bernstein 1970)

Crop plant	EC	Crop plant	EC
Barley	15.8	Rice (paddy)	6.2
Sugar beet	13.0	Maize	6.2
Cotton	12.0	Sesbania	5.8
Safflower	11.3	Broadbean (Vicia)	5.0
Wheat	10.0	Flax	4.8
Sorghum	9.0	Beans (Phaseolus)	2.5
Soybean	7.2		

Epstein and co-workers have suggested that there is a very great need to breed economic crops which are salt tolerant. Rush and Epstein (1976) argue that by generating strains of crops capable of coping with salinity, that, 'what is now a problem could become a vast opportunity for crop production by tapping the immense wealth of water and mineral nutrients of the oceans without the energy-costly process of industrial desalination'. In view of the enormous expenses of saline soils and the necessary increase in crop production to meet the world's expanding population breeding of salt tolerant crops is of utmost importance. However, the large amount of research, carried out mainly in the last decade into the molecular biology of salt tolerance has had disappointingly little success. This does not mean that it is impossible to create transgenic salt-tolerant crops, but it appears to be much more difficult than was believed 25 years ago. The reason for this is the catalogue of genes and attendant controls that have to be introduced. Moreover, the situation is complex since saline soils may considerably differ in their properties. Various plant species have developed tolerance strategies adapted to local conditions which may thus not be of general importance.

Much success in this respect could probably be achieved by unravelling the mechanisms for salt tolerance in *Spartina* (Wu and Seliskar 1998) a plant which grows in upper salt marshes and tolerates coastal seawater. *Spartina* is a member of the *Gramineae* so that the research findings could be used in breeding programmes in maize, wheat, barley and rice. The same interesting approach would be to investigate the salinity tolerance mechanisms of *Beta maritima* (Koyro and Huchzermeyer 1999) which is closely related to the sugar beet. A further possible strategy might be the introduction of gene coding for high yield properties into these halophytes so that crops with high yields and high salt tolerance are created.

Table 4.14 Effect of K^+ fertilization on leaf area (cm^2) of 2 year old Satsuma mandarins grafted onto 3 year old Poncirus trifoliata rootstocks. The differences in leaf area were significant at the 1% level (from Anac *et al.* 1997). Figures in brackets are relative values for each treatment, 100 signifies the maximum leaf area at lowest salinity for each treatment.

Salinity (dS/m)	K (g/tree)		
	0	70	156
0.65	25.95 (100)	27.60 (100)	33.61 (100)
2.00	19.95 (77)	25.17 (91)	29.35 (87)
3.50	15.63 (60)	22.14 (80)	24.95 (72)
5.00	12.65 (50)	24.78 (90)	23.17 (69)
6.50	10.19 (39)	16.86 (61)	21.57 (64)

Regardless of whether an unfavourably low K^+/Na^+ ratio in the cytoplasm is linked to Ca^{2+} deficiency under saline conditions (Cramer *et al.* 1985) or to less developed mechanisms to lower cytoplasmic Na^+, the simple supply of K^+ may alleviate salt stress effects on shoot growth by increasing the cytoplasmic K^+/Na^+ ratio. This could mean that K^+ fertilization on crops grown on saline soils may be a simple as well as an efficient tool for farmers to improve their crop yields, in particular in the arid and semiarid regions. Table 4.14 shows such an example for mandarin trees which clearly indicates that K^+ fertilization significantly increased leaf expansion (Anac *et al.* 1997). From the relative data shown in brackets it is evident that in the K^+ fertilized treatments growth depression by salinity was less than in the non K^+ fertilized treatments. As yet relatively few such field experiments have been conducted; however, this appears to be a promising approach for crop improvement in saline environments.

CHAPTER 5

PLANT GROWTH AND CROP PRODUCTION

5.1   Physiological Source and Sink Relationships

5.1.1   General

Higher plants have specialized organs characterized by specific metabolic processes. For growth and crop production one may distinguish between organs in which primary organic molecules are produced and organs in which organic matter is stored or consumed. In Crop Physiology the former are called *physiological sources* and the latter *physiological sinks*. Most important sources are green photosynthetic plant tissues in which chloroplasts import inorganic constituents to synthesize primary organic molecules such as sugars and amino acids (see page 149 and 174). Physiological sinks are meristematic tissues which use the primary molecules for growth and storage tissues such as fruits, seeds, stems, tubers and roots. The flow from the source to the sink is almost exclusively mediated by phloem transport. Partitioning of these primary molecules, the photosynthate, and their long distance transport from source to sink are essential processes for growth and crop production which is considered in the following sections.

5.1.2   Assimilate production and cellular partitoning in the physiological source

The typical physiological source of the higher plant is the expanding leaf showing maximal photosynthetic rates at around 70% of its final size (Turgeon and Webb 1975) and is characterized by the production and export of organic molecules. Young leaves which have not expanded act as sink organs but are transformed to source organs as the leaf ages. This sink-source transition starts at the leaf tip and progresses towards the base being usually complete when the leaf is around 50% expanded. A considerable proportion of triose phosphates synthesized and released by the chloroplasts in the source tissue is used for the synthesis of organic compounds. Triose phosphates are thus precursors of sucrose which is synthesized in the cytosol. In most plant species this is quantitatively the most important organic constituent which participates in phloem transport. In addition some organic carbon is transported in the form of amino acids (see Table 4.5). The chloroplast may thus be considered as the source of organic material. This is shown in Figure 5.1

Transitory starch is synthesized during the daytime and deposited inside the chloroplasts. In some species, like grasses fructans and sucrose rather than starch act as the storage product. Starch is synthesized from the triose phosphates which are generated by the Calvin cycle. The chloroplast imports inorganic phosphate, CO_2, NH_4^+, NO_2^- and SO_4^{2-} and exports triose phosphate and amino acids. The transport of both inorganic and organic phosphates across the membrane is brought about by the so-called phosphate translocator, which couples the uptake of inorganic phosphate with the export of triose phosphates and 3-phosphoglycerate (Heldt *et al.* 1977). The phosphate translocator acts as a strict antiporter which under physiological conditions translocates the import of inorganic and export of organic phosphates in a 1:1 stoichiometry (Flügge 1999). These photosynthates exported into the cytosol are used for the synthesis of sucrose and the released phosphate is imported into the chloroplasts to produce ATP. For a comprehensive account of the relevant pathways for starch synthesis in the chloroplast and sucrose synthesis in the cytosol the reader is directed to relevant reviews (Preiss 1982, 1991, Beck and Ziegler 1989, Beck 1993). In this context only the major factors controlling the assimilate partitioning between starch in the chloroplast and sucrose in the cytoplasm are considered.

One key factor regulating the partitioning are the relative concentrations of inorganic phosphate and triose phosphate. A low concentration of inorganic phosphate in the cytosol lowers the export of triose phosphate and thus promotes starch synthesis in the chloroplast. On the other hand a high concentration of inorganic phosphate in the cytosol inhibits starch synthesis and favours export of triose phosphate which is then synthesized to sucrose in the cytosol. Several enzymes are regulated by inorganic phosphate and triose phosphate. The chloroplastic enzyme ADP-glucose pyrophosphorylase is the key enzyme that regulates starch synthesis from glucose 1-phosphate and is inhibited by inorganic phosphate (Pi) and stimulated by 3-phosphoglycerate (3-PGA) (Preiss and Levi 1979). A high 3-PGA/Pi ratio is present in the light and the enzyme is 'switched on'. According to Heldt *et al.* (1977) highest rates of starch synthesis are obtained with a 3-PGA/Pi ratio of about 1.7. In the dark the ratio is low and thus starch synthesis is impaired.

$$\text{Glucose 1-phosphate} + \text{ATP} = \text{ADP glucose} + \text{PPi}$$
$$\text{(ADP glucose pyrophosphorylase reaction)}$$

Starch degradation in the chloroplast is probably associated with an increase in the concentrations of hexosephosphates, DHAP, and 3-PGA. Triose phosphates thus formed can be released into the cytosol. Starch degradation in the chloroplast is probably brought about by a phosphorylase according to the reaction

$$(\alpha\text{-Glucan})_n + \text{Pi} = (\alpha\text{-Glucan})_{n-1} + \text{glucose 1-phosphate}$$
$$\text{(Phosphorylase reaction)}$$

The equilibrium of this reaction depends on the level of Pi and glucose 1-phosphate, high levels of Pi favouring starch degradation. Also starch can be degraded by the action of amylase and probably, glucose rather than triose phosphate is exported from the chloroplast during night (Trethewey and ap Rees 1994). Control of sucrose synthesis is thus presumably different from that during the day.

Fructose-2,6-bisphosphate exerts central regulatory control of sucrose synthesis in the cytosol (Stitt 1990). By inhibiting cytosolic fructose -1,6-bisphosphate phosphatase, sucrose synthesis is decreased during the night. In the dark a low ratio of 3-PGA/Pi promotes the formation of fructose-2,6-bisphosphate and thus sucrose synthesis is decreased. Reciprocal conditions prevail in the light. This is an important regulatory step since it regulates the flow of organic carbon either into sucrose formation or organic carbon to feed glycolysis after cleavage of fructose-1,6 bisphosphate into triose phosphate. Thus glycolysis is favoured in the dark which is important since glycolysis is the sole metabolic source for providing energy during the night.

245

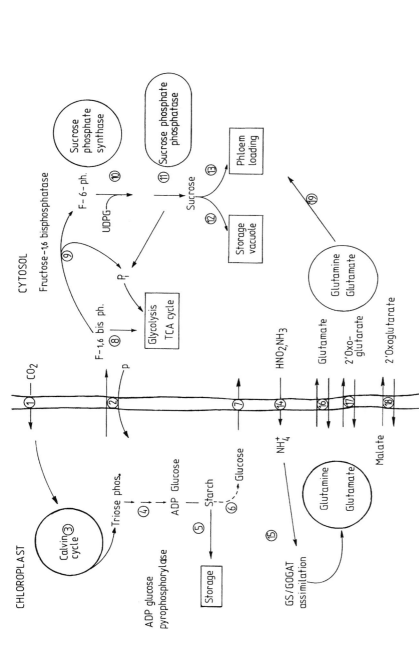

Figure 5.1 Assimilate partitioning between chloroplast and cytosol in the physiological source. 1) CO_2 difffusion across the chloroplast envelope. 2) Triose phosphate/phosphate translocator. 3) Calvin cycle. 4) ADP glucose pyrophosphorylase reaction. 5) Transient starch storage in the chloroplast stroma. 6) Starch breakdown. 7) Glucose release into the cytosol during night. 8) Carbohydrate channeling into glycolysis and TCA cycle. 9) Fructose-1,6 bisphosphatase. 10) Sucrose phosphate synthase. 11) Sucrose phosphate phosphatase reaction. 12) Transient sucrose storage in the vacuole. 13) Phloem loading of sucrose. 14) Diffusion of HNO_2 and NH_3 across the chloroplast envelope. 15) GS/GOGAT reaction producing glutamine and glutamate. 16) glutamine/glutamate translocator. 17) Glutamate/2-oxoglutarate translocator. 18) Malate/2-oxoglutarate translocator. 19) Phloem loading of amino acids.

$$\text{Fructose-1,6-bisphosphate} + H_2O = \text{Fructose 6-phosphate} + Pi$$
(Fructose 1,6-bisphosphate phosphatase reaction)

A further important enzyme for sucrose formation is sucrose phosphate synthase. In this enzymic reaction UDP-glucose and fructose 6-phosphate combine to yield UDP (uridine diphosphate) and sucrose 6-phosphate which in a final irreversible step is dephosphorylated and sucrose is synthesized. This latter reaction is catalyzed by sucrose phosphate phosphatase. Sucrose phosphate synthase is inhibited by inorganic phosphate and stimulated by glucose 6-phosphate and thus sucrose synthesis is also depressed in the dark. The activity of the sucrose phosphate phosphatase is stimulated in the light, possibly by increasing the *de novo* protein synthesis of this enzyme (Huber *et al.* 1994).

$$\text{UDP glucose} + \text{fructose 6-phosphate} = \text{UDP} + \text{sucrose 6-phosphate}$$
(Sucrose phosphate synthase reaction)

$$\text{Sucrose 6-phosphate} + H_2O = \text{sucrose} + Pi$$
(Sucrose phosphate phosphatase reaction)

Such cellular control between the chloroplast and the cytosol guarantees a continuous provision of organic carbon in the form of sucrose for the growing plant. Sucrose synthesis is promoted during the day as is the case for starch formation in the chloroplasts, in particular under optimal photosynthetic conditions. When sucrose supply is higher than plant demand, triose phosphate accumulates in the cytosol which is then used for starch synthesis. This kind of regulation is important for permanent assimilate supply and provides protection against major disturbance of the Calvin cycle. Only when triose phosphate is provided for metabolic processes, is inorganic phosphate released which is essential for the functioning of the Calvin cycle and thus also for photosynthesis. Apart from maintenance of CO_2 fixation, simultaneously in most species assimilates are stored in the form of transitory starch which is then degraded in the following night. Also, some sucrose is stored in the mesophyll vacuole during the day which is also released at night for phloem transport (Kaiser and Heber 1984). The export of organic carbon at night is of particular importance for meristematic tissues which frequently show higher growth rates at this time for reasons of a better water balance (page 203). On days with low temperature and low light intensity the rate of starch synthesis is low which is of relevance for crops harvested late in autumn such as vine and sugar beet. Sunny days in autumn are well known to improve the quality of these crop species. Thus chloroplastic starch as well as vacuolar sucrose in the leaf mesophyll have a buffer function which can be fully utilized on days of optimal photosynthesis.

As well as sucrose amino acids are also produced in the leaf mesophyll and are translocated by phloem transport to physiological sinks (see Table 4.5). Sucrose is important for the storage of carbohydrates, whereas amino acids are essential for protein synthesis in various physiological sinks such as seeds of leguminous species and cereal grains. As the result of nitrate reduction in the cytosol coupled to the assimilation of NH_3 via the GS/GOGAT cycle in the chloroplasts (Figure 5.1) followed by transamination the whole spectrum of amino acids including cysteine and methionine are synthesized in the cytosol of the mesophyll cell. Glutamate in exchange for α-oxoglutarate and glutamine in exchange for glutamate are channelled from the chloroplast into the cytosol. Together with carbon skeletons mainly in the form of 3-phosphoglycerate which derives from the Calvin Cycle the various amino acids are synthesized in the cytosol.

In contrast to sucrose and amino acids produced in the leaf mesophyll and translocated *via* the phloem, the synthesis of fatty acids occurs in the plastids of each cell, in the chloroplasts as well as in the chromoplasts and leucoplasts. Fatty acids are thus not be translocated in the phloem and lipids are synthesized in all cells. In the physiological sink the synthesis of lipids depends on the incoming carbon skeletons, *e.g.* from carbohydrates supplied by phloem transport. The composition of the phloem sap is shown in Table 4.5 (page 212).

The export of photosynthates from the chloroplast not only represents the provision of organic carbon but also the export of energy. The energy is present in triose phosphates and also in malate which may be produced in the chloroplast by the reduction of oxaloacetate (see page 155). These constituents may be incorporated into the glycolytic pathway and the citrate cycle and thus contribute to the energetic requirement of the cytoplasm.

5.1.3   Assimilate transport and phloem loading in the physiological source

The assimilates move from the mesophyll cells through the bundle sheath cells (typically developed in C_4 species; Plate 3.1) and phloem parenchyma cells to the sieve elements in the smallest veins of the leaf. The transport pathway is not yet completely clear. The pathway covers a distance of about two or three cell diameters. The assimilates may move through the symplast *via* plasmodesmata (symplastic path) to the sieve cell-companion cell complex of the phloem (see page 248). They may also travel in the apoplast (apoplastic path) to the sieve cell the complex where they are loaded into companion cell phloem (Figure 5.2). Both paths are utilized in different species and even in one species, one or both may be operative. The experimental evidence of Geiger *et al.* (1974) obtained with sugar beet leaves, indicates that the apoplast of the mesophyll tissue is an essential part of the pathway photosynthates follow when transported from the mesophyll cells towards the sieve cell-companion cell complex. Similarly, in tobacco and potato leaves the assimilates travel through the apoplast prior to phloem loading.

Table 5.1 shows the range of sucrose concentrations found in the various tissues of sugar beet (Kursanov 1974). It is supposed that sucrose leaks into the apoplast and diffuses here along a concentration gradient towards the minor veins. The transport of assimilates from the mesophyll apoplast into the sieve elements and companion cells is the apoplastic phloem loading step. The fact that sucrose is at a higher concentration in the sieve element-companion cell complex than in the surrounding mesophyll (almost 100 times higher) indicates that sucrose is transported into the phloem system against its chemical potential gradient. Due to this concentration gradient the solute potential of the mesophyll is around -1.3 MPa and of the sieve element-companion cell complex around -3 MPa. Thus experiments of Sovonick *et al.* (1974) and Smith and Milburn (1980) have shown that ATP is required for phloem loading with sucrose. Analogous results have been obtained for phloem loading with amino acids by Servaites *et al.* (1979).

248

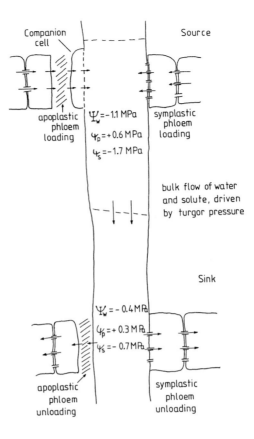

Figure 5.2 Diagram of possible pathways of phloem loading (physiological source) and phloem unloading (physiological sink). The loading as well as the unloading step can be apoplastic or symplastic. Phloem loading with photosynthate at the source induces water uptake into the phloem along the water potential gradient. Unloading of photosynthate is followed by water release and thus water movement from source to sink is established. Water potential data from Nobel (1991).

Table 5.1   Sucrose concentrations in various tissues of sugar beets
(Kursanov 1974)

	Sucrose concentration (mol/m³)
Mesophyll	3—3.5
Minor veins	20—25
Major veins	50—80
Phloem sap	200—300
Storage tissue of the root	400—600

Apoplastic phloem loading supposedly occurs *via* a H⁺-cotransport (Figure 5.3). This is a secondary transport mechanism (page 248) for uptake of sucrose and also of amino acids which is driven not directly by ATP hydrolysis, but indirectly by the H⁺ gradient established by the plasma membrane H⁺-ATPase (Bush 1993, Riesmeier *et al.* 1994, Lohaus *et al.* 1995). The prerequisite for efficient uptake is a low apoplastic pH because a steep H⁺ gradient drives the uptake of assimilates. It has been shown that with a high apoplastic pH sucrose uptake is reduced. Research of Mengel and Haeder (1977) has shown that the positive effect of K⁺ on phloem transport results from the direct influence of K⁺ on the phloem loading step. Potassium ion uptake into the companion cell-sieve element decreases the electrical potential thereby promoting the driving force for H⁺ extrusion into the apoplast. In addition, K⁺ directly stimulates the activity of the H⁺-ATPase (Briskin and Poole 1983). Van Bel and Van Erven (1979) have shown that K⁺ particularly promotes the uptake of sucrose and of glutamine at high apoplastic pH and they suggest that the cotransport mechanism depends on apoplastic pH. At low apoplastic pH a H⁺-cotransport is operative and protons are recycled, whereas at high apoplastic pH K⁺-cotransport is dominant. Phloem loading may also be dependent on the turgor pressure of the sieve elements. It is suggested by Geiger (1979) that a high turgor depresses and a low turgor promotes the process of loading. The sucrose concentration in the apoplast may also influence loading; the higher the apoplastic sucrose concentration the lower the chemical potential gradient to favour sucrose H⁺ cotransport.

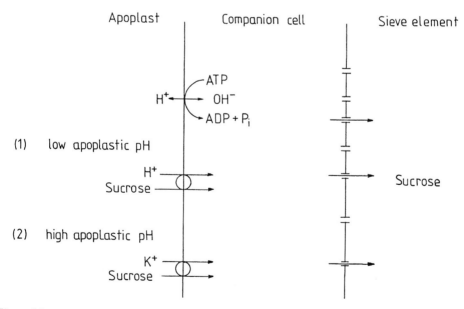

Figure 5.3   Apoplastic phloem loading: H⁺/sucrose cotransport at low apoplastic pH and K⁺/sucrose cotransport at high apoplastic pH. The plasmalemma H⁺-ATPase pumps H⁺ out of the cell into the apoplast, thus establishing a steep H⁺-gradient which drives sucrose uptake into the sieve elements through a sucrose H⁺ cotransporter. In case the gradient is not as steep, at high apoplastic pH K⁺ may drive uptake along the electrical gradient.

Molecular studies in this field also support a sucrose H^+-cotransport. Proton pumps and a sucrose -H^+cotransporter called SUC2 have been localized in the plasma membrane of companion cells in *Arabidopsis* (De Witt and Sussman 1995). Another cotransporter for sucrose called SUT1 and present in the plasmlemma of sieve elements has been found in potato, tomato and tobacco (Kuhn *et al*. 1997).

As mentioned on page 247 open plasmodesmata are a prerequisite for symplastic transport. However, the presence of plasmodesmata does not necessarily mean that the phloem loading step is symplastic. In many species plasmodesmata have been found between sieve element-companion cell complexes and neighbouring cells (Gamalei 1989). In the case of exclusive apoplastic loading few symplastic connections are present. Species that transport sugars symplastically show abundant connections and instead of sucrose other sugars *e.g.* raffinose and/or stachyose are translocated. It is an open question as to how sugars concentrate in the sieve element in case of diffusion through plasmodesmata. According to Van Bel (1992) sucrose diffuses from the bundle sheath cells to the companion cells *via* plasmodesmata. In the companion cells which are closely associated with the sieve elements *e.g.* stachyose is synthesized from sucrose and galactose and thus the diffusion gradient for sucrose is maintained. It is suggested that the plasmodesmata between bundle sheath and companion cells are able to exclude larger molecules than sucrose.

5.1.4   Phloem tissue

The phloem tissue contains sieve cells (sieve elements), companion cells and parenchyma cells. Sieve cells differ from the sieve elements in so far as they are still developing and not so highly specialized as the sieve elements. In comparison with the sieve elements, sieve cells contain more mitochondria, have a smaller diameter (2.5 μm) and, unlike the sieve element, are located in the minor vein system of leaves. Plate 5.1 shows a cross section of a minor vein of *Tagetes patula* from the work of Evert (1980). The vein comprises a tracheary element (xylem), a vascular parenchyma cell, a sieve cell, phloem parenchyma cells and companion cells. These latter cells do in fact 'accompany' the sieve cells and are derived from the same meristematic cells from which the sieve cells are developed. Each sieve cell is associated with one or more companion cells (Plate 5.1). Abundant symplastic connections between both cell types indicate their close functional relationship. They may therefore be considered as a functional unit, the so-called sieve element-companion cell complex (Geiger 1975). Companion cells as well sieve cells act as cellular sites for apoplastic phloem loading (Figure 5.3). Thus Trip (1969) observed that in sugar beet leaves newly synthesized assimilates are predominantly accumulated in the companion cells. The companion cells possess a large number of mitochondria and thus may provide the sieve elements with energy. (For further details, in particular about the different types of companion cells the reader is directed to botanical texts).

The sieve element is a highly specialised long stretched cell and the plasma membrane of a sieve element is continuous with that of the subsequent sieve element. Many cellular organelles occurring in other cells are lacking including cell nucleus, golgi bodies, ribosomes, microfilaments and tonoplasts. Mitochondria, plastids and endoplasmatic reticulum are modified. The elongated sieve cells contain lateral sieve

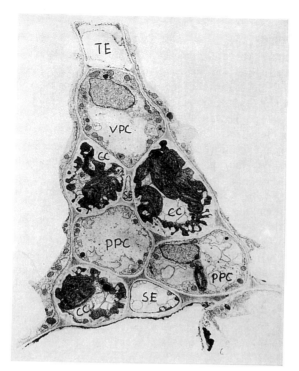

Plate 5.1 Cross section of a minor vein from leaf of *Tagetes patula* (Photo: Evert). TE = tracheary element (xylem), VPC = vascular parenchyma cell, CC = companion cell, SE = sieve cell, PPC = phloem parenchyma cell.

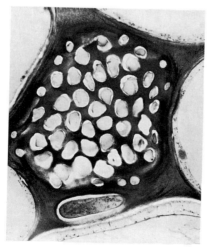

Plate 5.2 Sieve plate in a soya bean petiole which was quickly frozen *in situ*. The functional condition of this sieve tube was established by microautoradiograph (Photo: Fisher).

areas and porous sieve plates at their ends which separate the sieve elements from one another (Figure 5.2). The pores in the sieve plates between the sieve elements are open which is a prerequisite for transport. Translocation through living sieve elements of faba bean was visualized by applying a phloem mobile fluorescent dye which could be observed by the use of confocal laser scanning microscopy (Knoblauch and Van Bell 1998). Fisher (1975) showed that in the petiole tissue of soya bean about 70% of the sieve plate pores were essentially free from obstruction. Plate 5.2 shows an example of this investigation. The light area of the plate show the open pores. In the upper plant organs the sieve elements are present in the stem cortex, the petioles and in the major vein system of leaves. The roots contain the sieve elements in the central cylinder.

The minor vein system which comprises sieve cells and companion cells may be compared with a rather fine net mesh (Plate 5.3). In sugar beet leaves for example the total length of this fine vein system amounts to about 70 cm per cm² leaf area (Geiger and Cataldo 1969). This total length is about 10 times greater than that of the corresponding major veins. This demonstrates the importance of the minor vein system in the collection of photosynthates. The mean distance of the mesophyll cells from these fine veins is only about 70 μm or the length of about two cells. Photosynthates are thus easily accessible to the companion cells. Assimilates are transported from these minor veins to the major veins. Minor veins may therefore be regarded as 'contributory streams' to the major vein system.

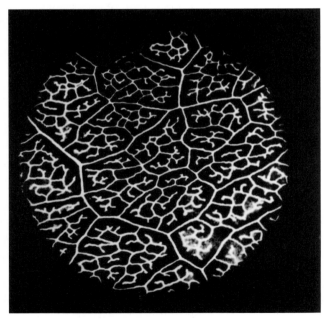

Plate 5.3   Autoradiograph showing the minor vein system of a sugar beet leaf (magnification × 10). The autoradiograph was obtained after treating the leaf with ^{14}C labelled sucrose and finally removing the free space sucrose by rinsing (Photo: Geiger).

5.1.5   Mechanism of phloem transport

It is 70 years ago since Münch (1930) proposed the 'pressure flow hypothesis' to account for phloem transport as a mass flow phenomenon. Stated simply, this is as follows: Phloem loading of solutes increases the pressure in the sieve tubes and phloem unloading decreases the pressure. Thus a pressure gradient is established which is responsible for a mass flow action. The process of loading and unloading of phloem affects the water status of the sieve tubes. Loading results in a decrease in solute potential. Hence the accumulation of sucrose in sieve tubes is associated with water uptake by osmosis. The sieve tube thus takes on a rather high turgor, which may amount to about +1.0 MPa or even higher (Wright and Fisher 1980). The reverse situation occurs when sieve tubes are unloaded. Loss of sucrose results in a decrease of the solute concentration (corresponding to an increase in the solute potential). As the water potential rises water leaves the phloem along the water potential gradient into the adjacent sink and a lower pressure in the sieve element results. If one imagines two sieve tubes in the same phloem strand, in which one sieve tube is being loaded and the other unloaded, it is easy to envisage mass flow along a pressure gradient (Figure 5.2), provided that the sieve plates do not present too great a resistance to the mass flow process. As has already been discussed, however, evidence now suggests that normal functioning sieve pores are open.

The mechanism of phloem transport has been discussed very thoroughly by Geiger (1975). He concludes that it is mainly a mass flow phenomenon driven by a turgor gradient in which phloem loading exerts the 'push' of the mechanism and unloading the 'pull'. The sieve element turgor at the source is generally higher than at the sink and the turgor difference sufficient for driving mass flow from source to sink through the phloem has been calculated to be at around +0.1 to +0.5 MPa (Fischer 1978). Solutes present in the bulk flow move at the same rate as water. The velocity of phloem transport is around 50 to 100 cm/h (Cronshaw 1981) and thus much faster than diffusion (100 cm per 32 years). As calculated by Nobel (1991) the water in the phloem moves along the turgor pressure gradient and against the water potential gradient from source to sink (Figure 5.2). Since water movement is a bulk flow and not driven by osmosis this is in accord with thermodynamic laws. Sieve tubes are open and there are no membranes to be crossed along the phloem transport pathway so that the higher water potential due to a less negative solute potential in the sink sieve element compared with the source phloem cannot reverse the movement of water (from sink to source). As mentioned, the driving force is the turgor pressure gradient which is set up by partly active transport mechanisms involved in phloem loading and phloem unloading. Thus despite 70 years of research, the Münch hypothesis — with modifications — is generally held by a number of eminent plant physiologists to be the most acceptable mechanism to account for phloem translocation.

In the literature numerous researchers have reported that K^+ promotes phloem transport and as mentioned on page 497 is probably related to the beneficial effect on phloem loading. According to Lang (1983) high K^+ uptake by the sieve tube-companion cell complex induces osmotic water flow into the complex to «push» the phloem mass flow. Release of K^+ from the phloem tissue on the other hand leads to water release and thus contributes to the «pull mechanism».

Phloem loading needs energy as has been already mentioned. Energy may also be needed for the release and consumption of photosynthates in so-called 'physiological sinks'. Whether the transport of solutes itself is also an energy consuming process is not yet clear. Consistent with the concept of the pressure-flow hypothesis there should be no direct energy requirement in order to drive translocation. Effects of temperature on the rate of translocation in the phloem cannot be used as an argument in favour of energy involvement because of the induced changes in viscosity of the phloem sap. Such changes can be considerable as the phloem sap contains large concentrations of sucrose. Low temperature (0°C) may also result in blockage of the sieve plates by plasmatic material (Giaquinta and Geiger 1973).

In summary, the experimental evidence shows that translocation is not directly driven by energy, that pores are open and that a pressure gradient is established, all of which is consistent with the pressure-flow hypothesis. In contrast to xylem transport, phloem transport can occur from root to shoot and *vice versa*. In accord with the Münch hypothesis bidirectional transport occurs in different sieve elements and has never been observed in a single phloem tube.

5.1.6   Phloem unloading and assimilate movement into the physiological sink

Different physiological sinks are present in plants including meristematic tissues (root and shoot tips and young leaves), vegetative storage organs (roots and stems) and generative storage organs (fruits and seeds). Phloem unloading and assimilate movement is therefore quite different (Figure 5.2). The release of assimilates from the sieve element represents the unloading step. After unloading the assimilates move to the final sink tissue which has been termed the post-unloading path. In the final step assimilates are metabolized to produce energy for growth or they are stored as starch, soluble sugars, oil or proteins.

As in the source, the assimilates move from the phloem through the symplast *via* plasmodesmata (symplastic path) or they may follow the apoplast at some point(s). The apoplastic unloading step represents assimilate flux across the plasma membrane of the sieve element into the adjacent apoplast. In young dicotyledonous leaves (*e.g.* sugar beet) the whole unloading path seems to be symplastic (Oparka and Van Bel 1992). In young monocot leaves (*e.g.* maize) few plasmodesmata have been observed, thus assimilate movement probably also occurs in the apoplast (Evert and Russin 1993).

Assimilates in growing root tips probably travel through the symplast (Oparka *et al.* 1995). In storage organs a part of the phloem unloading path is apoplastic. In developing seeds (wheat, barley, rice) assimilates move symplastically out of the phloem through the maternal tissue before entering the apoplast at the maternal/embryo parenchyma interface (Figure 5.4). The apoplastic step is necessary because of the absence of plasmodesmata between the maternal and filial tissues in developing seeds. The symplastic path may be prevalent in vegetative starch storing organs. In potato tubers numerous plasmodesmata form a symplastic path from the sieve elements to the places of starch deposition (Oparka 1986). Storage organs that accumulate osmotically active compounds to high concentrations, around 500 to 900 mol/m^3 sucrose in the taproot of sugar beet and around 500 mol/m^3 sucrose in the stem of sugarcane possess an apoplastic unloading step (Patrick 1990). In the case of symplastic unloading for

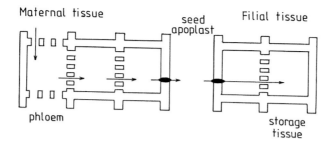

Figure 5.4 Phloem unloading pathway throughout the developing seed (modified after Patrick and Offler 1995).

reasons of high solute potentials assimilates can move in the direction opposite to that of the unloading path.

Symplastic movement of assimilates through plasmodesmata in the starch storing sinks occurs along a steep concentration gradient from phloem to storage cells and enables diffusion and mass flow. In these organs organic carbon is stored in polymers like starch or protein or is used for respiration and meristematic growth. In the case of apoplastic assimilate movement coupled with vacuolar accumulation of sucrose, changes in apoplastic solute concentrations, cell wall elasticity and membrane hydraulic conductivity found during the development of sugar cane (Moore and Cosgrove 1991) may regulate solute and turgor potentials in order to maintain simultaneously cell growth and storage of high sugar concentrations in the vacuole.

When apoplastic unloading or apoplastic paths are involved at some points assimilate movement becomes more complicated. Membranes have to be crossed and in several cases transporters which are dependent on metabolic energy play a role. Apart from active uptake, metabolic conversions in the apoplast may help to maintain the concentration gradient along the apoplastic path. Sucrose cleavage into hexose moieties in the apoplast by acid invertase can lower the apoplastic sucrose concentration (Eschrich 1989). The enzyme is present in the apoplast and has a high activity at low apoplastic pH. As sucrose moves into the apoplast it is supposed that it is thus split by invertase and cannot therefore be reabsorbed by the phloem.

In developing maize kernels such sucrose cleavage has been found (Porter *et al.* 1987), but some sucrose can cross the apoplast unchanged. In other species as in wheat, sucrose cleavage does not precede its uptake by sink cells (Jenner 1974). Movement of sucrose into the seed apoplast and transport of sucrose from the apoplast into the cereal endosperm is achieved in different ways for the various grain crops. In most legumes entry of sucrose into the apoplast as well as uptake into the embryo are carrier-mediated and energy-dependent. In cereals like wheat, sucrose release is not energy-dependent and occurs by facilitated diffusion, but the uptake into the endosperm is carrier-mediated and active (Patrick and Offler 1995). The uptake mechanism is generally a sugar H^+-cotransport. The subsequent assimilate movement in the endosperm then occurs through the plasmodesmata to the final sites of storage (Cook and Oparka 1983, Mc Donald *et al.* 1995). The subject has been reviewed in detail by Oparka and Van Bel (1992), Patrick and Offler (1995) and Patrick (1997).

5.1.7   Storage processes in physiological sinks

Apart for their importance in human nutrition, storage products in particular starch, as well as oils and sugars are becoming increasingly used in the non-food sector. Transgenic manipulations in starch and oil synthesis, have been successful in recent years (*e.g.* Murphy 1994, Müller-Röber and Koßmann 1994). For further successful development of new transgenic crop species the mechanisms of storage and control have to be elucidated. In the following section the main aspects of storage processes for carbohydrates are described. The processes involved in storage of proteins and oils are not considered here. Proteins are stored in specialized compartments (related to vacuoles and ER vesicles) which are called protein bodies and are enclosed by a unit membrane. Protein bodies are abundant in seeds. In potato tubers proteins may be deposited in vacuoles. Some recent relevant reviews which give some insight into these complex process of synthesis and deposition of storage proteins have been provided by Müntz *et al.* (1993), Staswick (1994), and Shewry *et al.* (1995). Storage lipids are synthesized in the ER system and are deposited in so-called oleosomes, lipid bodies which are surrounded by a half-unit membrane (see for details: Huang 1993).

In storage sinks such as in the sugar beet taproot and in sugar cane, sugar is stored at high concentrations in the vacuole. Energy is thus required for accumulation which takes place against a concentration gradient. A sucrose H^+-antiport has been suggested as an accumulating mechanism (Briskin *et al.* 1985, Getz *et al.* 1991). Vacuolar proton pumps move H^+ into the vacuole. The reflux of H^+ from the vacuole drives the sucrose import *via* a proton/sucrose antiport (Figure 5.5).

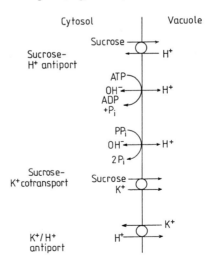

Figure 5.5   Sucrose deposition in the vacuole *via* sucrose H^+-antiport with H^+ ATPase and H^+ PPase as driving forces. Also K^+/sucrose cotransport coupled with a H^+/K^+ antiport may play a role in sucrose uptake.

Sucrose transport may also be directly influenced by K^+. According to Saftner and Wyse (1980) a K^+/H^+ antiport drives K^+ out of the vacuole back into the cytoplasm where a K^+ concentration around 100 mol/m^3 is required for efficient sucrose uptake into the vacuole. A K^+ cotransport is suggested which is driven by the inwardly directed concentration gradient of K^+. Potassium nutrition favours the storage of starch, sugars and proteins as observed in numerous crops (Mengel 1999) and apart from favouring photosynthetic efficiency and phloem loading K^+ may have also direct impact on storage. This is also shown for example for protein synthesis in grains (Table 5.2). Apart from increased assimilate translocation into the developing grain with higher K^+ supply (Seçer 1978), K^+ may directly affect protein synthesis. Wyn Jones and Pollard (1983) have suggested that the transcription and translation processes require K^+.

Table 5.2   Effect of K^+ on the incorporation of ^{15}N into grain proteins of wheat. K_1 and $K_2 = 0.3$ and 1.0 mM K^+ in the nutrient solution, respectively (Seçer 1980).

	K_1	K_2
Albumin, mg ^{15}N/kg	42.4	67.0
Globulin, mg ^{15}N/kg	36.4	49.2
Prolamin, mg ^{15}N/kg	108.0	151.0
Glutelin, mg ^{15}N/kg	130.0	194.0
K-concentration mg K/DM	4.9	5.0

Sucrose import into vacuoles is also of importance for storage of fructans. In the Jerusalem artichoke (*Helianthus tuberosus*) sucrose is probably also transported by a H^+ antiport mechanism (Frehner *et al.* 1987). Fructan is synthesized in the vacuole. Inulin is a fructan type with β-2,1 glycosidic bonds which as far as is known occurs only in dicotyledons while the levan type of fructans with β-2,6 glycosidic bonds prevails in grasses. According to Edelmann and Jefford (1968) two enzymes are involved in fructan synthesis: firstly, sucrose: sucrosefructosyltransferase (SST) which transfers a fructose moiety from one sucrose molecule to another and can be regarded as a priming reaction and secondly, fructan: fructanfructosyltransferase (FFT) which transfers fructose units to the growing fructan chain.

$$\text{Sucrose} + \text{Sucrose} = \text{Fructan} + \text{Glucose} \quad \text{(SST-reaction)}$$
$$\text{Fructan}_n + \text{Fructan}_m = \text{Fructan}_{n-1} + \text{Fructan}_{m+1} \quad \text{(FFT-reaction)}$$

In different vegetative (leaves, stems, roots) as well as generative organs (cereal grains) fructans represent a transient storage form and are remobilized during the vegetative period for growth and final storage as fructans *e.g.* in the Jerusalem artichoke tuber (Feuerle 1992) or for growth and starch deposition in the cereal grains (Houseley and Daughtry 1987). Fructans are also involved in leaf cell expansion of grasses. Schnyder *et al.* (1988) found highest fructan concentrations in the elongation zone of grass leaves which were higher during the day than during the night suggesting that fructans were stored during the day and consumed at night. For further details see Pollock and Cairns (1991).

Starch synthesis and deposition in storage organs occurs in amyloplasts as shown in Plate 5.4. In contrast to chloroplasts, hexose phosphate rather than triose phosphate is taken up and used for starch synthesis in amyloplasts. Glucose 1-phosphate (G1P) is the main sugar form which is translocated into amyloplasts from wheat endosperm (Tyson and ap Rees 1988, Tetlow *et al.* 1994) and potato (Kosegarten and Mengel 1994, Naeem *et al.* 1997) and then metabolized to starch *via* ADP glucose pyrophosphorylase and starch synthase. The high specifity of the glucose 1-phosphate translocator is shown in Table 5.3. From the inhibitory effects it is suggested that the translocator binds glucose and only phosphate groups in the C1 position which finally leads to the uptake of G1P.

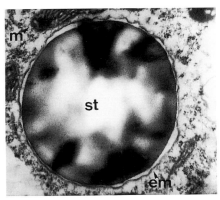

Plate 5.4 Typical amyloplast from starch storing tissue of *Solanum tuberosum* (Photo: Kosegarten). em: envelope membrane, m: mitochondrion, st: starch grain.

Table 5.3 Effect of various metabolites on the uptake rate of G1P by isolated amyloplasts from *Solanum tuberosum*. Metabolite concentrations were 0.5 mM, if not noted otherwise, pH was 5.7 (from Kosegarten and Mengel 1994).

Treatments	glucose 1-phosphate uptake rate (pmol × 10^6amyloplasts^{-1} × 25 min^{-1})
Experiment A	
G1P (control)	10.19
+ glucose 6-phosphate	10.03
+ fructose 6-phosphate	10.06
+ fructose 1-phosphate	5.73
+ ADP glucose	3.75
+ fructose 1,6-bisphosphate	10.18
+ 3-phosphoglycerate	11.28
+ dihydroxyacetone phosphate	10.83
+ inorganic phosphate	5.75
Experiment B	
G1P (control)	11.14
+ ß-D-fructose (2.5 mM)	11.53
+ α-D-glucose (2.5 mM)	6.90
+ α-D-glucose	8.29
+ sucrose (2.5 mM)	10.18
+ dihydroxyacetone	11.50

The major enzymes of starch synthesis in amyloplasts are: ADP glucose pyrophosphorylase, starch synthase and branching enzyme. In contrast to leaves, ADP glucose pyrophosphorylase in storage organs shows limited or no allosteric regulation (Kleczkowski *et al.* 1993). It is suggested by Dujardin *et al.* (1995) that the rate of starch deposition is primarily regulated at the transcriptional level of ADP glucose pyrophosphorylase. As yet it is not clear whether the rate of glucose 1-phosphate absorption has a direct effect on starch yield (Kosegarten and Mengel 1998). For further details about starch synthesis and related enzymes the reader is directed to relevant reviews: (Preiss 1991, Morell *et al.* 1995, Nelson and Pan 1995, Smith *et al.* 1997, Kosegarten and Mengel 1998).

5.1.8   Assimilate partitioning between physiological source and sink

That large amounts of assimilates should be produced in the physiological source (Gifford and Evans 1981) and that these assimilates should be utilized by a high sink strength are important factors in determining high yields in crop production. A high sink strength is a particularly important issue for high yield production. Sink strength is determined by sink size and sink activity (Warren Wilson 1972, Patrick 1988). Sink size is basically a result of the growth of the sink organ (cell division and cell enlargement) and sink activity mainly comprises the metabolic activity of assimilate uptake and conversion. However, the final amount of products stored in the physiological sink is dependent on several processes along the whole path of assimilate distribution, so-called partitioning within the plant (Patrick 1988). This section considers carbohydrate partitioning and an overview of the processes involved in partitioning between the physiological source and sink is given.

A central question is whether photosynthesis is a key limiting process in crop production. In general, an increase in photosynthetic rate results in an increased rate of translocation from the physiological source. Throughout plant development various sink organs (page 254) may compete for the photosynthate translocated *via* the phloem thus determining the assimilate distribution between the various sinks. For example, during the vegetative phase, shoot and root growth compete for assimilates and both these needs have to be balanced in order to produce a large shoot for a high photosynthetic productivity and a well developed root system to acquire sufficient water and nutrients (Geiger *et al.* 1996). In the generative phase of cereals the developing ear is by far the strongest sink without major sink competition (Roeb and Führ 1990). These examples indicate the enormous impact of sink competition on assimilate translocation and not least on crop yield.

The photosynthetic rate of many crop species often decreases during day despite of adequate light and CO_2. Schäfer *et al.* (1984) reported that in Central Europe under field conditions winter wheat in the stage of elongation and grain filling reached maximum net photosynthetic rates in the morning about three hours before solar irradiation and temperature were at maximum and independent of differential water supply. Gifford and Evans (1981) reported that net photosynthetic rates per unit leaf area of modern high yielding cultivars differed little from their wild homologous. The most important progress in improving crop yield has been achieved by an improved assimilate transport to more efficient vegetative or reproductive storage sinks. The so-called harvest index, the ratio of commercial or edible yield such as tubers and grains to total shoot yield is around 0.80 in potato tubers (Inoue and Tanaka 1978) and around 0.56 in modern wheat cultivars (Gifford *et al.* 1984, Mehrhoff and Kühbauch 1990). In wheat cultivars a high grain number per ear (Gifford and Evans 1981) and an efficient and complete use of fructans transiently stored in the cereal stem (Mehrhoff and Kühbauch 1990) contributes to a high grain yield in the modern cultivars.

Thus control and regulation of assimilate partitioning is of major importance in obtaining high edible proportions of crop yield. Several experiments influencing the source-sink relationship have shown that the photosynthetic rate increases when sink demand increases and decreases when sink strength is low. Short and long-term regulation mechanisms have to be distinguished. In principle, short-term regulations in the source leaf balance cytosolic sucrose synthesis on the one hand and chloroplastic starch synthesis on the other during light/dark periods to provide the growing plant continuously with organic carbon. In the source leaf, assimilate partitioning is regulated by several key enzymes: by sucrose phosphate synthase and fructose-1,6-bisphosphatase which are involved in sucrose synthesis and by ADP glucose pyrophosphorylase which regulates starch synthesis (page 245). Usually, by increasing sink demand the rate of photosynthesis increases, probably *via* increased sucrose phosphate synthase activity. By removing sucrose, inorganic phosphate is released which can be imported into the chloroplast regenerating Calvin Cycle intermediates and thus driving triose phosphate export into the cytosol. Here, glucose 6 phosphate is synthesized which stimulates the activity of sucrose phosphate synthase and thus provides sucrose for translocation to meet the increased sink demand. Under these conditions less starch is stored in chloroplasts during the day. However, carbon diversion from starch during the day is limited and it has been shown by Chatterton and Silvius (1979) that the shorter the day lengths the more is starch stored for mobilization and translocation for the longer night period. At night, starch degradation in the chloroplast provides organic carbon for continued phloem transport. When there is a low sink demand for sucrose, more starch is stored in the chloroplasts and more sucrose is released during the night (Fondy and Geiger 1982).

Long-term changes occur in the source leaf when source-sink relationships have been permanently changed. In potato, photosynthesis and assimilate translocation to the tubers and roots increases considerably after tuber setting (Moorby 1968). Photosynthetic activity declines immediately after the removal of potato tubers (Burt 1964) without affecting the growth rate of the remaining tubers (Engels and Marschner 1987). Several mechanisms may be responsible for inhibition of photosynthesis. As mentioned, sucrose synthesis in the source leaf is usually impaired at low sink demand and as shown by Riesmeier *et al.* (1993) chloroplastic starch synthesis is not sufficient to provide adequate inorganic phosphate levels. Thus, less inorganic phosphate may be available to drive ATP synthesis in the chloroplast and thus CO_2 fixation decreases. In transgenic potato plants it has been shown that photosynthesis is inhibited by the accumulation of carbohydrates in the leaf as a consequence of suppressed phloem transport (Heineke *et al.* 1992). This is in accord with the observation that high sugar levels decrease the transcription rate and gene expression for many photosynthetic enzymes (Krapp and Stitt 1995, Koch 1996).

In potato, photosynthesis may have a more direct impact on assimilate supply and thus on starch production in tubers (Engels and Marschner 1987) than in cereal crops. Reduction of photosynthesis during grain filling by shading (Judel and Mengel 1982) or defoliation (Jenner and Rathjen 1972) does not affect the grain yield. In contrast to cereal crops, however, restricting photosynthesis in potato may limit starch production in tubers (source-limitation) and it is tempting to speculate that this is because of the higher harvest index in potato and that the storage capacity has not reached its final maximum. Generally, the higher the sink size the higher should be the carbon import into the storage sink. Changes in sink activity are more complex, including phloem unloading and competing processes between growth and storage of the physiological sink. In potato, phloem unloading and assimilate movement towards starch storage is symplastic, however, in cereals assimilates must pass the apoplast on their path from unloading to starch synthesis. It is supposed by Kosegarten and Mengel (1998) that these latter processes rather than photosynthesis may thus control starch synthesis in cereals (sink-limitation).

The activity of the sucrose-cleaving enzymes acid invertase and sucrose synthase (page 262) may play a crucial role in the control of assimilate partitioning in the physiological sink and may influence yield quantity and quality. Cleavage of sucrose in the apoplast by acid invertase in some species (page 255) leads to glucose and fructose and displays optimum activity at a low apoplastic pH (Eschrich 1980). On cleavage the apoplastic sucrose concentration decreases and sucrose may continuously leak out of the phloem (Eschrich 1989). In storage organs depending entirely on symplastic assimilate movement, invertase located in the vacuole and sucrose synthase may maintain a concentration gradient to enable diffusion as well as mass flow along the solute gradient through the plasmodesmata. Also symplastic assimilate movement from the phloem to the maternal tissue of grains is driven by a steep sucrose concentration gradient (Fischer and Wang 1995). A strong sink, for example characterized by a high activity of these enzymes may deplete phloem sucrose concentration at the site of unloading and thus by increasing the sieve element turgor pressure gradient, assimilate translocation towards the sink organ is increased (Schulz 1994). The activity of invertase and sucrose synthase largely depends on the developmental stage of the tissue. Invertase is presumably involved in early growth and expansion before storage processes have begun, while sucrose synthase is coupled to storage of starch (Singh et al. 1991, Quick and Schaffer 1996). Geiger et al. (1996) have shown that gene expression of these enzymes changes with sink development and these enzymes probably may control import of organic carbon. It is argued by Koch (1996) that abundant import of organic carbon towards the sink organ favours genes for growth and storage processes while poor carbohydrate supply favours genes for photosynthesis and assimilate provision. Thus, sucrose is not only a carbon skeleton for storage and growth, but also may act as a signal regulating gene expression.

Sucrose synthase in the physiological sink usually catalyzes sucrose breakdown and generates UDP glucose and fructose in the presence of UDP (step 1). Since sink growth overlaps with starch storage, UDP glucose may be used for cell wall formation as well as for the synthesis of glucose 1-phosphate (step 2) which is channelled into starch synthesis (Kosegarten and Mengel 1998).

step 1: Sucrose + UDP = UDP glucose + fructose
(sucrose synthase reaction)

step 2: UDP glucose + PPi = Glucose 1-phosphate + UTP
(UDP glucose pyrophosphorylase reaction)

It is often observed that growth of the storage organ may dilute the concentration of stored material, *e.g.* starch in potato tubers (Engels and Marschner 1986). Thus the incoming sucrose is channelled into different metabolic pathways which may influence both the crop yield and its quality. Apart from influencing storage by competing growth processes of the storage organ, plant growth and storage of metabolites proceed concurrently over a long period and may thus compete for assimilates (Kosegarten and Mengel 1998). This is well known for potato and tomato. In several studies it has been shown that under conditions which limit plant growth, carbohydrates are channeled into storage processes. This has been shown under N-deficiency (Hehl and Mengel 1973) and also recently under a mild restriction in water supply where there were no visible symptoms of drought stress (Table 5.4) (Veit-Köhler *et al.* 1999).

Table 5.4   Effect of different water supply on plant growth and storage of various metabolites in mature fruits of *Lycopersicon esculentum*. Means ±SD (n = 5). Significant differences are denoted by: $^*P \leq 0.05$ (from Veit-Köhler *et al.* 1999).

parameter	70% water capacity			50% water capacity		
growth						
vegetative plant matter (g/plant)	836.6	±	69.4	809.2	±	23.5
number of fruit settings	6.0	±	0.0	4.6	±	0.6*
average fruit weight per plant (g/fruit)	53.9	±	7.9	54.3	±	17.9
total fruit yield (kg/plant)	2.3	±	0.8	1.6	±	0.7
marketable fruit yield (kg/plant)	1.6	±	0.4	1.4	±	0.4
storage in mature fruits						
glucose plus fructose (mg/ml)	46.67	±	0.98	50.86	±	1.37*
vitamin C (mg/ml)	0.27	±	0.03	0.35	±	0.03*
aroma volatiles (rel. peak areas):						
hexanal	1756.0	±	84.9	2120.5	±	258.5*
(Z)-3-hexenal	1097.8	±	73.1	1771.1	±	250.4*
(E)-2-hexenal	465.5	±	50.0	566.4	±	45.4*

This latter example shows that by small reduction in water supply (50% water capacity), plant growth in tomato, and in particular the number of fruit settings were depressed. The treatment with the lower water supply also favoured the translocation of carbohydrates into the growing fruit. However, the higher translocation of carbohydrates did not promote fruit growth and thus did not result in a higher edible yield, but in a significant improvement of fruit quality. It has been shown that apart from sugars the synthesis of a wide spectrum of metabolites in the tomato fruit, including vitamin C and aroma volatiles was favoured by a lower supply of water. Since not all fruits of the plants with higher water supply matured, the marketable yields of both treatments were similar similar of treatments (Table 5.4). Thus, only by a small reduction in water supply, a high proportion of fruits with a high quality can be harvested (Veit-Köhler *et al.* 1999). It is tempting to speculate that for the synthesis of secondary metabolites, the supply of energy in the storage organ is of importance. Increased K^+ fertilization favouring photosynthetic efficiency and assimilate translocation in the phloem has led to increased concentrations of vitamin C in mature tomato fruits (Anac and Colakoglu 1995). Vitamin C synthesis is closely linked to carbohydrate metabolism with glucose as a precursor (Wheeler *et al.* 1998) and it is suggested that sugar accumulation promotes vitamin C synthesis. Hexanal, hexenal and other aroma volatiles are important compounds for the taste and odour of many fruits and vegetables. Also, the levels of such aroma volatiles can obviously be increased at a high energy status in the tomato fruit (Table 5.4); however, much research in this field is needed to understand the pathways and regulation of production of secondary substances during fruit development.

5.2   Essential Growth Stages and Yield Components

5.2.1   General

The life cycle of a plant starts with germination. In this process the soil embedded seed requires an optimum temperature as well as a supply of water and oxygen, and the presence of favourable endogenous factors within the seed itself. These endogenous factors are mainly phytohormones such as abscisic acid, gibberellic acid, and indole acetic acid (Chapter 5.2.2). Germination begins with the uptake of water. The swollen seed provides the necessary conditions for respiration. With the uptake of oxygen the seed reserves of carbohydrates, fats and sometimes even proteins are oxidized to CO_2 and H_2O, and energy is converted into ATP and NADH. This form of energy is essential for growth processes. The storage proteins in seeds are mainly hydrolyzed and the resulting amino acids used for the synthesis of enzyme proteins and nucleic acids. Both are essential components in the formation of meristematic cells and for cell division, the process which initiates growth.

The first plant parts to be developed are the roots. This means that at an early stage of development of the plant, the organ is formed which is responsible for the uptake of water and nutrients. Shoot growth then starts and as soon as the shoot breaks through the soil surface, the synthesis of chlorophyll is induced by light. From this point onwards two further growth factors come into play, namely, light and the CO_2 of the atmosphere. The importance of these two factors increases the more the seed is depleted of its storage material. Young leaves are not self-sufficient. They must be supplied with carbohydrates and amino acids. In seedlings this supply of organic material is provided by storage compounds present in the seeds. However, with the onset of the vegetative stage which is characterized by the rapid development of leaves stems and roots, the sources of organic material for growth shifts from the seeds to the leaves. Photosynthates assimilated in the older leaves provide the source for younger tissues. Young leaves import carbohydrates until they have reached about one third of their final size. The net import of amino acids into young leaves continues longer and even until they are fully developed. Mature leaves export about 50% of their photosynthates. The remainder is needed for leaf metabolism itself and is mainly respired. As in the seed, carbohydrates are oxidized to CO_2 and water. In the process the energy released from the carbohydrates is converted to ATP.

The development of buds after the dormant phase is analogous to that described above. Some species store plant nutrients, such as organic N and carbohydrates in their roots and stems which are mobilized in spring by rising outer temperatures and provide energy and nutrients to the buds for the development of leaves and shoots. This is the case in the vine which over winter stores arginine and K^+ in the wood to be remobilized in spring for the rapid growth of new shoots and leaves (Kliewer and Cook 1971). Nitrate may even serve as a storage N compound in vine roots (Löhnertz et. al.1989a). This function of nitrate as a storage form for N is clearly shown in Figure 5.6 where root nitrate concentration of the vine root is shown in relation to the developmental stages of the crop.

Nitrate concentration increases steeply from the onset of growth until anthesis and then declines sharply to almost zero at the *veraison* stage, the time at which the berries are a few mm in diameter. The period between anthesis and veraison is characterized by a huge increase in foliage for which the stored nitrate provides the nitrogen source. This observation is in accord with the findings of Tagliavini *et al.* (1999) from experiments with nectarine trees. These authors found that late nitrogen application contributed more to the formation of storage N than did application at an earlier stage when the plant was still growing vigorously. During winter most of the N was stored in fine and coarse roots more than in the trunk and twigs. According to Millard (1996) who has reviewed the topic of internal cycling of nitrogen in plants, N storage and mobilization has been found to be a major source of nitrogen used in seasonal growth of woody plants. Nitrogen is accumulated in storage organs (wood, bark) before the dormant period (defoliation, leaf senescence) which generally begins in winter but also occurs in periods of drought in regions of Metiterranean climate. Besides the seasonal N storage, short term storage of N also occurs during priods when N is not required for growth. specific wood and bark proteins are known which are synthesized and stored in summer and autumn and remobilized in spring for growth. In many plant species N is mainly stored in the foliage in the form of proteins of which Rubisco (page 148) is the most important (Chapin *et al.* 1990). The amount of remobilized N in spring depends on the amount of N stored and not on N availability in the soil in spring. The amount of N stored prior to defoliation reflects the N nutritional status of plants during summer.

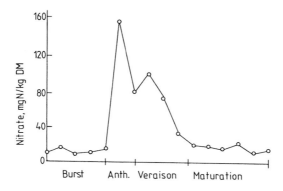

Figure 5.6 Nitrate concentrations in roots of vine during the entire growth peroid. Important developmental stages are: Bud burst, Anthesis, Veraison, Maturation (Löhnertz et al. 1989a).

In higher plants the vegetative stage is followed by the reproductive stage. This begins with flower initiation and it is succeeded after pollination or anthesis, by the maturation stage. In determinate plant species such as cereals, the vegetative and reproductive stages are quite distinct but for indeterminate species such as tomatoes they overlap. As plant growth progresses from the vegetative stage to maturity, photosynthates are more and more directed away from younger meristematic tissues towards storage tissues.

In the process of development light plays not only a role in providing energy, as developmental processes are frequently under the control of light with distinct wavelengths which are absorbed by phytochromes and triggered by processes such as stem elongation, leaf expansion, seed germination and flower initiation. In many cases this specific light effect is coupled with the action of phytohormones.

5.2.2   Phytohormones

These are organic molecules synthesized in the plant in only very low concentrations which control or influence developmental processes. In some cases they function in cells where they are formed, in other they are transported from one part of the plant to another to transfer physiological signals. Over short distances they are transported *via* the apoplast. Long distance transport occurs mainly in the phloem and xylem to target tissues where they may react with receptors in the plasma membrane to transfer information. A characteristic feature of the phytohormones in the synergistic or antagonistic effect that one group of hormones may have on another. The classical plant hormones are:

Auxins, indole acetic acid (IAA), and derivates of IAA
Gibberellins
Cytokinins
Abscisic Acid (ABA)
Ethylene

According to recent findings the brassinosteroids, an additional group of phytohormones may also be added to this list (Clouse and Sasse 1998).

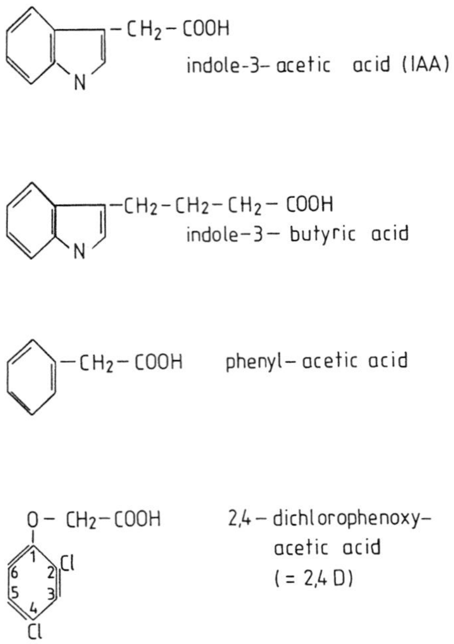

indole-3-acetic acid (IAA)

indole-3-butyric acid

phenyl-acetic acid

2,4-dichlorophenoxy-
acetic acid
(= 2,4 D)

Figure 5.7 Molecular structure of auxins.

Auxins are indispensable for growth promoting elongation of both, shoots and roots. High auxin concentrations are found in vigorouosly growing tissues whereas low concentrations are present in dormant tissues. The most widely occurring natural auxin in plants is indole-3-acetic acid (IAA). In addition there are other compounds with an analogous struture possessing auxin activity. For most of these but not all, the indole ring is the essential moiety conferring auxin activity (Figure 5.7). Such compounds include indole-3-butyric acid, phenyl acetic acid in which a phenyl group replaces the indole ring, and 4-chloro-indole acetic acid (Kende and Zeevaart 1997).

2,4-dichlorophenyl-acetic acid is a synthetic auxin with growth hormone properties which is used as a herbicide.The precursor of indole-derived auxin synthesis is the amino acid tryptophan which provides the indole ring. Auxin and nitrogen metabolism are thus interrelated. Auxins occur in free and in conjugated forms the latter being esterified and bound to hydroxyl groups of amino acids, peptides, and carbohydrates. The conjugated forms are biologically inactive and provide storage forms in seeds and hormonal homeostasis and are activated by enzymic hydrolyzation.

Auxins are synthesized in meristematic tissues from where they are transported. The transport across the plasmalemma into the apoplast is believed to take place by proton cotransport (Luetzelschwab *et al.* 1989). Long distance transport usually takes place from apex to base *i.e.* in a basipetal direction giving rise to a gradient in auxin concentration of great importance to the growth response. The gradient is responsible for the phenomenon of *apical dominance* (Jensen *et al.* 1998) in which the apical bud inhibits the development of lateral buds so that the growth of the main shoot depresses the development of lateral shoots. This may be accounted for by the high concentrations of auxins between the internodes below the apex which stimulate the formation of ethylene another growth inhibitor to prevent the burst of the lateral buds. This controlling effect of auxins is exploited in the use of synthetic auxins such as 2,4D as herbicides. Concentration gradients of auxins may also occur within plant organs themselves, as observed by Marschner *et al.* (1984) in potato tubers. IAA concentration increased from the basal part of the tuber to the apex whereas for the ABA (Abcisic acid, see page 273) concentration the reverse was found.

A most important process in which auxins are involved is the stimulation of the plasmalemma H^+ pump. Schubert and Matzke (1985) observed that the net release of H^+ from protoplasts was increased by the addition of IAA but depressed by ABA. This effect of a IAA promoting and ABA retarding the plasmalemma H^+ pump is a typical example of an antagonistic relationship between phytohormones. It is still a matter of controversy as to the means by which IAA stimulates the pump. According to Gabathuler and Cleland (1985) IAA decreases the K_m value of the H^+ ATPase and hence raises the affinity of the enzyme (ATPase) to the substrate (ATP). Ruck *et al.* (1993) reported that the effect of the auxin was to hyperpolarise the plasmamembrane. Felle *et al.* (1991) experimenting with coleoptiles of *Zea mays* reported that hyperpolarisation of the plasmamembrane took place about eight minutes after the addition of IAA. However, addition of IAA to cell suspensions of maize resulted in depolarization associated with a strong import of positive charge which according to the authors was due to a proton cotransport of anions into the cytosol. This example shows that auxins may also promote the uptake of nutrients. Protons pumped into the cell wall by the ATPase are of utmost importance for the growing process.

According to the acid-growth theory of Hager *et al.* (1971) low pH in the apoplast is a prerequisite for cell expansion in the activation of hydrolizing enzymes cleaving cell wall bonds (Fry 1989) and also for substitution of Ca^{2+} by H^+ and thus breaking Ca^{2+}-bridges. By this cleavage of cell wall bonds the strength of the cell wall is weakened and can be stretched as occurs in cell expansion followed by cell division (Jones 1994). This close relationship between the auxin supply and growth as expressed by the elongation of *Avena coleoptile* was already appreciated in the early experiments of Went and Thiman (1937 quoted by Jensen *et al.* 1998). Indeed growth of coleoptiles was formerly used widely as a biotest for assessing auxin concentration.

Meristematic cells are strong sinks for metabolites the uptake of which is mainly dependent on proton cotransport as shown in Figure 5.8. From this it is clear that the stimulation of the H^+ pump by auxins has an impact on the supply of the meristematic cells with metabolites and therefore on growth. Sugars thus imported into the cytosol supply the cell with energy. Similarly amino acids provide building blocks for proteins and nucleic acid synthesis. The import of K^+ is required amongst other things (see page 499) for building up cytosolic K^+ concentration needed for protein synthesis (Wyn Jones and Pollard 1983). Evidence for the uptake of K^+ into the cytosol of *Avena* coleoptiles by facilitated diffusion as a result of auxin supply hyperpolarising the plasmalemma has been obtained by Senn and Goldsmith (1988). Figure 5.8 also shows an auxin binding protein in the plasmalemma where the auxin binds and initiates a reaction sequence (see below). Much has still to be learned about the metabolism of auxins, their synthesis and degradation, and also the precise sites at which auxins they function. In recent years, however, proteins have been found which bind auxins very specifically and it is therefore supposed that these so called *auxin binding proteins* are target molecules for auxins. These auxin binding proteins are involved in reaction chains initiated by auxins (Jones 1994). Interestingly such auxin binding proteins have been found in the endoplasma reticulum (ER), the plasmalemma, the nucleus, in the cytosol, and also in low concentration in the apoplast. Jones (1994) suggests in a useful review article that the apoplast auxin binding protein may interact with a plasmamembrane proteins by modulating channel or pump activities of the H^+ ATPase. Binding of auxin to the protein (*target molecule*) may also initiate a chain of actions coupled with secondary messengers, such as guanosine triphosphate (GTP), Ca^{2+}, and phosphoinositol. According to Blatt and Thiele (1993) the effect of auxin on the plasmalemma H^+ ATPase enzyme is required for distinct biochemical reaction sequences. Some enzymes are also predicted to bind to auxins. Estruch *et al.* (1991) found a 60 kDa protein, a glucosidase, which had a low auxin, but a high cytokinin affinity. The cytokinin is O-linked to the glucose and in this state it is not biologically active. It is activated, however, by the binding of auxin to glucosidase which cleaves the glucose to release the cytokinin. Such a process may be of central importance for the interplay between both phytohormones, auxin and cytokinin.

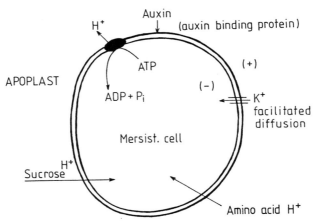

Figure 5.8 Uptake of amino acids and sucrose by H^+ cotransport and K^+ by facilitated diffusion brought about by the plasmalemma H^+ pump and the auxin binding protein initiating the reaction sequence.

Auxins may be oxidized by IAA oxidase which is activated by Mn^{2+} and uv light. Oxidation products are indole aldehyde and 3-methylenoxindol. The proces is inhibited by phenolic compounds and coumarin. Formerly it was assumed that the oxidative decarboxylation was the main process of catabolic breakdown of auxins. There is now evidence that in the normal catabolic reaction sequence the indole is oxidized to an oxiindole with subsequent glycosylation of the carboxylic group (Normanly *et al.* 1995).

Auxins are typical growth substances which exert a unique influence on meristematic cells where all organic growth begins.

The term *Gibberellin* originates from the fungus *Gibberella fujikuroi* which lives in paddy rice soils and which may stimulate the rice culm elongation by the secretion of gibberellins. Today more than 120 gibberellic acids (GA) are known although of these only a few are active (Kende and Zeevaart 1997, Evans 1999). The basic structure of gibberellins is the gibban ring (Figure 5.9). In the biosynthetic pathway of gibberellins the kaurene ring is an important intermediate, the precursor of which is geranyl-geranyl diphosphate (Graebe 1987). This means that the synthesis of gibberellins is related to lipid metabolism and particularly to acetyl-CoA. The synthetic pathway for gibberellins may differ slightly between various organisms but for higher plants the hydroxylation in C-13 is an essential step (see Figure 5.9).

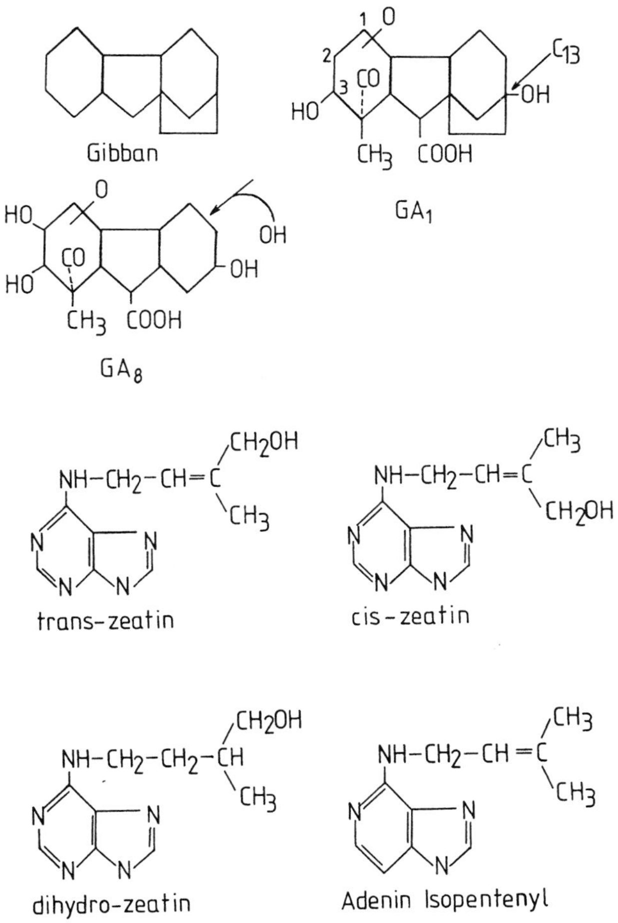

Figure 5.9 Molecular structure of various gibberellins and cytokinins.

The most important gibberellins are gibberellic acid 1 (GA-1) and gibberellic acid 4 (GA-4). Activity requires a β-hydroxyl at the C-3 position as shown in the formula (Figure 5.9). Recently the gene has been identified which codes for the corresponding 3-β hydroxylase which hydroxylates GA-9 and GA-20 to the active forms GA-1 and GA-4, respectively (Williams et al. 1998). Deactivation of active forms occurs by a further hydroxylation at C-2 in which the GA-1 is converted to GA-8 and GA-4 to GA-34 (Figure 5.4). Gibberellins are synthesized in plastids. Generally gibberellin concentration is higher in seeds than in vegetative organs although according to present knowledge gibberellins have no direct function in seeds (Graebe 1987). Young shoots and roots need gibberellins for elongation and seeds may provide gibberellins to these organs.

A major function of gibberellins is the inducton of cell expansion and stem elongation. It is supposed that gibberellins similar to auxins stimulate the plasmalemma proton pump. The resulting effect of gibberellic acid in combination with K^+ is shown in Table 5.5.

Table 5.5   Effect of gibberellin in combination with K^+ on the growth rate of sunflower internodes (data from Dela Guardia and Benlloch 1980)

Concentration of K^+, mol/m^3	Amount of GA-3, μg	Growth rate
0.0	0	3.8
0.5	0	6.0
5.0	0	19.0
0.0	100	29.2
5.0	100	41.4
5.0	100	56.6

This spectacular effect of GA on stem elogation presumably results from cell expansion associated with the formation of vacuoles promoted by GA. The additive effect of K^+ on the growth of internodes results from the effect of K^+ in increasing cell turgor, high cell tugor being required for cell expansion (page 219). Gibberellins enhance the growth of potato stolons and the tuberization of potato (Xu et al. 1998). As shown in Figure 5.10, tuberization frequency was suppressed with increasing concentrations of gibberellins.

Dwarf plants have shorter internodes than their taller analogues. These shorter internodes can be increased in length by the application of gibberellins since in most dwarf plant types one or more steps in the synthetic pathway of active GAs are impaired. Some plant species do not develop normally under short day conditions. In these species long days are neccessary for the synthesis of active GA forms and under long day conditions plant stems elongate and generative organs are developed (Graebe 1987). In a valuable review article Evans (1999) discusses the importance of the molecular structure of gibberellins, especially the location of hydroxyl groups, in relation to their physiological function.

More recent research has shown that gibberellins may delay leaf senescence as found for *Alstroemeria hybrida* (an ornamental plant) by Jordi et al. (1993). Kappers et al. (1998)

reported that leaf senescence in alstroemeria was delayed by gibberellins and red light. The effects were independent of each other as can be seen from Figure 5.11 (Kappers *et al.* 1998). Increasing concentrations of GA-4 led to an increase in chlorophyll concentrations indicative of a delay of senescence. This effect was found both in the dark treatment and in the treatment with the application of red light. Red light application, however, resulted in a higher level of chlorphyll. Interestingly the duration of red light application was only 10 min/day which shows that relatively short exposure to red light was suffcient to initiate chlorophyll production.

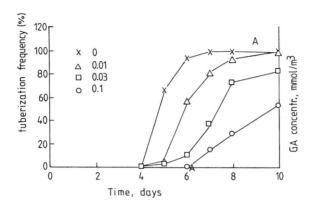

Figure 5.10　Tuber formation frequency as related to concentration of exogenously supplied gibberellins (after Xu *et al.* 1998).

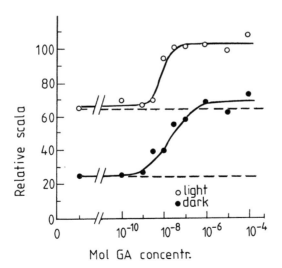

Figure 5.11　Effect of increasing GA-24 concentrations on relative chlorophhyll concentrations in leaves of *Astroemeria* hybrid subjected dark and light treatment (after Kappers *et al.* 1998).

Mengel *et al.* (1985) reported that wheat plants grown at low light intensity had significantly lower gibberellin concentrations in their grains as compared with control plants whereas the concentration of other phytohormones in the grains such as cytokinins, auxins and ABA was virtually unaffected. This specific effect of low light intensity in decreasing gibberellin concentration was associated with a single grain weight about half that of the grain weight obtained in the control. It is a matter of speculation, however, as to whether the low gibberellin concentration was a direct cause of the poor grain growth. Chlorocholine chloride (CCC) is an inhibitor of gibberellin synthesis and is used as a growth regulator for reducing the culm length of cereals (Linser *et al.* 1961).

The basic structure of *Cytokinins* is the adenine ring in which one H in the C6-bound NH_2 is substituted by isopentenyl or isopentenyl derivates. The structure of some naturally occurring cytokinins are shown in Figure 5.9 (Binns 1994). The synthetic compound 6-benzylamino purine (kinetin) has frequently been used in experimental work as a cytokinin substitute (Green 1983). The basic building block of isopentenyl moiety and its derivatives is acetyl CoA. Hence the synthesis of cytokinins is related to the N and lipid metabolism. As yet the precise pathway of cytokinin biosynthesis is unknown (Binns 1994). Cytokinins are a further important group of growth homones, promoting elongation and frequently acting in combination with other factors including other phytohomones. They depress apical dominance and thus have an antagonistic effect on auxins and promote the growth of laterals (quoted after Binns 1994). In studies of developing young rootlets developing from tomato cuttings Maldiney *et al.* (1986) found that in the beginning of the radicle formation IAA concentrations were high and declined thereafter. This first step of root formation, the reactivation of the pericyclic cell and the formation of root primordia was associated with a high IAA/Cytokinin ratio, whereas the second step of root formation, root elongation, was characterized by low concentrations of IAA and cytokinins and by a low IAA/cytokinin ratio so that at this this stage the cytokinins were of greater importance. Phytohormone concentrations were in the range of 50 to 250 µg for IAA, 100 to 2000 µg for ABA and zero to 125 µg for cytokinins per g fresh weight. Considering that the elongation is associated with the formation of vacuoles the cytosolic hormone concentrations and particularly that of the cytokinins must of course be much higher. From this it follows that especially in the phase of root elongation, cytokinins play an essential role. This is in accordance with earlier research work of Green (1983) who reported that growth rates of excised cucumber cotyledons was much increased by the addition of cytokinin (6-benzyl-aminopurine). Cytokinin promoted growth was associated with a decreasing amount of sucrose in the cotyledons as shown in Figure 5.12. This means that growth processes initiated by cytokinin consume energy. In this experiment of Green (1983) cytokinin induced the formation of ethylene.

A further important effect of cytokinins is the delay of leaf senescence (Thimann 1980). In this context there is evidence that cytokinins are involved in transcriptional regulation of gene expression (Lu *et al.* 1992). Frequently, however, cytokinin effects are related to other influences such as light or stress. For example in the transcriptional regulation of the gene encoding for nitrate reductase, light and nitrogen nutritional status are also involved besides cytokinin. Cytokinins are also essential for the control of the cell cycle

(Binns 1994). From this follows that cytokinins function at central metabolic sites. Cytokinins are synthesized in shoots and roots but generally roots are the more important location for synthesis.

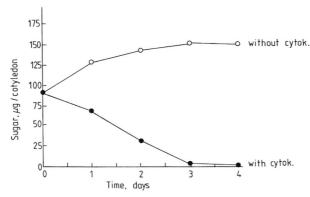

Figure 5.12   Sucrose content in cucumber cotyledons with and without cytokinin application (modified after Green 1983).

Like the other groups of phytohormones (auxins, gibberellins) the active form of cytokinins can also be inactivated by glycosylation. N-cytokinin glycosides are very stable and the formation of N-glycosides supposedly withdraws cytokinins from the active pool (Letham and Palmi 1983). In contrast, the O-glycosides are deglycosylated readily and may be involved in homeostatic control (maintenance of a constant level) of active cytokinin concentration. The effects of cytokinin are induced by increased concentration although too high a concentration may have a suppressing influence.

Abscisic acid (ABA) is rather a growth inhibiting than a growth promoting phytohormone. In many cases it is the natural antagonist of auxins, gibberellins and cytokinins. Its molecular structure (Figure 5.13) is characterized by a ionon ring showing its relationship to carotenoids, the epoxy ionon ring being essential for ABA synthesis (Kende and Zeevaart 1997).

The name abscisic acid is derived from leaf abscission which is induced by ABA. Drying out of soils stimulates the ABA synthesis in root tips and the translocation of ABA from roots to shoots (Davies and Zhang 1991). This occurs even under mild water stress with a soil water potential of -200 to -300 kPa. ABA transport is *via* the xylem and ABA concentration in the xylem sap, generally in the range of 10 μmol/m$^+$, may rise rapidly with mild water stress by one or two orders of magnitude. ABA arriving at sites of transpiration in the epidermal apoplast adjacent to guard cells initiates stomatal closure. Salt stress also considerably increases ABA levels in plant tissues. ABA is therefore also called a stress phytohormone, ABA transport from roots to upper plant parts being promoted under stress conditions and ABA functioning as a signal to modulate protection mechanisms against the stress. Proline and ethylene synthesis are stimulated by ABA, both compounds having a protective function under water and saline stress. Ludewig *et al.* (1988) concluded from their results with decapitated sunflowers that ABA increases

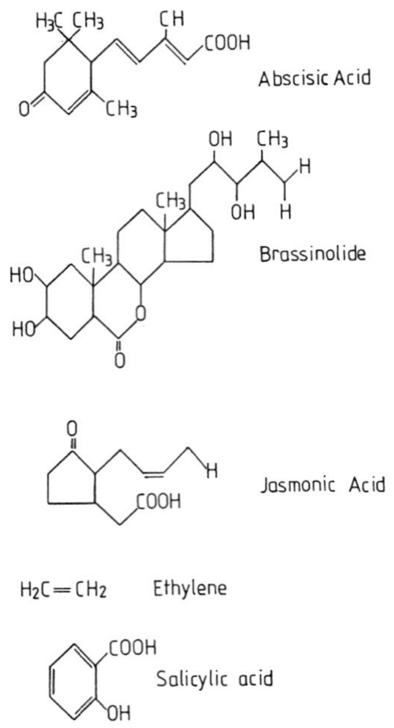

Figure 5.13 Molecular structure of abscisic acid, ethylene, brassinosteroids, jasmonic acid and salicylic acid.

membrane permeability for water thus facilitating the transport of water across membranes and also the uptake of water from the soil. Combined with the closure of stomata this mechanism improves or maintains water status under stress conditions. Cold acclimation has also been associated with an increase of ABA concentration in barley leaves with a level about four fold higher than the control being recorded (Bravo *et al.* 1998). The mechnism by which ABA improves freezing tolerance is not yet understood but it could relate to the maintainance of a favourable water status or the induction of the expression of specific genes which code for substances protective against freezing. In this context the water replacement hypothesis is of interest where stabilization of hydrogen bonds is achieved between OH-groups of sugars and polar groups of phospholipids and proteins. In these aggregations the sugars replace the water molecules (Crowe *et al.* 1993). ABA synthesis is also stimulated by insufficient nutrient supply as found for K^+ by Haeder and Beringer (1981).

An important finding is that a decrease in soil water availability is associated with an increase in xylem sap pH (Schurr *et al.* 1992) and pH increases in xylem sap from 5 to pH 8 may be obtained (Wilkinson *et al.* 1998). Generally xylem sap pH is in the range of 5.0 to 5.5 (Kosegarten *et al.* 1998). Presumably the apoplast pH must also be affected under such high xylem pH conditions with adverse effects on proton cotransport processes across the plasmamembrane (see page 126). pH in xylem sap and apoplast liquid depends on the activity of the plasmalemma proton pump (Hoffmann and Kosegarten 1995) which is known to be stimulated by auxin. It would seem that ABA also functions here as an auxin antagonist as shown by Schubert and Matzke (1985). The fall off in proton pumping associated with a depolarization of the plasmamembrane brought about by ABA induces the efflux of K^+ out of the cell and guard cells which decrease in turgor. Additionally ABA may cause a blocking of plasmalemma K^+ channels (Ding and Pickard 1991). High pH in the apoplast of growing tissue may stop growth completely because of the lack of sufficient H^+ to obtain an optimum cell wall extension (Kutschera and Schopfer 1986). This finding supports results of Marschner et al. (1984) who reported a negative correlation between the ABA concentration in potato tubers and their growth rate. Presumably inadequate cell wall extension at high ABA concentration is the reason why ABA acts as a growth inhibitor and why it is such a strong antagonist to gibberellins and auxins which stimulate the proton pump. Since leaf conductance for gases is lowered by ABA through stomatal closure, the uptake of CO_2 is suppressed and therefore also the formation of photosynthates.

ABA is not only synthesized in the roots but also in chloroplasts where it is partially sequestered by a so called anion trap (Davies and Zhang 1991). In the undissociated form ABA can readily penetrate biological membranes because of its lipophilic ionon ring. The uptake of ABA from the apoplast which generally has a low pH is therefore supposedly a passive process and high uptake rates are the reason why ABA concentrations in the apoplast are low. Partial deprotonation of ABA may occur in the cytosol and particularly in the chloroplasts as a result of the high pH in the stroma. In this anionic form rediffusion is restricted so that the anions are trapped. This ABA trap may regulate ABA activity. In addition ABA may be rapidly deactivated by oxidation to phaseic acid or by conjugation to the ester form (Kende and Zeevaart (1997). According to Müller et al. (1989) soil fungi and algae produce ABA and release it into the rhizosphere from where it can be taken up by plant roots. ABA induces the dormant phase of plant organs and its concentration in seeds and dormant buds is particularly high and falls off gradually with an increasing tendency to germination and bud burst, respectively. Concomitant with this decrease in ABA concentration, levels of gibberellins and auxins increase. ABA stimulates the tuberisation of potato tubers and is counteracted in this function by gibberellins which promote stolon elongation (Xu et al. 1998).

Brassinosteroids comprise a group of about 60 naturally occurring polyhydroxy-steroids with brassinolide the most prominent one (Figure 5.13). It was at first isolated from rape (*Brassica napus*) pollen (Grove et al. 1979). Brassinosteroids are growth promoting natural products found at low levels in pollen, seeds and young vegetative tissues throughout the Plant Kingdom. They are essential for normal plant growth and according to Clouse and Sasse (1998) must be considered along with auxins, gibberellins cytokinins, ABA and ethylene as a defined group of phytohormones. Biosynthesis is from isopentenyl diphosphate with campesterol and campestol as important intermediates. Brassinosteroids retard leaf abscission, enhance stress resistance and are involved in cell division and cell elongation presumably by enhancing cell wall extension. Pollen and immature seeds have the highest concentration of brassinosteroids which is in a range of 1 to 100 ng/g fresh weight. The concentration in leaves is about 100 to 1000 times lower. A useful review article on brassinosteroids has been published by Clouse and Sassse (1998).

Ethylene is formed from methionine with S-adenosyl-L-methionine and cyclic l aminocyclopropane- 1-carboxylic acid (ACC) as intermediates in the synthetic pathway and ACC synthase as the key enzyme for synthesis (Kende 1993). The enzyme has been found in the pericarp of various fruits where ethylene is involved in the fruit maturation. The activity of the ACC synthase is enhanced by stress wounding and auxin. Ethylene induces ABA synthesis and thus may induce massive leaf abscission under salt stress (Gomez-Cadenas *et al.* 1998). Auxin initiates the synthesis of ethylene which blocks the the formation of lateral buds in the effect of apical dominance (see page 267). According to Green (1983) cytokinin also promotes the synthesis of ethylene with Ca^{2+} functioning as synergist in this process. Catabolic breakdown of ethylene is brought about by the formation of ethylene oxide and ethylene glycol (Kende and Zeevaart 1997). Ethylene stimulates flowering and fruit maturation. The specific biochemical reactions in which ethylene is involved are as yet poorly understood.

Recent research has shown that *Jasmonic Acid* is an important phytohormone. Together with the brassinosteroids it belongs to the non-traditional phytohormones. In nature two forms of jasmonic acid (JA) have been found, (-) JA and its isomer (+) -7-iso JA. Jasmonates are widespread in the plant kingdom and have various physiological activities (Sembdner and Parthier 1993). The molecular structure of JA is shown in Figure 5.13.

The synthetic pathway of jasmonic acid begins with linolenic acid. Structural groups of the molecule required for biological activity are the acetyl C-3 chain, the pentenyl chain at C-7, and the keto group at C-6. Jasmonic acid is known to promote leaf senescence and associated chlorophyll and Rubisco degradation. On the other hand the finding that JA levels are high in zones of cell division, young leaves, and reproductive tissues suggests that JA plays a direct role in growth processes (Creelman and Mullet 1997). According to Koda (1992) JA is effective in tuber formation of potato, of yam (*Dioscorea*), and possibly also of tubers of *Helianthus tuberosus*. Jasmonic acid induces the synthesis of proteins which may serve as transient storage proteins and proteins which are involved in the defence of plant cells against pathogens, herbivores and also physical and chemical stress. Jasmonic acid supposedly stimulates seed germination by counteracting the effect of ABA (Creelman and Mullet 1997).

The *Oligosaacharins* are also included in the non-traditional phytohormones. They act as elicitors in pathogen defence mechanisms.

Salicylic acid (Figure 5.13) is known for inducing resistance to bacterial, fungal, and viral attack. Resistance develops at a distance from the site of attack and it has been shown that the signal for inducing the resistance is tranferred by salicylic acid (Gaffney *et al.* 1993).

5.2.3   Growth rate and nutrient supply

The yield of a crop may be considered in biological as well as agricultural terms. Biological yield has been defined as the total production of plant material by a crop whereas the economic yield or commercial yield takes into account only those plant organs for which particular crops are cultivated and harvested. Obvious examples include cotton bolls, grains, seeds, tobacco leaves, potato tubers, etc. For a number of crop plants such as forage crops and some vegetables, the amount of plant material produced above ground during the vegetative growth stage is equivalent to the economic yield. For most crop plants, however, this is not the case and plant development during the vegetative stage controls both the biological and economic yields. This dependence on the vegetative growth stage lies in the fact that during this period green plant tissues are formed which provide photosynthates for seeds or other storage tissues. Holliday (1976) denotes the economic or commercial yield (Yc) as the product of the total yield (Yt) and the *crop index* (Ic) = *harvest index*. This latter term is always < 1, and denotes the proportion of harvested crop, *e.g.* grains, of the total yield:

$$Yc = Yt \times Ic.$$

As a result of plant breeding, modern cultivars often have higher crop indices than older ones. The crop index of modern wheat cultivars is thus about 35 to 40%, whereas older cultivars have indices in the range of 23 to 30% (Austin *et al.* 1993).

Vegetative growth consists mainly of the growth and formation of new leaves, stems and roots. As meristematic tissues are charaterized by high protein synthesis, photosynthates transported to these sites are used predominantly in the synthesis of nucleic acids and proteins. It is for this reason that during the vegetative stage, the N nutrition of the plant controls the growth rate to a large extent. A high rate of growth only occurs when abundant N is available as shown in Table 5.6. In the treatment with the low level of N nutrition (0.5 g N/pot), the dry matter yield was much lower than in the treatment with sufficient N supply (2.0 g N/pot). The plants with the low N supply accumulated carbohydrates, particularly starch and fructans in leaves and stems, whereas their concentration of crude protein was much depressed. This clearly demonstrates that where N-nutrition is inadequate, photosynthates can only be utilized to a limited extent in the synthesis of organic N compounds. The remainder is stored in the form of starch and fructans. In an analogous way insufficient K^+ (Scherer *et al.* 1982) or phosphate supply (Freden *et al.* 1989) may retard vegetative growth associated with an accumulation of carbohydrates in leaves.

Table 5.6   Effect of N supply on the yield and organic constituents of young ryegrass (*Lolium perenne*); (Hehl and Mengel 1972).

	supply (g N/pot)	
	0.5	2.0
Yield, g dry matter/pot	14.9	26.0
Crude protein (g/kg DM)	123	264
Sucrose (g/kg DM)	77	63
Fructans (g/kg DM)	100	10
Starch (g/kg DM)	61	14
Cellulose (g/kg DM)	144	176

The example presented above demonstrates that for optimum growth of plants there must be a balance between the rate of photosynthate production and the rate of N assimilation. Under conditions where high photosynthetic activity can occur (high light intensity, optimum temperature, absence of water stress), the level of N and other plant nutrients must also be high and *vice versa*. In plant species in which the rate of CO_2 fixation is high and particularly species which assimilate CO_2 *via* the C-4 pathway the N demand is considerable when growth conditions are optimum (see Figure 5.15 on page 284).

The level of N-nutrition required for optimum growth during the vegetative period must also be balanced by the presence of other plant nutrients in adequate amounts. The synthesis of organic N compounds depends on a number of inorganic ions, including Mg^{2+} for the formation of chlorophyll and for the activation of numerous enzymes; phosphate for the synthesis of nucleic acids. Protein synthesis depends on the K^+ concentration in the cytosol (Wyn Jones and Pollard 1983). Potassium is important for growth and elongation probably in its function as an osmoticum and may react synergistically with indole acetic acid (Cocucci and Dalla Rosa 1980). The synergistic effect of gibberellic acid and K^+ on the elongation of sunflower stems is shown in Table 5.5 (Dela Guardia and Benlloch 1980).

Generally meristematic growth and leaf expansion depend much on water supply (Durand *et al.* 1997) since these processes are most sensitive to water stress (page 219). According to Kosegarten *et al.* (1998) the adequate provision of Fe to leaf primordia is essential for the development of young leaves. Growth and development at this early stage are very much dependent on temperature. Brouwer et al. (1973) reported that the rate of leaf appearance of young maize plants was mainly controlled by temperature, whereas light intensity had hardly any effect. Cell division and cell expansion are more sensitive to low temperature than is photosynthesis (Woolhouse (1981). Temperature increases tend to stimulate growth which in turn results in a dilution of carbohydrates and chlorophyll. This occurs particularly when the light intensity is low (cloudy weather). Under such conditions large pale-green leaves with long stems may be produced (Warren-Wilson (1969). On the other hand, high light intensity and lower temperatures especially during the night, result in smaller, more fungal disease-resistant plants with higher concentrations of chlorophyll and carbohydrates.

Unlike animals, plants cannot move from place to place to seek out better sources of nutrients including water. They must therefore expand the surfaces involved in the uptake and capture of nutrients and sunlight through the elongation of branches and stems, leaf expansion and the formation and branching of the root system including root hairs (Taylor 1997).

5.2.4   Grain crops

Grain yield depends on three main yield components: number of ears per ha, number of grains per ear and the single grain weight. The number of ears per ha in a cereal crop depends on seed density and tillering capability. Tillering capability is genetically controlled, but is also much dependent on environmental factors. Short day conditions associated with high light intensities, low temperatures and ample nitrogen supply favour tillering (Evans *et al.* 1975). These environmental conditions considerably influence phytohormone activity. Long day conditions and high temperatures are associated with a high auxin (IAA) production in the apex of the young primary cereal shoot (Michael and Beringer 1980). IAA or related compounds induce the production of ethylene, which inhibits the growth of lateral buds and thus the formation of tillers. This apical dominance occurs very strongly in monoculm varieties. The stimulating effect of N nutrition on tillering is thus partially the result of the favourable effect of N on cytokinin synthesis. In the stage of ear formation nitrogen has a beneficial influence on the initiation of spikelets during ear development and hence on grain number per ear (Ewert and Honermeier 1999).

Under the climatic conditions of Central Europe, maximum grain yields of wheat (6 to 7 Mg grain/ha[-1]) are obtained, when the number of ears is about 500 to 600 m^2. Such a high plant density is not always desirable. In more arid climates, a lower density often produces higher grain yields, as lower plant densities do not require so much water. In critical growing phases such as heading and flowering, the possibility of water stress is thus reduced. According to Day and Intalap (1970) water stress at heading decreases both the number of ears per unit area and grains per ear. Water stress during flowering accelerates the maturation process thus leading to smaller grains and lower grain yields. This observation is consistent with the findings of Pelton (1969) made under the arid conditions of South-Western Saskatchewan in the U.S.A., that lower plant densities were associated with longer ears, larger grains and thus higher grain yields of wheat. Because of a lack of water the importance of rain fed agriculture will increase in the near future. For the optimum use of the available water, crops should be well supplied with nitrogen as shown by Ryan *et al.* (1998) for wheat under the arid conditions of Morocco at fertilizer N rates of 40 kg N/ha.

Grain yield is influenced by various environmental factors of which water supply is one of the most important. Christen *et al.* (1995) applied water stress at three stages of wheat growth: 1. from stem elongation to flag leaf formation, 2. from flag leaf stage to ear emergence, and 3. from ear emergence to anthesis. At each of these stages soil water potential was reduced. A major impact of the water stress on grain yield was found in the phase from ear emergence to anthesis when the grain yield of the secondary tillers rather than that of the main shoot was reduced. Water stress mainly affected the number of kernels per ear and to a lesser extent the single grain weight. Insufficient water supply particularly reduced the apex length of the secondary shoots and this was obviously the cause of the decrease in number of grains per ear. Drought, *i.e.* intense water stress, lasting only for a few days in the grain filling stage has a disasterous influence on grain yield. In barley the grain number per ear and particularly the single grain weight were much depressed as was the green surface area of plants and the starch production (Savin and Nicolas 1996). Elevated temperature may also reduce the growth period and the duration of grain filling giving rise to low grain yields and the harvest indices, the effect being more pronounced the higher the temperature (Batts *et al.* 1997). The longer the period of ear formation the more spikelets can be developed and the greater is the chance of producing long ears containing a large number of grains (Evans *et al.* 1975). However, as the duration of ear development is inversely related to the duration of grain filling, a high number of grains per ear is often associated with a low single grain weight. Grains are developed within the spikelets which are initiated at an early stage in cereal growth (Plate 5.5). During subsequent development, a considerable proportion of spikelets degenerate. In the case of rice, degeneration can account for as many as half of the original number of spikelets (Yoshida 1972). Degeneration is promoted when N-nutrition is inadequate and by low light intensity or low temperatures at the time of spikelet development (Fuchs 1975).

The third important component of cereal yield, the grain size or single grain weight, is genetically controlled as well as depending on the environmental factors which influence the process of grain filling during maturation. In the filling process the ears or grains act as a physiological sink. The strength of this sink depends greatly on the number of endosperm cells developed (Höfner et al. 1984). According to Schacherer and Beringer (1984) K^+ has a beneficial influence on the development of endosperm cells and hence on the single grain weight of cereals. Endosperm development is restricted under dry weather conditions as reported by Tuberosa et al. (1992) for barley. The source for the grain sink are the leaves and to a much lesser extent hulls and awns. After the onset of flowering, photosynthates are used more and more for the grain filling process. This is shown in Figure 5.14 in which the distribution of photosynthates in a wheat plant is shown for the various stages from flowering until maturation (Stoy 1973). The arrows indicate the direction in which the photosynthates are translocated. The plant parts shown in black are the most important for assimilation. In the later stage, the flag leaf in particular provides photosynthates for grain filling. In awned wheat varieties photosynthates from the flag leaf contribute to about 70% of the total grain filling, whereas in varieties without awns about 80% may originate from the flag leaf. The remainder of assimilates mainly comes from the ear itself.

Not all the photosynthates required for grain filling are photosynthesized during the reproductive period (after flowering). Some of the carbohydrates are synthesized before flowering and stored in stems and leaves during the vegetative growth stage. The proportion of stored assimilates used in grain filling can vary from 0 to 40% for rice, 5 to 20% for wheat, 12 to 15% for maize and be in the region of 20% for barley (Yoshida 1972). These values depend to a large extent on photosynthetic activity after flowering and also on the conditions for the long distance transport of photosynthate. In water stressed plants long distance transport is much restricted and non-structural carbohydrates deposited in the culm are hardly used for grain filling after pollination as found by Phelong and Siddiqui (1991) under arid conditions in Australia. With optimum water supply, however, the carbohydrates stored in the culm contributed to 20% to grain filling. The most important carbohydrates stored transiently in the culms and leaves of grasses and cereals are the fructans (Pollock and Cairns 1991) which are deposited in the vacuoles during stem elongation. After anthesis fructans are mobilized for grain filling and converted to sucrose which is translocated via the phloem into the endosperm cells of the developing grains. According to Duffus and Binnie (1990) sucrose concentrations in the vacuoles of young wheat caryopses are high to protect the embryo from desiccation.

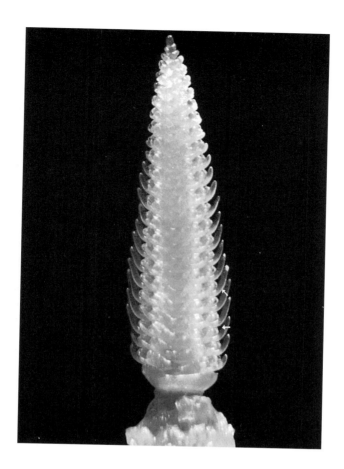

Plate 5.5 Young developing ear of barley in the 6th leaf stage at which spikelet degeneration frequently occurs. (Photo: Ruckenbauer).

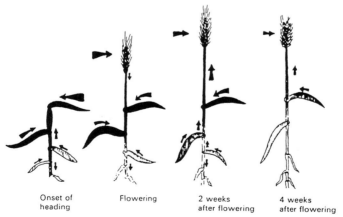

| Onset of heading | Flowering | 2 weeks after flowering | 4 weeks after flowering |

Figure 5.14 Assimilation and distribution of photosynthates at different stages in growth (after Stoy 1972).

A high rate of grain filling is obtained if the level of N-nutrition is high during the grain filling period and the K^+ status of plants is optimum. A high N concentration in leaves also implies a high concentration of chlorophyll and of enzymes involved in CO_2 assimilation, especially Rubisco. In crops well supplied with N, senescence of the flag leaf is delayed and respiration losses are low. Table 5.7 shows the influence of dropping the level of N nutrition during the grain filling period on the grain yield of spring wheat (Forster 1973a). The nutrient solution in the control treatment was maintained at a NO_3^- concentration of 6.2 mol/m^3 whereas in the comparative treatment it was lowered to 1.2 mol/m^3 NO_3^-. This lower level of N-nutrition during the grain filling period resulted in a yield reduction, which was largely accounted for by the smaller grains, obtained in this treatment. The concentration of crude protein in the grain was also lowered. The yield reduction, however, could not be explained solely by the lower crude protein concentration in the grains. This amounted to only 1.2 g per 1000 grains, whereas the yield depression (difference between the thousand grain weights) was 2.8 g per 1000 grains (TGW).

Table 5.7   Effect of N supply in the grain filling stage on the yield of spring wheat (Forster 1973a)

mol/m^3 NO_3^- nutrient solution	Grain yield g/16 plants	TGW	Crude protein mg/g
6.2	51.4 (100)	38.2 (100)	206 (100)
1.2	46.6 (100)	35.4 (100)	189 (100)

1000 grains of the 6.2 treatment contained 7.9 g crude protein. 1000 grains of the 1.2 treatment contained 6.7 g crude protein: Difference = 1.2 g

Most usually, grain filling is not primarily limited by photosynthate supply to developing grains (Jenner and Ratjen 1975), provided that photosynthate transport is not restricted by adverse conditons such as drought (Phelong and Siddique 1990) or even mild water stress (Tuberosa *et al.* 1992).

During the grain filling process the activities of phytohormone species in the grain attain a maximum value at different times between anthesis and maturation. Cytokinins which reach a peak at about 1 week after anthesis are believed to control the formation of grain endosperm cells and thus have a marked influence on grain size. This view is consistent with findings of Aufhammer and Solanski (1976) with spring wheat and Herzog and Geisler (1977) with spring barley, who observed that cytokinin application increased grain yield exclusively by increasing single grain weight. The peaks of gibberellic acid (GA) and indole acetic acid (IAA) occur about 4 and 5 weeks, respectively after anthesis. These two types of phytohormones also probably promote grain growth, while abscisic acid (ABA) has the reverse effect. Its peak coincides with the peak of grain fresh weight. This finding may be interpreted as an inducing effect of ABA in the maturation process which is in accordance with the well known effect of ABA in promoting senescence. A useful review on growth and development of wheat in relation to temperature was published by Porter and Gawith (1999).

In contrast to the other cereals, grain yields of rice can only be improved to a very limited degree by increasing grain size. This is because grain growth is physically restricted by the size of the hulls (Yoshida 1972). If the nutrient status of the plant is adequate it is not the rate of photosynthesis (physiological source) that restricts growth, but rather the physiological sink which is the limiting factor in rice yields. The strength of the physiological sink is dependent on the number of grains per unit area (Tanaka 1973). This number can be raised either by increasing the plant density or by increasing the number of grains per panicle and panicles per plant. The number of panicles per plant is initiated fairly late in tillering, and the number of spikelets per panicle is determined about 10 days before flowering. These critical stages are important for the grain yield of rice and only under optimum nutritional and weather conditions are a high number of fertile spikelets developed. When this occurs a strong sink capacity results which requires an abundant supply of carbohydrates during the grain filling period which is only attained when the N concentration in the leaves is high (Schnier *et al.* 1990a). Such high N concentrations are indicative of high concentrations of chlorophyll and enzymes involved in CO_2 assimilation. Shading at this stage by too dense crop stands affect photosynthesis (Schnier *et al.* 1990b) and may lead to an increased number of empty hulls.

Tillering in rice is favoured by low temperature, high light intensity and an abundant N supply. Spikelet degeneration is low under conditions of high light intensity and *vice versa*. Grain filling is improved by adequate O_2 supply to the roots during the grain filling stage (Murata and Matsushima 1975), as O_2 retards root senescence and hence allows the roots to supply the upper plant parts with cytokinins. The distribution of available photosynthates is affected by both the number of panicles and number of spikelets per panicle. If the rate of assimilate production is limiting, a high number of panicles per unit area is often accompanied by a reduction in the number of grains per panicle, and an increase in the number of empty hulls.

The growth rate after flowering also has a bearing on the grain yield of rice. If the maturation period is shortened, as can occur if the crop is deficient in N, a decrease in grain yield results. After flowering, rice especially needs an abundant supply of N and K, and for obtaining top yields in the range of 10 Mg grain/ha, a late dressing of N and K is often applied (Tanaka 1972). Table 5.8 shows the model yield components of a high yielding rice culture (Toriyama 1974). In the wet season where light intensity is low, solar radiation is often the yield limiting factor, and this considerably reduces the number of grains per unit area.

Flooded rice is unique in its efficient uptake and assimilation of NH_4^+ being adapted to soil conditions in which NH_4^+ prevails and NO_3^- levels are extremely low because of denitrification (see page 349). At neutral soil pH with the presence of algae in the flooding water the pH of the water can attain pH levels as high as 10 during the day (Mikkelsen *et al.* 1978) with substantial quantities of NH_3 present. The uptake of this NH_3 is not controlled and NH_3 may therfore rapidly penetrate the plasmalemma of the epidermal root cells. In the case of plants other than rice this can induce toxicity as reported by Kosegarten *et al.* (1997) for maize. For rice, however, these authors were able to demonstrate that NH_3 was not toxic because of the presence of a highly efficient glutamate synthase which was capable of assimilating the high rates of NH_3

284

entering the cytosol. Also under conditions of NH$_3$ influx into root cells, a normal low pH was maintained in the vacuole, in contrast to maize where the pH rose (Wilson *et al.* 1998).

Table 5.8   The model yield components of a high yielding rice cultivar grown in different climatic regions (Toriyama 1974)

Component	Monsoon climate		Temperate climate (Japan)
	Wet season	Dry season	
Panicle No per m^2	250	375	4000
Grain per panicle	100	100	80
Total number of grains, m^{-2}	25000	37500	32000
Filled grain in %	85	85	85
1000 grain weight, g	29	29	27
Computed grain weigt, Mg/ha	6.6	9.2	7.5

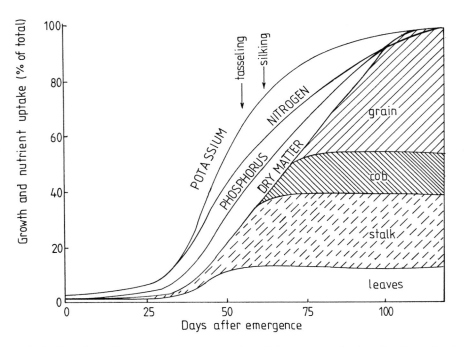

Figure 5.15   Potassium, nitrogen and phosphorus uptake and dry matter production of maize during the growing season (Iowa State University, after Nelson 1968).

Plate 5.6 Cob deformation in maize as a consequence of inadequate K⁺ nutrition (Photo: *Kali und Salz AG, Bern*).

The origin of maize is Mexico and Central America and it is therefore adapted to a semi arid climate with high light intensity. The numerous cultivars of maize available today mean that the crop can be grown under various climatic conditions provided that the temperature is not too low. *Zea mays is* the most prominent C-4 species of all the crops. As a C-4 plant it uses CO_2 with high efficiency due to its low CO_2 compensation point and does not suffer loss of CO_2 by photorespiration (see page 159). CO_2 supply is therefore only seldom a limiting factor in contrast to the C-3 crop species (Edwards 1986). For this reason maize has an enormous growth rate and its incremental gains in organic carbon per day are considerable. This also means that nutrient requirements per day are high in order to meet the inherent yield potential (Figure 5.15). For maximum yields high light intensities are required, higher than that for C-3 species (Fageria *et al.* 1991). Under such optimum conditions including sufficient water, maximum grain yields as high as 20 Mg/ha may be obtained. Such a high yield potential requires high fertilizer rates, especially nitrogen and potassium. Insufficient K⁺ supply may lead to cob deformation as shown in Plate 5.6. Boulaine (1989) reported that in France the average maize grain yield was raised form 2.5 to 6.0 Mg/ha over the years from 1956—1986 as a result of mineral fertilizer application and the use of improved cultivars. Nutrient uptake and dry matter production of maize are shown in Figure 5.15 (Nelson 1968). Maximum grain yields are only obtained if water stress is avoided (Shimshi 1969, Decau and Pujol 1973).

Maize is sensitive to low temperatures and this handicap frequently affects the growth of the crop in cool and wet springs in temperate climates. The young maize plants are stunted and chlorophyll formation is depressed as evident from the light green and yellow leaves. The main reason for this poor growth is the sensitivity of the pyruvate: phosphate dikinase to low temperatures (Edwards 1986). This enzyme is responsible for the regeneration of phospho-enol-pyruvate (PEP), the CO_2 acceptor in C-4 species (see page 155). If its activity is restricted the production of organic carbon is suppressed and the supply of growing tissues with sugars and amino acids is insufficient. Under temperate climatic conditions maize can mature too late, if the weather is cool. For this reason late N applications should not be recommended in regions where maturation may be a problem, as late N supply delays senescence.

Other important C-4 crops are Sorghum (*Sorghum bicolor*) and sugar cane (*Saccharum robustum and S. officinarum*). Sorghum is a crop well suited to semi-arid climatic conditions because of its inherent drought resistance. In the initial stage of development the rate of shoot growth is low in comparison with the simultaneous development of an intense root system with adventitious roots capable of utilizing soil water with high efficiency. After the development of the root system shoot growth starts vigorously. Sorghum may also adjust to drought conditions by cutting down on water loss from tranpiration. This is achieved by decreasing the solute potential in the leaves (*i.e.* increasing the concentration of osmotica, see page 226) thus maintaining the cell turgor at a low water potential (Ackerson *et al.* 1980). The crop can be cultivated over a wide range of soil pH from pH 5 to 8.5 (Fageria 1991). Sorghum is the crop species particularly well suited to rain-fed agriculture and will increase in its importance in future.

Sugarcane is adapted to tropical and subtropical climates and can be cultivated on a wide variety of soils over a pH range from 4 to 9. Because of high growth rates sugarcane needs large quantities of nutrients which are mainly in form of high rates of fertilizer application. This is particularly true for K^+ since the stalks which are removed at harvest contain high amounts of K^+ (Jones 1985). Sucrose is mainly stored in the stalks of sugar cane which at technical maturation contains about 70% water. About 50% of the dry weight is sucrose. Cool temperatures, high solar radiation and moderate N supply promote maturation and are favourable for sucrose storage.

5.2.5 Root crops

Yield physiology of root crops differs in several respects from that of cereals. The main difference between both groups of crops is the fact that in root crops a marked competition for carbohydrates is often observed between vegetative growth and the filling of storage tissues. Photosynthetic activity has a direct impact on tuber filling since in potato tubers numerous plasmodesmata form a symplastic path from the sieve element to the storage parenchyma of tubers (Oparka 1986). Sink filling may therefore frequently be the limiting step in yield production (Kosegarten and Mengel 1998). For potatoes and simlar root crops yield components constitute the number of plants per ha, the number of tubers per plant, and the tuber size.

Tuber initiation is induced by plant hormones (see page 288 and Figure 5.5). The question has been discussed by Krauss (1980). Abscisic acid (ABA) promotes initiation whereas gibberellins (GA) have an adverse effect. A continuous N supply results in a low ABA/GA ratio with regrowth of the tubers occurring (Plate 7), *i.e.* tuber growth ceasing and one or more stolons being formed on the tuber apex. Interrupting N supply increases the ABA concentration dramatically and thus gives rise to tuber initiation. This effect of N supply is sensitive enough to produce chain-like tubers (Plate 5.7) and may happen when after a period of regrowth the N supply stops and a second phase of tuber initiation is induced. In practice this reversible initiation of tuber growth by a high level of N nutrition often occurs at a late stage in tuber growth. This causes tuber malformation and the production of knobbly tubers (Plate 5.7). The inducing effect of ABA on tuber initiation was demonstrated by Krauss and Marschner (1976) by treating stolons with ABA. Treatment of tubers with chlorocholine chloride (CCC), which is known to be an inhibitor of GA synthesis, had an analogous effect as ABA on tuber initiation.

Long days restrict tuber initiation whereas short days with low night temperatures promote initiation. This is understandable as long days conditions are associated with high GA and low ABA levels in the plant, whereas for short days the reverse is true. For long days even at the end of the night, leaves still contain much starch because of the poor sink activity of the tubers (Lorenzen and Ewing 1992). Hence tuber growth depends not only on photosynthesis as a source of photosynthates but also on the transport and the rate of translocation of these photosynthates from the leaves to the tubers which relates to sink metabolism. The quantity of photosynthate produced is a function of leaf area per plant to bring about CO_2 fixation. The area of leaf per plant primarily depends on plant development during the vegetative growth stage (from germination until flowering). Vigorous growth is obtained if in addition to favourable climatic conditions the plant is adequately supplied with water and nutrients and especially with N and K^+. Nitrogen is required for the development of the foliage, and K^+ for efficient transformation of solar energy into chemical energy (Peoples and Koch 1979).

For tuber growth and the process of filling with starch, N nutrition is of particular importance for unlike cereals, enhanced N nutrition after flowering can stimulate vegetative growth and the initiation of new leaves. Photosynthates can thus be diverted from the filling of storage tissues to promoting vegetative growth. This is demonstrated for potatoes in the data of Table 5.9 obtained by Krauss and Marschner (1971). At the highest level of NO_3^- nutrition (7.0 m M NO_3^-) N uptake was greatly increased and tuber growth considerably depressed.

Table 5.9   Growth rate of potato tubers in relation to nitrate supply (Krauss and Marschner 1971)

Nitrate concentration mol/m^3	Nitrate uptake m mol/day	Tuber growth cm^3/day
1.5	1.18	3.24
3.5	2.10	4.06
7.0	6.04	0.44
—	—	3.98

Plate 5.7 Effect of N supply on the tuber formation of potatoes (Photo Krauss) — Upper part: Malformation of tubers as a result of disturbances in tuber growth caused by nitrogen — In between: Secondary growth after alternating N supply — Lower part: Regrowth of tubers following abundant N supply.

During the early stages of development root crops should be well supplied with N in order to develop the vegetative plant organs needed for photosynthesis. After flowering, however, the N supply to root crops should decline. This later stage should be characterized primarily by the synthesis of carbohydrates and their translocation to the tubers. Figure 5.16 demonstrates this pattern in potato development. Before flowering, leaf and stem material are predominantly produced. After this time, however, there is a rapid decline in the yield of leaf material and a steep increase in the dry matter production of tubers. Generally tuber initiation starts at flowering, but in modern cultivars tuber setting occurs before flower buds are developed.

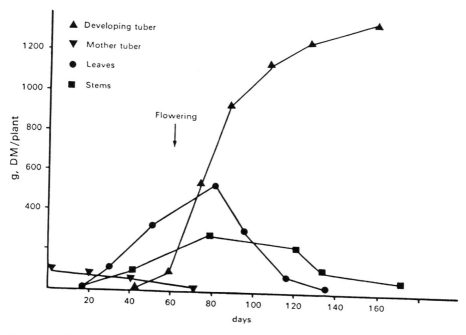

Figure 5.16 Changes in the dry matter yield of various parts of a potato plant during the growth period.

If tubers are excised, CO_2 assimilation is drastically depressed by a feedback mechanism as has been shown by Basu *et al.* (1999). In their experiments potato plants with tubers had a CO_2 assimilation rate of 20.7 µmol CO_2/m²/s whereas in potato plants in which the tubers were excised the CO_2 assimilation rate was only 1/3 of this value. The plants with excised tubers accumulated high amounts of starch in the chloroplasts and elevated concentrations of sucrose and hexoses were found in the leaves. There is evidence that a high level of hexoses in leaves promotes hexose phosphorylation with the effect that the cytosolic pool of inorganic phosphate is depleted. This depletion depresses the release of photosynthate in the form of triose phospahte from the chloroplast to cytosol which occurs as a counter exchange (antiport) for inorganic P import. Starch therefore accumulates in the chloroplast and CO_2 assimilation is depressed. This kind of a feedback mechanism is not only true for potato plants but for numerous crop species as reported by Geiger *et al.* (1992).

Basu *et al.* (1999) have reported diurnal changes in the photosynthetic rate of CO_2 assimilation of potato plants. In the morning an increase occurred until about 14.00 hrs when a peak of 20 µmol CO_2/m²/s was attained. After this time the photosynthetic rate strongly declined to a level of about 7 µmol CO_2/m²/s at 17 hrs. The authors suppose that the high sugar concentrations in leaves in the afternoon and that the feeedback mechanism described above were the reason for the fall in CO_2 assimilation at this time.

Cassava (*Manihot esculenta*) is a tropical crop grown for its tubers which are rich in starch. It is much adapted to poor soils and is able to tolerate soil acidity. Fertilizer application, especially phosphate, frequently low in tropical soils, may improve crop yields. Cassava extracts appreciable amounts of K^+ from soils and causes severe depletion of available K^+ in soils (Fageria *et al.* 1991). Cassava starch is an important diet for the endogenous population in the regions of the world in which it is grown.

Root crops of the beet family, differ from potatoes in yield physiology in so far as they are biennials, and are dormant between the vegetative and reproductive stages. Before going into the stage of winter dormancy this type of crop accumulates carbohydrate in the storage tissue. The economic yield of these crops thus closely relates to this accumulation process. For sugar beet (*Beta vulgaris*) this depends on the number of plants per unit area, the size of roots and their sugar concentration. The number of plants per unit area, or plant density, will be discussed in more detail below. Root size is dependent to a large extent on nutrient and water supply during the early growth of the crop. Vigorous leaf growth during this stage and the development of a large leaf area per plant is essential for voluminous roots. The more quickly in the growth period the leaves are able to form a complete canopy over the soil, the better are the chances of a good yield. Satisfactory leaf growth depends very much on a high level of N nutrition during the early stages of plant development. In the later phases of plant growth, however, generally beginning at the end of July or in the beginning of August in Central Europe, the level of N supply to the plants should decline. If this does not occur, photosynthates are diverted from filling root tissue with sugar and are utilized to a marked extent for the growth of new leaves. The data of Table 5.10 illustrate this relationship (Forster 1970). In one treatment of this solution culture experiment, the N concentration of the nutrient solution was reduced to one-third of the original concentration at 6 weeks prior to harvesting. This had no major effect on the root yield, but the leaf production was drastically reduced. The reduction in N supply resulted in a considerable increase in the sugar concentration of roots and hence an improvement of sugar yield by more than 30%. This example demonstrates, that in the final stages of sugar beet development, photosynthesis and the translocation of photosynthates towards the roots should be the main processes and not leaf growth. The same observations have been made in field trials (Bronner 1974). The relationship between N nutrition and sugar concentration in sugar beet is also dependent on phytohormone activity. Kursanov (1974) reported that abundant N supply during the late stage of sugar beet growth enhances the auxin (IAA) level which in turn, is believed to promote root growth and to delay sucrose storage. A high rate of both CO_2 fixation and translocation are particularly necessary when voluminous roots have been developed, for considerable amounts of sugar are required in the filling process. This explains why high root yields are sometimes accompanied by low sugar concentrations and also why plants well supplied with K^+ are often high in sugar concentration (Draycott *et al.* 1970).

Table 5.10 Effect of lowering the N supply in the late stage of growth on the yield and sugar concentration of sugar beet (Forster 1970)

	Roots	Leaves	Sugar	Sugar yield
		g/plant	mg/g	g/plant
Full N	957	426	164	93
1/3 N	955	360	190	125

Sugar beet is generally cultivated in temperate climates on fertile soils. It is for this reason that the N fertilizer rate may be much lower than the total N taken up by the crop, the latter being in a range of 200 to 300 kg N/ha. On these fertile soils much organic N is mineralized particularly because the crop stands for a long period of time in the field. Besides mineralizable soil N, interlayer NH_4^+ may also contribute to the N supply of sugar beet on particular soils (Marzadori *et al.* 1994). Trials of Heyn and Brüne (1990) have shown that on average 60 kg N/ha gave optimum sugar yields. Since the N nutrition of the crop must be precisely controlled as a prerequisite for obtaining high yielding sugar beet with a high quality, soil analysis for available N is indispensable. This has been demonstrated in 124 field trials carried out by Machet and Hebert (1983) in Northern France in which mineral N was determined using the N_{min} method (page 101) and by Nemeth *et al.* (1991) in some thousand field trials carried out in Germany and Austria. Nemeth *et al.* (1991) determined mineral N and mineralisable organic N using the EUF method (page 98). In both the experiments of Machet and Hebert (1983) and the field trials of Nemeth *et al.* (1991) the recommended rates of N application from soil analyses increased the root quality and improved financial net return to the farmers as compared with "rule of thumb" fertilizer practice.

Sucrose is stored in the vacuoles of roots and there are powerful sucrose uptake systems which translocate the sucrose across the tonoplast and which involve K^+ (Briskin *et al.* 1985, Saftner *et al.* 1983). It is for this reason that sugar beet require high K^+ fertilizer application rates (Loué 1989). According to Lindhauer (1989) Na^+ may subtitute for K^+ in osmotic functions but not in specific functions in the roots.

5.2.6   Fruit crops

Yield production of perennials such as grapes and fruit trees is also characterized by vegetative growth, mainly the production of branches and leaves and by filling of fruits during the regenerative period. Frequently branch and leaf growth are characterized by high growth rates which is very typical for vine (*Vitis vinifera*) as shown in Figure 5.17 (Huglin 1986).

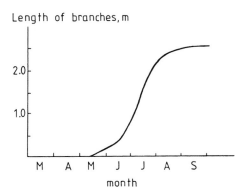

Figure 5.17 Growth of vine branches (*Vitis vinifera*) throughout the growth period under Central European climate conditions, cv Pinot gris (modified after Huglin 1986).

Growth of branches is associated with the formation and growth of leaves so that during the relative short period of about two months most of the plant nutrients, particularly nitrogen and potassium must be available. Some of these nutrients are stored in the wood of roots and stems and are mobilized with the onset of branch and leaf development (Löhnertz *et al.* 1989b).

In grapes, leaf position is important in the filling process. At the onset of fruit development, leaves in the direct vicinity of the fruits are the main contributors to fruit growth. As fruit development proceeds more and more of the leaves above the grape bunch provide photosynthates for fruit filling. Grape weights and sugar contents are higher the more leaf material is available to supply the grape bunches. A large leaf area is therefore important in grape production. In the early stages of fruit development this appears to be of significance in increasing the weight of grapes. Later its effect is mainly in increasing sugar concentration. The filling process also depends on the photosynthetic efficiency of leaves, which is not only controlled by light intensity and temperature but also by mineral nutrition. Sucrose along with smaller amounts of fructose, glucose and organic acids are the most important photosynthates translocated from the leaves to the grapes. The predominance of sugar or organic acid synthesis depends considerably on temperature. According to investigations of Kliewer (1964) with *Vitis vinifera*, the synthesis of organic acids is enhanced at lower temperatures, whereas sugars are synthesized to a much greater extent under warmer conditions. This may partly explain why temperature effects during berry filling have such a significant bearing on the taste and quality of wine.

According to Huglin (1986) vine berry growth is characterized by four stages. During the first stage — about 20 days after anthesis — seed growth dominates and growth rates are higher than that of the pericarp. During the second stage the growth rate of the pericarp is at least as high as that of the seeds. Throughout these first two stages berries still behave like leaves characterized by active stomata and by transpiration. During the second stage auxin concentration in the berries increases until the third stage which is called v*eraison*. At this stage there is virtually no growth, and veraison is characterized by metabolic changes. The berries become lucid and in red vine varieties the synthesis of anthocyans starts. Veraison is followed by the stage of maturation. During the first part of this stage berries have a high growth rate mainly as a result of the uptake of water and sugars. During the last part of maturation of the berries growth is practically at a standstill but still import of sugars, mainly sucrose occurs which is immediately hydrolyzed to form glucose and fructose. ABA concentration in vine berries increases during berry development and attains the highest concentration at maturation (Düring 1973).

The development of other fruit crops such as apples, pears and peaches is basically similar to that described above for grapes. A main factor in fruit yield is the quantity of fruits per tree or plant. Fruit setting is related to phytohormone activity and probably also to nutrition in a more indirect way. This problem is of particular significance for perennial fruit crops, which tend to bear fruits only every second season. This phenomenon, called alternance, is often observed in apple trees. Weller (1971) reported that in years with a low fruit yield, apple trees develop abundant fine roots during the late summer and in the beginning of autumn. Improved fruit setting is observed in these trees in the following year. The relationships between root development, phytohormone synthesis and flower setting, are by no means clear. Whether plant nutrients are involved in these relationships has yet to be firmly established.

The development of tomato fruits is in some way comparable with the growth of grapes with the exception that tomatoes are indeterminate plants, bearing flowers and fruits at the same time. Leaves located in the vicinity of the fruit truss provide the major source of photosynthates in fruit filling (Khan and Sagar 1967). The filling process and also the number of fruits per plant depend on the nutritional status of the plant (Forster 1973b).

Plantation crops such as oil palm, coconut, rubber trees, bananas and pineapple are not so strictly bound to seasons. Their growth and yield depend considerably on an ample supply of plant nutrients. Fremond and Ouvrier (1971) reported that the onset of fruiting of coconut palms was considerably earlier provided that the young plants were abundantly supplied with nutrients.

5.2.7 Growth regulators

Growth regulators or bioregulators are synthetically produced chemical compounds which can be applied to plants to modulate the activity of the phytohormones.

Table 5.11 Synergists, antagonists and inhibitors of phytohormones

Analogous sub.	Synergists	Antagonists and Inhibitors
Auxins		
Arylalkanes	Monophenols	Coumarins
carboxylic acids	Brassins	Triiodobencoic acid
Aryloxyalkane	—	Naphylphtalmic acid
carboxylic acids		
Derivates of benzoic acid	—	Chloroflurenol
Gibberellins		
Cyclic adenosine	Catecholamines	Onium compounds
monophosphate		quaternary N compounds
		Sulphonium and Phosphonium derivates
		Pyrimidine derivates
		Succinic acid N-dimethyl-hydrazide
Cytokinins		
Benzyladenine	—	Pyrrolo- and Pyrazolo pyrimidines
Furfuryladine		
Abscisic acid		
Xanthoxin	Farnesol	Fusicoccin
Phaseic acid		
Ethylene		
1-aminocyclopropane	Auxins	Aminoethoxy-vinylglycine
1-carboxylic acid	Glyoxime	—
2-choroethylphosphonic acid		

Table 5.11 shows some examples of the antagonistic, synergistic and inhibitory behaviour of these growth substances in relation to the naturally occurring phytohormones. The aim of growth regulator activity is often to influence vegetative growth thereby contolling culm length or shoot/root ratio, but it may also be focused on flower and fruit formation. Molecular structure of some important growth regulators is shown in Figure 5.18.

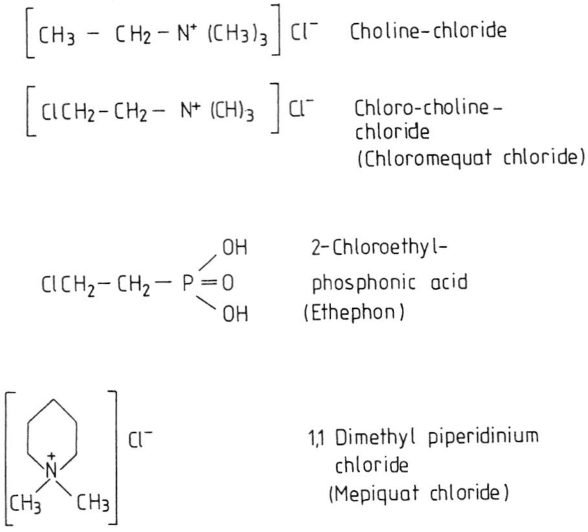

$$\left[CH_3 - CH_2 - N^+ (CH_3)_3 \right] Cl^-$$ Choline-chloride

$$\left[ClCH_2 - CH_2 - N^+ (CH)_3 \right] Cl^-$$ Chloro-choline-chloride
(Chloromequat chloride)

$$ClCH_2 - CH_2 - P \overset{OH}{\underset{OH}{=}} O$$ 2-Chloroethyl-phosphonic acid
(Ethephon)

1,1 Dimethyl piperidinium chloride
(Mepiquat chloride)

Figure 5.18 Important growth regulators.

Most of these compounds are used in cereal cultivation for reducing culm length and thus to protect the crop against lodging. About 40 years ago Tolbert (1960) found that chlorocholine chloride (CCC) depressed shoot growth in wheat. The application of CCC for reducing the culm length of wheat and thus improving its resistance against lodging was first introduced by Linser *et al.* (1961). It was found that the basal nodes were shortened and the diameter of CCC treated culms was thicker and the reduction of straw length resulted in a substantial increase in resistance to lodging. More recent results were found by Sanvicente *et al.* (1999) applying a combination of various growth regulators including CCC to winter barley. If was shown that the growth regulators increased both the stiffness of the culm and the dry weight per unit stem because of modifications of the supporting tissues. Besides CCC, of which is known commercially as *Cycocel* or *Chloromequat*, mixtures of CCC with other growth regulators are common especially choline chloride. Stahli *et al.* (1995) reported that the application of CCC together with choline chloride to wheat enlarged the surface of flag leaf, and increased its chlorophyll concentration and rate of CO_2 assimilation. This effect was enhanced when the mixture also included the herbicide *Imazaquin* which at low rates of application improves the effect of growth regulators by increasing their translocation from the place of treatment to other plant parts (Arissian *et al.* 1991).

In recent years other growth regulators and mixtures of growth regulators have been developed in order to improve efficiency and for use in cereal crops other than wheat. For barley and rye a mixture of 2 chloroethyl phosphonic acid (= *Ethephon*) and 1,1 dimethyl piperidinium chloride (*Mepiquat chloride*) known as *Terpal* is often applied by spraying at the end of tillering. For oats a mixture of CCC and choline chloride is used. The above cited growth regulators when applied under practical conditions in the field generally reduce the straw yield but have no direct impact on grain yield with the exception of cases where lodging is prevented. Here the treatment is beneficial as lodging may result in severe grain losses and also affects the quality of the grain. For this reason it is important to know whether there is a risk of lodging. Gate *et al.* (1996) have developed a model to predict this risk. The main factors of the model are the number of tillers with at least three well developed leaves at the stage when the ear is about 1 cm long, the length of the second internode and the length of the ear at a given time in relation to the sum of temperature which is the sum of average day temperature of each day beginning with ear emergence (Gate *et al.* 1996). Risk of lodging increases with the density of the stand and the length of the second internode. The model does not take into accout the nitrogen nutrition since it is assumed that N fertilization is adjusted to the availability of N in the soil and therefore avoiding excessive nitrogen is avoided. Excessive N supply by fertilizer and residual available N in soil or both may severely raise the hazard of lodging. On the other hand the application of culm-shortening growth regulators allows the application of somewhat higher nitrogen fertilizer rates since it reduces the risk of lodging. Resistance to lodging is also dependent on the use of cultivars with a low rate of culm elongation (Gate *et al.* 1996). The hazard of cereal lodging is particularly high in temperate climatic conditions with frequent rainfalls and also with thunderstorms during the growth period.

The application of 2-chloroethyl phosphonic acid, also known as *Ethephon*, is used in rubber production (Ribaillier and Auzac 1970). This delays the formation of the wound callus and thus extents the period of latex flow. Like CCC, the 2-chloroethyl phosphonic acid blocks the synthesis of gibberellins (Sadeghian and Kühn 1976). As shown in Table 5.11 quarternary N compounds such as choline chloride, chlorocholine chloride, and dimethyl piperidinium chloride as well as phosphonium derivatives (2-chloroethyl phosphonic acid) are typical gibberellin inhibitors and hence retard stem elongation.

Dimethyl-piperidinium chloride plays a role in cotton production. It is known that on fertile soils and on well watered soils cotton (*Gossypium hirsifum*) produces excessive vegetative growth which does not contribute to boll production (Reddy *et al.* 1990). Dimethyl-piperidinium chloride, also known as *Pix* or *Mepiquat chloride* reduces plant height and the number of internodes (Reddy *et al.* 1990). Its effect depends on temperature and may be diminished by day night temperatures higher than 30/20°C. According to Kerby *et al.* (1986) dimethyl piperidinium chloride favours the early boll setting at lower internodes while the late boll setting is somewhat decreased. For this reason a favourable effect of its application can be expected where the length of the season is a constraint on yield potential. The effectiveness of dimethylpiperidinium chloride application on vegetative growth decreases following its time of application (Reddy *et al.* 1990).

5.3 Nutrition and Yield Response

5.3.1 General

Since the famous experiment carried out by the Dutchman Jean Baptiste van Helmont (1577-1644) about 400 years ago it has been known that the soil contributes to only a small extent to the weight of material synthesized by plants. Van Helmont planted a willow in a pot and found that after a growth period of 5 years that the weight of soil in the pot had hardly changed whereas the willow had gained about 160 lb. Van Hemont suggested that the production of wood, bark and leaves resulted from water taken up from the soil. The British chemist Joseph Priestley (1733-1804) was the first to discover that plants and animals take up oxygen and release CO_2. This process of respiration occurring in the dark was interpreted by another Dutchman, J. lngenhousz (1730-1799) who also recognised the uptake of CO_2 and the release of CO_2 by plants in the light. Theodore de Saussure (1767-1845) made the first quantitative measurement of these gas exchanges of plants and showed that the only direct carbon source of this assimilation was atmospheric CO_2 and not soil humus. He also found that plants were unable to grow in pure water and needed the presence of substances from the soil (Drews 1998). At the beginning of the 19. century it was the currently held new opinion that plants lived from the humus of the soil, a concept propagated by the German Albrecht Thaer (1752-1828) who also introduced the regular application of farm yard manure which had a very beneficial impact on German agriculture (Boulaine 1989). Carl Sprengel was one of the first to suggest that mineral elements in the soil are the real plant nutrients. This idea was also propagated by the French Jean Baptiste Boussingault (1802—1887) who already at that time made a distinction between available and non-available soil nutrients. He found that soils may produce more nitrogen than is taken up by the crop and in contrast to Liebig he held the opinion that it was the soil and not the plants that may fix atmospheric nitrogen (Boulaine 1987). Justus Liebig's (1803-1883) famous book *The Application of Organic Chemistry to Agriculture and Physiology* (1840) was based on the current knowlege of those days. Liebig's particular merit is that he already saw these major processes as a natural cycle in which the plant is the great generator of organic matter by means of inorganic nutrients. This organic matter he regarded mainly as representing food for man and animals. Excrements of men and animals as well as their dead bodies, when incorporated into the soil, are decomposed to inorganic matter which can again be used by plants as nutrients. Liebig was also one of the first to appreciate the thermodynamic aspects of this cycle. He had received an interesting article of the physicist Julius Rober Mayer (1840) in which the maintenance of energy, the first Law of Thermodynamics, had been described. The article had been rejected for publication by *Poggendorfs Annalen* a famous German Journal for Physics, but Liebig immediately realized the enormous new insight of this concept to natural processes and in seeing that solar energy is transformed into chemical energy by the plant which can then be used by man and animals. This famous article of Julius Rober Mayer was published in *Liebig's Annals*.

We now know that as well as water, CO_2 and mineral nutrients determine the production of plant material. Carbon dioxide assimilation is the primary process involved in yield formation. It is for this reason that the rate of CO_2 assimilation and factors which influence it are of paramount interest. Photosynthates can be utilized for the plant's vegetative growth, for the synthesis of storage material and for respiration. The proportion of photosynthates directed towards these three sinks depends on the physiological age of the plant (Warren-Wilson 1969). In young seedlings structural growth dominates and therefore more than half of the photosynthates assimilated are used for growth. In mature plants, however, the major fraction of photosynthates is used in the synthesis of storage material (Table 5.12). During the main grain filling period of wheat about 80% of photosynthates are transported to the grains are used as metabolites for the synthesis of organic compounds, mainly starch. The remainder are respired (Evans and Rawson 1970). the same is also true for other plants. Tomato fruits for example respire 30% of the photosynthates they receive. Similar values (30 to 40%) were obtained for ryegrass by Alberda (1977). According to Peterson and Zelitch (1982) about 45% of the assimilates are respired in tobacco.

Table 5.12 Rates of utilization of photosynthates by three types of sinks at three stages of development (Warren-Wilson 1972)

	mg photosynthate/g dry matter/day		
	Structural growth	Storage	Respiration
Germinating seedling	20	5 *	10
Young vegetative plant	15	5	8
Mature plant	5	10	4

* assumes that seed reserves are not part of an embryo, otherwise a negative value results.

5.3.2 Net assimilation rate and leaf area index

In a crop stand CO_2 is continuously being fixed by photosynthesis and released by respiration. The net amount of C assimilated (net assimilation) is measured by the excess C gained from photosynthesis over that lost by respiration. The net assimilation rate (NAR) is often used to express the rate at which dry matter is accrued and is defined as net assimilation per unit leaf area.

In the process of respiration, molecular oxygen is taken up by plants and assimilates are oxidized to CO_2 and H_2O. In dark or mitochondrial respiration the oxidation process is associated by ATP synthesis. Green cells of C-3 species are also capable of light induced respiration (photorespiration) but in this process no ATP is produced (see page 150). Photorespiration is virtually absent in C-4 species such as maize, sorghum and sugar cane. This important difference between C-3 and C-4 species means that the CO_2 compensation point — the CO_2 concentration at the leaf surface at which CO_2 assimilation and rates of CO_2 released by respiration are equal — differs between

C-3 and C-4 species. For the latter it is in the order of 5 to 10 μL/L whereas C-3 plants are normally unable to diminish the CO_2 concentration at the leaf surface to less than 50 μL/L (Krenzer *et al.* 1975). The lower compensation point of C-4 plants is of importance in crop physiology, for these species may depress the CO_2 concentration to very low level in the leaf tissue due to intense CO_2 assimilation and still show a net assimilation. At such low CO_2 concentrations at the leaf surface — about 1/50 of the normal CO_2 concentration in the atmosphere — C-3 species have no net assimilation and therefore may not utilize favourable conditions such as light intensity and temperature for photosynthesis. Schäfer *et al.* (1984) reported that wheat and sugar beet under field conditions on sunny days and well supplied with nutrients and water, attained their maximum CO_2 assimilation several hours before noon and hence before the highest light intensity was attained. Such depressions of CO_2 assimilation as shown by Schäfer *et al.* (1984) in C-3 species are not found in C-4 species. This is one of the major reasons why tropical grasses (C-4 species) can grow at such an enormous rate under high light intensity and high temperature conditions.

Both a cyanide-resistant and a cyanide sensitive-respiration have been identified in plant mitochondria (Solomos 1977). The cyanide-sensitive respiration is highly efficient in ATP production whereas the resistant form is only poor in efficiency in converting energy released during respiration into ATP. The physiological importance of cyanide-resistant respiration has yet to be elucidated. This type of respiration is especially high in voluminous storage tissue such as tubers and roots (Lambers 1979).

Both CO_2 assimilation and respiration rates increase with temperature. The relationship between CO_2 assimilation rate and temperature is characterized by an asymptotic curve whereas the plot between respiration rate and temperature is hyperbolic (Figure 5.19). A temperature point therefore occurs at which assimilation is equal to respiration (mitochondrial and photorespiration). At this temperature net assimilation is zero and no net growth occurs.

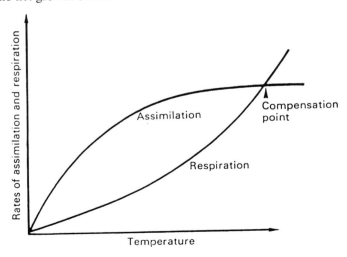

Figure 5.19 Rates of CO_2 assimilation and respiration in relation to temperature.

The higher the density of plants in a particular crop stand, the more mutual competition comes into play: competition for water, competition for nutrients, competition for light. Secondary effects may also be induced by competition. Mutual shading resulting from a high plant density may, for example, increase the susceptibility of the crop to fungal diseases and lodging. In very dense stands competition for light is often the limiting growth factor provided that water and mineral nutrients are present in adequate amounts, as is usual in fertile soils. Under such conditions the photosynthetic rate is decreased in shaded leaves whereas the rate of respiration is increased. As the crop density is increased and mutual shading is intensified, the net assimilation rate (NAR) is thus decreased.

Generally the density of a crop population is expressed in terms of leaf aea index (LAI). This is defined as the leaf area of the crop per unit soil area, on which the crop is growing (Watson 1952). An LAI of 4 for a given crop means that plants growing on an area of 1 m^2 soil have a leaf surface area of 4 m^2. The optimum LAI differs between crops. Some LAI values for different crops are given in Table 5.13. For cereals an LAI of 6 to 8 is usually recommended. Sugar beet requires an LAI of 3.2 to 3.7 growing under the light conditions of Central Europe but at higher light intensities as in Southern Europe, optimum LAI values for sugar beet are in the region of about 6. This example shows that a higher LAI value is acceptable, if the light intensity is enhanced. Crop type may also be important. This is evident in Table 5.13 in the case of rice. Modern rice cultivars are characterized by short culms and erect small leaves. These types minimize mutual shading and thus allow a better light absorption (Tanaka 1972). The view that mutual shading results in enhanced respiration has been questioned by Evans (1975). He suggests that old leaves respire at a low rate, and they die if the CO_2 loss by respiration exceeds CO_2 assimilation.

Crop nutrition and water supply also has an impact on the optimum LAI. When the LAI is optimum, excess N nutrition enhances leaf growth and the grain yields of rice are depressed because of higher mutual shading (Tanaka 1973). High levels of phosphate and K^+ nutrition cannot counteract this negative effect of N. Schnier et al. (1990a) found in field trials in the Philippines that the CO_2 assimilation rate of rice depended mainly on the N concentration in leaves and the LAI. The relationship found is shown in the equation below:

$$CO_2 \text{ ass.-rate} = a + b \ln (\%N)$$

where

$a = 28.5 \text{ LAI} - 13.6 \text{ LAI}^{1.28}$

$b = 3.49 \text{ LAI}^{1.23}$, %N = N concentration in leaf in %

N% = % N concentration in leaves

From this equation it is clear that LAI has an optimum value because the 2nd term for a has a negative sign. It is also evident that the N concentration in leaves is exclusively and positively related to CO_2 assimilation. The N concentration is indicative of the enzymes involved in photosynthesis. It should be borne in mind that high rates of N application not only increase the N concentrations in leaves but also promote leaf growth and may therefore affect CO_2 assimilation through mutual shading because of a greater leaf area. According to Schnier et al. (1990a) grain yield of rice depended mainly on the LAI during the vegetative stage and leaf N concentration during the reproductive stage.

Leaf growth and therefore also LAI depends much on water supply during the vegetative stage of development. Meinke *et al.* (1997) reported that under arid conditions the LAI of maize attained in the non-irrigated plots was only between 1 and 2. In this non-irrigated treatments the maximum LAI was reached at a late stage, namely 100 days after sowing in contrast to irrigated plants which attained their maximum LAI 75 days after sowing. In the irrigated plots the maximum LAI differed between 3 and 7 and increased with fertilizer N rates.

Table 5.13   Optimum LAI-values of various crops

Soybeans	3.2
Maize	5.0
Sugar beet	3.2 - 6.0
Wheat	6.0 - 8.8
Rice (new varieties)	7.0
Rice (local varieties)	4.0

5.3.3   Carbon dioxide assimilation, CO_2 concentration, and light intensity

Carbon dioxide (CO_2) concentrations in the atmosphere have increased considerably since the beginnning of industrialization in Europe as shown in Figure 5.20. This increase has resulted mainly from the use of fossil energy as well as from microbial oxidation of organic soil matter in forest soils after deforestation. This is particularly true for the felling of large areas of tropical forests which amounted in recent years to almost 1% per year of the total tropical forest area (Archibold 1995). According to Kimball (1983) the atmospheric CO_2 concentration will double by the year 2025 if consumption of fossil energy proceeds at the present rate. This may have a dramatic impact on the climate, the consequences of which are as yet unknown.

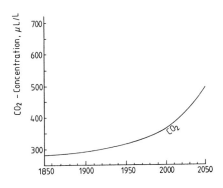

Figure 5.20   Increase of global atmospheric CO_2 concentration (after Fabian 1997).

Ingenhousz (1730-1790) was the first to observe that plants take up CO_2 and release O_2 under light conditions. Theodore de Saussure (1767-1845) measured the uptake of CO_2 and the release of O_2 in relation to the production of plant dry matter and light intensity. His work provided the first evidence that CO_2 is the sole source for carbon for plants and not, as believed in those days, the humus in the soil. It is clear that atmospheric CO_2 concentration exerts a major influence on the assimilation rate of CO_2 and hence on yield formation. Recent field trials carried out under the aegis of the European Stress Physiology and Climate Experiment program (ESPACE program) conducted at eight different sites throughout Europe with a total of 716 trials showed an average grain yield increase of 35% when the CO_2 concentration was doubled (Bender *et al.* 1999). The yield-increasing effect differed much between the various sites and the relative yield increase ranged between 11 and 121%. From this finding one may conclude that the yield increase due to a higher CO_2 concentration will only be attained if other growth factors are adequately available. The actual CO_2 concentration of the atmosphere is in the range of 300 to 350 µL/L. In a crop stand it may vary from this level because of crop assimilation and respiration. In a stand of sugar cane for example Chang-Chi Chu (1968) found values lower than 300 µL/L during the day but higher than 440 µL/L at night due to the absence of CO_2 assimilation and an intense respiration of the crop stand at night. According to Kimball (1983) a doubling of the atmospheric CO_2 concentration increases crop yields by a factor of 1.4. Mitchell *et al.* (1999) reported that increasing the CO_2 concentration form 350 to 650 µL/L raised the flag leaf photosynthesis of spring wheat by about 50%.

As a consequence of selection and plant breeding, yields of plant parts such as roots, tubers, fruits, seeds, and stems harvested by man have increased at least ten fold as compared with wild species. Interestingly enough, however, the carbon exchange rate per unit leaf area (net CO_2 assimilation) has not changed much by domestication (Gifford and Evans 1981). Total production, however, is not only a question of of a momentary net carbon exchange rate (CER) but also depends on the duration of a high CER. Strong physiological sinks such as tubers, seeds and grains of modern cultivars have a beneficial impact on CER, because there is no major feed-back from sink to source signalling a surplus of assimilates in the sink. Net carbon exchange depends much on the stomatal resistance which increases under water stress and hence affects the entry of CO_2 into the leaf tissue. Mythilli and Nair (1996) found in chickpeas (*Cicer arietinum*) that CER was at a maximum during seed filling and decreased with seed maturation. CER was negatively correlated with the stomatal resistance.

Figure 5.21 shows the effect of increasing the CO_2 concentration on the CO_2 assimilation rate (mg CO_2/Jm² leaf area/s) in *Trifolium pratense* growing under different light intensities. Under poor light conditions (50 J/m²/s) the rate of CO_2 assimilation was only slightly increased by raising the CO_2 concentration. From the three dimensional diagram it can be seen clearly, however, that by raising the CO_2 concentration at higher light intensities the rate of CO_2 assimilation was dramatically increased (Warren-Wilson 1969). This relationship, shown in Figure 5.21, is an excellent example of an interaction of two factors. In this example two factors, CO_2 concentration and light intensity interact to enhance the rate of CO_2 assimilation.

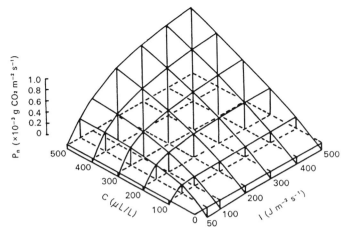

Figure 5.21 Effect of light intensity and CO_2 concentration on the net photosynthetic rate per unit leaf area of clover leaves (after Warren- Wilson 1969). C = CO_2 concentration in $\mu L/L$ — I = light intensity in Joule, $J/m^2/s$; 1 J m^2/s = 1 W/m^2 = 238 lux = 22.2 foot candle — Pn = net photosynthetic rate per unit leaf area (m^2) — Net photosynthesis means total C assimilation minus C loss by respiration.

From this example, it is evident that increasing one of the factors will only have a major effect provided that an adequate amount or intensity of the other factor is available. This is a general relationship which not only applies to light intensity and CO_2 concentration but also to other factors which influence growth such as the soil moisture regime, the level of N nutrition or the rate of supply of other plant nutrients. An analogous example is shown in Figure 5.22 where the net CO_2 assimilation rate per day is plotted against the daily irradiance. The lower curve represents the ambient (normal) CO_2 concentration, and the upper line to a doubling of the CO_2 concentration (Schapendonk *et al.* 1997).

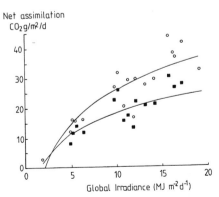

Figure 5.22 Relationship between daily irradiance ($MJ/m^2/day$) and daily CO_2 assimilation (g $CO_2/J/m^2/day$) of ryegrass. Closed symbols relate to the ambient (350 $\mu L/L$) and open symbols to the doubled CO_2 concentration (after Schapendonk *et al.* 1997).

This experiment of Schapendonk *et al.* (1997) was carried out over two years and the variance in light intensity relates to the varying light intensities and day lengths. Also Schapendonk *et al.* (1997) found that doubling the CO_2 concentration raised the net carbon assimilation of ryegrass by a factor of 1.28 (1994) and 1.38 (1995). The higher CO_2 efficiency in 1995 was due to more favourable weather conditions in that year. Besides the enhancing effect of the increased CO_2 concentration on the net CO_2 assimilation, *water use efficiency* (WUE) was increased as was also the release of organic carbon into the soil. The higher WUE (unit dry matter/unit water) resulted because transpiration was little affected as the increased CO_2 concentration raised the net CO_2 assimilation and thereby dry matter production. From the curves in Figure 5.23 it can be seen that the curve for the lower CO_2 concentration is flatter than that of the higher one indicating that the efficiency of light is greater at the higher CO_2 concentration.

The light intensity at which the plateau of CO_2 assimilation is obtained, the so called *light saturation*, depends on various factors. Shade plant species which generally live under trees have a much lower light saturation level (20-50 J/m^2/s) than species living in full light with a saturation level of 200 to 300 J/m^2/s. From Figure 5.23 it can be seen that light saturation for tobacco red clover and turnips is low as compared with wheat and particularly maize and sugar cane (Stoy 1973). These two latter species are C-4 plants and therefore use CO_2 with a much higher efficiency (page 159) and for this reason higher light intensities can still be used to enhance CO_2 assimilation.

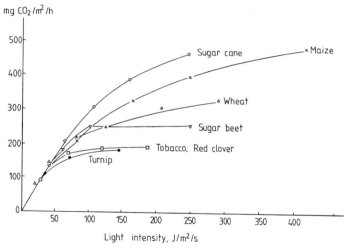

Figure 5.23 Effect of increasing light intensities on the CO_2 assimilation of various crop species (Stoy 1973).

The favourable effect of high light intensities or CO_2 concentration or the combined effects of both on CO_2 assimilation has been observed by numerous researchers. Brun and Cooper (1967) experimenting with soya beans found that when atmospheric CO_2 concentration was 300 μL/L, CO_2 assimilation levelled off at a light intensity of about 100 J/m^2/s. For rice, higher light intensities of about 300 to 400 J/m^2/s may be regarded as optimum (Yoshida 1972). An idea of what is meant by these light intensity values can be appreciated from the normal values which occur in Central Europe. On a clear day at noon in summer, light intensities as high as 300 J/m²/s may be reached. Under cloudy conditions the value is about 80 J/m²/s.

From the absorption spectrum for white light shown in Figure 3.16 (p) it can be seen that the photosynthetic active pigments mainly absorb blue and red light but no green light. Therefore the energy present in the green light is not used by the plant and the photosynthetic active radration (PAR) is lower than the total energy in the light spectrum from 350 to 700 nm. However, the conversion of the photosynthetic active radiation into chemical energy by plants is low and amounts to about 5%, the remaining 95% being converted to heat (Loomis *et al.* 1971). The available energy for annual crops is even lower than 5% because a considerable period of time is needed for these crops to grow to attain maximum leaf expansion. Leaf senescence in the canopy may also occur before the growing season ends. According to Holliday (1976), in an advanced agricultural system with intensive cropping, plants of the C-3 type utilize about 2.7% and C-4 plants 4% of the available radiation energy in the production of plant dry matter.

As described on page 140 it is a photon which initiates photosynthesis regardless of whether the photon is at a high energy (blue light) or a lower energy level (red light). For this reason it is appropriate to relate photosynthesis to the amount of photosynthetic photons hitting the surface of green leaves. The photon amount is expressed in mol and accordingly to the Loschmidt number, 6.023×10^{23} photons are 1 photon mol (= 1 Einstein). Figure 5.24 shows the effect of light intensity expressed as *photosynthetic photon flux density (PPFD)* on CO_2 asimilation (Loreto and Sharkey 1990).

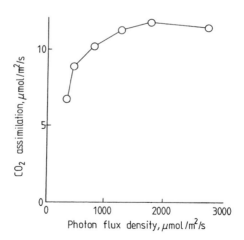

Figure 5.24 Effect of photosynthetic photon flux density on the CO_2 assimilation of *Quercus rubra* leaves (modified after Loreto and Sharkey 1990).

In the term *photosynthetic photon, flux*, the word *photosynthetic* implies that the photon contributes to the expulsion of electrons from the photosynthetic systems (see page 140) which are mainly photons of the red and blue light. The energy for the expulsion of one electron is one quantum. The quanta requirement for the net assimilation of one molecule CO_2 ranges between 12 and 22 depending on the CO_2 concentration and the photorespiration (Lippert 1993). The higher the photorespiration, the higher the quanta requirement for the assimilation of one molecule of CO_2.

High light intensities lead to an increased production of oxygen radicals since the high rates of electrons expelled in high photosynthetic flux density may be scavenged by O_2 thus forming the superoxide and hydroperoxyl radicals as considered on page 143 (Cadenas 1989). These radicals destroy photosynthetic pigments and hence depress photosynthesis and may even lead to plant death. Plant species differ in their capability to tolerate high photosynthetic photon flux densities because of the presence of carotenes and ascorbate which are able to neutralize the radicals. In these detoxification processes the ascorbate cycle plays an important role (Foyer *et al.* 1991). Shade plant species are only poorly equipped with such detoxification pathways and therefore cannot grow at high light intensities.

5.3.4   Yield curves

It has already been mentioned that improving conditions for growth by altering one growth factor can be without effect if another growth factor is limiting. This relationship which is known as the *Law of the Minimum* was discovered by Sprengel at the beginning of the nineteenth century and its significance for crop production was propagated especially by Justus Liebig. It is still of utmost importance in crop production to know which, if any, growth factor is limiting under a given set of growing conditions. Alleviating the limiting growth factor results in a yield increase. This relationship is roughly reflected in an asymptotic curve. In other words, as the growth factor is increased to improve yield, the yield increments become smaller. This pattern of yield increase is shown in Figure 5.25 where N is used to represent the growth factor. The highest yield increment results from the first unit of N applied and with successive applications of N units the yield increments become progressively smaller (diminishing response curve). Mitscherlich studied this relationship in numerous pot and field experiments and concluded that the yield increase brought about by a unit growth factor was proportional to the quantity of yield still required to attain maximum yield level. This is described in mathematical terms as follows:

$$dy/dx = k (A - y)$$

where y = yield, A = maximum yield, x = growth factor, k = constant. On integration, the equation below is obtained:

$$\ln (A - y) = c - kx$$

The term c is an integration constant which comprises all the invariable terms, except k. If x = zero, y is also equal to zero, so that in this case the following equation is obtained:

$$\ln A = c$$

Substituting c for ln A the following equation results:

$$\ln (A - y) = \ln A - kx$$

Converting the natural logarithms into common logarithms, the final form of the Mitscherlich-equation results:

$$\log (A - y) = \log A - cx$$

The term c in this case is porportional to k and results from the conversion of natural to common logarithms:

$$c = k \, 0.434$$

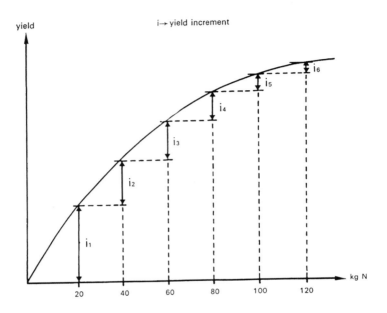

Figure 5.25 Response curve showing diminshing increments.

The Mitscherlich-equation is often written non-logarithmically in the following form:

$$y = A(1 - 10^{-cx})$$

From this equation it is clear that the Mitscherlich curve is an exponential function. The higher the x value the smaller the term 10^{-cx} and thus y approaches A asymptotically. Such curves attain a plateau but they have no maximum. Hence they are similar to Michaelis-Menten curves which describe the velocity of an enzymic reaction against the concentration of the substrate acted upon by the enzyme. In the case of the Mitscherlich growth curve, numerous reactions are integrated and in the ideal case these reactions yield the ideal curve. The rise of the curve is steeper the higher the c-value and for the three major plant nutrients N, K, and P, the curve is steepest for P and flattest for N provided that the nutrients are expressed in the same units *e.g.* kg N, K and P, respectively (von Boguslawski 1958).

Under practical conditions yield curves frequently do not attain a plateau asymptotically but at higher and excessive rates of a growth factor (x) the curve declines because of yield depressions. These depressions may occur for different reasons in contrast to the rising part of the curve where gradually the lack of a growth factor is overcome. Excessive N nutrition may induce lodging of cereals or may favour the infestation of fungi and thus depress crop yield. Excessive application of K^+ may depress the uptake of Mg^{2+} and thus affect growth and yield production and excessive phosphorus may lead to the precipitation of Zn in soils and therefore depress crop yield due to insufficient Zn nutrition. Mitscherlich (1950) attempted to describe the phenomenon of yield depression by excessive applictation rates of a plant nutrient in mathematical terms by introducing a depression factor k. This is zero when no depression occurs and increases with extent of the depression until 0.05. However since the cause of the depressions could differ so too could the shape of the depression curves. Von Boguslawski and Schneider (1964) elaborated the following formula in which the the yield depression is integrated:

$$y = A \ 10^{-z(\log x + i/m+i)\exp n}$$

where

A = maximum yield, x = growth factor, *e.g.* plant nutrient, m = growth factor at which the maximum yield was attained, z = constant, n = depression term
i = the distance between the intersection of the curve with the abscissa and the origin (zero).
This distance represents the proportion of the growth factor originating from the soil and not applied *e.g.* available N in the soil (Figure 5.26).
At the maximum point is x = m and therefore the term $(x + i)/(m + i) = 1$. The equation then reads:

$$y = A \ 10^{-z\log 1} = A \ 10^0 = A$$
$$\log 1 = O \text{ and } -z \ O = O \text{ and } 10^0 = 1$$

If the $(x + i) > (m + i)$ the quotient $(x + i) / (m + i)$ is >1. In this case the exponent with the negative sign increases too and y therefore decreases. The same is true for the increase of *n* which in the equation of v. Boguslawski and Schneider (1964) was between 1 and 2. The term z may also be changed which in the work of v. Boguslawski and Schneider (1964) was between 1 and 2. By the changing n and z the curve can be adjusted to sets of points from data obtained from trials. Using this adjustment the quantitative terms of i and m are obtained, the former indicating the amout of nutrient originating from the soil, the letter m indicating the amount of fertilizer (x) at which the maximum yield is obtained. This adjustment can now easily be made with the aid of modern computer techniques. In this way the maximum of the curve can be obtained which is of utmost importance for the evaluation of field trials with increasing rates of fertilizer. Therefore yield curves are not only of academic importance but are of oustanding practical significance. From the shape of the curve in Figure 5.26 it is clear that there was only a slight depression. The graph also shows the term i which represents the available nutrient in the soil and is therefore also of interest. This term i is analogous to the term b of Mitscherlich (1950) also denoting the amout of available nutrient in the soil.

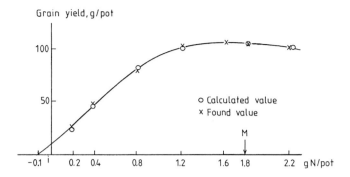

Figure 5.26 Yield curve according to the equation of v. Boguslawski and Schneider (1964) from data obtained in pot experiments (n = 2.0; z = 1.90). x = experimental data, o predicted data.

In addition to the equation of Von Boguslawski and Schneider (1964) data obtained from fertilizer trials with increasing rates of a plant nutrient can also be treated as a quadratic function in order to obtain the yield maximum.

$$y = a + bx - cx^2 \text{ (quadratic equation)}$$

Boyd (1970) in evaluating numerous fertilizer field trials came to the conclusion that in many cases the points obtained can be allocated to two straight lines one with a slope which reflects the increasing yields and one a horizontal line which reflects the plateau (Figure 5.27). Acccording to Boyd (1970) this relative simple rectilinear approach provides precise advice to farmers. From the results shown in Figure 5.27 it is evident that the yield plateau and the yield obtained in the plots without N fertilizer differed considerably because of crop rotation and weather conditions.

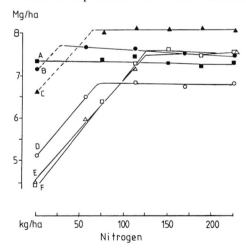

Figure 5.27 Rectilinear appoach to obtain the yield maximum. Effect of N fertilizer application on sugar yield (Boyd 1972). The shown six rectilinear lines relate to different years or group of years.

Using a modern computer it is possible to treat huge amounts of data and fit them to an equation. This is the approach used by George (1983) in assessing the yield maxium of N fertilizer field trials and he described in a useful publication. His concept is based on the Mitscherlich equation but in addition comprises a term for the yield depression. The equation used is as follows:

$$y = a + br^x + cx$$

where y = yield, x = fertilizer rate, a b and c are constants, a >1, b < 1, r > 0 and < 1, and c < 0 which means that the term c is negative and therefore the depression will be the higher the higher the value of x. According to numerous field trials r is assumed to be 0.99. For fitting numerous reseach data to this equation and for assessing the yield maximum a futher type of equation is established derived from the one above:

$$y_1 = a + \Delta a + br^{x + \Delta x} + c(x + \Delta x)$$

This equation brings about a horizontal shift by Δa and a vertical shift by Δx. George(1983) in evaluating numerous N fertilizer field experiments with increasing N rates and different blocks with additional treatments such as straw, green manure application, clover ley, and others fitted the yield data of these to the basic curve (curve of FYM without Δ increments) by altering Δa and Δx according to the data obtained in the various blocks. The resulting curve for all these experiments with FYM as the basis curve is shown in Figure 5.28.

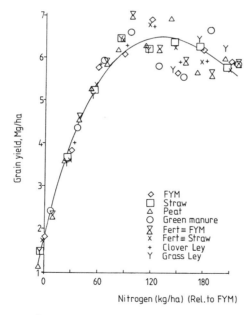

Figure 5.28 Yield curve according to the equation of George (1983) with a term c for yield depressions. Grain yield obtained from increasing nitrogen rates and the basis curve obtained for the block with FYM. The points for yields in the other blocks (straw, peat, green manure etc.) were obtained by adding increments for a and x (Δa and Δx) according to the equations of George (1983).

310

This curve shows a precise maximum. Although the yields differed between the various blocks, the nitrogen level for the maximum yield is fairly constant. Differences in yield were related largely to years while differences for the nitrogen level were were related largely to cropping systems.

An important reason why the shape of yield curves may differ considerably is the amount of available nutrient present in the soil. This amount is indicated by the term a in the equation of George (1983). When considering the available nutrients in the soil and expressing the yield in relative terms (maximum yield = 100%) ideal yield curves may be obtained as was recently shown by Everaarts and De Moel (1998) for cabbage yields obtained by increasing N fertilizer rates. Here the abscissa of the graph was the fertilizer N + mineral N present in a soil depth of 0 to 90 cm in spring. Analogous results were reported by Schön *et al.* (1985) for lowland rice fertilized with N (Figure 5.29). In this case the relative grain yield was plotted against the initial exchangeable ammoniun + fertilizer N. Interestingly in this investigation the plot of relative grain yield against, fertilizer N, *i.e.* omitting to take the available N present in the soil into account a yield curve was not obtained.

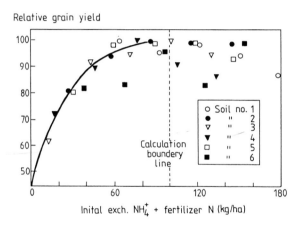

Figure 5.29   Relative grain yield of lowland rice *versus* fertilizerN + exchangeable NH_4-N at the beginning of the crop season (after Schön *et al.* 1985).

From Figure 5.29 it is also evident that excessive N rates (>100 kg N/ha) showed no clear yield pattern but rather a scatter of points which, as indicated above, may be caused by various factors inducing yield depressions.

Water supply to crops is a very important factor for the yield level and the shape of yield curves. This was shown in long-term field trials of Thompson and Whitney (1998) for sorhum and winter wheat. Figure 5.30 shows the grain yield of sorghum (*Sorghum bicolor*) against N fertilizer rates at different soil moisture levels. Soil moisture level was assessed as high when at planting time the moist soil depth was >136 cm. The corresponding values for medium moisture was a soil depth of 90 to 135 cm and low soil moisture at a moist soil depth of <90 cm. The trials were carried out in the semi-arid climate of the West Central Great Plains (USA) where rainfall during the growing season is low. In the treatment with high soil moisture, the N fertilizer produced a marked increase in grain yield, and the yield curve resembled a Mitscherlich-type curve. In the low soil moisture treatment the yield level was very low, the yield curve linear and the yield increase due to N fertilizer low.

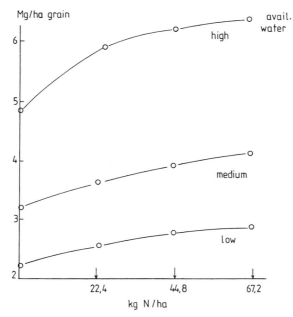

Figure 5.30 Effect of N fertilizer on the grain yield of sorghum at three different moisture levels. Average data from 1975 to 1996 (after Thompson and Whitney 1998).

This investigation also demonstrates that under semi-arid conditions with low rainfall during the growing season a knowledge of the amount of stored water in the soil profile is of high importance in the assessment of a reliable fertilizer recommendations. In this context it should also be emphasized that long-term field trials representative for the soils and climate of an agricultural area are of great value. This not only applies to N fertilizer application but even more to treatments of P and K (Munk and Rex 1990).

The maximum agricultural yield of a crop in terms of grains, tubers, seeds per ha, may not be the same as the maximum economical yield. As well as the benefits of increased yield from fertilizer application the input costs including fertilizer, its handling together with harvesting costs of the crop have to be taken into account in assessing the rate of fertilizer application required to obtain optimum profit. The relationship between N fertilizer application, yield in terms of grain and in terms of currency and costs and yield maximum and yield optimum are shown in Figure 5.31 which relate to the economic and agricultural conditions in West and Central Europe. The graph shown below is analogous to the graph derived by Mitscherlich (1950) for calculating economic yield.

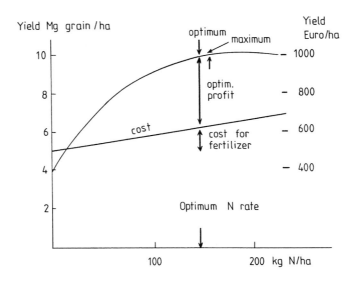

Figure 5.31 Effect of N fertilizer on grain yield and currency yield and fixed costs in relation to the yield maximum and the economic maximum. Fixed costs relating to fertilization comprise the cost for the fertilizer + additional costs such as fertilizer distribution and handling and costs for harvesting the yield increment resulting from fertilizer application. In the example presented the fertilizer costs were multiplied by 1.5 assuming that the additional costs including labour etc. amouted to 50% of the cost of the fertilizer.

The fixed cost for the treatment without N fertilizer indicated on the ordinate, is relatively high and in the example here surpass the yield of the unfertilized treatment. The slope for the increase of the fixed costs results from the cost for fertilizer, its handling and distribution plus the cost for harvesting the yield increment obtained by the fertilizer application. This slope is relatively flat which means that costs for fertilizers are reasonable in relation to the additional costs. The maximum distance between the straight line for the costs and the yield curve represents the economical optimum. In this example it is slightly below the yield maximum which was obtained at a fertilizer rate of 160 kg N/ha. For optimum profit a fertilizer rate of 150 kg N/ha was required. At this point the economic yield increase in terms of currency amounted to about 370 Euro while the cost for fertilizer plus application was 150 Euro. This example shows that fertilizer application can be very profitable. The profit is higher the lower the yield of the treatment without fertilizer and *vice versa*. It is for this reason that a knowledge of the level of available N in the soil profile at the beginning of the season is of utmost importance for the efficient application of fertilizers.

5.3.5   Modelling soil/plant relationships and crop production processes

By use of modern computers it has become possible to simulate complex dynamic systems to establish models which can be used to predict how changes of input to such sytems can affect the systems themselves. A model reflects a simplified picture of a mulitcausal system taking into account the most important factors which have an impact on the system. Such systems often comprise competing and sometimes interacting processes with the presence of different pools and flow rates between one pool and another. Changes of the system are described and converted into mathematical relationships so called algorithms. By running these algorithms changing the variables the best fit of the model to the system can be obtained so that the model can be used for prediction purpose. In recent years numerous models have been developed for use in plant nutrition and crop production such as models for nutrient and water flow in the soil (Kersebaum and Richter 1991, Moldrup *et al.* 1992, Funk *et al.* 1995, Lenzi and Luzio 1997) and for N and C turnover in soils (Molina *et al.* 1983, Jenkinson *et al.* 1994, Manguiat *et al.* 1994, Yang and Jansen 1997). In this section two models are described, one dealing with nutrient availablity in the mineralization of organic N in soils, the other forage production of ryegrass.

For fertilizer application and plant nutrition modelling of N mineralization is of particular importance. Manguiat *et al.* (1994) propose the following exponential model for assessing the N mineralization in paddy soils:

$$N_t = N_o (1 - e^{-kt})$$

N_t = concentration of mineralized N at the time t, t = time, N_o = N mineralization potential (= total mineralizable N),
k = mineralization constant

From this equation it is clear that at time zero N_t equals zero because then $e^o = 1$. If t is infinite the term $e^{-kt} = 0$ and $N_t = N_o$. Hence mineralization rates decrease with time exponentially and asymptotically approch zero. In this equation it is assumed that the N mineralization potential is a pool uninfluenced by other processes. This is actually not the case since the pool gains N from organic matter which becomes prone to mineralization. An inflow and an outflow of N from the pool is therefore given, the net decrease in N of the pool being generally lower than the the quantity of N mineralized (Mengel *et al.* 1999). The model of Molina *et al.* (1983) and the improved model of Nicolardot *et al.* (1994) take into account the most important processes involved in N minreralization and the N flow between various pools. The N flow of this model, NCSOIL (nitrogen carbon dynamics in soils) is shown in Figure 5.32. The source of N inflow is the N in organic residues mainly originating directly or indirectly from plants and the outflow of N is inorganic N (NH_4^+) the main source of which is pool II.

Physically or chemically stabilized organic N comprises N compounds tightly bound to clay minerals or incorporated into humic matter (Schulten and Leinweber (2000). Such bound N is scarcely affected by microbial enzymes and the half life of the N in Pool III is several hundred years in contrast to the half-life of the N in Pool I and II which is measured in days. The N compounds here are not irreversibly bound but can be released physically or chemically by particular conditions for example by drying or oxidation.

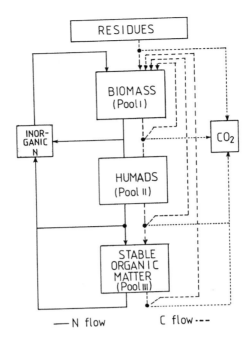

Figure 5.32 Scheme of the N flow model in soils of the NCSOIL model of Molina *et al.* (1983), modified after Nicolardot *et al.* (1994). Pool I = biomass = organic N of living microorganisms involved in the decomposition of organic matter. Pool (II) = organic N derived from biomass not yet stable = humads. Pool (III) = stable organic matter.

Nitrogen input comes mainly from plant residues from which the N, mostly proteins and nucleic acids, is generally quickly metabolized *i.e.* mineralized or incorporated into the microbial biomass. Some N losses may occur, as volatile N released into the gas phase or N transferred to a very stable humus-like form resistant to mineralization. As shown in Figure 5.32 there are several pools which are directly or indirectly involved in N mineralization. If the size of the pools and the reaction constants are known the flow rates between the pools can be calculated assuming that the reactions are first order reactions according to the equation for Nt shown above. Reaction constants differ for the various pools and are in the order of 0.33 for Pool (I) and of 0.16 for Pool (II) (Appel 1998b). According to Appel (1998b) Pool (II) is of utmost importance for net N mineralization. It must be borne in mind that these pools are not spatially separated in the soil and their quantification is not easy. Recently Appel (1998b) measuring net N mineralization in incubation experiments with 23 different agricultural soils found that the organic nitrogen (Norg) extracted by the $CaCl_2$ solution as recommended by Houba *et al.* (1986) was indicative of Pool (II) of the NCSOIL model. Parameter optimization provided evidence that the model must be supplied with organic N of a low C/N ratio of 3.5 indicative of amino acids and one may conclude that these represent the major N source for mineralization and that Pool (II) mainly consists of dead microbial biomass.

Using this approach Appel (1998b) found a close correlation between the measured and the precicted net N mineralization according to the model ($r^2 = 0.716$). About 80% of the mineralized N originated from Pool (II) and the remaining 20% from Pool (I) which represents living microbial biomass. The N extracted by the $CaCl_2$ solution represented only a small proportion of the total N in Pool (II), but this small proportion of about 3% was significantly correlated with the N in Pool (II) so that the quantity of N in Pool (II) could be assessed. This amoutd to 12 to 31% of the total soil N. Twenty three parameters are considered for running the NCSOIL model the most important of which are soil moisture, soil temperature and various reaction constants. The model therefore can be adjusted to the particular conditions of a site.

Various processes of crop production have also been modelled. Ten Berge *et al.* (1999) edited a monograph *Application of Rice Modelling*. Dauzat and Eroy (1997) modelled the light regime of maize and mung bean grown under coconut trees. Manschadi *et al.* (1998) simulated the growth and development of the root system of faba beans (*Vicia faba L.*). The model comprises weather data, soil data, management data and genetic data of the crop. The flow diagram of the various input parameters is shown in Figure 5.33. A major parameter in this flow diagram is the *Faba Bean Model* (FAGS) which provides the daily dry matter alloction to roots. To run a model easily it is very important to know which parameters have the strongest input. Meinke *et al.* (1997) reported that for modelling grain yield of spring wheat under the subtropical climate conditions of Queensland (Australia) water and nitrogen supply were the most important parameters.

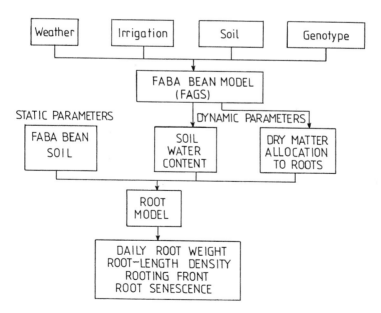

Figure 5.33   Flow diagram illustrating the required input parameters for the simulation of root growth (after Manschadi *et al.* 1998).

The simulation model for the productivity *of Lolium perenne* of Schapendonk *et al.* (1998) is discussed more in detail in order to show the relationships of physiological proceses to relevant parameters and their quantification. This model is run with separated algorithms for source and sink related processes. Repeated defoliation such as cutting or grazing leads to temporary shortage of assimilates, alternated with periods with assimilate surplus when the leaf area (LAI) is high and the foliage can intercept almost all the incoming photosynthetic active radiation (PAR). New leaves can only develop when carbohydratres are remobilized from a storage pool to compensate for photosynthetic rates after defoliation. The initial LAI in spring is fixed at 0.1 and initial carbohydrate in the stubble fixed at 20 g/m^2. The LAI after defoliation was obtained by a calibration curve and fixed at a value of 0.5. The cutting dates may be fixed by the model user. The light use efficiency (LUE = dry matter production per unit PAR, g/MJ) is rather constant for the same climate but may differ under various climatic conditions. It is assumed that the assimilate flow from the source and inflow of assimilates into the sink are semi-independent which means that both processes are not completely independent because of feedback mechanisms. High assimilate concentration in the sink thus may affect photosynthesis. The growth rate (Δ W = difference between two dates of measuring) equals either Δ W-source or Δ W-sink; in the first case photsynthesis, in the latter case metabolization of assimilates in the sink is the limiting process. The source is mainly radiation dependent, the sink depends much on temperature. The light profile in the canopy is calculated on the basis of LAI and the light extinction coefficient. The efficiency of conversion of absorbed light into carbon assimilation is variable in time and depends on radiation intensity, temperature and water availability. If water and nutrient supply are optimum the LUE (light use efficiency) depends on the intensity of photosynthetic active radiation (PAR), and LUE generally is higher at low than at high light intensity. Metabolic intensity of the sink is defined as the temperature-driven rates of leaf area increase. This depends on temperature as shown in Figure 5.34 from the work of Schapendonk *et al.* (1998). Besides leaf elongation the formation of new tillers also requires assimilates. The tillering is calculated from the proportion of buds in the new lesf axils. The supply of buds with assimilates is called site filling which decreases with increasing LAI and internal shading. LAI duration depends on longevity and tiller survival. Tiller death and leaf senescence are promoted by internal shading. As the LAI increases deeper layers of the sward become shaded with a critical LAI for internal shading of 4. Tillering rate is estimated as the ratio between the calculated amount of stored carbohydrates and an estimated maximum amount of available carbohydrates. Carbohydrate demand is generally higher than photoysnthetic supply during the first days after defoliation.

Under source limiting conditions the leaf area development is no longer determined by temperature-dependent leaf elongation rates but by the amount of assimilates partitioned to the leaves. Simulated growth starts when the 10 day moving average of daily temperature is above a critical level which was 3°C for the northern and 5°C for the southern countries. For model calibration, several parameters are used: the remaining LAI after cutting, the base temperature after cutting, the base temperature for the start of photosynthesis and development, and the daily temperature for optimum light use efficiency (LUE). The model was validated with an FAO database from 35 sites in

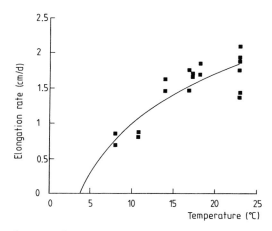

Figure 5.34 Leaf elongation rates of grass leaves as a function of daily temperature averaged over five days after cutting (after Schapendonk *et al.* 1998).

Europe. The average error between the predicted and observed yield was 14% for the irrigated grassland and 19% for the non irrigated grassland.

5.4 Nutrition and Plant Quality

5.4.1 General

The quality of plant products is not so easily defined and measured as yield. Quality standards depend very much on the purpose for which a plant is used. For example, very different qualities are looked for in grains of malting barley than in those used for animal feeds. The same is true for potato tubers, as to whether they are used for starch production or for human consumption. There are also other examples of this kind. Many quality factors such as flavour or taste are very difficult to measure and as these factors are also very much subjective, the absolute assessment of quality is often very difficult. For this reason the following section is restricted to considering some basic relationships between nutrition and the synthesis of organic compounds.

The major factors controlling crop quality are fixed genetically. Thus potato tuber proteins differ considerably between potato cultivars, whereas environmental factors including nutrition are scarcely able to influence protein pattern. On the other hand, exogenous factors can considerably influence the concentrations of organic compounds in plant organs of relevance to the produce quality. This is the case for example for both sugar beet and grapes where high light intensity during the filling period is desirable to obtain quality crops rich in sugar. Nutritional effects are dependent on the influence of particular nutrients on biochemical or physiological processes. The concentration of carbohydrates or sugars in storage tissues, grains and seeds thus relate directly or indirectly to the supply of plants with K^+ or nitrogen.

5.4.2   Root crops

Potassium promotes CO_2 assimilation and the translocation of carbohydrates from the leaves to the tubers of potatoes. This is the reason why the starch concentration in the tubers is high in potatoes well supplied with K^+ although too high a supply of K^+, may depress the starch concentration as reported by Beringer et al. (1983). Haeder et al. (1973) reported that at three weeks after flowering that in potatoes with an optimum K^+ supply about two thirds of the photosynthates produced in the leaves had been translocated to the tubers whereas in contrast with suboptimal K^+ nutrition only about half of the produced photosynthates were translocated to the tubers. It has been known for many years that application of potassium sulphate is preferential to muriate of potash (KCl), giving a higher starch concentration in the tubers (Terman 1950). This is due to the fact, that chloride has a detrimental influence on the translocation of carbohydrates from leaves to tubers (Haeder 1975). The quality of potato tubers is not only a matter of starch concentration. For tubers used in starch production there should be a high degree of esterification between phosphate and the hydroxyl groups of the starch. More highly esterified starches are more viscous and of better quality. Experiments of Görlitz (1966) and of Effmert (1967) thus showed that P fertilization not only increased the P concentration in potato tubers but also improved starch quality. In tubers used for human consumption ‚blackening' of tubers is often a problem. Blackening originates from a complex formed by iron and chlorogenic acid. Citric acid inhibits this complex formation, probably by chelating the Fe. As the concentration of citric acid in potato tubers is positively correlated with the K^+ concentration (Macklon and DeKock (1967), heavy potash dressing usually reduces the suceptibility of potato tubers to blackening (Vertregt 1968).

This relationship between the occurrence of black spot and the K^+ concentration in potato tubers is shown in Figure 5.35 from the work of Vertregt (1968). Of the tubers with a K^+ concentration < 500 mmol K^+/kg DM more than 50% were found to be suffering from black spot, whereas of those tubers with a K^+ concentratiion > 600 mmol K^+/kg DM less than 20% showed the disease. Figure 5.35 also shows a clear negative correlation between the K^+ concentration and the dry matter content of the tubers. The sensitivity of potato tubers to damage caused by mechanical harvesting or by transportation is also influenced by nutrition. Pätzhold and Dambroth (1964) reported that higher levels of phosphate application reduced the sensitivity of potato tubers to mechanical damage.

Potato tubers containing high concentrations of glycoalkaloids concentrations may be toxic (Sinden and Deahl 1976). The two main alkaloids found were α chacoine and α solanine; the concentration of the latter being somewhat lower. Very large differences in alkaloid concentration were found between the various parts of the tuber with highest concentrations in the eye area 1640 mg, in the peel 1000 mg, in the vascular bundle zone 80 mg per kg fresh matter. Only traces of alkaloids were found in the medulla of the tubers. Potato storage did not influence the glycoalkaloid concentration of the tubers.

Tubers of other crop species with carbohydrates as the main storage material, such as sweet potatoes (*Ipomoea batata*), cassava (= manioc), yam (*Dioscorea* species) and topinambur (*Helianthus tuberosus*), respond in a similar way to nutrition. Obigbesan (1973)

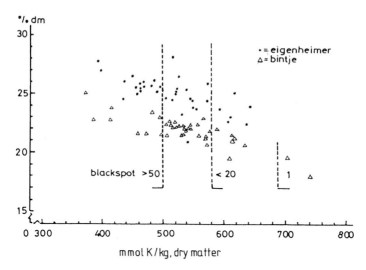

Figure 5.35 Relationship between K⁺ concentration in potato tubers, appearance of black spot, and dry matter in tubers (Vertregt 1968).

found that in tubers of cassava not only was the starch concentration enhanced by K⁺, but the concentration of the poisonous cyanide was also decreased.

The quality of sugar beet depends primarily on sugar concentration, but is also affected by the concentration of soluble amino compounds, and by the concentration of minerals, in particular K⁺ and Na⁺. The presence of soluble amino compounds and minerals disturbs crystallization during sugar refining and thus affects the sugar output. Increasing the K⁺ nutrition to an adequate level is generally accompanied by an increase in the sugar concentration and a decrease in the concentration of soluble amino compounds (Draycott *et al.* 1970). On the other hand, high K⁺ levels at least slightly increase the K⁺ concentration in roots, although the main response is in increasing the K⁺ concentration of the leaves. An increase of K⁺ in sugar beet roots is usually accompanied by a decrease of the Na⁺ concentration (Von Boguslawski and Schildbach 1969).

Nitrogen nutrition is of paramount importance for sugar beet quality. During the first period of beet growth abundant N supply is essential for satisfactory yields, but during the later stages particularly at root filling N supply should decline (Bronner 1974). Schmitt and Brauer (1975) found in field trials that N fertilizer rates exceeding the optimum depressed root yield and quality. Sugar beet roots grown at too high a level of N nutrition during the last months prior to harvest are generally characterized by low sugar concentrations and high concentrations of amino compounds and minerals. This results because the storage tissue is retained in a juvenile stage (Forster 1970, see page 290). Late application of N and the supply of N forms, which are not directly available, as for example anhydrous NH_3, should therefore not be recommended for sugar beet cultivation. Application of Na$^+$ may improve sugar yield and quality of sugar beets (Judel and Kühn 1975).

Sugar beet quality is not only dependent on mineral nutrition but is also influenced by environmental factors. In years with low rainfall and high light intensity, low root yields are generally harvested. Such roots are usally high in sugar as well as in amino acids and minerals. This pattern is typical for conditions of water stress, which result in poor root growth and thus in an accumulation of sugars, amino compounds and minerals in the roots. Best sugar yields are obtained when high light intensity conditions prevail during the last weeks of the sugar beet growth period and when adequate water is available for the crop. Optimum fertilizer application in combination with irrigation produces maximum root yields of satisfactory quality (v. Boguslawski and Schildbach 1969). In Central Europe under these conditions yields of as much as 80 tonnes roots/ha or 12 Mg sugar/ha can be harvested.

The problems of sugar cane quality are similar to those of sugar beet. Sugar cane grown on saline sites is often poor in quality, being low in sugar and containing very high amounts of minerals and amino compounds.

5.4.3 Grain crops

In cereals used for bread, grain baking quality is of particular interest. Wheat cultivars grown under the arid conditions of Canada or of South-Eastern Europe are partially of the hard wheat type possessing high baking quality properties. Cultivars grown under the more humid conditions of Western and NorthWestern Europe, are often poorer in baking quality, but they are characterized by a higher grain yield potential. In the past there were large differences between these two types. New high yielding cultivars of wheat are now available, however, which have very satisfactory baking qualities. In order to utilize this quality potential, the growing crop must be adequately supplied with nutrients and in particular with N. The endosperm of wheat grains contains two types of storage proteins, the gliadins and the glutenins. The gliadins have a low molecular weight of about 35 kDa, do not possess disulphide bridges and have a poor biological and baking quality (Payne 1987). The glutenins on the other hand are large heterogenous molecules which comprise 19 subunits linked together by disulphide bridges formed by cysteine (Shewry 1992). The elasticity of the glutenins largely controls baking quality which differs markedly between wheat cultivars (Payne 1987). According to Ewart (1978) a high degree of glutenin polymerization increases dough tenacity and thus improves of baking quality.

Nitrogen supplied to cereals at flowering increases the protein content of the grains substantially and thus improves baking quality. Split application of nitrogen fertilizer, one application in early spring, another after tillering, the another at anthesis is now commmon practice in areas where water is generally not a growth limiting factor. Ample N supply increases the protein concentration as well as the concentrations of gliadin and glutenin in grains thus improving baking quality (Triboi et al. 2000). Under more arid conditions, however, baking quality is often not improved by late N application, as observed by McNeal et al. (1963) using five hard red spring wheat varieties. According to field trials of Primost (1968) and Schäfer and Siebold (1972) wheat must be adequately supplied with K^+ in order to obtain the beneficial effect of late N application on baking quality.

The feeding quality of proteins is mainly determined by the concentration of crude protein and the proportion of essential amino acids in proteins. These are amino acids which cannot be synthesized by humans or animals, and for this reason must be supplied as constituents of the mammalian diet (Table 5.14). Grains of cereals and maize are particularly low in lysine and therefore are rather poor in protein quality. Late N application increases the concentration of crude protein in the grains, but the individual grain proteins are affected to a different degree. Albumin and globulin, the main proteins of the embryo and rich in essential amino acids, are hardly affected, whereas the concentrations of glutenins and especially of gliadins are increased (Michael and Blume 1960). Mitchell *et al.* (1952) found a similar relationship in maize grains; N application mainly increased the concentration of zein (the storage protein analogous to gliadin in wheat and hordein in barley).

Table 5.14   Essential amino acids in human nutrition

Valine	Threonine	Phenylalanine
Leucine	Methionine	Tryptophan
Isoleucin	Lysine	

Zein, gliadin and hordein are very poor in lysine, so that by increasing their concentration the nutritional value of the grains is reduced. This increase in the gliadin proportion occurs especially in the final stages of grain maturation (Sonntag and Michael 1973) whereas the glutenin concentration is only slightly increased during the final stage of the grain filling period (Mengel *et al.* 1981). Late N application to wheat, barley and maize thus usually increases the concentration of grain crude protein, but the nutritional value of the grain is reduced. Analogous results have been obtained for the proteins of rye grains (Bayzer and Mayr 1967).

With oats and rice the situation is different. Here high or late N fertilizer dressings predominantly increase the concentration of glutenin, a grain protein with a moderately high proportion of lysine. For these crops therefore an increase in grain proteins by a late N dressing is not detrimental to the nutritional value. This finding has been confirmed in feeding experiments using pigs and rats. It was observed that the protein quality of barley and wheat grains was impaired, but that of oat grains was improved by high N application (Brune *et al.* 1968).

From the foregoing discussion it can be seen that poor quality grain proteins of wheat, barley and maize can be little improved by plant nutrition. As these grains play a major role in human nutrition, especially in the developing countries, the improvement of grain proteins is a challenging target for plant breeders. One important advance in this direction was made by Mertz *et al.* (1964). These workers obtained strains of maize and barley with relatively high concentrations of lysine in the grain, because the storage proteins consist mainly of glutenin and not of zein. Increasing the level of N nutrition to these strains therefore mainly resulted in an increase of glutenin rather than prolamin concentration (Decau and Pollacsek 1970, Sonntag and Michael 1973). Breteler (1976) compared the effects of increasing fertilizer rates on a normal maize cultivar with a lysine-rich cultivar (opaque maize) in field experiments. The lysine rich cultivar gave a higher grain yield than the normal one but maturation of the lysine-rich cultivar was delayed by three weeks. At all N fertilizer levels the lysine-rich cultivar gave higher lysine and tryptophan concentrations in grains, hence these grains had a higher biological value. Generally, however, lysine rich cereals are not so high yielding as conventional cereal cultivars. This means that any increase in lysine concentration in the grains is offset by a lower grain production. It is for this reason that these lysine-rich cultivars have not been accepted into practical farming on a large scale. It is easier to produce essential amino acids by using other crops e.g. grain legumes.

In this context it is emphasized that the non-protein nitrogen in potato tubers is of a poor biological quality in contrast to the protein in potato tubers. Such protein is obtained by coagulation in the process of starch production from potato tubers. Burgstaller and Huber (1980) found that this potato protein had the same beneficial feeding effect as soya protein. The concentration of essential amino acids in some feeding stuffs are shown in Table 5.15 from the work of Burgstaller and Huber (1980).

Table 5.15 Concentration of essential amino acids in some feeding crude proteins (Burgstaller and Huber 1980)

	Potato	Fish-meal	Skim milk	Soya
	g amino acid/kg crude protein			
Lysine	84	83	77	64
Methionine	23	28	26	15
Cysteine	19	43	47	40
Threonine	58	43	47	40
Isoleucine	57	43	58	47

A close relationship appears to occur between the protein concentrations in cereal grains and the concentration of vitamins of the vitamin B group (thiamine, riboflavin, nicotinic acid). Late N application generally enhances the vitamin B concentrations of grains Jahn-Deesbach and May (1972) and thus improves their nutritional value. Vitamin B_1 mainly occurs in the aleurone layer and in the scutellum of the grains. The vitamin B concentration of flours is therefore lower in the higher milled grades.

In grains used for malting purposes and in particular barley, low protein and high starch concentrations are required. The grains should be large, as larger grains are generally richer in carbohydrates and the germination rate is higher. Grain protein concentration should not exceed 115 mg/g dry grain. It has been known for many years that crops adequately supplied with phosphate and K^+ produce better quality grains for malting barley, whereas high rates of N application impair quality due to an increase in the grain protein concentration. This has been confirmed in field trials by Schildbach (1972). Small scale malting tests were carried out by the same author in which it was found that several components important for beer quality were improved by phosphate and K$^+$ application. A high single grain weight of cereals is generally associated wth a high starch and a low crude protein proportion in grains. In this respect K$^+$ nutrition is of particular importance for filling the grains with starch and increasing the single grain weight (Burkhart and Amberger 1977). Drought after anthesis may severely affect the single grain weight of malting barley and thus impair the malting quality as was found in greehouse experiments by Savin and Nicolas (1996). The nitrogen amount per grain was was not influenced by high temperature or drought and N concentration of the grains was increased. Analogous effects were found by Mengel *et al.* (1985) for wheat when grown at low light intensity. Simlar results were reported for rice by Tashiro and Wardlaw (1991). Highest single grain weight was obtained at a day night temperature of 27/22°C. Above this threshold single grain weight decreased and at day/night temperature of 39/34°C was only half of that attained with the optimum temperature. High temperature depressed starch formation more than protein formation in the grains. Investigations of Bertholdsson (1999) have shown that barley cultivars characterized by late heading, many tillers, and many seeds per ear generally have relatively low grain protein concentrations even when grown with a fairly high N supply.

Oat grains contain a particular polysaccharide which plays a role in human nutrition. This compound is a ß-glucan with a (1-4) and (1-3) ß glycosidic bond. This glucan is located in the cell walls of the endosperm. According to Ganßmann (1993) it increases the viscosity of the small intestine content and thus causes a delayed reabsortion and a higher excretion of bile salts. Therfore more cholesterol is required for the synthesis of bile salts in the liver. This additional demand for cholesterol is provided by absorption from the blood and it is for this reason that this particular ß glucan of oats depresses the blood cholesterol level. The oil concentration in cereal grains is relatively low (20 to 50 g/kg) with a high proportion of grain oil being located in the embryo. Welch (1978) found a positive correlation between the protein and oil concentrations of 86 barley genotypes. No relationship was observed, however, between the oil concentration of barley and baking quality. More than 50% of the fatty acid of barley grain oil consists of linoleic acid, as shown in Table 5.16.

Table 5.16 Fatty acid concentration in oil of barley grain (Welch 1978)

	g acids/kg grain
Palmitic acid	124—287
Stearic acid	6—18
Oleic acid	104—169
Linoleic acid	524—583
Linolenic acid	45—73

5.4.4 Oil crops

In oil crops like grain crops, there is a marked competition for photosynthates between different metabolic sinks. This was already shown clearly in the results of Schmalfuss (1963) in an experiment with flax. Increasing the level of N nutrition enhanced the concentration of crude protein from about 220 to 280 g/kg seed, but lowered the lipid concentration considerably. Analogous results were reported by Appelquist (1968) for rape (*Brassica napus*). The most important data of this investigation are shown in Table 5.17. At the low level of N nutrition the oil concentration was higher but both the seed yield and seed size were depressed. This beneficial effect of the lower level of N nutrition on the oil concentration probably resulted because the earlier senescence of leaves in this treatment reduced the rate of seed filling during seed maturation. Herrmann (1977) reported that high N fertilizer rates result in an increase of erucic acid in rape oil which is detrimental to quality. Since erucic acid-free rape cultivars are now available, however, this effect of N is of no major relevance. Oil concentration in seeds also depends on K^+ nutrition and is low at suboptimal K^+ supply as found by Mengel and Forster (1976) for a number of rape cultivars.

Table 5.17 Effect of nitrogen supply on yield and oil concentration of rape seeds (Appelquist 1968)

Nitrogen rate	Seed yield g/pot	Seed single weight mg/seed	Oil concentration g/kg dry matter
Low	10.0	3.0	468
High	18.6	3.6	417

It is well known that oil seeds grown at low temperatures are comparatively richer in unsaturated fatty acids than saturated fatty acids (Beringer1971). An example of this relationship is shown in Table 5.18 from the work of Ivanov (1929). The more unsaturated fatty acids have higher iodine numbers. Beringer and Saxena (1968) working with sunflowers, flax and oats confirmed this finding. They further showed that the concentration of tocopherol (vitamin E) in the seeds was increased by higher temperatures. The higher concentrations of unsaturated fatty acids found in oil seeds from plants grown at low temperatures can be explained by the greater accompanying O_2 pressure in these seeds which facilitates the oxidative desaturation of saturated to unsaturated fatty acids (Harris and James1969). At higher temperature conditions, the O_2 pressure in the seeds falls because of the higher respiration rate.

In the future, lipids and oils produced by crops are likely to gain in importance as renewable raw materials for technical purposes such as biofuels and lubricants. Fatty acids obtained by the hydrolyzation of sunflower and rape oils (triacyl glycerols) can be methylated and the ester produced can be combusted in Diesel engines with a efficiency of between 92 and 95% of technical Diesel fuel (Devel 1992). Other plants which appear suitable for breeding programmes for triacyl glycerols include *Cuphea* and *Euphorbia* species. Long-chain fatty acids are particularly desirable because of their high energy content. Lines of sunflower and safflower are known in which the oleic acid amounts to 80 to 90% of the total fatty acid (Röbbelen 1993).

Table 5.18   Effect of location on the iodine number in oil of flax seeds (Ivanov 1929)

Location	Degree of latitude	Iodine No.
Archangelsk	64	195—204
Leningrad	59	185—190
Moskow	55	178—182
Woronesh	51	170
Kuban-Odessa	45	163
Taschkent	41	154—158

5.4.5   Forage crops

The quality of forage crops such as herbage (grasses, clover, lucerne) used as pasture or as hay depends much on the digestibility of the fodder. Digestibility decreases as the concentration of crude fibre (cellulose, lignin) is increased. As these compounds are accumulated as plants age, and the concentration of crude protein decreases, the quality of old forage crops is generally poor. This is particularly true for grass, very young grass having a crude protein concentration of about 200 to 250 g/kg dry matter and crude fibre concentration of about 200 g/kg dry matter. In old grass the crude fibre content exceeds the content of crude protein considerably and may be in the order of 300 g/kg, while the crude protein concentration may only amount to 100 g/kg dry matter. Mild water stress may delay the senescence of forage and thus have a positive impact on forage quality as found by Halim *et al.* (1989) for alfalfa. In this experiment the digestibility of the forage was negatively correlated with the cell wall cellulose but positively correlated with the cell wall hemicellulose. Nitrogen supply increases the concentration of crude protein. This has been observed by numerous authors Hoogerkamp (1974). Table 5.19 from the data of Goswami and Willcox(1969) demonstrates the effect of increasing the rate of N on the various nitrogenous fractions in ryegrass. High rates of N fertilization enhanced the protein concentration substantially and the NO_3^- concentration appeared to reach a maximum at rates higher than about 400 kg N/ha. An increase in crude protein concentration is usually accompanied by a decrease in soluble carbohydrates and especially in polyfructosans (Nowakowski1962).

The digestibility of forage may also be influenced by other forms of fertilizer practice. Schmitt and Brauer (1979) thus reported that potassium application to grassland had a clear effect on the digestibility of various constituents in the resulting herbage. The digestibility quotients which were estimated using animal trials are shown in Table 5.20. The application of potassium, made in addition to the phosphate of the control treatment, improved the digestibility of crude protein, protein and lipids and depressed the digestibility of crude fibre. The meadow on which the experiment was carried out was a mixed sward and the K$^+$ application greatly increased the proportion of legumes at the expense of herbs. This shift in botanical composition exerted an important influence on the digestibility of the herbage.

Table 5.19 Effect of an increasing nitrogen supply on the various nitrogenous fractions of ryegrass (Goswami and Willcox 1969)

kg N/ha	Total N	Protein N	Free amino acid N	Nitrate- +nitrite N
		g N/kg dry matter		
0	13.2	9.8	1.6	0.4
55	15.3	11.0	1.6	0.4
110	18.9	12.6	2.1	0.6
220	16.9	17.5	3.1	1.7
440	37.3	20.6	5.6	3.5
880	39.3	23.4	5.9	4.1

Table 5.20 Effect of K fertilization on the digestibility quotients of various herbage fractions (data from Schmitt and Brauer 1979)

	% digestibility	
	P fertilizer	P + K fertilizer
Dry matter	60.0	60.3
Organic matter	63.9	62.9
Crude protein	54.1	61.3
Protein	48.0	54.9
Lipids	44.3	50.6
Crude fibre	64.5	61.6

Generally high N application leads to an increase in the proportion of grasses, whereas P and K fertilization favours the legumes. This relationship is shown in Table 5.21 from a long-term field trial on a loamy alluvial soil typical of many meadows in Central Europe (Schmitt and Brauer 1979). The N treatments on these sites (150-300 kg N/ha) enabled at least 3 cuts per year to be taken. Where only two cuts are harvested grass is often in a senescent stage when it is cut and this results in a poorer quality. How many cuts can be harvested in any one year frequently depends on soil moisture. In arid conditions, in particular, water can often be the limiting factor to forage growth. Under these conditions mixed swards of grasses, legumes, and herbs are more resistant to water shortage and other unfavourable influences. These meadows are often not fertiized with N in order to maintain a high proportion of legumes which fix the N required by the sward. Legumes on the other hand need an abundant supply of phosphate and K$^+$ otherwise they are replaced by herbs of a poorer nutritional quality. Competition in grass-legume associations, which has been considered by Haynes (1980a) in a valuable review paper, depends not only on mineral nutrition. Mutual shading may also affect the growth of the stand. Generally legumes suffer more from shading than do grasses. In this respect a new clover cultivar (Blanca), bred in Great Britain, is of particular interest as it is characterized by long petioles and is therefore less susceptible to shading. Even at a mineral nitrogen fertilizer rate as high as 400 kg N/ha this cultivar made up about 40% of the total sward (McEwen and Johnston 1984). In the plots which received no mineral N, the rate of biological N fixation of this ryegrass clover association was extremely high amounting to between 200 and 450 kg N/ha.

Table 5.21 Effect of fertilizer practice on the percentage proportion of grasses, legumes and herbs in a meadow. Location 'Beerfelden'. Soil derived from Bunter sandstone mixed with some loess (data from Schmitt and Brauer 1979)

Fertilizer	Grasses	Legumes	Herbs
	% proportion		
No fertilizer	65.8	6.3	27.9
P + K	65.9	22.5	11.6
P + K + N	80.5	9.3	10.2

Nutrient availability in soils is of high importance for forage yield and quality. Duru *et al.* (1994) investigated this relationship at four sites in the Pyrenees at an altitude between 1250 and 1300 m. In pastures with the highest nutrient availability the number of plant species was lowest. Of the 29 species in this plot only one leguminous species was present namely *Trifolium repens* but occupying a high proportion of the sward. For the next two sites following the decrease in soil nutritional status *Trifolium repens*, and *Medicago lupulina* were also present. In the location with the poorest soil fertility herbs dominated and of the grasses only *Festuca rubra* and *Trisetum flavescens* were found. Early growth was much more predominant in swards with a high nutritional status. The accumulated temperature required for an LAI = 2 was lower the better the phosphate availability in the soil. The favourable effect of a phosphate + K$^+$ application on forage production in relation to accumulated temperature is shown in Figure 5.36.

Nitrogen entry into soils as dry and wet deposition from the atmosphere may also have an impact on the botanical composition of non cultivated grassland. According to Wedin and Tilman (1996) nitrogen deposition from the atmosphere on to the land surface has increased about tenfold during the last 40 years to 5 to 25 kg N/ha in North America and to 50 to 60 kg N/ha in western Europe. In trials with N fertilizer

328

application on non-cultivated grassland in Minnesota shifts in the botanical composition were found. Native warm season grasses were replaced by low diversity mixtures of cool-season grasses. The proportion of C-4 grasses mainly *Schizachyrium scoparium* declined and the weedy Eurasian C-3 grass *Agropyron repens* became dominant. The shift was associated with a decrease in the C/N ratio of the biomass, increased N mineralization and nitrate accumulation in the soil as well as nitrate leaching (Wedin and Tilman 1996).

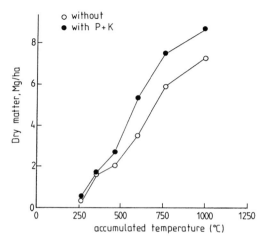

Figure 5.36 Effect of accumulated temperature on forage production as related to phosphate + K⁺ application. Filled circles with P+ K and open circles without P + K. Accumulated temperature = total of daily temperature above zero°C beginning on the 1. February (after Duru *et al.* 1994).

Forage crops contain energy (carbohydrates, fats, proteins) and the organic constituents (proteins) required for animal growth and the production of milk, eggs or wool. In addition to these organic constitutents, fodder supplies animals with the essential minerals such as P, S, Ca, Na, Mg, K, and heavy metals. In many cases, these minerals are present in forage crops in abundant quantities, but under conditions of intensive production a mineral shortage may occur. This is especially true in milk production, as dairy cows require considerable amounts of Na^+, Mg^{2+}, Ca^{2+} and phosphate. These minerals can be supplied as additives, but it is often also opportune to supply mineral contents in the fodder high enough to meet the normal requirement of the animals. For this reason pastures are often dressed with Na and Mg fertilizers, in order to increase the Mg^{2+} and Na^+ concentration in the herbage rather than to improve plant growth (Finger and Werk 1973). The Mg^{2+} nutrition of milking cows can be a problem, particularly in spring when animals are grazing young herbage. Frequently both the Mg^{2+} concentration, and its availability to the grazing animal are low in the young plant material. Generally 15 to 20% of the Mg^{2+} in the herbage consumed is taken up by the animals but in young herbage this available percentage of material already low in Mg can drop to values below 5% (Pulss and Hagemeister 1969). Under these conditions milking cows are undersupplied with Mg^{2+}, so that the Mg^+ concentration in the blood serum drops below the critical level (1.0 mg Mg/100 ml blood serum) and animals suffer from staggers or so-called grass tetany. The reason why the Mg^{2+} availability is poor in young grasses is not yet clear. High K^+ level in grass was often suggested as a cause for poor Mg^{2+} availability. Investigations of Kemp *et al.* (1961), however, showed that older herbage with high K^+ concentrations resulted in satisfactory Mg^{2+} resorption in milking cows. It is of particular interest that low Mg^{2+} availability is only observed in young grass and not in young legumes (Pulss and Hagemeister 1969).

5.4.6   Vegetables and fruits

Most of the plant nutrients such as P, K, Mg, Cl, S and the heavy metals are also essential elements for animals and man. In this respect nitrate is an exception; it is an important plant nutrient but it is not essential for the animal organism. Nitrate itself is not toxic but NO_2^- resulting from microbiological reduction of NO_3^- during storage or processing of plant material can be toxic as nitrite impairs oxygen transport of haemoglobin.

As yet it is not established whether NO_2^- can give rise to nitrosamine formation in the mammalian digestion tract. Carcinogenic nitrosamines are not synthesized in plants, even under conditions of a high nitrate supply (Hildebrandt 1979). According to Schlatter (1983), if there is any synthesis of nitrosamines from NO_2^- in the human digestion tract, the quantity produced is extremely low and in the order of 1/20000 of nitrosamines generally taken in with the diet. Nevertheless plants for direct consumption, such as herbage and vegetables should not have too high a nitrate concentration. In vegetables and particularly in spinach, a concentration of 2 mg NO_3^--N/g dry matter is regarded as a critical level. In forage crops concentratios of up to 4 mg NO_3^--N/g dry matter are acceptable.

Nitrate concentrations in forage crops and vegetables depend largely on the nitrate supply from the nutrient medium and the energy available for the NO_3^- reduction in the plant cells. For nitrate reduction NAD(P)H is required as shown on page 164. In green tissues these reducing equivalent are mainly provided by the oxidation glyceraldehyde 3-phosphate which is exported out of the chloroplast under light conditions. Under dark conditions, however, 3-phosphoglycerate is exported, which is unable to provide the reducing equivalents necessary for the reduction of $NAD(P)^+$. It is for this reason that nitrate reduction is depressed or even inhibited in green plant parts during night. This is evident in the nitrate concentration at the end of the night and during the day as can be seen in spinach shown in Table 3.9 from the work of Steingröver *et al.* (1982).

The nitrate concentrations decreased during day and it is also evident that nitrate concentrations were the highest in the petioles and lowest in the leaves due to the relatively high nitrate reductase activity in leaves. High nitrate concentrations occur in the vacuoles and in the xylem liquid particularly when plants are well supplied with nitrate. This nitrate is of osmotic importance (McIntyre 1997) and may also serve as a nitrogen reserve for periods of low N provision from the roots. When spinach plants grown in solution culture were deprived of the nitrate in the nutrient solution the high nitrate level in the upper plant parts fell to a very low level within a few days (Mengel 1984). The relative nitrate concentrations in plant parts of cauliflower are shown in Figure 5.37 from the work of Pimpini *et al.* (1970).

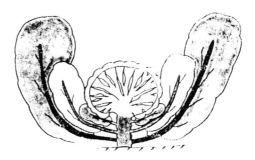

Figure 5.37 Relative nitrate concentrations in upper parts of cauliflower; nitrate concentrations are the higher the more shaded the plant part (after Pimpini *et al.* 1970).

Lowest nitrate concentrations were present in the younger leaves and in the head with highest concentration in the major veins, much higher than in the corresponding leaf blade.

Nitrate concentrations in plants depend also much on the level of available nitrate in the soil regardless of whether this originates from mineral fertilizer or from high nitrification rates as may occur after heavy slurry application. Table 5.22 shows nitrate concentration in lettuce grown in 14 commercial farms in Switzerland. Half of the farms were managed according to the conventional techniques using mineral fertilizer application; the other half of farms was along the lines of *organic farming* without mineral N fertilizer and mainly using farmyard manure as a N source. Generally organic farming operates at a lower level of nitrate in the soil as is evident from the lower NO_3^- concentrations in the heads of lettuce grown from May until the end of October. However, regardless of cultivation practice, nitrate concentrations in the heads of the lettuce were extremely high in the lettuce grown in the greenhouse in November. The high nitrate concentration results from the low light intensity in the greenhouse. The concentration is much higher than the tolerable level which is about 350 mg nitrate-N/kg fresh weight (Temperli *et al.* 1982).

A useful review paper dealing with nitrate in vegetables was published by Maynard *et al.* (1976).

The vitamin concentration in forage crops, vegetables and fruits is also a quality factor. Increasing of N supply may raise the carotene concentrations in tomatoes and carrots (Penningsfeld and Forchthammmer 1961). These authors also found that high levels of K^+ nutrition generally increased the concentration of vitamin C (L-ascorbate) in tomato fruits and carrot roots whereas higher N levels resulted in the reverse effect. Fritz *et al.* (1976) reported that the vitamin C concentration in tomato fruits ranged from 150 to 450 mg Vitamin C/kg fresh matter and was much dependent on light intensity which favours the synthesis of vitamin C. Anac and Colacoglu (1995) carrying out field trials in the Aegean region of Turkey observed a significant effect of K^+ fertilizer application on the vitamin C concentration in tomato fruits. According to these authors K^+ application to figs (*Ficus carica*) decreased the proportion of sunscaled and cracked fruits. Sunscale is characterized by a degradation of pigment due to higher

levels of oxygen radicals which are produced during e⁻ transort in chloroplasts at
high temperature and light intensity (see page 143). Vegetables and fruits may be
so damaged by sunscald so that they can not be sold on the market. Some species
such as tomatoes cucumber and pepper are rather resistant to sunscale because of
their high levels of carotenoids and SOD activity which detoxify the oxygen radicals
(Bowler *et al.* 1992).

Table 5.22 Nitrate-N concentrations in the heads of lettuce as related to growing season and farming
practice. Conventional farming mainly with mineral N fertilizer, organic farming without
mineral N fertiizer and mainly farm yard manure as N source (after Temperli *et al.* 1982).
ns = not significant, *** = significant $p < 0.01$

Growing season	Organic	Conventional
	mg nitrate-N/kg fresh weight	
Mai 1980	83	243 ***
June 1979	65	218 ***
June 1980	128	172 ***
Oct. 1979	226	280 ***
Oct. 1980	348	443 ***
Nov. 1979	816	809 ns
Nov. 1980	767	799 ns

The quality of fruits not only depends on the concentrations of organic constituents,
but also to a considerable extent on fruit size, colour, shape, flavour and taste. The latter
characteristics are also a question of maturation. Taste and aroma molecules are
frequently produced during the last phase of maturation.

In apple production it is of utmost importance to maintain an adequate Ca supply to the
fruits otherwise ‚bitter pit‘ occurs (see page 532). This disease is characterized by small
brown spots on and in the fruit resulting from a breakdown of fruit tissue (Plate 11.1).
Such fruits are generally low in Ca concentration (< 200 mg/kg Ca in the dry matter)
and may often be high in K^+ and Mg^{2+}. The Ca^{2+} concentration in apple fruits depends
considerably on transpiration conditions, as Ca^+ is moved to the fruits almost exclusively
in the transpiration stream. Apple fruits in which the transpiration rate is adequate are
generally well supplied with Ca^{2+}, provided that Ca^{2+} uptake by the roots is not impeded.
Under moisture stress conditions, however, leaves and fruits can compete for xylem sap,
and in the process the leaves are more effective. According to Wilkinson (1968) there may
even be a back-flow of xylem sap from the fruits to the leaves under very dry weather
conditions. Bünemann and Lüdders (1969) reported that bitter pit occurs more frequently
when fruit and shoot growth is high. It is known that Ca^{2+} moves preferentially
to actively growing vegetative tissues than to storage tissues. Shoot growth may
therefore compete with storage tissue for available Ca^{2+} in the plant and induce Ca^{2+}

deficiency in the fruits. This effect may be brought about by high levels of N nutrition. This may account for the well-known observation that Ca^{2+} disorders are stimulated by high levels of N fertilization.

Sprays of Ca salts can be used to alleviate *bitter pit*. Generally, however, they are not very successful, for if Ca^{2+} is to be useful it must be taken up through the fruits, for it is immobilized in the leaf. Application of gypsum or other Ca salts to the soil is a more efficient means of supplying Ca^{2+}, as most of the Ca^{2+} in apple fruits arrives there from the roots *via* the transpiration stream.

Shear (1975) cites a list of 35 such Ca^{2+} related disorders in fruits and vegetables. This demonstrates that not only apples but also other fruits and vegetables are affected in quality by conditions of localized Ca^{2+} deficiency within the plant. Seçer and Unal (1990) reported that Ca^{2+} had a favourable impact on yield and quality of melons. *Blossom end rot* of tomato fruits is also brought about by an inadequate local supply of Ca^{2+} and has many similarities to *bitter pit* (Shear (1975). The disease is more prevalent under high levels of K^+ nutrition. This relationship is shown in Table 5.23 (Forster 1973b). On the other hand, tomatoes should be well supplied with K^+ in order to prevent so-called *green back*. This disease is characterized by a delayed maturation of the fruit associated with low N/Ca and K/Ca ratios (Marcelle and Bodson1979). Often the tissue around the fruit stem remains green whilst the remainder becomes a yellowish-red colour. Tissue affected by *green back* is hard and tasteless, and for this reason such fruits are of poor quality. Several authors have observed that abundant K^+ supply prevents or at least reduces *green back* in tomato fruits (Winsor 1966, Forster 1973b, Forster and Venter 1975).

Table 5.23 Occurrence of greenback and blossom-end rot at tomato fruits in relation to the level of potassium nutrition (Forster 1973b). (Number of fruits/6 plants)

	1	3	9
	mol K^+/m^3 nutrient solution		
Greenback	82	2	0
Blossom-end rot	9	15	21
Healthy fruits	96	192	221
Total No. of fruits/6 plants	187	209	242

Wine quality is dependent on numerous components and in addition is also subjective and therefore not easy to describe but some main features of wine quality are considered below. The basis of a good quality wine is the must, the pressed juice of grapes, and there is a saying that half of the wine quality comes from the must and is therefore produced in the vineyard and the other half is made in the wine cellar, particularly by fermentation. A good must should have a high specific weight which should be between 1060 and 1130 g/L or even higher. Must with lower specific weights gives very poor wines and generally are not used for wine making. Specific weight is mainly depedent on sugar concentration with glucose and fructose being by far the most important sugars present. This sugar concentration reflects the concentration of alcohol (ethanol) which can be obtained by fermentation. Alcohol alone is not an indicator of wine quality. The sugar must concentration, however, also mirrors the synthetic conditions during the last weeks of vine berries maturation. If there is enough sunshine, cool nights and no deficiency of water numerous constituents besides sugar relevant to wine quality are present in the must. A considerable proportion of these, such as flavonoids and terpenes are synthetized in pathways of the secondary plant metabolism which in particular respond very sensitively to microclimatic and weather conditions and other factors such as the exposition of the vineyard, soil water, rooting depth of the soils and soil minerals and also the weather conditions during berry maturation. It is for this reason that the location of a vineyard plays such an important role for the type and quality of wine.

Table 5.24 shows the concentration of most important constituents in the must in the course of berry maturation (Peynaud 1981). It is evident that sugar concentration increases and that of acids decreases during maturation. The maturation process depends much on temperature and if there is a humid cool autumn full maturation is frequently not be attained and poor wines result. The sugar/acid ratio is an indicator for the degree of berry maturation and is used assess the time of grape harvesting. The reducing sugars are glucose and fructose. Sucrose concentration is low because it is immediately hydrolized after it is imported *via* the phloem into berries. Acids are essential components of must and wine since they give the wine its freshness. They should be well balanced with other compounds and not dominate. The most important acids are tartaric acid and malic acid and sometimes also a small proportion of citric acid is present. As shown in Table 5.24 the concentration of tartaric acid does not change very much during berry maturation while that of malic acid falls considerably. Acid concentration in berries depends on climate, weather and the particular wine cultivar, with cool humid conditions during maturation favouring the production of acids and slowing down the synthesis of sugars. The famous white vine cultivar *Riesling* is known to produce relative high amounts of acids in contrast to *Chasselas*.

Other important constituents in the must are the polyphenols which comprise the tannins and the flavonoids. The latter have flavane as basic structure and comprise the flavone characterized by a yellow colour and the anthocyanidins with a red to bluish colour. These constituents are generally present in must and wines in glycosylated form, the common name for the red colour in red wines being anthocyans which are a group of anthocyanidin glycosides. The most important anthocyanidin in red wines is malvidin. The synthesis of anthocyans requires a light stimulus and generally berries exposed to high light intensity produce more anthocyans which is of special importance for red wines.

Tannins comprise two groups of contsituents those derived from the catechins and hydrolysable tannins which yield gallic acids and polymeres of gallic acid after hydrolyzation. Tannins give the wine a deep long lasting taste if they are well developed during wine storage and are indispensable constituents of good red wines. Terpenes are a futher group of essential constituents in wines, most of them are mono-terpene alcohols and mono-terpenediols. About 50 of such compounds are known. They give wine its fine fragrant aroma (Rapp 1989). Fingerprinting terpenes in wine can be used to identify vine cultivars of its origin. Muscat-type wines are very rich in these terpenes giving wines with a typical spicy (Gewürz) character. Riesling type wines have a medium concentration of terpenes and the fine fragrant aroma of a good riesling results mostly from these terpenes. Silvaner-type wines are virtually free of terpenes. This shows that the terpene concentration in wine is predominatly controlled by grape genetics.

Table 5.24 Change in concentrations of sugars and acids during berry maturation; cultivar Cabernet Sauvignon grown in Pauillac, Bordeaux (after Peynaud 1981)

Date	Reducing sugars	Tartaric acid	Malic acid
		g/L	
28. August	88	9.7	16.7
4. Sept.	124	9.4	11.6
11. Sept.	136	7.5	8.7
18. Sept.	166	7.5	6.6
25. Sept.	188	7.3	5.2
2. Oct.	196	7.2	4.6
11. Oct.	204	6.8	3.9

Ferulic acid is synthesized in the berries. It is the starting molecule from which numerous aromatic molecules are synthesized. The mature berry contains about 400 such compounds, the wine about 800 showing that during fermentation numerous aromatic molecules are produced. These compounds are mainly aldehydes, esters, ketones, sugars, and sugar alcohols.

Fermentation of grapes from red wines starts in tanks or wooden vessels before the juice is expressed. The alcohol produced destroys the membranes so that the anthocyans located in the vacuoles of the berry skin tissue are released and together with some of the tannins present in the skin and in the seeds. The grapes are then pressed and the fermentation continues in the juice. Tannins give the wine a harsh and unpleasant taste. They therefore need to "mature". They originate not only from the must but also from the oak barrels in which the young wine is stored. During storage in the bottles the tannins are altered and some of them form a sediment. This formation of the sediment is an indication that the wine is mature and hence finally has obtained the aromatic fruity smell and the smooth, full and harmonious taste typical of good red wines.

In white wine making, the berries are immediately expressed and the juice is fermented. Fermentation means that the must is oxidized by yeast in the absence of O_2. There are about a dozen yeast species involved in fermentation with *Saccharomyces ellipsoideus*, *Kloeckera apiculata* and *Hanseniaspora uvarum* as the most important ones (Peynaud 1981). The must is an ideal nutrient medium for the yeast and at the beginning multiplication rate of new yeast cells proceeds exponentially. The N concentration is of crucial importance for the yeast multiplication. In this respect amino acids in the must are the most important source for the growth of the yeast. Grape juice quality for yeast growth depends most on the proteinogenic amino acids with highest concentrations of proline, arginine, glutamate, and serine (Bisson 1991). According to Prior (1997) arginine concentration in the must is correlated with mineral nitrogen concentration in the soil from which the grapes are harvested. Metabolization of methione may lead to methionol which can decompose to produce S containing volatile molecules with a very unpleasant smell. This may happen if the must is poor in amino acids so that the N of methionine is used as an N source by the yeast cells (Rauhut 1996). Hence the concentration of amino acids in the must is also a quality characteristic which depends on vine nutrition.

During fermentation heat is developed which may give rise to the loss of volatile aroma molecules and therefore impair the wine quality. This problem is particularly acute for the production of white wines under warmer climatic conditions. Today fermentation temperature can be controlled using modern technology with the effect that fresh, fruity wines with a full bouquet can be produced in warmer countries (Sautter 1999).

The moderate consumption of wine may have a beneficial impact on human health and diminish the risk of cardiovascular diseases (Francois 1990). In this respect the *French paradox* is often discussed that despite exposure to high degree risk factors such as a diet high in saturated fatty acids and cholesterol, hypertension, and smoking the rate of mortality from coronary heart diseases in France is the lowest of the 22 industrialized countries in the Western World. The consumption of fruit and vegetables in the Western world is inversely related to the mortality from coronary heart diseases, however, no protection is as significant as that from the moderate consumption of wine (Stockley 1997).

CHAPTER 6

FERTILIZER APPLICATION

6.1   Nutrient Cycling

6.1.1   General

During the growing period roots act as a sink collecting available nutrients to be utilized in the synthesis of organic plant constituents. After the termination of growth and the start of decay, the process is reversed and nutrients are released into the soil from the breakdown of plant debris. Not all nutrients taken up by plants over the growth period need necessarily be released back into the soil in the same year. In trees and vine, for example, an appreciable proportion of the annual uptake of some nitrogen and K^+ is retained in bark and wood over winter and remobilized in spring (Löhnertz et al. 1989b, Millard 1996, Tagliavini et al. 1999). On the other hand, some nutrients may be lost from growing plants by the leaching effect of rain. Senescent leaves are particularly susceptible to the loss of potassium, sodium, chloride, nitrate, and phosphate. The leaching of Ca^{2+} from plant leaves may also provide an important and beneficial effect of mobilizing Ca^{2+} since cycling of this nutrient within the plant is virtually absent (see page 522). The Ca^{2+} brought back into the soil may be taken up by plant roots and then can be translocated into younger leaves.

The turnover of nutrients in the 'soil-plant cycle', as described above, is controlled by a number of factors. In particular it is dependent on the intensity of weathering of soil particles, the nature of the soil parent material and the rate of leaching of plant nutrients from the upper soil layer. When the rate of leaching is high, plant nutrients may be leached out of the upper soil layers at a faster rate than they are taken up by plants. Soils in which this occurs have a negative nutrient balance and become more and more acidic due to the leaching of organic anions and bicarbonate (see page 58). Such soils become progressively poorer in available plant nutrients and show typical features associated with low soil pH conditions. These include low phosphate availability, high levels of soluble cationic Al species and Mn and, severe retardation or complete inhibition of N_2 fixation and nitrification (see page 400 and 401). The podzolic soils (*Spodosols*) are an example of soils with the features described above. These soils are found throughout the world and develop under conditions where the parent material is poor in plant nutrients and the leaching intensity is high. On the other hand when the nutrient balance is positive, very fertile soils can develop because of the continuous accumulation of plant nutrients. The chernozems (*Mollisols*) are typical of these natural fertile sites. Between these two extreme examples many other soil types develop, which are intermediate in their nutrient balance status. Of course not all soils with a positive nutrient balance are fertile. Toxic salt accumulations can also occur, as is the case in the saline and alkaline soils (see page 62).

337

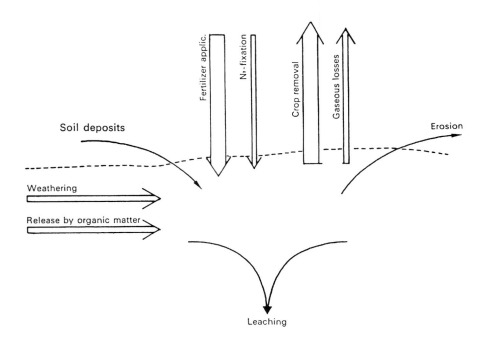

Figure 6.1   Nutrient gains and nutrient losses in soils.

The development of agriculture has lead to a disturbance in the nutrient balance in soils. In the older farming systems a high rate of recycling of plant nutrients was guaranteed. Animals were fattened by extensive grazing so that a large proportion of nutrients was returned to the soil in faeces and urine. Similarly most crops were consumed locally and mineral plant nutrients were brought back to the soil as manure. The main nutrient loss from the system therefore resulted from leaching, and to some extent this was overcome by the age-old practice of liming. With the onset of industrialization, however, the situation changed. Plant nutrients were removed from the cycle to an increasing extent in the form of crop and animal products to meet the food demands of the growing towns. Intensity and specialization in agriculture increased and farming techniques improved. All this resulted in a widening nutrient gap which was filled by the application of plant nutrients in the form of mineral fertilizers. By this means the nutrient balance between the input and removal of plant nutrient was redressed. This idea was put forward by Liebig (1841), when he wrote: "It must be borne in mind that as a principle of arable farming, what is taken from the soil must be returned to it in full measure". In addition the use of mineral fertilizers improved the nutrient status of many soils which previously had been poor in plant nutrients and of little agricultural value as was the case for agricultural development in European countries. Today huge areas in third-world countries are in a situation similar to that of Europe about 150 years ago. Sanchez and Leakey (1997) reported that most soils in Sub- Saharan Africa are infertile becaues of a lack of nutrients, particularly phosphate. Nutrient imbalance in this part of the world is particularly evident in soils of smallholders. On average this imbalance amounts to a loss of 700 kg N, 100 kg P and 450 kg K per ha over the last 30 years (Sanchez and Leakey 1997). Here input of plant nutrients into the soils provides a start for a gradual improvement of soil fertility and thus an improvement of soil and labour efficiency. Since mineral fertilizers are usually expensive in third world countries, cropping systems should be developed which minimize plant nutrient loss. with the integration of leguminous crops into the rotation this nitrogen source may contribute to balance the nutrient cycle of farming.

The major components of the nutrient balance in agricultural soils are shown in Figure 6.1. Soil nutrient gains arise from nutrient release in the soil by weathering of soil minerals such as micas, feldspars, and carbonates and by the mineralization of organic matter, the application of mineral and organic fertilizers, nitrogen fixation by N_2-fixing bacteria, and the supply of nutrients contain in rain and snow. The main losses result from crop removal, leaching, volatilization (NH_3, N_2, N_2O) and erosion. These processes are now considered in more detail.

6.1.2   Nutrient removal by crops

The quantity of nutrients which a crop takes up during growth depends much on crop species and yield. Both crop yield and nutrient uptake also depend considerably on crop cultivar. This is demostrated in the data of Table 6.1 which shows the nutrient uptake of a rice crop (Kemmler 1972). The new high yielding cultivar has higher nutrient requirements, because of its increased yield potential.

Table 6.1   Yield level and nutrient uptake of conventional local rice and the high yielding rice cultivar "TN1" (Kemmler 1972)

Cultivar	Grain yield, Mg/ha	N	P	K, kg/ha
Local	2.8	82	10	100
TN1	8.0	152	37	270

The nutrient uptake of various field crops is presented in Table 6.2. Comparative figures are given in Table 6.3 for nutrient removal by the most important plantation crops. Nutrient removal in these data is calculated on the basis of those plant parts which are generally harvested and thus removed from the field. The same also applies to the data of Table 6.4 which shows nutrient removal by fruit trees. In field crops only those plant parts of agricultural importance are usally removed from the field as for example grains, tubers or roots. This must be taken into account when calculating nutrient balances. The mature grains of cereal crops contain about 70% of total N and P of the aerial plant parts. Most K, however, occurs in the green plant parts, and as little as 25% of the total K is present in the grains. In contrast, the roots and tubers of root crops contain only about 30% of the total N absorbed, but as much as 70% of the total K.

Table 6.2   Quantities of nutrients removed by various crops (data calculated from Eakin (1972) and Kluge (1992); Mg = Mega g = 1000 kg

Crop	Yield Mg/ha	N kg/ha	P kg/ha	K kg/ha	Ca kg/ha	Mg kg/ha	S kg/ha	Cu g/ha	Mn g/ha	Zn g/ha
Barley (grain)	2.2	40	8	10	1	2	3	34	30	70
(straw)	2.5	17	3	30	9	2	5	11	360	60
Wheat (grain)	2.7	56	13	14	1	7	3	33	100	160
(straw)	3.8	22	3	33	7	4	6	11	180	56
Oats (grain)	2.9	55	10	14	2	3	6	34	134	56
(straw)	5.0	28	8	75	9	9	10	34	—	360
Maize (grain)	9.5	150	27	37	2	9	11	66	100	170
(straw)	10.0	110	19	135	29	22	16	55	1700	359
Sugar beet (roots)	50	90	22	125	—	—	—	—	—	—
(leaves)	30	84	10	120	—	—	—	—	—	—
Rape (seeds)	3	100	24	32	—	—	—	—	—	—
(straw)	6	84	20	240	—	—	—	—	—	—
Faba beans (seeds)	3	120	16	34	—	—	—	—	—	—
(straw)	3	45	20	62	—	—	—	—	—	—
Peas (seeds)	2	72	10	22	—	—	—	—	—	—
(straw)	2	30	2	42	—	—	—	—	—	—
Sunflower (seeds)	3	84	21	58	—	—	—	—	—	—
(straw)	4	60	4	144	—	—	—	—	—	—
Flax (seeds)	1	35	5	8	—	—	—	—	—	—
(straw)	1	8	1	12	—	—	—	—	—	—
Sugar cane	75	110	27	250	31	26	6	—	—	—
Tobacco (leaves)	2.2	83	8	110	83	20	15	33	600	80
Cotton (seed + lint)	1.7	45	11	14	2	4	3	66	120	350
Potatoes (tubers)	27	90	15	140	3	7	7	44	100	60
Tomatoes (fruit)	50	130	20	150	8	12	15	80	145	180
Cabbage	50	145	18	120	22	9	50	44	110	90

The data presented in the tables cited above provide only a general guide to the amounts of nutrients removed from the soil by crops. The figures can vary considerably. As well as the effects of yield differences, the use of new cultivars and modern techniques have also had a bearing on crop nutrient requirements and thus also on the amounts of nutrients taken up by the crops. Today under favourable climatic conditions,14 Mg maize grains, 12 Mg rice grains and 10 Mg wheat grains can be obtained per ha and season not thought possible twenty or thirty years ago. Such high crop yields require heavy fertilizer application.

Table 6.3   Nutrient removal of plantation crops (Cooke 1974)

	Yield/ha	N	P	K	Ca	Mg
				kg/ha		
Oil Palm	2.5 Mg of oil	162	30	217	36	38
Sugar cane	88 Mg of cane	45	25	121	—	—
Coconuts	1.4 Mg dry copra	62	17	56	6	12
Bananas	45 Mg of fruit	78	22	224	—	—
Rubber	1.1 Mg of dry rubber	7	1	4	—	—
Soyabeans	3.4 Mg of grain	210	22	60	—	—
Coffee	1 Mg of made coffee	38	8	50	—	—
Tea	1300 kg of dried leaves	60	5	30	6 3	

Table 6.4   Nutrient removal by fruit crops; medium yield, normal spacing (according to Jacob and von Uexküll (1963)

	N	P	K
		kg/ha/year	
Pome fruits	70	9	60
Stone fruits	85	9	65
Grapes	110	15	110
Oranges	170	23	120
Lemons	180	23	115

For a rough calculation of nutrient uptake by crops the following figures may be used:

1 Mg grains (cereals) contains: 20 to 25 kg N, 5 kg K, and 4 kg P

1 Mg dry matter of leaves and stems (forage, vegetables) contains: 20 kg N, 20 kg K, and 2 kg P

Data relating to nutrient removal, and fertilizer application rates in particular, are frequently given in the old oxide formulae, *e.g.* kg P_2O_5 or kg K_2O per ha. Recently it has become more common to express these values in terms of the element, *e.g.* kg P or kg K per ha. This modern nomenclature has been used throughout in this book as a means of expressing nutrient removal and leaching losses. The only exception to this is for fertilizer usage which is still often given in the old oxide terminology. The numerical difference between the two types of nomenclature is not large for K and to calculate the 'K_2O rates', the K data must be multiplied by a factor of 1.2. The respective factors for

P_2O_5, MgO and CaO are 2.29, 1.66 and 1.4. From these figures it is clear that differences in data between the two forms of terminology are greatest for phosphate.

6.1.3 Nutrient removal by leaching

The extent to which nutrients are transported down the soil profile varies considerably between soils. It depends mainly on the climate, soil type and the quantity of nutrients present in the soil in readily soluble form. Freely drained soils are very prone to nutrient removal by leaching. This is the case in the tropics and in the humid temperate climate particularly on pervious soils (van Wambeke 1991, Sims and Ellis 1983, Ballif 1992). An important parameter determining the downward movement of water in soil profiles is the *drainable porosity* which is described by the following equation:

$$\mu = P / \Delta H \quad (mm)$$

in which μ is the drainable porosity, P the amount of percolating water in mm, and ΔH the change in water table in mm. From this equation it is clear that the drainable porosity increases with the amount of percolating water and decreases with the rise of the water table. Water percolation and therefore movement of solutes in the water only occurs to a major extent when the water potential of soils is at or over field capacity.

The rate of leaching is generally measured using lysimeters in which the flow of water and nutrients can be followed through columns of soil similar to those of the soil profile. Although movement of water in natural soils differs from water movement in lysimeters, as these generally have an artificial soil profile, leaching rates obtained using lysimeters provide a useful guide to assessing the rate at which plant nutrients are lost from the soil. Under the climatic conditions of Central Europe with a rainfall of about 700 mm per year, about 25 to 50% of the rainwater passes through the soil profile to a depth of more than one metre. Higher water percolation figures are typical of sandy soils whilst the lower value is more usual in heavier soils. The quantities of nutrients transported along with the water into deeper soil layers are shown in Table 6.5 (Vömel 1965/66). The data presented are the maximum and minimum values observed over a period of eight years on arable soils in Germany and agree well with leaching figures obtained in England (Cooke 1972). The data show that the rate of leaching of nutrients is highest on light soils and the lowest on heavy soils where the rate of percolation is low.

Table 6.5 Rates of leaching of plant nutrients from soils of different texture (Vömel 1965/66)

Soil	Clay concentr.	N	K	Na	Ca	Mg
				kg/ha/year		
Sand	<3 g/kg	12—52	7—17	9—52	110—300	17—34
Sandy loam	16 g/kg	0—27	0—14	1—69	0—242	0—37
Loam	280 g/kg	9—44	3—8	11—45	21—176	9—61
Clay	390 g/kg	5—44	3—8	9—42	72—341	10—54

The leaching of plant nutrients is not only dependent on the rate of percolation and on the quantity of nutrients present in the upper soil layer. The tightness and the extent to which nutrients are bound to soil particles is also important. This can be seen from the data of Table 6.5. Although the concentrations of exchangeable Na^+ in these soils were much lower than the concentrations of exchangeable K^+, the leaching rates were higher for Na^+. Both cation species, Mg^{2+} and Na^+, are only weakly adsorbed by clay minerals (see page 19) and for this reason they are particularly susceptible to leaching. Calcium ions are also not very strongly adsorbed by inorganic colloids and, as Ca^{2+} is present in most mineral arable soils in comparatively high concentration in the soil solution (see Table 2.14), the rate of Ca^{2+} leaching is high. Potassium ions on the other hand, can be bound very tightly to 2:1 clay minerals (see page 19). For soils rich in these minerals, K^+ leaching rates are low. This is the case for numerous fertile soils derived from loess (*Mollisols*). Clay minerals of the kaolinitic types do not adsorb K^+ selectivily. High rates of K^+ leaching have therefore been observed in kaolinitic soils under tropical climate conditions (Pedro 1973). Also Ca^{2+} and Mg^{2+} leach readily under such conditions (Sims an Ellis 1983).

Of the major plant nutrients phosphate is leached at the lowest rate. Cooke and Williams (1970) found that on heavy arable soil, phosphate originating from an annual dressing of 33 kg/ha given over the past 100 years only penetrated about 20 cm below the plough layer. For grassland similar results were obtained. Usually the rate of phosphate removal by leaching from the upper soil layer in mineral soils is rather low and amounts to about 0 to 1.75 kg P/ha/year (Dam Kofoed and Lindhard 1968). In organic soils, however, higher phosphate leaching rates may occur as phosphate is less strongly bound to soil particles (Munk 1972). The application of excessively high rates of poultry litter (Mozaffari and Sims 1996) or pig slurry (Peters and Basta 1996) to soils also introduces extremely high rates of phosphate with the consequent hazard of phosphate leaching into deeper soil layers. This has been shown by Werner *et al.* (1988) and Leinweber (1996).

Nitrate in soils is most mobile since it is not bound to soil particles. High concentrations of NO_3^- can arise rapidly in soils not only by the application of fertilizer N but also by transformation of the N present in soil organic matter. Nitrate leaching rates may therefore be high depending on climate, weather and soil conditions. Under temperate climatic conditions nitrate leaching occurs mainly during winter and early spring. In summer evaporation is high enough to keep nitrate in the rooting profile. Only on sandy soils after prolonged high rainfall may NO_3^- leaching occur. Therfore fertilizer nitrogen is usually mainly applied in spring or summer and not before winter. Teske and Matzel (1976) in studying the N balance in lysimeter experiments with ^{15}N labelled urea, found that the N leaching rate ranged from 8.8 to 16.7 kg N ha/year. Of this amount only 12 to 15% originated from fertilizer N. Similar results were found by Kjellerup and Dam Kofoed (1983) with lysimeter experiments in which the fertilizer N (Nitrochalk) was also labelled with ^{15}N. On a long term basis N losses by leaching amounted to only 5% of the N fertilizer rate. High leaching losses, however, may occur when the N fertilizer rate is not adjusted to the N demand of the crop and the available N in soils. Müller *et al.* (1985)] studying different ecosystems in the valley of the Mosel found large N leaching rates in vineyards because of excessive N dressings. Nitrogen leaching rates from forest soils and agricultural soils were low and in the range of 7 to 17 kg N/ha/year whereas the rate from vineyards was 144 kg N/ha/year because of excessive nitrogen fertilizer application.

Field trials were carried out by Borin *et al.* (1997) in the Venetian Plain (Northeast Italy) with a well distributed rainfall throughout the year and an average annual rainfall of 950 mm. In this experiment nitrate leaching at four intensity levels were investigated. At the highest intensity the average annual N input was 447 kg N/ha which comprised mineral fertilizer and slurry. Here high nitrate leaching peaks were obtained. Nitrate leaching was particularly intense if slurry application in autumn was followed by rainfall. The second highest level of intensity received an average N application of 167 kg N/ha. Here nitrate leaching was much lower and on average not much higher than the amount found in the low input treatments with N inputs of 111 and 91 kg N/ha. On average over the total rotation nitrate leaching amounted to 42, 12, 12, and 7 kg nitrate N/ha/year respectively for the four intensity cropping systems. From this it is evident that the high surplus of N applied in the treatment with the highest intensity resulted in relatively modest nitrate leaching. The soil in this experiment was a silty clay with 400 g clay/kg and was therefore relatively impervious. Some of the surplus N may have been denitrified and some incorporated into the soil organic N pool. As reported by Kücke and Kleeberg (1997) N incorporated into this pool of organic N may later be mineralized to provide an indirect source of nitrate leaching. These authors also found on medium-textured loess-derived soils that nitrate leaching rates increased with N fertilizer application rates and were in the order of 100 kg nitrate N/ha after slurry application in autumn. On sandy soils nitrate leaching rates ranged from of 30 to 40 kg N/ha and were hardly influenced by mineral N fertilizer application.

Nitrate leaching may be considered from both agronomic and environmental aspects. From the agronomic aspect leaching of nitrate from the rooting profile of soils means a loss of a valuable plant nutrient whereas from the environmental aspect leached nitrate may reach the ground water from which drinking water may be obtained and therefore may act as a pollutant (Owen and Jürgens-Gschwind 1986). As considered above nitrate leaching is high if nitrogen fertilizer rates are excessive and not adjusted to the available nitrogen already present in the rooting profile at the beginning of the growth period. This reserve of available nitrogen present, mainly in the form of nitrate is closely related to crop yield (Wehrmann and Scharpf 1979, Blackmer *et al.* 1989). It is therefore of utmost importance to know how much available nitrogen is present in the rooting depth of soils. This may be measured by means of the "N_{min}-method" as described on page 101. Frequently in spring, however, there is little time available to take soil samples from deeper soil layers for determination of available nitrogen. Attempts have therfore been made to assess the available nitrogen to crops by means of simulation models. In such models it is assumed that the water present in the soil profile is at a water potential of field capacity or less and is replaced at deeper layers together with the solutes it contains by water brought on the soil surface, *e.g.* by rainfall (Smith *et al.* 1990). Flow models for nitrate leaching consider the flow rate in the profile, the nitrate concentration, the mineralizable organic nitrogen together with other parameters (Kersebaum and Richter 1991, Moldrup *et al.* 1992). As yet, however, the capacity of these models, to predict nitrate leaching is still unsatisfactory (Otter-Nacke and Kuhlmann 1991, Stockdale *et al.* 1997). A major problem is the estimation of the mineralizable organic nitrogen in soils (Thicke *et al.* 1993). According to Stockdale *et al.* (1997) a combination of field measurements and predictive modelling seems to be the way forward to achieve practical prediction for the use of fertilizer N by farmers. Engel *et al.* (1989) estimated the nitrate present in the soil profile (0—90 cm) in spring by means of a regression equation which takes into account weather data throughout the winter and the nitrate quantity present in a soil depth of 0—60 cm in spring. This model yields reliable results for available N but has the drawback that soil samples must be taken in spring and analyzed for nitrate.

The nitrate concentration in soils is decreased after harvest by the incorporation of straw into soils (Schmeer and Mengel 1984) because of the assimilation of mineral nitrogen including nitrate is assimilated by soil microorganisms decomposing the straw (Mengel 1996). By this means the rate of nitrate leaching is also lowered. A further means of decreasing nitrate leaching during winter is the cultivation of a so called catch crops which take up the available nitrate present in the rooting profile and thus protect it against leaching (Thorup-Kristensen 1994). In spring the catch crop is incorporated into the soil and the organic nitrogen mineralized from it during summer serves as nitrogen source for the main crop. A valuable review article on nitrate turnover in soils including nitrate leaching has been published by Smith *et al.* (1990).

Cover crops may reduce plant nutrient leaching to an marked extent. High leaching rates are often found in fallow soils in particular because in the absence of a nutrient demand by a growing crop, more nutrients are available for translocation. Similarly the percolation rate of water is also higher under fallow conditions because of the lack of a crop requirement. Differences in leaching between fallow and cropped land can be seen very clearly in the results of a lysimeter experiment carried out by Coppenet (1969) under the very humid conditions of the French Atlantic coast. The data of this experiment, presented in Table 6.6 are taken from observations made over a 12-year period. Leaching rates for all nutrients, with the exception of phosphate, were higher in the fallow treatment. The differences were particularly marked for N and K which are required and taken up by plant roots at high rates. Crop type can also affect the rate at which nutrients are leached from the soil. For widely spaced crops, such as grapes or maize, leaching losses are generally higher than for high density crops. Under grassland, N leaching is usually quite low although higher rates occur when leguminous species are present. This was shown by the work of Low and Armitage (1970) under English climatic conditions. Nitrogen leaching rates were especially high when the deterioration of the white clover sward began, as can be seen from the data of Table 6.7.

Soil cover is of importance for the recharge of ground water. Generally on arable land ground water recovery is higher than on grassland or forests because arable land frequently carries a thin or even no vegetation during autumn and winter, when ground water is mainly filled up by drainage water. Ground water recovery is particularly high on sandy soils due to their high rates of downward water movement (Strebel *et al.* 1985).

Table 6.6   Leaching rates of plant nutrients from a clay loam soil (18% clay) under fallow and cropped treatments (Coppenet 1969)

	Fallow	Cropped
	kg nutrient/ha/year	
N	142	62
P	0.3	0.3
K	46	24
Ca	310	230
Mg	24	18

Table 6.7   Nitrogen leaching rates in relation to soil cover (Low and Armitage 1970)

Period	White clover	Grass	Fallow
	kg N ha/year		
1952—1953	27	1.8	114
1953—1954	26	1.3	113
1954—1955	60 *	3.9	105
1956	131 **	2.0	41

* White clover dying out; **White clover removed

The movement of ions in solution involves the transport of cations and anions in equivalent amounts. In the soil profile the main cation species transported are Ca^{2+}, Mg^{2+}, Na^+ and K^+ and the major anions are NO_3^-, Cl^-, SO_4^{2-} and organic anions. In calcareous or alkaline soils HCO_3^- may also be leached at high rates. Leaching contributes to a varying degree to the loss of individual plant nutrients from the rooting depth of soils. For phosphate and K^+, leaching losses are in most cases less important than the removal by plant uptake. For Na^+, Ca^{2+} and SO_4^{2-}, however, leaching losses under humid conditions can amount to more than 50% of the total removal from the rooting depth of soils (Saalbach 1984).

The movement of ions into deeper soil layers may not only be regarded in terms of nutrient loss. Additionally, it is a process by which the accumulation of ions in the upper soil layer is prevented and hence also the development of salinity. This is an often forgotten advantage of regions of humid climate, for by appropriate fertilizer usage nutrient losses by leaching can be controlled. The retention of plant nutrients in the upper soil layers only becomes a major problem under tropical conditions where rainfall is high and soils are often very pervious. Under these conditions the use of slow-release fertilizers is likely to reduce leaching losses (see Chapter 6.2.4. on page 364).

6.1.4   Volatilization and denitrification

Nitrogen may be lost from the soil in the form of gases. Considerable losses may occur by the volatilization of NH_3. Losses generally increase with the pH of the soil according to the equilibrium

$$NH_4^+ \Leftrightarrow NH_3 + H^+ \text{ (pK = 9.21)}$$

Volatile losses of NH_3 may originate from NH_4^+ or urea containing fertilizers or both (Terman 1979) and may be relatively high on soils with neutral to alkaline pH in accord with the above equation. Additionally NH_3 volatilization may be high from soils with low CEC lacking specific adsorption sites for NH_4^+ (Fenn and Hossner 1985). Urea is more susceptible to volatilization than ammonium because the application of urea to soils is alkaline in effect as its amino groups may be protonated.

$$H_2N - CO - NH_2 + H^+ \Rightarrow H_2N - CO - NH_3^+$$
$$\text{Urea} \qquad\qquad \text{Protonated urea}$$

This reaction which may occur when urea comes in contact with the soil, gives rise to an alkaline zone around the urea granule favouring NH_3 volatilization. Ammonia is produced by urease an enzyme which splits urea into CO_2 and NH_3 as shown below.

$$H_2N - CO - NH_2 + H_2O \Rightarrow 2 NH_3 + CO_2$$

Urease is present in plant material and plant debris and therefore is more concentrated in soils the higher the content of organic matter or the higher the amount of plant residues remaining in the soil (Freney *et al.* 1992). These authors found that urea hydrolysis also depends on the water content of plant residues, hydrolysis being particularly high with moist plant material. Urea may also be hydrolyzed by a pure chemical process. However, this is rather slow and has therefore of no major significance as a source of volatile NH_3 losses. Much NH_3 may be lost after slurry application (Braschkat *et al.* 1997, see also page 357) and also on pastures where urea excreted by animals with the urine may give rise to the formation of NH_3. Ammonia emission mainly results from agriculture and in particular from intensive livestock farming (Isermann 1987). Table 6.8 shows the "agricultural NH_3 emission density" which is the total NH_3 originating from agriculture per unit surface of agricultural land of a country and is expressed as rate in kg N/ha/year.

Table 6.8   Agricultural NH$_3$ emission density of various European countries (Isermann 1987)

Country	NH$_3$ emission density (kg N/ha)
Netherlands	70
Belgium	55
Denmark	39
Norway	36
German Democratic Republic	32
Federal Republic of Germany	30
France	22
Great Britain	21
Ireland	20
Italy	20
Greece	10

Growing plants also release gaseous nitrogen, mainly NH$_3$ into the atmosphere a fact known since the beginning of the 20. century from the work of Wilfarth *et al.* (1905 quoted by Farquhar *et al.* 1983). Major gaseous N losses by plants are due to the release of NH$_3$ and amides, the latter mainly being volatilized by flowers. Ammonia is produced in the glycolate pathway (see page 152) and this may be a major source for the release by higher plants, particularly in senescent leaves unable to reassimilate NH$_3$.

According to investigations of Stutte *et al.* (1979) with soyabean carried out under field conditions, the volatile N release of the crop may amount to 45 kg N/ha/year under high temperature and transpiration conditions. Ammonia lost by plants mainly passes through the stomata.

Atmospheric NH$_3$ concentration may vary to a marked extent. Thus "clean air" contains about 3 to 5 μg NH$_3$/m^3 whereas for an ungrazed pasture the NH$_3$ concentration at the canopy base was 13.5 and above the canopy 1 μg NH$_3$/m^3. These results demonstrate that the NH$_3$ released by the soil was to a large extent absorbed by the grass.

The atmospheric NH$_3$ concentration profile for a crop stand at day and night is shown in Figure 6.2 from the work of Lemon and van Houtte (1980). It is obvious that during the day the atmospheric NH$_3$ concentration in the crop stand was much reduced because of the uptake by the crop. Absorption and loss of NH$_3$ occur mainly *via* the stomata and thus higher NH$_3$ concentrations were found in the crop stand during the night. These values were particularly high near the soil surface showing that the soil was a major source of NH$_3$.

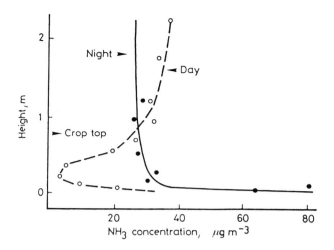

Figure 6.2   Vertical profile of atmospheric NH_3 concentration within and above a canopy of quackgrass (*Agropyron repens*) at night and day (modified after Lemon and Van Houtte 1980).

In the flooding water of paddy soils, high pH values may give rise to considerable losses of NH_3. Mikkelsen *et al.* (1978) found a diurnal change in the pH of flooding water. Highest pH values were found at noon, lowest at midnight and the authors suppose that the pH change results from the photosynthetic activity of algae living in the flooding water. On a neutral soil, maximum flood water pH values as high as 9 were observed at noon when considerable amounts of NH_3 were volatilized. Nitrogen losses from N fertilizers were highest when applied to the surface. With deeply placed urea the losses were low and amounted to less than 10 % of the fertilizer rate (Schnier *et al.* 1988). Volatile losses of NH_3 originating from urea may be reduced by the application of urease inhibitors to the soil which depress the activity of soil urease with the effect that the formation of NH_3 is delayed and the NH_3 produced has a good chance of being taken up by plant roots. In this way NH_3 losses can be reduced as reported by Watson *et al.* (1990) in experiments using N-(n-butyl) thiophosphoric triamide (NBPT) as inhibitor. Similarly Pedrazzini *et al.* (1987) found that the application of the urease inhibitor phenyl phosphorodiaminate (PPD) in pot experiments depressed soil urease activity efficiently and thus the formation of NH_3 in the soil was delayed.

Under anaerobic soil conditions, nitrate is often reduced to volatile nitrogen forms such as N_2 and N_2O (see page 48). This process, which is mediated by microbiological activity, is called denitrification. Denitrification is defined as the microbial reduction of NO_3^- or NO_2^- to dinitrogen (N_2) and nitrogen oxides mainly nitrous oxide (N_2O). Some species of bacteria which use O_2 as the terminal e^- acceptor in respiration under aerobic conditions adjust to anaerobic conditions by producing enzymes which transfer the e^- originating from respiration to other e^- acceptors such as NO_3^-, NO_2^- or others as shown in Figure 6.3.

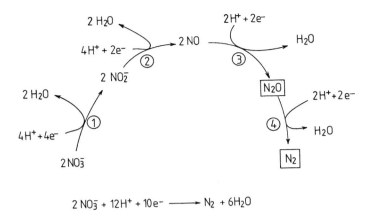

$$2 NO_3^- + 12H^+ + 10e^- \longrightarrow N_2 + 6H_2O$$

Figure 6.3   The various steps of the reduction of nitrate to nitrous oxide and dinitrogen in the process of denitrification. 1. Reduction to nitrite. 2. Reduction to NO. 3 Reduction to N_2O. 4. Reduction to N_2.

From the various steps shown in Figure 6.3 it is evident that NO_3^-, NO_2^-, NO, and N_2O may serve as e^- acceptors. Each step is catalyzed by a specific enzyme and also various bacteria may be involved in the reduction. The reduction of NO_3^- to NO_2^- is also known as dissimilatory nitrate reduction. Twenty-three genera of bacteria are now known which are capable of denitrifying NO_3^- and NO_2^-, the most important being *Pseudomonas, Alcaligenes, Azospirillum* and *Rhizobium* (Firestone 1982). Hence soils rich in *Rhizobium* bacteria may also possess a high denitrification potential which may be in the range of their N_2 fixation capacity (Casella 1988).

The volatile N molecules produced during this reduction process are NO, N_2O and N_2. Of these, N_2 by far is released in highest quantities while the release of NO is minimal. The ratio between N_2O and N_2 released is not constant. Generally the release of N_2O represents less than 10% of that of N_2 release.

Under field conditions Rolston et al. (1976) found that highest N_2O rates were released 8 days and highest N_2 rates were released 18 days after nitrate fertilizer application. Nitrous oxide (N_2O) is not only released in the process of denitrification but is also produced in the process of nitrification in which ammonium is the nitrogen source. Breitenbeck and Bremner (1986) reported that after the application of ammonium containing fertilizers the release rates of N_2O were in the order of 2 to 4 kg N/ha while in the treatment with nitrate fertilization only about 0.5 kg N/ha were released in the form of N_2O.

The production of N_2O is also of environmental importance (see page 395) since it is a greeenhouse gas and it is also involved in the destruction of ozone in the stratosphere (Crutzen 1981). From the denitrification reaction sequence shown in Figure 6.3 it is evident that the processes require e^- which ultimately originate from organic matter such as plant residues and dead microbial biomass (Gök and Otto 1988). The reaction sequence shown in Figure 6.3 also indicates that the reduction steps need H^+ and it is for this reason that denitrification increases soil pH. Denitrification also requires anaerobic conditions. In this respect the proportion of water filled soil pores is a better indicator of denitrification conditions than the bulk water content of soils as emphazised by Aulakh *et al.* (1992) in a valuable review paper on denitrification (see also Figure 6.20). Denitrification is localized within the soil at anaerobic microsites (von Rheinbaben 1990). Analogous to other soil microbial processes, denitrification rates depend much on soil temperature with high rates in a temperature range of 20 to 30°C; below 5°C virtually no denitrification occurs (Aulakh *et al.* 1992). Under the climatic conditions of Europe denitrification occurs mainly during summer and early autumn. Later in autumn and winter it is almost absent even if soils are saturated with water (Schneider and Haider 1992). The anaerobic conditions of flooded rice soils frequently associated with optimum temperatures represent a particular problem in paddy rice production, especially on poor and acid soils as reported by Snitwongse *et al.* (1988). In flooded soils denitrification losses may amount to about 50% of the fertilizer N application (Aulakh *et al.* 1992).

From an agronomic viewpoint the amounts of nitrogen lost by denitrification from soils merit particular attention. It must be emphazised that the quantitative measurement of relatively small amounts of N_2 released by denitrification relative to the huge quantity of atmospheric N_2 presents difficulties. According to Schneider and Haider (1992) denitrification loss in arable soils is in a range of 10 to 20 kg N/ha/year. This is in the range of losses reported by Benckiser *et al.* (1986) under similar environmental conditions. Duxbury and McConnaughei (1986) found denitrification losses on arable soils in the order order of 2 to 5 kg N/ha/year. Similar rates of denitrification losses were reported by (Moiser *et al.* 1986). Kowalenko and Cameron (1977) using ^{15}N labelled fertilizers to follow the fate of N found a total recovery (^{15}N in crop + soil) of 69% in the first year and 54% in the following year. The remaining non recovered 31% and 46% were assumed to be mainly denitrification losses. Riga *et al.* (1980) and also Kjellerup and Dam Kofoed (1983) also found that under field conditions a major amount of N loss can be attributed to denitrification.

Certainly these variations in denitrification losses result from varied soil conditions some more favourable to denitrification than others. Aulakh *et al.* (1992) state that on arable soils N denitrification loss oscillates between 1 and 98 kg N/ha/year. On grassland denitrification rates are low. Here mainly N_2O is released some of which may be dissolved in soil water depending on soil moisture. Nitrate losses by denitrification or leaching or both may be reduced by the application of nitrification inhibitors which inhibit the the oxidation of ammonium and thus the formation of nitrite and nitrate (see page 414).

6.1.5 Erosion

An additional nutrient loss in soil can be caused by erosion. Large quantities of soil can be permanently removed from agricultural land to rivers and lakes. As this material contains a high proportion of fine soil particles, considerable amounts of plant nutrients can be lost. The degree of erosion depends on a number of factors including the rainfall and its intensity, the relief of the land and, in particular, on the soil cover. For most soils which are not subjected to erosion to any great extent, losses of nutrients are relatively small and amount to only a few kg/ha of nutrients at the most. Sharpley

(1993) found an annual phosphate loss of 438 g available P/ha/year in the runoff of no tillage catchements while in the runoff from conventional tillage catchements the loss of available phosphate was 1288 g P/ha. Under fallow conditions the effects of erosion are more pronounced. On soils susceptible to erosion, particularly slopes, non-tillage or minimum tillage is recommended (Blevins *et al.* 1990). Lenzi and Luzio (1997) used a Non-Point source Pollution Model integrated with the Geographic Information System (GIS) to assess the runoff from a water shed in the Alpine area of Italy. The term "Non-Point" means that the sources of nutrients removed by water are diffuse and do not originate only from a few points. Sources from agricultural land are diffuse. The water shed investigated by Lenzi and Luzio (1997) comprised mainly vineyards, woodland and meadows with a total surface of 7700 ha. Runoff was high with high rainfall events. The quantities of plant nutrients removed with the runoff were 0.8 kg N and 0.16 kg P/ha/year. From an agronomic point of view these represent minor nutrient losses. However, in terms of environmental impact even small amounts can be of significance for P since particulate removal of P by erosion presents a physical mechansim by which P can be transferred from soil to water (Sharpley and Smith 1990). This is of particular relevance to the many regularly fertilized agricultural soils where a build of in P status has occurred. Phosphate can be accumulated in the top few cms of soil which is most prone to erosion (Haygarth *et al.* 1997). Diffuse leaking of P from this and other sources can undermine water quality by contributing to eutrophication (Foy and Withers 1995).

Water and wind erosion is a severe problem in the tropics where high rainfall intensities prevail and young soils (*Entisols*) with a high proportion of sand (Psamments) are frequent (van Wambeke 1991). After the clearance of tropical forest followed by arable agriculture deterioration of soil fertility is a great hazard (Picard 1994). The ploughed layer becomes more and more compact, yields decrease and with them the the protective vegetation cover. Intensive land use, such as overgrazing or the cultivation of arable crops such as peanuts, cotton and soya favour the erosive effect and finally result in desertification. Wind and water erosion first remove the fine soil particles such as organic matter and clay which not only means a loss of plant nutrients but also the potential of the soil to adsorb and store plant nutrients is lowered as well as the water holding capacity. According to an FAO study (Aldhous 1993) 11.3 million ha of tropical rain forest were lost each year during the 1970s and in the last two decades about 50% of the total tropical rain forest has been lost by deforestation (Skole and Tucker 1993). Restoration of such degraded land is expensive and in many cases an almost impossibility to achieve (Dayly 1995). This problem is also a challenge for worldwide agriculture where the aim is to produce food without major environmental damage and where crop yields still can be considerably increased without environmental pollution and the waste of resources.

6.1.6   Nutrient supply by precipitation and atmosphere

Plant nutrients contained in rainwater snow and fog may contribute to the nutrient supply of crops. An intensive investigation into the plant nutrient content of the precipitation from twelve different sites in Norway was carried out by Låg (1968).

The data from the sites with the highest and the lowest rainfall are shown in Table 6.9. The Vagamo site which is situated in the mountains is characterized by a rather low precipitation, containing very low amounts of plant nutrients. On the Lista site the pattern is quite different. Lista is on the Southern coast of Norway where the rainfall is high. The precipitation is considerably influenced by sea water and is relatively rich in inorganic ions, in particular in Na^+ and Cl^- and substantial amounts of Mg^{2+} and sulphate are also present in the rainwater at this site. The concentrations of Na^+, Cl^- and Mg^{2+} all decreased from the coastal districts to the interior. This observation could be of more general significance for tropical conditions when the natural distribution of oil palms and coconut palms is restricted to coastal areas. According to Ollagnier and Ochs (1971) these species have a particularly high requirement for Cl^- (see page 641).

The results from the two Norwegian sites demonstrate the great variability in the nutrient content of rainwater. It must be emphasized, however, that these are extreme sites and are therefore by no means representative. On average, the quantities of plant nutrients generally supplied in precipitation are somewhere between the values given in Table 6.9. Older data reported by Riehm and Quellmalz (1959) show that under the climatic conditions of Central Europe, amounts of 5 to 10 kg/ha/year for K^+, Na^+, Mg^{2+}, Cl^- and nitrogen in the form of $NO_3^- + NH_4^+$ are supplied by the atmosphere (dry deposition) and by rain (wet deposition) to vegetation and soils. These amounts are not high and contribute only to a small extent to crop production and with the exception of nitrogen are still applicable to today's condition. As shown in Table 6.8 appreciable amouts of NH_3 are released from intensive livestock farming. In recent years the NH_4^+ concentration in the rain water has increased considerably in areas with high slurry applications. Here NH_4^+ rates supplied by rain may be as high as 60 kg N/ha/year or even higher (Roelofs et al. 1988). According to Zobrist et al. (1987) NH_3 neutralizes sulphuric acid and nitric acid in fog droplets to form ammonium sulphate.

Table 6.9   Amounts of plant nutrients supplied by precipitation on two Norwegian sites (Låg 1968)

Location Precipitation	mm/year	S	Cl	NO_3^--N	NH_4^+-N	Na	K	Mg	Ca
					kg/ha/year				
Vagamo	294	0.9	0.4	0.1	0.1	0.4	0.3	0.2	1.2
Lista	1871	19.2	264	3.5	2.8	147	8.6	17.8	14.2

Nitrogen deposition originates not only from ammonium but also from nitrogen oxides (NO, NO_2). These are produced to a large extent by burning fuel including petrol (Isermann 1990) and the fuel burned in air traffic (Egli 1995). Nitrogen oxides dissolve in water droplets and forms acids according to the following equations:

$$2\,NO_2 + H_2O \Rightarrow HNO_3 + HNO_2$$

$$HNO_2 + \tfrac{1}{2}\,O_2 \Rightarrow HNO_3$$

Analogous reactions apply for sulphur oxides (SO_2, SO_3):

$$SO_2 + H_2O \Rightarrow H_2SO_3$$
$$H_2SO_3 + \tfrac{1}{2} O_2 \Rightarrow H_2SO_4$$
$$SO_3 + H_2O \Rightarrow H_2SO_4$$

Strong acids are formed by these reactions which, when brought back to the earth surface by rain, dew and fog contribute to soil acidification.

Quantities of sulphur supplied by wet and dry deposition may differ considerably between various sites and depend much on the extent of industrialization which was and still is in some areas an important source of sulphur to crops. In the last decade, however, in most of the industrialized countries the atmosphere has been cleaned up and SO_2 levels greatly reduced so as to contribute only to a few kg S/ha/year (Schnug 1991). Unsworth *et al.* (1985) reported that in Great Britain sulphur deposition amounted to an average about 3 to 4 kg S/ha/year. Sulphate is easily leached from arable soils (Ballif and Muller 1985) so that the removal of the atmosphere source may lead to an insufficient S supply particularly for crops such as rape which have high requirements.

Oxides of sulphur and nitrogen (NO_x, SO_x), mainly originating from the burning of fossil energy (Baldocchi 1993), may react with water droplets in the atmosphere to form acids as shown above. The acid fog so produced may have a pH of about 3 and is detrimental to the cuticle of the needles of spruce trees (Mengel *et al.* 1987).

6.2 Mineral Fertilizers, Manures and Fertilization Techniques

6.2.1 Mineral fertilizers

In many soils the rate of removal of plant nutrients by crop uptake, leaching, volatilization and denitrification is well in excess of nutrient release by weathering and mineralization. A negative nutrient balance thus results unless nutrients are applied in the form of fertilizers or manures to make up the difference. Generally, the more intensive the cropping system and the higher the yields, the greater must be the amounts of nutrients applied to the soil in order to maintain soil fertility. For most soils the use of inorganic fertilizers is thus almost essential and a wide range of fertilizers of different grades and nutrient ratios are now marketed. It is beyond the scope of this book to discuss these in detail. Only a brief survey of the most important fertilizer types will be given. The more common fertilizers are considered in the chapters following which deal with the individual plant nutrients.

Fertilizers contain those nutrients such as N, P, and K that are rapidly taken up and required in high quantities by crops. Nitrogen is mainly given in the form of nitrate, ammonium or urea. More specialized fertilizers contain nitrogen in a more insoluble form, such as urea formaldehyde and isobutylidene urea. These forms are slow release nitrogen sources. Phosphorus fertilizers generally contain P mainly in the form of ortho-phosphate. In a small number of phosphate fertilizers P is present as polyphosphates. An important criterium of P fertilizers is solubility. Superphosphate for example is very soluble in water; ground rock phosphates, on the other hand, are insoluble in water. Potassium is applied to soils mainly in the chloride or sulphate forms. Potassium nitrate and potassium polyphosphate play only a minor role. Sulphur fertilizers can be obtained in the form of sulphate in ammonium sulphate, superphosphate and potassium sulphate. In addition to supplying S, these fertilizers are also a source of nitrogen, phosphorus and potassium respectively. Calcium and magnesium are applied as sulphates or in the form of carbonates or oxides. These two latter compounds have an alkaline reaction and are thus mainly used to increase soil pH (see page 538).

Although most inorganic fertilizers, such as ammonium sulphate, calcium nitrate or potassium chloride, are salts which are neutral in reaction, they can affect soil pH by their physiological reaction. Fertilization with nitrate results in a pH increase of soils (Smiley 1974) because nitrate uptake by plant roots occurs mainly as a proton cotransport (Ullrich 1992) so that the protons pumped into the rhizosphere are recycled into the cytosol. If the rate of H^+ recycled into the cytosol is higher than the rate of H^+ pumped by the plasmalemma ATPase into the rhizosphere the soil pH will increase. With ammonium nutrition there is no major recycling of H^+ into the cytosol and it is for this reason that ammonium nutrition decreases soil pH. The same is true for the symbiotic nitrogen supply *i.e.* N_2 fixation by symbiotic living bacteria, since nitrate is hardly taken up (Hauter and Steffens (1985). The potassium fertilizers, potassium chloride and potassium sulphate, tend to be neutral in reaction.

Fertilizers containing only one of the three most important plant nutrients, nitrogen, phosphorus or potassium, are called straight fertilizers. Typical examples of this group are superphosphate (P), muriate of potash (K), ammonium nitrate (N) and a mixture of ammonium nitrate with calcium carbonate (N), called 'nitro-chalk'. Compound and mixed fertilizers contain two or three of the main plant nutrients N, P and K. NPK-fertilizers very commonly differ in their NPK-ratios. An NPK compound 15-15-15, for example, means that the ratio of $N:P_2O_5:K_2O$ is equal to 1:1:1 and that the concentration (grade) of these plant nutrients in the compound is 15% N, 15% P_2O_5 and 15% K_2O. As this example shows, the nutrient concentration of straight and compound fertilizers is generally expressed in terms of % P_2O_5 and % K_2O and not as a percentage of the element.

6.2.2   Organic manures and crop residues

Organic manures mainly originate from the wastes and residues of plant and animal life. They are rich in water and carbon compounds but are usually comparatively poor in plant nutrients. One of the most important organic manures is farmyard manure (FYM). This is a mixture of partially decomposed straw containing faeces and urine. In recent years there has been a decline in its use as modern methods of livestock management tend to use little or no straw for bedding, the basis of FYM. The production of livestock slurries has thus increased. The nutrient concentration of organic manures can vary widely depending much on their source and moisture content. Some mean values for a number of organic manures are listed in Table 6.10. The relative amounts of the major nutrients differ considerably. Farmyard manure is comparatively poor in phosphate as it contains a high proportion of straw. Sewage sludge is low in potassium as this is lost during preparation.

Table 6.10   Nutrient concentrations in organic manures

	Moisture	N	P	K	Ca	Mg
				mg/g fresh matter		
Farmyard manure	760	5.0	1.1	5.4	4.2	1.1
Cattle slurry	930	3.1	0.7	3.2	1.1	0.4
Pig slurry	970	2.0	1.0	2.0	—	—
Sewage sludge	550	8.3	2.2	0.4	0.7	

The nutrient concentration of slurries is often difficult to assess, since slurries can differ considerably in water content. For this reason Vetter and Klasink (1977) have proposed a calculation for the amount of plant nutrients 'produced' by farm animals on the basis of animal number. In this calculation 1 cow or 7 adult pigs or 100 hens are considered as one 'animal manure unit'. The unit is such that each animal group 'produces' about the same amount of plant nutrients. In Table 6.11 the amounts resulting in a one year cycle are shown. It can be seen that the quantity of nitrogen 'produced' by one unit is about the same for the three animal groups, but there are differences for K, P and Mg resulting from the type of food fed to the animals. Roughages and green fodder are rich in K and thus these materials give rise to relatively high K concentrations in slurries. On the other hand cereals are rich in P, and since they are fed to poultry and pigs in relatively high amounts, the slurries of these animals have a high P concentration. In areas of intensive animal husbandry huge amounts of manure are produced, which may lead to overfertilization of soils and thus result in pollution problems (Leinweber 1996, Mozaffari ans Sims 1996).

Table 6.11   Amounts of plant nutrients excreted by 'one animal manure unit' per year (Vetter and Klasink 1977)

	N	P	K	Mg
		kg/year		
Cattle (1 animal)	77	18	90	6.6
Pigs (7 animals)	75	29	34	5.4
Poultry (100 animals)	80	32	32	4.2

The value of organic manures cannot be assessed simply by analysis for the total quantity of plant nutrients. Nutrient availability to crops is highly important and this can only be determined by field trials. Most of the N in manures occurs in organic compounds. In slurries about half of the total N consists of NH_4^+ the remainder being organic N. NH_4^+ is readily available to the crop, whilst only about 25 % of the organic N is mineralized.

The remaining organic fraction is very stable as has been demonstrated by Amberger *et al.* (1982). Table 6.12 shows that the pH of slurries is >7 which means that the NH_4^+-N present is susceptible to volatilization which may range from a few % to 99% of the total NH_4^+-N present in slurries (Braschkat *et al.* 1997). Losses due to NH_3 volatilization decrease as the infiltration rate of slurry into soils is increased. Thus because of its lower viscosity and higher infiltration rate, ammonia losses from pig slurry are lower than from cattle slurry. Braschkat *et al.* (1997) studying NH_3 losses from grassland under field conditions reported that in addition to the dry matter conctent of slurries, solar radition had marked impact on NH_3 volatilization. Infiltration rates decreased with an increase in dry matter concentration. Solar radiation directly influences slurry temperature higher temperatures favouring NH_3 volatilization. According to these authors the effects of wind speed, air humidity, and air temperature on NH_3 volatilization were only of minor importance.

Table 6.12 Nitrogen concentrations, pH and dry matter in slurries. N concentrations are based on a dry weight of 100 g dry weight/kg (after Amberger *et al.* 1982)

Source	pH	Dry matter	Total N	NH_4^+	NO_3^-	Mineralizable N in % of total organic N
			in g/kg			
Cattle	7.5	62	4.4	1.9	trace	27
Pigs	7.2	49	8.6	5.7	trace	17
Poultry	7.3	193	7.1	4.8	trace	22

Injection of slurries into soils results in much lower volatile losses as compared with surface application and application with rainfall >10 mm promotes the infiltration rate so that virtually no NH_3 loss occurs. Mannheim *et al.* (1997) developed a computer model to predict the volatile NH_3 losses of slurries. The most important parameters of the model are: quantity of slurry applied in m^3/ha, soil surface in relation to infiltration rate (bare, covered with straw, grassland), type of slurry (pig or cattle), technique of application (injection braoadcast), and rainfall during application.

The organic matter concentration of slurries is relatively low and therefore is virtually of no significance for increasing soil organic carbon. Organic N in slurrries is, however, of relevance to the availability N of soils since some of the organic N can be rapidly mineralized (see Table 6.12). Barekzai *et al.* (1993) found that with increasing rates of slurry the easily mineralizable organic nitrogen in soils extracted by electroultrafiltration (EUF, see page 92) also increased. Thus regular slurry application may increase the pool of mineralizable nitrogen in soils. As was shown by Kücke and Kleeberg (1997) in field trials, this organic nitrogen, if mineralized in late suummer or autumn, may lead to substantial nitrate leaching by autumn and winter rainfall. High rates of slurry affect the continuity of grassland swards as well as their botanical composition as shown in Table 6.13.

Table 6.13 Effect of slurry rates applied per year during a period of four years on the continuity and the botanical composition of grassland sward (Barekzai 1992).

Slurry quantitiy applied m³/ha/year	Herbs %	Grasses %	Legumes %	Continuity % of gaps
Without slurry	42	44	16	25
30	29	68	4	30
60	19	81	—	35
90	8	92	—	38

Most of the organic nitrogen compounds in farmyard manure (FYM) are resistant to decomposition and only become slowly available over long periods. About one-third of the total N is easily released and available in the season of application (Cooke 1972). Rees *et al.* (1993) found that of a total of 129 kg N/ha applied as poultry manure only about 10 to 20% of this N was taken up by the following spring barley. The easily available N fraction of manure consists mainly of inorganic N (ammonium, nitrate), urea, and peptide N (Sluijsmans and Kolenbrander (1977). Regular FYM application, however, may increase the quantity of mineralizable nitrogen in soils considerably as shown by the long term field experiments in Rothamsted. Some of the most interesting results are depicted in Figure 6.4 (Johnston 1994).

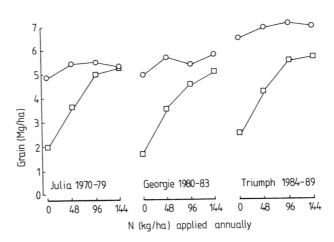

Figure 6.4 Effect of mineral N fertilizer on the grain yield of three spring barley varieties. In one treatment (○) the soil had received FYM regularly since 1852 while in the other treatment (☐) no FYM was applied. Organic matter concentration in the soil with FYM application was 46 g/kg and in the control soil 18 g/kg soil. The data represent means of several years (after Johnston 1994).

From the results it is clear that the long term regular FYM application must have greatly increased the soil in mineralizable organic nitrogen as the mineral N application gave no yield response. It is also evident that FYM application enriched the soil with organic matter. Similar results have been reported by Garz and Chaanin (1990) from long term field trials in Germany. As is shown in Table 6.14 application of FYM and also of straw increased the mineralizable N. This finding demonstrates the importance of organic matter in increasing the pool of mineralizable N and it also shows that the straw combined with mineral fertilizer produced even more mineralizable N (see last line of Table 6.14) than the farmyard manure. Mineral fertilizer alone, however, was unable to enrich the pool of mineralizable soil N. From the data it is also clear that by far the largest proportion of organic soil N is stable and very resistant to decomposition.

In most cases a higher level of organic matter in soils signifies an improvement of the physical soil parameters such as pore volume, soil structure, and water holding capacity of soils (Michael and Djurabi 1964, Asmus *et al.* 1987).

Table 6.14   Effect of long term FYM and NPK mineral fertilizer application on soil N fractions (after Garz and Chaanin 1990).

Treatment	Total N g/kg	mineralizable N* mg/kg	Stable N g/kg
No fertilizer	1.05	58.8	0.95
NPK, mineral	1.07	59.5	0.93
NPK + FYM	1.17	70.3	1.06
NPK + Straw	1.13	72.3	1.00

* According to the method of Stanford and Smith (1972)

Most of the organic matter brought into soils by FYM is decomposed and the resulting CO_2 released into the atmosphere. Of the total C applied in FYM only a very low proportion is found in the humates (Michael and Djurabi 1964).

Not only FYM but also other organic constituents especially plant residues such as straw and leaves contribute to the formation of organic matter in soils and frequently improve the physical properties of soils. This is of particular relevance for rainfed agriculture in the tropics where water conservation in soils is of utmost importance. Here mulching with green manure or straw may improve water use substantially and increase crop yields (Moitra 1996).

The rate of decomposition of organic matter in the soil depends much on organic matter type and especially on the proportion of less hydrolyzable substances such as cellulose and lignins. These are relatively high in FYM and low in young green manure, with straw being between the two. Figure 6.5 shows the decomposition of maize residues from the trials of Wagner and Broder (1993). It is obvious that after about two years only a low percentage of the total organic matter brought into the soil was still present.

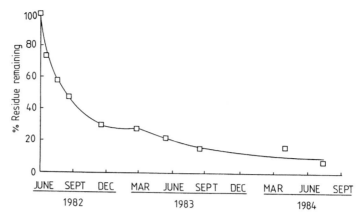

Figure 6.5 Pattern of decomposition of maize residues incorporated into soils during a period of about two years. Total amount of organic matter incorporated at the beginning = 100%. (after Wagner and Broder 1993).

In temperate zones, straw incorporation into soils is a common practice. It contributes substantially to the maintanance of the level of organic soil matter in a similar way to FYM (Yang and Janssen 1997). Under humid climatic conditions most of the straw incorporated into soils is decomposed over a one year period (Thomsen 1993). Crops profit from the plant nutrients in straw such as phosphate and especially K^+.

Straw has a marked impact on nitrogen turnover in soils. Since straw is relatively rich in decomposable organic carbon, soil microorganisms feeding on the straw need an addtitional nitrogen source (Mengel 1996). For this reason straw application may lead to a transitional biological nitrogen fixation that is to say the assimilation of inorganic nitrogen by soil microorganisms, a process known also as "immobilization" (Scherer and Mengel 1983). In particular cases crops and soil microorganisms may therefore compete for inorganic nitrogen including fertilizer nitrogen. Under these conditions an additional fertilizer application may be appropriate to ensure sufficient N supply to crops. This can be particularly important if soils are poor in mineralizable organic nitrogen (Scherer and Mengel 1983). On the other hand straw incorporation into soils in late summer or autumn leading to a biological fixation of nitrate may protect the nitrate from leaching by winter rainfall (Schmeer and Mengel 1984). The nitrogen N thus fixed may be mineralized and available for the crop in the following year. The potential for N immobilization depends on the N concentration in straw and according to Jensen (1996) is much higher in barley straw than in pea straw.

Biederbeck *et al.* (1980) in comparing straw application with straw burning over a period of 20 years under the climatic conditions of Saskatchewan, found that ploughing straw into the soil in spring improved soil structure and reduced the hazard of erosion as compared with straw burning. Straw burning on the other hand resulted in an increase of available phosphate and exchangeable NH_4^+ in the soil. The 'straw burning' treatments gave somewhat higher yields as compared with the straw application treatments, although the difference was not significant.

As already mentioned, besides straw other plant residues may also play a role in the maintenance of soil organic matter. Crop rotation is therefore of utmost importance in sustaining the levels of organic N and organic C in soils (see also Chapter 2.2.2 on page 34). In this context the favourable effect of perennial crops and of grass legume mixtures should be especially mentioned. Legumes also improve the nitrogen status of soils and may thus improve crop yields as found in the oldest field trial in USA, the Morrow Plots (Odell *et al.* 1984). Some interesting results of this long-term experiment are shown in Table 6.15. In the treatment with no fertilizer, the rotation with a leguminous crop yielded much higher grain yields than the permanent maize treatment.

Table 6.15 Grain yield of maize in relation of crop rotation and fertilizer application. Long term field trials in Illinois "Morrow plots". Grain yields shown are the average in the period 1967 to 1978. (after Odell *et al.* 1984)

	Permanent maize	Maize-Soya	Maize-Oats-Clover
		Mg grain/ha	
No fertilizer	1.54	2.38	2.97
Lime since 1955 + NPK	4.05	4.69	4.76
FYM since 1904 + lime + P	4.12	4.64	4.79
Since 1955 lime + NPK	4.27	4.91	4.82

The level of organic soil matter may also be elevated by the application of sewage sludge which generally contains much organic carbon as well as nitrogen and phosphate while its K^+ concentration is extremely low. Nearly all K^+ flowing into sewage plants is not sedimented but leached into the rivers. The nitrogen in the sewage sludge is almost completely in organic form and resistant to mineralization because the more easily mineralizable nitrogen has already been released during sewage processing. Nevertheless in the application of sewage sludge to the soil some N does become plant available. Steffens *et al.* (1987) testing the effect of sewage sludge application over several years in field trials found that the application to grassland yielded an increase of hydrolysable soil nitrogen whereas application to arable land was without effect on soil nitrogen. Application of sewage sludge from industrial plants recovering phosphate should also be considered in relation to phosphate recycling since phosphate is a limiting resource (Winteringham 1992). Sewage sludge may contain excessive concentrations of heavy metals and for this reason its application should be controlled by soil and sludge analysis for toxic elements. In a valuable review article Juste and Mench (1992) discuss the results of long term field trials in which sewage sludge was applied. Phytotoxicity due to sewage-borne metals was rarely observed on grain crops. Harmful effects, however, found on some legume species were explained by the detrimental influence on microbial activities in soils, especially on N_2 fixation. The sewage sludge application was found to increase plant growth in 65% of cases. Cadmium, Zn, and Ni were the most plant-available heavy metals whereas the availability of Cr and Pb was insignificant. The cumulative metal input into the soils was the major factor influencing metal concentrations in plant tissues. A trend of a progressive decrease in metal uptake by crops was observed in the time lag following sludge application.

Most organic matter applied to soils has a favourable influence on soil miroorganisms (Wagner and Broder 1993, Oberson et al. 1996) and their enzyme activities which is of importance for the turnover of soil organic matter (Martens et al. 1992).

6.2.3   Liquid fertilizers

Over the past two decades the application of fertilizer 'fluids' has become more common. The term 'fluid' includes 'liquids' in which the fertilizer is completely dissolved and 'suspensions' in which the fertilizer is present in the form of a suspension. Liquid fertilizers are generally easier to transport than solid ones and cause fewer labour problem in handling and application. A further advantage is their homogenity and the even distribution which can be achieved when they are applied to the soil. The two most important liquid N fertilizers are anhydrous ammonia and aqueous ammonia.

Anhydrous ammonia is the simplest liquid nitrogen fertilizer. It consists only of NH_3 which is present in a liquid form under pressure. Anhydrous ammonia is a high grade N-fertilizer with 82% N. This high concentration is of considerable advantage in terms of transport costs. On the other hand, the liquid under pressure requires special handling precautions and also suitable equipment for transportation and application. Its use is therefore often restricted. A special injection assembly is used to apply it into the soil at a depth of 15 to 20 cm to avoid loss of NH_3 by volatilization.

Aqueous ammonia is a solution containing about 25% NH_3. This solution is only under a very low pressure; it is therefore easier to handle and does not require the rather expensive application equipment which is needed for anhydrous ammonia. It must be remembered, however, that aqueous ammonia is a low grade fertilizer and contains only about $21-29\%$ N. Again it is also necessary to ensure that it is applied below the soil surface in order to avoid loss of NH_3 by volatilization.

Nitrogen solutions were developed by the Tennessee Valley Authority (TVA) in the USA. 'Low pressure' solutions are made up from urea, ammonia and ammonium nitrate and have about $30-40\%$ N. They have the advantage that they are more concentrated than aqueous ammonia and more easy to handle. 'No pressure' solutions made up directly from urea and ammonium nitrate contain less than 30% N. The use of both urea and ammonium nitrate in solution is based on the fact that a mixture of these fertilizers has a higher solubility than either of its individual components.

Phosphorus and N containing solutions have been in use in the USA since the early 1950's. These 'NP solutions' were first produced by neutralizing orthophosphoric acid with an ammonium salt to produce mono- and diammonium phosphates; the standard solutions resulting from this procedure are about $8-24-0$ (8% N, 24% P_2O_5, 0% K_2O). More recently by substitution of polyphosphates for orthophosphates it has been possible to manufacture and ship higher graded fertilizers. The basic component of these solutions is superphosphoric acid. This has a high P concentration and is made from orthophosphoric acid and a series of polyphosphoric acids. The proportion of each depends on the total P concentration; as this increases the proportion of longer chain acids goes up and the ortho acid content is reduced. This is illustrated in Figure 6.6 (Slack 1967). Superphosphoric acid is neutralized by the addition of NH_4^+ (ammoniation) and NP-solutions can be obtained with grades of 10—34 0 and 11—37—0, depending on the P-concentration of the superphosphoric acid used. These grades are based on a salting out temperature of 0°C. This is the temperature at which crystallization occurs. As high nitrogen grade fertilizers are generally required, NP solutions are supplemented by a mixture of dissolved urea plus ammonium nitrate (28—32% N). Solutions of varying NP ratios are thus obtained. Polyphosphates tend to chelate heavy metals and Mg. This is a further advantage of NP solutions based on superphosphoric acid, because impurities of Mg, Fe or other heavy metals do not cause phosphate precipitation. Unfortunately, the addition of KCl to NP solutions results in precipitation. In solutions containing nitrate, KNO_3 crystallizes out. According to Hignett (1971) nitrate containing NPK solutions have a maximum grade of 7—7—7 (7% N, 7% P_2O_5, 7% K_2O), whereas for solutions containing only urea as nitrogen component, a grade of 9—9—9 can be produced at a salting out temperature of 0°C. The preparation of suitable NPK solutions by the addition of KCl is thus difficult. High graded PK solutions can be obtained by the neutralization of superphosphoric acid with KOH. Potassium hydroxide, however, is rather expensive and the resulting high production costs present a substantial drawback. For this reason most farmers using liquid fertilizers in the form of N- and NP solutions apply potassium in the solid form.

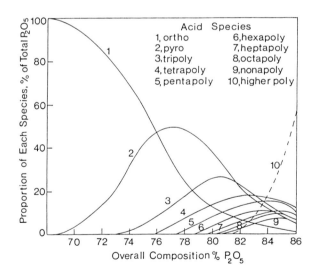

Figure 6.6 Equilibrium proportions of phosphoric acid species in the P_2O_5- H_2O- system at high P_2O_5 concentrations (after Slack 1967).

Some work has been carried out to develop suspensions in which the KCl is present as finely divided crystals, stabilized by the addition of clay (1 to 3% as suspending agent). According to Slack (1967) high grades in the range of 15—15—15, and $10-30-10$ can be obtained by this technique. Suspensions (slurries), however, are not easy to deal with and for this reason their application is limited. Liquid fertilizers are easier to handle and to apply than solid ones, provided that suitable equipment is available for application. Once brought into contact with the soil, liquids behave in the same way as comparative solid fertilizers, and generally no differences are observed in relation to growth and crop yields.

6.2.4 Controlled-release fertilizers

Controlled-release fertilizers (CRF) can be devided into coated fertilizers and fertilizers with a low solubility *per se*, the latter being mainly derivatives of urea. In both groups the actual solubility is low because they need to provide a controlled release of plant nutrients adapted to the demand of crops and to minimize losses by leaching. Of the urea derived non coated controlled-release fertilizers, urea formaldehyde polymers (UF), isobutylidenurea (IBDU), and crotonylidendiurea (CDU) are the most important on contact with the soil. These fertilizers are subjected to hydrolysis and when brought into the soil form urea as shown in Figure 6.7.

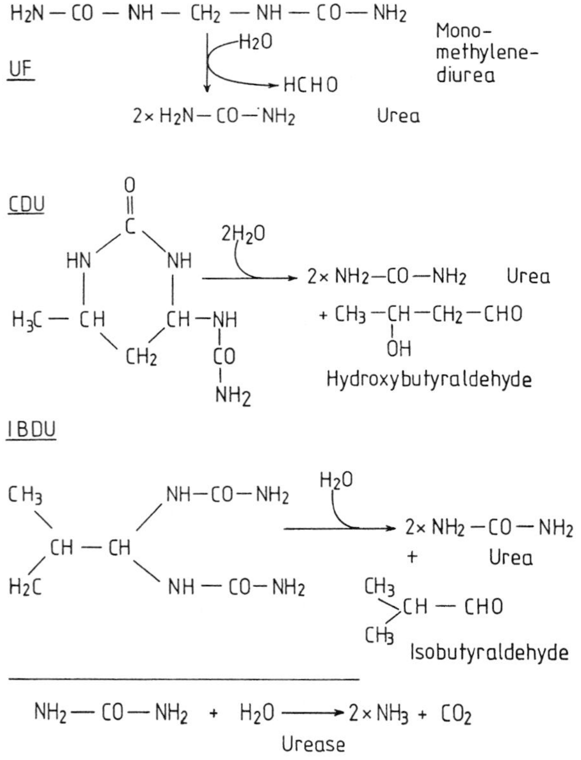

$$H_2N - CO - NH - CH_2 - NH - CO - NH_2$$

UF

Mono-methylene-diurea

$-H_2O$

$\longrightarrow$ HCHO

$2 \times H_2N - CO - NH_2$ Urea

CDU

$2H_2O$

$\longrightarrow 2 \times NH_2 - CO - NH_2$ Urea

$+ CH_3 - CH - CH_2 - CHO$

$\quad\quad\quad OH$

Hydroxybutyraldehyde

IBDU

H_2O

$\longrightarrow 2 \times NH_2 - CO - NH_2$

$+$ Urea

$\underset{CH_3}{\overset{CH_3}{>}}CH - CHO$

Isobutyraldehyde

$$NH_2 - CO - NH_2 + H_2O \xrightarrow{\quad\quad} 2 \times NH_3 + CO_2$$
$$\text{Urease}$$

Figure 6.7 Hydrolysis of urea derived non-coated controlled-release fertilizers. UF formaldehyde polymers, CDU crotonylidendiurea, IBDU isobutylidenurea.

Urea thus produced is quickly hydrolyzed to ammonium which in most cases is the substrate for nitrification. According to Hadas and Kafkafi (1974) the rate of microbial decompostion of non coated urea formaldehyde is controlled by temperature. At a temperature $<14°C$ the breakdown of the fertilizer is delayed in the first weeks after application.

Coated fertilizers may be mantled by different materials such as sulphur, plastics or waxes. The so called osmocotes are surrounded by a plastic mantle which allows the entry of water. As a result of the osmotic uptake of water, the mantle tears and the plant nutrients diffuse into the soil. Hence nutrient release from these fertilizers depends much on soil moisture and since plant growth also depends on soil moisture nutrient release may be adapted to plant growth. The decomposition of the sulphur coat depends on the oxidation of the sulphur which is mainly brought about by soil microorgamisms. In this case nutrient release depends on soil microbial activity.

The controlled-release fertilizers are expensive to produce and are thus only be applied under very specific conditions. Generally they are of no importance for arable and forage crops including grassland (Stählin 1967, Allen et al. 1971). Where environmental conditions are not easily regulated the use of controlled fertilizers is difficult to predict as emphasized by Miner et al. (1978). Nutrient release can even coincide with nutrient loss from the soil by leaching.

The main uses of controlled release fertilizers is in the production of high value crops grown in greenhouses, container plants and in some cases for lawns (Sharma 1979). San Valentin et al. (1978) were able to reduce K^+ leaching losses considerably by applying coated potassium fertilizer to tobacco grown on sandy soils. Of higher importance is minimizing nitrate leaching by controlled release fertilizers. Controlled release fertilizers are used in citrus cultivation on pervious soils where nitrate leaching losses can be very high (Dasberg et al. 1984, Embleton et al. 1986).

The efficiency of controlled-release fertilizers in reducing leaching losses is shown in Figure 6.8 from the work of Terman and Allen (1970) in which sulphur coated fertilizers were compared with urea and ammonium nitrate in a leaching experiment carried out in the greenhouse. In the treatments cropped with Bermuda Grass, all the leachates were low in N regardless of the fertilizer type applied. In the fallow treatments, however, there were marked differences between the amounts leached from the various fertilizers. Similar advantages of sulphur coated KCl were reported by Terman and Allen (1970).

Paramasivam and Alva (1997) tested various controlled-release fertilizers including a new type of resin-coated products (" Osmocote" and "Meister") in leaching experiments carried out in columns filled with a very pervious soil (97% sand). The resin-(wax) coated fertilizers showed the lowest nitrate leaching and retarded nitrification much more than did the S coated urea derived polymer. The problem of controlled-release fertilizer has been considered by Oertli (1980) in a useful review paper.

366

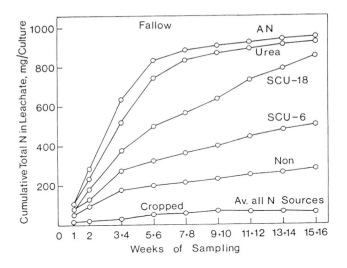

Figure 6.8 Cumulative average amounts of total N in leachates as affected by N source and cropping. AN = ammonium nitrate, SCU = sulphur coated urea, SCU-6 having a lower solubility than SCU-18, Cropped = average of all N sources cropped with grass (after Terman and Allen 1970).

6.2.5 Techniques of application

One of the most important aspects of fertilizer usage is to know when fertilizers should be applied. This depends primarily on the crop and on the mobility of the particular nutrients applied to the soil. Potassium and phosphates, which are hardly leached on medium to heavy textured soils are often applied in autumn and incorporated into the soil ready for crop growth in the spring. On the other hand, nitrogen fertilizers, which are generally susceptible to leaching, are applied in spring and also in the form of top dressing during vegetative growth. Under semi-humid climatic conditions some of the nitrogen required by a crop can be applied in autumn for the following spring crop. The N fertilizer is ploughed well into the soil, so that it is available in the deeper rooting zone in the following year. In many cases it is not opportune to apply all the nitrogen in one dressing but rather to split the total amount into two or several applications. This type of nitrogen treatment is particularly common in intensive cropping systems where crop yields are high and where large amounts of nutrients are applied. Both in rice production in East Asia and wheat production in Europe, split application is a well known fertilizer practice. Broadcast application of nitrogen fertilizer as urea or NH_4^+ to the flooding water of rice may lead to high losses by NH_3 volatilization. In such cases the deep placement of urea super granulates is recommmneded or the injection of a urea solution into the soil as proposed by Schnier *et al.* (1988). By these measures the volatilization of NH_3 is almost completely avoided.

Total nitrogen application rates for wheat are in the order of 100 to 160 kg N/ha. This can sometimes be split into three or four applications. The last application at the flowering stage is particularly important as it promotes grain filling and improves the grain protein content (see page 320). Nitrogen treatment of winter crops, *e.g.* winter wheat or winter rape, can be split into autumn and spring applications. This is practiced in the Mid-West (Indiana) of the USA where N application in autumn is necessary for good growth of winter wheat (Huber et al. 1980). Optimum N fertilizer rates depend much on other growth conditions, especially water supply and should be higher the better the growing conditions. Even in arid soils with rainfed agriculture, cereals may profit from small N fertilizer rates in the range of 40 kg N/ha as reported by Ryan *et al.* (1998) for Morocco. The beneficial effect of split N application on the annual yield of the grass crop has been reported by Brockman (1974). Split application of lime, K^+, and Mg^{2+} may also be opportune on pervious soils in the tropics where a considerable proportion of nutrients when applied at high rates may be leached by heavy rainfall.

Most fertilizer applied is in solid form and is broadcast or in other words distributed uniformly on the soil. For soils of poor nutrient status the application of fertilizer in a row or a band, can often yield better results. The same is true for soils which strongly fix plant nutrients. Broadcasting phosphate fertilizers allows maximum contact between the fertilizer and soil particles so that it promotes P fixation. If the fertilizer is applied in the form of a band (placed application), however, the extent of contact between the phosphate soil fixing particles is reduced. Soil fixation is thus depressed and if the band is near the seed, a zone of high phosphate concentration is accessible to developing plant roots. The effect of phosphate placement is particularly noticeable on soils where P is limiting (Reith 1972). Tomar and Soper (1981) reported that placing urea in the form of a band at a depth of 0.1 m into the soil improved N utilization by the crop considerably as compared with broadcast application. On potassium deficient soils K placement is also often superior to broadcast application. On sites deficient in either phosphate or potassium, fertilizer application should be made in the spring when the crop demand is high. Fertilizer application on phosphate or K^+ fixing soils months before the crop is sown may result in a considerable nutrient fixation as shown for K^+ by Karbachsch (1978) and for phosphate by Barekzai and Mengel (1985).

6.2.6   Foliar application

Leaves and other aerial plant organs are well able to take up nutrients in gaseous form such as CO_2, O_2, and SO_2 *via* the stomata. However, the uptake of nutrients in ionic form from solution is limited as the outer epidermal cells of the leaf are covered by the cuticle which occurs in all terrestrial plant species. The cuticle consist of two components cutin and wax. Cutin is a polymer built up by hydroxy fatty acids with $C_{18:1}$ and $C_{16:0}$ chains and due to the presence of hydroxyl groups it is hemihydrophylic. The waxes are a mixture of hydrophobic molecules such as very long-chain fatty acids, hydrocarbons, esters and alcohols. They are partially embedded in the cutin. In many plant species epicuticular crystalline wax structures overlay the cutin wax layer. Waxes represent the water proofing component of the cuticle (Post-Beittenmiller 1996). They shed rain water from the leaf surface and limit non stomatal water loss. Due to their hydrophobic nature waxes impede the diffusion of hydrophylic ions and frequently surfactants are neccessary to ensure diffusion of plant nutrients across the cuticle. From this it follows that leaves are not constructed for the uptake of hydrophilic ions dissolved in water. Nevertheless uptake of aqueous plant nutrients occurs at least in small quantities.

Those nutrients that are taken up by leaves can contribute to the nutrient needs of the plant. Nutrient uptake by the leaf is more effective the longer the nutrient solution remains in the form of a fine film on the leaf surface. Therefore on hot clear days when evaporation is high the water from the foliar spray can easily evaporate and salts accumulate on the leaf surface without being absorbed. This causes leaf scorching and burning. Such detrimental effects can be avoided by using solutions of low concentrations, about 20 to 50 g salt/L, and spraying on cool cloudy days or in the evening. In order to obtain thin surface films which adhere to the leaf surface and which provide a large surface contact, foliar spray solutions are often supplemented with surfactants which reduce the water tension.

Basically the process of nutrient uptake by leaf cells is the same as that of nutrient absorption by plant root cells, the main step being the transport across the plasmalemma. Analogous to the effect of nitrate and NH_4^+ nutrition on the pH of the nutrient medium during uptake by roots, the pH in the leaf apoplast is also increased by nitrate and depressed by NH_4^+ uptake (Hoffmann and Kosegarten 1995). This finding demonstrates that the basic uptake process for nitrogenous ions is the same for leaves and roots. Uptake rates depend on the nutrient concentration in the leaf apoplast which is largely controlled by the rate of diffusion of ions across the cuticle and by the nutrient reservoir present in the water film of the leaves.

Foliar application of plant nutrients can be very efficient under certain conditions, but it should be remembered that in general leaves are only able to take up a relatively small quantity of nutrients in comparison with plant demand. This is particularly the case for macronutrients, which are required by crops in high amounts. Foliar application is therefore not very common in practice. For nitrogen the most appropriate form of foliar application is urea, which is quickly taken up and metabolized by leaf cells. According to Franke (1967) urea increases the permeablity of the cuticle and thus favours diffusion conditions. Mathur et al. (1968) reported that spraying urea on cotton resulted in significant yield responses and that foliar application was superior to soil application. On soils with a low N availability, Walter et al. (1973) obtained a yield increase and an improvement of grape quality by foliar application of urea. Such effects can be found, but the results must not be generalized. Under conditions of practical farming it is often difficult to distinguish between foliar uptake and root uptake, for much of the urea sprayed on leaves may fall on to the soil and find its way into the plant through the roots.

Foliar application is particularly useful under conditions where nutrient uptake from the soil is restricted. This is often the case for the heavy metals such as Mn, Zn and Cu. These nutrients are frequently fixed by soil particles and for this reason are scarcely available to plant roots. On such sites, foliar application in the form of inorganic salts or chelates is a valuable tool in combating nutrient deficiencies (Tukey et al. 1962). As micronutrients are only required in small quantities, a foliar spray applied once or twice and correctly timed, is adequate to meet the demand of the crop.

Nutrient sprays are particularly used for fruit trees. These crops are often deep rooting so that fertilizer applied to the soil surface may be of little use and more readily available to the cover crop. Thus Cooke (1972) reported that leaf spraying with urea was an effective means of applying nitrogen to apple trees in grass-sown orchards where the trees often suffer from nitrogen deficiency. Physiological disorders, such as 'bitter pit' in apples, which results from Ca deficiency, can also be alleviated to some extent by spraying the fruits with a solution of a Ca salt. Several applications must be made (Schumacher and Frankenhauser (1968).

6.2.7   Nutrient ratios and recommendations

As already outlined in Chapter 5 (see page 305) the maximum effect of one particular plant nutrient can only be expected, if the supply of other plant nutrients is adequate. For this reason the ratio in which plant nutrients are applied in fertilizers is also important. This ratio depends on a number of factors including soil fertility status, crop species and crop management. If a soil is poor in one particular nutrient, as for example phosphate, fertilizers with a relatively high P content should be applied. For extreme deficiency in phosphate or potassium, the application of appropriate straight fertilizers is recommended in order to raise soil nutrient status to a satisfactory level. This is often the case for phosphate or potassium fixing soils.

On soils on which the availability of plant nutrients is satisfactory, crop nutrient uptake ratios correspond very closely to crop nutrient requirements. Thus to some extent nutrient uptake ratios can be used to calculate fertilizer application rates. The ratio of the amounts of N, P and K taken up by cereals is in the order of 1:0.3:0.8; the corresponding ratio for sugar beet and potatoes is 1:0.3:1.8. Hence for cereals, fertilizers containing nutrient ratios of about 1:0.5:1 are recommended, whereas sugar beet and potatoes require ratios with a higher proportion of potassium. The N, P and K uptake ratios for vegetative plant material such as grass, clover or lucerne are about 1:0.15:1.1. In clover and lucerne, N is mainly supplied by nitrogen fixation and for this reason only phosphate and potassium are generally applied in large amounts. The quantities of phosphate required by crops are higher than the amounts indicated by phosphate removal figures and the NPK uptake ratios because of the substantial quantity of fertilizer phosphate fixed in the soil. Phosphate fixation is particularly high in acid soils and may amount to 50% of the quantity of P applied. The corresponding value in neutral soils is 20% (Sturm and Isermann 1978, Jungk et al. 1993).

In calculating nutrient ratios suitable for mineral fertilizer application, farm practice must also be taken into account. Crop rotation is important as the residues from one crop can considerably influence the amount of nutrient required by a following crop. Wheat straw with a grain yield of 6 Mg/ha, for example, contains about 17 kg N/ha, 3 kg P/ha and 30 kg K/ha. If this is ploughed into the field, the crop following will have a useful source of K, but will need an additional N supply than normal, because part of the N application will have been utilized by the soil bacteria in decomposing the straw or even denitrified. Sugar beet leaves are a good source of plant nutrients, containing approximately 100 kg N/ha, 10 kg P/ha and 100 kg K/ha. These nutrients, especially P anf K, can largely be used by a following crop. Mineralization of organic nitrogen present in plant residues may provide inorganic nitrogen for the crop following.

The nitrate, however, produced in the process of mineralization may be lost by leaching or denitrification. Weather, climate, and soil conditions therefore have a great impact on the availablity of nitrogen originating from crop residues. It is thus particularly difficult to assess the value of residues of potatoes, rape, and some vegetables. Potato tubers and rape seed contain only low amounts of nitrogen. The larger part of their N is present in the the leaves and stems and these remain on the field (Aufhammer *et al.* 1994) and the organic nitrogen from these tissues is mineralized at least partially in late summer or autumn. The resulting nitrate may be leached by winter rainfall.

On grazing pastures a considerable proportion of plant nutrients can be recycled directly by animals. On permanent pastures grazing animals can return as much as 70 to 80% of the phosphate and potassium taken up by the sward. The nutrient ratio of fertilizers applied to pastures should therefore contain low proportions of these nutrients (Bergmann1969) in contrast to meadows where huge amounts of P and K may be removed in the hay. Nitrogen loss from pastures in the form of volatile NH_3 may be considerable and increases with the intensity of grazing (Isermann 1987).

The nutrient ratio of applied fertilizers also depends on the degree of agricultural intensification. Where yield levels are high and intensive cropping is being carried out, the proportion of plant nutrients originating from the soil is generally small. This in particular is important for potassium. In extensive agricultural systems potassium is very largely provided by the soil. According to Cooke (1974) potassium fertilizers are increasingly needed when agriculture is intensified.

Although methods for soil testing are well established and it is possible to use these methods for fertilizer recommendtions for crops (page 92), on a world wide-scale the proportion of agricultural land which is regularly tested for its plant nutrient status is low. In Germany it amounts only to about 8 to 10 %. There is no doubt that fertilizers can be used more efficiently if the nutrient status of soils is known. Not only can crop yield and quality be improved but appropriate fertilizer application can also raise the efficiency of other resources such as land, water, energy, and manpower. Low levels of available soil nutrients are frequently found in developing countries (Sanchez and Leakey 1997) whereas in developed countries, however, the reverse is more likely. Here there are soils enriched by plant nutrients and thus need less fertilizer than is frequently applied.

A first approach to applying adequate amounts of plant nutrients is to establish a balance sheet of plant nutrients taken away from the field by crop removal and by an estimate of nutrient losses caused by fixation, leaching, volatilization and denitrification. Producing such a balance sheet is relatively easy for phosphate and potassium provided that the soils are well supplied with these nutrients and therefore do not fix them to any great extent. For K^+ major losses are not to be expected except in highly pervious soils *via* leaching and in soils containing 2:1 clay minerals capable of K^+ fixation (see page 484). Estimation of available residual nitrogen is much more difficult since appreciable amouts of organic N mineralized in late summer or autumn may be leached by winter rainfall or denitrified under waterlogging soil conditions. In addition of the total organic N present in plant residues only a part is mineralized. Soil microorganisms which bring about mineralization also use the nitrogen present in the organic matter for growth and multiplication and it is only that nitrogen which is in surplus to their demand is released in the form of NH_4^+ and later nitrified (see page 408). Field trials of Paul (1994) *e.g.* have shown that of a total of about 200 kg organic N present in sugar beet tops only about 20% was mineralized during winter and the following growth period. Mineral nitrogen originating from plant residues may be estimated provided no major changes in a long practiced rotation occur. Such estimations are based on farmers' observations, experiences and some field trials. Thus the net amount of mineral N orginating from sugar beet leaves or plant residues of rape or potatotes have been estimated between of 30 to 50 kg N/ha under typical farming conditions on arable soils derived from loess in Germany (Mehl 1999).

Fertilizer recommendations for phosphate and K^+ based on soil analysis do not present a problem in most cases. Many field trials on a wide selection of soils using a variety of crops have been carried out to calibrate soil tests and crop responses to obtain the threshold values above which no substantial yield increase can be obtained. On arable soils derived from loess (Alfisols, Mollisols) in central and west Europe this critical K^+ level is in the range of 150 to 200 mg exchangeable K^+/kg soil and for phosphate of 60 to 80 mg P (CAL or DL-P, see page 95)/kg soil. Farmers should try to maintain these levels in their soils. These critical levels for K^+ and phosphate are for crop species with the highest demand for these nutrients. These high demanding crops include sugar beet, potatoes and rape but not cereals (see page 340). Under such agricultural conditions it is recommended to apply K^+ or phosphate or both only to the crop species with the highest demand for these nutrients in quantities so that the optimum available nutrient levels are maintained.

In practice this means that phosphate and K^+ are applied every 2. to 3. years. This practice is only recommended for soils in which no substantial leaching of K^+ and no major fixation of the applied phosphate occurs. A similar strategy can be used for the application of lime. Soils should be tested for their K^+, phosphate and lime status in intervals of three to four years.

The strategy for the recommendation of nitrogen fertilizers differs much from that of K^+, phosphate, and lime for several reasons. Phosphate and K^+ fertilizers even when applied at high rates are much buffered by the soil so that no major changes in the K^+ and phosphate concentration occurs in the soil solution. This is not true for nitrogen fertilizers, particularly not so for nitrate. Nitrogen fertilizer application rates must therefore be more adapted to the demands and the developmental stages of crops. On most soils nitrogen fertilizers have a rapid and marked impact on crop growth and crop quality. As is well known ample nitrogen nutrition during the grain filling stage may raise the protein concentration in cereal grains by improving the baking quality of wheat but be detrimental to the malting quality of barley. Too high a suply of nitrogen to sugar beet particularly during the last months of the growing period is also detrimental to the qualitiy of beet. Optimum N supply to crops may reduce the hazard of fungal infestation as shown by Huber *et al.* (1980) with take-all root and crown rot (*Ophiobolus graminis*) in wheat. In addition fertilizer nitrogen and soil nitrogen are subjected to various turnover processes as described above. These examples show that nitrogen fertilizer application, and especially rates and timing must be carried out with much care and for this reason in many cases a soil test on available nitrogen is neccessary.

A first approach to optimize N fertilizer rates is the estimation of the maximum yield potential of a location including soil and climate conditions. If the water supply because of soil or climatic conditions or both allows an average wheat grain yield of only 4 Mg/ha, the total nitrogen demand of the crop will be much lower than for a maximum yield level of 10 Mg grain/ha. This yield potential of locations is considered in the nitrogen fertilizer application concept proposed by Barekzai *et al.* (1992). Of further importance for the estimation of an optimum nitrogen fertilizer rate is the amount of available mineral nitrogen in the rooting profile at the beginning of the growth period. For aerobic soils this is mainly nitrate (Wehrmann and Scharpf 1986, Blackmer *et al.* 1989) and for flooded rice soils mainly exchangeable NH_4^+ (Schön *et al.* 1985). The most difficult parameter to estimate or to determine is the net mineralization of soil organic nitrogen over the growth period. It is the merit of Stanford and Smith (1972) for having begun the investigations of soils for their potential mineralising soil nitrogen. These workers found considerable differences between Soil Orders and particulary within a given Soil Order a finding indicative of the numerous factors which influence the nitrogen mineralization potential of a soil. Thicke *et al.* (1993) tested several soil nitrogen mineralization indices in field trials. Most of these indices were unsatisfactory. Best results were obained using the aerobic incubation method in which the soils were incubated for a period of one week. Incubation methods, however, are laborious and therefore not suitable for routine analysis. In this respect the extraction of organic nitrogen by EUF or a $CaCl_2$ solution deserves particular mention (see page 100). It has been shown that these methods provide a satisfactory indication of the available inorganic and mineralizable soil nitrogen under practical farming conditions and can be used as basis for a reliable nitrogen fertilizer recommendation (Ziegler *et al.* 1992, Barekzai *et al.* 1992, Fürstenfeld *et al.* 1994, Groot and Houba 1995).

A useful method for forecasting fertilizer requirements of vegetable crops growing on widely different soils was developed by Greenwood et al. (1974). These workers argue that of the more than 20 vegetable crops grown on widely different soils in the United Kingdom it is impracticable to carry out trials to cover more than a few possible combinations of crop and soil. They therefore developed an alternative 'short cut' method in which a model was first devised then calibrated against experimental data and used to predict response curves in different situations. Experiments were carried out to characterize the responses of 22 crops to N, P and K fertilizers so that the approach could be applied in practice. This modelling approach has proved to be valuable in NPK fertilizer prediction for vegetable crops in the United Kingdom (Greenwood *et al.* (1980).

In future, models may also play a major role in optimizing fertilizer rates. As yet the use of various current models to predict soil nitrate values in the rooting profile in spring without the integration of soil test data has not proved satisfactory (Otter-Nacke and Kuhlmann 1991). There is evidence, however, that the integration of a soil test nitrogen parameter such as C/N ratio, rotation or rainfall may improve such models considerably (Appel 1998, Mehl 1999, Stockdale *et al.* 1997).

6.2.8   Hydroponics soilless cultivation and fertigation

Terrestrical plants require soils not only as a source of plant nutrients from the soil solution but also as a means of anchorage of the entire plant by the root (see page 78) Nevertheless terrestrial plants may complete their life cycle when cultivated without soil in nutrient solution if some prerequistes are met. The German scientists Julius Sachs (1860) and Knop (1865) (both quoted by Asher and Edwards 1983) were able to demonstrate in the mid of the 19. century that higher plants can be cultivated in nutrient solutions without soil. This proved that plants are able to feed exclusively on inorganic materials without the need of soil or humates. From that time the growth of plants in solution culture or hydroponics has been an important tool in the study of mineral nutrition such as the study of the uptake of plant nutrients, the antagonstic and synergistic behaviour of plant nutrients, the effect of particular plant nutrients on growth and the synthesis of plant molecules of importance for crop quality, and other interesting physiological questions. Even today solution culture of plants is still an indispensible technique in plant science (Asher and Edwards 1983).

Over the years various nutrient solutions have been formulated by research workers (Hoagland and Arnon 1950, Johnson et al. 1957, Hewitt 1966, Lesaint and Coic 1983, Morard 1995). One of the best known nutrient solution is that of Hoagland (Hoagland and Arnon 1950) the composition of which is shown in Table 6.16.

Table 6.16   Composition of the Hoagland nutrient solution for the cultivation of higher plants Hoagland, after Asher 1978; Asher and Edwards 1983)

Macronutrients in mol/m³		Micronutrients in mmol/m³	
Nitrate	14	Iron	25
Ammonium	1	Boron	46
Potassium	6	Manganese	9
Calcium	4	Zinc	0.8
Magnesium	2	Copper	0.3
Phosphorus	1	Molybdenum	0.1
Sulphur	2	Chloride	18

There are major differences between soilless culture and culture of plants in soil. Nutrient solutions are not buffered, either in pH or in nutrient supply and plants grown in nutrient solutions are continuously and amply supplied with water. In order to overcome the problem of nutrient buffering, nutrient concentrations of solution cultures are generally much higher than in the soil solution (see Table 2.14). In the Hoagland nutrient solution the concentration of K^+ is 6 mol/m³ which is about 10 times higher than the K^+ concentration in the soil solution. There is an even greater difference between the phosphate concentration in nutrient solutions and the phosphate concentration in the soil solution. A further problem of solution culture is that concentrations of plant nutrients are depleted by plant uptake at different rates *e.g.* the nitrate concentration is much more rapidly depleted than the sulphate concentration. From this follows that the composition of the nutrient solution changes with time of cultivation and for some nutrients concentrations may fall greatly whereas for others even toxicity levels may be reached. In order to overcome this problem nutrient solutions should be changed frequently or the volume of solution per plant should be large (Parker and Norvell 1999). Ingestad and Lund (1979) propose that nutrients should be added to the solution in quantities according to the growth rate and plant demand. Asher and Edwards (1983) and recently Parker and Norvell (1999) who have considered the subject of nutrient solutions in a useful article recommend the use of flowing solution culture. In this system large volumes of nutrient solution are cycled between a resevoir and plant roots and plant nutrient concentrations and pH are strictly controlled in relation to plant nutrient uptake. Such systems require considerable technical expertise. Nutrient concentrations used in recycling systems are low and in the range of concentrations found in the soil solution as shown in Table 6.17. It is therefore possible to carry out meaningful experiments simulating soil solution conditions which are not possible using the usual high concentration of standard nutrient solutions. The use of flowing nutrient solutions has been treated by Wild *et al.* (1987) in a useful review article.

Table 6.17 Nutrient solution concentrations in flowing solution culture. According to Asher and Loneragan (1967 quoted after Asher and Edwards 1983).

in mmol/m^3			
Nitrate	750	Iron	2
Ammonium	100	Boron	3
Potassium	250	Manganese	1
Calcium	250	Zinc	0.5
Magnesium	100	Copper	0.1
Phosphorus	0.04–25	Molybdenum	0.02
Sulphur	100	Cobalt	0.04
Chlorine	100		

This solution also contains cobalt which is required for the nodulation of leguminous species.

The *Nutrient film Technique (N.F.T.)* is also a system involving recycling and control of the nutrient solution (Cooper 1976), and is used in practical horticulture. The nutrient solution flows in a thin film across a slight inclined plane to drive the solution in one direction. The roots are not completely immersed in the solution so that oxygen supply of roots is ensured (Morard 1995). This system is much used in commercial glasshouse production of salad crops.

One of the earlier problems associated with growing plants in hydroponics was the means of supplying Fe. Inorganic Fe^{2+} quickly oxidizes to Fe^{3+} and precipitates as $Fe(OH)_3$. Iron was therefore mainly given as Fe citrate. Now Fe chelates are used *e.g.*

Fe- EDTA (ethylene diamine tetra acetate) or Fe- EDDHA (ethylene diamine-di-o-hydroxyphenyl acetate). These compounds remain soluble over a wide pH range and serve as an efficient Fe source for plants. Besides the nutrients shown in Table 6.16 and Table 6.17 plant roots also need oxygen. Nutrient solutions must therefore be aerated.

The pH change in nutrient solutions is mainly related to the uptake of cation and anion species and especially the uptake of nitrate and ammonium. Nitrate is taken up in the form of proton cotransport, ammonium in most cases is not (see page 126). Hence protons pumped into the nutrient solution by plant roots are recycled back into the cytosol by the uptake of nitrate. If the rate of this H^+ recycling is higher than the H^+ release rate by plant roots the pH in the nutrient solution increases. Phosphate and sulphate are also taken up by means of a H^+ cotransport. However, in the vegetative stage of plant growth about 80% of total anion uptake can be accounted for as nitrate, so that uptake of nitrate plays the most important role in increasing nutrient solution pH provided the plants are exclusively supplied with nitrate as N source. If plants are supplied with ammonium, nutrient solution pH decreases since protons are not recycled back to the cytosol to any major extent. This effect of the N source on solution pH is shown in Figure 6.9 from the work of Mengel et $al.$ (1983). It should be noticed that in this experiment there was no significant influence of the nitrogen form on the growth of the maize.

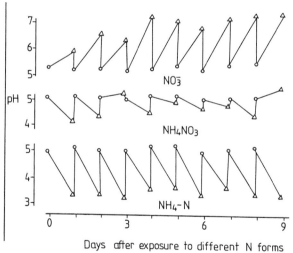

Figure 6.9 The effect of the nitrogen source on the pH in the nurtient solution in which maize was grown. The upper trait represents the treatment with nitrate, the trait in between the treatment with ammonium nitrate, and the lower trait the treatment with ammonium. The nutrient solution was replaced every 24 h and the circles denote the nutrient solution pH of the new solution, the triangles the pH after 24 h. (after Mengel et $al.$ 1983).

It is evident that the pH of the replenished nutrient solution was about 5 for all three nitrogen treatments: nitrate, ammonium nitrate, and ammonium. After a cultivation period of 24 h the pH of the nitrate solution was considerably increased particularly in the later days of the cultivation whereas the pH of the NH_4^+ solution was markedly depressed. Considering that the pH scale is a logarithmic one a depression of nutrient solution pH from about 5 to about 3.3 over 24 h is indicative of an enormous net release of H^+ of plant roots into the outer solution. Nutrient solutions supplied exclusively with ammonium as N source quickly run into acidity problems. In the treatment with ammonium nitrate the oscillation of pH was relatively small with the tendency of a pH increase with plant growth.

In the nutrient solution technique discussed above no substrate is used such as sand, gravel or rockwool. Generally plants are grown in pots or trays fillled with the nutrient solution and covered with a lid with holes through which the stems of the plants are mounted and wrapped with a piece of wool or plastic foam. In hydroponics in which a substrate is used this provides anchorage for the roots.

The solution culture techniques developed for scientific purposes formed the basis for using nutrient solutions for the commercial growth of crops. Today highly technical systems are employed in the supply of inorganic fertilizers to grow highly priced crops. These include vegetables and fruits (tomatoes, cucumbers, strawberries, melons) and ornamental plants. In many cases crops are cultivated on substrates such as gravel, rockwool, and sand combined with a fully recycled and automatically controlled nutrient solution. These hydroponic systems have the advantage that they can control the nutrient supply to crops precisely *i.e.* using relatively large amounts of nitrogen during the vegetative growth in order to encourage the early development then reduce the nitrogen supply during fruit formation in order to improve fruit quality. In properly handled hydroponic systems water and plant nutrients can be used with a high efficiency. Morard (1995) in a useful book on hydroponics emphasizes that soilless crop cultivation (hydroponic with or without a supporting substrate) can avoid soil pathogen infection, weeding problems and can be used where fertile soils are rare because of salinity, acidity or other adverse conditions. This kind of crop production is therefore of particular interest in arid areas where water is a limiting resource. In this context also the use of sewage effluents should be considered (Wallace *et al.* 1978). There are, however, indications that fruit quality of crops grown in hydroponics is poorer than fruits grown on soils (Morard 1995). This interesting question needs further research.

The largest area with commercially grown crops in solution culture is in Europe where the area of soilless crop cultivation has increased from about 500 to 3000 ha from 1981 to 1990 (Morard 1995). The global area of soilless cultivation of crops amounts to about 5000 ha most of which being found in highly developed countries as shown in Table 6.18.

Crop growth on poor soils in hot, arid climate needs irrigation. Under such conditions the conventional form of water supply by furrow or sprinkler irrigation is associated with a high water loss because of intense evaporation and hence the water efficiency is very poor.

Trickle irrigation system was developed to cope with such conditions (Kafkafi 1994). The trickle which is the end of the irrigation pipe releases water to the base of the stem of the crop plant, *e.g.* tomato. Water is discharged into the moist soil volume in quantities that at least cover the evapotranspiration but do not lead to a major water loss by leaching and so that the root system of each plant is surrounded by a "bulb- shape" soil volume of moist soil as shown schematically in Figure 6.10 (Kafkafi 1994).

Table 6.18 Countries with the largest area of hydroponics for the commercial production of vegetables, fruits and ornamental plants (Morard 1995).

Country	Area, ha
Netherlands	3000
France	1000
Great Britain	600
Japan	400
Belgium	300
South Africa	300

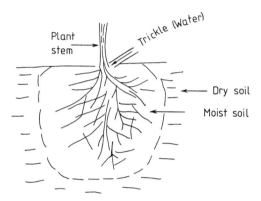

Figure 6.10 Scheme of a tricle with a bulb-shaped moist soil volume around roots.

The quantities of water discharged depend much on transpiration and evaporation and may be in a range of several L/h. The size of the bulb-shaped moist soil volume depends greatly on the size of the root system and water flow conditions. It is particularly high on sandy soils and is much smaller for loamy soils as shown by Bresler (1977) for the radial and vertical water distribution in a sandy soil (sand dunes of Sinai Desert) and a loam soil (loess-derived loam in Gilat). A useful paper on trickle irrigation was published by (Bresler 1977).

The term *Fertigation* means a combination of irrigation and fertilizer application in which plant nutrients are dissolved in the irrigation water. Bar-Yosef (1999), who discusses fertigation in a useful review distinguishes between macrofertigation and microfertigation. In the latter the the plant nutrients are dissolved in the water supplied by trickle irrigation (= drip irrigation). The advantages of microfertigation are the adaption of nutrient supply to the crop demand at various growth stages and to weather conditions, precise application and avoiding excess of fertilizer. Nitrate uptake of vegetables can be regulated where low nitrate concentrations in the crop are desired at harvest. On the other hand the technical equipment of fertigation requires additional costs for fertilizer injectors as well as shipping and storage of large volumes of dilute fertilizer solution.

It is obvious that such trickle irrigation systems especially on poor soils require a continuous nutrient supply to the crop in order to make the best use of water. The system is a kind of an open solution culture with the soil as substrate. Concentrations of nutrient solutions used are in the range of the concentrations for hydroponics (Table 6.16 and 6.17). In most cases micronutrients are not required with the exception of Fe which may be needed on calcareous soils.

The dosage of nutrient supply must be handled according to nutrient demand of crops in order to avoid deficiency or oversupply. Nitrogen may be given in the form of nitrate or ammonium or both. Excess ammonium may lead to ammonium toxicity which may particulary occur if roots are poor in nonstructural carbohydrates which are required in the assimilation of ammonium. This question of nitrate and ammonium application in fertigation has been discussed by Kafkafi (1994) in a useful article.

Fertigation is a highly sophisticated crop production system developed in Israel in order to make the best use of extremely poor soils and the limited water reserves. Its installation and maintainance is expensive and therefore it is only appropriate for cash crops with highly priced.

6.3   General Aspects of Fertilizer Application

6.3.1   Fertilizer use and agricultural production

The application of science to agriculture has had an enormous impact on agricultural production as can be seen from the data of Table 6.19 which shows yield per ha, fertilizer consumption and production per head and number of people supported by the production of 1 ha land (Siemes 1979) over the last 200 years in Germany. Whereas at the beginning of the 19th century 1 ha farmland scarcely produced food for 1 person, today 4.5 people can be fed from the production of 1 ha. It is also shown in Table 6.19 that the production per head has not changed much during this period because of the high population increase. In other countries similar dramatic increases in crop production have also occurred. This increase in productivity is an important achievement of scientific research and an essential one, for without it famine would be common even in regions now producing a surplus of agricultural products. Sturm (1992) reported that wheat grain yields in West Germany increased from 3.2 Mg to 9.0 Mg/ha in the period from 1950 to 1990. This dramatic increase was due to higher fertilization, especially with nitrogen, to the use of more efficient cultivars, and to the application of herbicides, fungicides, and insecticides.

In an excellent book "Feeding the Ten Billion- Plants and Population Growth" L.T. Evans (1998) gives an impressive historical view on the growth of human population paralleled by the development of agriculture. Figure 6.11 from this source shows the increase of world population, surface of arable land, fertilizer consumption and rice and wheat yields for the last 150 years.

Table 6.19 Mineral fertilizer consumption, yield per ha, production per head, and number of people supported by the production of 1 ha. Production expressed in grain «units» which is equivalent to 1 Mg cereal grains (after Siemes 1979).

Year	Number of people supported by 1 ha	Production per head, grain unit	Yield grain unit/ha	Fertilizer consump. kg NPK/ha
1800	0.8	0.91	0.73	—
1875	1.3	0.92	1.20	3.1
1900	1.6	1.14	1.84	15.6
1925	2.1	1.09	2.28	43.9
1950	3.3	0.91	2.98	101.9
1975	4.6	0.95	4.43	233.5
1978	4.5	1.03	4.46	255.8

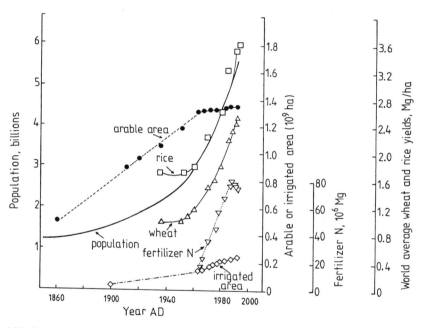

Figure 6.11 Increases in world population, arable area, the average yield of wheat and rice, the amount of N fertilizer used, and the irrigated area of the world (after Evans 1998).

Yield increases at the end of the 19th and at the beginning of the 20th centuries were also due to the introduction of forage legumes (red clover, lucerne) the growth of which were favoured by the application of lime and phosphate fertilizers (Schmitt 1958). The remarkable increase of maize grain yields in France over the years from 1956 to 1986 is shown in Figure 6.12 (Boulaine 1989). This impressive yield increase which is typical for other western European countries was also the result of the application of agrochemicals including fertilizers and the use of more efficient cultivars.

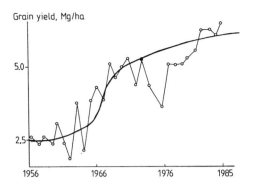

Figure 6.12 Effect of the application of agrochemicals including fertilizers and the use of more efficient cultivars on maize grain yield increase from 1956 to 1985 in France. The low yield in 1976 was due to extremely dry weather (Boulaine 1989).

Fertilizer usage plays a major role in the universal need to increase food production to meet the demands of the growing world population.

Uloro and Mengel (1996) reported a marked influence of inorganic fertilizer application on starch yield increase in *Ensete ventricosum*, a major staple food crop for millions of people in south and southwestern Ethiopia. From Figure 6.13 it is clear that the crop, which was grown on a typical soil of this area, responded to N, P, and K application by more than tripling the starch yield. Such yield effects are also beneficial to the working efficiency of farmers in Ethiopia.

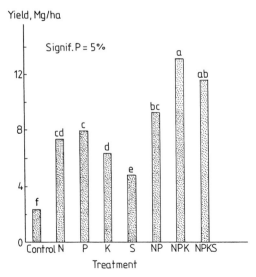

Figure 6.13 Effect of the application of various plant nutrients and plant nutrient combinations on starch yield of *Ensete ventricosum* grown on poor soils in Ethiopia (after Uloro and Mengel 1996). Columns with different letters indicate significant differences, p < 5%.

In order to obtain satisfactory fertilizer responses new cultivars are often required. This is particularly the case for rice and wheat. Local cultivars are often tall plants and are susceptible to lodging, especially if fertilized with nitrogen.

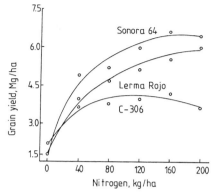

Figure 6.14 Response to nitrogen fertilizer of one old and two modern wheat cultivars (after Chandler 1970).

The newer dwarf cultivars are very resistent to lodging and high grain yields may be obtained when they are adequately supplied with plant nutrients. This is shown in the data of Figure 6.14 of Chandler (1970). The typical tall local Indian wheat cultivar C-306 only responded to a low rate of nitrogen application and maximum grain yields were obtained at about 80 kg N/ha. In contrast the short stiff straw Mexican cultivars Sonora 64 and Lerma Rojo gave a much greater response to N and about twice the grain yields were obtained with a nitrogen application rate of 160 kg N/ha.

The beneficial effect of fertilizer application on grain yield of modern rice cultivars is shown in Table 6.20 from numerous FAO field experiments carried out on various soil types in Sri Lanka (Braun 1989). Such yield increases are required in order to feed the growing world population (Neumann 1997).

Table 6.20 Effect of single applications of N,P, and K on rice grain yield in Sri Lanka, FAO trials (Braun 1989).

Soil Class	No.of sites	Control	N, 120 kg/ha	P, 35 kg/ha	K, 66 kg/ha
			yield in Mg/ha		
Regosol	9	2.3	10.1	8.5	12.2
Vertisol	25	3.2	20.9	7.1	14.6
Ferralsol	9	2.9	3.3	10.7	5.7
Luvisol	337	3.2	7.9	8.8	7.5
Fluvisol	157	3.3	8.6	8.2	8.0

There are numerous examples for various crops from different parts of the world which all show that application of the individual nutrients N, P, and K all give marked yield responses demonstrating that these nutrients were not sufficently available. The Ferralsols are particularly low in available phosphate. Highest responses are usually attained if all three nutrients had been applied together as was the case for *Ensete* shown in Figure 6.13.

6.3.2 Fertilizer application and energy consumption

In the USA on an average size farm, one man produces enough food for about 50 people. This high degree of efficiency has only been achieved by the use of a considerable amount of energy mainly in the form of petrol and electricity. The application of fertilizers, herbicides and pesticides also involves an indirect consumption of energy as the production of these materials requires energy. With modern production techniques the energy requirement for fertilizer production, particularly N fertilizers has been much reduced in recent years. The energy demand per kg nutrient in representative fertilizers are shown in Table 6.21. The values relate to modern production techniques and were previously much higher, particularly for N fertilizers.

Lewis and Tatchell (1979) carried out a thorough investigation into the energy consumption of agricultural production under the conditions of UK agriculture. Their main results are reported in Table 6.22 From these data it is clear that crop production has a positive energy balance since the energy output/input ratio is >1. The reverse is true for animal production. The table also shows that the application of mineral fertilizers improved the output/input ratio of crops. This finding is consistent with results of Pimentel *et al.* (1973) obtained for maize production in USA. It must be emphasized that the figures in Table 6.22 were calculated from data of only the edible plant parts *i.e.* grains or sugar or potato tubers. Other plant parts, which also contain energy, were not taken into account. The amount of energy present in cereal straw is usually higher than that contained in the grain which is considerable, for 1 Mg of barley or wheat grain contains about 15 GJ of energy. Considerably higher ratios would thus have been obtained if these non edible parts had been taken into account.

Table 6.21   Energy consumption for the production of plant nutrients in representative fertilizers (BASF).

Fertilizer	MJ per kg nutrient		
	N	P	K
Ammonia, 82% N	35	—	—
Urea, 46% N	42	—	—
Ammonium nitrate urea solution, 28% N	35	—	—
Nitrochalk, 27% N	31	—	—
Diammonium phosphate, 18% N, 20% P	36	14.4	—
Triple superphosphate, 20% P	—	14.0	—
Potassium chloride, 50% K	—	—	2.9
PK fertilizer, 20 + 17 + 0 (N + P + K)	—	15.5	3.8
NPK fertilizer, 15 + 15 + 15 (N + P + K)	32	.15.3	3.6

Table 6.22   Energy output/input ratio of crop and animal products (data from Lewis and Tatchell 1979)

	Total	Increment obtained by fertilizer application
Wheat (winter)	2.2	3.3
Barley (spring)	2.0	3.2
Sugar	2.5	2.1*
Potatoes	1.3	2.0
Milk	0.40	0.41
Beef**	0.21	0.22
Lamb**	0.20	0.18

* Low value because of poor weather, ** Total edible output

A particularly low energy input is required for the cultivation of legumes as these crops require very little N fertilizer. Thus, according to Gasser (1977), over a cultivation period of 3 years lucerne requires only about 9 GJ/ha but yields about 320 GJ/ha. This is an output/input ratio about 35. Animal grazing especially on leguminous swards is also low in energy demand (Leach 1976).

In discussing energy demand for agricultural production it is important to understand that animal and plant production require only a small amount of energy in comparison with the total consumption of modern society. In the UK for example only about 4% of the total energy consumption is needed for agriculture. Of this 4%, about 1% is required for power machinery and 1% for the production of fertilizers (White 1976). This comparison clearly indicates that energy saving should not primarily be directed to agricultural production. Indeed Greenwood (1981) has shown that on a world scale only a minute fraction of present fossil energy consumption would be needed to manufacture all the fertilizer to grow enough food for everyone.

It should be emphasized that crop production is one of the few production processes with a positive energy balance. It now seems likely that in order to meet future energy needs this acquisition of energy by plants will play an increasingly important role. Hall (1977) cites 5 plant species, eucalyptus trees, hibiscus shrubs, Napier grass (a tropical fodder grass) sugar cane, and cassava, which are considered to be suitable for *sun energy harvesting*. Species of *Euphorbiaccae* have also been considered as possible *energy crops*. These plants contain latex which consists mainly of hydrocarbons and can easily be processed. A further advantage of these species is that they have a low water requirement and can grow in rather arid regions (Calvin 1980).

6.3.3   Fertilizer application and the environment

Some major aspects of fertilizer application which merit attention in relation to pollution problems are pollution of drinking water, eutrophication of lakes and rivers, volatilization of NH_3 and nitrogen oxides, imbalanced quality of plant products and agricultural land as sinks for methane.

6.3.3.1 *Drinking water and fertilizers*

The main constituent of fertilizers which has an undesirable effect on the quality of drinking water is nitrate. The presence of other plant nutrients, such as phosphate, potassium and magnesium can improve the quality of drinking water, as these ions are also directly essential for human and animal nutrition. Nitrate itself is not toxic, but nitrite originating from the reduction of nitrate induces methaemoglobinemia in infants by inhibiting O_2 transport in the blood. Nitrite and amines can react to form nitrosamines which are cancinogenic. As yet it is not yet clear, however whether this reaction occurs in the human digestion tract.

Acceptable values for NO_3^- concentration in drinking water have been set at $50-100$ mg nitrate/L (WHO), a maximum of 50 mg nitrate/L (EU), and a maximum of 45 mg nitrate/L(US Public Health Service).

As nitrate fertilizers are being used in agriculture in increasing quantities on a global scale there has been considerable concern that some fertilizer nitrates may be leached and carried into the deeper aquifers supplying drinking water. It is not possible to generalize on the effects of fertilizer N on the nitrate concentration of ground water and drinking water. It is well known that the recovery of inorganic N fertilizers is incomplete and increasing rates of application have reduced average recovery. The extent to which leaching plays a role in this loss is not so clear. For arable crops $20-60\%$ of applied N is taken up. The comparative figure for grass is $40-80\%$. Both these values vary depending on the soil, the season and N fertilizer rates. On soils with a medium to high clay concentration water leaching rates during rainfall periods may be low and also the rate of nitrate leaching is only about 5% of the fertilizer amount given as reported for the climatic and agricultural conditions of the Mid-West USA (Parker 1972). Kjellerup and Dam Kofoed (1983) found in lysimeter experiments carried out under the humid conditions of Danmark that from the [15] N labelled fertilizer applied only 5% was leached out of the rooting profile and 58% was taken up by the crop as shown in Table 6.23.

Table 6.23 Nitrogen balance in a rotation of barley, rape wheat, and Festuca. Lysimeter experiments with ^{15}N labelled fertilizer (Kjellerup and Dam Kofoed (1983).

	% of fertilizer N
Uptake of the crop	58
Leaching	5
Soil	21
Unaccounted (volatile) losses	16

In similar lysimeter experiments also carried out with ^{15}N labelled fertilizers Teske and Matzel (1976) found an even lower percentage of leached nitrogen under the more continental climate conditions of eastern Germany. In both cases, however, nitrogen fertilizer rates were adjusted to the demand of crops. Under practical farming conditions this is often not the case, particularly in spring when the soil profile can contain relatively high quantities of mineral nitrogen, mainly as nitrate. This was shown by Baumgärtel *et al.* (1989) in numerous field trials conducted in northern Germany. As soon as the sum of residual mineral N + fertilizer N was higher than about 200 to 250 kg N/ha in the rooting profile residual nitrate after harvest increased from the relatively low level of about 30 kg N/ha to higher levels prone to leaching by winter rainfall. The relationship between the sum of mineral nitrogen in the rooting profile + fertilizer N and the grain yield of winter wheat is shown in Figure 6.15. It is evident that the maximum yield of about 9 Mg grain/ha is attained as soon as the sum of mineral nitrogen in the rooting depth + fertilizer N is 250 kg N/ha. The maximum yield is attained at a still tolerable level of residual nitrate in the rooting profile (0—90 cm depth). This is about 30 kg N/ha the quantity which is equal to the amount of N released from wheat roots into the soil during the growth period (Rroco and Mengel 2000).

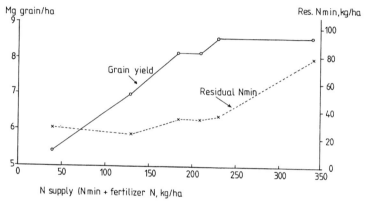

Figure 6.15 Residual nitrate in the rooting profile after harvest of winter wheat in relation to fertilizer N rate + mineral N in the rooting profile in the preceding spring related to wheat grain yield. Each point in the graph is the mean of 20 to 40 field trials (Baumgärtel *et al.* 1989).

Analogous results have been found for sugar beet in Dutch field trials (Prins *et al.* 1988). From this finding it is clear that for adequate N fertilizer rates it is pertinent to know the quantity of mineral nitrogen in the rooting profile at the beginning of the season and therefore in many cases a soil test for available nitrogen is needed. Such a strategy contributes considerably to avoiding a waste of nitrogen fertilizer and to reducing nitrate leaching into aquifers.

On sandy soils the problem is even more severe because of the highly pervious character of such soils which contain hardly any mineral nitrogen in spring under conditions of winter rainfall because of leaching. According to Thies et. al. (1977) fertilizer N applied to sandy soils in spring may easily be assimilated (immobilized) by soil microbes and hence become unavailable for the crop. Nitrogen so fixed may be mineralized in autumn after crop harvest and later leached by winter rainfall. According to Appel and Mengel (1992) this is the main reason why the apparent nitrogen fertilizer recovery is low on sandy soils. On the other hand sandy soils are the most important soils for drinking water regeneration in Central Europe (Strebel *et al.* 1985). Nitrate leaching in arable soils can be reduced by so called *catch crops* (Thorup-Khristensen 1994). These are crops grown in autumn after the main crop has been removed from the field with the purpose of taking up the available nitrate and the nitrate formed during autumn and even winter. Such crops are winter rye, winter rape, *Sinapis alba* and *Phacelia tenacetifolia*.

The application of organic fertilizers (farmyard manure, green manure, slurries) in autumn or winter may cause substantial nitrogen losses because of nitrate leaching and therefore may contribute to nitrate accumulation in aquifers. Borin *et al.* (1997) reported that nitrate leaching was particularly high if rainfall followed the liquid manure application. Even more severe are the effects of manure, slurry, and livestock waste when applied in quantities much higher than that required by crops (Owen and Jürgens-Gschwind 1986). Here phosphate may also accumulate in soils to very high levels (Leinweber 1996, Mozaffari and Sims 1996, Peters and Basta 1996). This accumulation indicating that nitrogen application must also have been in surplus. Most of this excessive nitrogen applied is lost either by leaching or by volatilization in the form of NH_3 or after denitrification in the form of N_2 and N_2O. Dinitrogen oxide (N_2O) is involved in the destruction of the ozone layer in the stratosphere (see page 394). Ammonia released from these intense livestock farms into the atmosphere is brought back into soils by precipitation where it my be oxidized to nitrate and thus also contribute to the nitrate in lakes, rivers and aquifers (Roelofs *et al.* 1985). The accumulation of available soil phosphate to very high levels after the repeated application of excessively high rates of slurry also means the waste of a very valuable limited resource (Mengel 1997). Nitrate leaching, N_2O release into the atmosphere and the loss of phophate because of too high rates of livetock manure and slurry application are a hazard to the environment. For this reason the application of organic fertilizers should be strictly controlled.

As well as agriculture, traffic also contributes to the entry of nitrogen oxides into the environment (Isermann 1990). These nitrogen oxides, mainly NO and NO_2, produced in combustion engines may react with water and O_2 to form nitric acid according to the following equation:

$$NO + \tfrac{1}{2} O_2 \Rightarrow NO_2$$

$$2NO_2 + H_2O \Rightarrow HNO_3 + HNO_2$$

$$HNO_2 + \tfrac{1}{2} O_2 \Rightarrow HNO_3$$

Nitrate leached into soils is not only a direct hazard for drinking water, it may also affect the regeneration of drinking water in soil profiles. Nitrate in water percolating through the soil profile may be denitrified on its way to the aquifer. The major electron donors for this reduction process are Fe^{II} and organic matter. The distribution of Fe^{II} and nitrate in such a percolating profile is shown in Figure 6.16 from the work of Lind and Pedersen (1976). In the upper 2 m layer virtually no Fe^{II} is present and nitrate prevails. In the transition zone nitrate concentration declines sharply associated with a steep increase in Fe^{II}. Denitrification decreases the amout of reducing equivalents which in this case is Fe^{II} and in the case of organic matter the oxidizable organic carbon. If soil profiles are loaded with high amounts of nitrate the oxidizable substrate, Fe^{II} and organic carbon or both may be depleted and the profile may lose its potential to regenerate drinking water with a tolerable nitrate concentration.

The vertical movement rate of the nitrate front in soil profiles depends on the permeability of soils and is in the range of 0.3 to 1.0 m per year under temperate climate conditions (Owen and Jürgens-Gschwind 1986).

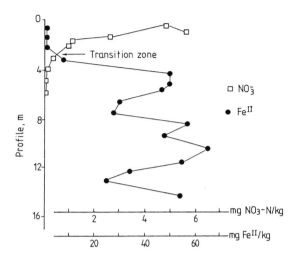

Figure 6.16   Distribution of nitrate and Fe^{II} in a deep profile of an arable soil. Modified after Lind and Pedersen (1976).

6.3.3.2 *Eutrophication and fertilizers*

Eutrophication or the promotion of the growth of plants, animals and microorganisms in lakes and rivers, is a natural process. If this is allowed to occur uninterrupted, it results in a progressively increasing deficiency of oxygen in the water. Thus organisms which live under anaerobic conditions are favoured more and more at the expense of aerobic organisms. Under these conditions organic material is not decomposed completely to H_2O and CO_2 but remains largely in a reduced form and accumulates. Besides this accumulation of organic compounds, metabolic end products of anaerobic microorganisms are produced, such as methane, ethylene, H_2S, butyric acid and other low molecular weight substances. These compounds are by and large toxic to aerobic living organisms. This is the main reason why the eutrophication of lakes and rivers has such a detrimental effect.

To understand the process of eutrophication the interrelated influences of the various kinds of organisms should be considered in more detail. The relationships between algae, photosynthetic bacteria and anaerobic bacteria living in lakes, where the circulation of water is mainly confined to the upper layer, is illustrated in Figure 6.17 (Stanier *et al.* 1995, p. 381). In the upper layer algae and other photosynthetically active green plants are present. Photosynthetic activity ensures that this layer is aerobic and enriched with dissolved oxygen. The boundary between the aerobic water layer and the deeper anaerobic water zone favours the growth of photosynthetic bacteria, for at this depth the light intensity is still high enough to maintain photosynthesis but the medium is anaerobic, as required by most photosynthetic bacteria. These organisms, the purple sulphur and green bacteria, are fed by the metabolic end products of the anaerobic microorganisms, which are present below mainly in the muddy sediment of the lake. The photosynthetic bacteria use these end products, such as H_2S, butyric acid or other fatty acids as electron donors in photosynthesis and thus decompose the compounds, which are toxic to green plants. This narrow band of photosynthetic bacteria acts as a filter and keeps the upper layers of the water free from toxic substances. If the balance between these organisms is disturbed, increased quantities of the toxic substances may reach the surface of the lake, and affect the growth and activity of green plants. Photosynthetic oxygen production is reduced and the surface of the lake gradually becomes anaerobic and the life of aerobic organisms including fish is endangered.

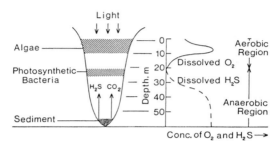

Figure 6.17 Oxygen and H_2S concentrations in a narrow deep water lake in relation to the growth of various organisms (after Stanier *et al.* 1995).

Such a disturbance in the biological balance can be induced by too vigorous a growth of algae. When large amounts of dead algal material sediment, the anaerobic microorganisms at the bottom of the lake are provided with an abundant source of food. Large amounts of toxic substances are then produced. If these are in excess of the capacity of the photosynthetic bacteria, the filtering effect of these organisms is then reduced or lost and toxic substances are able to reach the upper layers of the lake.

Frequently phosphate is the limiting factor in the growth of algae in lakes and streams, and increasing concentrations of phosphate in the water often run parallel with the degree of eutrophication. The eutrophic threshold level below which algal growth is limited is considered to be in the region of 10 μg/L P.

The rise in the use of phosphatic fertilizers has sometimes been blamed for the increase in eutrophication. As already discussed on page 343, however, phosphate is very tenaciously held by soil particles and is leached from agricultural land into lakes and rivers at only a very low rate. Leaching is not affected by phosphate fertilization, provided that phosphate is not apppplied at extremely high rates. However, this may be the case in areas with high livestock density, where very high quantities of phosphate may be brought to the field with the slurry (Leinweber 1996). Some phosphate in lakes and rivers originates from phosphate run-off from agricultural land (Sharpley 1993, Lenzi and Di Luzio 1997). But the quantities removed run-off are low and generally below 0.5 kg Pha.

By far the highest source of phosphorus, however, is not of agricultural origin and is that of detergents and urban wastes. Soil phosphates contribute only to about 4 to 5% to the total phosphate leached into surface water. Ryden et al. (1973) have provided a most useful detailed review on P in run-off and streams.

Nitrate is essential for algal growth. However as the critical concentration for growth of 0.3 mg N/L is below the usual rainwater concentration (0.7 mg N/L), the effects of NO_3^--leaching are not of major importance. Other plant nutrients do not limit the growth of aquatic plants to any extent.

6.3.3.3 Fertilizers - inorganic and organic

The quality of plant products can be considerably affected by plant nutrition (see Chapter 5), and the question is often asked whether any major differences in plant quality occur between plants supplied with inorganic or organic fertilizers. It is often believed by those who ought to know better, that for some reason inorganic fertilizers induce all manner of ills to man and beast, and should on no account be applied to the soil. Several points need to be clarified.

It must first be remembered that even in organic fertilizers such as farmyard manure, slurries and green manure, most plant nutrients, including potassium magnesium and phosphate, are present in an inorganic form. Other nutrients, in particular nitrogen and sulphur, are converted to inorganic forms by soil microorganisms before the absorption by plant roots takes place. Thus, although plants may be supplied with organic fertilizers, they nevertheless take up inorganic nutrients derived from these organic materials. This is the basic reason why there are usually no major differences between a crop supplied with organic or with inorganic fertilizers. Inorganic and organic fertilizers do, however, differ in the availability of the plant nutrients they contain. Nutrients in inorganic fertilizers are directly available to plant roots, whereas the nutrients of organic materials and especially organic nitrogen are of low availability. Only about one-third of the N of farmyard manure applied to the soil is available to a crop in the first year (Cooke 1972). This relatively slow release of nitrogen by organic fertilizers in comparison with inorganic N fertilizers can have some advantage in relation to crop quality and nitrate concentrations in plant material (see Table 5.22). Similar effects, however, can be obtained by inorganic N fertilizers by using a timed or split application.

One drawback to the use of organic fertilizers (manure, green manure) is their dependence on environmental factors for the mineralization of organic nitrogen by soil microorganisms (see page 409) whose metabolic activities are much dependent on soil temperature and soil moisture (Honeycutt 1991). When manure is ploughed into a soil in a dry spring scarcely any decomposition occurs. Very little nitrogen is thus available for the crop when it is most needed. If a wet summer or autumn follows this dry period, mineralization of organic nitrogen takes place very rapidly and provides nitrogen when it is no longer required by the crop. This may in part be leached and thus contribute to pollution (Kücke and Kleeberg 1997). In the case of sugar beet cultivation, a high rate of release of nitrogen from organic fertilizers late in the season may badly impair crop quality (see page 320). Temperli et al. (1982) tested the effect of mineral nitrogen fertilizer application as compared with organic N application, mainly in the form of farmyard manure, on the nitrate concentration in lettuce. The experiment comprised 7 farms in which N was applied as mineral N fertilizer and 7 farms in which organic N application mainly as farmyard manure was the N source (see Table 5.22 on page 331).

The lettuce was grown in the field, with exception of the lettuce cultivated in November which was grown in the greenhouse. The low light intensity prevailing here resulted in very high nitrate concentrations which did not differ between the plants grown with mineral or organic nitrogen. Lettuce grown in the field, however, showed significantly higher nitrate concentration when grown with mineral nitrogen as compared with organic nitrogen. This is mainly due to the lower quantities of nitrogen applied in organic farming systems compared with mineral N farming systems. This lower N rates used in organic farming was also reflected by lower yields, which in the present example was evident from the smaller lettuce heads harvested with organic farming. In farms growing vegetables soil nitrate concentration is frequently high (Wehrmann and Scharpf 1986). In such farming systems it is essential to analyse the soil for available nitrate and adjust nitrogen fertilizer rates to the level of available nitrogen already present in the soil. By this strategy excessive nitrate concentrations in vegetables can be avoided.

Critical nitrate concentrations in vegetables for humans are in the range of 400 to 500 mg nitrate-N/kg fresh matter and, as shown in Table 5.22, nitrate concentrations of lettuce grown in the field with mineral nitrogen are still tolerable while nitrate concentrations of lettuce grown under low light intensity in the greenhouse were too high regardless of whether grown with organic or inorganic nitrogen. Nitrate concentration is not uniformly distributed in plant organs. It is generally high in xylem containing tissues, because the nitrate concentration in the xylem sap is in an order of 20 to 30 mol/m^3 nitrate (Marschner et al. 1997). Pimpini et al. (1970) reported that the nitrate concentration in the ribs of cauliflower was several times higher than in the "flower" (see Figure 5.37 on page 330).

The favourable effect of farmyard manure on soil structure, soil water holding capacity and humus has already been considered on page 35. The humus content of soil appears to be much more dependent on particular cropping systems and crop rotation rather than the application of organic fertilizers (Bruin and Grootenhuis 1968). Cooke (1977) in discussing this question gives the following preferential sequence of crops and crop management treatments for increasing or maintaining the C content of soils: Clover/grass + FYM > Clover/grass grazing > Lucerne > Arable crops. It should be emphazised that inorganic fertilizers also increase the humus content of soils. Früchtenicht et al. (1978) reported that the application of mineral fertilizers over a period of 100 years deepened the humus enriched top layer of arable soils by a factor of 3 to 4. Odell et al. (1984) in evaluating a 100 years old field experiment in the USA (Morrow Plots, Illinois) found that organic N and organic C in soils had been increased significantly by liming and the application of mineral fertilizer without the application of farmyard manure. Cooke (1974) in referring to Rothamsted field experiments reported that "no damage to productivity need result from continuous application of large amounts of fertilizers". Some results of these experiments are shown in Table 6.24. The soil of this Broadbalk field from which the data were obtained is a silty clay loam which for the past 140 years has received annually the various treatments shown. Highest yields were obtained with the farmyard manure application, the NPK application and the farmyard manure + N application. Farmyard manure can thus to some extent replace the use of inorganic fertilizers. In practical farming terms, however, the complete substitution of inorganic fertilizers by farmyard manure is not a feasible proposition as not enough manure is generally available to maintain the nutrient balance. In addition farming without mineral fertilizers may result not only in yield depressions but also affect animal fertility. This has been shown by the Haughley Experiment with dairy cows carried out in East Anglia in England (Altherr 1972, Balfour 1975). In organic farming the nutrient cycle is not closed (Nolte and Werner 1994) so that in the long run nutrient losses must be made up in order to maintain soil fertility.

Table 6.24 Yield of wheat and potatoes grown in rotation on Broadbalk field (Rothamsted England) (Cooke 1974)

Treatment	1973	1972	1971	1970
	Wheat, Mg/ha of grains			
None	2.4	3.4	2.5	2.3
PK	3.1	4.2	2.6	2.5
PK + 144 kg mineral N/ha	3.9*	6.5	6.0	4.9
Farmyard manure	5.7	8.0	6.9	5.9
Farmyard manure + 96 kg/ha miner.N	4.3	6.9	4.9	5.6
	Potatoes, Mg/ha of tubers			
None	13.5	10.8	7.8	12.6
PK	21.7	16.2	9.6	19.1
PK + 192 kg/ha mineral N	42.9	38.8	46.6	41.8
Farmyard manure	47.1	40.2	36.2	43.8
Farmyard manure + 96 kg/ha	48.8	41.4	49.4	49.1

* Seriously lodged in 1973

These data show that highest yields were obtained with farmyard manure but it must be emphasized that for the production of such high rates of manure at least two " animal manure units" (see page 356) per ha are required. If all farmers were to practice such a system an enormous surplus of animal products (milk, meat, eggs) would be produced. The nitrogen exported from these farms and also the nitrogen leached from the soils would have to be replaced. This may be possible by growing forage legumes or by the purchase protein concentrates. In the case of legumes a substantial proportion of the farm land, about 30%, would to be cultivated with legumes and considerable amounts of lime would have to be applied to prevent the increasing soil acidity associated with biological N_2 fixation (see page 60). Finally it should be stressed that animal protein production is not very efficient. Proteins present in forage, seed, and grains are converted to animal proteins used for human nutrition with efficiencies ranging between 6 and 30% (Cooke 1975). Animal proteins in as bones, skin and hair are of no significance for human nutrition and frequently end up as an undesired waste product. A substantial part of protein nitrogen eaten by animals is excreted in the form of urea or uric acid and mostly is not completely recycled back as nitrogen fertilizer but leached as nitrate into the ground water or released in the form NH_3 into the atmosphere. From an ecological point of view it is more reasonable to produce plant proteins for human nutrition, such as high quality soya bean proteins. According to Mengel (1992) the production of one unit pork protein requires about 6 times more nitrogen than 1 unit soya bean protein.

6.3.3.4 *Greenhouse gases and the destrution of the stratospheric ozone*

Methane is a greenhouse gas. Its radiative absorption potential is 32 times higher than that of CO_2. For several decades its atmospheric concentration has increased steadily at a rate of 1% per year (Goulding *et al.* 1995). Sources of methane are of natural origin such as wetland or of anthropogenic origin such as mining of fossil energy and petrochemical industries. Methane is the end product of the microbial decomposition of organic carbon under anaerobic conditions. Therefore the main methane sources related to agriculture are flooded rice soils and enteric fermentation by ruminants with domestic animals being the most important source (Hütsch 1998a). Methane is produced and released by slurry stores and according to Sommer *et al.* (1996) also from non ruminant slurry, such as pig slurry where it may be produced when the slurry is applied to the soil. According to Boeckx and van Cleemput (1996) landfills can be a substantial source of methane and may still increase in importance. Covering landfills with a soil layer is a cheap and efficient option to reduce or even prevent methane release since soil bacteria oxidize methane to CO_2 under anaerobic soil conditions (Whalen *et al.* 1990). Thus soils can act as a biofilter for methane which is produced in deeper horizons under anaerobic conditions and oxidized in the upper soil layer before it reaches the atmosphere. According to Ambus and Christensen (1995) uptake and emision of methane from soils reached maximum rates when soils dried up, presumably because of better diffusion conditions in the soil medium.

The methane oxidation potential of aerobic soil microorganisms (methanotrophs) has a substantial impact on global methane turnover and is also influenced by vegetation, soil properties and fertilizer application. The rate of methane decomposition for various ecosystems decreases according to the following sequence (Goulding *et al.* 1995):

Forest soil > Cut grassland > Grazed grassland > Arable land

Hütsch *et al.* (1993) found that long term nitrogen fertilizer application had a significant impact on the methane oxidation potential of soils. This effect is depicted in Figure 6.18 in which the methane oxidation rates of various long term N fertilizer treatments of the Broadbalk field trial (Rothamsted) are shown. It is evident that increasing mineral fertilizer rates depressed the methane oxidation potential of soils and farmyard manure application had only a minor depressive effect. The reason for this finding is not yet understood. It is supposed that the continuous application of mineral fertilizer restricts the growth of methane oxidizing bacteria and that soils have only a limited capacity of methanotrophs (methane oxidizing bacteria) to grow *in situ* (King and Schnell 1994). Short term nitrogen fertilizer application has produced evidence that ammonium has a strong depressive effect on methane oxidation while nitrate is virtually without influence (Hütsch *et al.* 1994).

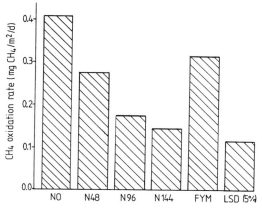

Figure 6.18 Effect of long term nitrogen fertilizer application on the methane oxidation potential of soils of the Broadbalk (Rothamsted) long term field experiment. N0 no fertilizer, N48, N96, N114 = 48, 96 144 kg mineral N/ha/year repectively. FYM = farmyard manure (after Hütsch *et al.* 1993).

This specific short term effect of ammonium on methane oxidation is related to methane monooxygenase which may oxidize methane and also ammonia. The reaction requires a pyrrolo-quinoline quinone as shown in the reaction sequence in Figure 6.19.

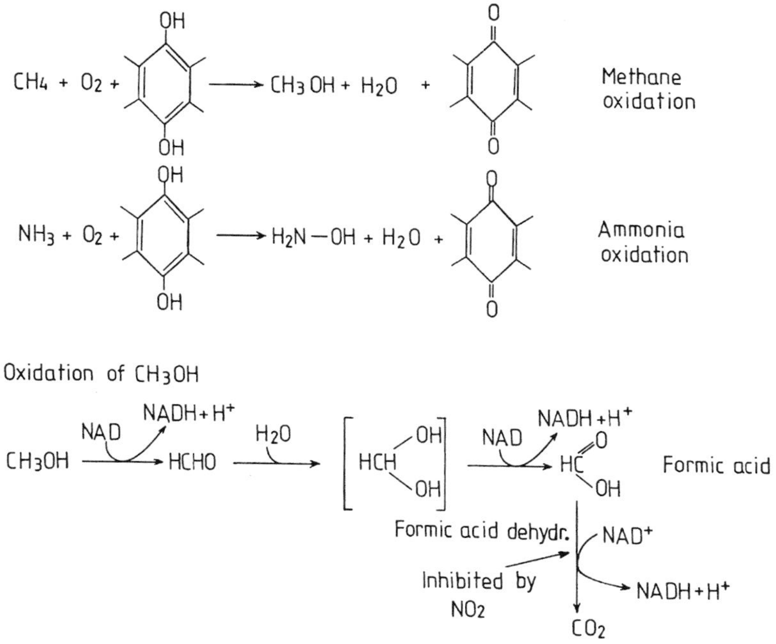

Figure 6.19 Reaction sequence of methane oxidation in soils with methane monoxygenase as the starting enzyme which may also oxidize ammonium (after King and Schnell 1994).

Ammonium and methane may thus compete for the same enzyme. In addition NADH produced in the process of methane oxidation to CO_2 inhibits methane oxidation to CO_2 (King and Schnell 1994). As can be seen in the Figure 6.19, NADH is produced in the oxidation of hydroxymethyl, of formaldehyde, and of formic acid.

Nitrous oxide (NO_2) is also a greenhouse gas and absorbs light in the infrared region. In addition it too is involved in the destruction of the ozone in the stratosphere. Nitrous oxide has a long lifespan as is shown in Table 6.25.

Table 6.25 Concentration and lifetime of some atmospheric gases related to the greenhouse effect and to the destruction of ozone in the stratosphere (Schönwiese 1988).

Gas	Current concentration	Lifetime
CO_2	347 mL/m^3	5—10 years
Ozone	30 μL/m^3	30—90 days
Chlorofluoro methane	0.4 μL/m^3	50—100 years
N_2O	0.3 μL/m^3	20—100 years
Methane	1.65 mL/m^3	4—7 years
NH_3	<1 mL/m^3	7—14 days

The main natural sources for N_2O production are:

Forest and other land > Wetland and waters > Agricultural land.

Main anthropogenic sources are:

Combustion and traffic > Animal manure > N fertilizer (Granli and Bockman 1994).

Nitrous oxide is an intermediate product in denitrification (see page 350) and in nitrification. In the latter process NOH is formed which may react to form N_2O according to the following reaction:

$$NOH + NOH \Rightarrow N_2O + H_2O$$

Global mineral N fertilizer consumption has increased from 32 Tg N/year to 77 Tg N/year from 1970 to the present time (Granli and Bockman 1994). Nitrification of ammonium and denitrification of nitrate in agricultural soils have also increased on a global scale and with this increase so too the release rates of N_2O from soils into the atmosphere. Farmyard manure application provides organic carbon and mineralizable nitrogen for soil bacteria and therefore favourable conditions for nitrification and denitrification which may lead to substantial rates of release of nitrous oxide from soils as found by Lessard et al. (1996). Both nitrification and denitrification rates in soils depend much on soil moisture. In very dry soils both processes are blocked; but with an increase in soil moisture nitrification associated with the formation of N_2O increases. This relationship is shown in Figure 6.20 (Davidson 1991). Maximum nitrification is obtained at about field capacity. At this soil moisture nitrification and denitrification overlap. Water filled pores beyond field capacity considerably promote denitrification and particularly the formation of N_2. Under waterlogging conditions the functioning of oxygen as e^- acceptor (see page 48) is at a minimum so that the N_2O produced is used as an e^- acceptor and thus the harmless N_2 is formed.

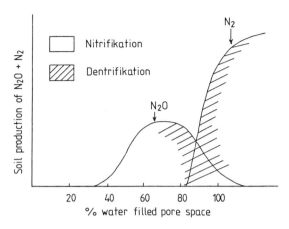

Figure 6.20 Relation between the proportion (%) of water filled pores to nitrification and denitrification the latter evidenced by the relase of N_2 and N_2O (Davidson 1991).

The N_2O/N_2 ratio of the two gases released by denitrification increases with increasing nitrate supply in the soil (Moisier *et al.* 1986). When nitrate levels are high in the soil relatively much oxygen in the nitrate is available and therefore the O in the N_2O is required less as an e^- acceptor for bacterial respiration resulting in high N_2O/N_2 ratios.

Nitrous oxide is not directly involved in the destruction of ozone but it can react with an O-atom (singlet oxygen) to produce NO according to the following equation:

$$N_2O + O \Rightarrow 2\ NO$$

This reaction product, NO, functions as catalysator in the destruction of ozone as shown below :

$$2\ O_3 + N + NO + hv \rightarrow 3\ O_2 + NO + N$$

Phosphate is a limiting factor in the growth of algae in lakes and streams, and increasing concentrations of phosphate in the water often run parallel with the degree of eutrophication. The aquatic environment is highly sensitive to very small changes in P supply and the eutrophic threshold value above which algal growth takes place is considered to be in the region of 10 μg P/L. For drinking purposes this concentration of P is allowable and very much greater with an EU guideline of 0.3 mg total P/L. The greatest sources of P pollution are detergents and urban wastes. In recent years, however, there has been increasing concern in the role of agriculture as a source of P transfer to the water. Since industrial and municipal discharges are progressively being controlled.

Vertical movement of P through the soil profile has generally been considered of little importance for most soils because of their high P fixation capacity (Cooke and Williams 1970). Values given for leaching are in the order of about 0.5 kg P/ha/annum. However, leaching rates can be much higher particularly in sandy soils which have become P saturated and which were treated with heavy application of animal manure (van Riemsdijk *et al.* 1987). There is evidence that in such acid soils P can be leached as humate-Fe-Al-P complexes (see page 456) Gerke (1994). Heckrath *et al.* (1995) reported the results of an experiment in which P was determined in the drainage water of a long-term field experiment on a heavy soil, a silty clay loam at Rothamsted, England, in which plots received no P, P in farmyard manure (about 40 kg P/ha or superphosphate (up to 35 kg P/ha) annually for over 150 years. The total P concentration in the drainage water was considered in relation to available P concentrations in the plow layer using the Olsen method (0,5 M $NaHCO_3$ extractable P). The results shown in Figure 9.8 for Olsen P values from $10-45$ mg/kg cover a range from below to excess of adequate recommended values for maximum crop yields (see page 473). In agronomical terms total P levels in the drainage water were very low. At higher levels of Olsen P, well in excess of recommended values, enhanced P losses occurred through surface runoff even on this heavy soil. These results draw attention to enhanced P, leaching through the soil which can arise on soils over fertilized with P as well as to the need to make P fertilizer application to soils in relation to P availabilty measurements in the soil.

There is intense interest in subject of phosphate pollution not only from the viewpoint of soils but also in relation to agriculture as a whole including animal production. Most valuable reviews have been produced by Foy and Withers (1995), SCOPE (Scientific Committee on Phosphates in Europe) 1997, and Tunney *et al.* (1997).

CHAPTER 7

NITROGEN

7.1 Nitrogen in the Soil and its Availability

7.1.1 General aspects

Nitrogen is one of the most widely distributed elements in nature. It is present in the atmosphere, the lithosphere, and the hydrosphere. As can be seen from Table 7.1 the atmosphere is the main reservoir for N (Delwiche 1983).

Table 7.1   Nitrogen quantities in various spheres (after Delwiche 1983)

	g atom N $\times$ 10^{12}
Atmosphere, N_2	2.8×10^8
Atmosphere, N_2O	1.3×10^2
Lithosphere, organic N	5.7×10^7
Lithosphere, inorganic N	1.4×10^7
Hydrosphere, organic N	2.4×10^4
Hydrosphere, inorganic N	7.1×10^3
Soils, organic N	1.25×10^4
Soils, inorganic N	1.15×10^4
Terrestrial plants	5.7×10^2
Terrestrial animals	1.5×10
Marine plants	1.4×10
Marine animals	1.4×10

The soil accounts for only a minute fraction of lithospheric N, and of this soil N, only a very small proportion is directly available to plants. This occurs mainly in the form of NO_3^- or NH_4^+. Nitrogen is a very mobile element circulating between the atmosphere, the soil and living organisms. Many factors and processes are involved in this N-turnover, some of which are physico-chemical, and others biological. The main outlines of the N cycle in nature are given in Figure 7.1.

7.1.2 Biological nitrogen fixation

The most important process by which the relatively stable atmospheric dinitrogen (N_2) is fixed and converted to NH_3 is termed *nitrogen fixation*. This is carried out by a number of different prokaryotic species. The biochemistry and importance of this process have

397

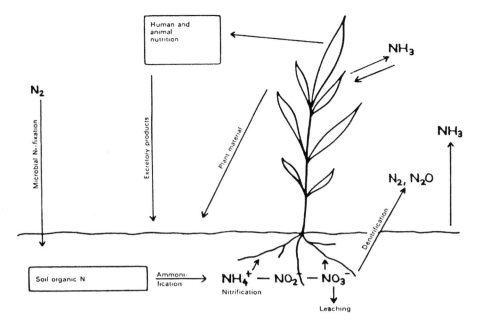

Figure 7.1 Nitrogen cycle in nature.

already been discussed on page 169. Nitrogen may also be fixed chemically by the Haber-Bosch Process in which N_2 and H_2 react together under high temperature and pressure conditions to yield ammonia.

$$N_2 + 3H_2 \Rightarrow 2NH_3$$

This is the basis of ammonium fertilizer production. The quantity of technical N_2 fixation currently amounts to 73×10^6 Mg per year (Brändlein 1987) which is substantial as compared with the biological nitrogen fixation which is 118×10^6 Mg/year (Delwiche 1983). Hence of the total of biologically and technically fixed nitrogen almost 40% can be attributed to the technical process. Most of the technically fixed nitrogen is used for the production of fertilizer and is therefore appplied to soils and thus affects the global nitrogen cycle (Mengel 1992).

Some atmospheric N_2 is fixed by electrical discharge (lightning) which results in the oxidation of N_2. The quantities gained in this way, however, are rather small and amount to only a few kg nitrate-N/ha/year under temperate climatic conditions. In the tropics the amount is higher although not usually in excess of about 10 kg nitrate-N/ha/year. Nitrogen oxides produced in the burning process of petrol of motor engines may also react to form nitric acid in the atmosphere and in soils. In areas associated with dense traffic this form of N may contribute substantially to the input of fixed N into soils and waters and constitute an environmental hazard (Isermann 1970).

The amount of N_2 fixed by bacteria may differ considerably from one site to another. This depends much on the presence of N_2 fixing microorganisms and further factors such as soil properties and climate and vegetation cover of soils. Table 7.2 shows the rates at which N is gained by biological fixation under different ecosystems (Hauck 1971). Fixation is rather low on arable land, but for pastures, forests and to a lesser extent for paddy, biological N_2 fixation provides an important source of N. Nitrogen fixation in rivers and lakes also appears to be rather high.

Table 7.2   Nitrogen gains from biological N_2 fixation (Hauck1971)

Ecosystem	Range in reported values kg N/ha/year
Arable land	7—28
Pasture (non-legume).	7—114
Pasture (grass-legume)	73—865
Forest	58—594
Paddy	13—99
Waters	70—250

Only prokaryotes are capable of fixing (assimilating) dinitrogen. Eleven of the 47 bacteria families and six of the eight Cyanophyceae families are able to fix N_2 (Werner 1980). Some of these species are free living N_2 fixers; others live symbiotically. The most important free living N_2 fixing bacteria are members of the genera *Azotobacter, Beijerinckia, Spirillum*, and *Enterobacter;* important N_2 fixers of the Cyanophyceae belong to the genera *Nostoc* and *Anabaena*. The extent to which these free-living bacteria are able to fix nitrogen depends much on their supply of organic carbon such as carbohydrates and lipids. It is for this reason that most N_2 fixing bacteria live in the rhizosphere or at the surface of plant roots or even in the intercellular spaces of roots, stems, and leaves because the direct contact with higher plants favours the supply of organic carbon. This physical and physiological direct contact between the bacteria and host plant is called "association". In this respect endobacteria are of particular interest. These microorganisms live in the intercellullar spaces of cell walls and have direct access to carbohydrates released into these spaces by adjacent cells. As found in recent investigations, a wide range of different N_2 fixing bacteria are present in the roots stems and leaves of sugar cane (Boddey *et al.* (1991) belonging to the genera *Erwinia, Azotobacter, Derixia, Azospirillum* and *Enterobacter. Acetobacter diazotrophicus*, found in large numbers in roots and stems of sugar cane, in particular merits attention (Calvacante and Döbereiner 1988) since it tolerates pH levels as low as 3 and its nitrogenase activity is hardly affected by ammonium or nitrate. The finding that sugar cane and other tropical C_4 species may grow well without N fertilizer is explained by the close association between the N_2 fixing bacteria and the C_4 species (Döbereiner 1997). Species such as maize, sorghum, and other tropical grasses are mainly infected by *A. lipoferum*, whereas C_3 species such as wheat, barley, oats, rye, and rice are mainly colonized by *A. brasilense*.

Boddey *et al.* (1991) reported that of 135 field trials carried out all over Brazil with sugar cane, in which N fertilizer was applied, only 19% of the trials showed significant N response, although N fertilizer rates were much lower than the N uptake of the crop. These results show that sugar cane has evolved an efficient association with N_2 fixing bacteria capable of fixing large amounts of N in the range of 100 kg N/ha and more. The N_2 fixing performance of free living bacteria in plant associations under temperate climatic conditions are much lower and in the range of 5 to 15 kg N/ha.

Inoculation of crops with *Azospirillum* species may improve the N nutrition of the host plant and nitrogen gains in the order of 20 to 80 kg N/ha may thus be obtained (Döbereiner 1983). As yet, however, inoculation with N_2 fixing bacteria are not always successful. Millet *et al.* (1984) for example reported that the inoculation of wheat with *A. brasilense* resulted in only a yield increase of 8%. The problem with inoculation is that the association between host plant and bacteria may be very specific. In this respect Döbereiner(1983) reported in a useful review paper that of 31 cultivars or ecotypes of *Paspalum notatum*, a tropical grass, only one cultivar was asscociated with the N_2 fixing bacterium *Azotobacter paspali*.

The N_2 fixing efficiency of these bacteria depends much on environmental conditions. Most of the bacteria discussed above are sensitive to low pH, to high O_2 concentrations and to inorganic soil nitrogen (ammonium, nitrate). Dinitrogen fixing bacteria need also to be free from enzymes which catalyze denitrification. Maximum N_2 fixing rates have been obtained at high soil temperatures (33°C). The potential N_2 fixing capacity of free living bacteria is thus highest in subtropical and tropical regions. Cyanophyceae species play a major role in paddy soils. The nitrogen fixing blue-green algae *Anabaena azollae* lives in association with the aquatic fern *Azolla pinniata*. Watanabe *et al.* (1977) reported that by cultivation of this *Azolla-Anabaena* association in paddy, considerable amounts of N_2 can be fixed and used by the rice crop. In some cases where fertilizer N was not applied, up to 30 kg N/ha were estimated to be fixed by blue-green algae in one harvest. The same workers were also able to obtain 22 harvests of *Azolla* with a total amount of 465 kg N/ha fixed per year (Watanabe *et al.* 1980).

As the level of soluble carbohydrates may often limit N_2 fixation in free living bacteria it seems possible that this may account for the evolution of some species of microorganisms which live symbiontically with higher plants. The most important symbiotic relationships in which NH_3 is provided by the bacterium and organic carbon by the host plant are the symbiosis with *Rhizobium*, *Bradyrhizobium* and *Actinomyces* species. *Rhizobium* and *Bradyrhizobium* live in symbiosis with leguminous species. Frankia species, formerly called *Actinomyces* live in symbiosis with numerous tree and shrub species. All species develop root nodules, those formed by *Actinomyces* are called actinorhizas (Quispel 1983). The best known such root nodules are those of alder, *e.g. Alnus glutinosa*. Actinorhizas are found in species of the order of Frankia and occur in more than 140 plant species belonging to 8 families which are not closely related to each other taxonomically (Betulaccae, Casuarinaccae, Myriaceae, Rosaceae, Elagneaceae, Rhamnaceae, Coriariaceae, Datiscaccae). Valuable articles on actinorhizal symbiosis have been published in the *Proceedings of the International Symposium on Frankia and Actinorhizal Plants* (Plant and Soil, Vol 87, 1985). The nitrogen fixing capacity of this kind of symbiosis may be considerable and plays a particular role in Agroforestry

(Franco and de Faria 1997). *Actinomyces alni* living in symbiosis with alder (*Alnus rugosa*) may fix about 150 kg N ha/per year. Such high rates of N_2 fixation can obviously have an important influence on the nitrogen economy of woodland soils as also shown in Table 7.2 for the forest ecosystem.

In agriculture *Rhizobium* and *Bradyrhizobium* species living symbiotically with legumes are the most important N_2 fixers. According to Polhill (1981) there are about 19,700 known legume species of which about 80% grow symbiotically with *Rhizobium* or *Bradyrhizobium*. About 200 of these legume species are used as crop plants. According to a more recent classification one should differentiate between *Rhizobium* and *Bradyrhizobium*. Both belong to the Rhizobiaceae family but are distinguished by their genetic properties and host specifity. In this respect *Bradyrhizobium* species are not so selective as *Rhizobiumm* species (Werner 1987). Pigeon pea (*Cajanus*) may be infected by *Rhizobium* as well as by *Bradyrhizobium*. These authors report that *Bradyrhizobium* yielded a much higher N_2 fixation rate than *Rhizobium* although *Rhizobium* grew faster than *Bradyrhizobium*. This difference is explained by a higher activity of the tricarboxylic acid cycle enzymes in *Bradyrhizobium* as compared with *Rhizobium*. *Bradyrhizobium* is able to store appreciable amounts of a heteropolysaccharides in the nodules (Streeter and Salminen 1993). Important *Rhizobium* species together with their host plants are listed in Table 7.3.

Table 7.3   Rhizobium species and their most important host plants. (After Werner 1987)

Bacteria species	Host species
Rhizobium leguminosarum	*Pisum sativum, Vicia hirsuta*
Rhizobium meliloti	*Medicago sativa*
Rhizobium loti	*Lotus corniculatus > Medicago sativa, Phaseolus vulgaris*
Rhizobium fredii	*Glycine soja* (wild type of soya) *> Glycine max*

> potential for nodule formation is lower.

Efficient N_2 fixation requires that the host plant has to be infected by the appropriate bacteria species. Infection of a host plant followed by nodulation with *Rhizobium* or *Bradyrhizobium* bacteria are compex processes. Infection begins by contact of the bacteria feeding on the root mucilage. Only a very small segment of the root which is below the root hair zone and above the calyptra is accessible to infection at sites where root hairs are formed. Recognition between bacterium and the root is supposedly achieved by the reactions of lectins (Long and Ehrhardt 1989), which are sugar binding proteins present in the cell wall of the root cell, with a lipochito-oligosaccharide located in the cell wall of the bacterium (Mylona *et al.* 1996). This recognition mechanism is very specific and is related to a fatty acyl bound to the non-reducing end of the oligosaccharide. In addition, the length of the oligosaccharide chain plays a role in the recognition. An infection thread is then formed by invagination of the root hair and it is through this thread that the bacteria migrate into the root cortex. At the outset the bacteria are enclosed by cell wall material of the infection thread. As time proceeds, however, the cell wall material is decomposed allowing the bacteria to enter cortex cells by Endocytosis. These infected cells grow, their organelles such as mitochondria, ribosomes, and ER multiply and the starch content of infected host cells decreases (Bosabalidis and Tsaftaris 1991). The bacterium is embedded in a vacuole of which the outer membrane, the peribacteroid membrane, is formed by the plasmalemma of the host cell. The peribacteroid membrane is the interface between the bacteroid and the host plant and plays an important role in controlling the exchange of metabolites. The bacterium itself is also enclosed within a membrane envelope. The space between both membranes is called intersymbiontic space which contains leghaemoglobin functioning as an O_2 carrier and O_2 buffer. The bacterium develops to a bacteroid which is much larger than the bacterium and its content in DNA is several fold greater than that of the bacterium. In Plate 7.1 the host cell of a soya root nodule is shown filled with numerous bacteroids which appear black in contrast to the infection vacuole which appears white. As can be seen, the infected host cell has a high turgor as compared with non infected neighbouring cells. The bacteroid is an organelle comparable with mitochondria or chloroplasts. The most important characteristic of the bacteroid is that it contains nitrogenase, the enzyme which brings about the assimilation of molecular nitrogen (see page 169).

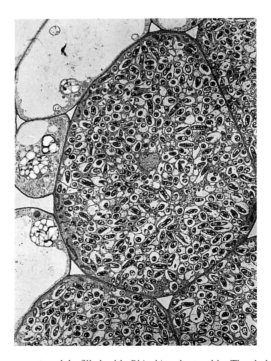

Plate 7.1 Cell of a soya root nodule filled with *Rhizobium* bacteroids. The dark centre of the organelle is the bacteroid imbedded in a small vacuole. The infected cells have a higher turgor than the non-infected ones (Photo Werner).

The host plant defends itself against the infecting bacteria and only about 1/10 of infections are successful leading to the formation of bacteroids. Not all cells of the cortex are infected. Thus cells enriched with bacteroids and others devoid of bacteroids occur simultaneously. Useful reviews on infection and nodulation have been published by Mylona *et al.* (1996).

The infection is followed by nodulation which is associated with the synthesis of nodulins which are enzyme proteins required to metabolize the fixed nitrogen (NH_3). These proteins also include leghaemoglobin which facilitates the O_2 transport to bacteroids in the infection vacuole. Interestingly the genetic code of the nodulins is located in the nucleus of the host plant while the signals for their expression come from the bacteria (Govers and Bisseling 1992). This shows the high degree of symbiosis between the higher plant and the bacteria. The upper part of Plate 7.2 shown nodulated roots of *Phaseolus* and *Orithopus* and the lower part shows a cross section of a root nodule of *Alnus glutinosa*. Bacteroids embedded in an infection vacuole can also be seen. Legume nodules are characterized by five different zones: the nodule meristem, followed by the prefixation zone, the inter zone, the nitrogen fixation zone, and the senescence zone (Mylona *et al.* 1996).

Of the legumes there are three genera which do not produce root nodules but stem nodules (see Plate 7.3). These genera are *Aeschynomene*, *Sesbania*, and *Neptunia* (Alazard and Duhoux 1987). *Aeschynomene afraspera* is an annual plant widely distributed in Africa on sites which are flooded during certain periods in the year. The plant has a height of 1.0 to 1.5 m and the stem inserted nodules still maintain the ability to fix dinitrogen when the lower part of the plant is flooded. Dinitrogen fixation rates measured using the acetylene test gave N_2 fixation rates for *Aeschynomene afraspera* which were higher than those for *Vicia*, *Vigna* and *Lupinus* (Alazard and Duhoux 1987). These stem nodulated legumes have a high potential as N suppliers of other crops as reported by Becker *et al.* (1995). The quantity of N_2 assimilated by *Rhizobium* and *Bradyrhizobium* bacteria depends to a large extent on nutritional conditions. Generally if the plant is infected by an efficient bacterial strain, growth conditions for the plant control the performance of the N_2 fixing bacteria. This means an ample supply of plant nutrients other than N and in particular with phosphate and K^+. Optimum pH and adequate water supply and photosynthetic energy are also required to give high amounts fixed N per unit area (Mengel 1994b). Many *Rhizobium* and *Bradyrhizobium* species are sensitive to low pH conditions in the soil medium. Schubert *et al.* (1990) reported that low soil pH restricted the multiplication of *Rhizobium* and led to a retardation of root nodule development and N_2 fixation. The depressing effect of low pH on nitrogenase activity, shown in Figure 7.2, was particularly great at pH 4.7 and still obvious at pH 5.4. Root nodule development (Scherer and Danzeisen 1980) and nitrogenase activity are also depressed by mineral N supply. The mechanism of the inhibition is not yet understood (Vessey and Waterer 1992). Streeter (1985) found that nitrate nutrition of soya led to an elevated level of ureides in the nodules and this author suggests that ureides may inhibit nitrogenase. Other reduced N compounds may have a similar effect. Mild water stress also depresses nitrogenase activity of *Vicia faba* but enough reduced N is still produced for the host plant (Plies-Balzer *et al.* 1995). A similar observation was made by Schubert *et al.* (1995) with water stressed *Medicago sativa*.

404

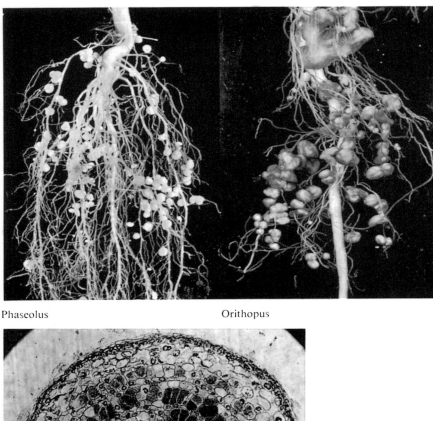

Phaseolus   Orithopus

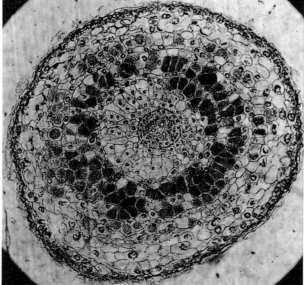

Cross section nodule
of alder

Plate 7.2 Nodulated roots of *Phaseolus vulgaris* (left) and *Orithopus sativa* (right). Below: transverse section through a young root nodule of *Alnus glutinosa*. (Photo Becking) Courtesy of Bayrische Landesanstalt für Acker- und Pflanzenbau.

Plate 7.3 Stem of *Sesbania rostrata* covered by numerous stem nodules (Photo; Mengel).

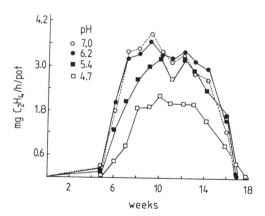

Figure 7.2 Nitrogenase activities throughout the growth period expressed as ethylene production by root nodules of *Vicia faba* in relation to soil pH (after Schubert *et al.* 1990).

These authors also suggest that the depression in nitrogenase activity may be dependent on a feedback control mechanism regulated by high concentrations of amino acids present in the nodules as a consequence of increased cycling of reduced N from the shoot because of lower shoot demand. Severe water stress of the host plants (*Vicia faba*) with leaf water potentials of about -1.4 MPa leads to a deformation of nodules and to a decrease of the leghaemoglobin concentration and N_2 reduction potential (Guerin *et al.* 1990).

Nodules are rich in soluble amino acids as compared with other plant organs. The supply of amino acids from the nodules closely relates to the life cycle of the host plant. In the first few days after infection of the roots of young legume plants, the bacteria is completely dependent on the host plant, and the amino acids synthesized are used for the growth of the bacteroids. In the later stages, however, most of the amino acids synthesized are transported to the host plant. For most leguminous species this transfer comes to an end rather abruptly on the termination of the flowering stage. Root nodules thus supply the plant directly with amino acids during the vegetative growing stage. The intensity of this supply depends to a large extent on the rate of photosynthesis and on the supply of root nodules with carbohydrates from the host plant.

7.1.3 The agronomic importance of biological dinitrogen fixation

Cultivation of leguminous crops for human and animal nutrition contributed much to the agricultural development in Central Europe in the 19th century (Kühbauch 1993). The growth of clover in particular as well as lupins on sandy soils cultivated with appropriate applications of lime, potassium, and phosphate as promoted by Liebig (1865) lead to an increase in biological N_2 fixation and was a new powerful N input into agricultural systems. Yields increased considerably, more forage and food was produced, more organic residues remained on the field and higher quantities of manure were were returned to the soil. By this means the levels of soil organic matter and available nutrients in soils were raised with an enormous benefit to the welfare of farmers (Schmitt 1958). In the 2. half of the 19. century biological N_2 fixation was the main N fertilizer source for a technical method of fixing of atmospheric nitrogen was still to be invented. The discovery by Hermann Hellriegel (Böhm 1986) that root nodules of legumes contain bacteria capable of fixing atmospheric nitrogen greatly promoted the cultivation of leguminous species.

The use of fertilizers in developing countries at the present time is in some aspects like that of Europe in the mid of the 19th century with various nutrients often inadequately supplied and correspondingly low crop yields. According to Franco and de Faria (1997) of the land cultivated for crops growth is limited by insufficient nutrient supply: on 80% of the land by phosphate, 35% by Mo and 5-10% by and S, K, Zn. In most tropical soils nitrogen is also insufficient for satisfactory crop yields. The restricted use of chemical fertilizers, especially N, in many developing countries results from their high costs, unavailability, and poor infrastructure for marketing (Dakora and Keya 1997). Biological N_2 fixation is therefore generally the cheapest and most efficient means of improving yields and soil fertility. Numerous leguminous crops such as grain legumes, fodder-pasture legumes, N_2 fixing shrubs and trees and tropical grasses with N_2 fixing endobacteria can be used as a nitrogen source for various cropping systems. All of these have a common requirement that other plant nutrients except N must be available in sufficient quantities.

Stem nodulating legume species growing in flooded soils are of particular interest for rice cultivation (Plate 7.3). Today 21 species of the genus *Aeschynomene*, three species of *Sesbania* and one species of *Neptunia* are known which form stem nodules (Ladha *et al.* 1992). Recent field trials of Becker *et al.* (1995) carried out in the International Rice Research Institute in the Philippines have shown that *Sesbania rostrata* is a suitable legume capable of providing a source of nitrogen for flooded rice equivalent to 90 kg N/ha in the form of urea. Satisfactory growth and high N_2 fixation rates of *Sesbania rostrata* were only obtained when phosphate was applied (Engels *et al.* 1995). Generally stem-nodulating legume species have a higher N_2 fixation potential than root-nodulating species. It appears that stem nodules are better supplied with organic carbon than root nodules since the green cortex of the nodules is photosynthetically active and contributes to the energy supply of the *Rhizobium* bacteroids. A valuable review article on stem nodulating legume species has been published by Ladha *et al.* (1992).

The amouts of N fixed by leguminous shrubs and trees is in the range of 100 to 600 kg N/ha/year (Dakora and Keya 1997). These plant species, mainly belonging to the genera *Accacia*, *Erythrina*, *Leucaena*, and *Mimosa* (Franco and de Faria 1997) may be used in agroforestry with alley cropping. The leaves are used as green manure while the wood produced may serve other purposes. According to Döbereiner (1997) it has recently been shown that roots, stems and leaves of palm trees are also colonized by N_2 fixing bacteria (*Herbaspirillum*, *Azospirillum*) raising the possibility that palm trees, too may be used as N source in agroforestry. Döbereiner (1997) reported that the central savannahs of Brasil, the Cerrados, were partly transformed into highly productive agricultural land by the cultivation of soybeans inoculated with *Rhizobium strains* selected for this region. Grasses may be colonized by various diazotrophs (N_2 fixing bacteria) such as *Herbaspirillum*, *Azospirillum*, *Burkholderia* species. These diazotrophics represent a potential source of endophytic bacteria which in future may play a possible part in supplying non leguminous species with N. The great benefit of the N_2 fixing endotrophic bacteria to the N economy of sugar cane is because the bacteria are mainly settled in the intracellular spaces and have direct access to photosynthates produced by the host plant (Boddey *et al.* 1991, Hecht-Buchholz 1998). This direct access is the reason why these bacteria are generally more efficient in N_2 fixation than are the N_2 fixing bacteria in the rhizosphere. A useful review article on the significance of the endophytic dinitrogen-fixing bacteria has been published by Hecht-Buchholz (1998).

Vigorous growth of legumes and the enrichment of soil organic matter by high biomass production transfer large amounts of phosphate from the inorganic to the organic pool and thus may render phosphate more available to crops and restrict phosphate fixation which is common on acid tropical soils (see page 471). In addition leguminous species such as white lupins (*Lupinus albus*) and red clover excrete citric acid into the rhizosphere and thus mobilize fixed phosphate (Gardner *et al.* 1983, Gerke *et al.* 1995). A negative aspect of legume cultivation, however, must be mentioned, namely the acidifying effect of legume roots. This occurs in plants fixing N_2 as a result of the net H^+ efflux from the roots (page 66). When little or no nitrate is taken up the H^+ pumped into the rhizosphere by the plasmalemma proton pump is not cycled back into the root cells via the H^+/nitrate cotransport (page 127) so that to a large extent the protons remain in the soil (Hauter and Steffens 1985). As a consequence of the proton pump and an absence of nitrate uptake N_2 fixing plants also accumulate substantial amounts of organic anions associated with high mineral cation uptake and concentration in the plant (Israel and Jackson 1978). Yan et al. (1996b) found that the quantity of H^+ excreted by roots of *Vicia faba* was approximately equivalent to the quantity of alkalinity in the upper plant parts. After incorporation of this plant material into the soil an equivalent rise of soil pH was found. Later soil pH declined as a result of nitrification. If the released nitrate is taken up by crops, however, it again raises soil pH because of H^+ cotransport during the uptake. This example shows that in legume cultivation there is no greater acidifying effect provided that the plant residues such as stems and leaves remain on the field. This is largely true for grain legumes but not for forage legumes. The latter thus generally have a stronger acidifying effect on the soil. Bromfield et al. (1983) for example in Australia reported marked long term effects on subterranean clover dominated pastures on soil pH depression and increase in exchangeable Mn leading to Mn toxicity. The physiologically alkaline effect of organic anions (see page 58) and nitrate can be affected by leaching since these anions may be leached out of the rooting depth of soils and their alkalizing effect on the top soil is lost. This particularly may be the case on poor soils in the tropics. It is for this reason that soil pH should be carefully controlled in legume cultivation. A valuable series of articles on the importance of biological dinitrogen fixation for sustainable agriculture in the tropics has been recently published in Vol. 28 No. 5/6 in the journal Soil. Biol. Biochem., 1997.

7.1.4   Ammonification and nitrification

Nearly all of the organic nitrogen present in soils originates from plants (Loll and Bollag 1983). Most of the soil nitrogen is very stable and resistant to biological decompostion. Of the various forms of organic nitrogen in soils, amino-N is the most important source for nitrogen mineralization. Amino-N is present in amino acids, peptides including the polypeptides and also in amino sugars, the latter mainly present in the cell walls of bacteria and fungi. In arable soils amino-N can make up about 30% of the total N present and may amount to 1700 to 2000 kg N/ha in the upper layer (10 cm) (Catroux and Schnitzer 1987). Of this huge amount only a small proportion is mineralized annually over a growth period. Mineralization which is brought about by microorganisms begins with the hydrolyzation of larger molecules into monomers such as polypeptides into amino acids and the polymers of microbial cell walls (muropeptides) into amino sugars. The hydrolysis of these N compounds which is catalyzed by many heterotrophic micoorganisms leads to the formation of NH_3 in the process of ammonification as shown in Figure 7.3

The microorganisms involved may use the monomers thus produced directly as a nutritional source for polymer synthesis such as cell wall material and enzymes.

When this occurs no mineral nitrogen is produced. Mineral nitrogen is only produced if the N needs of the microorganisms are satisfied in their requirement for organic carbon for synthetic or energetic purposes (Mengel 1996). Thus of the total N present in proteins and/or cell wall polypeptides of microorganisms, only a part is mineralized and this mineralized proportion depends on the C/N ratio of the organic matter which is subject to microbial decomposition. Generally the lower the C/N ratio the higher the proportion of N mineralized. This is illutrated by reference to Figure 7.4.

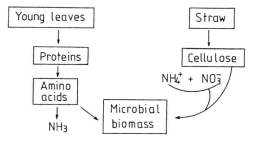

Figure 7.3 1. Hydrolyzation of peptide — 2. Oxidation of the resulting amino acid — 3. Deamination of the imino acid.

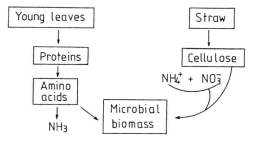

Figure 7.4 Mineralization of young leaves and straw. Microbial decomposition of leaves (narrow C/N ratio) produces NH_3, microbial decomposition of straw (wide C/N ratio) consumes mineral nitrogen required for the synthesis of microbial biomass.

In young leaves proteins make up 80% of the total N so that C/N ratio of the leaves is low. If the proteins in these leaves are broken down in the soil by peptidases excreted by microbes, the amino acids produced by hydrolyzation are partially used to cover energy demand and for this reason are deaminated. The resulting carboxylic acid is used as an energy source and the NH_3 is released into the soil medium. In the case of straw decomposition the reverse is the case and nitrogen is fixed rather than released. Cereal straw is rich in N-free organic carbon compounds particularly cellulose so that microbes breaking down the straw find a surplus of such compounds in relation to organic nitrogen. They therefore take up mineral nitrogen (ammonium, nitrate) from the immediate environment to assimilate these compounds, a process known as N immobilization (see page 430). The incorporation of straw into soils may therefore decrease the nitrate concentration in soils considerably (Gök and Ottow 1988) and thus also inorganic N flows into the microbial biomass.

Generally plant debris on and in soils is at first broken down by fungi which themselves are later decomposed by bacteria. These bacteria find a food which is enriched in organic N and fungal hyphae with a C/N ratio of about 10. In bacteria the C/N ratio may be as low as 3 to 5 (Jenkinson and Ladd 1981). The decomposition of microbial biomass therefore generally leads to the release of NH_3. Most heterotrophic soil mircoorganisms are able to produce NH_3. Important sources for NH_3 are peptides and muropeptides. Heterocyclic N as found in the N bases of nucleic acids is much more resistant to deamination (Mengel et al. 1999).

The biological oxidation of ammonia to nitrate is known as nitrification, a process which comprises various steps. Ammonia is first oxidized to NO_2^- then to hydroxylamine (NH_2OH) and finally to NO_3^-. The process is mediated by autotrophic bacteria which in contrast to heterotrophic bacteria obtain their energy from the oxidation of inorganic matter in the case here ammonia and nitrite. Two specialized groups of bacteria are involved, one in the oxidation of NH_4^+ to NO_2^- and the other in the oxidation of NO_2^- to NO_3^-. Several genera and species of ammonium and NO_2^- oxidizing autotrophs are known. Genera oxidizing ammonium include *Nitrosomonas*, *Nitrosolobus* and *Nitrosospira* (Paul and Clark 1996). All have been isolated from a variety of soils including soils from the long term field experiments in Rothamsted, England and from acid tea soils from Bangladesh and Sri Lanka (Walker 1976). Interestingly *Nitrosomonas* was only found in soils which had received farmyard manure and other animal excreta whereas *Nitrosolobus* was ubiquitous. These observations strongly indicate that in many soils *Nitrosolobus* plays a much more significant part in nitrification than is generally recognized, and is more important than *Nitrosomonas* (Bhuija and Walker 1977). Nitrite produced by the ammonium oxidizing autotrophs is rapidly oxidized to nitrate by *Nitrobacter* species. Both ammonium oxidizers and nitrite oxidizers are obligately aerobic and in most habitats they are closely associated so that the nitrite produced is rapidly oxidized and does not accumulate. In waterlogged soils the oxidation of NH_4^+ is restricted because of a lack of oxygen required for the oxidation as can be seen from the equations below. In addition, the nitrifying bacteria prefer more neutral to slightly acid pH conditions. Therefore in acid soils nitrification is usally substantially depressed and can even be completely inhibited. This is illustrated in Table 7.4 (Munk 1958).

The oxidation of NH_3 to NO_2^- is brought about by two steps the first being catalyzed by an NH_3 mono-oxygenase with hydroxyl amine as the oxidized product:

$$NH_3 + O_2 \Rightarrow NH_2(OH) + \tfrac{1}{2} O_2$$

$$NH_2OH + H_2O \rightarrow HNO_2 + 4e^- + 4H^+$$

Nitrite (HNO_2^-) oxidation follows the equation:

$$HNO_2 + H_2O \Rightarrow HNO_3 + 2e^- + 2H^+$$

In the oxidation of hydroxylamine and nitrite, each water molecule is split into $2e^-$ and $2H^+$ a process which is analogous to the photolysis of water (see page 140). Similarly to photosynthesis the e^- produced can be used to build up an electrochemical potential across a membrane to drive the synthesis of ATP. Both oxidations, that of NH_3 and that of HNO_2, lead to acid formation namely nitrous acid (HNO_2) and nitric acid (HNO_3). The formation and accumulation of the strong nitric acid is the cause of the acidification of soils during nitrification.

Table 7.4 Rate of nitrification of NH_3 in relation of soil pH. Total N added was 20 *mg in form of* ammonium sulphate (after Munk 1958).

Incubation duration, days	pH 4.4 mg nitrate N produced	pH 6.0 mg nitrate N produced
14	1.78	8.0
21	2.30	12.0
35	4.72	21.4

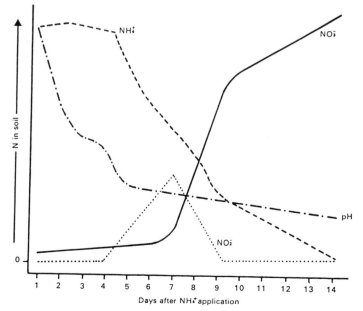

Figure 7.5 Relationship between microbial NH_4^+ oxidation, nitrite and nitrate formation and soil pH (after Duisberg and Buehrer 1954).

412

The relationships of NH_3 oxidation, pH shift nitrite and nitrate formation are well demonstrated in the classical experiment of Duisberg and Buehrer (1954) the main results of which are presented in Figure 7.5. During the incubation period of 14 days nearly all the NH_3 was oxidized to NO_3^- with a concomitant drop in soil pH. After one week of incubation, a peak in the NO_2^- concentration occurred, which later disappeared due to the activity of the *Nitrobacter*. Generally nitrite does not accumulate in soils because it is readily oxidized by *Nitrobacter*. Both ammonia and nitrite oxidizers, obviously function 'in series'. Ammonia is thus rather rapidly converted to nitrate, provided that suitable conditions for nitrifying bacteria are present in the soil, as was the case in the example presented in Figure 7.5. The soil used was a fertile calcareous sandy loam with a pH of 7.8 and the experiment was carried out in the laboratory with optimum soil water and temperature conditions. In the field, nitrification often occurs at lower rates, for as already mentioned, in soils of low pH and in waterlogged soils, nitrification is restricted or even completely inhibited. Under these conditions the soil may thus accumulate ammonium. Nitrification is also depressed in dry soils. Temperature has a marked effect on ammonification and nitrification. According to the findings of Beck (1983) nitrification attains an optimum at 26°C whilst the optimum for ammonification is as high as 50°C. This difference results in a characteristic distribution of nitrate and ammonium in soil in relation to temperature as shown in Figure 7.6. Thus in tropical soils even under neutral pH conditions ammonium may accumulate as the result of the low rate of nitrification.

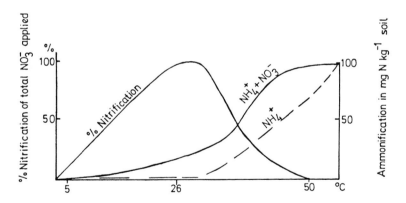

Figure 7.6   Effect of incubation temperature on % nitrification, NH_4^+ concentration, and $NH_4^+ + NO_3^-$ concentration in an arable soil (after Beck 1983).

Heterotrophic microorganisms may also produce nitrate. The characteristics of heterotrophic organisms is that they possess mitochondria and that the e^- transfer during respiration originates from the H atoms of NAD(P)H which acts as the H donor being split into H^+ and e^-. Some of these heterotrophs may use NH_3 as e^- donor as shown in the following sequence (modified after Paul and Clark 1996):

$$NH_3 + O_2 \Rightarrow NH_2(OH) + \tfrac{1}{2} O_2$$

$$NH_2(OH) \Rightarrow NOH + 2H^+ + 2e^-$$

$$NOH + H_2O \Rightarrow HNO_2 + 2H^+ + 2e^-$$

$$HNO_2 + H_2O \Rightarrow HNO_3 + 2H+ + 2e-$$

This sequence also shows that NH_3 may be oxidized by heterotrophic soil microorganisms. NOH formed in this reaction sequence may give rise to the formation of volatile N_2O, presumable according to the reaction:

$$NOH + HNO \Rightarrow N_2O + H_2O$$

The nitrous oxide (N_2O) thus produced is volatile and may therefore be lost from the soil. Breitenbeck and Bremner (1986) reported that after the application of anhydrous NH_3 to a soil much more N_2O was released into the atmosphere than with nitrate fertilizer application. Nitrogen losses in the form of N_2O after NH_3 application were in the range of 2 to 4 kg N/ha at a fertilizer rate of 180 kg N/ha. Such losses from an agronomic point of view are low. It should be kept in mind, however, that N_2O is a greenhouse gas. In soils N mineralization and the immobilization (assimilation of inorganic N) may occur simultaneously. The difference between both (N mineralization − N immobilization) is the net N mineralization, which is of particular agronomic importance because it is the quantity of N which can be supplied to the crop by the soil. Net nitrogen mineralization rates throughout a growing season can differ considerably and depend on soils crops and cropping systems. In sandy soils N mineralization is generally faster than in medium-textured soils (Jenkinson and Rayner 1977) because of the lower amounts of clay minerals which can adsorb organic nitrogen and thus protect it from microbial decomposition. Under central European climate conditions about 50 to 90 kg N/ha may be mineralized in arable soils under cereals over one growing season (Barekzai et al. 1992). For sugar beet because of the longer growth period during summer the amounts of mineralized N may be even higher. In field trials under fallow conditions Winner et al. (1976) found a net N mineralization of 150 kg N/ha over the growth season for a number of fertile soils in Germany. Very high rates of mineralization rates have also be found in grassland soils (Berg and Rosswall 1985).

Ammonification of organic soil N followed by nitrification is an important step in the global nitrogen cycle because the nitrate formed may be denitrified to N_2 and thus recycled back into the atmosphere (see page 349). By this process the huge atmospheric N pool is refilled and the process may to some extent balance the N_2 taken out of the atmosphere by biological dinitrogen fixation as shown in Table 7.5 (Werner 1980). From the data it is clear that the assessments of denitrification and biological fixation are still imprecise.

Table 7.5   Global turnover between soil N and atmospheric N in 10^6 Mg/year (Data from Werner 1980)

	Gain		Loss
Industrial production	46	Denitrification	200−300
Biological[1] fixation	100−200	NH_3 volatilization	165
NO_3^-/NO_2^- precipitation	60		
NH_3 precipitation	140		

7.1.5   Nitrification inhibitors

In order to avoid major nitrate losses as a result of denitrification or leaching, nitrification inhibitors have been developed in recent years. These inhibitors block the oxidation of NH_3 to NO_2^- brought about by species of *Nitrosomonas*, *Nitrosocystus*, and *Nitrosospira* (Bhuja and Walker 1977). Some of the most important inhibitors and their formulae are listed below:

Nitrapyrin (2-chloro-6-[trichloromethyl] pyridine) = N-serve
ST (2-sulphanilamide thiazole)
Terrazole (5-ethoxy-3-trichloromethyl-1,2,4 thiadiazole)
AM = 2-amino-4-chloro-6-trimethylpyrimidine
KN_3 (potassium azide)
Dichlorophenyl succinic amide
Dicyandiamide (DCD)
Ca carbide (C_2Ca)

The formula of some of these inhibitors are shown below in Figure 7.7.

$$H_2N-C-NH-C{\equiv}N \qquad \text{Dicyandiamide}$$ 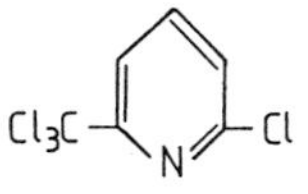
$$\overset{\|}{NH}$$

Cl_3C — [ring, N] — Cl 2- chloro -6- (trichloromethylpyridine)
(= Nitrapyrin)

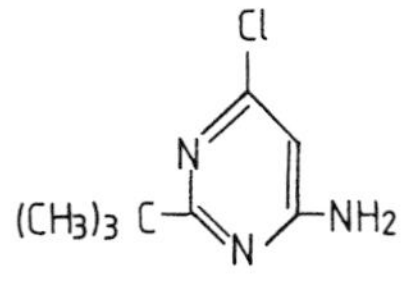

Cl

$(CH_3)_3\,C$ — [ring, N, N] — NH_2 AM= 2 amino-4-chloro-
6 trimethylpyrimidine

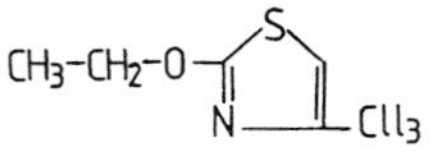

CH_3-CH_2-O — [ring, S, N] — Cll_3 Terrazole = 5-ethoxy-3-tri-
chloro methyl-1,2- 4 thiadizole

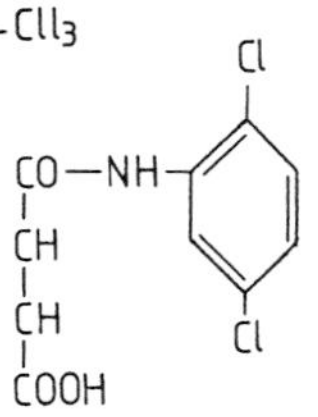

Cl

CO—NH— [ring]
CH
CH
COOH   Cl   2,5-dichlorophenyl-
succinic acid amide

Figure 7.7 Formula of most important nitrification inhibitors.

Nitrapyrin is the most thoroughly investigated nitrification inhibitor (Huber *et al.* 1977). The inhibitors block the microbial oxidation of ammonium to nitrite and thus also the formation of nitrate. The loss of N by leaching or denitrification is therefore prevented. Touchton *et al.* (1978) in studying the behaviour of nitrapyrin in different soils, found that the chemical is rather immobile in the soil and is especially bound to organic matter. The degradation of nitrapyrin depends on the degree to which it is adsorbed by soil colloids. Thus the degradation rate was higher in a sandy soil, low in organic matter, than in a clay loam rich in humus. Nitrapyrin degradation proceeds at a higher rate in soils of neutral pH than in more acid soils.

According to Touchton *et al.* (1978) the half life period of nitrapyrin degradation is about 4 weeks, but this may differ depending on soil conditions and soil microbial activity. Soil accumulations of nitrapyrin and its principal metabolic product (6 chloro-picolinic acid) should be avoided, as both products can be taken up by plants and may be toxic to them. A typical nitrapyrin degradation curve is shown in Figure 7.8 from the work of Touchton *et al.* (1978).

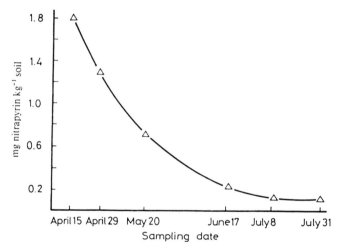

Figure 7.8 Decline in soil nitrapyrin concentration following application (after Touchton *et al.* 1978).

Keerthisinghe *et al.* (1993) who tested the inhibiting effect of nitrapyrin in comparison with wax-coated Ca carbide under the conditons of flooded rice culture found that the wax- coated Ca carbide was more effective than nitrapyrin. Ca carbide reacts with water to form $Ca(OH)_2$ and acetylene which inhibits nitrification and the wax coat ensures a slow release and therefore a long lasting effect.

AM (= 2-amino-4-chloro-6-trimethylpyrimidine) is soluble in water and liquid NH_3 but not in organic solvents. It is applied in quantities of 5 to 6 kg ha and can be adsorbed by soil colloids (Slangen and Kerkhoff 1984). Another nitrification inhibitor is dicyandiamide (Blaise *et al.* 1997) which may also be produced in the soil during the degradation of calcium cyanamide (Rathsack 1978). Pyrite inhibits nitrification and in addition has an acidifying effect on the soil and thus may depress NH_3 volatilization. The quanties of pyrite, however required for an effective inhibition are high and in the range of several 100 kg/ha (Blaise *et al.* 1997). Natural inhibitors have been found such as Neem which occurs in the seeds of *Azadiracta indica* and Karanjin which is present in seeds, leaves, and bark of *Pongamia glabra*. Such natural inhibitors may well become important in developing countries where the costs of agrochemicals are high (Slangen and Kerkhoff 1984).

Nitrification inhibitors are mainly applied in autumn together with ammonium fertilizers. Numerous field experiments carried out in Indiana (USA) over a 5 year period have shown favourable results of nitrapyrin when applied together with ammonium or urea in autumn. This autumn applied N gave the same grain yields as a split N application (autumn and spring) without nitrapyrin. Nitrapyrin application not only reduced N losses but also resulted in a more uniform N supply to the plant roots and higher grain protein contents were obtained (Huber *et al.* 1980).

7.1.6 Ammonium fixation

In contrast to NO_3^- which is rather mobile in the soil and hardly adsorbed by soil particles, NH_4^+ is adsorbed to negatively charged clay minerals because of its cationic properties. The planar bound NH_4^+ can be easily exchanged (page 17) and plays no major role in aerobic soils since NH_4^+ is quickly nitrified. Ammonium like K^+ can be bound in the interlayers of 2:1 clay minerals, such as illites, vermiculites and smectites and thus is largely protected against nitrification (Guo *et al.* 1983). The mechanism of this interlayer fixation is described in more detail on page 484. Ammonium and K^+ may compete for the same selective binding sites. Bartlett and Simpson (1967) reported, that the fixation of fertilizer K^+ was lowered by NH_4^+ application, and the reverse is also true that K^+ restricts the fixation of NH_4^+ (Dou and Steffens 1995). Not all soils contain interlayer NH_4^+. It is present in soils with 2:1 clay minerals especially vermiculite, smectites, and to a lesser degree illites (Scherer and Weimar 1994, Scherer and Ahrens 1996, Steffens and Sparks 1999). For arable soils the concentration of fixed NH_4^+ may be in a range of 20 to 1000 mg NH_4^+-N/kg soil which is equivalent to a soil depth of 30 cm of about 60 to about 3000 kg N/ha (Smith *et al.* 1994). Since deeper soil layers also contain fixed NH_4^+ (van Praag *et al.* 1980) the total amount of fixed NH_4^+ N in the rooting profile may be even much higher.

Most of fixed NH_4^+ is not plant-available and one may distinguish between native interlayer NH_4^+ which may become available only after weathering and decomposition of clay minerals and recently fixed NH_4^+. The latter originates from fertilizer NH_4^+ and NH_4^+ produced by mineralization of organic N and is plant-available (Norman and Gilmour 1987). Feng (1995) investigating 15 typical paddy rice soils from China found that plants only decrease the interlayer NH_4^+ to a certain level below which is assumed to be native interlayer NH_4^+. The maximum amount of NH_4^+ which can be fixed above this threshold level represents the potential of plant-available fixed NH_4^+ and in Feng's investigation amounted to between 22 and 156 mg NH_4^+-N/kg soil. Field and pot trials carried out on arable aerobic soils are also in accord with a given level of fixed NH_4^+ below which the NH_4^+ is not plant-available (Li *et al.* 1990, Scherer and Weimar 1994). According to Feng (1995) there is a correlation between the native interlayer NH_4^+ values and the illite concentrations in soils so that presumably illite may be a major mineral in which native NH_4^+ is present.

The release of interlayer NH_4^+ is brought about by plant roots which may deplete the concentration of NH_4^+ and K^+ near the surface of NH_4^+ fixing minerals to a very low level (Smith *et al.* 1994, Scherer and Ahrens 1996) and thus induce the release associated with an expansion of the interlayer and swelling of the mineral. For the release it is important that the K^+ concentration near the surface of the mineral is also low (Smith *et al.* 1994). It is for this reason that nitrifying bacteria such as *Nitrosomonas*, *Nitrosolobulus*, and *Nitrospira* generally do not induce a release of interlayer NH_4^+ since they require much more NH_4^+ as an energy source as compared with their demand for K^+. They therefore do not decrease the K^+ concentration of the soil solution. Heterotrophic bacteria may profit from interlayer NH_4^+ particularly when their growth is boosted by the addition of easily digestible carbohydrates (Breitenbeck and Paramasivam 1995, Scherer and Werner 1996). The kinetics of release follows the Elovich function (Steffens and Sparks 1997) which means that the release rate decreases the more the pool of available interlayer NH_4^+ is depleted.

The change of interlayer NH_4^+ (non exchangeable NH_4^+) throughout a growth period can follow a typical characteristic as depicted Figure 7.9 from the work of Mengel and Scherer (1981) in which the change of interlayer NH_4^+ level was investigated under oats from early spring until the end of September at three rooting depths. During early spring, the level of fixed NH_4^+ was lowered in the two upper profile layers, 0 to 30 and 30 to 60 cm. Later the same occurred at the deeper soil layer (60 to 90 cm) whereas the level of fixed NH_4^+ in the two upper soil layers rose. The significant decrease in the upper soil layer in autumn may be due to the growth of heterotrophic bacteria. The changes in the NH_4^+ uptake mirror plant uptake and refilling of fixing positions by NH_4^+ produced by mineralization. As can be seen from the data in Figure 7.9 the quantities thus fixed or released are considerable. The fixed NH_4^+ is largely protected from nitrification and therefore also indirectly from denitrification and leaching in the form of nitrate.

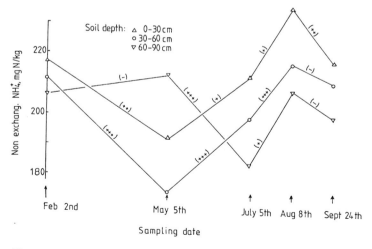

Figure 7.9   Change of the level of fixed NH_4^+ on an alluvial soil grown with oats in three soil depths throughout a growing period. (-) non significant, (+) significant at $p < 0.05$, (++) significant at $p < 0.01$, (+++) significant at $p < 0.001$ difference between two dates (after Mengel and Scherer 1981).

Nitrogen supply of crops by fixed NH_4^+ is of particular importance for flooded rice soils since here NH_4^+ is the most important inorganic nitrogen form which can be directly taken up by plants. Appreciable amounts of NH_4^+ originating from the fixed pool can be taken up by rice as shown by Keerthisinghe *et al.* (1984) in field experiments in the Philippines and by Schnier *et al.* (1987) in pot experiments with ^{15}N labelled NH_4^+ provided that the soils contained NH_4^+ fixing minerals. Release of fixed NH_4^+ in flooded soil occurs relatively quickly. As shown recently by Schneiders and Scherer (1998) a decreasing redox potential in flooded soils promotes NH_4^+ fixation and an increasing redox potential favours the release of fixed NH_4^+. Hence rice roots excreting O_2 into the rhizosphere provide conditions which promote NH_4^+ release. This effect presumably is related to the oxidation and reduction of Fe in fixing minerals. A useful review article on fixation and release of interlayer NH_4^+ has been published by Scherer (1993).

7.1.7   Nitrogen of the soil solution

The most important N containing solutes in the soil solution are nitrate, low molecular organic N compounds and NH_4^+. Ammonium is particularly high in anaerobic soils (flooded soils) and frequently is very low in aerobic soils because of rapid nitrification of NH_4^+. Heavy application of slurry may also lead to a transient increase of soluble NH_4^+ (Barekzai *et al.* 1988). The concentration of N-containing solutes, especially that of nitrate may change quickly due to processes such as uptake by plant roots and microorganisms, leaching and denitrification. Conditions favouring nitrification result in an increase in the NO_3^- concentration in the soil solution. Thus in spring when the temperature rises and aeration of the soil increases, the NO_3^- concentration in the soil solution is also raised. When crop demand is high, NO_3^- is rapidly taken up by plant roots associated with a depletion of nitrate in the soil solution (Page and Talibudeen 1977). White and Greenham (1967) reported that in orchards under a grass cover, only low amounts of NO_3^- were found whereas in a similar fallow soil the NO_3^- concentration in the soil increased until summer, and decreased later when the NO_3^- was leached into deeper soil layers by the summer rains. Nitrate levels in the soil solution can be as high as 20 to 30 mol/m^3 after nitrogen fertilizer application. In fertile soils it normally ranges from 2 and 20 mol/m^3 depending on the rate of mineralization and the uptake by plants. Nitrogen fertilizer application in spring to winter wheat resulted in a steep increase of nitrate in soils; the nitrate level, however, rapidly declined and was almost zero at the flowering stage of wheat (Gutser and Teicher 1976). Usually the NO_3^- concentration of the soil solution is of direct importance for the N uptake of the crop. In dry periods NO_3^- may accumulate in the upper soil layer (Page and Talibudeen 1977).

7.2   Nitrogen in Physiology

7.2.1   General

Dry plant material contains about 20 to 44 mg N/kg. This appears rather low in comparison with the C concentration which is in the order of about 400 mg/kg dry matter.

Nevertheless N is an indispensable elementary constituent of numerous organic compounds of general importance, amino acids, proteins, nucleic acids and compounds of secondary plant metabolism such as the alkaloids. Higher plants are major contributors to the large amount of N which is continuously being converted from the inorganic to the organic form. The most important sources involved in this conversion are NH_4^+ and NO_3^-.

7.2.2   Uptake

In contrast to other plant nutrients nitrogen may be taken up in the form of a cation as NH_4^+ or an anion as NO_3^-. Uptake rate of both depends mainly on the availability of these ions in the nutrient medium in general one NH_4^+ may substitute for NO_3^- and *vice versa*. Uptake rates are determined mainly by the physiological need of the plant and not so much on whether the source is a cation or an anion (Mengel *et al.* 1983). Many species take up NH_4^+ preferentially to NO_3^- including those of forest trees *e.g. Picea abies* (Marschner *et al.* 1991) and grasses. This effect of the form of N supply on the N uptake is demonstrated in the most interesting results of Breteler and Smit (1974) shown in Table 3.3 on page 132. It is evident that the application of NO_3^- increased the nutrient solution pH while NH_4^+ had the reverse effect. In the treatment with 4 mM NO_3^- + 1 mM NH_4^+, however, the pH of the nutrient solution was markedly depressed although the NO_3^- concentration was four times higher than the concentration of NH_4^+. This is indicative of the preferential uptake of NH_4^+. Similar results have been found recently by Rroco and Mengel (2000) who grew spring wheat in a nutrient solution with NH_4NO_3 as N source. The plants took first up the NH_4^+ and when this was depleted to a low level they then took up NO_3^-. Ammonium uptake was associated with a pH decrease in the nutrient solution and NO_3^- uptake with a rise of the pH.

As already discussed on page 127 the uptake of NO_3^- is mainly a H^+/NO_3^- cotransport (Ullrich 1992) with the H^+ pumped out of the cell by the plasmalemma proton pump being recycled back into the cytosol. Hence nitrate uptake is associated with an pH increase in the outer medium. With NH_4^+ nutrition recyling of H^+ back into the cytosol is restricted and the H^+ pumped out of the cell remain mainly outside and hence the pH is depressed. This is particularly evident when plants are grown in solution culture which is generally not H^+ buffered. In soils the pH change is not so spectacular due to the H^+ buffer power of soils. The uptake of NO_3^- and NH_4^+ can, however, induce large changes in the pH of the rhizosphere of up to 2 pH units higher or lower, respectively than the pH of the bulk soil (Smiley 1974, Marschner *et al.* 1991). Nitrate is the dominant form in most agricultural soils and it is this form which is usually taken up in greatest amounts by crops with exceptions such as lowland rice which is adapted to NH_4^+ uptake (see page 50)

Uptake of NH_4^+ may be brought about by facilitated diffusion since it may lead to a sharp depolarisation of the plasma membrane (Ullrich 1992). Hence uptake of NH_4^+ is analogous to that of K^+ brought about by the electropotential difference and cation selective channels (see page 127) with the difference that K^+ may accumulate in the cytosol to high levels in the range of 100 mol/m^3. This is not the case for NH_4^+ which is immediately assimilated by the GOGAT pathway (see page 174) and the cytosolic NH_4^+ concentration is in a range of 2 to 30 mol/m^3 as reported by Feng *et al.* (1999).

The uptake of ^{15}N labelled NH_4^+ was associated with an efflux of non labelled NH_4^+ which originated from a soluble N fraction in the root. The efflux rate was about 10% of the influx rate. Besides this passive uptake of NH_4^+ the possibility of H^+/NH_4^+ cotransport has been discussed (Schachtman and Schroeder 1994). This form of transport might be of importance when the NH_4^+ concentration in the soil solution is extremely low and of relevance in the acquisition of interlayer NH_4^+ in the soil by roots. Wang et al. (1994) studying the NH_4^+ uptake of rice seedlings in relation to the NH_4^+ concentration of the nutrient medium found that the uptake was biphasic with a curve up to 1 mol NH_4^+/m^3 followed by a linear relationship at higher concentration. This uptake behaviour supports the concept that in the lower concentration range NH_4^+ uptake is mediated by a specific transporter system (high affinity transporter) whereas at higher concentration it is mainly a facilitated diffusion through NH_4^+ specific channels. Analogous results have been reported by Kronzucker et al. (1996). These authors reported an NH_4^+ uptake system in roots of Picea glauca operating at 2.5 to 20 mmol/m^3 NH_4^+ in the uptake solution with a K_M of 20 to 40 mmol/m^3. The low- affinity transporter was linear over an concentration range of 0.5 to 50 mol/m^3. Plants with a low NH_4^+ supply prior to the uptake experiments showed higher NH_4^+ uptake than plants well supplied with NH_4^+ (Wang et al. 1994).

Ammonium uptake depends much on the carbohydrate status in the roots as was already reported by Kirkby and Hughes (1970) and by Michael et al. (1970). This observation is confirmed by recent research of Feng et al. (1999) who found a diurnal change carbohydrate concentration in maize roots which declined during the night and rose during the day. Assimilation of NH_4^+ in the roots requires carbohydrates translocated from shoots to roots provide the C skeletons and the energy (ATP and NADPH) for the NH_4^+ assimilation process. If the carbohydrate supply is insufficient due to high temperature and respirative decomposition of carbohydrates plants may die as was reported by Ganmore-Neumann and Kafkafi (1983) for strawberries. Most of the NH_4^+ taken up by roots is assimilated in the roots and translocated in the form of amino acids and amides to the shoots.

According to Oscarson and Larsson (1986) the net uptake of NO_3^- results from nitrate influx and efflux. The efflux of nitrate is particularly high if the plasmalemma ATPase is inhibited (Matzke and Mengel 1993). In this case the protons required for the cotransport of nitrate are lacking. The nitrate uptake process is regulated by other factors including the internal N nutritional status of the plant (Siddiqi et al. 1989, Imsande and Touraine 1994). Aslam et al. (1992) also found that nitrate uptake rates depend on the activity of nitrate reductase. In seedlings which had not been supplied with nitrate and which therefore had only a very low nitrate reductase activity, a high affinity uptake system for nitrate was found to be operating. The V_{max}, however, was very low. In contrast in seedlings with the induced nitrate reductase a high affinity transporter system for nitrate with a K_M of 0.7 mmol/m^3 was present but the V_{max} was about four times higher than in the non-induced plants. In induced seedlings a linear uptake system of NO_3^- was also found up to a concentration of 500 mmol/m^3 nitrate. The investigations quoted above show that plants possess sensitive N uptake systems which may be adjusted to N demand and environmental conditions.

Most plant species need much N during the vegetative stage whereas in the generative phase the N uptake rates by roots are low and the retranslocation of organic N is high. Also the uptake rate of nitrate is dependent on the energy status of plants (Rufty et al. 1989). In contrast to NH_4^+, nitrate can be transported at high rates into upper plant parts and can be stored at high concentrations in the vacuole (McIntyre 1997).

An important difference between NO_3^- uptake and NH_4^+ uptake is in their sensitivity to pH. NH_4^+ uptake takes place best in a neutral medium and it is depressed as the pH falls. The converse is true for NO_3^-- absorption, a more rapid uptake occurring at low pH values because more protons are available for the proton cotransport of nitrate. At low pH levels in the the nutrient solutions as brought about by NH_4^+ nutrition root growth is often depressed. According to Yan et al. (1992) in experiments with Faba beans this comes about because H^+ ions rediffuse into the cytosol to stress the cytosolic H^+ buffer system by decreasing the malate concentration which is needed for the neutralization of H^+. Plant roots may adapt to low pH conditions by the alteration of the K_M and the V_{max} of the plamalemma ATPase (Yan et al. 1998). Nitrate uptake can be depressed by NH_4^+, as shown by Minotti et al. (1969) and Blondel and Blanc (1973) using young wheat plants and by Kronzucker et al. (1999) with rice. Ammonium uptake on the other hand is not affected by NO_3^- as found for rice by Mengel and Viro (1978).

Gaseous ammonia may also be absorbed by the upper plant parts via the stomata (see page 348). This net uptake depends on the partial pressure of NH_3 in the atmosphere. Farquhar et al. (1980) reported that in Phaseolus vulgaris net NH_3 uptake was zero at the low atmospheric partial pressure of 0.25 MPa at 26°C. Increasing the partial pressures of NH_3 increased the net uptake and lowering the partial pressure resulted in a loss of NH_3 from the plant. The source of this released NH_3 is not yet clear. According to experiments of Hooker et al. (1980) with wheat plants, NH_3 loss was higher at more advanced than at an earlier growth stage. These authors suggest therefore that the NH_3 release results from protein decomposition in senescent leaves. Since NH_3 release is highly temperature dependent, photorespiration may also be a potential source of NH_3 (see page 152).

Ammonium in solution can be toxic to plant growth. The toxicity results mainly from ammonia (NH_3) which affects plant growth and metabolism at low concentration levels at which NH_4^+ is not harmful. The distribution of NH_4^+ and NH_3 in aqueous solutions is described by the equilibrium relationship:

$$NH_3 \text{ (aq)} + H^+ \Rightarrow NH_4^+$$

$$NH_3 \text{ (aq)} \Rightarrow NH_3 \text{ dissolved in water}$$

This equation shows that the NH_3 (aq) concentration depends very much on the pH of the medium and at a pH of <6 virtually all NH_3 is protonated. NH_4^+ shifts over to NH_3 at high pH and this can be very toxic. The toxicity of NH_4-N is therefore much dependent on the pH and especially on the pH in the apoplast which is much buffered by H^+ pumps (Kosegarten et al. 1999a) This H^+ pump activity may therefore be the contol in the sensitivity of various plant species to NH_3 toxicity.

The relationship between NH_3 toxicity and pH is illustrated in Figure 7.10 in which NH_4-N activity is plotted against pH (Bennett 1974). The incipient toxic concentration of NH_3 (aq) is about 0.15 mol/m^3 and the lethal concentration about 6.0 mol/m^3.

The combinations of pH and NH_4^+ activity ($NH_4^+ + NH_3$) which produce these levels are shown in the figure.

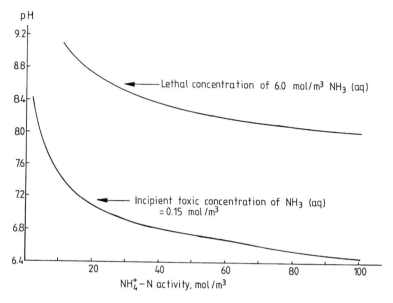

Figure 7.10   Lethal concentration and incipient toxic concentration of NH_3 (aq) in relation to the pH and NH_4^+ activity in the solution (after Bennett1974).

It is clear that the toxic effects of NH_4^+/NH_3 resulting from NH_3 (aq) are more likely to occur at higher pH levels. NH_3 (aq) particularly affects root growth. Bennett and Adams (1970) reported that the roots of young cotton seedlings were injured by NH_3 (aq) concentrations as low as 0.2 mol/m^3. The germination of seeds can also be impaired even by low NH_3 concentrations (Barker *et al.* 1970). Recent research by Kosegarten *et al.* (1997) has shown by means of a sensitive pH measurement in the cytosol and vacuole (Wilson *et al.* 1998) that shifting to a high nutrient solution pH in the presence of NH_4^+/NH_3 in the solution leads to an immediate increase of the cytosolic pH because of the uncontrolled NH_3 penetration of the plasma membrane and the vacuole. In the case of maize this high cytosolic pH inhibited glutamine synthetase while in the case of rice the cytosolic pH was downregulated so that the activity of the glutamine synthetase was not affected. Rice is obviously adapted to such high pH conditions which may occur in flooded rice soils covered by algae (Mikkelsen *et al.* 1978). These results show that inhibition of the glutamine synthetase by high cytosolic pH is the primary cause of NH_3 toxicity. This inhibition of glutamine synthetase means that the whole N metabolism is blocked with disasterous consequences for growth. The above findings are in accord with the work of Gerendas *et al.* (1990) who showed by means of NMR that a high pH in the nutrient solution in the presence of NH_4^+/NH_3 increases the pH in both the cytosol and vacuole because of the uncontrolled penetration of NH_3 into these compartments.

Wilson *et al.* (1998) found even at a low pH of 5.0, at which no NH_3 is present in the nutrient solution, that in maize roots the vacuolar pH increased to a relatively high level and could not be restored whereas in rice roots at the same external pH 5 there was only a minor vacuolar pH increase. Because of the high pH in the cytosol some NH_4^+ is deprotonated and NH_3 may penetrate the tonoplast and thus raise the vacuolar pH. This example not only shows that plant species may differ in their sensitivity to NH_3 toxicity but also that at low pH in the nutrient solution at which no NH_3 is present, NH_3 toxicity may occur if the NH_4^+ concentration of the nutrient solution is high. In the experiments of Wilson *et al.* (1998) the NH_4^+ concentration in the nutrient solution was 2 mol/m^3. A high vacuolar pH implies a decrease of the proton gradient across the tonoplast with a consequent depression in anion accumulation in the vacuole and other metabolic processes (page 129).

Highest N uptake rates were observed by Blondel and Blanc (1973) when both N forms, NH_4^+ and NO_3^-, were present in the nutrient solution. Similar observations were made by Cox and Reisenauer (1973) from dilute nutrient culture experiments. This beneficial effect of NH_4^+ when supplied together with NO_3^- has been recently confirmed by Kronzucker *et al.* (1999) in experiments with rice which showed that in the presence of nitrate, uptake and assimilation of NH_4^+ was promoted.

Urea is generally converted into NH_4^+ by urease in the soil (page 364). It can, however, also be absorbed directly by plants although the rate of absorption appears to be low in comparison with that of NO_3^-. Table 7.6 shows that for sunflowers grown in water culture, higher dry matter yields were obtained in a 2 mol m^3 NO_3^- than in a 8 mol/m^3 urea treatment. The supply of urea not only resulted in a lower uptake of N, but also in a disturbance of the protein metabolism and a pronounced accumulation of asparagine occurred (Kirkby and Mengel 1970). Hentschel (1970) studying the uptake of NH_4^+ and urea with labelled N also reported that urea was absorbed at a lower rate than NH_4^+ by *Phaseolus vulgaris*. A number of reviews have dealt with the uptake of different forms of nitrogen and their effects on growth and metabolism (Raven and Smith 1976, Haynes and Goh 1978, Kirkby 1981). A valuable and comprehensive review on the uptake and utilization of nitrogen by plants has been published by Engels and Marschner (1995).

Table 7.6 Influence of the form and level of nitrogen nutrition on the yield, nitrogen percentage and N-uptake by sunflower plants (Kirkby and Mengel 1970)

N concentration in the nutrient solution mol/m^3	Yield g DM/24 plants	N concentration mg N/g DM	Total N uptake mg N/24 plants
2 Nitrate	25.4	58.0	1,473
2 Urea	11.6	33.0	384
4 Urea	14.9	56.7	696
8 Urea	17.1	67.9	987

7.2.3 Nitrogen fractions

Nitrogen turnover in plants in characterized by three main steps. The first step consists of the conversion of inorganic N into organic N compounds of low molecular weight. The details and the various processes involved in this step have already been discussed on page 162 and 174. In the second step, the synthesis of high molecular weight N compounds takes place. These compounds include proteins and nucleic acids. Low molecular weight organic N compounds and particularly amino acids serve as building blocks for these synthetic reactions. The third step represents the breakdown of the N containing macromolecules by hydrolyzing enzymes. These three steps in N turnover represent the pathways between the three main N fractions involved in N metabolism: inorganic N, low molecular weight organic N compounds, and macromolecular organic N compounds.

The concentration of soluble amino acids in leaves mirrors the physiological state of plant tissues. This is shown in Table 7.7 from the work of Prior (1997) with leaves of vine plants (*Vitis vinifera*).

Table 7.7 Concentration of soluble amino acids in leaves of vine plants at different physiological age (after Prior 1997)

Amino acid	Five leaf stage	Beginnning berry maturation	Full berry maturation
		mg amino N/kg leaf dry matter	
Glutamate	3600	40	60
Glutamine	5600	2	80
Aspartate	1200	2	50
Alanine	800	40	40
Residual amino N	2200	40	160

The *Five leaf stage* as indicated in Table 7.7 represents a young vine shoot in spring with with five young rapidly developed leaves supplied with storage amino acids mainly glutamine as is evidenced by the very high concentration of soluble amino acids with glutamine dominating. At this stage protein synthesis is the principal metabolic process in the leaves and the various amino acids required fot this process are provided mainly by transamination. It is for this reason that the concentration of residual amino N is also high. The physiological state at the beginning of berry maturation is quite different. The leaf is no longer growing. Its main task is photosynthesis so that the concentration of soluble amino acids is very low. At full berry maturation leaf protein degradation has already begun so that the concentration of soluble amino acids has increased, particularly that of glutamine which is the major amino acid translocated from the leaves into the bark and wood. Interestingly, the concentration of the residual soluble amino acid concentration is also high at full berry maturation stage and originates from protein hydrolyzation.

Increasing the level of N nutrition results in an increase in all fractions but the extent to which this occurs differs between fractions. There is much experimental evidence to show that the concentration of soluble amino compounds (free amino acids, amines, amides) is considerably increased whereas the concentration of protein is only raised to a limited extent by high applications of fertilizer N as already shown in Table 5.19 (page 326) from the work of Goswami and Willcox (1969).

In green plant material, protein N is by far the largest N fraction and amounts to about 80 to 85% of the total N. The N of the nucleic acids makes up about 10% and the soluble amino N about 5% of the total N present in plant material. Many crops are cultivated essentially to produce plant proteins. In vegetative plant material the proteins are mainly enzyme proteins, whereas in seeds and grains the major protein fraction is made up of storage proteins. In relation to protein function it is necessary to distinguish between enzyme proteins, storage proteins and structural proteins. Nitrogen is also an essential constituent of various coenzymes and compounds of secondary metabolism especially alkaloids.

7.2.4 Translocation

Nitrogen taken up by plant roots is translocated in the xylem to the upper plant parts. The form in which N translocation occurs depends on the N uptake source and root metabolism. Ammonium taken up by roots is almost completely assimilated in the root tissue and the N is translocated to a great extent in the form of amino acids to the upper plant parts. In N_2 fixing legume species of temperate climates, glutamine and asparagine are the major N compounds translocated from the nodules *via* the xylem to the upper plant parts whereas in tropical leguminous species this role is played by allantoin and allantoic acid (Pate 1980). Nitrate-N can be translocated unaltered from the roots to shoots and leaves but this depends on the nitrate reduction potential of the root (see page 166). Nitrate and amino acids are thus the main forms in which N is translocated in the vascular system of higher plants. Generally in the xylem sap 70 to 80% of amino acids present are rich in N, with a N/C ratio greater than 0.4. It is believed that the function of these nitrogen-rich molecules (glutamine. asparagine) is to transport N with a minimum amount of C (Pate 1980). In the phloem amino acids are the predominant form of N-transport occurring in very much higher concentrations than in the xylem. Nitrate can be present but only in very low concentrations (Jeschke *et al.* 1997). For many plant species in which NO_3^- is assimilated predominantly in the shoot, the translocation of amino acids *via* the phloem from shoot to root is essential to cover the growth demand of the roots for nitrogen. This transport of amino acids in the phloem can also act as a feedback signal to control the uptake of NO_3^-. When plants are growing rapidly and there is a high shoot demand for N, a lower proportion of N is cycled back to the roots as amino acids. When shoot demand is low, however, there is a relative increase in cycling of N in the phloem which represses uptake and transport of N (Engels and Marschner 1995). These findings are in accord with the observations of Touraine *et al.* (1992) that feeding amino acids *via* soyabean seedling cotyledons resulted in increased amino acid concentrations in the phloem sap and a decrease in uptake of NO_3^- by the roots.

Nitrogen translocation is an important process in plant life. Young leaves are supplied with amino acids until they have reached maturity When the supply of N from the root medium is inadequate, N from older leaves is mobilized to feed the younger plant organs. For this reason, plants suffering from N deficiency first show deficiency symptoms in the older leaves. In such leaves protein has been hydrolyzed (proteolysis) and the resulting amino acids have been redistributed to the younger tips and leaves. Proteolysis results in a collapse of the chloroplasts and thus in a decline of the chlorophyll content. Hence yellowing of older leaves is a first symptom of inadequate N nutrition. A most useful review on nitrogen metabolism of higher plants has been published by Blevins (1989).

7.2.5   Nitrogen deficiency symptoms

Visual diagnosis of nutrient deficiency provides a valuable means of assessing the nutritional conditions of a crop. It is practiced successfully only by experts, as it requires much experience. Visual symptoms are only the consequence of metabolic disturbance, and different causes can lead to very similar syndromes. This is also the case for N deficiency. It is beyond the scope of this book to describe N deficiency symptoms in detail for different crops. Useful monographs giving exact descriptions and illustrated by coloured plates have been published by Bergmann (1992), Wallace (1961) and Chapman (1966) for various crops and by Baule and Fricker (1970) for forest trees.

Nitrogen deficiency is characterized by a poor growth rate. The plants remain small, the stems have a spindly appearance, the leaves are small and the older ones often fall prematurely. Root growth is affected and in particular branching is restricted. The root/shoot ratio, however, is usually increased by N deficiency. Nitrogen deficiency results in the collapse of chloroplasts and also in a disturbance of chloroplast development Hence leaves deficient in N show chlorosis which is generally rather evenly distributed over the whole leaf. Necrosis of leaves or parts of the leaf occurs at a rather late and severe stage in the deficiency. In this respect N deficiency differs fundamentally from K^+ and Mg^{2+} deficiencies, where the symptoms also begin in the older leaves but where chlorotic and necrotic spots appear at a rather early stage. Deficiency symptoms of Fe, Ca, S are also similar to N deficiency being characterized by yellowish and pale leaves. In these deficiencies, however, the symptoms occur first in the younger leaves. These more general observations may be used to serve as a first means of distinguishing between these various nutrient deficiencies.

Plants suffering from N deficiency mature earlier, and the vegetative growth stage is often shortened. This early senescence probably relates to the effect of the N supply on the synthesis and translocation of cytokinins (see page 272). Nitrogen deficiency in cereals is characterized by poor tillering; the number of ears per unit area and also the number of grains per ear are reduced. The grains are small, but often relatively high in protein concentration, due to a decrease in the import of carbohydrate into the grains, during the later stages of grain filling.

7.3   Nitrogen Fertilizer Application and Crop Production

7.3.1   General

There is general agreement, that of all the nutrient amendments made to soils, N fertilizer application has had and still has by far the most important effects in terms of increasing crop production. This is true for very different crops growing under the most widely varying conditions throughout the world. In cereals for example in a worldwide study there is a very close correlation between grain yield/ha and the average rate of fertilizer N applied in each nation when there are less than 160 kg N/ha applied (Greenwood and Walker 1990). Numerous field experiments carried out in the past have shown that for many soils, N is the most important growth limiting factor.

7.3.2   Response to nitrogen application

Although crops usually respond to N fertilizers, this is not always the case. Response to N depends on soil conditions, the particular crop species and principally on the amount of available N in the soil and the amount of N which will became available during the growth period. The response is generally poorer the higher the level of available N in the soil (Wehrmann and Scharpf 1979). In the absence of a response, residual N and/or the rate of N release by microbial mineralization of soil organic N is probably adequate to meet the demands of the crop. In extended field trials Mølle and Jessen (1968) observed many years ago that on sandy soils under the humid climatic conditions of Northern Denmark, N application rates of 90 to 135 kg N/ha for barley resulted in optimum economic returns. On peat soils, rich in organic N, however, rates of 45 kg N/ha were sufficient for optimum yields. High release rates of soil N may also be expected, if a grass sward is ploughed. The same is true for crops following a leguminous crop in the rotation. In both these examples N application is generally unneccessary. The cultivation of soils in the tropical rain zone also results in a high release of soil N in the first years of cropping. During this period the soils do not respond to N application (Agble 1974). In following years, however, N application becomes increasingly important as the organic N of the soil is gradually exhausted.

Drainage of organic soils may lead to an explosive mineralization of organic N as reported by van Diest (1977). In order to increase the nutrient holding capacity of these organic soils the water table was lowered from 25 to 75 cm below the soil surface. This resulted in N mineralization rates of about 1000 kg N/ha about half of which was taken up by the grass. The remaining N was leached or denitrified.

In most cases the assessment of the net-N mineralization potential is critical. A reliable method for the determination of the available mineral N and the net-N mineralization during the growth of the crop is still needed as already discussed on page 101. Mehl (1999) studied the net-N mineralization in numerous farmers' fields under winter cereals over several years and found that in farms with livestock an additional N mineralization of 20 kg N/ha could be accounted for. For the precrops sugar beet, potato, rape and fallow 30 kg N/ha and for farms with field vegetables an additional 50 kg N/ha produced by N mineralization were taken into account.

This example shows that for the assessment of a precise N fertilizer rate for winter cereals the farming system precropping was of higher importance than other soil characteristics such as texture and soil pH which showed no major effect in the investigation (Mehl 1999).

Optimum utilization of fertilizer N is only attained if other plant nutrients do not limit growth as was shown for K^+ and phosphate in an older field experiment of Gartner (1969). Without applications of phosphate and fertilizer potassium, the yield response to increasing levels of nitrogen was smaller than when adequate amounts of P and K were applied. The most important factor controlling the effect of nitrogen fertilizer on crops is water supply. This has already been convincingly shown in the field trials of Thompson and Whitney (1998; see Figure 5.26). In studying water and nitrogen fertilizer use by spring wheat in a semi arid environment, Meinke *et al.* (1997) found that nitrogen shortage restricted leaf expansion, leaf number and tillering and thus reduced the leaf area index. When water supply is insufficient, fertilizer nitrogen is inadequately taken up by the crop and some of it may be lost by denitrification as reported by Corbeels *et al.* (1998) in a wheat/sunflower cropping sequence under semi arid Mediterranean conditions. Field trials in Tunisia carried out by Latiri-Souki *et al.* (1998) over several years showed that N fertilizer increased grain yield with and without irrigation. Plants fertilized with N extracted water from deeper layers from the soil profile. Nitrogen application increased the LAI which in turn raised the total accumulated photosynthetic radiation (PAR) to values from 410 to 930 MJ/m^2. Since irrigation water is expensive in arid and semi arid regions, Oweis *et al.* (1999) propose that it be used at rates of about 2/3 of the maximum requirement together with low rates of N fertilizer so that the highest economic return can be made from both these commodities.

Efficiency in N fertilizer usage has been much improved by the introduction of modern cultivars as was shown by Austin *et al.* (1993) in comparison between the modern wherat cutivar *Brimestone* and an old cultivar *Square Head's Master*. Both cultivars produced the same amount of biomass but the modern cultivar had a higher harvest index producing much more grain and less straw and was thus more efficient at converting fertilizer N into grain. Analogous observations have been made in other crop species such as rice, maize and sorghum. The efficiency of nitrogen fertilizer usage as defined above is much dependent on factors such as water supply and the presence of other plant nutrients in the soil. The nitrogen status of the soil is also particularly important. For soils high in available nitrogen, N fertilizer efficiency is low and *vice versa*. According to Crasswell and Godwin (1984), who have considered this question in a useful review paper, one may distinguish between agronomic efficiency, apparent nitrogen recovery and the physiological efficiency of fertilizer N:

$$\text{Agronomic Efficiency} = \frac{\text{Grain yield}_F - \text{Grain Yield}_C}{\text{Fertilizer N applied}} \text{ kg/kg}$$

$$\text{Apparent Nitrogen Recovery} = \frac{\text{N uptake}_F - \text{Nuptake}_C}{\text{Fertizer N applied}} \times 100\%$$

$$\text{Physiological Efficiency} = \frac{\text{Grain Yield}_F - \text{Grain Yield}_C}{\text{N uptake}_F - \text{Nuptake}_C}$$

In these equations the indices F and C denote «fertilized crop» and «unfertilized control» respectively. High «agronomic efficiency» is obtained if the yield increment per unit N applied is high. This is generally the case when the soil is low in available N and the rates of N application are not too high. In Table 7.8 agronomic efficiencies of N fertilizer are shown which were obtained for winter cereals grown on representative agricultural soils from farmers' fields in Hessia, Germany (Kané 2000). Nitrogen fertilizer rates were adjusted to the mineral nitrogen present in the rooting profile in spring (N-min method, see page 101).

Table 7.8 Agronomic efficiencies of N fertilizer for winter cereals grown on farmers field and fertilized according to the Nmin method (After Kane 2000).

Location, Crop,	Year	kg N/ha	Grain Yield, Mg/ha	Agron. Effic.
Giessen, barley	89/90	80	5.48	6.55
Wernborn, wheat	89/90	160	7.06	9.89
Hassenh., barley	89/90	90	6.61	30.3
Bruchk., wheat	98/90	180	4.00	8.94
Wernborn, wheat	90/91	160	6.83	30.82
Giessen, wheat	91/92	100	6.61	4.00
Hassenh, wheat	91/92	160	8.28	7.70
Ossenheim, wheat	91/92	180	5.00	7.32

Depending on the prices of N fertilizer and grain the Agronomic Efficiency should be at least 5. As can be seen from the data in Table 7.8 this target was obtained with one exception; on the soil Giessen (wheat) the Agronomic Efficiency was only 4.0. On this soil the N fertilizer gave no yield increase. However, the crude protein concentration in the grains from the corresponding plot untreated with N was 153 mg N/kg in the dry matter. This is a high value and obviously sufficient N was available in the unfertilized soil. The soil of this *Giessen* location is a deep- rooting alluvial soil rich in interlayer NH_4^+ from which the crop presumably drew much nitrogen. Very high efficiencies of about 30 were obtained in two cases. Here the grain yields in the unfertilized plots were very low. In general the results show that N fertilization is profitable provided that the N rates are well assessed.

A satisfactory % recovery of N fertilizers is attained if the fertilizer applied is not lost (leaching, denitrification) or fixed, but mainly taken up by the crop. Generally recovery levels of about 50% are obtained in practical farming, but much higher and lower values have also been found (Craswell and Godwin1984). High physiological efficiency of N usage in cereal crops is achieved when a high proportion of the N taken up is used in grain formation. For the production of 1 t wheat grains about 30 kg N are required, 18 kg are present in the grains, 6 kg in the roots and 6 kg in the straw. Highest physiological efficiency thus occurs, when 33 kg grain are produced for every kg N taken up by the crop.

Within the past decade there has been enormous development in the study of modelling to simulate N turnover in the soil-crop system for predictive use in crop N management and many models have been proposed. One such model, The Rothamsted Nitrogen Turnover Model, (Bradbury et al. 1993), uses readily available input parameters to describe N turnover, in terms of uptake of mineral N by the crop, mineralization-immobilization turnover of organic matter and gains and losses of N from the soil-crop system. SUNDIAL (SimUlation of Nitrogen Dynamics In Arable Land) is a user- friendly, PC- based version of this model (Smith et al. 1996) and can be used for example by farmers and agricultural advisers by entering a particular farm management strategy to simulate N turnover and predict nitrate losses. In a most valuable review, Greenwood (2001) discusses his model N_ABLE which simulates the response of crops to N and which is based on fundamentally derived equations for groups of processes that dominate in plant nutrition. The equations include ones for the decline in critical %N with increase in plant mass, for the dependence of growth rate on sub-optimal %N, for the development of root systems and their ability to extract nitrate from soil. These have been combined with those for soil processes into the model which calculates daily increments in N-uptake, growth, changes in the distribution of water and nitrate down the soil profile and the amounts of N leached out of the profile. It requires only readily available inputs including, highest expected yield, times of planting and harvest, fertilizer level, mineral N status down the profile and crop and weather. The model has been calibrated for different crops and its validity tested against the results of field experiments. User-friendly versions have had an impact on growing vegetable crops in the UK. An internet version can be run interactively at http://www.qpais.co.uk/nable/nitrogen.htm. The model enables simulations to be made for 22 different crops, grown on different soils using either actual daily weather records or best estimates for daily weather for any of 134 sites throughout the world. It takes usually less than 2 minutes to run and various tabular and graphical outputs are offered.

7.3.3 Nitrogen fertilizer application rates

The amount of N that should be applied to a crop depends largely on the particular crop species and on the prevalent soil conditions, especially on the available N in the soil. Generally the quantity of N taken up by a good crop over the growth period serves as a guideline in assessing the appropriate rate of N application (kg N/ha). These quantities are listed for several crops in Table 6.2 (page 340). When the rate of inorganic N release from soil organic matter is high, lower N application rates need to be applied. On the other hand, for poor soils low in N, the N application rate should be in excess of the total amount of N uptake.

Straw application to soils generally increases the immobilization of fertilizer N and may lead to a depression in crop yields (Amberger and Aigner 1969), although the effect can be overcome by an additional N application of about 10 kg N per Mg straw applied. In paddy soils incorporation of straw into the soil was not found to affect rice yields (Williams et al. 1972).

Optimum rates of N application also depend on winter rainfall, as observed in field experiments, carried out over a number of years in the Netherlands (Van der Paauw 1962b).

In mild winters with heavy rainfall, considerable amounts of N are leached out of the soil. In order to maintain yield levels this loss must be compensated for by increased application of fertilizer N in spring.

In the long run, N removal from the soil be it as plant products, leaching or denitrification, must be balanced by an adequate return of N fertilizer. Large amounts of N can be fixed by the *Rhizobium* legume associations. For this reason, forage legumes such as lucerne or clover and in most cases also seed legumes are not usually fertilized with N. Indeed N application may even depress yields, as N fertilizer N inhibits N fixation and favours the growth of weeds. In grass legume associations, N application favours grasses which are able to compete with legumes for plant nutrients and other growth factors. Mixed grass clover swards should therefore not be treated with fertilizer N, if the proportion of legumes is adequate to meet the demands of the sward. When the proportion of legumes in the sward is very low, however, N fertilizer treatment is usually recommended. In Denmark Dam Kofoed and Sondergaard Klausen (1969) found that rates of 150 to 300 kg N/ha/per year gave best results on grassland where the proportion of clover was very low or absent. Where conditions favour the growth of legumes, high forage yields can be obtained without N fertilizers. This is the case for large grassland areas of New Zealand. High forage yields without N fertilizer can also be obtained in Europe with modern clover cultivars (McEwen and Johnston 1984).

Excessive N application may be detrimental to crops. As already mentioned (see page 290) a high level of N nutrition during the last months of sugar beet growth reduces the quality of the roots. For cereals lodging may result. In order to reduce susceptibility to lodging, it has become a common practice in Central Europe to apply chemicals, which shorten cereal straw length (see chapter 5.2.7). Excess N can stimulate various fungal crop diseases (Trolldenier 1969). Examples of this kind are brown rust (*Puccinia hordei*) on barley, brown leaf spot (*Helminthosporium oryzae*) on rice (Hak 1974) and *Fusarium graminearum* on wheat (Bunescu *et al.* 1972). Disease may be especially severe, if the supply of K and P to the crop is low. On the other hand two diseases of maize *Pseudomonas syringae* (chocolate spot) and *Helminthosporium turcicum* (Northern corn leaf blight) appear to be reduced by high N application rates (Karlen *et al.* 1973).

7.3.4   Nitrogen fertilizers

The most common straight N fertilizers are listed in Table 7.9. For most N fertilizers NO_3^- and NH_4^+ are the N carriers. This is also the case for mixed and compound fertilizers. Ammonium is partially adsorbed on soil colloids and its uptake rate is usually therefore lower than that of NO_3^- under field conditions. For this reason most crops do not respond as quickly to NH_4^+ fertilizers as to NO_3^- application. Nitrate fertilizers are known to produce a rapid response in the plant. In most cases, however, the difference between both types of N fertilizers plays only a minor role. Thus Huppert and Buchner (1953) in evaluating numerous field experiments carried out in Germany, found that there were no major differences in yield response whether crops had been dressed with nitrate or ammonium. Only on more acid soils was NO_3^- superior to NH_4^+. Widdowson *et al.* (1967) reported that $Ca(NO_3)_2$ gave larger grain yields of barley than $(NH_4)_2 SO_4$ in three quarters of all the experiments carried out on light or medium textured soils in England. The yield differences, however, were not great. Nitrogen losses by denitrification and leaching may be higher with NO_3^- than with NH_4^+ application. In Table 7.10 the N recovery in a three years' rotation is shown from the work of Riga *et al.* (1980). The experiment compares the N recovery from 100 kg of labelled N fertilizer (^{15}N) supplied either as NO_3^- or NH_4^+ in the first year of the rotation. The nitrogen was given in a split application either as 2 or as 3 equal dressings. As can be seen from the table, N losses (unrecovered N) were higher with NO_3^- than with NH_4^+ application and with 2 rather than with 3 dressings. NH_4^+ application resulted in a higher amount of N incorporated into the soil. About half of the N applied was taken up by wheat in the first year while uptake by the following crops, 2nd year oats and 3rd year two harvests of forage maize, was comparatively low.

Table 7.9   Major nitrogen fertilizers

Fertilizer type	Formula	%N
Ammonium sulphate	$(NH_4)_2SO_4$	21
Ammonium chloride	NH_4Cl	26
Ammonium nitrate	NH_4NO_3	35
Nitrochalk	$NH_4NO_3 + Ca\,CO_3$	21
Ammonium nitrate sulphate	$NH_4NO_3\,(NH_4)_2SO_4$	26
Potassium nitrate	KNO_3	14
Urea	$CO\,(NH_2)_2$	46
Calcium cyanamide	$CaCN_2$	21
Anhydrous ammonia	NH_3	82

Table 7.10   Fate of labelled N applied in the form of NO_3^- or NH_4^+ in a split application experiment with 2 or 3 equal dressings given in the first year of a crop rotation experiment; N fertilizer rate = 100 kg N/ha (data from Riga *et al.* 1980)

Fertilizer form and number of dressings	Wheat 1. year	Oats 2. year N uptake	2 × Maize 3. year	N in soil at the end of the exp.	N loss (unrecov. N)
			kg N/ha		
$3 \times NO_3^-$	57.4	1.31	2.02	8.1	31.2
$2 \times NO_3^-$	46.7	1.27	1.54	8.2	42.3
$3 \times NH_4^+$	64.1	1.70	2.74	13.4	18.5
$2 \times NH_4^+$	45.6	1.70	2.74	14.1	35.9

Ammonium containing fertilizers should not be mixed with calcareous unfertilizers *e.g.* ammonium sulphate with $CaCO_3$ since the resulting alkaline pH will lead to a deprotonation of NH_4^+ and NH_3 is released.

High NH_3 losses may occur on alkaline and calcareous soils because of NH_3 volatilization (see page 51). Such soils should therefore be dressed with NO_3^- and not with urea or NH_4^+ containing fertilizers. On paddy soils N losses can be high as a result of denitrification (see page 50). These soils should therefore not receive nitrate-containing fertilizers. Here urea and ammonium fertilizers are recommended. They should be worked well into the soil (Schnier et al. 1988).

Ammonium sulphate used to be a very important fertilizer, but its relative consumption has declined more and more in recent years. To some extent it has been replaced by urea which has a higher N concentration and is easily handled in solution form (see page 362). Ammonium nitrate is an explosive salt and for this reason its direct use as fertilizer is prohibited in some countries. This rather high graded N fertilizer is now often used in the preparation of liquid fertilizers. Because of the risks of fires, however, NH_4NO_3 is often mixed with limestone. This mixture is safe and easy to handle and called nitrochalk. It is a well known straight N fertilizer, which due to its content of limestone prevents or delays acidification of the soil. Ammonium chloride is of only minor importance. According to Cooke (1972) it is suitable for paddy soils, where the use of $(NH_4)_2SO_4$ leads to the production of undesirable sulphide. Ammonium nitrate sulphate is a double salt of ammonium nitrate and ammonium sulphate made by neutralizing nitric and sulphuric acids with NH_3. The product is easy to handle and to store. Its consumption, however, has decreased in the recent years. Potassium nitrate contains 44% K_2O (= 35.2% K) in addition to N. This fertilizer is often used as foliar spray. The same is true for urea, which is also used to a large extent in liquid fertilizers (see page 362).

Calcium cyanamide contains N in the amide and in cyanide forms. This fertilizer has a darkish colour as it also contains some C, which is formed during the production process. Calcium cyanamide is soluble in water. As shown in Figure 7.11 Ca cyanamide is converted in the soil to urea, which is again split into NH_3 and CO_2.

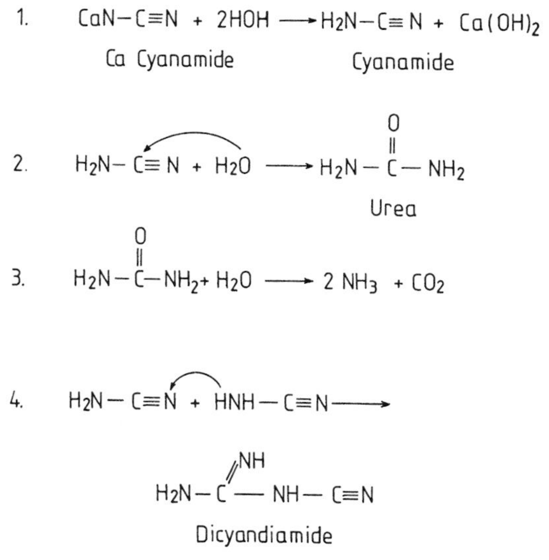

1. $CaN-C\equiv N + 2HOH \longrightarrow H_2N-C\equiv N + Ca(OH)_2$

    Ca Cyanamide   Cyanamide

2. $H_2N-C\equiv N + H_2O \longrightarrow H_2N-\overset{\overset{\textstyle O}{\|}}{C}-NH_2$

                                  Urea

3. $H_2N-\overset{\overset{\textstyle O}{\|}}{C}-NH_2 + H_2O \longrightarrow 2 NH_3 + CO_2$

4. $H_2N-C\equiv N + HNH-C\equiv N \longrightarrow$

                     $H_2N-\overset{\overset{\textstyle NH}{\|}}{C}-NH-C\equiv N$

                     Dicyandiamide

Figure 7.11 Decomposition of Ca cyanamide in soils. — 1. Ca cyanamide hydrolyzes into cyanamide and Ca hydroxide — 2. Cyanamide reacts with water to form urea — 3. Urea reacts with water forming ammonia and carbon dioxide — 4. Two cyanamide molecules react to form dicyandiamide

The conversion of Ca cyanamide in soils needs water. For this reason the response of plants to calcium cyanamide application is delayed in dry soil conditions. During the conversion process in the soil intermediate toxic products can be formed. Generally calcium cyanamide is applied before sowing and worked well into the soil. The intermediate toxic substances are good weed killers and nitrification inhibitors. Cyanamide is a slow reacting N fertilizer. Dicyandiamide formed during the degradation of calcium cyanamide (Figure 7.11) retards the ammonification of urea and also the oxidation of NH_3 since it is an nitrification inhibitor (see page 414) The N effect of the Ca cyanamide fertilizer is thus delayed (Rathsack 1978). During the conversion process $Ca(OH)_2$ is also formed, which has a favourable effect on pH and soil structure in acidic soils.

The application of anhydrous NH_3 has already been discussed on page 362. The main drawback of this high graded N fertilizer is the special equipment required for its transport and application. Numerous field trials have shown that anhydrous NH_3 produces very much similar responses to solid N fertilizers (Dam Kofoed et al. 1967). On heavier textured soils and under more continental climatic conditions, anhydrous NH_3 can also be applied in autumn without the risk of N loss by leaching (Korensky and Neuberg 1968). On these soils the risk of leaching losses only occur, if the NH_4^+ is oxidized to NO_3^- to any great extent. This microbial oxidation depends largely on the prevalent soil temperatures in autumn and winter. Thus under climatic conditions with mild winters and high rainfall the application of anhydrous NH_3 or other NH_4^+ fertilizers can lead to substantial N losses due to the leaching of NO_3^- (Dam Kofoed et al. 1967). This loss of N during winter is probably the main reason why anhydrous NH_3 applied in autumn often results in lower yields than are obtained following a spring application (Roussel et al. 1966).

8.1   Soil Sulphur

Sulphur occurs in the soil both in inorganic and organic forms but in most soils organically bound S provides the major S reservoir. In most instances in soils of humid and semi humid areas the range of total S is from 100 to 1000 mg S/kg, a range that is similar to that of total P (Syers *et al.* 1987). Organic S concentration generally decreases with soil depth and approaches zero in soil depth >1.5 m. In peat soils organic S may amount to almost 100% of the total S. The C:N:S ratio of soil organic matter is approximately 125:10:1.2. Sulphur in soil organic matter can be divided into 2 fractions, carbon bonded S, and non carbon bonded S. These fractions are distinguished by treatment with hydriodic acid since S which is not bonded to C can be reduced to H_2S. This includes compounds with a C-O-S linkage, an organic ester sulphate or a C-N-S linkage as sulphamate. Carbon bound-S is reduced and is present in cysteine and methionine in which the S atom has accepted an e^- from each of both binding partners. The sulphate esters can be readily hydrolyzed by soil sulphatases (Martens *et al.* 1992) and the S present therefore becomes rapidly available to plant uptake. By contrast, however, mineralization of reduced S in organic matter proceeds at a much slower rate (Maynard *et al.* 1985, Nguyen and Goh 1992). The reverse process of mineralization of organic S, that of immobilization of inorganic S, can also occur by microbial assimilation (Pirela and Tabatabai (1988). These authors found that under ryegrass the concentration of organic S increased. The biochemical pathway of assimilation involves the reduction of SO_4^{2-} to sulphide which combines with serine to form cysteine (Scott 1985) so that there are clear similarities with sulphate reduction by higher plants (see page 178). Immobilization of SO_4^{2-} in soils is driven primarily by microbial activity and the conversion of S into the microbial biomass and is thus greatly favoured by the addition of C sources as for example cereal straw with a wide C:S ratio (Wu *et al.* 1993).

The inorganic forms of S in soil consist mainly of sulphate. In arid regions soils may accumulate high amounts of salts such as $CaSO_4$, $MgSO_4$ and Na_2SO_4. Sulphate may be present dissolved in the soil solution or adsorbed on soil colloids both sulphate fractions being in dynamic equilibrium. Sulphate like phosphate is adsorbed to sesquioxides and clay minerals, although the binding strength for sulphate is not as strong as that for phosphate. Sulphate adsorption capacity follows the order: Al_2O_3 > kaolinite > bauxite > peat > limonite > haematite > hydrated aluminium > goethite. According to Turner and Kramer (1991) sulphate may be adsorbed as mononuclear and binuclear complexes as shown in Figure 8.1. The adsorption is by ligand exchange with surface hydroxyl groups by which sulphate replaces OH^- coordinated to Fe or Al to form a covalent bond with the surface. As shown in the Figure 8.1, in the mononuclear complex two sulphates bind to one Fe^{III} while in the binuclear complex one sulphate replaces two OH_2^+ to bind to two Fe^{III} nuclei. At low pH the mononuclear complex dominates.

This binding is very strong, stronger than the adsorption of water and therefore the mononuclear binding of sulphate is considered as irreversible. The Figure 8.1 also shows SO_4^{2-} adsorbed electrostatically (non specifically) to sites bearing a positive charge as for NO_3^- and Cl^- adsorption. There is now some evidence that this form of adsorption of SO_4^{2-}, a weaker form of adsorption, is more important than was formally recognised (Marsh *et al.* 1987). As shown in Figure 8.10 on page 449 the pH dependance of non specific sulphate adsorption differs much between soil types.

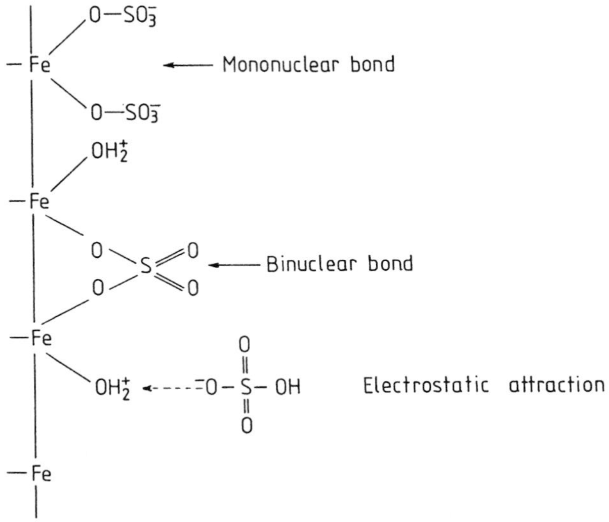

Figure 8.1 Adsorption of sulphate to goethite. The adsorption takes place by electrostatic attraction of sulphate by a protonated OH group of the goethite followed by the formation of binuclear and mononuclear complexes.

Sulphate adsorption capacity of clay minerals follows the sequence: kaolinite > illite > bentonite. Adsorption strength for sulphate decreases as soil pH increases. Martini and Mutters (1984) found significant correlations between soil pH and the concentration of exchangeable sulphate in Coastal Plain soils and Piedmont soils of South Carolina. There was also a clear relationship between the clay concentration of soils and the concentration of exchangeable sulphate; exchangeable sulphate increasing with soil clay concentration.

Under waterlogged conditions, inorganic S occurs in reduced forms such as FeS, FeS_2 (pyrites) and H_2S. Total soil levels depend on organic matter contents and also on climatic conditions. In soils under humid conditions high amounts of SO_4^{2-} are leached whereas in arid soils SO_4^{2-} accumulates in the upper profile. Soils of the temperate regions are generally higher in S the more organic matter they contain.

Microbial mineralization of organic S in soils under anaerobic conditions leads to the formation of H_2S and if aerobic conditions follow, the H_2S formed readily undergoes autoxidation to SO_4^{2-}. In anaerobic media, however, H_2S is oxidized to elemental S by chemotrophic sulphur bacteria (*Beggiatoa, Thiothrix*). The same bacteria can also oxidize S to H_2SO_4 in the presence of oxygen. Elemental S is also oxidized by chemotrophic bacteria of the genus *Thiobacillus*. The overall process may be expressed as follows:

$$2H_2S + O_2 \Rightarrow 2H_2O + 2S + 510 \text{ kJ}$$
$$2S + 3O_2 + 2H_2O \Rightarrow 2H_2 SO_4 + 1180 \text{ kJ}$$

$$\text{net: } 2H_2 S + 4O_2 \Rightarrow 2H_2SO_4 + 1690 \text{ kJ}$$

The oxidation of S thus results in the formation of H_2SO_4. A consequent increase in soil acidity thus occurs. In this way the sulphate acid soils are formed as decribed by Ahmad and Wilson (1992) for the Carribean region. Because of the extremely low pH, these soils are very rich in exchangeable Al and therefore very infertile and are difficult to ameliorate. Oxidation of elemental S is generally a microbial process in arable soils but may also proceed abiotically (Wainwright 1984). The same process also accounts for the acidification resulting from the addition of elemental S to soils. This treatment is sometimes used to depress the pH of alkaline soils (Jolivet 1993). In a similar way to the reactions described above, FeS can be oxidized biologically and chemically to elemental S (Schoen and Rye 1971) according to the equation:

$$FeS + H_2O + 1/2 O_2 \Rightarrow Fe (OH)_2 + S$$

Under reducing soil conditions (waterlogged soils, paddy) H_2S is the most important end product of anaerobic S degradation. Organic sulphides, such as methyl- and butyl sulphides are also formed, which like H_2S are both characterized by an unpleasant odour. Photosynthetic green and purple bacteria can oxidize H_2S to S by utilizing the H of the H_2S as e⁻ for photosynthetic e⁻ transport. When this process is restricted H_2S may accumulate to toxic levels and thus impair plant growth. To some extent, the detrimental effect of H_2S can be alleviated by the addition of ferrous salts which form the sparingly soluble FeS. Sulphate reduction under anaerobic conditions is mainly brought about by bacteria of the genus *Desulfovibrio* (Ponnamperuma 1972). These bacteria utilize the oxygen of the SO_4^{2-} as a terminal e⁻ acceptor.

The processes of S conversion in soils discussed above are shown in Figure 8.2. As most soils are aerobic the formation of SO_4^{2-} is favoured and the concentration of SO_4^{2-} in the soil solution represents the dynamic equilibrium of processes leading to the formation or removal. As this ion is relatively mobile in soils, some SO_4^{2-} is usually lost by leaching. Sulphate is the most important form in which plants absorb S from the soil medium. In the plant itself a part of the absorbed sulphate is reduced and converted to an organic form (see page 178). Organic S present in the remnants of dead plant material in the soil is also involved in the S cycle as shown in Figure 8.2. Under reducing conditions H_2S is produced. Some H_2S can be released into the atmosphere and is thus lost from the soil system. Most useful reviews dealing with the S cycle and S turnover in soils have been presented by McGrath *et al.* (1994) and Zhao *et al.* (1996).

438

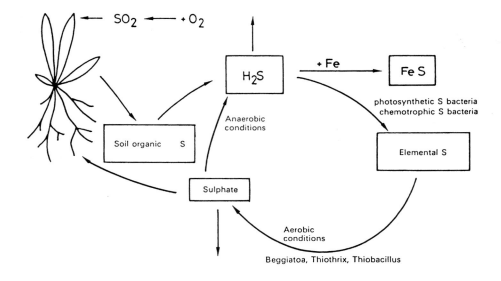

Figure 8.2 Sulphur cycle in nature.

8.2   Sulphur in Physiology

8.2.1   Uptake and translocation

The most important S source for plants is SO_4^{2-}. Its uptake across the plasmalemma is H^+ cotransport in which at least 3 H^+ are involved per 1 SO_4^{2-} (Clarkson *et al.* 1993, Tanner and Caspari 1996). Hawkesford *et al.* (1995) reported that S- starved tomato plants had a higher plasmamembrane S transporter activity and also a higher ATP: sulphurylase activity, the latter enzyme being essential for the assimilation of sulphate (see page 177). According to Leustek and Saito (1999) most of sulphate transporters which have been analyzed are encoded by a gene family. High affinity transporters are mainly located in the plasmalemma of roots with a K_M of about 10 mmol/m^3 and low affinity transporters with a K_M of about 100 mol/m^3 in the plasmalemma of leaves. Sulphate transport across the tonoplast is driven by the electrical gradient between the vacuole and the cytosol. Plants may also take up H_2S and SO_2 from the atmosphere and plant canopies act as a strong sink for SO_2 deposition (Baldochini 1993). Other plant nutrients scarcely affect the absorption of SO_4^{2-} by plant cells. Selenate, however, which is closely chemically related to SO_4^{2-}, depresses SO_4^{2-} uptake substantially (Leggett and Epstein1956) indicating that both ion species probably compete for the same uptake system. The sulphate transport protein, located in the plamalemma requires a pH gradient with the pH in the outer medium lower than in the cytosol (Clarkson *et al.* 1993).

Sulphate transport from the cytosol into the vacuole is a "downhill transport" which means that the SO_4^{2-} follows the electrochemical gradient built up by tonoplast H^+ pumps (see page 129). This is in line with investigations of Lin (1981) who found that SO_4^{2-} uptake by protoplasts of maize roots was promoted by a low pH in the outer medium. As yet it is not yet clear whether the SO_4^{2-} stored in the vacuole can be remobilized for metabolic purposes.

Sulphate taken up by the roots is distributed within plants by xylem and phloem transport (Rennenberg 1984). Phloem transport of S from older to younger leaves, however, can often be poor and this is the reason for the frequent appearance of S deficiency first in younger leaves when S supply from the root is unable to meet their demands (Bouma 1967). Bell et al. (1995) sugggested that the SO_4^{2-} stored in vacuoles of mesophyll cells is released only under conditions of prolonged S stress and is too slow to support new growth. However, S is mobile in the phloem (see page 212) and indeed since the reduction of sulphate is to a large extent a light dependent process (see page 178) occuring predominantly in the chloroplast of mature leaves (Brunold 1990) roots are dependent on the transport of reduced S from the shoot. Rennenberg and coworkers have identified the tripeptide glutathione (GSH) as main long distance transport form of reduced S (Rennenberg et al. 1979) and also as an inhibitor of SO_4^{2-} uptake by roots (Rennenberg et al. 1988). The very convincing experimental findings of this group with tobacco plants support the hypothesis that GSH in the phloem can act as a shoot to root signal which provides a very sensitive control of SO_4^{2-} uptake by roots and hence the S nutrition of plants (Herschbach and Rennenberg 1994). Low GSH concentrations in the phloem sap favour the synthesis of SO_4^{2-} transporters (Leustek and Saito 1999). These findings are also very much in accordance with the well known observation that the root sulphate transporters are strongly repressed in the presence of an adequate S supply.

8.2.2   Metabolic functions of sulphur

The assimilation (reduction) of SO_4^{2-} has already been described on page 178. Normally, reduced-S is rapidly incorporated into an organic molecule, the first stable organic S compound being cysteine (Leustek and Saito 1999). The SH (sulphydryl or thiol) group from cysteine can be transferred to phosphohomoserine to form cystathionine which breaks down to produce homocysteine. This compound in turn, can be converted to methionine by a CH_3 group transfer. According to Wilson et al. (1978) plants may also produce H_2S from SO_4^{2-} if very high amounts of sulphate are supplied. In such cases of oversupply of sulphate O-acetylserine becomes the limiting factor and therefore the reduced S cannot be assimilated and hydrogen sulphide is formed and emitted into the air as found by Rennenberg (1983).

Cysteine and methionine are the most important S containing amino acids in plants, where they both occur as free acids and as building blocks of proteins. One of the main functions of S in proteins or polypeptides is in the formation of disulphide bonds between polypeptide chains. The synthesis of the dipeptide cystine from two cysteine molecules brings about the formation of a disulphide bond (S-S-bond) from two SH groups (Figure 8.3).

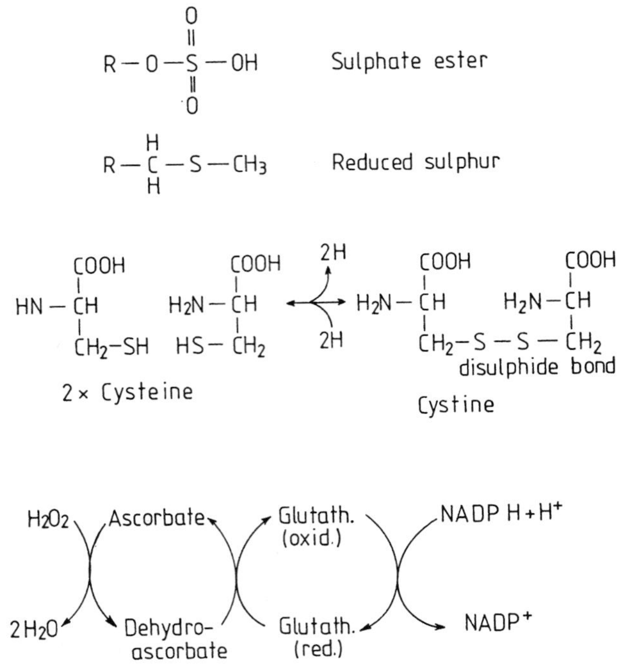

$$R-O-\overset{\overset{\displaystyle O}{\|}}{\underset{\|}{S}}-OH \qquad \text{Sulphate ester}$$

$$R-\overset{\overset{\displaystyle H}{|}}{\underset{|}{C}}-S-CH_3 \qquad \text{Reduced sulphur}$$

$$\underset{\text{2}\times\text{Cysteine}}{\overset{\overset{\displaystyle COOH}{|}}{\underset{\underset{\displaystyle CH_2-SH}{|}}{HN-CH}} \qquad \overset{\overset{\displaystyle COOH}{|}}{\underset{\underset{\displaystyle HS-CH_2}{|}}{H_2N-CH}}} \quad \underset{2H}{\overset{2H}{\rightleftharpoons}} \quad \underset{\text{Cystine}}{\overset{\overset{\displaystyle COOH}{|}}{\underset{\underset{\displaystyle CH_2-S-S-CH_2}{|}}{H_2N-CH}} \quad \overset{\overset{\displaystyle COOH}{|}}{H_2N-CH}}$$

disulphide bond

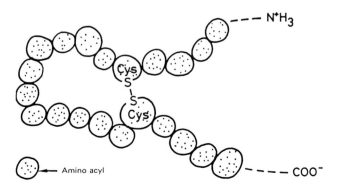

Figure 8.3 Sulphate ester, reduced organic S, oxidation of cysteine to cystine and *vice versa.* — Lower line: Halliwell-Asada reaction pathway.

The disulphide bond can serve as a covalent cross linkage between two polypeptide chains or between two points on a single chain. It thus stabilizes the polypeptide structure (see Figure 8.4). The formation of disulphide bonds in polypeptides and proteins is an essential function of S in biochemistry, as these S-S bridges contribute to the conformation of enzyme proteins. It is also important in the baking quality of flour

Figure 8.4 S-S bridge of a polypeptide chain.

for breadmaking because the sulphydryl and disulphide groups of cysteine and cystine in grain storage proteins such as glutenin are essential to the viscoelasticity of the dough. A further essential function of SH groups in metabolism is their direct participation in enzyme reactions, although not all free SH groups in enzymes are active.

As shown in Figure 8.3 cystine is formed by the oxidation (release of H) of two molecules of cysteine. The whole reaction serves as a redox system which can take up or release H atoms depending on prevailing metabolic conditions. Under reducing conditions (excess of H or reduced coenzymes) the equilibrium is shifted in favour of cysteine whereas under oxidizing conditions cystine is formed. The system thus functions as either an H donor or a H acceptor. The glutathione redoxsystem is analogous to the cysteine/cystine system. Glutathione is a tripeptide consisting of γ-glutamyl (= residue of glutamate), cysteinyl (= residue of cysteine) and glycine moieties (Figure 8.5). The reactive group of the system is the SH group of the cysteinyl moiety which forms an S-S bridge with the SH group of another glutathione molecule. Because of its higher water solubility, the glutathione redox system plays a more important role in metabolism than the cystine/cysteine redoxsystem. Reduced glutathione serves as a sulphhydryl buffer which maintains the cysteine residues of proteins in a reduced form. The oxidized form of glutathione can be reduced by NADPH (see Figure 8.5). The reduced form of glutathione also acts a storage and transport form of reduced S as already discussed (Herschbach and Rennenberg 1994). An analogous tripeptdide to glutathione is γ-glutamyl- cysteinyl- serine which is found particularly in grasses including rice and has the same functions as glutathione (Klapheck et al. 1992).

The glutathione redox system is an essential part of the *Halliwell-Asada pathway* shown in Figure 8.3 which in conjunction with an ascorbate dehydrogenase provides an important mechanism in the detoxification of oxygen radicals. This reaction pathway detoxifies H_2O_2 produced by superoxide dismutase in chloroplasts (Bowler et al. 1992).

The reactive sites of *thioredoxins* also contain cysteine molecules. Thioredoxins are ubiquitous proteins with a molecular weight of about 12 kDa possessing a well conserved reactive site structure with an amino acid sequence of:

Trp- Cys- Gly- Pro-Cys

The SH groups of both cysteine molecules are involved in the redox process and form an S-S bridge when oxidized. Thioredoxins function as regulator proteins and activate various enzymes, especially those involved in sugar metabolism in plants (Schürmann 1993). They also play a role in sulphate reduction (see page178) According to Kobrehel et al. (1992) a thioredoxin reduces gliadin and glutenin storage proteins in wheat endosperm before these proteins are hydrolyzed and therefore has a signal function in grain germination.

Cysteine is also an essential amino acids in *phytochelatins*. These sulphur rich polypeptides chelate heavy metals, the primary structure being characterized by the amino acid sequence

(glutamyl-cytseinyl-)$_n$ glycine

which is the glutathione sequence. Alanine may substitute for glycine in some cases; n ranges between 2 and 11. The main function of these compounds is to complex

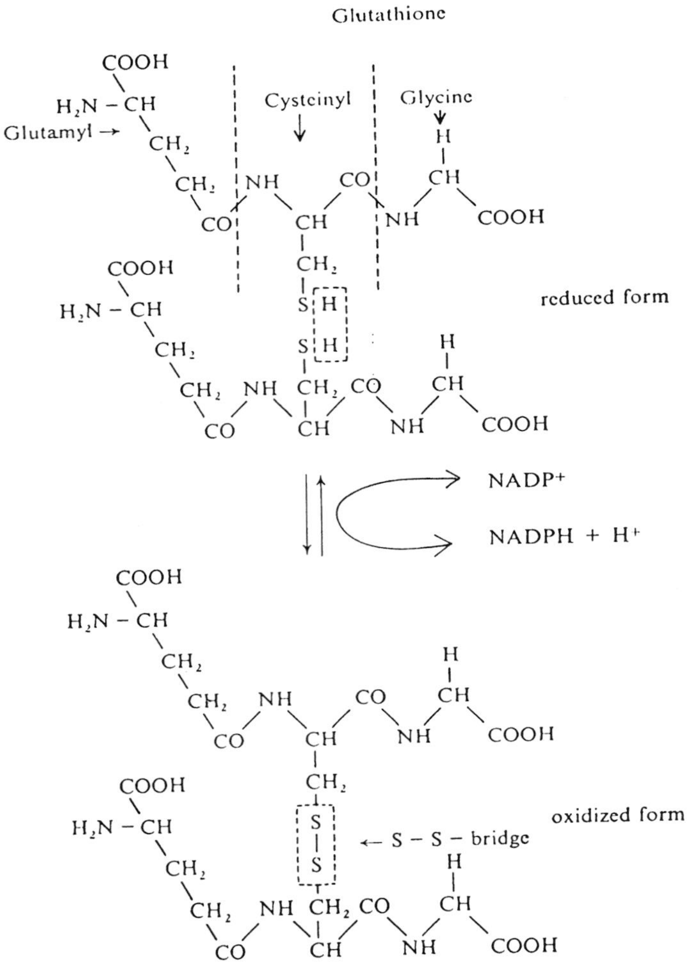

Glutathione

Figure 8.5 Oxidation and reduction of gluthathione by means of NADP+ or NADPH + H+, repectively.

heavy metals particularly Cd^{2+} (Rauser 1993) which is complexed by 4 S atoms of a cysteinyl moiety two forming ionic and two forming coordinative bonds. Cd^{2+} induces the synthesis of such phytochelatins. The affinity of the SH groups to heavy metals following the sequence shown below (Grill *et al.* 1987):

$$Cd > Pb > Zn > Sb > Ag > Ni > Hg$$

These cysteinyl-rich polypeptides play an essential role in the detoxifcation of heavy metals in plant and animal metabolism.

The SH group of lipoic acid also participates in redox reactions in a similar way to the glutathione redox system (Figure 8.6). Lipoic acid is a coenzyme which is involved in oxidative decarboxylation of of 2 oxo acids.

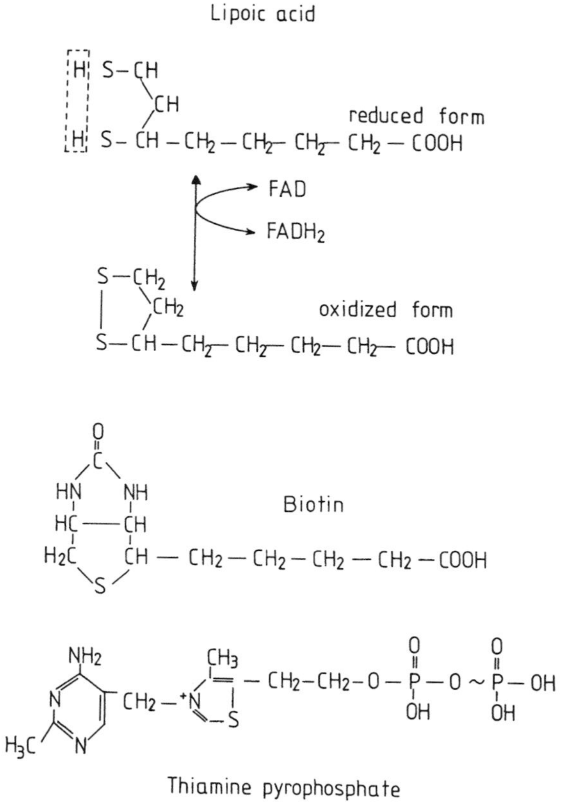

Lipoic acid

Figure 8.6 Oxidation and reduction of lipoic acid, and formula of biotin, and thiamine pyrophosphate

Another important group of S containing compounds are the ferredoxins (a type of non haem iron sulphur protein). These low molecular weight proteins contain a high proportion of cysteine units, and an equal number of S and Fe atoms in addition to the S contained in the cysteine and methionine units of the protein chain. The S and Fe atoms linked together in the form of a cluster which is bound to the protein chain *via* the S atoms of cysteine (see Figure 3.19). This configuration confers a highly negative redox potential, the most negative known in any biological compound.

Sulphur is a constituent of CoA and of the vitamins biotin and thiamine. Biotin is involved in the carboxylation of acetyl-CoA. Sulphur is an essential element of the thiazole ring which is a component of the vitamin thiamine (Figure 8.7). Thiamine may occur as the free vitamin or as thiamine pyrophosphate. In contrast to animal tissues, plants contain thiamine very largely in the free vitamin form. Little is known of the role of thiamine in plants. Thiamine pyrophosphate acts as a coenzyme in the decarboxylation of pyruvate to acetaldehyde and the oxidation of 2'oxoacids. The basis of these reactions is the ability of the thiazole ring of thiamine pyrophosphate to bind and activate aldehyde groups.

In coenzyme A (CoA) the active site of the molecule is the SH group (Figure 8.7). It can react with organic acids according to the equation:

$$R\text{-}COOH + HS\text{-}CoA \Rightarrow R\text{-}CO\text{ -}S\text{ -}CoA + H_2O$$

In this way the SH group becomes esterified with the acyl group of an organic acid. Coenzyme A thus serves as a carrier of acyl groups.

$$CH_3\text{-}CO\text{-}S\text{-}CoA \;(= \text{Acetyl CoA})$$

Acetyl CoA is formed when CoA reacts with acetic acid. This is an important example of an activated acid of this type and plays a very key role in fatty acid and lipid metabolism.

Many plant species contain small amounts of volatile S compounds. These are mainly di- or polysulphides. They occur in onions where they are responsible for their lachrymatory effect, *i.e.* bringing tears to the eyes. The main component in garlic oil is diallyldisulphide shown below.

$$CH_2 = CH \text{ - } CH_2 \text{ - } S \text{ - } S \text{ - } CH_2 \text{ - } CH = CH_2$$

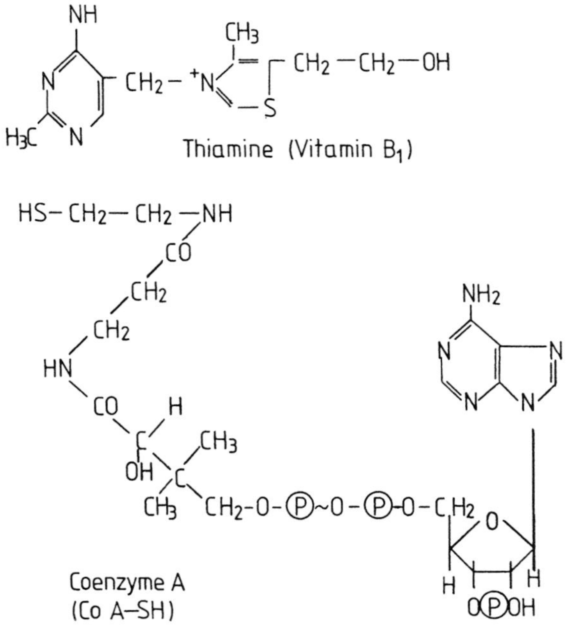

Figure 8.7 Thiamine (Vitamin B$_1$) and Coenzyme A (CoA-SH)

The mustard oils, the *glucosinolates* are derived from amino acids. The main synthetic pathway is outlined below in Figure 8.8.

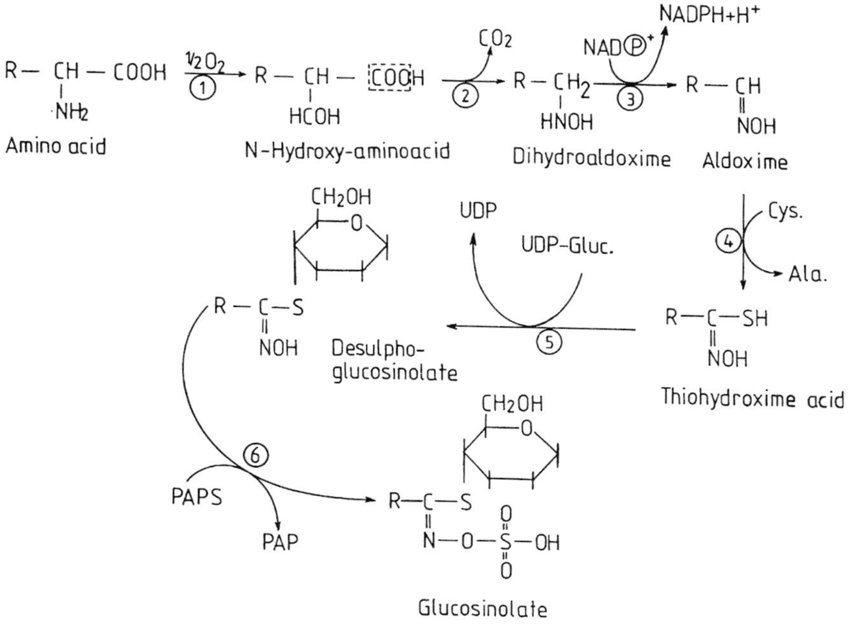

Figure 8.8   Putative pathway for the synthesis of glucosinolates (Glendening and Poulton 1988).
— 1. The amino acid is oxygenated and a N-hydroxyamino acid is formed. — 2. The N- hydroxyamino acid is decarboxylated and a dihydroaldoxime is formed. — 3.The dihydroaldoxime is oxidized and an aldoxime is formed. — 4. A cysteine tranfers the SH group to the aldoxime and a thiohydroximic acid is formed. — 5. UDP-Glucose transfers its glucosyl to the thiohydroximic acid and a desulpho- glucosinolate is formed. — 6. Phospho-Adenosine-Phospho-Sulphate (PAPS) transfers its sulphuryl group to the desulphoglucosinolate and the glucosinolate is formed.

The term R in the general formula for glucosinolates is most usually not the residue of a simple amino acid but is a product of the alteration of amino acids such as gluconapin which is $CH_2 = CH - CH_2-CH_2-$. In rape 2-hydroxy-3-butenyl (CH_2-CH-CHOH-CH_2-) dominates. Indolmethyl (= Glucobrassicine) also functions as the radical R in glucosinolates. Glucosinolates which largely account for the high S concentrations generally found in the Cruciferae may be hydrolyzed by myrosinase by which process sulphate, glucose and HCNS is produced (Schnug 1993). Sulphate thus produced may be used as nutrient source if the S supply is insufficient (Schnug 1989). Glucose can be used in carbohydrate metabolism and HCNS may protect the plant against microbial and insect attack (Schnug 1991). Modern plant breeding has developed rape cultivars which are almost devoid of glucosinolates.

The total S concentration in plant tissues is in the order of 2 to 5 mg S/g dry matter. With increasing S supply the concentration of sulphate S increases whereas the concentration of organic S remains constant after having attained a certain level. This relationship is shown in Figure 8.9 from the data of Deloch (1960) for sunflowers. These results demonstrate that S taken up in excess of the demand of the plant for the synthesis of organic S compounds is stored as SO_4^{2-}. In plant species capable of synthesizing glucosinolates, these rather than SO_4^{2-}, are the primary storage form. According to Marquard et al. (1968) the concentration of glucosinolates in these plant species depends very closely on the S supply. Plants well supplied with S are high in glucosinolate concentration. On soils with a sufficiently high level of available S application of S fertilizer can enhance glucosinolate concentration in the rape seeds without increasing seed yield or the oil content of seeds as reported by Fismes et al. (2000).

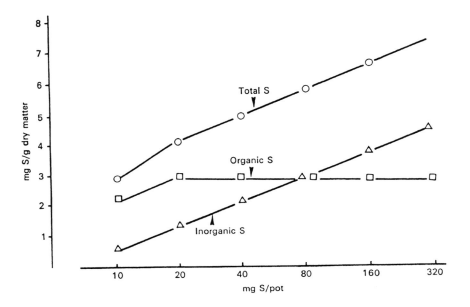

Figure 8.9 Influence of an increasing sulphate supply on the sulphate-S and organic S concentration in sunflower leaves (after Deloch 1960).

With the exception of plant species containing S glycosides, the bulk of organic S consists of protein S in the form of cysteinyl- and methionyl-residues. As proteins have a defined composition, the N/S ratio of proteins varies only slightly and is in the order of 30/1 to 40/1 (Dijkshoorn and van Wijk 1967). Somewhat similar N/S ratios were found by Rendig et al. (1976) in the protein of young maize plants. Chloroplast proteins and proteins associated with nucleic acids, however, have lower N/S ratios, as these proteins are comparatively rich in S (see Table 8.1).

Table 8.1 N/S ratio of various plant proteins (data from Dijkshoorn and van Wijk1967)

Protein	N/S ratio
Gliadin (grains)	33
Albumin (grains)	28
Globulin (grains)	67
Chloroplast proteins	15
Nucleoproteins	18

8.2.3 Sulphur deficiency and toxicity

As S is an essential constituent of proteins, S deficiency results in an inhibition in protein synthesis. The S- containing amino acids (cysteine, methionine) which are essential building blocks of protein are deficient and thus proteins cannot by synthesized. For this reason non-S containing amino acids accumulate in S deficient plant tissues (Linser *et al.* 1964). According to the findings of Rendig *et al.* (1976) the accumulation of amide N in S deficient maize plants is associated with low levels of sugars. These low sugar concentrations result from the poor photosynthetic activity of chlorotic S deficient plants. From the foregoing discussion it is not surprising that in S deficient plants protein concentration is low. This holds not only for vegetative plant material, but also for cereal grains. It has been known for many years that the grains of S deficient cereals contained less methionine and cysteine than found in cereals well supplied with S (Coic *et al.* 1963, Eppendorfer 1968). The ratio ,organic N/organic S' is therefore considerably higher in S deficient plant tissues (70/1 to 80/1) as compared with normal plant tissues. This ratio can serve as a guide to indicate whether or not plants are adequately supplied with S. Another feature of S deficient tissues is the accumulation of NO_3^- and amides.

In field crops sulphur deficiency and nitrogen deficiency are sometimes difficult to distinguish. In this instance leaf analysis can be invaluable. In S deficient plants the SO_4^{2-} concentrations are low and soluble amino N concentrations high. This contrasts markedly with N deficiency where soluble N levels are depressed and SO_4^{2-} concentrations are unaffected. In plants suffering from S deficiency the growth rate is reduced. Generally the growth of the shoots is more affected than root growth. Frequently the plants are rigid and brittle and the stems remain thin. In *Cruciferae* the lateral extension of the leaf lamina is restricted and the leaves are rather narrow. According to Schnug (1989), in rape S deficiency symptoms appear at first in the younger leaves in the form of reddish colouration and necrosis beginning at the margins. Finally the entire leaf becomes necrotic. The fact that the S deficiency appears first in the younger leaves shows that older plant tissues cannot contribute to any extent to the S supply of younger leaves, which are obviously mainly dependent on S taken up by the roots. Inflorescences of rape which are normally yellow may turn white under S deficiency conditions. The critical S concentration in fully developed rape leaves is 6 mg/g dry matter

(Schnug 1989). Concentrations below this level indicate that the S supply of the plant is insuffcient.

Plants are comparatively insensitive to high SO_4^{2-} concentrations in the nutrient medium. Only in cases where SO_4^{2-} concentrations are in the order of 50 mM as for example in some saline soils, is plant growth adversely affected. The symptoms, a reduction in growth rate and a dark green colour of the leaves, are not specific for S excess and are more typical of salt affected plants (see page 233).

High SO_2 concentrations in the atmosphere may be toxic to plants. According to Saalbach (1984) the critical SO_2 level for annual plants is 120 μg SO_2/m^3. For forest trees and other perennial species it is about half this value. SO_2 concentrations in the atmosphere are in the range of 10 to 40μg/m^3. In industrial areas, however, concentrations several times higher than these normal levels have been registered. Sulphur dioxide absorbed by the leaves dissolves in the moist surfaces of the mesophyll cells in the stomatal cavities, the resulting sulphurous acid dissociates giving rise to H^+, HSO_3^- and SO_3^{2-} (Silvius et al. 1975). From these findings it appears that a major reason for the toxic effect of SO_2 is that the anions HSO_3^- and SO_3^{2-} can accumulate and uncouple photophosphorylation (see page 147).

In industrial regions, high levels of SO_2 in the atmosphere have resulted in the eradication of certain lichen species. SO_2 toxicity in plants is characterized by necrotic symptoms in leaves. SO_2 is absorbed by small fog droplets in which it may be oxidized by reacting with ozone and water to sulfuric acid according to the equation below (Sigg et al. 1987):

$$SO_2 + O_3 + H_2O \Rightarrow H_2SO_4 + O_2$$

In this reaction a strong acid is produced which together with HNO_3 also present in fog droplets may produce a pH in the droplet liquid as low as 3 which leads to a destruction of the epicuticular waxes of the needles of spruce trees (Picea abies) with detrimental effects on the water retention by the needles under conditions of water stress (Esch and Mengel 1998).

8.3 Sulphur in Crop Nutrition

8.3.1 Sulphur balance

Although the S concentration in plants is of the same order as that of P, application of S has not generally played such an important role as P fertilization. The reason for this is that S in soil organic matter is often fairly accesssible and that SO_4^{2-} is much more weakly bound to soil particles than is phosphate. This is particularly so in the soil pH range required for the growth of most crops as is demonstrated in Figure 8.10 from the findings of Marsh et al. (1987). Another factor of benefit at least until recent years has been the substantial contribution made by atmospheric S deposition to the S nutrition of crops, in areas affected by industry. Also ground water if it can be utilized can provide a rich source of sulphate as reported by Bloem (1998).

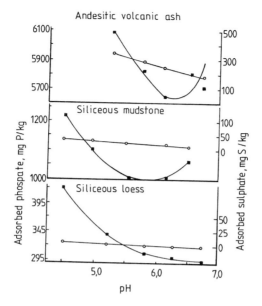

○ Sulphate
● Phosphate

Figure 8.10   Effect of soil pH on sulphate and phosphate adsorption for three contrasting soils (after Marsh *et al.* 1987.

On the other hand large amounts of SO_4^{2-} can be leached when rainfall is high and this constitutes the main loss of S from agricultural soils. The ability of soils to retain SO_4^{2-} depends on their sulphate adsorption capacity which can often be low even on medium textured soils. Sulphate leaching is also stimulated by common agricultural practices such as liming which releases a large proportion of adsorbed sulphate as a consequence of an increase in soil pH (see Figure 8.10, Curtin and Smillie 1986). Similarly phosphate fertilization enhances the potential for SO_4-leaching by inducing SO_4^{2-} exchange into the soil solution (Bolan *et al.* 1988). Leaching losses can represent a large proportion of S inputs. In lysimeter experiments in the UK annual leaching losses under grass ranged from 11. 8 kg S ha[-1] (21% of the input) on a heavy soil to 30.2 kg S ha[-1] (52% of the S input) on a sandy soil (Bristow and Garwood 1984). Amounts of SO_4-leached can often far exceed the quantities taken up by crops. In long term lysimeter experiments. Jürgens-Gschwind and Jung (1979) reported strong negative S balances on sandy soils with annual leaching losses of about 100 kg S/ha. Similar amounts of leached S were found on the pervious chalk soils in the Champagne (France) in a sugar beet/wheat rotation by Baliff and Muller (1985). Under natural vegetation cover the amounts of S leached were much lower. The amount of available sulphate in the rooting depth of soils in spring depends much on winter rainfall, high precipitation depleting the available sulphate and *vice versa* (Bloem 1998).

Appreciable amounts of SO_2 come into soils by dry and wet deposition (Baldocchi 1993) which contribute to soil acidification Hallbäcken (1992) since the SO_2 may be oxidized to SO_3 which together with water forms H_2SO_4. The quantities of S originating from the atmosphere depend to a large extent on atmospheric pollution of SO_2. Dry deposition usually describes uptake of SO_2 by soil and plants whereas the amount of S deposited in rainfall is referred to a wet deposition. In areas affected by atmospheric pollution of SO_2 both dry and wet depositions are important but in non industrial regions of high rainfall a very high proportion of the S deposition can be in rainfall.

Because of the concerns over the effects of acid rain particularly on poorly buffered natural ecosystems, concerted efforts have been made in industrial countries throughout the world to decrease total emissions of SO_2. In the UK they have decreased from 6. 4×10^6 tonnes in 1970 to 2. 9×10^6 tonnes in 1995 and other European countries have achieved even more dramatic decreases over the same period (McGrath et al. 1996). This fall has been accompanied by corresponding marked reductions in atmospheric SO_2 concentrations in some cases to <10 ug/m³. This decrease occurring as it has done simultaneously with a reduction in the application of sulphate containing fertilizers has lead to an increased incidence of S deficiency in arable crops (McGrath et al. 1996) and in oilseed rape in particular (Schnug 1991). Many areas in Europe and North America are now at risk of S deficiency because of this atmospheric clean up which has taken place over the past 30 years. In the UK for example three quarters of the land area receives less than 20 kg S/ha from the atmosphere which is a smaller amount than is required by most agricultural crops if S leaching is taken into account (p. 449).

It is difficult to assess the S status of soils in the field because of fluctuating amounts of available S in the soil throughout the growth of the crop as a result of variations in the rates of inputs of S from the atmosphere, mineralization of organic S and losses due to leaching. McGrath and Zhao (1995) have therefore proposed a quantitative model to assess the risk of S deficiency in cereals in Britain, by compiling a risk index from over 6000 soil data points which take in account the S inputs and losses as discussed above. This model predicts that if the target for reduction in SO_2 emission by the year 2003 is met soils in Britain will be at risk from S deficiency, 23% at high risk and a further 27% at medium risk. These findings are of relevance not only to the UK but wherever SO_2 concentrations in the atmosphere are decreasing. The Sulphur Institute in Washington (USA) estimates an annual S fertilizer deficit worldwide increasing from a current value of 7.5 million tonnes to over 10.0 million tonnes by 2010 (Ceccotti 1996).

8.3.2   Sulphur application

In regions far from the sea and industry, inadequate S supply and S deficiency of crops can be common. About 70 years ago Storey and Leach (1933) observed S deficiency in tea plants in Malawi, and the deficiency became known as ‚tea yellows'. In various regions of the tropics and subtropics sulphur deficiency in crops has been observed and this for different reasons. Many areas are remote from SO_2 releasing industries. Crops such as peanuts remove appreciable amounts of S from the field and soil reserves are depleted. Many tropical soils are poor in organic matter and therefore also poor in organic S as reported by Pasricha and Fox (1993) in a useful review article.

Sulphur deficiency has now been widely recognized in many parts of the world including Africa, Australia, New Zealand and the U.S.A. Sulphur application to crops is thus becoming increasingly common and essential. Walker and Adams (1958) carrying out field experiments on S deficient soils in New Zealand observed competition for S between grasses and clover. In the treatment without S nearly all the available SO_4^{2-} was taken up by grasses and N fixation by associated clover was negligible. Sulphur dressings of about 17 kg S/ha combined with adequate N fertilizer resulted in remarkably good clover growth, and yields of dry matter and N recovered by the sward. The highest S treatment also resulted in extremely high N uptake by the herbage which was about three times higher than the N fertilizer rate applied. This experiment demonstrates the essential role of S in promoting growth and N fixation by leguminous plants. Scherer and Lange (1996) found in pot experiments that SO_4^{2-} application increased the nodule weight of *Vicia faba* and *Pisum sativum* and the nodule number per pot in *Medicago sativa* and *Trifolium pratense*.

The interrelationships between S and N are of particular importance to S application because crop response to S fertilization frequently depends on the amount of N fertilizer applied. If S supply is marginal then N application can induce S deficiency. Also greater responses to S application are obtained when abundant amounts of N are applied (Zhao *et al.* 1999).

The total S requirement of different crops depends on plant matter production and also on the crop species. Crops with a high production of organic matter such as sugar cane, maize and Bermuda grass have a high demand for S which is in the order of 30 to 40 kg S/ha/season. (see Table 6.2). A high S requirement is also characteristic of protein rich crops (lucerne, clover) and particularly of the Cruciferae. Thus the requirement of rape is about 3 times higher than that of cereals. The S uptake of glucosinolate-free rape cultivars is about 15 kg S/ha and thus about half as much as glucosinolate containing cultivars (Schnug 1991) which shows that Cruciferae need an appreciable amount of S for the synthesis of glucosinolate oils. For this reason glucosinolate containing cultivars respond most sensitively to an inadequate S supply. The S requirement of different crops is also reflected in the S concentration of their seeds and grains, as is shown in Table 8.2 (Deloch 1960).

Table 8.2   Sulphur concentrations in grains and seeds of various crops (Deloch 1960)

Gramineae		Leguminosae		Cruciferae	
			mg S/g dry matter		
Barley	1.8	Broad beans	2.4	Rape (cv with glucosinolate)	10.0
Oats	1.8	Bush beans	2.4	White mustard	14.0
Wheat	1.7	Peas	2.7	Oil radish	17.0
Maize	1.7	Soya	3.2	Black mustard	10.0

As a consequence of the enormous decreases in atmospheric S inputs in industrialized countries more attention has been given to the effects of this limiting S supply on crop growth, particularly in the *Brassica* crops because of their high S demand (20-40 kg/ha) and the cereals because of their importance and widespread cultivation. In the case of cereals when supply falls below the 15-20 kg/ha required for optimum growth, reproductive growth is more sensitive to a lack of S than is vegetative growth. In addition, to effects on yield the small grained S deficient plants have grains of low sulphur status which accumulate low S containing storage proteins with a high grain N/S ratio. Changes in protein composition are associated with a reduction in breadmaking quality and particularly loaf volume (Byers *et al.* 1987). These relationship have been discussed by Zhao *et al.* (1999) in an excellent review article of the effect of S nutrition on the growth and quality of wheat.

The most important S containing fertilizers are gypsum, superphosphate, ammonium sulphate, potassium sulphate, and sulphate of potash magnesia. According to Fismes *et al.* (2000) ammonium thiosulphate $(NH_4)_2S_2O_3$ may also be used as S fertilizer with the advantage that it has a narrower N/S ratio than ammonium sulphate. Single superphosphate was formerly available as a S source (12—14% S) but for the past 20 years manufactures have mainly produced triple superphosphate which contains no S. Sulphur coated fertilizers (see page 365) also contribute to the S supply of plants. Dressings of gypsum $(CaSO_4\ 2H_2O)$ are often used in cases where soils are absolutely deficient in S. The rates generally applied are in the range of 10 to 50 kg S/ha. In regions with high rainfall spring application is recommended in order to avoid leaching by winter rains. Elemental S may be applied on alkaline soil. Its oxidation produces sulphuric acid and hence a maraked decrease in soil pH. Elemental S should be applied on the soil surface and not incoporated into the soil so that oxygen has a good access. The sulphate produced by oxidation can be taken up by plants (Jolivet 1993).

CHAPTER 9

PHOSPHORUS

9.1   Soil Phosphorus

9.1.1   Phosphorus fractions and phosphate minerals

Phosphorus in soils occurs almost exclusively in the form of orthophosphate with total P concentrations usually in the range of 500—800 mg/kg dry soil. Quite a substantial amount of this P is associated with organic matter and in mineral soils the proportion of organic P lies between 20 and 80% of the total P.

From the viewpoint of plant nutrition soil P can be considered in terms of 'pools' of varying accessibility to plants. Phosphate in the soil solution is completely accessible but this makes up only a minute fraction of the total soil P. The bulk of soil P is virtually inaccessible. More than 90% of total P is present as insoluble and fixed forms including primary phosphate minerals, humus P, insoluble phosphate of Ca, Fe and Al and P fixed by hydrous oxides and silicate minerals. This fraction can be described as non labile. A proportion of insoluble phosphate is more accessible than that of the bulk reserves. In this labile fraction solid phosphate is present in phoshate precipitations and is also held on soil surfaces. The labile phosphate is in rapid equilibrium with soil solution phosphate. Removal of phosphate from the soil solution by plant roots disturbs the equilibrium between the soil solution, P concentration and the labile pool at the solid soil phase which leads to a release of P into the soil solution.

These three fractions are shown schematically in Figure 9.1. As well as showing the accessibility of the phosphate pools to the plant it can also be seen that after addition of phosphate to the solution, substantial amounts of P can be fixed in the non labile fraction. The non labile fraction is also a source of very slow release of phosphate. This concept of pools of accessibility within the soil is of great importance in relation to understanding the use of P fertilizers in crop production.

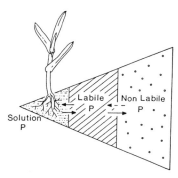

Figure 9.1   Schematic representation of the 3 important phosphate soil fractions for plant nutrition.

These three pools relate to the P compounds present in the soil. Soil P compounds not in the soil solution or the weakly adsorbed labile fraction can be considered in four main groups:

1. Sparingly soluble phosphates of Ca in calcareous and alkaline soils, of Fe and Al in acidic soils.
2. Strongly adsorbed phosphates by Fe and Al hydrous oxides
3. Occluded phosphate by Fe and Al compounds.
4. Organic phosphorus.

On neutral to calcareous soils the concentration of phosphate in soil solution is governed mainly by the formation and dissolution of calcium phosphates which in turn depends on soil pH and the Ca^{2+} concentration of the soil solution. The lower the Ca/P ratios of the Ca phosphates the higher their solubility in water. The equilibria of Ca phosphates from the soluble calcium dihydrogen phosphate $Ca(H_2PO_4)_2$ to the highly insoluble hydroxyapatite is shown below.

$$Ca(H_2PO_4)_2 + Ca^{2+} \Leftrightarrow 2CaHPO_4 + 2H^+$$

Ca dihydrogen phosphate $\qquad$ Ca monohydrogen phosphate

$$3\ CaHPO_4 + Ca^{2+} \Leftrightarrow Ca_4H(PO_4)_3 + 2H^+$$
$$\text{Ca octophosphate}$$

$$Ca_4H(PO_4)_3 + Ca^{2+} + H_2O \Leftrightarrow Ca_5(PO_4)_3OH + 2H^+$$
$$\text{Hydroxyapatite}$$

From these equilibria it is clear that H^+ promotes the solubilty of Ca phosphates in the soil and Ca^{2+} has the reverse effect. Thus hydroxyapaptite has a very low solubililty in water. The analogues of hydroxyapatite, chloroapatite and fluoroapatite are even less water soluble.

$$Ca_5(PO_4)_3Cl \qquad \text{Chloroapatite}$$
$$Ca_5(PO_4)_3F \qquad \text{Fluoroapatite}$$

The Ca phosphates shown above may be present in different crystalline forms, with or without water of crystallisation (Lindsay *et al.* 1989). In the upper layers of calcareous and alkaline agricultural soils amorphous Ca phosphates generally dominate. In most soils Ca^{2+} concentration in the soil solution is high enough for any $Ca(H_2PO_4)_2$ added to the soil to be transformed quickly into $CaHPO_4$. This is the fate of $Ca(H_2PO_4)_2$ which is present in various phosphate fertilizers. The $CaHPO_4$ formed although of low solubility in water provides a readily available source of phosphate to plants and the low solubility means that phosphate is hardly leached. As shown in the equations above apatite is only formed when the Ca^{2+} concentration in the soil solution is high. This is particularly the case in calcareous soils *i.e.* soils rich in $CaCO_3$ which have a high pH and a high Ca^{2+} concentration in soil solution.

In neutral and acid soils phosphate adsorption is the dominant process affecting phosphate availability in plants. Specific phosphate adsorption is brought about by ligand exchange (see page 23) in which the OH⁻ on the adsorbing surface is replaced by phosphate as shown below for

$$|= Fe\text{-}OH + H_2PO_4^- \Leftrightarrow |=Fe\text{-}H_2PO_4 + OH^-$$

From this equation it follows that phosphate adsorption is stronger the lower the OH⁻ concentration *i.e.* the lower the soil pH. It is for this reason that the adsorbed phosphate fraction is dominant in acid soils. For most acid soils simply by raising the soil pH *e.g.* by liming the solubility of adsorbed phosphate can be greatly increased (Haynes 1982). As well as this mono nuclear adsorption bi nuclear adsorption also can occur (see page 23) which binds the phosphate even more strongly to the surface of the adsorbing particle. As for mono nuclear adsorption these reactions are favoured by low soil pH.

Phosphate adsorbing surfaces include Fe oxides/hydroxides, Al hydroxides, allophanes, clay minerals, organic Fe complexes and calcite (Parfitt 1978). Iron oxides adsorb more strongly than the clay minerals per unit weight. Adsorption depends not only on the type of adsorbing minerals but also on their specific surfaces. Thus freshly precipitated material has a higher phosphate adsorption capacity than more crystallinic material. This effect of adsorption strength increasing with the degree of the amorphous nature of compounds is shown in Table 9.1 from the work of Burnham and Lopez-Hernandes (1982).

Table 9.1 Phosphate adorption capacity of various adsorbents. The adsorption index is a measure of the phosphate adsorption per unit weight of the adorbents (after Burnham and Lopez-Hernandes 1982)

Adsorbents	Adsorption index
Precipitated amorphous Al(OH)₃	1236
Precipitated FeOOH	846
Fe oxyhydrate	453
Aged Fe oxyhydrate	111
Laterite	21
Crystalline goethite (FeOOH)	0
Crystalline gibbsite Al(OH)₃	0
Calcite (CaCO₃)	46

Since soils may differ considerably in concentration of adsorbing minerals and in their degree of crystallinity, phosphate adsorption capacities of soil types may also differ considerably (Pagel and van Huay 1976).

The effect of pH on phosphate adsorption is shown in Figure 8.10 from the work of Marsh *et al.* 1987. Phosphate adsorption declines with increasing pH until a pH of 6–6.5 is attained.

Adsorption of phosphate to soil particles is frequently not an ideal adsorption process but rather a combination of adsorption and precipitation (Larsen (1967b). Thus Ca carbonates adsorb phosphate which is then slowly converted into apatites (Parfitt 1978). In this way some phosphate of the labile pool is continously being rendered immobile and so transferred to the non-labile phosphate fraction. This process of 'phosphate ageing' is especially rapid in acid soils with a high adsorption capacity.

Phosphate can also be adsorbed to humic- Fe(Al) surfaces through a ligand exchange mechanism as reported by Gerke and Hermann (1992). A possible mechanism for the binding of phosphate to a humate surface is shown below for Fe. An analogous phosphate adsorption mechanism applies for Al.

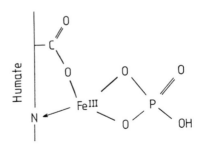

Figure 9.2 Upper part: Fe-hydroxy skin covering the phosphate adsorbed to Al oxide/hydroxide. — Lower part: Phosphate adsorption by humate-Fe-surface.

Phosphate is bound *via* Fe or Al bridges to the surface of the humate. Gerke and Hermann (1992) have shown that the adsorption capacity of the humate Fe surfaces is about tenfold higher than that of amorphous Fe oxides. The high adsorption capacity of these Fe and Al complexes may help to explain the positive correlation between organic C and phosphate adsorption in soils found by some workers. It now appears that humate Fe(Al)P complexes are much more important in phosphate availability and possibly phosphate mobility in the soil than was previously recognised (Gerke and Hermann 1992). In a study of sandy soils Gerke (1992) found that up to 50% of the total soil solution P was present in the form of such complexes. A main reason why the significance of these humicFe(Al)P complexes has not previously been appreciated is the lack of stability of these complexes under the acid conditions usually used for extraction and determination of soil phosphates (see page 456).

Occluded phosphate is a particular form of adsorbed phosphate which is trapped and therefore not directly available to roots. The principle of phosphate occlusion is shown in the upper part of Figure 9.2. Phosphate is adsorbed to the surface of Al hydroxide and bound by an overlying layer of Fe^{III} hydroxide. In this structure the phosphate bridges the Al- with the Fe^{III} hydroxide so that the surface of the Al phosphate particle is enveloped by an Fe^{III} hydroxide skin. Under anaerobic conditions the Fe^{III} oxide skin of occluded phosphate can be broken down by the reduction of Fe^{III} to soluble Fe^{2+}, resulting in a release of phosphate (Sah and Mikkelsen 1986). This may happen particularly in paddy soils under anaerobic conditions where the concentration of dissolved Fe^{2+} may rise considerably (Ponnamperuma 1972) causing Fe toxicity in rice (Ottow *et al.* 1991).

Soil phosphates of minor importance are strengite, variscite and vivianite. According to Lindsay *et al.* (1989) the solubility of strengite and variscite increases with soil pH. Stable forms exist only under acid conditions and can be transformed into vivianite in submerged soils by the reduction of Fe^{III} to Fe^{II}.

$$\text{Variscite Al } H_2PO_4(OH)_2$$

$$\text{Strengite } Fe^{III}H_2PO_4(OH)_2$$

$$\text{Vivianite } Fe^{II}{}_3(PO_4)_2$$

The fourth soil phosphate fraction is *organic phosphate*. Most of the organic soil phosphates are present in the form of the inositol phosphate ester while the proportion of phospholipids and nucleic acid in soils is small due to the fact that the two groups of phosphate esters are quickly dephosphorylated by microbial phosphatases. Plants only produce myoinositol hexaphosphate. In soils, however, different isomers of hexaphosphates occur with myo-inositol the dominating inositol phosphate. Other inositol phosphates occur to a lesser degree such as the di-tri- and tetraphosphates of inositol. Some of these organic phosphates are produced by higher plants, most, however, are synthesized by microorganisms (Dalal 1977). Inositol phosphates are prone to adsorption because they carry anionic phosphate groups.

The ultimate process by which organic phosphates are rendered available is by cleavage of inorganic phosphate by means of a phosphatase reaction. The principle of this reaction is a hydrolysis as shown in Figure 9.3.

458

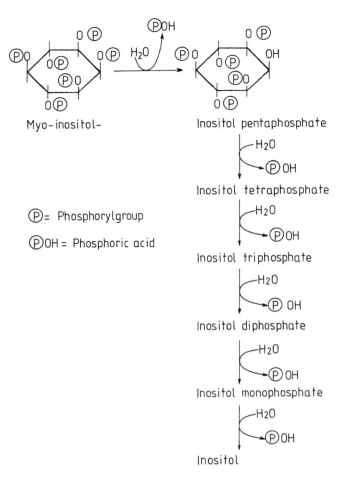

Myo-inositol-

Inositol pentaphosphate

P = Phosphorylgroup

POH = Phosphoric acid

Inositol pentaphosphate
  │─H₂O
  │
  ↓──►P OH
Inositol tetraphosphate
  │─H₂O
  │
  ↓──►P OH
Inositol triphosphate
  │─H₂O
  │
  ↓──►P OH
Inositol diphosphate
  │─H₂O
  │
  ↓──►P OH
Inositol monophosphate
  │─H₂O
  │
  ↓──►P OH
Inositol

Figure 9.3 Hydrolysis of myo-inositol phosphate to inorganic phosphate and inositol catalyzed by phosphatases.

The enzyme phosphatase is produced by the roots of higher plants as well as by numerous microorganisms (*Aspergillus*, *Penicillium*, *Mucor*, *Rhizopus*, *Bacillus*, *Pseudomonas*). Phosphatases are phosphate esterases which hydrolyze inositol phosphates, nucleic acids and phosphoglycerates. According to Sentenac *et al.* (1980) root cell walls are high in phosphatase activity. In intact root systems growing in sterile nutrient solution containing low molecular weight organic P compounds including lecithin, Na phytate and Na glycerophosphate are hydrolyzed at higher rates than necessary to meet the plant demand (Tarafdar and Claassen 1988). Similar findings have been reported by Dou and Steffens (1993). These authors found that acid phosphatase in soils was stimulated by the addition of inorganic and especially of organic phosphates. Phosphatase activity and phosphate turnover are much higher in the rhizosphere relative to the bulk soil (Helal and Sauerbeck 1984, Tarafdar and Jungk 1987).

Another influence microorganisms can have on phosphate availability is the release of acids and chelating agents which can act upon insoluble or adsorbed soil phosphate. Such microorganisms include *Aspergillus niger*, strains of *Escherichia freundi*, some *Pencillium* species and *Pseudomonas* species (Subra Rao 1974, Kucey *et al.* 1989).

Microbial activity depends much on temperature and is highest in the range of 30 to 45°C. It is for this reason that organic phosphates are of particular importance to plant nutrition under tropical climatic conditions as compared with temperate conditions. Dalal (1977) in a review paper written more than 30 years ago but still highly informative today, emphasizes that the soil solution also contains appreciable amounts of organic phosphates. Microorganisms can also immobilize phosphate by uptake, and there is a balance in the soil between immobilization and mineralization. The breakdown of organic residues low in P in particular can induce biological immobilization of P. The importance of organic phosphates as a source of phosphate for plant nutrition should not be underestimated. In the soil solution organic phosphate may make up 20 to 70% of total P (Ron Vaz *et al.* 1993) and in the rhizosphere it may be as much as 80 to 90% of total P (Helal and Dressler 1989).

The above four groups of soil phosphates contribute to plant nutrition since they may release inorganic phosphate into the soil solution which is directly available to plant roots. Ca phosphates provide soluble phosphate by dissolution, adsorbed phosphate by desorption, organic phosphate by dephosphorylation and occluded phosphate by the reduction of Fe^{III} thus breaking the enveloping skin of Fe^{III} oxides/hydroxides. These sources may be considered as the major components of the labile pool of soil phosphate and sources of phosphate in soil solution as shown in Figure 9.4. The contribution of these components differ between soils and are particularly dependent on soil pH.

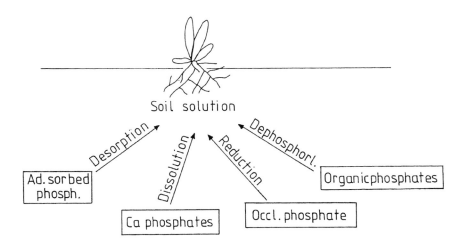

Figure 9.4 The most important soil phosphate fractions releasing phosphate into the soil solution and supplying plants with phosphate.

9.1.2   Phosphorus in solution and plant root interactions

The amount of phosphate present in the soil solution is very low in comparison with adsorbed phosphates or Ca phosphates. Adsorbed phosphate exceeds the phosphate of the soil solution by a factor of 10^2 to 10^3. The phosphate concentration of the soil solution itself is very dilute and in fertile arable soils is about 10^{-2} to 10^{-3} mol/m^3 (Hossner *et al.* 1973).

   The most important inorganic phosphate ions in soil solution are HPO_4^{2-} and $H_2PO_4^-$. The ratio of these two ion species in soil solution is pH dependent. High H^+ concentrations shift the equilibrium to the more protonated form according to the equation:

$$HPO_4^{2-} + H^+ \Leftrightarrow H_2PO_4^-$$

Figure 9.5 shows that at a pH 5, HPO_4^{2-} is almost absent whereas at pH 7 both phosphate species are present in fairly equal proportions.

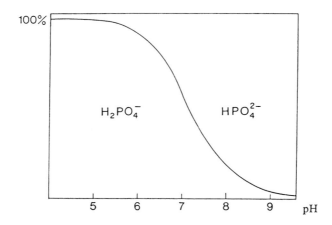

Figure 9.5   Ratio between $H_2PO_4^-$ and HPO_4^{2-} in relation to pH.

   As plant roots push their way through the soil they come in contact with the phosphate of the soil solution. The thinner the roots and the higher the root length density the better can the phosphate be intercepted. Root hairs and root hair length in particular are also important in this respect (Föhse *et al.* 1991). Provided that the roots have a high demand for phosphate — and this is generally the case for growing plants — phosphate is absorbed by the roots at a high rate and the soil solution in the direct root vicinity is depleted of phosphate. This depletion creates a gradient between the phosphate concentration near the root surface and the phosphate concentration in the bulk soil (Jungk and Claassen 1997), the concentration gradient regulating the rate of phosphate diffusion towards the plant root (see page 69). The importance of phosphate diffusion in supplying phosphate to plants has been demonstrated by Bhat and Nye (1974) in experiments showing that the degree of phosphate depletion around onion roots corresponds fairly well with phosphate diffusion calculations. Mass flow can also play a part in the transport of phosphate towards plant roots (see page 66) but usually its contribution is minimal since the phosphate concentration of the soil solution is so low (Barber 1995).

The importance of root colonization by mycorrhiza for the exploitation of soil phosphates has already been discussed on page 88. Root infection by endotrophic mycorrhizal fungi can stimulate plant growth by increasing the rate of phosphate uptake. An example of the dramatic effects that can be obtained is shown in Plate 9.1 from the work of Sanders and Tinker (1973). Onion roots which were infected with an endotrophic mycorrhizal fungi took up phosphate at a considerably higher rate per unit length of root than non infected roots. This effect may be brought about not only by the increased absorbing surface of the root fungus association which allows greater exploitation for soil phosphate but also several other factors may be involved. The smaller diameter of the hyphae than the roots allows them to penetrate soil pores of small diameter thereby increasing the volume of accessible soil. The hyphae too may take up phosphate more effectively and from lower phosphate concentration in the soil than the root is able to. Mycorrhizal roots may also be able to use sources of P that are not available to the roots possibly because of localized alterations in pH or production of organic anions as chelating agents or enhanced phosphatase activity. These possibilities have been discussed by Smith and Read (1997).

Mycorrhizal plants are often more tolerant to drought (Smith and Read 1997) which may be of particular importance for the accessibility of soil phosphate under dry soil conditions. High levels of available soil phosphate are well known to depress the development of mycorrhiza (Mosse1973). This probably relates to the enhanced exudation of sugars and especially amino acids from the roots of P deficient as compared with plants well supplied with phosphate (Graham et al. 1981). The colonisation of roots with vesicular-arbuscular mycorrhiza closely reflects this release of organic C. As the mycorrhizae take up phosphate more effectively it also indicates that the development of a mycorrhizal association is an adaptive response to P deficiency.

The influence of root exudates on the solubility of phosphate in the root vicinity has attracted considerable attention. It is established that relatively large amounts of organic C assimilated in photosynthesis are transferred from the roots into the soil (see page 84). Barber and Martin (1966) for example found that about 20% of the photosynthate of wheat seedlings was released into the soil and a significant fraction of the material consisted of chelating acids. Such organic chelating agents can of course exchange with surface bonded phosphate thereby releasing phosphate for plant uptake (Gerke 1994). This possibly explains the findings of Brewster et al. (1976) who observed phosphorus depletion zones around rape roots which were much deeper and wider than predicted from diffusion calculations (see page 69).

a   b

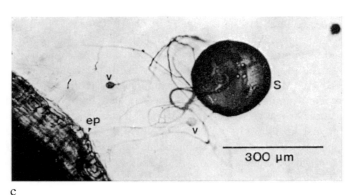

c

Plate 9.1   Root infection with endotrophic mycorrhizal fungi and phosphate uptake. — a) Response of onion growing on a P deficient soil to P fertilization and inoculation with endotrophic mycorrhizal fungi. Left, phosphate fertilizer. Centre, inoculation with endotrophic mycorrhizgl fungi. Right control. — b) Endomycorrhizal infection of onion by *Glomus mosseae*. The root cortex is full of hyphae, some of which bear vesicles (V). An external hypha (eh) is attached, entering the root at the point arrowed. — c) A germinated resting spore (S) of the endomycorrhizal fungus *Glomus macrocarpa*. One of the hyphae produced on germination has entered a nearby clover root (at point ep) to establish an internal infection. External vesicles (v) are seen. (Photos: Sanders)

Rhizosphere pH is of particular importance in the P nutrition of plants differing from the bulk soil by up 2 units depending on plant and soil factors (Marschner 1995).

The most important factor controlling rhizosphere pH is the source of nitrogen taken up, *i.e.* nitrate, ammonium or symbiotically fixed N. Nitrate uptake involves the recyling of H^+ into the cytosol due to H^+/nitrate cotransport (see page 127) and since nitrate may be taken up at a high rate the soil pH at the root surface may increase considerably (Smiley 1974). On the other hand with ammonium or symbiotically fixed N the rhizosphere becomes acidic because of the net H^+ efflux from the plasmalemma located ATPase. Under these conditions P uptake is stimulated since the lower pH shifts the equilibrium of the species from $H_2PO_4^{-2}$ to $H_2PO_4^-$ which is taken up very much more rapidly (Li and Barber 1991). In studying the dynamics of P in the rhizosphere of ryegrass Gahoonia *et al.* (1992) were able to show another benefit of soil acidification induced by ammonium nutrition in the release of phosphate from a Luvisol of pH 7.4 by solubilizing Ca-phosphates. In an Oxisol of pH 4.8 ammonium nutrition was ineffective whereas nitrate nutrition favoured the desorption of phosphate as a consequence of the increase in rhizosphere pH.

Legumes dependent on biological N_2 fixation also in the absence of nitrate uptake lack a means of recycling H^+ into the cytosol and therefore acidify the rhizosphere. According to de Swart and van Diest (1987) for different legume species about 0.4 to 0.8 mol of H^+ were released per mol N_2 fixed. Plants fixing N_2 are thus better able than nitrate fed plants to utilize P from rock phosphate or other sparingly soluble Ca phosphates in neutral and alkaline soils. The benefit of this acidification on P uptake and growth of *alfalfa* was reported by Aguilars and van Diest (1981).

Rhizosphere acidification is a widespread response to P deficiency. Depending on plant species acidification is brought about by proton or organic acid release. In P deficient tomato plants net H^+ efflux results as a consequence of depressed nitrate uptake (Heuwinkel *et al.* 1992). A number of species such as rape (Hoffland *et al.* 1989), leguminous species (Ohwaki and Hirata 1992) release organic acids particularly citric acid. Pigeon pea exudes pisidic acid (a derivate of tartrate) and particularly if insufficiently supplied with phosphate. Pisidic acid is a strong chelator of Fe^{III}. Thus Ae *et al.* (1990) reported that pigeon pea plants are able to mobilize Fe containing soil phosphates by the excretion of pisidic acid by roots which chelates the Fe and therefore breaks up the Fe phosphate complex. The roots of rice plants have also been found to excrete high amounts of citric acid if insufficiently supplied with phosphate. Citrate may also chelate Fe^{III} and thus contribute to the breakdown of adsorbed and/or occluded soil phosphate. Gerke *et al.* (1995) have shown that citrate released by red clover roots mobilizes adsorbed phosphate. For this reason red clover is less responsive than ryegrass to fertilizer P.

The example of proteoid roots as occur in white lupin which excrete high rates and amounts of citric acid in very confined zones has already been discussed on page 87. In these clusters of short rootlets covered in a dense mat of root hairs, the citric acid released is able to chelate Ca, Fe and Al thus mobilizing P from sparingly soluble compounds in both acidic (Gardner and Parbery 1982) and calcareous soils (Dinkelacker *et al.* (1989). Substantial amounts of citric acid were shown to be released by white lupin on a calcareous P deficient soil the release amounting to 23% of the net photosynthesis. This chelate release from roots and root hairs not only provides a highly efficient means of solubilizing phosphate but also because of the cluster root formation a very effective system for the plant to utilize the phosphate released. In addition proteoid rootlets showed a high reducing capacity reducing Fe^{III} to Fe^{2+} and thus also producing soluble inorganic phosphate. A useful review of phosphate dynamics in the rhizosphere has been published by Marschner (1995).

9.2   Phosphorus in Physiology

9.2.1   Absorption and translocation

Plant roots are capable of absorbing phosphate from solutions of very low phosphate concentrations of only a few $mmol/m^3$. Generally the phosphate concentration in the cytosol of plant cells is in the range of 5 to 8 mol/m^3 (Lauer et al. 1989) and therefore about a thousand fold greater than in the soil solution. Thus phosphate is taken up by plant cells against a very steep concentration gradient. Uptake is active as already discussed on page 123 and is mediated by H^+ cotransport (Ullrich-Eberius et al. 1981, Tanner and Caspari 1996). Kinetic data suggest the presence of two types of transporters with different affinities for P_i, one with a high affinity with a K_M of 3 to 5 $mmol/m^3$ and the other a low affinity system with a K_M of 50 to 330 $mmol/m^3$. Molecular studies have confirmed the presence of multiple genes encoding phosphate transporters that are differentially expressed (Schachtman et al. 1998). Some are strongly upregulated when phosphate supply is inadequate and under such conditions high affinity phosphate membrane transporters are either activated or derespressed (Leggewie et al. 1997). Other phosphate uptake systems are constitutive which means they are present a priori and are not affected by changes in phosphate concentration in the nutrient medium. Sentenac and Grignon (1985) reported that the pH in the apoplast controls the $H_2PO_4^-$ the uptake rate. At the same $H_2PO_4^-$ concentration in the apoplast (2.50 $mmol/m^3$), phosphate uptake at pH 4 was three times greater than at pH 6. The results are consistent with the view that a protonated transporter functions in phosphate uptake. It is doubtful whether organic P compounds are taken up by plant roots to any great extent. According to investigations by Roux (1968), P present in polyphosphates was only taken up by young barley plants after hydrolysis to the orthophosphate form.

Recent studies using ^{31}P NMR spectroscopy have provided quantitative information on the distribution of phosphate in cells (Ratcliffe 1994). The phosphate concentration in the cytosol is maintained at fairly constant concentrations in the range of 5 to 8 mol/m^3 (Lauer et al. 1989) regardless of the external phosphate concentration in the nutrient medium except under severe P deficiency (Lee and Ratcliffe 1993). By contrast the vacuolar phosphate concentration, which represents surplus uptake and storage of phosphate can vary widely. Under phosphate starvation it may be almost undetectable (Schachtman 1998 et al.) whereas at high P status values as high as 25 mol/m^3 can be attained (Lee and Ratcliffe 1993).

Phosphate is readily mobile in plants and can be translocated in upward or downward directions. Phosphate taken up by roots is translocated via the xylem mainly to fast growing young laminae where it is required for leaf expansion and growth (Fredeen et *al.* 1989). In a study of P transport and assimilation in intact castor bean plants Jeschke *et al.* (1997) observed that young leaves are supplied not only by phosphate taken up by the roots but also by phosphate from somewhat older leaves. This is particularly so when phosphate uptake by the roots is insufficient to meet the demand of the young leaves. The retranslocated phosphate from hydrolysis of organic P is supplied *via* the phloem. Phosphorus remobilized from mature leaves is also transported *via* the phloem but is directed to the roots. Inorganic P is present in phloem sap in substantial concentrations and makes up most of the total P (Hall and Baker 1972, Jeschke *et al.* 1997).

Drew and Saker (1984) suggest that the concentration of phosphate cycled in the phloem from shoot to root acts as a feedback signal to regulate P uptake. High concentrations of phosphate in the phloem induced by low demand in the shoot repress uptake. Conversely low concentrations associated with high shoot demand stimulate uptake (Marschner *et al.* 1997). Regulation appears to be dependent on the vacuolar phosphate concentration of root cells (Lee and Ratcliffe 1993) and can be delayed under rapid and widely fluctuating P nutritional regimes as can occur in plants grown in nutrient solution. This explains that when phosphate is resupplied to plants that are P deficient, there is a rapid and uncontrolled uptake of phosphate which induces symptoms of P toxicity in the shoots (Clarkson and Scattergood 1982). Zinc deficiency can also similarly lead to P toxicity through restriction of phosphate transport from shoot to root and thus impairment of the signal which controls phosphate uptake (Marschner and Cakmak 1986).

9.2.2   Phosphate fractions and metabolic functions

Phosphate in the plant occurs in inorganic form as orthophosphate and to a minor extent as pyrophosphate. The organic forms of phosphate are compounds in which the orthophosphate is esterified with hydroxyl groups of sugars and alcohols or bound by a pyrophosphate bond to another phosphate group. A typical example of a phosphate ester is fructose-6-phosphate shown in Figure 9.6. Such organic phosphates (phosphorylated sugars and alcohols) are mainly intermediary compounds of metabolism. Phosphate is also bound to lipophilic compounds particularly the phospholipids such as diacylglycerol 3-phosphate (phosphatidic acid) or as phosphosphingosine. Lecithin a phosphadityl derivate shown in Figure 9.6 is a typical example of this kind. As can be seen from the formula, P is bound in a diester linkage, one bond linking the choline and one bond the phosphadityl moiety. Such compounds possess hydrophobic and hydrophylic properties in the fatty acid radical and phosphate groups, respectively. Compounds of this type, *e.g.* lecithin and phosphatidyl ethanolamine, are essential components of biological membranes (see page 114). The phosphoryl group is an esssential building block of biological membranes since two adjacent phosphoryl groups bind together with Ca^{2+} to form a bridge which stabilizes the membrane (see Figure 3.4).

The most important compound in which phosphate groups are linked by pyrophosphate bonds is adenosine triphosphate (ATP). The formula and some properties of this coenzyme have already been described on page 145. The pyrophosphate bond is an energy rich bond which on hydrolysis releases 30 kJ/mol. The energy absorbed during photosynthesis, or released during respiration or anaerobic carbohydrate breakdown is can be conveyed to various endergonic processes as active ion uptake and the synthesis of numerous organic compounds. In these processes there is usually an initial phosphorylation (kinase) reaction. This involves the transfer of the phosphoryl group from ATP to another compound, as shown in the Figure 9.6.

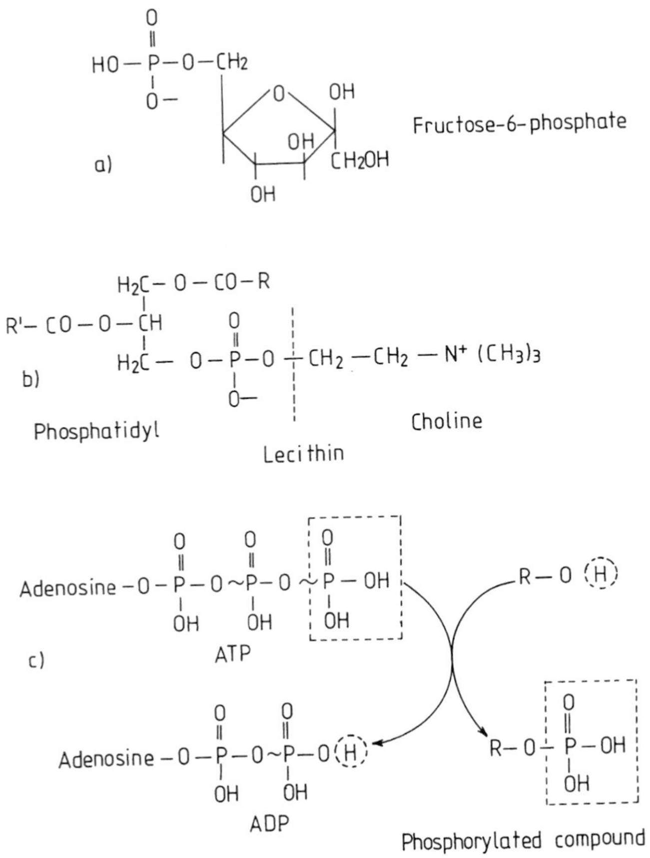

Figure 9.6 a) Phosphorylated metabolite, fructose-6 phosphate — b) Phosphatidyl derivate, lecithin — c) Phosphorylation of a metabolite by ATP.

In this reaction the phosphorylated compound is loaded with energy (= priming reaction) and is thus enabled to participate in further metabolic processes.

It appears that the unique function of phosphate in metabolism is its formation of pyrophosphate bonds which allow energy transfer. Uridine triphosphate (UTP), cytidine triphosphate (CTP) and guanosine triphosphate (GTP) are analogous compounds to ATP. ATP is required for the synthesis of glucans inclucding amylose and amylopectin (starch) with ADP-glucose as glucosyl donor. In an analogous way UTP is required for the synthesis of sucrose, cellulose and callose with UPP-glucose as glucosyl donor.

CTP is needed for the synthesis of phospholipids. All these nucleotide triphosphates (ATP, GTP, UTP, and CTP) are also essential building blocks for the synthesis of nucleic acids (DNA and RNA).

The structures of DNA and RNA are shown in Figure 9.7. The phosphate group in nucleic acids bridges the ribose (RNA) or deoxyribose respectively with another ribose or deoxiribose. DNA is the carrier of genetic information and the RNAs function in protein synthesis. This shows the universal and essential role of phosphate not only in plants but also in all other living organisms.

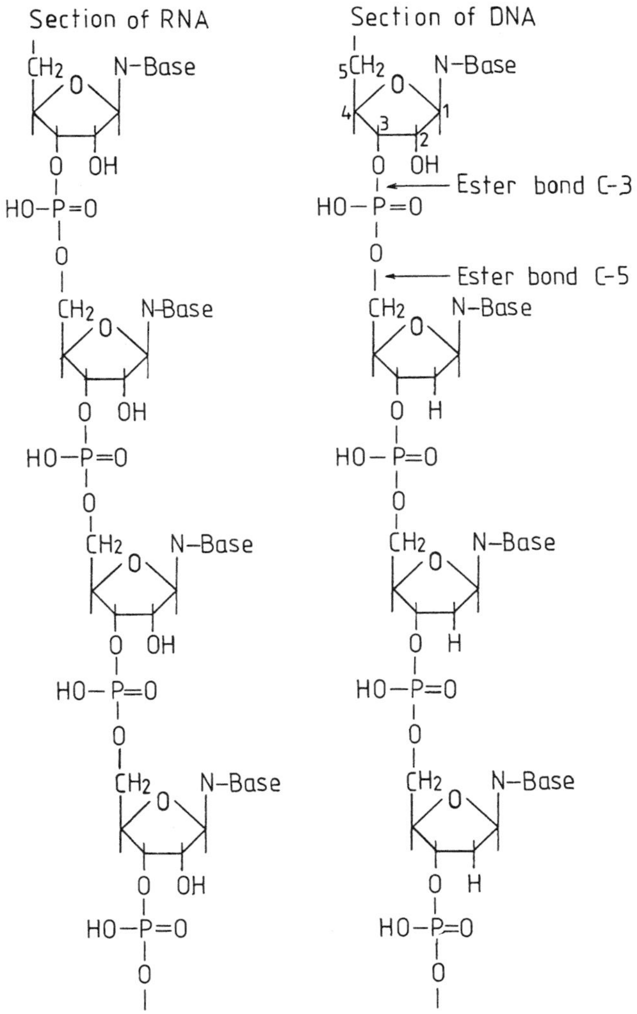

Figure 9.7 Section of DNA and RNA, showing the phosphate ester bond at C_3 and C_5 of the ribose or deoxyribose, respectively.

Recent research has shown that sections of RNA possess enzymatic properties as is the case in ribosomal RNA for catalyzing the translation process that is to say the translation of the mRNA information on polypeptide synthesis in the ribosome (Noller 1991). Such enzymes are called ribozymes. They represent a new class of enzymes which are made up entirely from RNA. All known ribozymes are metalloenzymes which mainly require divalent cations, particulary Mg^{2+}. Most of the ribozymes are relatively small molecules comprising about 100 nucleotides. They catalyze the cleavage of diester bonds and and also peptidyl transfer at the ribosome (Pyle 1993).

One of the most important groups of phosphorylated metabolites in plant metabolism are the triosephosphates (dihydroxyacetone phosphate, glyceraldehyde 3-phosphate). These compounds produced in photosynthesis are exported from the chloroplast into the cytosol *via* an antiporter (Geiger and Servaites 1994) which simultaneously mediates in the import of inorganic phosphate (page 243). Low concentrations of inorganic phosphate in the cytosol prevent the export of triosephosphates which leads to starch synthesis and accumulation in the chloroplast. Inorganic phosphate in the cytosol has a regulatory function by influencing the activity of various enzymes. One of these is 6, phosphofructo-2 kinase which catalyzes the synthesis of fructose-2,6 phosphate. This is a signal metabolite in all eurkaryotic cells (Stitt 1990) which activates phosphofructo kinase and hence the synthesis of 1,6 fructose bisphosphate, the metabolite which opens the glycolytic pathway to initiate carbohydrate degradation.

The phosphate reserve in seeds and fruits is phytin (anion phytate) which is synonymous with inositol hexaphospate (see page 458) The counter cation species of phytate are Ca^{2+} and to a lesser degree Mg^{2+}. In vegetative tissue the phosphate storage form is mainly the inorganic phosphate of the vacuoles. In fungi this role is played by polyphosphates. In phosphate-deficient plants it is especially the level of inorganic phosphate of stems and leaves and the phytate in seeds and fruits that are decreased. This relationship is shown in Table 9.2 from the old but still valid data of Michael (1939).

Table 9.2. Effect of P supply on the concentration of various P forms in spinach leaves and oat grains (Michael 1939).

P supply	Phospholipid	Nucleic acid	Phytate	Inorganic
		P in mg/g dry matter		
		Oat grains		
Inadequate	0.22	2.1	0.05	0.5
Adequate	0.22	2.4	0.5	1.3
		Spinach leaves		
Inadequate	1.1	0.9	-	2.2
Adequate	1.1	0.9	-	18.0

This example shows that even under P deficiency, P in the membrane and the nucleic acid P are not mobilized for the P supply of growing tissues. Phytate concentrations in seeds responds sensitively to phosphate supply as reported by Lickfett *et al.* (1999) for rape seed, the phytate proportion of the total seed P rising from 20 up to 75% with increasing phosphate supply.

9.2.3 Phosphorus deficiency

Plants suffering from P deficiency are small and stunted and have a rigid erect appearance. Young P-deficient plants have a bluish green colour in the early stages of growth. In cereals tillering is decreased. Fruit trees show reduced growth rates of new shoots and flower initiation is impaired. The formation of fruits and seeds is especially depressed in plants suffering from P deficiency. Thus not only low yields but also poor quality fruits and seeds are obtained from P deficient crops.

Generally the symptoms of P deficiency appear in the older leaves which are often of a darkish green colour. The stems of many annual plant species suffering from P deficiency are characterized by a reddish colouration originating from an enhanced formation of anthocyanins. The leaves of P deficient plants are frequently tinged with brownish colour and senesce prematurely. A detailed account of the appearance of P deficiency in crop plants is given by Bergmann (1992).

One of the earliest symptoms of P deficiency is a specific inhibition of leaf expansion and leaf surface area (Fredeen *et al.* 1989). This is attributed to a restriction in the delivery of water to the growing leaves curtailing leaf expansion, as a result of a reduction in hydraulic conductance in the roots (Radin 1990). The photosynthetic CO_2 assimilation capacity of the leaves is also decreased. Inadequate supply of phosphate prevents the export of triosephosphates from chloroplasts (Walker 1980) and therefore the synthesis of sucrose. Phosphate deficiency also decreases ATP supply and the rates of synthesis of RuBP needed for carboxylation (Rao and Terry 1989). The effects of P deficiency on photosynthesis, however, are less severe than on growth. In the source leaves in particular chlorophyll concentrations can even increase (Rao and Terry 1989) allowing some photosynthetic activity to be maintained.

By contrast to the effects on shoot growth, root growth is much less inhibited. Under P deficiency the roots act as a dominant sink for photosynthates (Freden *et al.* 1989) and P (Jeschke *et al.* 1997) from mature leaves. Dry weight shoot/root ratios are therefore typically decreased by the increased partitioning of carbohydrates to the root where sucrose can accumulate (Khamis *et al.* 1990). This response of the plant to P deficiency in favour of root growth allows increased exploitation of the soil in the acquisition of phosphate and this is enhanced by accompanying morphological and physiological changes occurring in the roots. These include the formation of longer and more slender roots (Anghinoni and Barber 1980) increase in the number and length of root hairs (Föhse and Jungk 1983), release of organic acids to chelate insoluble phosphates (Gerke *et al.* 1994) release of protons (Heuwinkel *et al.* 1992) stimulated phosphatase activity at the root surface (Tarafdar and Jungk 1987) and increased degree of mycorrhizal infection.

The P concentratioms of P deficient plants are generally low with about 1 to 2 mg P/g dry matter or less. Cereals and herbage adequately supplied with phosphate have P concentrations of about 3 to 4 mg P/g dry matter, during the vegetative growth stage. Generally the P concentration is higher in younger plants or plant organs than in older ones. Thus the P concentration in mature straw of cereals is rather low (1.0 to 1.5 mg P/g dry matter), whereas in seeds and grains, P concentrations in the range of 4 to 5 mg P/g dry matter are found. This example shows that during grain and seed formation a considerable amount of P is translocated from leaves and stems towards the seeds or grains.

Phosphorus deficient tissues show an increase in phosphatase activity as a consequence of higher turnover rates of P and remobilization of phosphate (Smith and Chevalier 1984). According to O'Connell and Grove (1985) for eucalyptus seedlings this is a more sensitive parameter than total tissue P concentrations as an indicator of growth limitation by P.

9.3   Phosphorus in Crop Nutrition

9.3.1   Phosphorus availability and crop requirements

From the early days of applying mineral fertilizers to soils, phosphate fertilization has always been important. Indeed vast areas of potentially good land are still agriculturally poor because of P deficiency. It must be remembered that in the soil phosphate can readily be rendered unavailable to plant roots, and that P is the most immobile of the major plant nutrients in the soil.

From the discussion on soil phosphorus it is clear that in general phosphate availability to plants can be assessed by measuring the phosphate concentration in the soil solution and the ability of the soil to maintain the soil solution concentration (phosphate buffer capacity). Concentrations of about 100 mmol/m^3 phosphate in the soil solution are considered as high and represent a high level of available soil phosphate. Phosphate concentrations of about 1 mmol/m^3 in the soil solution are generally too low to supply crops adequately with P. In addition to the phosphate concentration of the soil solution, the phosphate buffer power of the soil plays a crucial role in determining the rate of P supply to crops (Nair1996). Optimum soil solution phosphate concentrations probably differ for individual crops, cropping systems, and particular sites.

The quantity of P present in soil solution, even in soils with a fairly high level of available phosphate, is only in the range of 0.3 to 3 kg P/ha. As rapidly growing crops take up phosphate quantities of about 1 kg P/ha/day, it is clear that the soil solution phosphate must he replenished several times per day by mobilization of phosphate from the labile pool. The quantity of this fraction present in the top soil layer (20 cm) is in the range of 150 to 500 kg P/ha. The rate of desorption is higher in soils with a higher P buffer capacity. For this reason such soils are better able to buffer the phosphate concentration of the soil solution during the growing season.The optimum P concentration of the soil solution may thus be low if the phosphate buffer capacity is high and *vice versa* (see page 76). This relationship has been used in investigations of van Noordwijk *et al.* (1990) who calculated P fertilizer requirements of crops in relation to the P-concentration of the soil solution and the phosphate buffer capacity.

Soils which are prone to strong phosphate fixation (adsorption to sesquioxides and clay minerals) often require extremely high phosphate fertilizer applications in order to alleviate the effects of fixation. Increasingly higher rates must be applied, the steeper the phosphate buffer curve. This is particularly the case in oxisols and ultisoils developed under tropical conditions (van Wambeke 1991). In these strongly phosphate fixing soils, pH correction is also recommended, since phosphate adsorption is especially strong at low pH (see page 455). Liming can therefore be an efficient measure to increase the available soil phosphate by rendering strongly adsorbed phosphate to less strongly adsorption (Sims and Ellis 1983). The positive effect of liming on the recovery of soil and fertilizer phosphate was shown by Sturm and Isermann (1978) by evaluating numerous long-term field trials carried out in Germany (Table 9.3).

Table 9.3 Percentage recovery of fertilizer phosphate from long-term field trials on representative agricultural soils to the lime status of soils. Recovery = P uptake of the crop and change in DL-soluble phosphate in the soils. (Sturm and Isermann 1978).

Lime status	% Recovery
Arable soils, very well supplied with lime	80
Arable soils, well suppplied with lime	70
Arable soils, moderately supplied with lime	65
Arable soils, poorly supplied with lime	60
Arable soils, poorly supplied with lime and dry locations	50
Grassland	80

Similar recovery percentages of fertilizer phosphate were obtained by Jungk *et al.* (1993) on Luvisoils with a neutral to alkaline pH. Maintaining an optimum soil pH is not only of agronomic interest but also an economic one since phosphate is a limited resource with mining reserves of only about another hundred years (Mengel 1997). About 90% of the mined phosphate is used for the production of fertilizers. Phosphate fertilizer and soil phosphate therefore should be used efficiently.

The need for phosphate fertilizer differs much between the developing and the highly developed industrial countries. In highly developed countries phosphate fertilizers have been applied for many years and soils are frequently enriched in available phosphate. Mallarino (1995) in evaluating field trials from 1980 to 1990 in the Midwest of USA found that of the 41 sites only 13 responded to P fertilizer application by increasing maize yield so that on about 2/3 of the sites the soil contained adequate amounts of available phosphate. The P nutritional status of the maize was reflected by the P concentrations in ear leaves at silking with a threshold value of 2.3 to 2.5 mg P/g leaf dry matter above which yield increase was not obtained.

In Germany too it has been found that many soils are well supplied with available phosphate. In developing countries it can be quite different. Sanchez and Leakey (1997) reported that in sub-Saharan Africa the per capita food production continues to decrease and the soils of the smallholders become more and more infertile because of a negative balance of plant nutrients, including phosphate. The beneficial effect of inorganic and organic P fertilizers on such soils in Kenya was described by Jama *et al.* (1997). The authors emphasize that the P + N fertilization gave a considerable net benefit in terms of currency in the range of 50 to 150 USD/ha per season. Field trials carried out in Kenya on acid soils, poor in available P showed that the natural fallow-maize-rotation and particularly a Sesbania fallow-maize-rotation with Sesbania as a N_2 fixing species (see page 403) gave higher grain yields as compared with the maize monoculture. Treatments with additional P fertilizer (triple superphosphate) resulted in further yield increases. The authors state that "improved fallows have a potential to provide nutrients to crops, but they are unlikely to eliminate the need for P fertilizers on P deficient soils". In this investigation net benefits of about 200 USD/ha were also obtained. These examples show that appropriate fertilizer application can improve the welfare of the local population. In highly developed agricultural systems the level of available phosphate is frequently high and the net benefit of P fertilizer may be marginal. Field trials with a wheat-maize rotation carried out over 10 years on the slightly alkaline Mollisol in the Vojvodina (Jugoslavia) showed that yield increases for wheat were greater than those for maize. The reason for this difference was that maize was better able to exploit the soil for P (Leskošek *et al.* 1972). Agronomic efficiencies (kg grain/kg fertilizer P) for the application of 40, 67, and 80 kg P/ha were 40, 26, and 22 respectively for wheat but only 16, 7.0, and 6.1 for maize so that the net benefit of P fertilization of maize if any was low.

In order to recommend an appropriate P fertilizer application a soil test for available phosphate is indispensable in most cases. This is not only true for soils with insufficient available phosphate but also for soils with excess available phosphate as shown in Figure 9.8 from the relationship between P availability as determined by the Olsen method and grass yields (Poulton *et al.* 1997). Such tests may save money and also avoid a waste of the limited raw material, phosphate. High to very high levels of available soil phosphate are particularly found in areas where organic manure, slurries and waste products are applied to soils in too high amounts (Leinweber 1996, Mozaffari and Sims 1996, Peters and Basta 1996, see also page 386). The soils are enormously enriched in phosphorus to levels which are several times higher than required for a good crop stands. Such high P concentrations may lead to phosphate leaching. These amounts may be small in agronomic terms but have an accumulating effect in soil leachates and present an environmental hazard (see page 388).

The rates generally applied to arable crops range from 20 to 80 kg P/ha according to crop species and available soil phosphates. On soils with a high phosphate adsorption capacity, however, rates of 100 to 200 kg P/ha may be applied (Jama *et al.* 1998). Crops with a high growth rate, producing large quantities of organic material, require a higher level of available soil P than slow-growing crops. A high level of available phosphate in soils is required by maize, lucerne, intensive grass production, potatoes and sugar beet. In addition all intensive systems of arable cropping have a relatively high demand for phosphate. Phosphorus is particularly important for leguminous plants possibly by its influence on the activity of the *Rhizobium* bacteria. For mixed swards

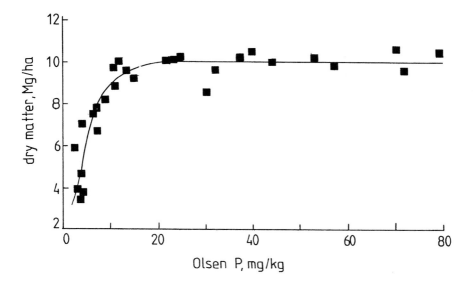

Figure 9.8 Relationship between "Olsen P" (see page 95) and grass yield (after Poulton *et al.* 1997).

it is therefore important that soil P levels should be kept high in order to maintain the leguminous species. If the P supply to cereals is inadequate during the early stages of development, a reduction in the number of ears per unit area results and hence a depression in crop yield. Adequate P supply may counteract mild water stress affecting tillering and leaf growth of wheat (Gutierrez-Boem and Thomas 1998). Phosphate is of particular importance for fruit setting fruit quality and resistance to diseases. Insufficient phosphate nutrition may delay fruit maturation. Ample phosphate supply to potatoes enhances the phosphate esterification of starch in potato tubers and thus improves starch quality (see page 318).

9.3.2 Phosphate fertilizers

The straight phosphate fertilizers which are used at the present time differ in chemical composition and solubility as can be seen from Table 9.4. Superphosphate is produced by the treatment of ground rock phosphate with sulphuric acid. The process yields a mixture of $Ca(H_2PO_4)_2$ and gypsum ($CaSO_4$). In the manufacture of triple superphosphate, phosphoric acid is used instead of sulphuric acid and the resulting product is $Ca(H_2PO_4)_2$. Mono and diammonium phosphates are made by adding NH_3 to phosphoric acid. Basic slags $[Ca_3(PO_4)_2 \cdot CaO + CaO \cdot SiO_2]$ are a by-product of the steel industry. In this process P originating from P containing ores is bound to CaO and silicates during smelting.

Table 9.4   Straight phosphate fertilizers

Name	Chemical composition	Solubility	Contentration of P_2O_5 in %
Superphosphate	$Ca(H_2O_4)_2 + CaSO_4$	water sol.	18−22
Triple superphosphate	$Ca(H_2PO_4)_2$	water sol.	46−47
Monoammonium phosphate	$NH_4H_2PO_4$	water sol.	48−50
Diammonium phosphate	$(NH_4)_2HPO_4$	water sol.	54
Basic slag (Thomas slag)	$Ca_3P_2O_8 \cdot CaO+CaO \cdot SiO_2$	citric acid sol.	10−22
Sinterphosphate (Rhenania type)	$CaNaPO_4 \ Ca_2SiO_4$	NH_4 citrate sol.	25−29
Ground rock phosphate	Apatite	soluble in	29
Fused Mg phosphate	Ca-Mg phosphate	citric acid	20

For this reason basic slags also contain Ca as oxide and silicates. In addition Mg and some heavy metals (Fe, Zn, Cu) are also present. Sintered phosphate (Rhenania-type) is produced by a disintegration of rock phosphate with Na_2CO_3 and silica in a rotary kiln at about 1250°C. The main constituents of this P fertilizer are $CaNaPO_4$ and Ca_2SiO_4 in a mixed crystalline structure. Other more recently developed fertilizers based on superphosphoric acid (polyphosphates) have already been considered on page 363.

Water soluble P fertilizers, basic slags, and sinter phosphate are suitable P fertilizers for most soil types. Rock phosphates differ widely in their fertilizer value depending on their provenance. Hard crystalline apatites are very insoluble and thus almost useless as fertilizer materials. Only ground soft rock phosphates may be used as fertilizers under particular conditions. Soft rock phosphates consist mainly of apatite with only a minor proportion of Fe^{III}- and Al phosphate. They must be ground in order to obtain a solublity in soils sufficient for fertilizer purposes. The solubility of soft rock phosphates depends much on origin. According to Anderson *et al.* (1985) solubility and availability in relation to provenance decreases according to the following sequence: Carolina > Gafsa > Sechura > Central Florida > Huila > Tennessee. The solubilty of soft rock phosphates is related to the degree to which the phosphate is substituted by carbonate in the crystalline structure of apatites. Such carbonate-apatites are are called francolite. Their solublity rises with increasing proportion of carbonate (Khasawneh and Doll 1978).

Besides the physical and chemical properties of the various soft rock phosphates, soil conditions, climate, and crops may have decisive impact on their solubility and therefore also on their fertilizing efficiency. Basically dissolution depends on the solubility product of dissolved phosphate, H^+, and Ca^{2+} in the soil solution acting upon the rockphosphate, according to the equation shown on page 454. Protons promote and Ca^{2+} and phosphate depress the dissolution. As shown by Anderson *et al.* (1985) phosphate plays an important role in this respect and therefore the solubility of soft rock phosphates is low in soils well supplied with phosphate because their phosphate concentration in the soil solution is relatvely high. Also soil pH is of fundamental importance for the use of soft rock phosphates as P fertilizer. As already documented in the older literature from numerous field experiments (van der Paauw 1965, Cooke 1966) soft rock phosphates are satisfactory fertilizers on soils with a pH < 5. There may be exceptions particularly where mycorrhizal fungi promote the dissolution and acquisition of the phosphate by crops as shown by (Mengel 1997). High soil moisture and therefore weather and climatic conditions may also improve the dissolution of soft phosphates, because high soil moisture favours phosphate diffusion in soils (Bhadoria *et al.* 1991). This means that the phosphate concentration adjacent to the fertilizer particle in the soil is lowered and depending on the solublity product, the dissolution of the rock phosphate promoted. Even under humid conditions, however, with high annual rainfall during the vegetation period, soft rock phosphate fertilizers may be of little value. In long-term grassland experiments carried out at 8 different sites in the humid area of Slovenia with a high rainfall Leskošek *et al.* (1973) found that on soils with a high pH superphosphate was superior to basic slag and soft rockphosphate gave no response. Over the pH range of 4.5 to 6.0 basic slag gave the highest forage yield followed superphosphate and then rock phosphate. The soil pH was influenced by the form of P fertilizer. It was slightly increased by basic slag slightly decreased by superphosphate, and rock phosphate was without effect. This example shows that the various forms of P fertilizer have an influence on soil pH.

Under temperate climatic conditions soft rock phosphates are inappropriate fertilizers for arable soils since they are generally only effective on acid soils. As phosphate recovery is poor on such soils (see Table 9.3) it is more useful to lime these soils in order to improve phosphate utilization. The idea that rock phosphates dissolve the longer they are in soils with pH > 5 has not been confirmed by relevant investigations of Wildhagen *et al.* (1983) and Steffens (1987). Similar results have been found by Bolland *et al.* (1988) in field trials on lateritic soils in Western Australia. In a valuable review article Bolland and Gilkes (1990) emphasize that the efficiency of rock phosphates was much lower than that of superphosphate tested under the arid conditions of Western Australia. This was true for the provenances Carolina phosphate, Queensland phosphate and also for Calciphos, a calcinated Christmas Island rock phosphate which in addition to apatites also contains Fe^{III} and Al phosphates. The term calcination implies that this hard rock phosphate has been dehydroxylated at a temperature at 500°C by which the crystalline structure has been destroyed. After application to the soil, however, rehydroxylation occurs to reduce the solubility.

Partially acidulated rock phosphates are rock phosphates which have been treated with sulphuric or phosphoric acids in amounts so that a part of the apatite has been transformed into more soluble phosphate form mainly $CaHPO_4$ (Resseler and Werner 1989). This provides an available source of phosphate to plant roots. The residual apatitic phosphate, however, has the same poor or even poorer solubility as the original rock phosphate (Resseler and Werner 1989). On soils in which the rock phosphates are only poorly soluble so also is the residual apatitic phosphate of partially acidulated rock phosphate (Bolland and Gilkes 1990).

Under tropical conditions, where soluble phosphates are easily leached from sandy acid soils, the application of rock phosphates is useful. According to Bolland (1996) on the humic sandy podsols in South Western Australia with an annual rainfall of >800 mm, phosphate is leached into deeper soil layers. This is particularly true for water soluble phosphate fertilizers such as superphosphate. In field trials superphosphate was compared with the partially acidulated Ecophos and Coastal Superphosphate. The effect of these fertilizers differed from season to season depending on rainfall and phosphate leaching. Under high leaching conditions superphosphate was inferior and *vice versa*. Bolland (1996) concludes from these results that Ecophos is a suitable alternative phosphate fertilizer to superphosphate and Coastal Superphosphate. The latter contains elemental sulphur which, when oxidized in the soil produces H^+ thus promoting the dissolution of rock phosphate.

Favourable effects of rock phosphate on the growth of wheat, soya beans and maize have been observed in India (Mandal 1975). Maloth and Prasad (1976) growing cowpeas (*Vigna sinensis*) on an alkali soil (pH 8.4) found that 86 kg P/ha in the form of soft rock phosphate gave the same yield increase as 43 kg P/ha in the form of superphosphate. On this particular soil, cereals did not respond to rock phosphate indicating that the effect of this fertilizer also depends on crop species. Leguminous species with roots excreting H^+ may cope better with the dissolution of rock phophates than cereals. This may also be true for crop species such as rape which excrete citric and malic acids (Hoffland *et al.* 1989) or red clover which excretes citric acid (Gerke *et al.* 1995). Haque *et al.* (1999) tested two rock phosphate provenances, Minjingu rock phosphate from a sedimentary deposit in Tansania and Chilembwe rock phosphate from an igneous deposit in Zambia, both in a non acidulated and in a partially acidulated form. *Trifolium quartinianum* was grown in four consecutive cuts on an acid Vertisol at 2370 m altitude above sea level and high rainfall in Ethiopia. The soil was very low in available phosphate. The response of phosphate application on the cumulative yield of clover is shown in Figure 9.9. With increasing phosphate rates increasing quantities of dry matter yield were obtained and even with the highest rate applied, 80 kg P/ha, the plateau of the yield curve had still not be attained. Interestingly the application of triple superphosphate gave the lowest, and the soft rock phosphate of Minjingu provenance the highest yield. A rock phosphate from the igneous deposit (Chilembwe) gave a very poor yield response which is typical for ignous rock phosphates (data not shown). The marked yield effect of the soft non acidulated rock phosphate from Minjingu demonstrates that on such soils and under such climatic conditions soft rock phosphates are good fertilizers and superior to water soluble fertilizers since they are much cheaper. This aspect is of particular importance for developing countries. The example also shows the enormous yield increase obtained with the highest phosphate rate. Apparent phosphate recovery was surprisingly high and amounted to almost 70% for the non acidulated rock phosphate and to 54 % for superphosphate at the highest phosphate level (Figure 9.9).This shows that there was no major phosphate fixation. The greater recovery of the rock phosphate presumably was due to the high rainfall conditions which leached the more soluble phosphate in the superphosphate. High yield levels of quality forage have a beneficial impact on farm animal production and on the farmer's income. For the farmer it is important to develop a farming system that keeps as much phosphate in the farming cycle as possible.

An interesting concept comparing the effectiveness of water soluble and non-water soluble phosphate fertilizers has been published by Chien *et al.* (1990).

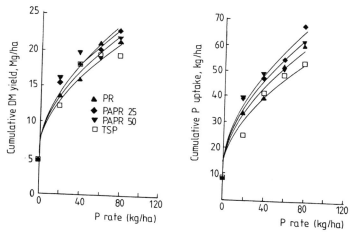

Figure 9.9   Cumulative yield and cumulative P uptake by clover during four consecutive cuts, one cut per year, related to the phosphate fertilizer rates given at the beginning of the field trial. PR = phosphate rock, provenance Minjingu, PAPR25 = phosphate rock acidulated at 25%, PAPR50% = phosphate rock acidulated at 50%. TPS = triple superphosphate — Left: Cumulative yield. Right: cumulative P uptake (after Haque *et al.* 1999).

Although the solubility of phosphate in the various forms of compound fertilizers does not differ as much as that in straight phosphate fertilizers, there are differences in phosphate solubility and effectiviness. These become especially evident when phosphate containing compound fertilizers are applied to soils with a low phosphate status. From numerous field experiments carried out in India it was found that the P response was highest with superphosphate followed by ODDA nitrophosphate and least with PEC nitrophosphate The abbreviations 'ODDA' and 'PEC' refer to different technological procedures.

9.3.3   Phosphate application

Only a relatively small amount, less than 15% of the P fertilizer applied to a soil, is usually recovered by a crop grown immediately after application (Greenwood 1981). The remainder is converted to insoluble form in the non labile fraction. Recovery of this P by crops in subsequent years becomes less and less. In order to keep fixation to a minimum and to ensure efficient uptake of P this can be achieved by application in granular form or as bands within the soil in close proximity to the roots. Phosphate is thus concentrated in a limited soil volume saturating the soil phosphate adsorption capacity and increasing the soluble phosphate in the soil. This is particularly important in high fixing soils or soils low in phosphate. The beneficial effect of placed applications as compared with broadcast treatments has been reported by a number of authors (Prummel 1957, Ryan 1962, Reith 1972; see also page 367).

Another important principle of application is that the root surface needs to absorb very high rates of phosphate in the early stage of growth if high yields are to be obtained (see Table 2.16 on page 83). Placement of fertilizer P as a starter application can thus be extremely effective (Stone 1998). Phosphate fertilizers can be applied at any time of the year, provided the phosphate fixation capacity of the soil is low. When this is the case, water soluble phosphate materials should be applied preferentially in spring. On acid strongly phosphate adsorbing soils the recovery of superphosphate was significantly reduced within a few months as reported by Barekzai and Mengel (1985). On such soils phosphate fertilizer should be applied just before sowing. This reduces the fixation of the fertilizer P to a minimum as it allows the crop the best opportunity to compete with the soil for P utilization. In soils in which available phosphate levels are adequate, application of phosphate can be made every second year without running any risk of yield depressions (Prausse 1968). The rate of phosphate application, however, should be in the same order as the sum of the corresponding two single treatments. As the mobility of P is low there is very slow movement down the profile and P tends to accumulate close to the soil surface. Plant roots thus proliferate in the rich P surface zone where the bulk of soil water that is potentially available to plants is also stored. Under moisture stress soil P can become unavailable (Ozanne 1980). The benefits of incorporated phosphate fertilizers into the soil and particular low water soluble fertilizers, can be appreciated from the results of Sewell and Ozanne (1970). Banding fertilizer at a 15 cm soil depth induced a threefold increase in root development in this banded zone. The plant were thus better protected against water stress.

The use of mechanistic modelling to calculate crop response to P fertilizer for field crops is not so advanced as for N or K application even though the underlying principles are the same and the transport of phosphate from soil to plant roots is well understood (see page 67). One relatively simple approach used in the Netherlands (van Noordwijk et al. 1990), where the P_w (see page 95) value based on water extraction of soil P is used as a basis for P fertilizer recommendations, has been to model P transfer in the soil from P adsorption isotherms to crop uptake requirements. P_w values have been calculated which are needed for adequate P uptakes by various crops. Other models to predict P uptake have been developed which take into account the transport of P in the soil via diffusion (see page 67) as discussed by Jungk and Claassen (1997). When P supply was ample (100 kg P ha^{-1}) good agreement between predicted and measured values was found throughout the development of the crop. In the low P treatment, however, in which no P had been applied to the soil for 20 years the model substantially underestimated the P uptake. The same observation was made by Brewster et al. (1976) who found an even higher underestimation in P deficient plants. The reason for this underestimation in the low P treatment is that other sources of P must have become available to the plants which were not accounted for by the model. This is in accordance with the various mechanisms by which the roots of P deficient plants can increase P acquisition from the soil as already described. In soils high in phosphate as occur in many well fertilized arable soils models based on calculation of diffusion of P through the soil may also be inappropriate as nutrient transport is unlikely to limit P uptake (Barraclough 1986). Advances are still to be made in modelling P utilization by crops. So far models have been tested on only a few field sites and none has predicted effects of added fertilizer on crop yield. There is also a need for models to take into account root development during growth as well as feedback effects by the plant and particularly changes in P demand as the critical internal P concentrations decline as plants increase in size. In the long term, modelling may be expected to refine but not replace the recommendations based on soil chemical analysis for P fertilization.

CHAPTER 10

POTASSIUM

10.1 Soil Potassium

10.1.1 Potassium minerals and potassium release

The average K^+ content of the earth's crust is in the order of about 23 g/kg. By far the greatest part of this K^+ is bound in primary minerals and in the secondary clay minerals which largely make up the clay fraction of the soil of particle size less than 2 μm.

The clay content of a soil is to some extent dependent on the soil parent material, but is also considerably affected by pedogenesis. Old highly weathered soils are often low in K^+ (Portela 1993). Thus highly weathered sandy soils contrast markedly with younger soils either derived from volcanic material (Graham and Fox 1971) or with a silt fraction containing appreciable amounts of K^+ bearing minerals as is the case for Aridisols in Pakistan (Mengel and Rahmatullah 1994). Loess- derived Luvisols also contain high amounts of K^+ important for the nutrition of plants (Mengel *et al.* 1998).

The K^+ concentration of soils also depends on the type of clay minerals present. In an extensive investigation comprising more than 1000 sites in Central Europe, Laves (1978) found that soil K^+ content was closely related to the content of illites and Al-chlorites and to a lesser degree to the content of smectites. Highest K contents were found with high chlorite and illite contents associated with low levels of smectites as occurred in the brown soils of the mountainous areas. The alluvial soils in which the reverse was the case and smectites were dominant were lower in K. The loess soils were intermediate between the two. Organic soils are frequently low in clay and K^+ contents. The level of K in organic soils is in the region of 0.3 g/kg although the figure may vary somewhat between different classes of organic soils.

The main source of K^+ for plants growing under natural conditions comes from the weathering of K^+ bearing minerals. The most important of these minerals are listed in Table 10.1. The basic structural element of the feldspars is the tetrahedral three-dimensional framework. K^+ is located in the interstices of the Si-Al-O framework of the crystal lattice and held tightly by covalent bonds (Sparks 1987). The weathering of feldspars begins at the surface of the particle. Potassium is initially released by water and weak acids at a more rapid rate than other constituents. Earlier reports that weathering of feldspars produces a Si-Al-O residual envelope around the unweathered core have not been confirmed (Sparks 1987). According to Sparks (1987) and to more recent investigations of Wulff *et al.* (1998) feldspars in the sand fraction may provide an important and continuous supply of K^+ to crops. In the lithosphere feldpars are the dominant minerals and their weathering is an important process of global dimension as emphasized by Sparks (1987) in a valuable review article.

481

Minerals of the mica type and also the secondary minerals of the 2:1 layer silicates differ fundamentally in structure from the feldspars. For this reason they also differ in their properties of releasing and binding K^+. The micas are phyllosilicates and consist of unit layers each composed of two Si-Al-O tetrahedral sheets between which an M-O, OH octahedral sheet is located, where M is usually an Al^{3+}, Fe^{2+}, Fe^{3+}, or Mg^{2+} (Fanning et al. 1989). Potassium ions occupy the approximately hexagonal spaces between the unit layers and as a consequence the distance between unit layers is relatively small e.g. 1.0 nm in micas. The replacement of non hydrated interlayer K^+ by hydrated cation species such as Na^+, Mg^+ or Ca^{2+} expands the mineral increasing in the distance between the unit layers, e.g. to 1.4 nm in vermiculite (see Figure 10.1).

Table 10.1   Potassium concentration of some primary and secondary clay minerals (Scheffer and Schachtschabel 1982)

	concentration in g/kg
Alkali feldspars	32—120
Ca-Na feldspars	0—24
Muscovite (K mica)	60—90
Biotite (Mg mica)	36—80
Illite	32—56
Vermiculite	0—16
Chlorite	0—8
Montmorillonite	0—4

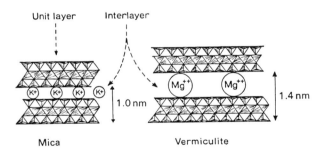

Figure 10.1   Unit layer and interlayer of mica and vermiculite.

Generally lattice bound K^+ is vulnerable to weathering and can diffuse out of the mineral in exchange for other cation species. This interlayer release is promoted by a low K^+ concentration adjacent to the K^+ bearing mineral. Threshold concentrations below which K^+ release taken place are 50 to 400 $mmol/m^3$ for biotite, and 2.5 $mmol/m^3$ for muscovite and illite. These values indicate the much higher rate of weathering of biotite than that of muscovite (Martin and Sparks 1985). During K^+ uptake plants also greatly reduce the K^+ concentration in the immediate vicinity of the roots and this can induce release of K^+ from minerals (Kuchenbuch and Jungk 1984). As shown by Hinsinger and Jaillard (1993) roots of ryegrass grown in close proximity to phlogopite (mica) induced the release of K^+ from the mica which in turn was converted to vermiculite. According to recent results of Rahmatullah and Mengel (2000), it is the lowering of the K^+ concentration *per se* and not of H^+ released by roots which initiates the release of interlayer K^+. As adsorption sites of the interlayer are made free by the release of K^+ they become occupied by other cation species. The adsorption of these cations Na^+, Ca^+ and Mg^{2+} in hydrated form results in swelling of the interlayer (Figure 10.1). This is the case of typical frayed edge or wedge zone formation in weathering micas which are indicative of the dissolution of the silicate layer following the release of interlayer K^+ (Rich1968). Initially the process is an accelerating one as the resulting widening gap between the two layers of the mineral favours the diffusion of the replaced K^+ out of the mineral.

According to Farmer and Wilson (1970) this kind of weathering converts micas to the secondary 2.1 clay minerals, illite and vermiculite, the reaction sequence being as follows: Micas (about 10% K) → hydromicas (6 - 8% K) → illite (4—6 % K) → transition minerals (3% K) → vermiculite or montmorillonite (<2% K). Intensive cropping of soils without the application of K^+ fertilizer may lead to a degradation of illites as has been shown by Tributh *et al.* (1987). Illite has a dioctahedral structure and has similarities to muscovite and mica. In comparison with mica, however, the crystalline structure of illite is not so well defined, it contains H_2O and has a lower K^+ concentration (Aktar 1994). High H^+ concentrations as occur at about pH 3 adjacent to the interlayer may not only induce the release of interlayer K^+ but destroy the mineral since as well as K^+, other cations such as Mg^+ and Al cation species and originating from the octhedral centres are also released (Feigenbaum *et al.* 1981). Figure 10.2 shows a model of expandible phyllosilicate with interlayers saturated with different cation species.

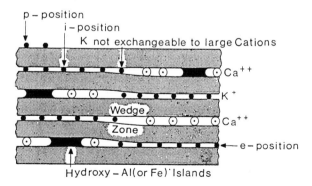

Figure 10.2   Model of an expandible layer silicate with interlayers, wedge zone, p-, e-, and i-position (after Rich 1968).

The rate of K^+ release by weathering not only depends on the K^+ concentration of minerals but is also dependent by even slight structural differences between minerals. Thus biotite a ferromagnesian mineral although generally lower in K^+ concentration than muscovite, releases K^+ at higher rates because the K^+ in muscovite is more closely located to the negatively charged oxygen in the interlayer and hence is more strongly bound (Martin and Sparks 1985). Calcium feldspars are degraded at a higher rate than K feldspars (Duthion 1966).

10.1.2 Potassium fixation

As the depletion of K^+ from the interlayers of minerals continues, the rate of release becomes progressively slower. Interlayer sites become depleted of K^+ although they still retain a very high K^+ selectivity with respect to divalent cations (Raman and Jackson 1964). Addition of K^+ to such minerals results in a strong K^+ adsorption to these positions causing a contraction of the mineral (Graham and Lopez 1969).This process is called K^+ fixation. The contraction results in a reduction in the unit distance to about 1 nm. Zones of vermiculite structure in weathered biotite and muscovite characterized by a 1.4 nm layer distance thus regain the 1 nm spacing of mica when treated with K^+ (see Figure 10.1).

The degree to which K^+ fixation occurs depends on a number of factors including the charge density of the mineral, the extent of the wedge zone, the moisture content, the concentration of K^+ and the nature and concentration of competing cation species in the surrounding medium. Fixation tends to be high when the negative charge per unit silicate layer (= charge density) is high. When this occurs K^+ is very strongly held by the negatively charged layers. If the wedge zone is confined to the edge of the particle only small amounts of K^+ can be fixed. However, if the zone penetrates deeply into the mineral, considerable amounts of K^+ can be withheld. Some minerals such as weathered micas, vermiculites and illites fix K^+ under both moist and dry conditions, whereas smectites only fix K^+ under dry conditions. For this reason fixation is frequently higher under dry than moist soil conditions (Schroeder 1955). As the NH_4^+ is very similar to K^+ in ionic radius, this too can be fixed by expanded 2:1 clay minerals (Bartlett and Simpson 1967). Ammonium can also exchange for fixed K^+. The same holds true for H^+ (Rich and Black 1964).Thus both ion species, NH_4^+ and H^+, can compete with K^+ for K^+ fixing binding sites. This means that K^+ fixation is generally not so important in restricting K^+ availability to plants on acids soils (pH < 4.5) as on limed agricultural soils. The fixing power of the 2:1 clay minerals usually follows the order vermiculite > illite > smectite. Reduction of Fe^{III} located in octahedra increases the K^+fixation potential because the negative charge density in the interlayers is increased (Chen et al. 1990).

Potassium fixation is of considerable importance in agricultural practice (see page 508) as the quantities of fertilizer K^+ rendered unavailable in this way can be very high. Generally the K^+ fixation capacity is higher in the deeper soil layers because of greater clay concentrations

10.1.3   Potassium adsorption and mobility

Soil clay concentration is not only of importance for K^+ release and fixation, it also considerably influences the mobility of K^+ in the soils. Potassium ions are adsorbed by clay minerals to binding sites which differ in selectivity. The older literature distinguishes between three different types of adsorption sites for the 2:1 clay minerals such as illites, vermiculites and weathered micas as shown in Figure 10.2 (Schouwenburg and Schuffelen1963). These are sites at the planar surfaces (p-position), at the edges of the layers (e-position), and in the interlayer space (i-position). The specificity of these three binding sites for K^+ differs considerably. As has been already outlined on page 22, this specificity of K^+ binding in relation to other cations can be expressed in quantitative terms by the Gapon coefficient which is higher the greater the specificity of the binding site for K^+. According to Schuffelen (1971) the three different K^+ binding sites of illite have the following Gapon coefficents

p-position: 2.21 $(mol/m^3)^{-1/2}$; e-position: 102 $(mol/m^3)^{-1/2}$; i-position: infinite $(mol/m^3)^{-1/2}$

The Gapon coefficient may be calculated from molar or millimolar concentrations. As numerical values can differ between the two forms of expression, units should always be stated. The coefficients shown above are based on mol/m^3 concentrations and they refer to K^+/Mg^{2+} exchange, thus comparing the specificity of the binding site for K^+ with that of Mg^{2+}. Values of the same order of magnitude have also been reported by Duthion (1966) for K^+/Ca^{2+} exchange. The considerable differences between the Gapon coefficient for the three K^+ binding sites demonstrate the immense distinction between sites for K^+ selectivity. The binding selectivities for K^+ by organic matter and kaolinitic clays are similar to the p-position sites. Here the K^+ bond is relatively weak so that K^+ adsorbed may be easily replaced by other cations, and particularly by Ca^{2+} and Mg^{2+}. The i-position has the greatest specificity for K^+. These binding sites account for K^+ fixation in soils.

   Gapon coefficients of the p- and e-position are not so clear-cut as shown above but may overlap. This is in agreement with experimetal results of Jardine and Sparks (1984) who found in their K^+/Ca^{2+} exchange experiments that there are only two clear adsorption sites for K^+: One easily acessible for hydrated K^+, the other accesible with some difficulty binding non-hydrated K^+. From this follows a clear distinction between K^+ adsorbed in the hydrated form and K^+ adsorbed in the non-hydrated form, the latter located at sites of the interlayer. This is of relevance for the availability of K^+ to plant roots as shown below.

   The importance of the clay concentration in soils and the type of clay minerals on K^+ solublity, K^+ buffer power, exchangeable (adsorbed) K^+, and K^+ fixation was convincingly shown by Sharpley (1990) who investigated the most important K^+ interrelationships between these parameters, of 102 different arable soils from a wide variety of taxonomic groups in Nebraska. The soils were divided into three groups depending on clay types: Soils with mainly kaolinitic clays, soils with mainly smectitic clays, and soils with mainly mixed clays containing smectitic and kaolinitic clays. The impact of the clay concentration in soils and of the clay types on the K^+ solubility in soils is shown in Figure 10.3.

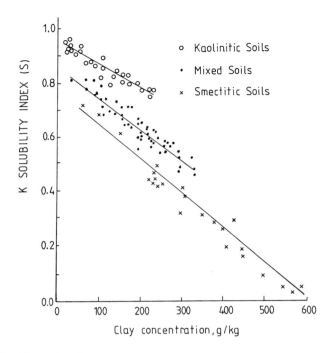

Figure 10.3 Solubility indices for K$^+$ related to clay concentration and clay types (after Sharpley 1990).

The soil group with the smectitic clay minerals, including those mainly with interlayer K$^+$ showed the lowest K$^+$solubility which decreased linearly with increasing clay concentration of the soils. Highest solubility of K$^+$ was found in soils with mainly kaolinitic clays and the soil group with the mixed clays (kaolinitic + smectitic) was intermediate between the other two in relation to K$^+$ solublity. This result confirms earlier investigations of Nemeth *et al.* (1970) carried out with arable soils from Germany. In the investigation of Sharpley (1990) K$^+$ fixation was greatest by far in the the soil group with smectitic (2:1) clay minerals in which the K$^+$ release rate was potentially high in agreement with the high K$^+$ buffer power of this group. Potassium fixation potential, K$^+$ adsorption potential, and K$^+$ buffer power were lowest in the kaolinitic soil group owing to the low cation exchange capacity (CEC). This soil group showed the highest proportion of soluble K$^+$ and therefore also a relatively high K$^+$ concentration in the soil solution. This finding highlights the potential hazard of K$^+$ leaching in soils in which kaolinitic clay minerals dominate.

10.1.4 Potassium fractions

Soil K$^+$ can be divided in 3 fractions: K$^+$ as a structural element of soil minerals including interlayer K$^+$, K$^+$ adsorbed in exchangeable form to soil colloids such as clay minerals and organic matter, and K$^+$ present in the soil solution. By far the greatest fraction of soil

K^+ is located in mineral structure. The exchangeable K^+ represents only about 5 % of the total soil K^+ (Table 10.2) and is defined as the K^+ which is exchanged by a 1000 mol/m³ NH_4CL or 1000 mol/m³ NH_4 acetate solution. This extraction comprises practically all K^+ adsorbed in hydrated form, the K^+ in the soil solution, and depending on the K^+ bearing minerals also a minor proportion of the non-hydrated K^+ (interlayer K^+). Plants, however, not only take up this "exchangeable K^+" but also a K fraction which is not exchangeable with 1000 mol/m³ NH_4^+. This so called nonexchangeable K^+ released during cropping originates mainly from interlayer K^+ and also from feldspars and has been determined by extracting the soil with 1000 mol/m³ HNO_3 (Pratt 1965). In Table 10.2 the quantities of exchangeable K^+ ($CaCl_2$ extract, extracting only hydrated K^+), nonexchangeable K^+, mineral K^+, and total soil K^+ are shown for a sandy loam and a loamy sand from the work of Martin and and Sparks (1983).

Table 10.2   K fractions of a sandy loam and a loamy sand soil (Martin and Sparks 1983)

	Exchang. K^+ (CaCl₂)	Nonexchang. K^+ (HNO₃)	Mineral K^+	Total K^+
		in mol K kg⁻¹ soil		
Sandy loam	1.72	2.20	37.6	41.5
Loamy sand	1.15	2.09	31.3	34.5

As can be seen from the figures the non exchangeable K^+ was not much higher than the exchangeable K^+ and it includes only a very low proportion of the interlayer K^+. Although the nonexchangeable K^+ fraction is of relevance for the nutrition of plants, its quantitative assessment is difficult because it is not the quantity of K^+ extracted by the HNO_3 solution that is indicative of K^+ availability but rather the rate at which K^+ is released from the interlayers (Mengel and Uhlenbecker 1993). This release rate follows Elovich kinetics (Havlin *et al.* 1985). Mengel and Uhlenbecker (1993) and Mengel *et al.* (1998) showed by means of repeated K^+ extraction with electroultrafiltration (EUF) (see page 99) that the b value of the Elovich function was indicative of the release rate of interlayer K^+ and correlated with the K^+ uptake of ryegrass.

y = a + b ln x   Elovich equation
y = cumulative extracted K^+
a = intercept on the ordinate
x = number of repeated extractions

Using the results of this EUF extraction together with the findings of pot experiments to test uptake of K^+ it was shown that ryegrass took up high amounts of interlayer K^+ originating from the silt fraction of loess-derived Luvisols. These amounts were in the same range as K^+ released from the clay fraction of these soils. Hence in these soils representative of many fertile soils a considerable amount of interlayer K^+ accessible to crops originates from the silt fraction which makes up 60 to 80% of the total soil mass (Mengel *et al.* 1998).

The main relationships between the structural K^+ in minerals, the adsorbed K^+ and the K^+ in the soil solution from which it is drawn by plant roots are shown in Figure 10.4.

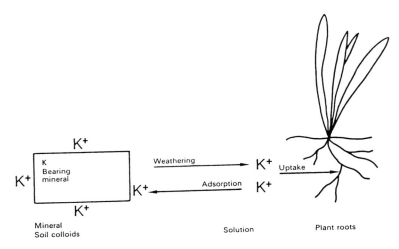

Figure 10.4 Potassium relationships between K^+ bearing minerals, K^+ in soil solution, and K^+ uptake by roots.

Potassium released by minerals into the soil solution can then be taken up directly by plant roots or be adsorbed by soil colloids. An equilibrium may thus be set up between adsorbed K^+ and the free K^+ in soil solution. The K^+ level in the soil solution resulting from this equilibrium depends much on the selectivity of the adsorption sites. If there are many adsorption sites for K^+, the concentration of K^+ in the soil solution tends to be low and *vice versa*. If there are no interlayer binding sites for K^+ as may be the case for organic soil and sandy soils the K^+ concentration in the soil solution is hardly buffered and may change quickly depending on K^+ uptake by crops, K^+ fertilizer application, and K^+ leaching as shown by Wulff *et al.* (1998; see Figure 10.5).

Potassium concentration in the soil solution and K^+ buffer power largely control the K^+ diffusion rate towards the plant roots and thus the K^+ uptake of crops (see also page 76). The more the exchangeable K^+ fraction is exhausted the greater is the contribution of the nonexchangeable K^+ to plant supply. Soils derived from loess (Luvisol) are known to release considerable amounts of interlayer K^+ and a substantial proportion of K^+ taken up by crops on these soils may originate from the nonexchangeable fraction. Since most soil tests do not take the nonexchangeable K^+ fraction into account, they may relate only poorly to K^+ responses by crops as has been observed by Kuhlmann and Wehrmann (1984) for loess- derived soils in Germany. According to Steffens and Mengel (1979) grasses are especially able to utilize interlayer K^+.

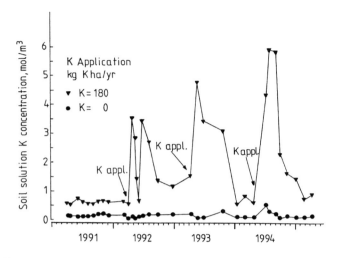

Figure 10.5 Effect of K fertilizer on the K^+ concentration in the soil solution of a sandy soil (after Wulff *et al.* 1998).

10.2 Potassium in Physiology

10.2.1 Uptake and translocation

Potassium is an essential element for all living organisms. In plant physiology it is the most important cation species not only in regard to its concentration in plant tissues but also with respect to its physiological and biochemical functions. A main feature of K^+ is the high rate and efficient means by which it is taken up and translocated throughout the plant. Accumulation by roots achieved by plasmamembrane transporters which are able to adjust quickly to the wide range of K^+ in soil solution (Maathius and Sanders 1997). This rapid rate of K^+ uptake is due to various K^+ uptake systems, mainly specific channnels in the plasmalemma and tonoplast of plant tissues (page 129). Potassium channels are structurally the simplest in the superfamily of cation channels and are characterized by a short section in the pore region which is responsible for the K^+ selectivity (Hille 1996). Both outwardly and inwardly directing channels are present in which the K^+ flux follows the electrical potential gradient (Colombo and Cerana 1991, Pottosin and Andjus 1994). A variety of different channels occur including Ca^{2+} or Na^+ activated K^+ channels.

For K^+ two main groups of transport systems in the root plasma membrane can be distinguished; high-affinity transporters and low-affinity transporters, the former being very selective for K^+ and characterized by a low K_M in the range of a few $mmol/m^3$ and the latter less selective and characteized by K_M of about 1 mol/m^3 (Fox and Guerinot 1998). The K^+ uptake mechanism of both groups is different; for high-affinity transport, uptake proceeds *via* K$^+$ H$^+$ symport whereby H$^+$ is transported together with K$^+$ across the plasmalemma into the cytosol (Schachtman and Schroeder 1994, Rubio *et al.* 1995, Schachtman *et al.* 1991). According to Fernando *et al.* (1992), and Schachtman and Schroeder (1994) the synthesis of such high-affinity K$^+$ transporters is induced when K$^+$ supply is insufficient and is an adaptation to low-availability of K$^+$. Both K$^+$ transport systems are coupled to the activity of the plasmalemma H$^+$ pump which builds up the electrochemical gradient required for the transport across the membrane and in the case of the high-affinity transport also for the provision of H$^+$ for the cotransport. The low-affinity transport is a facilitated diffusion of K$^+$ through selective channels into the cytosol (Fox and Guerinot 1998). Whereas the high-affinity transport system is charaterized by saturation kinetics, the functioning the low-affinity transporters is reflected by a biphasic curve with a linear increase at the higher K$^+$ concentration range in accordance with what would be expected of a facilitated diffusion. According to Maathuis and Sanders (1997) the low affinity transporters come into action at K$^+$ concentrations in the nutrient solution of >0.1 mol/m^3 and and they play an increasing role on total K$^+$ uptake as the K$^+$ concentration of the outer solution increases. At concentrations >1.0 mol/m^3 K$^+$, most of K$^+$ is taken up by the low-affinity system (facilitated diffusion along specific channels). Both inwardly and outwardly directed K$^+$ channels occur by which uptake and retention of K$^+$ are regulated (Maahuis and Sanders 1997). The basic mechanisms of the two groups of K$^+$ transporters are shown in Figure 10.6 from a valuable review of Maathuis and Sanders (1996) on mechanisms of K$^+$ absorption.

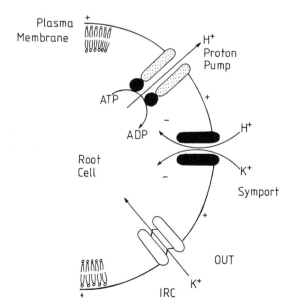

Figure 10.6 High affinity K transporter (H$^+$/K$^+$ cotransport = symport) and low affinity K transporter following the electrochemical gradient, IRC = inwardly rectifying channel. Both transporters are driven by the proton pump (after Maathuis and Sanders 1996).

The low affinity K^+ transport channels allow diffusion along the electrochemical gradient. These low-affinity channels have a K_M of about 1 mol/m^3 K^+ which means they may play a role when the K^+ concentration in the soil solution is high as is the case when K fertilizer is applied to soils with a low K^+ buffer power (see Figure 10.4). The high-affinity transporters are of importance for the mobilization of interlayer K^+ which needs a low K^+ concentration adjacent to the K^+ bearing mineral (see page 68). Voltage-gated channels for cation species such a K^+, Ca^{2+} and Na^+ are present in the membrane which are controlled by the voltage across the membrane and therefore are dependent on the activity of the membrane proton pump (Catterall 1995). Such channels, which bring about membrane transport for different cation species, may be responsible for cation antagonism (see page 134). According to Bouteau et al. (1996) outwardly directed K^+ channels are present which allow an efflux of K^+. The blocking of these channels by Ca^{2+} may be responsible for the so called Viets effect known for very many years in which Ca^{2+} promotes the net uptake of K^+ (Viets 1944, Mengel and Helal 1967). This diversity of K^+ uptake systems allows an efficient control of K^+ uptake and K^+ distribution in plant tissues according to needs the plant (Glass 1983).

Of particular interest is the K^+ transport across the plasmalemma of guard cells (Talbott and Zeiger 1996). According to Roth-Bejerano and Neyidat (1987) the opening and closing of K^+ channels through the plasmalemma of guard cells is induced by the reception of red light from phytochromes. Kim et al. (1993) reported that the leaflet movement Samanea saman results from a change in turgor of the extensor and flexor cells. The leaflets open in the morning when K^+ together with Cl$^-$ streams into the flexor cells. In the evening when the leaflets close the reverse process takes place and the K^+ and Cl$^-$ stream out of the flexor cells into the extensor cells. In these ion transports K^+ channels were found to open in the flexor cells when they were closed in extensor cells and vice versa. Diurnal rhythms of K^+ uptake have also been reported by Le Bot and Kirkby (1992) and by Macduff and Dhanoa (1996). Potassium uptake rates were highest at noon and lowest at midnight. The mechanism of this rhythm is not yet completely understood. According to Macduff and Dhanoa (1996) energy supply is not the controlling factor.

The distribution of K^+ within the cell is strictly controlled whereby a constantly high K^+ concentration is maintained in the cytosol whereas the K^+ in the vacuole may alter depending on K^+ supply to the tissue (Leigh and Wyn Jones 1986). This relationship of K^+ concentration in the cytosol depending on K^+ supply is shown in Table 10.3 (Fernando et al. 1992).

Table 10.3   Effect of increasing K^+ supply on the K^+ concentration in the vacuole and cytosol (after Fernando et al. 1992).

K^+ in the nutrient solution mol/m^3	Vacuole mol/m^3	Cytosol mol/m^3
0.01	21	131
0.1	61	140
1.2	85	144

The K⁺ concentration of the outer solution (nutrient solution) was varied by a factor of 100 while the responding K⁺ concentration in the vacuole of barley root tissues was changed by a factor of about 4 and the cytosolic K⁺ concentration remained almost constant. High K⁺ concentrations in the cytosol are required for enzymic activities as is shown below. Potassium in the vacuole and in the cytosol also functions as an important osmoticum. Potassium concentrations in leaf apoplast are about 10 mol/m³ and depend on K⁺ supply (Long and Widders 1990). Mühling and Läuchli (1999) found a diurnal change in the K⁺ concentration in the apoplast of *Vicia faba* leaves as shown in Figure 10.7. The build up of such relatively high K⁺ concentrations must have been dependent on the low-affinity K⁺ transporters.

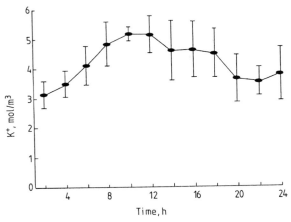

Figure 10.7   K⁺ concentration in the apoplast of *Vicia faba* in the course of one day (after Mühling and Läuchli 1999).

Potassium is highly mobile in plants and long-distance transport takes place in xylem and phloem. Concentration in the exudation sap of decapitated plants range from 10 to 30 mol/m³ (Kirkby *et al.* 1981). These concentrations in exudation sap, however, are not identical with the K⁺ concentration in the xylem of intact plants as shown in Table 4.4 (page 209). The latter are generally much lower and in a range of 5 to 10 mol K⁺/m³ (Schurr and Schulze 1995). Nevertheless, of all inorganic ion species present in the xylem sap, K⁺ generally has the highest concentration. The phloem sap K⁺ concentration is in a range of 50 to 150 mol/m³ (Marschner *et al.* 1997). Transport of K⁺ in the xylem is exclusively from root to shoot whereas in the phloem transport may be directed towards various sink organs both in the shoot (apex, seeds, fruits) including storage roots (beets) and tubers. Potassium and other nutrients which are mobile in the xylem and phloem can be recycled in the plant, *i.e.* transport from root to shoot and back to roots and from here again to shoots.

Recycling has been studied only in only a few species where it is possible to obtain both xylem and phloem sap from growing plants, such as castor oil bean (*Ricinus communis*) and white lupin (*Lupinus albis*) (Pate 1975, Kirkby and Armstrong 1981, Jeschke 1997). From this and other work it has become clear that cycling of K⁺ and other nutrients may serve several well defined functions as summarized by Marschner *et al.* (1997). In the first place K^+, by acting as a counter ion for NO_3^- transport in the xylem from the root faciletates the movement of NO_3^- to the shoot which for many plant species is the major site for nitrate reduction and assimilation (see page 166). Secondly, after NO_3^- reduction in the shoot, K^+ again acts as counter ion from shoot to root, but this time for assimilation products such as amino acids and organic anions, especially malate. A third function that recycling of K^+ may serve is to communicate the K^+ requirement of the growing shoot to the root. This suggestion seems most reasonable as there is good evidence that in K^+ — replete plants, growth and K^+ uptake are in balance (Kochian and Lucas 1988) and even with interruption of K^+ supply, acclimatory changes can take place to maintain adequate K^+ in the phloem from shoot to root to act as a signal to regulate K^+ uptake. It is suggested that at high shoot demand, a lower proportion of K^+ cycles back to the root as a feedback signal for enhanced uptake and xylem transport. The results of White (1997) from investigations of the regulation of K^+ influx into roots of rye seedlings are in accord with the concept of K^+ supply from the shoot regulating K^+ uptake by the root.

10.2.2 Meristematic growth

A basic function of K^+ is its impact on proton pumps. About 20 years ago Briskin and Poole (1983) reported that the plasmalemma ATPase is stimulated by K^+ which favours the process of H^+ pumping by the dephosphorylation of the ATPase subunit. This effect might be expected to be due to the cytosolic K^+ concentration. There are, however, findings which ascribe an activating effect on the ATPase to the K^+ concentration at the outer side (apoplast). Mengel and Schubert (1985) found that in presence of a $3\ mol/m^3$ K^+ concentration in the outer medium the net release of H^+ was four times higher as compared with the absence of K^+ in the outer medium. The authors explain this phenomenon as a membrane-depolarizing effect of K^+ because K^+ uptake lowers the electrical gradient across the membrane and hence less energy is required in the up-hill transport per unit H^+ out of the cytosol into the outer medium. Berkowitz and Peters (1993) as well as Shingles and McCarty (1994) found that K^+ stimulated the ATPase in the inner envelope of chloroplasts. The activating effect of K^+ on ATPase is not only of relevance to meristematic growth, but also to electron transport in photosynthesis and to phloem loading.

The growth process is initiated by a plasmalemma located ATPase which pumps H^+ out of the cytosol into the apoplast. The acidification of the apoplast results in a loosening of cell wall material and in the activation of hydrolysing enzymes (Hager *et al.* 1971). This loosening of cell wall material is a prerequisite for cell expansion. As already discussed above the H^+ release rate depends much on the presence of K^+ because of the depolarization of the plasmalemma. Potassium also functions in another way in plant growth which is a dominating process in meristematic growth. It is required particularly in polypeptide synthesis in the ribosomes and is probably involved in several steps of the translation process including the binding of tRNA to ribosomes (Wyn Jones and Pollard 1983). The effects of K^+ on protein synthesis are discussed on p. 499.

Phytohormones, which are involved in the growth of meristematic tissues, are enhanced in effect by K^+. Cocucci and Dalla Rosa (1980) found a synergistic effect between K^+ and indole acetic acid on the growth of maize coleoptiles. Similar results have been reported by Dela Gardia and Benlloch (1980) for the influence of gibberellic acid on the growth of young sunflowers. Green (1983) has also reported that the effect of cytokinins on the growth of cucumber cotyledons was much enhanced by K^+. All the above examples indicate that K^+ plays a crucial role in meristematic growth.

10.2.3   Water regime

Potassium is of utmost importance for the water status of plants. Uptake of water in cells and tissues is frequently the consequence of active K^+ uptake (Läuchli and Pflüger1978). Because of its high concentration in the cytosol and under ample K^+ supply also in the vacuole K^+ is an important osmoticum which contributes to water retention by plants and to high efficiency of water use (Mengel and Forster 1973). The importance of K^+ for water retention in wheat seedlings is shown in Table 10.4 from the work of Scherer et al. (1982). In the treatment with no K^+ in the nutrient solution the water retention by the shoots (water/g dry matter) was significantly lower throughout the experimental period of one month. The lower water retention of the young plants deprived of K^+ caused that the water status of the plants fell below an optimum for physiological and biochemical processes so that lower growth rates were obtained. Cell expansion depends on turgor and optimum turgor require sufficient K^+ (Mengel and Arneke1982). The turgor threshold required for cell expansion is about 0.7 MPa which is not attained if the K^+ supply is insufficient. Measurements of Pfeiffenschneider and Beringer (1989) in the medulla of mature *Daucus carota* storage roots showed that both the cell volume and the elastic modulus of the cell walls were much increased in the treatment with the higher K^+ supply.

Table 10.4   Effect of K^+ nutrition on water retention of young wheat seedlings (Scherer *et al.* 1982)

Date	18. 11.	21. 11.	24. 11.	27. 11.	30. 11.	3. 12.	7. 12.	10. 12.
				g H_2O/g dry weight				
K_o	7.89	7.42	6.76	6.50	6.18	5.55	5.20	4.75
K_1	8.67	7.65	7.31	7.36	7.10	6.66	6.35	5.76

K_o → nutrient solution without K^+.
K_1 → nutrient solution with K^+
All data of the K_1 treatment are significantly different ($p < 0.001$) from K_o.

The high water efficiency (production of unit dry matter per unit water transpired) at optimum K^+ nutrition results from more closely controlled opening and closure of stomata (Brag 1972). As already described above K^+ is the major osmoticum in guard cells

(Talbott and Zeiger 1996). Especially at sunrise when water status is usually adequate a rapid and precise opening of the stomata depends on a rapid influx of K^+ into the guard cells so that photosynthesis can begin. Photosynthesis depends much on leaf water potential and declines considerably with decreasing water potenial. Pier and Berkowitz (1987) showed that the presence of an ample K^+ supply very much counteracted this photosynthetic decline. Towards the end of the day some of the K^+ in the guard cells may be substituted in function by sucrose (Talbott and Zeiger 1996).

Convincing evidence for the function of K^+ in stomatal opening and closing was provided by electron probe analysis studies of Humble and Raschke (1971). Using this technique it was possible to measure the K^+ concentration in guard cells of open and closed stomata. Figure 10.8 shows the result of this experiment. It can be seen that the K^+ concentration in the open state is considerably higher than that of guard cells from closed stomata. Under light conditions the guard cells produce abundant ATP in photosynthetic phosphorylation, providing sufficient energy not only for the plasmalemma H^+ (Humble and Hsiao 1970) but also for the activation of guard cell K^+ channels (Wu and Assmann 1995). Potassium is accumulated in the guard cells in considerable concentrations and the resulting high turgor pressure causes the opening of the stomata. Figure 10.8 also shows that the inorganic anions (chloride, phosphate) do not accompany the K^+ uptake to any great extent. The major anion charge balancing the accumulated K^+ is malate. Depending on species and conditions, however, also chloride may function to a major extent as counter ion for K^+ (Talbott and Zeiger 1996).

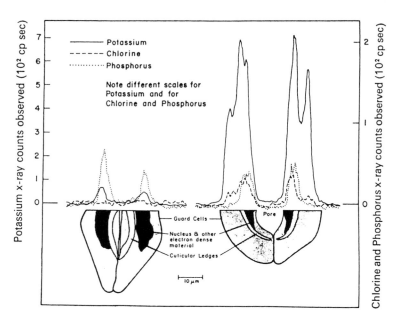

Figure 10.8 Ion concentrations in guard cells with closed and open stomata (after Humble and Raschke 1971).

It appears that in most plant species this opening/closing mechanism is absolutely dependent on the specific K⁺ uptake process. Other univalent cations are thus generally unable to replace K⁺ in this specific function (Trolldenier 1971b), except in a few plant species, *e.g. Kalanchoe marmorata* where Na⁺ is effective. From the discussion above it is evident that K⁺ has biophysical and biochemical functions, the former mainly osmotic functions, the latter enzyme activations. Barraclough and Leigh (1993) investigated critical plant K concentrations in ryegrass in relation to these different functions. Expressed on a tissue water basis the biophysical function needed 126 mol/m³ K⁺ at low Na⁺ supply but only 82 mol/m³ at high Na⁺ supply. The biochemical function required 46 mol/m³ K⁺ and was independent of Na⁺ supply. Critical K⁺ concentrations in the aerial plant material decrease with age because the proportion of growing tissue decreases and structural tissue increases as is evidenced by the fall in N concentration in vegetative plant organs (Greenwood and Stone 1998).

10.2.4 Photosynthesis and translocation of photosynthates

The older literature provides ample experimental evidence that K⁺ has a favourable influence on the efficiency of photosynthesis and not only on CO_2 assimilation but also on the transformation of light energy into chemical energy. Weller and Höfner (1974) and also Overnell (1975) reported that K⁺ stimulates the photosynthetic O_2 production which means it has a promoting impact on the e⁻ flux through the photosynthetic e⁻ transport chain (page 146) and therefore also on the reduction of NADP⁺ and the generation of ATP. This is in agreement with results of Pflüger and Mengel (1972) who found in chloroplasts of different species that K⁺ stimulated photophosphorylation and particuarly increased the photosynthetic reduction capacity. Thus higher rates of NADPH and ATP are provided for various synthetic processes including CO_2 assimilation. More recent results of Wu and Berkowitz (1992) obtained with isolated chloroplasts from *Spinacia oleracea* show that the K⁺ influx across the inner chloroplast membrane into the stroma through a K⁺ selective channel was associated with a flux of H⁺ through a separate channel driven by a H⁺ pump. It was shown that the elevated stromal K⁺ concentration stimulated e⁻ flow in the photosynthetic e⁻ transport chain by means of O_2 produced by photolysis. According to Berkowitz and Peters (1993) were it not for a system to pump H⁺ out of the illuminated chloroplasts, the increase of stromal pH as a consequence of H⁺ release into the thylakoid lumen (page 146), would quickly dissipate.

The favourable influence of an optimum K⁺ nutrition on photosynthesis has been found by several authors. The results of Peoples and Koch (1979) are of particular interest because they also relate to respiration. The investigations of these workers revealed that K⁺ promoted the *de novo* synthesis of the enzyme ribulose bisphosphate carboxylase (Rubisco). Potassium also decreased the diffusive resistance for CO_2 in the mesophyll. Some of the most important data of this investigation are given in Table 10.5 showing that the increase in CO_2 assimilation was paralleled by an increase in photorespiration and a decrease in dark respiration. This demonstrates that in the case of an optimum K⁺ nutrition less carbohydrate was respired and thus more could be used for growth and crop production.

Table 10.5   Effect of K+ of CO₂-assimilation, photorespiration, and dark respiration (Peoples and Koch 1979).

K+ in leaves mg K+/g DM	CO₂-assimilation mg dm⁻² h⁻¹	Photorespiration dpm dm⁻² h⁻¹	Dark respiration mg dm⁻² h⁻¹
12.8	11.9	4.00	7.56
19.8	21.7	5.87	3.34
38.4	34.0	8.96	3.06

Demmig and Gimmler (1983) found that isolated chloroplasts in the light took up considerable amounts of K+. High K+ concentrations (100 mol/m³ which is in the range of cytosolic K+ concentration) in the outer medium induced a broader pH optimum for the enzyme RuBP carboxylase as compared with low K+ concentrations (10 mol/m³). Thus at the higher K+ concentration the rate of CO₂ assimilation was almost twice the rate occurring at the low K+ concentration. This effect of K+ concentration on the pH optima for RuBP carboxylase is shown in Figure 10.9.

Numerous authors have shown that K+ enhances the translocation of photosynthates. According to Mengel and Haeder (1977) this promoting effect of K+ on phloem transport is related to phloem loading and not primarily to processes in the physiological sink. This is confirmed by data of van Bel and van Erven (1979) who found that K+ favours the uptake of sucrose and glutamine by sieve cells particularly at high pH. Such high pH levels may occur if protons are recycled back *via* H+ cotransport of sucrose and glutamine (Figure 10.10). Van Bel and van Erven (1979) therefore suppose that the form of cotransport is dependent on the apoplastic pH, with K+ cotransport at a high pH and H+ cotransport at low pH.

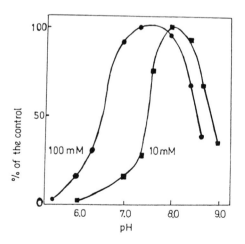

Figure 10.9   pH-optimum of ribulosebisphosphate carboxylase at a K+ concentration of 10 and 100 mol/m³ (after Demmig and Gimmler 1983).

APOPLAST

Sieve tube companion cell complex

High pH

K⁺ sucrose

Low pH

H⁺ sucrose

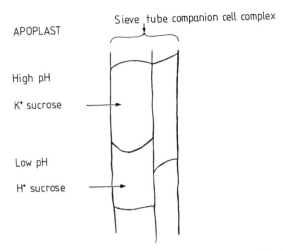

Figure 10.10 Sucrose uptake by the sieve tube companion cell complex: High pH as K⁺/sucrose cotransport; low pH as H⁺/sucrose cotransport.

From this concept K⁺ cotransport should be particularly effective at high rates of phloem loading.

Potassium not only promotes the translocation of newly synthesized photosynthates but has also a beneficial effect on the mobilization of stored material. Thus Koch and Mengel (1977) as well as Seçer (1978) found that in wheat K⁺ increased the mobilization of proteins stored in leaves and stems and also promoted the translocation of amino acids towards the grains. Hartt (1970) as well as Mengel and Viro (1974) showed that the effect of K⁺ on photosynthate translocation is a direct one and not a consequence of a higher CO_2 assimilation rate.

10.2.5 Enzyme activation

The main function of K⁺ in biochemistry is its activation of various enzyme systems. The problem was treated by Evans and Sorger (1966) in a still informative review article.

Figure 10.11 shows a typical example of an enzyme activated by univalent cations. The enzyme, a starch synthetase from sweet corn, was most strongly activated by K⁺, followed by Rb⁺, Cs⁺ and NH₄⁺. The activation by Na⁺ was very poor and Li⁺ had hardly any effect. With several enzymes, activation by NH₄⁺ or Rb⁺ is as efficient as the activation by K⁺. This is of theoretical interest as the sizes of these three ion species are almost the same when hydrated. Activation thus seems to be related to the size of the activating ion. In growing plants, however, K⁺ can not be replaced as an activator of enzymes by NH₄⁺ and Rb⁺ as these ions are toxic at the concentrations required (El-Sheikh and Ulrich 1970, Morard 1973).

In vitro experiments have shown that maximum K⁺ activation is obtained within a concentration range of 40 to 80 mol/m³ K⁺. *In vivo* activation may occur at the same

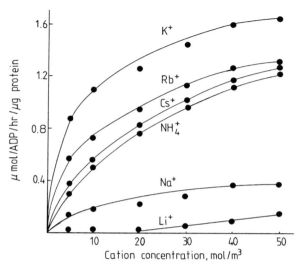

Figure 10.11 Effect of univalent cation species on the activity of starch synthetase isolated from sweet corn (after Nitsos and Evans 1969).

range or even higher as shown in Figure 10.9 for ribulosebisphosphate carboxylase. Polypeptide synthesis in the ribosomes also requires such a high K^+ concentration as suggested by Wyn Jones and Pollard (1983) and hence K^+ seems to be essential for protein synthesis. This is also supported by the data of Koch and Mengel (1972, 1974) which show that subotimal K^+ supply to tobacco restricted the incorporation of N into proteins. Figure 10.12 shows this relationship, with insufficient K^+ supply there was a relative enrichment of ^{15}N label in the nitrate and amino acid fractions indicating a repression of nitrate reduction. A further indication that K^+ is involved in protein synthesis comes from a water culture study of the effect of NaCl salinity on the incorporation of labelled N into the protein fraction of young barley plants (Helal and Mengel 1979). Salinity depressed the K^+ concentration in the shoots and also led to an accumulation of labelled inorganic N in shoots and to a marked depression of labelled N in the protein fraction of shoots and roots. This depressing effect of salinity on protein synthesis was completely counteracted by increasing the K^+ supply and K^+ uptake by the plants. This experiment provides another indication of the importance of an adequate K^+ concentration in the plant tissue and particularly in the cytosol for protein synthesis.

The means by which K^+ can bring about conformational changes of enzymes was already considered in the older literature (Evans and Sorger 1966) but it is only recently that more precise details have become known. According to Miller (1993) K^+ can be centred in an oxygen cage in which six oxygen atoms are oriented to the centre of the cage (Figure 10.13). In the case of the enzyme dialkyl-glycine carboxylase these oxygen atoms are provided by three carbonyls of an amino acid, one from a water molecule, one from serine and one from the carboxylic group of aspartate. The structure thus formed resembles an octahedron with one oxygen atom at each of the six corners

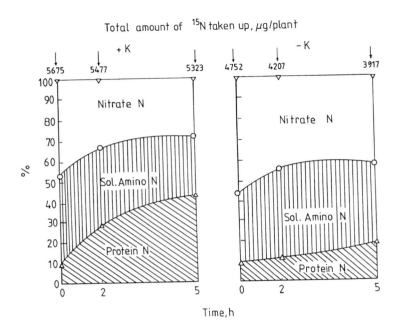

Figure 10.12 Effect of K⁺ supply on the incorporation of ¹⁵N labelled nitrate into the soluble amino acid and into the protein fraction. Numbers indicated by arrows represent the amount of ¹⁵N in the tissue (after Koch and Mengel 1974).

and K⁺ in the centre (octahedra see Figure 2.9). This K⁺ binding is very selective relative to Na⁺ because it requires dehydration of the cation in accordance with the lower hydration energy of K⁺ as compared with Na⁺. If Na⁺ binds to this site the conformation changes from the octahedral geometry to a distorted trigonal bipyramidal form and access of the substrate to the enzymic reaction site is blocked. Na⁺ and also Li⁺ therefore have a strong inhibitory effect on enzymes activated by K⁺ (Toney *et al.* 1993). The difference between K⁺/Na⁺ binding is analogous to that in the interlayers of K⁺ fixing minerals (see page 484). The activation of enzymes by univalent cation species, as shown in Figure 10.11, indicates that those cation species with a low hydration energy (K⁺, Rb⁺, Cs⁺, NH₄⁺) are capable of activation in contrast to Na⁺ and Li⁺. It is supposed that in most cases of K⁺ activated enzymes, the change in conformation is brought about by the formation of an octahedra with K⁺ bound coordinatively to six O atoms (Toney *et al.* 1993).

Severe K⁺ deficiency leads to the synthesis of toxic amines such as putrescine and agmatine (Smith and Sinclair 1967). Arginine is the precursor for the synthesis of these polyamines as shown in Figure 10.14.

Synthesis is promoted by low pH in the cytosol which may possibly result from less H⁺ pumping across the plasmalemma because of a lack of K⁺. At low cellular pH putrescine may be positively charged to counterbalance the cation deficit caused by a lack of K⁺. It is is now established that other nutritional regimes which generate

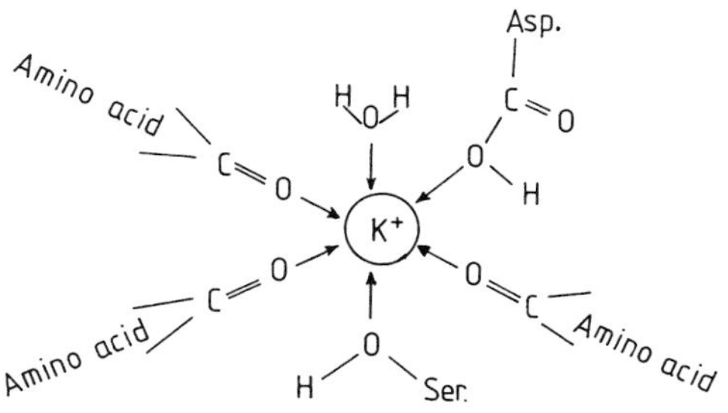

Figure 10.13 Oxygen atoms oriented to the K+ centre (modified after Miller 1993).

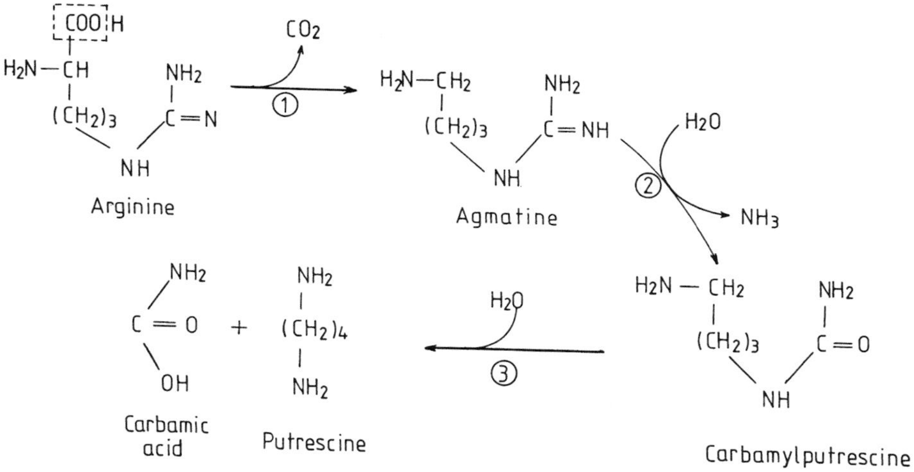

Figure 10.14 Synthesis of putrescine. 1. Decarboxylation of arginine. 2. Deamination of agmatine. 3. Hydrolysis of carbamyl putrescine

H^+ in plant tissues including NH_4^+ nutrition, exposure to SO_2 and acid feeding also increase endogenous concentrations of putrescine (Priebe *et al.* 1978, Young and Galston 1983).

10.2.6 Replacement of potassium by sodium

The question of whether Na^+ can replace K^+ in physiological processes in the plant is not only of academic interest but also of practical importance in relation to fertilizer usage (see page 509). In less specific processes such as raising cell turgor some replacement is possible as shown for *Lolium* by Barraclough and Leigh (1993). The extent to which substitution can occur, however, depends much on the uptake potential for Na^+. As is known to date Na^+ uptake across the plasmalemma is mainly a facilitated diffusion through low-affinity channels which also allow the passage of other cation species (Fox and Guerinot 1998). Net uptake of Na^+ also depends on the active extrusion of Na^+ into the outer medium (Schubert and Läuchli 1988) which may play a particular role in the resistance of plant species to salinity (see page 237).

The net uptake of Na^+ differs considerably between plant species (Marschner 1971). Table 10.6 shows the Na^+ uptake potential of various crops. For the "high and medium Na^+ species", the favourable effect of Na^+ is important on plant growth. This is particularly the case for the *Beta* species (El-Sheikh and Ulrich 1970). In these species Na^+ contributes to the osmotic potential of the cell and thus has a positive effect on the water regime of plants.

Table 10.6 Uptake potential of various crops for sodium (data from Marschner 1971).

High	Medium	Low	Very low
Fodder beet	Cabbage	Barley	Buckwheat
Sugar beet	Coconut	Flax	Maize
Mangold	Cotton	Millet	Rye
Spinach	Lupins	Rape	Soya
Swiss chard	Oats	Wheat	Swede
Table beet	Potato		
	Rubber		
	Turnips		

The beneficial effects of Na^+ on plant growth are particularly observable, when the K^+ supply is inadequate (Hylton *et al.* 1967, Amin and Joham 1968). Table 10.7 shows such an example for rice (Yoshida and Castaneda 1969). In the lower range of K^+ concentrations, Na^+ increased grain yields, whereas at the higher K^+ concentration Na^+ induced a slight yield depression. Effect and uptake of Na^+ may not only differ between species but also plant organs of the same species as reported by Lindhauer (1989) for sugar beet. Here Na^+ was well able to substiute for K^+ in the leaves but not in the storage roots. The main results of this investigation are given in Table 10.8 showing the beneficial effect of Na^+ on leaf growth but not on the growth of storage roots (Lindhauer 1989). Under field conditions Na^+ deficiency of crop plants has not yet been observed. Wooley (1957), however, succeeded in inducing Na^+ deficiency symptoms in tomatoes under glasshouse conditions. According to Brownell and Crossland (1972), Na^+ is an essential nutrient for some C_4 species. where it is supposed to maintain the integrity of chloroplasts and thus allow the uptake of pyruvate (Brownell and Bielig 1996). As shown on page 2 this is an essential process in C_4 species because the pyruvate is then phosphorylated to PEP which is the primary CO_2 acceptor. The Na^+ concentrations required for this beneficial effect are low and therefore Na^+ is considered as a micronutrient in some C_4 species (Brownell and Crossland 1972). The effect of Na^+ on plant growth has been considered in papers by Marschner (1971) and Jennings (1976).

Table 10.7 Effect of increasing potassium concentrations in the grain yield of rice, in the presence or absence of a high Na level in the nutrient solution (Yoshida and Castaneda 1969).

K concentration	Grain yield, g/pot	
mol/m^3	- Na$^+$	+ 43 mol/m^3 Na$^+$
0.025	4.6	11.0
0.050	6.9	19.9
0.125	26.4	46.6
0.250	63.3	67.3
1.25	67.5	75.9
2.50	90.8	87.6
5.00	103.6	92.6

Table 10.8 Effect of the Na$^+$/K$^+$ ratio in the nutrient solution on growth of sugar beet leaves and storage roots (after Lindhauer 1989)

Nutrient solution, mol/m^3	4.5 K$^+$, 0.5 Na$^+$	0.5 K$^+$, 4.5 Na$^+$	0.5 K$^+$, 0.5 Na$^+$
Leaves, g d.m.	50.3	65.2	58.0
Roots, g d.m.	107	83	81

Responses to Na$^+$ may differ between genotypes. In experiments with 3 genotypes of sugar beet Marschner *et al.* (1981] observed that substitution of half the K$^+$ in the nutrient medium by Na$^+$ increased the dry weight of the plants and the sucrose concentration in the storage tissue in all three genotypes. However, when 95% of the K$^+$ was replaced by Na$^+$ widely varying responses were obtained.

10.2.7 Potassium deficiency

Potassium deficiency does not immediately result in visible symptoms. At first there is only a reduction in growth rate (hidden hunger), and only later do chlorosis and necrosis occur. These symptoms generally begin in the older leaves, due to the fact that these leaves supply the younger ones with K$^+$. According to investigations of Pissarek (1973) with rape and of Morard (1973) with sorghum, K$^+$ deficiency symptoms are first seen in the 2nd and 3rd oldest leaf and not in the oldest ones. In most plant species chlorosis and necrosis begins in the margins and tips of the leaves (maize, cereals, fruit trees) (Plate 10.1), but in some species such as clover irregularly distributed necrotic spots occur on the leaves (see Plate 10.2).

Plants suffering from K$^+$ deficiency show a decrease in turgor, and under water stress they easily become flaccid. Resistance to drought is therefore poor (Pissarek1973) and the affected plants show increased susceptibility to frost damage, fungal attack, and saline conditions. Photosynthate export in the phloem is dependent on the cycling of K^+ and Mg^{2+} through source leaves. In K deficient source leaves therefore both phloem loading of photosynthates and subsequent solute volume flow through the sieve tubes are depressed and sucrose is accumulated in the leaves (Cakmak *et al.* 1994).

Abnormal development of tissues and cell organelles is observed in K^+ deficient plants. According to Pissarek (1973) inadequate K^+ supply resulted in a reduced growth rate of the cambium in stems of rape. The formation of xylem and phloem tissue was restricted whereas the cortical tissue was affected only to a minor extent. Lignification of the vascular bundles is generally impaired by K^+ deficiency. This effect probably makes K^+ deficient crops more prone to lodging. K^+ deficiency also results in a collapse of chloroplasts (Pissarek 1973) and mitochondria (Kursanov and Vyskrebentzeva 1966).The development of the cuticle is much retarded when K^+ nutrition is low. This is illustrated in Plate 10.3, which shows an electron micrograph of the cuticle of a cotton leaf well supplied with K^+(left side) compared with the cuticle of a similar leaf suffering from K^+ deficiency (right side). Potassium deficiency symptoms of various plant species have been described in detail by Bergmann (1992).

10.3   Potassium in Crop Nutrition

10.3.1   Crop requirements and response

For thousands of years naturally occurring soil K^+ was virtually the only K^+ source for plants so that soils low in available K^+ were infertile. Nowadays, inadequate soil K^+ levels are frequently corrected by the use of K^+ fertilizers. In intensive cropping systems in particular, high application rates are employed. In order to maintain the fertility level of a soil, the amount of K^+ taken up by crops (kg K/ha) and that lost by leaching should at least be balanced by K^+ fertilization. It is only in highly developed agricultural systems with top level yields, however, that the amount of K^+ returned to the soil is equal or in excess of that removed by crops. This means that in many cases soil K reserves are being depleted (Sanchez and Leakey 1997).

The quantities of K^+ removed from the soil by various crops are listed in Table 6.2. The table gives some indication of the rates at which K^+ should be applied, the normal range being from about 40 to 250 kg K^+/ha/yr. The quantity of K^+ removed from a soil depends very much on the yield level and also on the rate of K^+ leaching. The figures in Table 6.2 therefore provide only a rough working basis for fertilizer recommendations, as uptake is highly dependent on the K^+ availability. Where availability is poor such as in K^+ fixing soils or in soils low in exchangeable K^+ and with a high K^+ buffer power, K^+ uptake by the crop from the soil is also low and unsatisfactory yields are obtained. Under such conditions fertilizer recommendations based on an average total K^+ uptake for a particular crop do not meet the needs of the crop nor raise the soil fertility status. The reverse situation occurs when soils are high in available K^+. Here fertilizer applications based on crop uptake can result in a higher K^+ uptake than is needed for maximum yield and hence also in a waste of fertilizer.

Plate 10.1 K[+] deficiency symptoms in older leaves of rape. (Photo Pissarek)

Plate 10.2 K[+] deficiency symptoms in white clover.

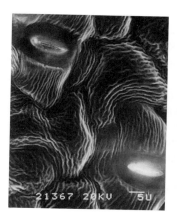

Plate 10.3 Electron micrograph of the cuticle of a cotton leaf well supplied with K[+] (left hand side) and poorly supplied with K[+] (right hand side). Magnification 1500 x. (Courtesy of Sunxi and Rao Li-Hua)

The level of available K^+ in the soil which may be considered optimum, cannot be expressed in general terms, since it depends on the crop as well as climate and soil conditions. Von Braunschweig (1978) carrying out numerous field trials in Germany found that the lactate soluble K^+ level needed to be higher for soils rich in clay. Loué (1979) in evaluating more than 300 field trials carried out in France reported that K^+ application resulted in marked grain yield responses if the concentration of exchangeable K^+ was <80 mg/kg soil. On medium textured soils with levels of >160 mg/kg exchangeable K^+ only small increases in grain yield were obtained following K^+ fertilizer application. Leskošek (1978) carrying out numerous grassland trials found that K^+ fertilizer application gave significant yield increases on soils low in available K^+ and yield responses decreased exponentially with an increase of available soil K^+ and practically approached zero at 130 mg EUF extractable K^+/kg soil which is equivalent to about 180 mg exchangeable K^+/kg soil. Potassium yield response decreased linearly with an increase of K^+ concentration in the grass and was negligible at a K^+ concentration of 30 mg K/g dry matter.

Generally grasses and cereals respond less favourably to K^+ application than dicots, especially potatoes and legumes (van der Paauw 1958, Schön et al. 1976). These latter crops thus require higher levels of available soil K^+. For soils rich in non exchangeable K^+, this source may be utilized by crops to some degree (Singh and Brar 1977, George et al. 1979). This is even true for sandy soils containing K^+ bearing minerals in the sand fraction (Wulff et al. 1998). If, however, these sources are also depleted crop yields gradually decline (v. Boguslawski and Lach 1971).

In order to evaluate K^+ requirements of crops it is also important to consider the total length of the growing period. Tomatoes and sugar beet for example take up about the same amount of K^+ as sugar cane, per unit area per year. In the two former crops, however, the growth period is only about 120 days whereas in sugar cane the growth period may extend over the whole year (Nelson 1968). This indicates that crops with the same total uptake may have a very different K^+ demand (= uptake/unit time). Requirement also varies depending on the stage of growth, the highest uptake rate often being in the vegetative stage. In potatoes for example 50% of the total K^+ is absorbed in the first third of the growth period. In the cereals too, K^+ is particularly needed during vegetative growth, and K^+ application during the reproductive stage hardly affects grain yield (Chapman and Keay 1971).

A further important factor in determining K^+ uptake by crops is the type of rooting system and its extent. This is well demonstrated in the competition which occurs between legumes and grasses for K^+ uptake. In the field when these species are growing together, K^+ uptake is considerably higher by grasses, and under low level K^+ conditions this can lead to the disappearance of legumes from the sward. This competition was studied by Lambert and Linck (1964) using intact root systems of lucerne and oats supplied with $^{42}K^+$. When whole root systems were used 91% of the labelled K^+ was found in oats and only 9% in the lucerne. However, when the $^{42}K^+$ was applied separately to intact root segments of lucerne and oats more labelled K^+ appeared in the lucerne than in oats. These results indicate that root morphology (root length, number of root hairs) and probably also the individual K^+ uptake potential (high-affinity K^+ transporters, see page 127) of crop species appear to be important factors influencing competition between plant species for K^+.

In order to to improve K⁺ fertilizer practice Greenwood and Karpinets (1997a) have developed a mechanistic model that can be calibrated for widely different crops. One of the major principles on which it is based is that the critical % of K needed for maximum growth rate declines with increasing plant mass per unit area (as is also a main feature of the similar N model, see page 430). The model calculates throughout the growth of the crop, the potential increase in dry weight, the increments in root length and the % K of the plant mass. It also calculates the maximum amount of K⁺ that could be transported to the root surfaces and modifies the potential uptake by feedback of plant K⁺ status on root absorption (see page 214). The inputs are relatively easily obtained. These include soil characteristics (especially those relate to K⁺, *i.e.* amounts of fixed soil K⁺, and the velocity constants for fixation and release of soil-K), cultural practices, crop species, and weather. In a second paper Greenwood and Karpinets (1997b) reported tests of the validity of the model against results of multi level K⁺ fertilizer experiments and reasonably good agreement between measured and simulated values were obtained. Simulations with the model indicated that in central England no response of 10 crops to K fertilizer would be likely on soils containing more than 170 mg K/kg NH_4NO_3-extractable K/kg soil with a clay concentration of between 150 and 450 g/kg.

The response to K⁺ fertilizer by crops depends to a considerable extent on the level of N nutrition. Generally the better the crop is supplied with N the greater the yield increase due to K⁺ (Heathcote 1972,) and as shown in numerous field trials with field crops by Loué 1989). On the other hand applied N is only fully utilized for crop production when K⁺ supply is adequate. In moving from an extensive to an intensive cropping system responses to K⁺ fertilizers are frequently not observed in the first years of application (Anderson 1973). This is particularly true in more arid regions where little or no K⁺ losses occur by leaching. Potassium reserves accumulated in the upper soil layers are often sufficient to supply crop needs in the first few years of intensive cropping. However, as soon as these reserves are exhausted because of the higher K⁺ requirements of increased crop yields and continuous cropping, responses due to K⁺ fertilization may be expected (Stephens 1969).

10.3.2   Deficient soils and fixation

Potassium deficiency occurs commonly on a number of different soil types. It may appear where K⁺ has been leached as for example on light sandy acid soils low in K⁺ on highly leached lateritic soils (Anderson1973). Organic soils and peats have a low capacity in storing K⁺ and the adsorbed K⁺ may be easily exchanged and leached. Occasionally, too, K⁺ is deficient on soils which have been heavily cropped and in soils which fix K⁺ into a non exchangeable form (Ulrich and Ohki 1966). For these soil types, studies in continuous cropping without K⁺ supplementation are of particular interest. On organic soils and sandy soils poor in K⁺ bearing minerals, yield levels drop rapidly year by year as exchangeable K⁺ is depleted. Finally very poor harvests are obtained. With soils richer in K⁺ bearing minerals, yields do not fall so rapidly. Moreover as environmental factors also cause yield fluctuations from year to year the slight yield depressions resulting from K⁺ depletion are more difficult to observe. As pointed out on page 506,however, the K⁺ bearing minerals do not provide an inexhaustible K⁺ source and with time the rate of release of non exchangeable K^+ declines. In soils where K^+ originates mainly from the interlayers of 2:1 minerals the depletion of K^+ from these minerals enhances potential K^+ fixation. The more the mineral is depleted the higher becomes the fixation capacity. Eventually K^+ release becomes extremely low and drastic yield depressions occur. In order to obtain satisfactory yields from such soils high fertilizer rates are required because of the fixation of K^+ by the expanded clay minerals (Table 10.9).

For soils in which vermiculite is the dominant clay mineral enormous amounts of K^+ may be fixed. In such a sandy clay loam soil from Michigan in the USA, Doll and Lucas (1973) reported that this was the case of about 92% of the applied K^+ fertilizer. Tomato production was increased by applications of up to 1600 kg K/ha. Similar reports have been made in other parts of the USA as well as in the Danube valley in Germany (Schäfer and Siebold 1972, Burkart and Amberger 1978).

Typical results of K^+ fertilizer applications to K^+ fixing soils are shown in Table 10.9 from field trials of Kovacevic and Vukadinovic (1992). The soil was a calcareous silty loam with vermiculite, smectitic clay minerals and illite in the clay fraction and an exchangeable K^+ level of 2 mmol/kg (80 mg K/kg soil). Even with a fertilizer rate of 2200 kg K/ha the level of exchangeable K^+ determined was only to 3.3 mmol/kg soil whereas if no K^+ had been fixed in this value should have risen to about 20 mmol/kg soil. This result demonstrates the enormous K^+ fixation potential of this soil. It is for this reason that only heavy fertilizer rates gave optimum grain yields of maize and raised the K^+ concentration in the leaves to a level indicative of adequate K^+ supply. Plants insufficiently supplied with K^+ tended to lodge and some plants even died before maturation.

Table 10.9 Effect of K^+ fertilizer rates on yield of maize and K^+ concentration in leaves grown on a K^+ fixing soil (Kovacevic and Vukadinovic 1992)

Fertilizer kg K/ha	Leaf K mg K/g d.m.	Grain yield Mg/ha	% H_2O in grain	% Lodging
125	6.4	1.75	31.5	42
275	7.8	2.57	28.7	21
460	8.6	4.66	28.6	18
650	10.3	6.95	29.2	20
835	14.3	7.76	29.7	5
1580	17.1	8.98	29.7	2
2200	18.6	8.88	29.3	2
LSD, 5%	1.0	0.65	1.5	

10.3.3   Disease

Potassium not only influences crop production by enhancing growth assimilate transport and the synthesis of various chemical compounds relevant to crop quality. It is also highly important in raising the disease resistance of many crop species. In maize for example stalk rot and lodging are usually more severe when soil K^+ is low in relation to other nutrients (Krüger 1976). The beneficial effect of K^+ in preventing lodging of maize (Table 10.9) is also true for other cereal crops (Trolldenier 1969). In wheat too a lower incidence of powdery mildew caused by the fungus *Erysiphe graminis* has been observed in plots treated with additional potassium (Glynne 1959). Some other crop diseases less frequent in plants well supplied with K^+ include: brown spot in rice caused by *Ophiobolus miyabeanus*, brown rust in barley infected with *Puccinia hordei* and *Fusarium* wilt in bananas resulting from *Fusarium oxysporum* (Goss 1968). According to the observations of Baule (1969) forest trees adequately supplied with K^+ are also more resistant to fungal diseases. The nature of the action of K^+ in controlling the severity of plant diseases is still not understood. It may relate in part to the effect of K^+ in promoting the development of thicker outer walls in epidermal cells thus preventing disease attack (Trolldenier and Zehler 1976). In addition as already indicated, plant metabolism is very much influenced by K^+. Since plants well supplied with K^+ have a higher energy status than those with a suboptimal K^+ nutrition (Peoples and Koch 1979) the well supplied plants have more energy *e.g.* ATP and NADPH for the synthesis of phytoalexins and hence enable the host plant to counteract the attack of pathogens. The effect of K^+ on disease resistance in crop plants has been well reviewed by Goss (1968).

10.3.4 Fertilizers and application

Potassium is supplied to crops as a straight fertilizer or in the form of compounds. Most K^+ fertilizer can be grouped in chlorides and sulphate (Table 10.10). The most widely used and cheapest potash fertilizer is potassium chloride (KCl) which is known commercially as muriate of potash.

Table 10.10 Important potassium fertilizers

Name	Formula	K	K_2O^*	Mg	N	S	P
Muriate of potash	KCl	50	60	-	-	-	-
Sulphate of potash	K_2SO_4	43	52	-	-	18	-
Sulphate of potash - Mg	$K_2SO_4 \cdot MgSO_4$	18	22	11	-	21	-
Kainit	$MgSO_4+KCl+NaCl$	10	12	3.6	-	4.8	-
Potassium nitrate	KNO_3	37	44	-	13	-	-
Potassium metaphosphate	KPO_3	33	40	-	-	-	27

* Expressed as K_2O

Besides this high grade form (50% K) lower grade KCl fertilizer types (41% K and 33% K, or 58% K_2O and 40% K_2O) are also on the market. These forms contain substantial amounts of NaCl and are therefore suited to natrophilic crops (sugar beet, cabbage, oats). Potassium sulphate (sulphate of potash) contains a somewhat lower K^+ concentration and is more expensive due to higher production costs. Potassium nitrate is also a low-grade N fertilizer and potassium metaphosphate also a P fertilizer in which the P concentration is almost as high as that of K^+. Both fertilizers, potassuim nitrate and potassium metaphosphate are of relatively minor importance. Another K fertilizer is sulphate of potash-magnesium. It has a relatively high Mg^{2+} concentration and is useful where Mg^{2+} is required in addition to K^+. Magnesium kainite is a low-grade K fertilizer with a high concentration of NaCl.

With the exceptions of potassium metaphosphate and potassium silicate all potash fertilizers are soluble in water. They are therefore very similar in availability so that differences between these fertilizers result from accompanying anions. In some cases the application of sulphur, magnesium, or sodium may be agronomically beneficial and an appropriate fertilizer should be chosen. Some crops are sensitive to high amounts of chloride. These chlorophobic species include tobacco, grapes, fruit trees, cotton, sugar cane, potatoes, tomatoes, strawberries, cucumber and onions. It is preferential to treat these crops with potassium sulphate. For potatoes, the use of sulphate rather than chloride generally results in higher starch concentrations. The effect of KCl as compared with K_2SO_4 on growth and yield of grapes (*Vitis vinifera*) was well demonstrated by Edelbauer (1976) in a solution culture experiment. The most interesting data of this investigation are shown in Table 10.11 The depressive effect of chloride also had a negative influence on the sugar concentration of the juice whereas the acid concentration in the juice was hardly affected by the varying chloride/sulphate nutrition (Edelbauer 1977). Haeder (1975) found that chloride affects the translocation of photosynthates from the upper plant parts to the tubers. Under field conditions the depressive effect of chloride on gowth and yield of choride-sensitive crops is not as severe as shown in the Table 10.11. because chloride may be diluted and leached into deeper soils layers under humid conditions while K^+ is bound to soil colloids.

Table 10.11 Effect of a varying chloride/sulphate nutrition on grape yield (Edelbauer 1978)

Nutrient solution		Grape yield g/pot	Weight of cluster, g	No. of clusters/pot
KCl, me/L	K_2SO_4, me/L			
4.0	-	111	58.9	1.89
2.5	1.5	149	70.5	2.10
1.0	3.0	252	84.2	3.00
-	4.0	254	91.3	2.78

Most field crops are not sensitive to chloride and for this reason are generally treated with muriate of potash. Oil palms and coconuts even appear to have a chloride requirement (von Uexküll (1972). Potassium nitrate is mainly used for spraying on leaves of fruit trees and horticultural crops. Potassium metaphosphate and potassium silicates are used in cases where it is desirable that solubility should be low in order to prevent high concentrations in the root vicinity. Because of the high prices of these low soluble K^+ fertilizers, they are only used occasionally for horticultural crops. Tokunaga (1991) reported that a potassium silicate produced from ash and dolomite had a low K^+ solubility but still released enough K^+ for an optimum nutrition of rice and the Si released was beneficial for rice growth. Cucumbers dressed with this K silicate gave a somewhat higher yield than the fertilization with KCl.

Potassium fertilizers are applied broadcast and only in soils with a low level of available K^+ or with a high K^+ fixation capacity is banded application recommended. Using this technique the K^+ fixation capacity of a restricted soil volume can be saturated and within this zone excess K^+ is available for uptake. In experiments with maize Welch et al. (1966) observed that responses to banded K fertilizer were as much as 4 times greater than a broadcast treatment. It is also opportune to apply K fertilizers to K^+ fixing soils just before sowing the crop and also later as a top dressing in order to reduce the time of contact between fertilizer K^+ and the K^+ fixing minerals. The longer the time of contact the more fertilizer K^+ is fixed.

On fine textured soils the vertical movement of K^+ in the soil profile is restricted. This may affect the supply of fertilizer K^+ to the roots of some crop species. For example Budig (1970) found that grapes suffered from K^+ deficiency although the upper soil layer was rich in available K^+. However, the deeper soil layers (40 to 60 cm), from which the grape roots mainly drew their nutrients, were depleted of K^+. For such crops deep application of fertilizer K^+ is recommended. Accumulation of K^+ in the upper soil layer may also occur by practice of zero tillage over several years (Mackay et al. 1987, Hütsch and Steffens 1992).

High K^+ losses due to leaching occur only on sandy soils (Wulff et al. 1998), organic soils and soils with kaolinite as the main clay mineral (see page 342). These soils should have a low level of available K^+ (Wulff et al. 1998) and be treated with fertilizer K^+ just before the crop is sown or planted in order to avoid excessive K^+ losses by leaching in rainy periods (winter rainfall, rainy season). For some crops split applications are even recommended under high leaching conditions.

A comprehensive book (about 1200 pages) on all aspects of «Potassium in Agriculture» was edited by Munson (1985).

CHAPTER 11

CALCIUM

11.1 Soil Calcium

11.1.1 Occurrence in soils, weathering and leaching

The mean Ca concentration of the earth's crust amounts to about 36.4 g/kg. It is thus higher than that of most other plant nutrients (see Table 11.1). Calcium in the soil occurs in various primary minerals. These include the Ca bearing Al-silicates such as feldspars and amphiboles, Ca phosphates, and Ca carbonates. The latter are particularly important in calcareous soils (Donner and Lynn 1989) and are usually present as calcite ($CaCO_3$) or dolomite [$CaMg(CO_3)_2$]. Calcium is present in the various forms of Ca phosphates such a apatites, octo-Ca-phosphate and $CaHPO_4$ (see page 454). The Ca concentration of different soil types varies widely depending mainly on parent material and the degree to which weathering and leaching have influenced soil development. Soils derived from limestone or chalk, and young marsh (polder) soils are usually rich in Ca with a high content of $CaCO_3$ being in the range of 100 to 700 g/kg Ca. Even the surface layer of soils developed on limestone, however, can be low in Ca when leaching is excessive. Old soils, highly weathered and leached under humid conditions, are generally low in Ca. Two typical examples of such soils are the Spodosols of the temperate zone and the Oxisols and Ultisols of the humid tropics. In more arid conditions high Ca contents in the upper soil layer may occur in the form of an accumulation of gypsum ($CaSO_4 \cdot 2H_2O$).

Calcium bearing minerals play an important role in pedogenesis. Thus soils derived from Ca containing parent material such as basalt and dolerite generally contain higher amounts of secondary clay minerals. Soils developed from calcite, dolomite and chalk mainly belong to the rendzina group of soils (Mollisols). These shallow soils contain appreciable amounts of $CaCO_3$ and for this reason are alkaline in pH. A high soil pH and the presence of Ca^{2+} favours the formation of Ca-humate complexes and accounts for the dark colour of the Mollisols.

The weathering of the primary Ca bearing minerals depends considerably on the formation of H^+ in the soil. Hydrogen ions and probably also chelating agents can release the Ca^{2+} from the lattice structure of minerals thus causing a dissolution of the mineral. These weathering processes along with the release of Ca^{2+} from exchange sites of soil colloids by H^+ account for the considerable amounts of Ca^{2+} leached under humid climatic conditions. The rate of leaching of Ca^{2+} increases with the annual rainfall and with the content of Ca bearing minerals in the soil. Quantities of Ca^{2+} leached under temperate conditions are in the range of 200 to 300 kg Ca/ha (see page 343). Hallbäcken (1992) reported that in Sweden during the period from 1927 until 1984 the concentration of exchangeable cations, mainly Ca^{2+} and Mg^{2+} in forest soils was considerably reduced. This decrease was about twice as much as would have been expected by the uptake of the forest trees. The difference results form anthropogenic input of acids, mainly sulphuric acid.

Table 11.1 Mean chemical composition of the earth crust to a depth of 16 km

Element	weight g/kg	volume mL/L	Element	weight g/kg	volume mL/L
O	464.6	917.7	H	1.4	0.6
Si	276.1	8.0	P	1.2	
Al	80.7	7.6	C	0.9	0.1
Fe	50.6	6.8	Mn	0.9	
Ca	36.4	14.8	S	0.6	
Na	27.5	16.0	Cl	0.5	0.4
K	25.8	21.4	Br	0.4	
Mg	20.7	5.6	F	0.3	
Ti	6.2	-	All other elements	5.2	1.0

The weathering of carbonates is much dependent on the production of CO_2 in the soil. Calcite ($CaCO_3$) is a relatively insoluble mineral, the solubility only being in the order of 10 to 15 mg Ca/L (about 0.3 mol/m^3). In the presence of CO_2, however, $Ca(HCO_3)_2$ is formed which is much more water soluble (Donner and Lynn 1989).

$$CaCO_3 + CO_2 + H_2O \Rightarrow Ca(HCO_3)_2 \Rightarrow Ca^{2+} + 2\ HCO_3^-$$

Calcareous soils thus have a high concentration of HCO_3^- in the soil solution which may induce Fe chlorosis (see page 566). In such soils the formation of $Ca(HCO_3)_2$ (Monohydrogen Ca carbonate) is also an important process making Ca^{2+} soluble and hence prone to leaching. Also in soils not containing $CaCO_3$, Ca^{2+} and HCO_3^- (monohydrogencarbonate = bicarbonate) may be leached where HCO_3^- can also be produced from the dissociation of carbonic acid contained in rain water or by the reaction H_2O and CO_2 the latter mainly produced by respiration.

$$CO_2 + H_2O \Leftrightarrow H^+ + HCO_3^-$$

The reaction is dependent on pH and on the partial pressure of CO_2 which may be much higher in the soil than in the atmosphere. Both high pH and high CO_2 partial pressure favour the formation of the $HCO_3^- + H^+$. Hydrogen ions thus produced may exchange Ca^{2+} from soil colloids. These reactions to some extent may account for why applications of organic material to soils may promote Ca^{2+} leaching since during organic matter decomposition there is an increase in CO_2 release and hence also in $H^+ + HCO_3^-$ formation in the soil.

In quantitative terms, however, NO_3^- formation appears to play a more important role than HCO_3^- in balancing Ca^{2+} and also Mg^{2+} in the leaching process in many soils (Larsen and Widdowson 1968). This is particularly true for soils with a pH < 7.

Under natural conditions NO_3^- originates largely from soil organic N mineralization and nitrification. Nitrification produces H^+ as desribed on page 411 which may react with Ca bearing minerals or exchange Ca^{2+} from soil colloids. The process of nitrification therefore exerts a major influence on soil acidification and the leaching of Ca^{2+}. Under conditions where nitrification occurs the application of NH_4-fertilizers has the same effect in increasing soil acidification and Ca^{2+} leaching. Thus under humid conditions in general for every 100 kg N as $(NH_4)_2 SO_4$ added to the soil about 45 kg Ca^{2+} are removed in drainage water (Russell 1973). Some soils rich in organic S may also acidify as a result of the formation of H_2SO_4 which is particlur true for sulphate acid soils (Ahmad and Wilson 1992, see page 437).

Roots, themselves, are capable of excreting H^+ and thus may contribute to soil acidification. Net H^+ efflux of roots is particularly high when plants are fed with NH_4^+ as explained on page 375. Also under N_2 fixing legumes soils pH drops (see Figures 2.22 and 3.14). As shown by an experiment of Mundel and Krell (1978) carried out over a 17 year period on an alluvial meadow soil with a low H^+ buffer power, soil pH fell during this period from 7.0 to 5 or 4 in the plot with the high N treatments. Although N was applied as nitrochalk, a fertilizer which contains $CaCO_3$, and is thus alkaline, the soil pH and base saturation of the soil colloids decreased. This was related to the rate of fertilizer application as can be seen from Figure 11.1. Increasing the rate of N application decreased the pH of the soil. It is assumed that this fall in soil pH was mainly due to an enhanced nitrification of the applied NH_4^+. Also the loss of considerable amounts of potential bases such as organic anions (see page 57) are removed with the harvested biomass.

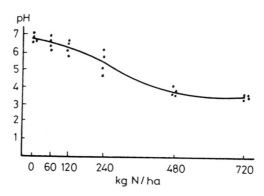

Figure 11.1 pH decrease in the upper soil layer during a 17 years period in relation to the N fertilizer rate (after Mundel and Krell 1978).

Under humid conditions acidification proceeds even without the major impacts of man. This has been illustrated by soil development occurring during a 250-yr primary successional sequence of vegetation along the Tanana River in Alaska (Marion *et al.* 1993). Over this period these soils lost about 75% of their original Ca content and the profile upper 10 cm acidified by 2-3 pH units.

In recent years soil acidification has increased on a global scale because of acid precipitation. The burning of fossil fuels leads to the formation of acids such as H_2SO_4, H_2SO_3, HNO_2, and HNO_3 which are returned from the atmosphere to the earth in precipitation (see page 353). Acid rains with a pH as low as 2.4 have been recorded. On contact with the soil these acid rains rapidly break down soil minerals as well as depressing soil pH and must therefore be considered as a serious ecological hazard (Likens *et al.* 1996).

During pedogenesis under humid climatic conditions the leaching of Ca as nitrate and bicarbonate and thus soil acidification have contributed considerably to soil degradation. This detrimental process which gradually renders fertile soils more and more infertile is still continuing in many parts of the world (Williams 1980, Bromfield *et al.* 1983). Acid soils not only possess a poor soil structure but are often low in Ca and Mg and contain high amounts of soluble Al and Mn, which are often toxic to crops (see page 54).

In the history of man it is only relatively recently that this gradual soil degradation has been prevented by the application of alkaline materials to the soil. The application of limestone was already practiced by the Romans. In Europe it was introduced in the 19th. century and has maintained and even improved the fertility status of many cultivated soils. It appears that the application of alkaline material to soils is an essential means of maintaining the soil pH at an optimum level. Limestone is the most important material used for this purpose. The question of liming is discussed more thoroughly on page 533.

Besides the Ca bearing minerals a substantial amount of Ca^{2+} is adsorbed to organic and inorganic soil colloids and is here of particular importance for soil structure. As already outlined on page 28, Ca^{2+} promotes the flocculation of soil colloids and thus improves soil structure and the stability of soil particles. For soils in which 2:1 clay minerals dominate, about 80% of the soil cation exchange capacity should be saturated with Ca^{2+} in order to maintain a satisfactory soil structure. For soils with kaolinite and sesquioxides as the most important clay minerals a lower percentage of Ca^{2+} saturation in the order of about 20% is recommended (Broyer and Stout 1959). The adsorption sites of the inorganic soil colloids are not very selective for Ca^{2+}. As the electrostatic charge of Ca^{2+} is high due to its divalency and rather thin hydration shell, Ca^{2+} is relatively strongly adsorbed to different kinds of clay minerals in the soil. The adsorption bond of Ca^{2+} to organic colloids and especially to the humates is more specific. Thus in chernozems (Mollisols) and calcareous peat soils, both of which soil types contain $CaCO_3$, the humic acids are mainly present in the form of Ca humates.

Calcium adsorbed to soil colloids equilibrates with the Ca^{2+} of the soil solution. According to the investigations of Nemeth *et al.* (1970) on a number of soil types, there is a fairly linear relationship between the exchangeable Ca^{2+} and the Ca^{2+} of the soil solution under equilibrium conditions. However, it is important to recognize that the effects of soil pH on Ca^{2+} availability in soil solution are closely related to soil pH as summarised by McLaughlin and Wimmer (1999) as follows. At pH levels above 5.0 increasing acidity causes both increased mineral weathering and increased availability of Ca^{2+} within the soil solution. This is effected by replacement of Ca^{2+} on soil cation-exchange sites by H^+. Both natural and human influences that acidify soils induce the release of the exchangeable Ca^{2+} into the soil solution. As soil pH drops below 5.0, cationic Al species are dissolved more rapidly than Ca^{2+}, because release of cations is proportional to charge, resulting in in higher amounts of cationic Al species in the soil solution. A further fall in pH below 4.5 in some forest soils can rapidly dissolve significant reserves of Ca-oxalate which can represent half of the exchangeable Ca^{2+} in the forest floor.

11.1.2 Ecological aspects

Soils differ very widely in their pH and Ca contents, particularly in uncultivated soils. During evolution plant species have adapted to these varying pH and Ca conditions. For this reason marked differences in tolerance occur between plant species and even varieties of a single species. In this respect plant species may be divided into calcicoles and calcifuges. The calcicoles are typical of the flora observed on calcareous soils whereas the calcifuge species grow on acidic soils poor in Ca. Fairly clear differences occur in Ca metabolism of these two groups. Many of the calcicole species contain high levels of intracellular Ca and high concentrations of malate, whereas the calcifuges are normally low in soluble Ca. Species and even cultivars may differ considerably in their capability of precipitating Ca^{2+}. Mainly in the form of Ca oxalate although other Ca containing crystals may also be formed (Bangerth 1979).

The Ca concentration in plants is to a large extent genetically controlled and is little affected by the Ca^{2+} supply in the root medium, provided that the Ca availability is adequate for normal plant growth. This has been very well demonstrated by Loneragan and Snowball (1969), who compared the Ca concentrations of 18 different species grown in nutrient solution, with those grown in the field. As can be seen from Figure 11.2, there was little difference in the Ca concentration of a particular species whether the plants had been grown in soil or in solution culture. Additionally the concentrations of Ca of the monocotyledons was much lower than that of the dicotyledons, a point which is considered later.

The level of Ca in the soil, however, is not the only factor of importance in the calcicole-calcifuge question, for calcareous and acid soils differ in other respects. Calcareous soils are higher in pH and carbonate content. They are richer in nutrients, the level of soluble heavy metals is usually lower and in addition the activity of the nitrifying and nitrogen fixing bacteria is higher. Thus as well as the effects of Ca levels *per se* all these other factors have a bearing on the ecology of plants growing on these soils. An example showing the significance of increased heavy metal solubility under acid conditions is shown from the work of Rorison (1960a and b). It was observed that the inability of the calcicole *Scabiosa columbaria* to grow under acid conditions originated largely from its Al intolerance. Plants grew poorly in soil at pH 4.8 but in nutrient solutions there was little difference in growth between pH 4.8 and 7.6. When Ca^{2+} levels were increased without inducing a pH change, by addition of $CaSO_4$, no improvement in plant growth was observed. Addition of Al^{3+} to the low pH solution culture treatment, however, depressed growth in a similar way to the plants growing in soil at the same pH.

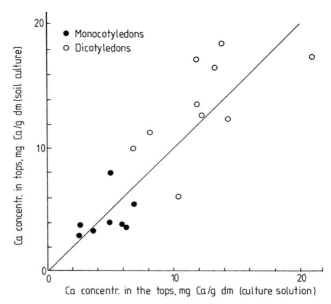

Figure 11.2 Relation between the Ca concentration in the tops of 18 plant species grown in soil to the Ca concentration in the tops of the same plant species grown in culture solution (after Loneragan and Snowball (1969).

Another important aspect in the calcicole-calcifuge problem is the ability of different ecotypes to utilise Fe. Hutchinson (1967) tested 135 different plant species in their susceptibility to lime-induced chlorosis by high $CaCO_3$ levels in the soil. It was observed that plant species originating from acid soils were more susceptible than those from calcareous sites. Lime-induced chlorosis is a physiological disorder which is characterised by a light green to yellow colour of the youngest leaves and which results from the surplus of HCO_3^- in calcareous soils (Mengel 1994a). Hutchinson (1967) too observed that the susceptibility of the ecotypes from acid soils to lime-induced chlorosis was not due to the lack of uptake of Fe by roots but rather in the inability of the roots to metabolise Fe. Lime-induced chlorosis can often be a severe problem in growing crops on calcareous soils (see page 566).

McLaughlin and Wimmer (1999) summarise three processes that in the long-term lead to decreasing Ca availability for plants. One is the incorporation of Ca into the plant biomass, where it is temporarily unavailable to the system (in the case of the woody biomass in forest ecosystems) or continuously removed from the site through the harvested yield (in the case of agricultural systems). Secondly, there is the loss of available Ca from the soil through the leaching as increasing amounts of exchangeable Ca^{2+} from soil reserves are replaced by H^+ and cationic Al species on the soil-exchange complex. Last but not least, reduced availability of Ca to the roots can be attributed to competitve interference from cationic Al species and other cations such as Mn^{2+} and Fe^{2+} that increase in the soil solution as pH levels drop to values lower than 5.0.

Ecological aspects of plant nutrition have been very well reviewed by Epstein (1972a). A useful volume dealing with ecological aspects of mineral nutrition has also been edited by Rorison (1969). McLaughlin and Wimmer (1999) have also reviewed ecological aspects of available Ca in forestry.

11.2   Calcium in Physiology

11.2.1   Uptake and translocation

Higher plants often contain Ca in appreciable amounts and generally in the order of about 5-30 mg Ca/g dry matter. These high Ca concentrations, however, mainly result from the high Ca^{2+} levels in the soil solution rather than from the efficiency of the Ca^{2+} uptake mechanism of root cells. Generally the Ca^{2+} concentration of the soil solution is about 10 times higher than that of K^+. The uptake rate of Ca^{2+}, however, is usually lower than that of K^+. This low Ca^{2+} uptake potential occurs because Ca^{2+} can be absorbed only by young root tips in which the cell walls of the endodermis are still unsuberized (Clarkson and Sanderson 1978). The uptake of Ca^{2+} can also be competitively depressed by the presence of other cations such as K^+ and NH_4^+ which are rapidly taken up by roots (see page 134).

The amount of Ca^{2+} absorbed by the plant depends on the concentration in the root medium and is also genetically controlled. This has been demonstrated by Clarkson (1965) who compared a calcicole and a calcifuge species of *Agrostis* in their response to Ca^{2+}. Both species were grown at pH 4.5 in nutrient solution with increasing concentrations of Ca^{2+}. *Agrostis setacea* (the calcifuge species) was little affected in growth by the additional Ca^{2+} supply, whereas *Agrostis stolonifera* (the calcicole species) responded considerably to the higher Ca treatments. The same pattern was observed in Ca concentration in shoots as shown in Figure 11.3.

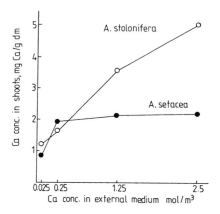

Figure 11.3   Ca concentration in the shoots of *Agrostis* species as a function of Ca^{2+} in the external medium (after Clarkson 1965).

Hence, the Ca^{2+} concentrations found in plant tissues may vary considerably. In general the lower Ca^{2+} concentrations are found in monocotyledons than in dicotyledons as already mentioned. The uptake of Ca by both plant groups as represented by ryegrass and tomato was investigated by Loneragan and Snowball (1969) using flowing nutrient solution in which very low but constant supplies of Ca^{2+} could be maintained. The results for these two contrasting plant species are shown in Table 11.2.

Table 11.2   The effect of calcium concentration in nutrient solution (mmol/m³) on relative growth rates of plants and calcium concentration in the shoots (Loneragan *et al.* 1968 and Loneragan and Snowball 1969).

	Ca supply in mmol/m³				
Species	0.8	2.5	10	100	1000
	Relative growth rate (%)				
Ryegrass (*Lolium perenne*)	42	100	94	94	93
Tomato (Lycopersicon esculentum)	3	19	52	100	80
	mg Ca g/dry matter				
Ryegrass (*Lolium perenne*)	0.6	0.7	1.5	3.7	10.8
Tomato (Lycopersicon esculentum)	2.1	1.3	3.0	12.9	24.0

Maximum growth rates of ryegrass (a monocotyledon) were obtained at 2.5 mmol/m³ Ca^{2+} as compared with 100 mmol/m³ Ca^{2+} for optimal growth of tomato (a dicotyledon). The corresponding Ca concentrations were also much lower in the ryegrass than in the tomato plant thus indicating a higher demand by the tomato plant. The reason for this higher demand of dicotyledons appears to relate to the higher cation exchange capacities of the tissues of these plants indicative of the greater presence of carboxylic groups in the cell walls. Additionally the tomato plant is able to precipitate large amounts of Ca as Ca oxalate in its tissues. This example also shows that Ca^{2+} can be supplied at high concentrations and can reach more than 20 mg/g dry weight of plant tissues. Hence, the concentration of Ca is one of the highest of the inorganic elements in plant material. Interestingly despite the high concentrations of Ca in plant cells the distribution of Ca within cells is most uneven. The high concentrations measured in mol/m³ that may occur outside the cytosol *e.g.* in the apoplast, the vacuole and mitochondria are highly toxic to cytosolic processes within the cytosol. The Ca^{2+} concentration of the cytosol is extremely low and in the range of 10^{-3} to 10^{-5} mol/m³ (Marmé 1983). In the alga *Chara* Williamson and Ashley (1982) have reported cytosolic concentrations of Ca^{2+} in the order of 10^{-4} to 10^{-3} mol/m³. The maintenance of this low cytosolic Ca^{2+} concentration is of vital importance for the plant cell (Hanson 1984) for there is now evidence that Ca^{2+} may inhibit various enzymes located in the cytoplasm

(Gavalas and Manetas 1980). The same is also true for chloroplasts. The low Ca^{2+} concentration has not only to be maintained to prevent inhibition of enzyme activity, however, but also the precipitation of inorganic phosphate as Ca-phosphate and the competition with Mg^{2+} for binding sites (Hepler and Wayne 1985). The levels for apoplastic Ca^{2+} for terrestrial plants are typically in the order of 10^{-1} mol/m³ (Bush 1995). Calcium ions in contrast to K^+ and phosphate are not transported effectively by the symplast and this may be caused by the necessity of maintaining low levels of Ca^{2+} in the cytosol.

The typical pathway for uptake of Ca^{2+} involves initial movement into the free space of the root apoplast and then further movement through apoplastic pathways. This pattern of movement through the cell walls avoids trans-membrane passage and the toxicity problems that contact with the cytoplasm would entail (Haynes 1980a). The movement of Ca^{2+} from the cortex to the stele of the roots is therefore restricted to the apoplastic or free space pathway which is only accessible in non suberized young roots. As roots age the endodermis becomes suberized and this explains, why the uptake of Ca^{2+} (and Mg^{2+}) is restricted to an area just behind the root tips. This pattern of uptake of Ca^{2+} by plant roots has been extensively studied by Clarkson and co-workers using Ca-45 as a radioactive tracer. The main results of this work are discussed by Russell and Clarkson (1976). By contrast K^+ and phosphate were found to be absorbed along the whole length of the root as movement takes place *via* the symplastic pathway. In contrast to the apoplast the continuity of the symplast is maintained through the endodermal walls by plasmodesmata (Figure 4.6).

The passage of Ca^{2+} through the plasmalemma is presumably achieved by facilitated diffusion following the electrochemical potential between the apoplast and the cytosol (Hedrich and Schroeder 1989). In intracellular transport Ca^{2+} and Mn^{2+} have the same transporter (Fox and Guerinot 1998) and competion for the transporter between both cation species may occur. Intracellular transport involves transport into the vacuole and also into mitochondria. In order to maintain the low Ca^{2+} concentration in the cytosol, Ca^{2+} entering the cytosol is continuously pumped back into the apoplast by a selective Ca^{2+} pump (Serrano 1989) or transported into the vacuole and mitochondria by Ca^{2+} transporters. In contrast to K^+ and phosphate, Ca^{2+} is translocated only poorly in the phloem sap (see page 212). Its translocation is mainly restricted to the xylem sap and follows strictly the unidirectional transpiration stream. In addition to transpiration, however, root pressure can also play a role in the upward translocation. This is particularly true at night when the transpiration rate is low. The importance of root pressure in the transport of Ca^{2+} in cabbage was recognised by Palzkill and Tibbitts (1977) who showed that the low transpiring inner head leaves were unable to obtain adequate amounts of Ca^{2+} unless root pressure occurred. High transpiration favoured Ca^{2+} transport to the outer leaves at the expense of the inner leaves.

The movement of Ca^{2+} in the xylem vessels, however, can not be explained simply in terms of mass flow as Ca^{2+} is absorbed by adjacent cells and also adsorbed to indiffusible anions in the xylem walls. According to Biddulph *et al.* (1961) the xylem cylinder of bean stems operates as an exchange column for Ca^{2+} and according to their investigations mass flow in the vessels was inadequate to explain the rapid deep-seated exchanges observed for this [45]Ca tracer. This conclusion is supported by results of

Isermann (1970), van de Geijn and Petit (1979) and Momoshima and Bondietti (1990) who found that the adsorbed Ca^{2+} in the xylem tissue can be exchanged by other cation species and that such an exchange favours the upward translocation of Ca^{2+}. Kuhn et al. (1995) investigated this effect in Norway spruce using stable isotope labelling and found that occupancy of more basipetal binding sites in the cell wall by other cations, such as cationic Al species, facilitated an apical movement of Ca^{2+} through the xylem. Thus, although net Ca^{2+} uptake was reduced by lower pH and the presence of cationic Al species in solution, the upward movement of Ca^{2+} in the stem was enhanced. This example stresses the relevance of exchangeable Ca^{2+} in the cell wall as a reservoir for Ca^{2+} supply to acropetal growing centres (McLaughlin and Wimmer 1999).

The importance of exchange reactions in Ca^{2+} movement is particularly clear from studies on individual plant organs where the correlation between intensity of transpiration and uptake of Ca^{2+} is often much less close than for the plant as a whole. In leaves for example the influx of Ca^{2+} sharply declines after leaf maturity even though a constant transpiration rate is maintained (Koontz and Foote 1966). The same holds true for the influx of Ca^{2+} into fruits. In growing plants there is evidence that Ca^{2+} is translocated preferentially towards the shoot apex even though the transpiration rate here is much lower than in the older leaves. It now seems likely that this preferential movement is induced by the auxin indole acetic acid (IAA), which is synthesised in the shoot apex. It is believed that during growth an IAA stimulated proton efflux pump in the elongation zones of the shoot apex increases the formation of new cation exchange sites so that the growing tip becomes a centre for Ca^{2+} accumulation. This relationship was investigated by Marschner and Ossenberg-Neuhaus (1977) using the IAA transport inhibitor 2, 3, 5 tri iodobenzoic acid (TIBA). Their results indicate a causal connection between TIBA-induced inhibition of IAA transport and the inhibitory effect of TIBA on Ca^{2+} translocation into the shoot apex. Bangerth (1979) suggested that the basipetal IAA transport forces Ca^{2+} to be translocated acropetally.

The rate of downward translocation of Ca^{2+} is very low due to the fact that Ca^{2+} is transported in only very small concentrations in the phloem (Wiersum 1979). Rios and Pearson (1964) observed that the downward transport of Ca^{2+} in cotton plants was inadequate to support root growth in the Ca^{2+} deficient nutrient solution portion of a split root medium. Once Ca^{2+} is deposited in older leaves it cannot be mobilized to the growing tips. This has been convincingly shown by Loneragan and Snowball (1969) using autoradiographs. Plants were not able to utilise Ca^{2+} from older leaves for the growth of meristematic tissues, even when Ca^{2+} deficiency symptoms were observed in the growing tips. Another example demonstrating poor Ca transport in the phloem has been presented by Marschner and Richter (1974). These workers supplied root segments of intact young maize seedlings with labelled Ca^{2+}. It was found that this Ca^{2+} was translocated exclusively to the upper plant organs and not to the root tips. Some data of this experiment are shown in Figure 11.4.

The reason why Ca^{2+} is present in the phloem sap in only very minute concentrations is not really known. Marschner (1974) supposes that the extremely low levels of Ca^{2+} in the phloem sap result from an accumulation of Ca in the cells surrounding the phloem. He also suggests the possibility of Ca^{2+} specific pumps, located in the membranes of the sieve elements which could be responsible for removing Ca^{2+}. As a result of low Ca^{2+}

concentration in the phloem, all plant organs which are largely provided with nutrients by the phloem sap are rather low in Ca. On the other hand the K⁺ concentrations of these organs are relatively high, because K⁺ is present in the phloem sap in abundant quantities (Table 4.5). This relationship is particularly evident when the Ca and K concentrations of leaves are compared with those of fruits and storage tissues. The poor supply of Ca²⁺ to fruits and storage organs can result in Ca deficiency in these tissues (see page 531).

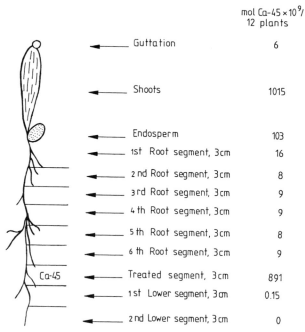

Figure 11.4 Distribution of ⁴⁵Ca in a maize seedling following application to a specific section of the root (after Marschner and Richter 1974).

11.2.2 Calcium forms and contents

Calcium occurs in plant tissues as free Ca²⁺, as Ca²⁺ adsorbed to indiffusible ions such as carboxylic, phosphorylic and phenolic hydroxyl groups. It serves as counter cation for organic and inorganic anions. It is also precipitated as Ca oxalate, carbonate and phosphate. These compounds often occur as deposits in cell vacuoles. In seeds, Ca is present predominantly as the salt of the inositol hexaphosphosphate (phytate). As already indicated Ca²⁺ in the cell wall is associated with the indiffusible carboxylic groups of the pectins and saturates most of these sites. Calcium ions are also bound to membranes where they may bridge anionic groups and thus stabilize the membrane (Figure 3.4).

The interrelationship between 'free' and 'bound' Ca is of importance in fruit ripening. This process may be considered as a special case of senescence and is associated with an increase in ethylene production. Ethylene synthesis is regulated by an enzyme system in the cell wall membrane complex (*i.e.* outside the cytoplasm). The production of ethylene together with an increase in membrane permeability as a result of a fall in physiologically active Ca^{2+} may be considered as essential steps in fruit ripening. In addition fruit ripening requires the removal of Ca^{2+} from the middle lamella. This is correlated with an increase in the activity of polygalacturonase, the enzyme responsible for dissolving the pectates of the middle lamella and softening the tissue. The involvement of Ca in ripening is shown in Table 11.3 which compares the Ca concentrations of two cultivars of tomato, a normal cultivar (Rutgers) and a non ripening mutant cultivar (rin) at three stages after anthesis. The table shows in the normal "Rutgers" cultivar that during ripening the total Ca remained fairly constant but there was a marked shift from the bound to soluble fraction. In the case of the non ripening mutant "rin" the total Ca concentration increased dramatically as did the bound fraction. This finding is in accordance with the observation that fruit ripening is depressed by Ca treatment and stimulated by Ca deficiency (Faust and Shear 1969). One may speculate that the low Ca^{2+} concentration in phloem sap maintains the Ca at a low level in fruits and storage organs so that maturation can take place.

In plants which are poorly supplied with Ca^{2+} a high proportion of the plant Ca (>50%) may occur in the cell wall fraction or as oxalate (Mostafa and Ulrich 1976, Armstrong and Kirkby 1979). Under such conditions essential functions of Ca^{2+} may be impaired such as in membrane stabilization. In a study of genotypes of tobacco Brumagen and Hiatt (1966) were thus able to show that differences in response to Ca deficiency were at least in part related to differences in oxalate formation.

Table 11.3 Calcium concentration in non ripening rin and Normal Rutgers pericarp tomato tissue at different stages of fruit development (Poovaiah 1979)

Days after anthesis	Soluble Ca, μg Ca/g DM		Bound Ca, μg Ca/g DM	
	rin	Rutgers	rin	Rutgers
40	299	349	530	562
50	412	602	667	246
60	492	622	1357	291

The Ca concentration differs not only between the different tissues and organs at the whole plant level, but there are also considerable variations in Ca concentrations within the cells. Hence, steep electrochemical gradients for Ca^{2+} occur across the the membranes of different organelles. These gradients have been proposed to be an essential prerequisite for the diverse functions of Ca^{2+} in regulating physiological and structural processes in plants. Particularly steep gradients are present across the plamamembrane and the tonoplast. The endoplasmic reticulum and the mitochondria are also relatively rich in Ca^{2+} (see Figure 11.5). Gradients across these organelles and the cytosol are normally relatively stable in unstimulated cells, although they are established by a dynamic balance between influx and efflux of Ca^{2+} across each of the cellular membranes (Bush 1995).

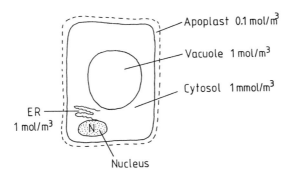

Apoplast 0.1 mol/m³

Vacuole 1 mol/m³

Cytosol 1 mmol/m³

ER
1 mol/m³

Nucleus

Figure 11.5 Typical Ca^{2+} concentrations in various compartmnets of fully expanded plant cells.

Most of the water soluble Ca^{2+} of plant tissue is normally localized in the vacuole and is electrochemically balanced with organic and inorganic anions such as malate, nitrate and chloride. The concentration of vacuolar Ca^{2+} is approximately $1 mol/m^3$ which is several orders of magnitude higher than that in the cytosol which is in the $mmol/m^3$ range. Hence, the vacuole represents as a storage compartment for Ca^{2+}. Elevated Ca^{2+} levels were also proposed for the endoplasmatic reticulum and, like the vacuole, is thought to function as a Ca^{2+} store. High Ca concentrations are also found in the middle lamella of the cell wall (Westermark *et al.* 1986), where crosslinking of acidic pectins by Ca^{2+} bridges is an important factor in wall rigidification. While, as mentioned above, the Ca^{2+} concentration in the cytosol is extremely low the Ca^{2+} concentration at the exterior surface of the plasma membrane is much higher. This provides a steep gradient for Ca^{2+} across the plasma membrane with an equivalent electrical potential of -150 to -200 mV (Bush 1995).

The maintenance of low Ca^{2+} concentration in the cytosol is achieved by Ca^{2+} pumps (Ca^{2+} ATPase) located in the plasmalemma and are analogous to the H^+ pumps (Serrano 1989) in the plasmalemma (see page 118). Both pumps are driven by the hydrolysis of ATP and pump their respective cations Ca^{2+} or H^+ against a steep gradient out of the cytosol into the apoplast. The Ca^{2+} pump is a type of ATPase in which the formation of a phosphorylated intermediate during transport is Ca^{2+} dependent. Vanadate inhibits this phosphorylation and thus the transport activity. The second major class of Ca^{2+} transporters found in the vacuole are the Ca^{2+}/nH^+ antiporters (Barkla and Pantoja 1996). The antiporters are secondary transporters which do not directly require ATP for transport and can use the proton gradient to drive Ca^{2+} transport across the membrane. The affinity of the antiporter for Ca^{2+} is relatively low compared with the Ca^{2+} ATPases and the stoichiometry of the antiporter transport is greater than two H^+ per Ca^{2+}. Bush and Wang (1995) compared both Ca^{2+} transporters in wheat aleurone cells, the K_m for Ca^{2+} of an antiporter associated with the tonoplast was 100 times greater than that for a Ca^{2+} ATPase associated with the same membrane. However, the antiporter operated at a much higher maximal rate than did the ATPase, so that at a concentration of 1 mmol Ca^{2+}/m^3, both transported similar amounts of Ca^{2+}. Function and properties of the active Ca^{2+} transport by plant cell membranes has been considered by Evans *et al.* (1991) in a useful review paper.

The steep electrochemical gradients for Ca^{2+} that occur within the cell requires not only the activity of Ca^{2+} transporters to maintain the gradient, but also the existence of influx transporters into the cytosol in order to provide a tightly regulated transport. This is mediated by Ca^{2+} channels located in the membranes. Principally three types of Ca^{2+} channels have been identified in plants, namely voltage-, second-messenger- and stretch-operated channels. The ion channels require different stimuli to be open gated for Ca^{2+} and they differ in their selectivity for Ca^{2+}. In sugar beet tonoplasts for example two types of voltage-operated channels were found. One was open gated at highly negative vacuolar potentials (Pantoja *et al.* 1992) and the other was gated open at potentials that are positive inside the vacuole (Johannes *et al.* 1992). The function of voltage-operated Ca^{2+} channels are well known from animal cells. As reviewed by Catterall (1995) voltage gated ion channels are responsible for the generation of conducted electrical signals in neurons and other excitable cells. They are particularly important for the transmembrane transport of Na^+, K^+ and Ca^{2+}. The term voltage gated means that the voltage across the membrane (membrane polarization) controls the permeability of the channels. The ion conductance brought about by these channels is highly selective and remarkably efficient. Open channels for the three cation species conduct the cations at rates approaching their rates at free diffusion in water. The structure of these cation channels is basically the same: four homologous transmembrane domains containing six transmembrane-helices surround a central pore. Opening of the pore is thought to be driven by a voltage-driven conformational change of the membrane. Toxins, drugs and inorganic cations are blockers of the voltage-gated ion channels. Small changes in the amino acid sequences of the membrane protein may have dramatic effects on the selectivity. Sodium ion and Ca^{2+} channels have similar overall structures but differ remarkably in ion selectivity. In the presence of Ca^{2+}, Na^+ cannot pass a Ca^{2+} channel while in the absence of divalent cations a Na^+ passage through the Ca^{2+} channel is possible.

The second-messenger-operated channels are also well known from animal cells. One well-established mechanism for the release of internal stores of Ca^{2+} is through activation of innositol-triphosphate (IP_3) operated channels. As reviewed by Coté and Crain (1993) the ability of IP_3 to alter cellular Ca^{2+} in higher plants has been demonstrated, although the physiological relevance of this channel for intact plants has not yet been established.

Stretch-operated channels are activated by tension and are in contrast to the other channels less selective for Ca^{2+}. These type of channels have been identified in the plasma membrane of several plants and are supposedly involved in turgor regulation, thigmotropic responses, aluminium toxicity, and responses to temperature and hormones (Bush 1995).

11.2.3 Biochemical functions

Calcium plays a key role in plant growth and development. A wide variety of diverse bio-chemical functions are initiated by or associated with changes in cellular Ca. The unique chemical binding properties makes Ca also most important for stability and integrity of biological membranes and tissues. The necessity of Ca^{2+} for plant growth can easily be demonstrated by interrupting Ca^{2+} supply to the roots. Their growth rate is immediately reduced and after some days the root tips become brown and gradually die.

Calcium is required for cell elongation and cell division (Burström 1968). Cell extension requires the loosening of the cell wall, a process in which the auxin induced acidification of the apoplast plays a role by replacing Ca^{2+} from the cross-links of the peptic chain of the cell wall (Cleland et al. 1990). There is evidence that the auxin induced H^+ pumping (ATPase) of meristematic cells is related to the presence of Ca^{2+} (Marmé 1983).

Calcium ions are essential for newly synthesised membranes because of both, its activation of enzymes involved in structural biosynthesis (Brummel and McLachlan 1989) and its stabilising effect through the formation of Ca^{2+} bridging within the more polar regions of the membrane (Minorsky 1985). According to Caldwell and Haug (1982) Ca^{2+} is adsorbed to negatively charged phosphate groups of membrane lipids (see Figure 3.4) and it is probably in this way that it restricts the permeability of membranes to hydrophilic solutes. Replacement of Ca^{2+} from the plasmamembrane by other cations such as Na^+ or cationic Al species has been suggested as a main factor involved in salinity stress (Cramer et al. 1985). The authors examined the root hairs of intact cotton plants and monitored the binding of Ca^{2+} on the plasma membrane by using a fluorescence dye. Membrane bound Ca^{2+} was considerably reduced with increasing Na^+ supply. A significant efflux of K^+ occurred at high Na^+ concentrations in the root medium (225 mol/m³). This efflux was remarkably reduced in the presence of Ca^{2+} in the root medium. These results suggested that in case of Na^+ salinity Ca^{2+} may be exchanged from the plasmalemma resulting in higher permeability and higher efflux of K^+ out of the cells. Also for the Al toxicity, the exchange of Ca^{2+} from the plasmamembrane by cationic Al has been suggested to play role (Rengel 1992).

Microelectron probe studies of Roland and Bessoles (1968) have revealed that Ca is located especially in the border zone between the cytoplasm and cell walls indicating high Ca contencentrations at the outer plasmalemma. Calcium ions can be removed from membranes by treatment with EDTA. This treatment increases membrane permeability to such an extent that inorganic and organic compounds can diffuse out of the cell and considerable damage may result (van Steveninck 1965). Impairment of membrane permeability by Ca deficiency, like the effect of EDTA, influences the retention of diffusible cellular compounds (Dickinson 1967). McLaughlin and Wimmer (1999) reviewed a number of studies dealing with Ca^{2+} losses from foliar leaching by rainfall. They concluded that the exchange of Ca^{2+} by the H^+ of in acid precipitation results in relevant losses of cations from the canopy in forest ecosystems.

As a result of Ca^{2+} exchange the plasmalemma become leaky and as the deficiency progresses there is a general disintegration of membrane structure (Marinos 1962). In whole plants, the disorder occurs first in meristematic tissues such as root tips, growing points of the upper plant parts and storage organs. Brown melanin compounds resulting from polyphenol oxidation are associated with the deficient tissues. DeKock *et al.* (1975) claim that in tissues containing adequate amounts of Ca, this oxidation is inhibited by the chelation of the phenolic compounds by Ca^{2+}.

The role of Ca^{2+} in membrane stability is not only of importance in ion uptake (see page 527) but also in other metabolic processes. Typical features of senescence for example, are similar to these of Ca deficiency and can be retarded by Ca^{2+}. The features include a breakdown in the compartmentation of the cell and an increase in respiration following the leakage of endogenous respiratory substrates from the vacuole to the respiratory enzymes in the cytoplasm (Bangerth *et al.* 1972). Poovaiah and Leopold (1973) have demonstrated that senescence in maize leaves can be deferred by the addition of either Ca^{2+} or cytokinin and that the effect of these substances is additive. This role of Ca^{2+} in retarding senescence has also been clearly indicated from the findings of Poovaiah (1979). Abscission of leaf blades of kidney bean caused by senescence of the pulvinar tissue was dramatically delayed by high Ca^{2+} concentrations in the nutrient medium. Another aspect of the importance of Ca^{2+} in membrane stability has been discussed by Marschner (1978). He suggests that the low Ca concentration in storage organs induces a high membrane permeability and allows solute diffusion in these tissues. This is obviously of importance in fruits and storage organs which accumulate large amounts of sugars from the phloem.

Calcium as second messenger is involved in a large number of cellular functions that are regulated in plant cells by changes in cytosolic Ca^{2+} concentrations, such as ionic balance, gene expression, carbohydrate metabolism, mitosis and secretion (Bush 1995). One may question how it is that a single ion species and even a relatively immobile one is able to regulate such numerous and complex processes. The key to understand this lies in the function of Ca^{2+} as a so called secondary messenger. An environmental stimuli, wounding for example, creates a Ca^{2+} impulse which can be measured as a transient increase of the cytosolic Ca^{2+} level (Davies 1987). The term second messenger implies that this Ca^{2+} impulse does not directly produce a biological response in the form of activating an enzyme for example, but affects a secondary receptor. This secondary receptor is then capable of producing a biological response, *e.g.* by activating callose production at the plasma membrane cell wall interface (Lerchl *et al.* 1989). In this example the activated enzyme represents the final target protein for the Ca^{2+} signal. Signalling *via* a secondary receptor is capable of inducing biological responses of high specificity. For such regulatory actions of Ca^{2+} to occur three conditions must be satisfied. Firstly a homeostatic system must exist that regulates Ca^{2+} concentrations in the various cell compartments. In eucaryotic cells this is always provided by the diverse ion channels and Ca^{2+} transporter systems. Secondly, a secondary receptor is necessary to mediate the signal transduction into a wide array of cellular responses. Thirdly, the final target proteins are required that are specifically activated by the secondary receptor molecules. As a consequence of the new techniques that are now available much progress has been made during the last decade in molecular characterization of target proteins and in identifying the structure and function of the secondary receptors in the Ca^{2+} signal transduction pathway. *Calmodulin* and calmodulin-like proteins have been found to be most important in eukariotic cells to act as second receptor molecules in the Ca^{2+} signal pathway. Calmodulin is ubiquitous in animal and plant cells and is thus an essential protein in eukaryotic cells. Calmodulin is a polypeptide consisting of 148 amino acids, is heat stable and insensitive to pH changes. It is able of binding four Ca^{2+} forming a compact structure by change of conformation (see Figure 11.6) and displacing a hydrophobic section of the polypeptide chain. Calmodulin acts by binding to short peptide sequences within target proteins, thereby inducing structural changes, which alter their activities in response to changes in intracellular Ca^{2+} concentration (Zielinsky 1998). Although calmodulin has no activity on its own, the role of messenger is exerted through its capacity to activate other enzymes. Calmodulin is ideally suited to serve as a Ca^{2+} activated signal because of its high solubility, low moleqular weight, and the fact that it changes in charge and conformation on binding with Ca^{2+} (Kauss 1987). Numerous isoforms of calmodulin and of calmodulin-like proteins have been identified in plants, and there is a growing list of targets of calmodulin, suggesting that a complex Ca-based regulatory network controls a wide variety of responses to the environment (Snedden and Fromm 1998). An impressive number of enzymes are activated in the way described by calmodulin, including cyclic nucleotide phosphodiesterase, adenylate cyclase, membrane bound Ca^{2+}-ATPase, NAD-kinase. According to Veleuthambi and Poovaiah (1984) calmodulin activates the phosphorylation of soluble and membrane bound proteins. As Ca^{2+}-ATPase brings about active extrusion of Ca^{2+} through the plasmalemma any increase in Ca^{2+} concentration in the cytosol directly induces Ca^{2+} extrusion. Since the electrochemical gradient for Ca^{2+} between the apoplast and the cytosol favours the passive influx of Ca^{2+} into the cytosol, it is supposed that there is a permanent extrusion of Ca^{2+} into the apoplast or vacuole or both. Results of Mitsui et al. (1984) provide evidence that calmodulin is involved in the synthesis of α-amylase and its secretion into the scutellum of rice seeds. Aluminium is believed to interfere with calmodulin and thus antagonize the Ca^{2+} effect (Siegel and Haug 1983). Recent work of Zhang and Rengel (1999) has shown that cationic Al species elicited an increase of Ca^{2+} concentration in the apical root cells of wheat. Aluminum toxicity as described on page 661 is likely to result from this Ca/Al interaction.

Bush (1995) reviewed in detail three complex and intensively studied stimulus-response couples to demonstrate the involvement of Ca^{2+} in signal transduction pathways, namely abscisic acid and cell volume, gibberellic acid and secretion, and red light and gene expression. The author noted the importance of being able to understand how information is encoded in a Ca^{2+} change to dictate and specify a particular response. There are two currently prevalent hypotheses to explain specificity: a spatio-temporal model, in which the timing and location of the Ca^{2+} change encode for specificity, and a multiple-messenger model, in which specificity is encoded by the simultaneous action of several intercellular mediators. These models are not mutually exclusive. How Ca^{2+}-induced signals are generated and transmitted in plants and how such signals are translated into physiological responses has been discussed by Trewavas and Malho (1997), McAinsh and Hetherington (1998), Snedden and Fromm (1998). Timing and specificity of signals and their relationship to whole-plant physiology are summarised by McLaughlin and Wimmer (1999). Some types of Ca-signals occur as oscillations, with both a specific frequency and amplitude. This provides an easy analogy to encoding in radio waves, which can be AM (amplitude- modulated) or FM (frequency-

530

modulated), or perhaps both, in plant systems (Berridge 1997). Another aspect of discussion is the analogy to computer-based progammable translation systems. Thus, the funcionally more-complex responses to environmental stimuli that lead to plant growth and development can be explained by coordination between genetic 'ROM' (read-only memory) and environment-stimulated 'RAM' (random-access memory) (McLaughlin and Wimmer 1999).

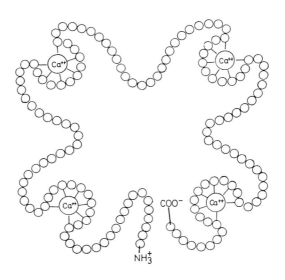

Figure 11.6 Calmodulin - a polypeptide strand consisting of 148 amino acids. Ca^{2+} is bound at 4 sites.

In some cases Ca^{2+} stimulates enzymes without the help of calmodulin or calmodulin-like proteins. This was suggested for example in mitrochondria where the Ca^{2+} concentration is normally much higher than in the cytosol (Marmé 1983). One such enzyme which occurs in the mitochondria is glutamate dehydrogenase which brings about the deamination of amino acids. Calcium dependent protein kinases have also been found to be directly stimulated by Ca^{2+} (Cohen 1989, Roberts and Harmon 1992, Fallon *et al*. 1993).

There is evidence that the response of plant tissues to low temperature may be mediated by the passive influx of Ca^{2+} from apoplast to cytosol. Pickard (1984) obtained a dramatic rise in voltage transients in auxin depleted stem tissue of pea when chilled for a period of only one minute. It is claimed that low temperature activates voltage dependent cation channels so that the permeability of the plasmamembrane to Ca^{2+} increases and Ca^{2+} moves passively from apoplast to cytosol. Minorsky (1985) has proposed that Ca^{2+} plays a primary role as a physiological transducer in chilling injury to plants.

11.2.4 Calcium deficiency and disorders

As already outlined above Ca deficiency is characterized by a reduction in growth of meristematic tissues. The deficiency can be first observed in the growing tips and youngest leaves. These become deformed and chlorotic and at a more advanced stage necrosis occurs at the leaf margins. The affected tissues become soft due to a dissolution of the cell walls. Brown substances occur which accumulate in intracellular spaces and also in the vascular tissue where they can affect the transport mechanism (Bussler 1963).

Absolute Ca deficiency as described above occurs relatively seldom as most mineral soils are rich in available Ca. Indirect Ca deficiency resulting from an undersupply of Ca to fruit and storage tissues, however, is an often observed disorder. Shear (1975) cites a list of 35 such Ca related disorders in fruits and vegetables. In apple the disease is called bitter pit as the whole of the surface of the apple is pitted with small brown necrotic spots (Plate 11.1). In tomato the disease is known as blossom-end rot and is characterized by a cellular breakdown at the distal end of the fruit (Plate 11.1). A similar Ca deficiency disorder occurs in water melon. Ca deficiencies in vegetables, such as blackheart of celery, internal browning of *Brassica oleracea* (Brussels sprouts) blossom end rot of pepper and cavity spot of carrots have been described by Maynard (1979) in a useful review paper. All these tissues are mainly supplied with Ca^{2+} by the transpiration stream. If the xylem sap is low in Ca^{2+} or the rate of transpiration of the fruits is poor, as occurs under humid conditions, inadequate levels of Ca^{2+} may be supplied to the fruits and deficiency symptoms may result. Calcium ion translocation in the xylem sap may be depressed by NH_4-nutrition, soil water stress and high salt concentrations in the soil. These factors have therefore been found to favour the occurrence of blossom-end rot in tomatoes.

Calcium appears only to be transported from the soil solution to the upper plant parts *via* root tips (Russell and Clarkson 1976). Any factor which prevents the growth of new roots (poor aeration, low temperatures etc.) may therefore be expected to prevent Ca^{2+} uptake and thus induce deficiency. This may account for the observation that Ca related disorders often occur on soils adequately supplied with Ca, and that the weather appears to be a controlling factor (Scaife and Clarkson 1978, Bangerth 1979, Kirkby 1979).

The importance of maintaining an adequate level of Ca^{2+} in the xylem sap is very clear from the results of Chiu and Bould (1976). These workers observed that Ca stress during the fruiting stage of tomatoes caused serious blossom-end rot of fruits although older leaves had high Ca concentrations. This oservation shows that Ca^{2+} absorbed by the plants before fruiting and deposited in the leaves could subsequently not be mobilized for fruit development. The results provide further evidence of the immobility of Ca in the phloem. In addition they also indicate that leaf analysis for Ca is not a reliable index for predicting Ca deficiency in fruits. The occurrence of the physiological disorders described above depends very much on the Ca levels in the fruits. In apples for example a close negative correlation has been found between the occurrence of bitter pit and Ca concentration (Sharples 1968). Calcium supply to fruit trees is therefore a well established means in horticulture of avoiding diseases of fruits as a result of Ca deficiency. Because of the restricted mobility of Ca^{2+} not only liming but also more importantly Ca-containing sprays are used (Rease and Drake1996).

Fruits and storage tissues growing in the soil, such as peanuts, potatoes and celery bulbs are not supplied by the transpiration stream and for this reason Ca^{2+} must be absorbed directly from the soil medium. According to investigations of Skelton and Shear (1971) growth and yield of fruits of peanuts (*Arachis hypogaea*) depend considerably on the Ca availability of the soil. Inadequate Ca supply to celery causes black heart. Seçer and Unal (1990) reported that the Ca concentration in petioles of melons (*Cucumis melo* L.) was posively correlated with fruit yield and fruit quality.

Under the humid tropical conditions soils are often cation-depleted. At such settings Ca^{2+} can be the primary nutrient limiting tree growth (Cuevas and Medina 1988). This view is supported by the study of Reich *et al.* (1995) who analysed a number of different woody species in an oligotrophic Amazonian rain forest. They found lower levels of Ca, N and P occurring with later successional species than with early ones. Although photosynthetic rates were correlated with N and P in early successional species, at the lower levels of Ca, N and P occurring with later successional species, leaf photosynthetic rates were closely correlated with leaf Ca concentrations ($r^2 = 0.80$) and not with leaf concentrations of N or P. The role of Ca^{2+} in plant physiology and its importance for crop production have been reviewed by Foy (1974), by Marschner (1974), by Bangerth (1979), by Hepler and Wayne (1985) and by Bush (1995). The importance of Ca in soil-plant relationships in tropical and subtropical conditions is considered by Malavolta *et al.* (1979) and by McLaughlin and Wimmer (1999).

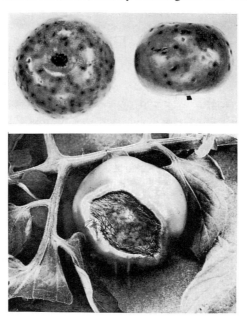

Plate 11.1 Upper part, bitter pit in apples, by courtesy of *U.S. Dep. of Agric.* Beltsville (Photo: Shear). Lower part, severe blossom end rot in tomato by courtesy of the Macaulay Land Use Research Institute, Aberdeen, Scotland. (Photo: DeKock)

11.2.5   Strontium

Calcium and Sr are closely related chemically and in plants they show similar behaviour. The uptake and distribution of both elements in plants are thus alike, but not identical (Hutchin and Vaughan 1968). For example, in experiments with *Pisum sativum* Myttenaere (1964) found that Sr^{2+} is deposited to a greater extent than Ca^{2+} in cell walls. What is of major interest is that Sr^{2+} cannot substitute for Ca^{2+} in physiological processes. High Sr concentrations in plant tissues are toxic. Toxicity symptoms appear first in older leaves, which become brown and necrotic. The uptake of Sr^{2+} is restricted by Ca^{2+}. Liming can thus reduce the uptake of Sr^{2+} (Reissig 1962). On the other hand plants low in Ca absorb Sr^{2+} at a higher rate (Balcar *et al.* 1969).

11.3   Liming and Calcium in Crop Nutrition

11.3.1   The pH effect and the calcium effect

The application of liming materials such as $CaCO_3$, CaO or $Ca(OH)_2$ to the soil has several effects. It supplies Ca^{2+} to crops, it may improve soil structure, it neutralizes soil acidity and reduces the level of toxic Al species in soils. Most inorganic soils contain high enough levels of Ca^{2+} in soil solution and their exchange sites are well enough saturated with Ca^{2+}, to adequately meet crop demands. It thus appears that liming is primarily a means of improving soil structure and pH. On acid peat soils, however, as are frequently used as a substrate in horticulture, the natural Ca content can be so low that plants suffer from Ca deficiency. Here the application of Ca containing fertilizers is advisable.

As already outlined on page 57 acidification of soils and Ca^{2+} loss by leaching run parallel under humid climate conditions. Thus the alkaline reaction of the liming material is needed for neutralizing soil H^+. In principle the pH increase brought about by liming material could also be induced by other compounds of alkaline nature, such as K_2CO_3 or Na_2CO_3. These compounds, however, are of no importance for the pH improvement of acid soils under practical conditions. The effects of soil pH on crop yields may differ considerably for various crops (see Table 2.10, page 57). In a 50 years old field experiment carried out on a sandy soil in Germany, Köhn (1976) found the following order of sensitivity to yield production in the 'no liming' treatment: fodder beets > barley > oats > wheat > potatoes. Rye in contrast gave the highest yields in the non limed plots.

The main purpose of liming acid soils is to reduce soluble Al by precipitation. This was clearly shown by in pot experiments by Hütsch and Steffens (1986) as can be seen from some of the results given in Table 11.4.

Table 11.4 Effect of liming on soil pH, soluble cationic Al species anionic Al compounds and grain yield of spring barley. Soil Al was extracted by electroultrafiltration (EUF, see p. 99) which differentiates between cationic Al species and anionic Al compounds, the latter are mainly organic Al complexes. The CaO quantities applied corresponded to 1/2, 1 and 2 times of the equivalent amount of exchangeable Al (data from Hütsch and Steffens 1986).

Treatment	pH	Cationic Al, mg/kg soil	Anionic Al, mg/kg soil	Grain yield, g/pot
without CaO	3.9	14.9	18.4	0
½ CaO	4.3	6.1	14.7	2.5
1 CaO	4.7	1.2	17.4	23.2
2 CaO	5.7	1.1	15.3	29.2

The application of CaO increased the pH and had a marked impact on soluble cationic Al which was so high in the treatment without CaO that no grain was produced. Interestingly the amount of EUF extractable anionic Al was virtually uninfluenced by the CaO treatment and was without toxic effect. This result is in agreement with data of Mühling *et al.* (1988).

The soil used in the above quoted experiments was rich in Al/Fe oxides/hydroxides and therefore resembled the Oxisols of the tropics. These acid tropical soils may contain high concentrations of soluble Al. Sims and Ellis (1983) in studying the effect of limestone application to an Al rich Ultisol found that liming reduced the exchangeable Al considerably and also increased phosphate availability and uptake. The most important soil parameters influenced by liming are shown in Table 11.5 from this work. Two rates of liming and a control were compared, the lower rate of liming which brought about the neutralisation of exchangeable Al, being one third of that of the higher rate of lime application. Highest phosphate uptake was obtained at the lower rate of lime application. Liming had a clear influence on available Al, exchangeable cations, soil pH, and phosphate availability. The authors suggest that such acid soils should be limed to a level which results in Al neutralisation.

Haynes (1984) also reported a decrease in extractable Al by applying $Ca(OH)_2$. Ca silicate, however, did not show this effect. The influence of pH and liming on phosphate adsorption has already been considered on page 455. A useful papers dealing with the effects of liming on phosphate availability have been published by Haynes (1982) and Kerschberger (1987).

On soils in which a pH increase is not desirable, but where Ca^{2+} is needed for amending soil structure, neutral Ca salts should be applied. Such soils are mainly salt affected soils, characterized by neutral to alkaline pH values and by an excess of Na^+. This is mainly adsorbed to soil colloids (see page 62). The application of neutral Ca salts results in a replacement of the adsorbed Na^+ by Ca^{2+}, thus increasing the Ca^{2+} saturation of soil colloids and inducing flocculation. The most important neutral Ca salt

Table 11.5 Effect of liming an Ultisol on various soil parameters and on phosphate uptake of oats grown in pots. (Data from Sims and Ellis 1983)

Parameter	Without lime	Lime (CaCO₃)	
		5.4 Mg/ha	17.2 Mg/ha
Extr. Al, mol/kg	8.57	3.17	1.23
Exchang. Al, mol/kg	11.2	3.17	0
Al sat. %	74	9.8	0
pH, KCl	3.37	3.61	6.76
Exchangeable Ca, mmol/kg	1.85	18.8	54.4
Exchangeable Mg, mmol/kg	0.85	0.7	0.1
Exchangeable K, mmol/kg	1.6	1.5	1.0
Available P (Bray) mg/kg	0.55	1.00	1.06
P uptake, mg P/pot	4.76	7.02	5.93

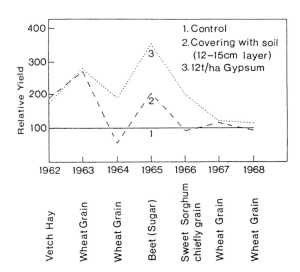

Figure 11.7 Effect of various amelioration measures applied to a saline soil on the relative yields of various crops (after Raikov 1971).

used for salt affected soils is gypsum ($CaSO_4 \cdot 2H_2O$). The quantities applied depend on the degree of salinization. Generally amounts of 5 to 10 Mg gypsum/ha are applied (Shainberg et al. 1989). The effect of two treatments on a saline soil are shown in Figure 11.7 (Raikov 1971). In one treatment gypsum was applied and in the other the saline soil was covered with a non saline soil layer. Crop yields obtained over a period of several years were compared with a control treatment, the relative yields of which are represented by '100' (line parallel to the x-axis in Figure 11.7). The treatment with gypsum resulted in considerable yield increases, in excess of those resulting from the more expensive 'soil covering treatment'. The gypsum treatment also improved soil structure, and the soil of the treated plots became darker in comparison with the control treatment. In addition the soil was easier to cultivate and did not compact crack or retain surface water.

On sandy infertile soils, in particular heavy gypsum applications (i.e. more than 5 Mg/ha) are detrimantal to crop growth because of the preferential exchange of Mg^{2+} which allows Mg to be leached from the upper part of the soil profile (Syed-Omar and Sumner 1991). Sumner (1993) referred to many other workers who found such downward movement of Mg after gypsum application. Care should therefore be taken after gypsum application to provide sufficient Mg for plant nutrition.

The application of gypsum is not referred to as liming but as the Ca^{2+} effect is similar to that of lime both treatments of liming and gypsum application are closely related. However in contrast to lime, gypsum is more readily dissolved and this is important in some cases of amelioration. Two types of gypsum, namely, mined and industrial by-product materials, are used. The by-product from the manufacture of phosphoric acid (phosphogypsum) is the most important gypsum on a world-wide basis. Because of its P concentration, this by-product contributes to some extent to P nutrition when applied with high rates. Generally the mined gypsum materials have been found to dissolve slower than the by-products. The speed with which gypsum dissolves is of importance if used to ameliorate acid subsoils. Acid subsoils are usually deficient in Ca and contain toxic levels of Al. Root proliferation below the plough layer is often severely limited resulting in great susceptibility to drought stress (Sumner 1993). The area that is affected by subsoil acidity is estimated to over 1 billion ha. While acid topsoils can be readily ameliorated by the use of limestone, lime is of little immediate benefit for subsoils because it does not readily move down the soil profile. In a number of studies Sumner (1993) illustrated that gypsum treatment increased exchangeable Ca^{2+} and decreased exchangeable Al to a soil depth of at least 90 cm, which allowed a greater proliferation of roots in the subsoil and the extraction of water by the crop. As a result the crop was better able to withstand periodic droughts during the dry season, so that increases in yield were obtained. Several mechanisms supposedly result in a decrease of exchangeable Al. One presumably particularly important is called the "self-liming effect" proposed by Reeve and Sumner (1972). The mechanism involves ligand exchange of OH^- by SO_4^{2-} on sesquioxide surfaces by which $Ca(OH)_2$ is formed which may react with Al^{3+} forming the insoluble $Al(OH)_3$, as shown in the following equations.

$$Fe^{III} Al(OH)_6 + CaSO_4 \rightarrow Fe^{III} Al(OH)_4SO_4 + Ca(OH)_2$$

$$2 Al^{3+} + 3 Ca(OH)_2 \Rightarrow 2Al(OH)_3 + 3 Ca^{2+}$$

11.3.2 Liming materials

Table 11.6 shows the most important liming materials. The carbonates are simply ground limestone or chalk. Burning Ca carbonate at 1100°C results in thermal dissociation:

$$CaCO_3 + H_2O \Rightarrow CaO + CO_2$$

Table 11.6 Liming materials

Liming Material	Formula	Neutralizing value in CaO
Chalk or limestone	$CaCO_3$	50% CaO
Slaked lime	$Ca(OH)_2$	70% CaO
Burnt lime	CaO	85% CaO

'Burnt lime' is produced in this way. The Ca oxide formed (CaO) readily reacts with water thus forming Ca hydroxide (hydrated lime or slaked lime).

$$CaO + H_2O \Rightarrow Ca(OH)_2$$

On contact with CO_2 this forms $CaCO_3$.

$$Ca(OH)_2 + CO_2 \Rightarrow CaCO_3 + H_2O$$

Thus when hydrated lime is exposed to the atmosphere for a long period of time it is gradually converted to Ca carbonate by atmospheric CO_2. Frequently liming materials also contain substantial amounts of Mg as well as Ca. Burnt magnesium lime for example consists mainly of CaO and MgO, and contains more than 5.5% Mg. Ground magnesium limestone is a mixture of $CaCO_3$ and $MgCO_3$ with an Mg content of 3% Mg and more. These Mg containing materials are particularly used for liming Mg deficient acid soils.

The value of liming material depends on their 'neutralizing value', which is expressed in terms of equivalents of CaO. 100 kg of $CaCO_3$ have the same neutralizing effect as 50 kg CaO. Thus the neutralizing value of 100 kg $CaCO_3$ is 50, whereas 100 kg $Ca(OH)_2$ has a neutralizing value of 70. As the neutralizing value is related to the quantity of carbonate or oxide present in the liming material $MgCO_3$ or MgO also contribute to the neutralizing efficiency. Generally all oxides, carbonates, and even silicates are alkaline in reaction. For this reason Ca silicates present in basic slags and sinter phosphates (see page 474) have a H+ neutralizing effect and are thus of some importance in controlling soil pH. In long-term field experiments Schmitt and Brauer (1969) found, that in plots supplied regularly with basic slag, the pH was only slightly depressed, whereas in plots treated with equivalent amounts of other P fertilizers the soil pH fell from 6.5 to 5.4 over a ten year period. Similar results have been obtained by Roscoe (1960). Some waste products are also used as liming materials. These are mainly carbonates. The waste product of sugar factories has a neutralising value of about 20.

11.3.3 Lime application and reaction in the soil

Because of the high solubilities of CaO and $Ca(OH)_2$ both these compounds are quick acting in comparison with $CaCO_3$. Thus when a rapid change in soil pH is required or where soil reactions are slow, as in cold and wet soils, the application of CaO or slaked lime [$Ca(OH)_2$] is recommended.

$$CaO + H_2O \Rightarrow Ca(OH)_2; \; Ca(OH)_2 + 2\,H^+ \Rightarrow Ca^{2+} + 2\,H_2O$$

$CaCO_3$ reacts more slowly. Under strong acid conditions it dissolves relatively quickly by neutralizing soil H^+:

$$CaCO_3 + 2\,H^+ \Rightarrow Ca^{2+} + H_2O + CO_2$$

Under weak acid or even neutral conditions the presence of CO_2 favours the dissolution of $CaCO_3$ by forming Ca bicarbonate (monohydrogen carbonate) which in turn neutralizes soil H^+:

$$CaCO_3 + CO_2 + H_2O \Rightarrow Ca(HCO_3)_2 \Rightarrow Ca^{2+} + 2\,HCO_3^-$$

$$2\,HCO_3^- + 2\,H^+ \Rightarrow 2\,H_2O + 2\,CO_2$$

Ca silicates present in sinter phosphates and in by products from the steel industry are even slower in their neutralising reaction than $CaCO_3$.

$$CaSiO_3 + 2\,H^+ \Rightarrow Ca^{2+} + SiO_2 + H_2O$$

The rate of dissolution of liming materials also depends on particle size. Finely ground material reacts more rapidly than coarse material due to its larger surface area (Barrows *et al.* 1968).

Generally the application rates of liming materials are in the order of about 3 to 4 Mg/ha CaO or 4 to 6 Mg/ha $CaCO_3$ supplied over a 3 to 5 year cycle. The quantities required not only depend on soil pH (actual acidity), but also on the content of H^+ adsorbed to soil colloids (potential acidity). This relationship between the amount of limestone required in order to obtain given pH levels in different soils is shown in Figure 11.8 (Peech 1961). It is evident that in order to correct the pH of heavy acid soils (high cation exchange capacity = CEC) particularly large amounts of limestone must be applied.

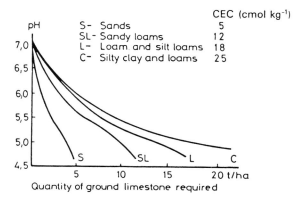

Figure 11.8 Relationship between the quantity of limestone required and the pH increase for various soil classes. (modified after Peech 1961)

Lime can be applied at any time in the year provided that soil moisture allows the soil to be worked. Lime should not be applied with NH_4^+ containing fertilizers, as the pH shift converts NH_4^+ to NH_3 which is partially lost by volatilization.

$$NH_4^+ + OH^- \Rightarrow NH_3 + H_2O$$

Soil classes and types differ in the optimum pH level at which they should be cultivated. Generally as the clay content of a soil increases, the optimum pH also rises.

Raising the pH of acid soils is also a means of providing more suitable conditions for soil bacteria. This may influence various processes such as microbial N_2 fixation, denitrification of NO_3^- and mineralization of organic soil N. Kuntze and Bartels (1975) reported that too low a soil pH resulted in N deficiency of the herbage as a consequence of inhibited N mineralization of the organic matter in peat soils. A pH shift away from acid conditions is therefore frequently accompanied by an enhanced rate in the decomposition of soil organic matter by microorganisms. On sandy soils, organic matter plays an essential role in water retention. For this reason the pH of these soils should not be too high, in order to avoid excessive organic matter decomposition. Soil pH also has a pronounced influence on the availability of various soil nutrients (see page 54).

Liming not only plays a role in the amelioration of agricultural land it is also of importance for the recultivation of waste heaps. Davison and Jefferies (1966) in experimenting with substrate material from coal mine waste heaps found that liming in combined with mineral fertilizer application resulted in very high responses in the growth of *Agrostis tenuis*. The authors claim that the pH increase enhanced fixation of heavy metals which occur in excess in this waste material and which are phytotoxic under low pH conditions.

The rate of Mg^{2+} removal from soils depends considerably on the amount of Mg containing minerals in the soil, their rate of weathering, and the intensity of leaching, as well as on the Mg^{2+} uptake by the vegetation and its removal from the field by crop products. In many soils the release of Mg^{2+} by weathering is able to balance the removal by leaching because in comparison with K^+ uptake rates of Mg^{2+} by crops are generally low. Frequently on sandy soils loss by leaching predominates. In such soils the subsoil often contains higher levels of Mg than in the upper part of the profile. In an observation of 63 uncultivated Swedish profiles Wiklander (1958) found an average Mg^{2+} saturation of the CEC in the upper soil layers (0-20 cm) of 17% (pH 5.4) whilst in the subsoil (40—50 cm) it was 29% (pH 5.8).

The level of Mg in soils depends to a large extent on soil type. Highly leached and weathered soils such as podzols and lateritic soils are generally low in Mg. On the other hand soils formed in depression sites, where leached nutrients may accumulate as in marsh soils or gleyed soils, tend to be high in Mg. The same applies to soils which are not leached such as the Solonchak (Aridisols) and Solonez (Mollisols) in which Mg salts usually occur. Parent material too plays a role and generally soils developed on Mg rich rocks such as basalt, peridotite and dolomite are well supplied with Mg.

Serpentines are 1:1 structured silicate minerals and are therefore similar to kaolinite (see page 31) but with Fe^{2+}, Fe^{3+}, Mg^{2+} and Ni^{2+} substituted for octahedral Al^{3+} (Dixon 1989). Soils derived from serpentine contain serpentine concentrations in the range of 200 to 900 g/kg according to their degree of weathering with Entisols showing the highest concentration (Bonifacio and Barberis 1999). Such soils are very high in available Mg^{2+} and have a low Ca/Mg ratio on the exchange complex. Even Ca deficiency of crops may occur on them (Epstein 1972b). In addition high levels of toxic heavy metals especially Ni and Cr may be present. Serpentine soils support a special vegetation adapted to these unusual conditions (Rodenkirchen and Roberts 1993). Serpentine soils have been discussed by Krause (1958) and Epstein (1972b).

12.2   Magnesium in Physiology

12.2.1   Uptake and translocation

Generally the concentration of Mg^{2+} in the soil solution is higher than that of K^+ but the uptake rate of Mg^{2+} by root cells is much lower than the uptake rate of K^+. This lower ability of roots to take up Mg^{2+} in comparison with K^+ is probably not only restricted to root tissue but also holds for other plant parts as well. The reason for this behaviour is not yet clear. One may speculate that the low uptake potential reflects the lack of a special uptake mechanism transporting Mg^{2+} across the plasma membrane. Fox and Guerinot (1998), who have treated cation transport in a useful review article have not considered Mg^+ transport across membranes. Since the Mg^{2+} concentration in the soil solution is relatively high one may conclude that plants have not developed an efficient Mg^{2+} uptake system and that Mg^{2+} flux across the plasmalemma may be a facilitated diffusion through more or less specific channels following the electrochemial gradient for Mg^{2+}. There is also not much known about the Mg^{2+} flux across the tonoplast.

Since the Mg^{2+} concentrations in the cytosol and in the vacuole are in a range of a few mol/m^3 and do not differ very much it is feasible that here too, the Mg^{2+} transport across the tonoplast is mediated by facilitated diffusion.

In this transport, cation competition may play a major role and the uptake of Mg^{2+} can be greatly depressed by an excess of other cation species, especially of K^+ and NH_4^+ which are taken up at high rates and may compete with Mg^{2+} for the negatively charged cytosol. This competition can lead to Mg^{2+} deficiency in plants. Not only the uptake but also the translocation of Mg^{2+} from the roots to the upper plant parts can be restricted by K^+ and Ca^{2+} (Schimansky1981).

The data of Grimme et $al.$ (1974) show that high Mg concentrations may occur in plants supplied with a low level of K^+ nutrition. These higher Mg concentrations cannot be explained simply in terms of a "concentration effect" resulting from a lower rate of growth but probably originate directly from enhanced Mg^{2+} uptake in the presence of low K^+ concentrations in the soil solution. This observation agrees well with the findings of Leggett and Gilbert (1969) who reported that the Mg^{2+} uptake by soya beans was especially high when the nutrient solution was free of K^+. Similar observations have also been reported by Hall (1971) showing very much elevated Mg levels in Ca-deficient tomato tissues. Magnesium uptake of plants is frequently low on acid soils. This results not only from the low Mg^{2+} availability associated with acid conditions but is also related directly to the low pH level. According to experiments of Grimme (1983) it is not so much the H^+ concentration as the increased level of cationic Al species which depresses the uptake of Mg^{2+}. On the other hand Mg^{2+} may reduce the uptake on Mn^{2+}. Löhnis (1960) showed in a number of plant species that it is possible to prevent the appearance of Mn toxicity by increasing the Mg^{2+} supply. Evidence of decreased Mn^{2+} uptake by Mg^{2+} has also been found by Maas et $al.$ (1969).

Although high levels of K^+ nutrition often depress total Mg^{2+} uptake, increasing K^+ supply affects the Mg concentration of different plant organs to a varying extent. As can be seen from Table 12.1 increasing K^+ supply reduced the Mg concentration in tomato leaves and roots considerably. The Mg concentration in the fruits, however, was somewhat increased by the higher levels of K^+ in the nutrient solution. This observation was obviously not accidental as it was found at all six harvests of the tomatoes (Viro1973). Linser and Herwig (1968) also reported that increased K^+ supply resulted in higher Mg concentrations in the seeds of flax. Similar findings have been reported by Addiscott (1974) for potatoes. It thus appears that K^+ promotes the translocation of Mg^{2+} towards fruits and storage tissues. In contrast to Ca^{2+}, Mg^{2+} is very mobile in the phloem and can be translocated from older to younger leaves or to the apex (Steucek and Koontz 1970, Schimansky1973). The same is true for K^+. As fruit and storage tissues are highly dependent on the phloem transport for their mineral supply they are thus higher in K and Mg than in Ca (see Table 12.1).

Table 12.1   Effect of an increasing K⁺ supply on the cation concentration in various organs of the tomato plant (Viro 1973)

Treatment	K	Na	Ca	Mg
mol K/m³ nutrient solution	in mg/g dry matter			
Leaves				
2	5.0	4.0	47	6.1
10	33	1.9	42	2.7
20	42	1.8	33	1.5
Roots				
2	2.0	3.6	39	3.3
10	22	2.5	32	3.1
20	24	1.3	33	2.6
Fruits				
2	16	1.0	0.9	0.7
10	25	0.7	0.8	0.8
20	27	0.6	0.7	0.9

12.2.2   Biochemical functions

In plant tissues a high proportion of the total Mg, often over 70%, is diffusible and associated with inorganic and organic anions such as malate and citrate. Magnesium is also counter cation of indiffusible anions such as carboxyl and phosphoryl groups (Kirkby and Mengel 1967). Cereal grains contain Mg as the salt of inositol hexaphosphoric acid (Mg phytate). The most well known role of Mg is its occurrence at the centre of the chlorophyll molecule (see Figure 1.3). The fraction of the total plant Mg associated with chlorophyll, however, is relatively small and only in the order of 15 to 20%.

A major function of Mg^{2+} is the formation of the bridge it brings about between ATP and an enzyme as shown in Figure 1.2. This bridging is of general importance because phosphorylation is catalyzed. In all organisms phosphorylation is a basic mechanism by which energy is tranferred and enzymic processes are enabled. A well known example of this kind is the phosphorylation of the Fe complex of the nitrogenase (Figure 3.38) by which energy is transferred and the e⁻ flow through nitrogenase induced. Phosphorylation occurs mainly on hydroxyl groups of amino acid residues which brings about a conformation change of the protein. The cytosolic Mg^{2+} required for the activation is in the range of several mol/m³ (Cohen 1989). Dephosphorylation catalyzed by phosphatases (kinase reactions) is also activated by Mg^{2+}. According to

Cohen (1989) dephosphorylation is a major mechansim for the control of intracellular events in eukaryotic cells. Most ATPases (H⁺ pumps) require Mg^{2+} for activation as is also the case for V-pyrophosphatase located in the tonoplast (Barkla and Pantoja 1996). In this reaction Mg^{2+} probably functions as an allosteric modulator (Rea and Poole 1993). Recent investigations of Pei *et al.* (1999) have shown that the cytosolic Mg^{2+} concentration controls the ion flux across the tonoplast *via* fast vacuolar channels in guard cells. Here Mg^{2+} ensures that fast channels do not function as a continuous vacuolar K⁺ leak, which would prohibit stomatal opening.

A key reaction of Mg^{2+} in photosynthesis is the activation of ribulose bisphosphate carboxylase (Rubisco). As has been already described on page 146, light triggers the import of Mg^{2+} into the stroma of the chloroplast in exchange for H⁺ thus providing optimum conditions for the carboxylase reaction. This relationship is illustrated in Figure 12.1 from Walker (1974). When the chloroplasts are illuminated H⁺ are driven into the inner space of the thylakoid and the thus established electrochemical gradient across the thylakoid membrane induces a contra flow of Mg^{2+} into the stroma. This process provides the conditions for the Rubisco activation, *i.e.* high pH and a high Mg^{2+} concentration in the stroma. The favourable effect of Mg^{2+} on CO_2 assimilation and subsequent processes such as sugar and starch production are probably the consequence of this activation of ribulose bisphosphate carboxylase. According to Barber (1982) Mg^{2+} is the most important cation species neutralizing the indiffusible anions of the thylakoid membrane.

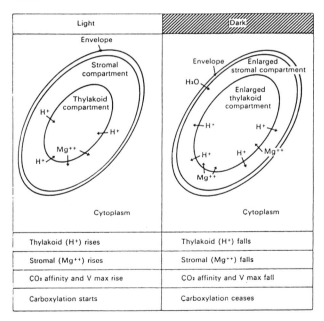

Figure 12.1 Mg^{2+} and H⁺ flow across the thylaoid membrane at light and dark providing optimum activation conditions for the activation of RuBP carboxylase- high pH and high Mg^{2+} concentration in the stroma- at light and blocking the carboxylation at night because of low pH and low Mg^{2+}concentration in the stroma (modified after Walker1974).

A further fundamental function of Mg^{2+} in all organisms is its essentiality for the structure and conformation of nucleic acids (Travers 1989). This is true for DNA and r RNA (ribosomal nucleic acid). Ribozymes are an important new class of enzymes which are made up entirely from RNA and in contrast to the usual enzymes not from amino acids (Pyle 1993). Ribozymes are located in the ribosomes and may bind to other RNA strands and are able to cleave the phosphodiester bond. All ribozymes are metalloenzymes which require divalent cations, particularly Mg^{2+}. In the ribozymes high Mg^{2+} concentrations are required for binding the tRNA strand to the ribosome (Ahsen and Noller 1995). From this it is obvious that an undersupply with Mg^{2+} will have a detrimental impact on polypeptide synthesis and thus on protein formation.

This conclusion is in agreement with earlier findings of Haeder and Mengel (1969) who reported that when plants are Mg deficient the proportion of protein N decreases and that of non-protein N increases. The effect is probably caused by dissociation of the ribosomes into their sub-units in the absence of Mg^{2+} (Watson1965). Magnesium ions appear to stabilize the ribosomal particles in the configuration necessary for protein synthesis and is believed to have a similar stabilizing effect in the matrix of the nucleus.

12.2.3 Magnesium deficiency

Magnesium deficiency symptoms differ between plant species although some general characteristics are apparent. Magnesium ions are mobile in the plant and deficiency always begins in the older leaves and then moves to the younger leaves as Mg^{2+} is phloem-mobile and translocated from older to younger leaves when supply is insufficient. Interveinal yellowing or chlorosis occurs and in extreme cases the intercostal areas become necrotic. Plate 12.1 shows Mg deficiency symptoms in a sugar beet leaf. This appearance is typical for a number of other dicotyledenous plants including grapes, field beans, bush beans, potatoes and tomatoes. In sugar beet the deficiency can often be mistaken for virus yellows. Individual leaves suffering from Mg deficiency are stiff and brittle and the intercostal veins are twisted. Magnesium- deficient leaves often fall prematurely. In cereals and the monocots in general the appearance of Mg deficiency is different. As in the case of the dicots, the water and carbohydrate metabolism of the plant is also affected and deficiency begins in the older leaves. With the cereals, however, the base of the leaf first shows small dark green spots of chlorophyll accumulation which are apparent against the pale yellow background colour of the leaf. In more advanced stages of deficiency the leaves become more chlorotic and striped. Necrosis occurs particularly at the tips of the leaf. The symptoms are the same for wheat, oats, rye and also for maize in the early stages. As the plants age, however, the leaves of maize take on a more spotted appearance.

An adequate supply of Mg depresses shoot and root growth. Assimilates accumulate in Mg deficient source leaves although there is controversy as to the cause of the accumulation. Fischer *et al.* (1998) reported increases in concentration of carbohydrates in both source and sink leaves of Mg deficient spinach plants and thus argue that a lack of carbohydrate supply can not have been the cause of growth restriction. These workers suggest that assimilate accumulation occurs in source leaves because of a lower sink demand and not because of impaired phloem loading through lack of Mg as proposed by Cakmak *et al.* (1994). It should also be born in mind as discussed above, that a primary effect of a lack of Mg^{2+} is to disturb polypetide synthesis in the ribosomes thereby restricting protein production and growth.

In Mg deficient chloroplasts the lack of Mg^{2+} to activate RuBP carboxylase impairs CO_2 fixation and leads to a shift to reductive O_2 activation, *i.e.* RuBP oxygenase (see page 151). Thus in Mg deficient leaves superoxide radicals are favoured which in turn enhance the activities of antioxidants such as ascorbate and H_2O_2 scavenging enzymes. Magnesium deficient leaves are thus highly photosensitive and photooxidation of the thylokoid membranes appears to be a major contributory factor to Mg deficiency chlorosis.

As might be expected marked differences occur in chloroplast structure caused by Mg^{2+}. In chloroplasts of *Phaseolus vulgaris* the grana are reduced in numbers, are irregular in shape and granal compartmentation is reduced or absent. In some cases starch grains are accumulated (Thomson and Weier 1962). Chevalier and Huguet (1975) studying the effect of Mg deficiency on the ultrastructure of chloroplasts of apple leaves found that inadequate Mg^{2+} supply resulted in a deformation of the lamellar structure. Plants suffering from Mg^+ deficiency are not only lower in chlorophyll, the concentration of carotinoids is also decreased. Plastoquinone and tocopherol concentrations are little affected (Basynski *et al.* 1980). In mitochondria Mg^{2+} deficiency restricts the development of cristae.These symptoms of ultrastructural disorganization precede visual Mg deficiency symptoms. *In vivo* concentrations of chlorophyll and thus also Mg concentrations are considerably higher in chloroplasts than in the cell as a whole (Hewitt and Smith 1975). It is therefore not surprising that chlorosis is often a first symptom of Mg deficiency. Pozuelo *et al.* (1984) reported that Mg deficiency led to an increase in suberization in the endodermis and hypodermis of maize roots.

In leaf tissue the threshold value for the occurrence of deficiency symptoms is in the region of about 2 mg Mg/g dry matter, although this is dependent on a number of factors including plant species.

Plate 12.1 Mg deficiency symptoms in a sugar beet leaf.

In grapes a syndrome is observed which is related to Mg nutrition and which is called in the French literature "dessechment de la rafle" which means drying out of the petioles of a grape which occurs during the veraison (Delas 1992). The petioles of the berries, the main axis of the grape and its ramification become necrotic.The necrosis appears in the peripheral cell layers and is associated with plasmolysis. The drying out, however, is never total and the vessels still allow some solute transport. The disease very much affects grape yield and quality. There are some indications that the disease is related to the Mg^{2+} supply to the vine. Vines with Mg deficiency on the leaves as well as vines which have received heavy K fertilizer application are particulary prone to the drying out of the grape petioles. Petioles thus affected are low in Mg and Ca but high in K concentrations. Magnesium application to the grapes cures the disease (Delas 1985). The effect of spraying the grapes with a Mg sulphate solution on the percentage of grapes affected is shown in Table 12.2.

Table 12.2   Effect of spraying grapes with a Mg sulphate solution on the percentage of grapes affected by the " dessechment de la rafle" (Drying out of grape petioles, after Delas 1985)

Treatment	% of grapes with the disease
Control	50.2
3 times application of a 12% MgSO$_4$ solution	7.3

Critical Magnesium concentrations in plant tisues and Mg deficiency of numerous crop species are precisely described and shown in coloured pictures in a useful mongraph by Bergmann (1994). Problems of Mg deficiencies and Mg functions in important crop species are treated by Huguet and Coppenet (1994) in a useful booklet.

12.3   Magnesium in Crop Nutrition

12.3.1   Crop requirements and critical levels

The amounts of Mg^{2+} taken up by some important crop plants is shown in Table 6.2. For arable crops the average uptake is in the region of about 10-25 kg Mg/ha/yr and generally the uptake of root crops is about double that of cereals. Sugar beet, potatoes, fruit and glasshouse crops are particularly prone to Mg deficiency. In recent years the importance of Mg as a fertilizer has increased. Previously Mg was applied unwittingly as an impurity along with other fertilizers. The high purity of fertilizers used at the present day, however, means that this source of Mg application to the soil no longer exists. Increased crop yields resulting from higher applications of non-Mg containing fertilizers have also placed a greater demand on the soil for Mg. High levels of K^+ or NH_4^+ as already mentioned restrict Mg^{2+} uptake by plants. For these reasons Mg deficiency in crop plants is becoming more frequent and Mg applications are now common. Deficiencies occur particularly in highly leached organic acid soils or on sandy soils which have been given heavy dressings of lime. In some cases Mg deficiency occurs on soils high in K. The presence of high concentrations of K^+, NH_4^+ or Ca^{2+} or combinations of these ions restrict the uptake of Mg^{2+}. In acid soils, cationic Al species are believed to depress Mg^{2+} uptake markedly (Grimme 1983) while on calcareous soils the same effect is brought about by an excess of soluble Ca^{2+}.

Numerous experiments have been carried out to investigate Mg relationships between plant and soil under Mg deficient conditions. Prince *et al.* (1947) concluded that, if Mg^{2+} constitutes less than 6% of the CEC, crops are likely to show response to Mg fertilizer application. In experiments with sugar beet Tinker (1967) obtained responses in soils with CECs of 5 to 10 cmol/kg soil when they contained less than 0.2 cmol/kg exchangeable Mg^{2+} (2 to 4% of the CEC = 24 to 48 mg/kg exchangeable Mg^{2+}).

From a survey of 60 field experiments Draycott and Durrant (1971) have suggested a limit of 35 mg/kg soil of exchangeable soil Mg^{2+} and 0.4 mg Mg/g dry matter of the leaf as critical levels above which no further yield increases result from Mg-fertilizer. In an advisory paper published in the UK agricultural advisory service Mg application is recommended for all crops growing on soils with less than 25 mg/kg exchangeable Mg^{2+} and to susceptible crops when it is less than 50 mg/kg (MAAF 1979). Above these levels Mg applications are only necessary when K levels are high, or when hypomagne-saemia may occur in animals or when glasshouse and fruit crops are being grown. These findings agree fairly well with the data reported in a review by Doll and Lucas (1973). Best Mg fertilizer responses are generally found on light sandy soils. Thus Dam Kofoed and Hojmark (1971) reported that on sandy soils in Denmark, fodder beet, swedes, and potatoes all gave high yield responses to Mg application. The effect was less spectacular in cereals. Slight Mg deficiency in cereals during vegetative growth does not always result in a reduction in yield. Grain yield depressions do occur, however, when Mg deficiency symptoms are present on the flag leaves or ears (Pissarek 1979). The relationship between the Mg concentrations in shoots of oats at the time of culm elongation and final grain yield are shown in Figure 12.2.

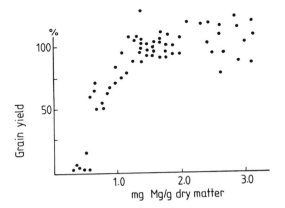

Figure 12.2 Relationship between the relative grain yield of oats at maturation and the Mg concentration in the shoots at the onset of culm elongation (modified after Pissarek 1979).

Uebel (1996) reported that young beech (*Fagus silvatica*) established under pine (*Pinus silvestris*) growing on a sandy soil poor in available Mg^{2+} and K^+ showed typical deficiency symptoms of Mg and K in the treatment not fertilized with KMg fertilizer. The deficiency was aggravated by the atmospheric deposition of nitrogen. Young beech plants which were fertilized in the first and second year after planting with a total of 120 kg K/ha and 60 kg Mg/ha showed no deficiency symptoms, were more resistant to water stress and late frost in spring and thus had less breakdown (loss). The Mg fertilizer application significantly increased the Mg concentration in the leaves from about 1 to 2 mg Mg/g dry matter. Altherr and Evers (1975) obtained growth responses to Mg fertilization by spruce, grown on a soil on bunter sandstone in Germany. Magnesium deficiency in spruce and fir grown on acid granite and phyllite sites was also reported by Zech and Popp (1983). The chlorotic needles were characterized by yellow tips. The Mg concentrations of these yellow needles were 0.25-0.27 mg Mg/g dry matter, whereas those of the green needles were about twice as high. Application of sulphate of potash magnesia to the soil brought about regreening of the affected leaves. The damage in forest trees now widely spread in Central Europe and attributed to acid rain appears to be related to the lower concentrations of K and Mg in the needles of the affected trees (Huettl 1984). Zöttl and Mies (1983) observed that spruce needles exposed to the light were prone to yellowing while needles growing in the shade remained green. In the yellow needles the chloroplasts were severely damaged. Needle analysis revealed that the yellow needles had significantly lower Mg and Zn concentrations while the concentrations of other elements were not much affected (see Table 12.3). The authors hold the view that in needles exposed to sunlight, photooxidants may destroy the chloroplasts and the chlorophyll, which in turn supposedly increases the leaching of Mg^{2+} from the needles. Spruce trees with yellow needles low in Mg occurred especially on soils low in exchangeable Mg^{2+}. Zöttl and Mies (1983) suggest that on such sites the Mg^+ leached from the needles by rain cannot be fully compensated for by Mg^{2+} uptake.

Table 12.3 Concentrations of some elements in four year old green and yellow needles of *Picea abies* (after Zöttl and Mies1983)

	Green	Yellow		Green	Yellow
	mg/g DM			μg/g DM	
N	10.2	10.0	Mn	515	784
P	2.31	2.45	Fe	99	84
K	8.1	9.6	Zn	40	24*
Ca	3.19	3.39	Cu	2.9	4.4
Mg	0.64	0.22*	Al	486	471
S	1.74	1.56	Pb Cd	1.53 0.09	1.86 0.07

* significant difference to the green needles.

12.3.2 Magnesium fertilizers

The major Mg fertilizers used and their approximate Mg concentrations are shown in Table 12.4. Magnesium is supplied in most cases as carbonate, oxide or a sulphate. In general sulphate fertilizers are more rapidly effective than carbonate fertilizers but are also more expensive. Applications of dolomitic limestone are particularly useful on acid soils which need regular liming. Decomposition of the dolomite is also assisted by low soil pH. On more neutral soils $MgSO_4$ *e.g.* kieserite is more appropriate particularly on arable land where high levels of Mg are rapidly required.

Table 12.4   Mineral Mg fertilizers

	Mg, g/kg
Magnesian limestone (Mg carbonate)	30—120
Ground burnt magnesian lime (Mg oxide)	6—200
Kieserite ($MgSO_4 \cdot H_2O$)	160
Epsom salts ($MgSO_4 \cdot 7\ H_2O$)	96
Sulphate of potash magnesia ($K_2SO_4 \cdot MgSO_4$)	66
Magnesite ($MgCO_3$)	270

The various forms of $MgSO_4$ differ considerably in solubility. Epsom salts $MgSO_4 \cdot 7\ H_2O$ although more expensive is more soluble than kieserite ($MgSO_4 \cdot H_2O$). This has a practical significance, for, as has been pointed out by Cooke (1972), whilst 500 kg/ha $MgSO_4$ as kieserite applied to soil may be needed to prevent Mg deficiency in tomatoes, the trouble can be controlled by spraying 35 kg/ha of Epsom salt ($MgSO_4 \cdot 7\ H_2O$) dissolved in 400 L water applied on a number of occasions over the growing season. Fertilizers containing only smaller percentages of Mg such as kainite, basic slags and some PK and NPK fertilizers, are useful in maintaining the Mg level of the soil. In cases, however, where Mg deficiency is suspected higher graded Mg fertilizers are preferred (see Table 12.4). Yamauchi and Winslow (1989) cured discoloration of rice grain grown as upland rice on Ultisos in Nigeria by Mg sulphate application. The application of Mg particularly in combination with Si increased the grain yield considerably. Another Mg fertilizer not included in the sulphate or carbonate group is magnesium ammonium phosphate. This is a sparingly soluble salt used in horticulture particularly for young valuable plants sensitive to other forms of Mg treatment. The relative values of different Mg fertilizers as well as the use of other Mg sources including farm yard manure, basic slags, and liming materials have been thoroughly discussed by Cooke (1972) and by Darre and Gilleron (1992). The rates of Mg application on sandy soils are in the range of 50 to 100 kg Mg/ha. These rates have resulted in substantial yield increases of various arable crops on sandy soils in Denmark (Dam Kofoed and Hojmark 1971). Potatoes in particular generally show a marked response to Mg treatment (Jung and Dressel 1969). Magnesium application is also important for pastures in relation to animal nutrition. Intensive grassland management frequently results in herbage with low concentration of available Mg which does not meet the demand of milking cows so that animals suffer from grass tetany or staggers (hypomagnesaemia). The problem of hypomagnesaemia and the functions of Mg in the animal metabolism have been treated by Rayssiguier (1992) in an interesting article.

CHAPTER 13

IRON

13.1 Soil Iron

Iron makes up about 5% by weight of the earth's crust and is invariably present in all soils (see Table 11.1). The greatest part of soil Fe usually occurs in the crystal lattices of numerous minerals. The primary minerals in which Fe is present include the ferromagnesian silicates such as olivine, augite, hornblende and biotite. In biotite and illite Fe is located in the centre of the octahedra and may be di-or trivalent (Fanning *et al.* 1989). By weathering of these minerals Fe oxides such as goethite, haematite, and ferrihydrate are formed. The solubility of these Fe^{III} oxides is extremely low (Schwertmann 1991). Goethite (α-FeOOH) is the most widespread Fe mineral in soils. Together with other Fe oxides it greatly influences soil colour which is between yellowish brown and brown (Allen and Hajek 1989). Haematite (α-Fe_2O_3) is also a widespread Fe oxide in soils particularly in well-drained tropical soils. In Oxisols and Ultisols it represents a signifcant proportion of the clay fraction. Ferrihydrate ($HFe_5O_8 \cdot 4H_2O$) has a high specific surface and for this reason may be of importance as an Fe source for plants. As a consequence of their high stability Fe oxides accumulate during oxidative weathering as hydrous oxides in the clay fraction. Thus in soils at an advanced stage of oxidative weathering as is the case of lateritic soils, these oxides together with Al oxides and kaolinite predominate in the profile. According to Chen and Barak (1982) the solubility of Fe oxides/hydroxides decreases in the following sequence: $Fe(OH)_3$ amorphous > $Fe(OH)_3$ in soils > γ-Fe_2O_3 maghaemite > γ-FeOOH lepidocrocite > α-Fe_2O_3 haematite > α-FeOOH goethite.

The concentration of soluble Fe in soils is extremely low in comparison with the total Fe concentration. Soluble inorganic forms include Fe^{3+}, $Fe(OH)^{2+}$, $Fe(OH)_2^+$ and Fe^{2+}. In well aerated soils, however, Fe^{2+} contributes little to the total soluble inorganic Fe except under high soil pH conditions. Iron solubility is largely controlled by the solubility of the hydrous Fe^{III} oxides. These give rise to Fe^{3+} and its hydrolysis species (Lindsay 1991):

$$Fe^{3+} + 3OH^- \Leftrightarrow Fe(OH)_3 \text{ (solid)}$$

The equilibrium is very much in favour of $Fe(OH)_3$ precipitation and is highly pH dependent, the activity of Fe^{3+} falling with increasing pH. At higher pH levels Fe^{3+} activity in solution decreases 1000 fold for each pH unit rise (see Figure 13.1). In this high pH range (pH 7 to 9) $Fe(OH)_2^+$, $Fe(OH)_3$ and $Fe(OH)_4^-$ are formed. The solubility level reaches a minimum in the pH range between 7.4 to 8.5 (Lindsay and Schwab 1982). Acid soils are thus relatively higher in soluble inorganic Fe than calcareous soils where levels can be extremely low.

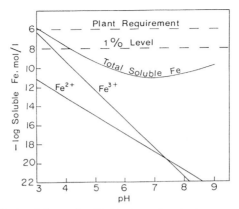

Figure 13.1 Solubility of Fe in relation to the pH. The dotted lines indicate 100% and 1% plant requirement (after Lindsay 1974).

When soils are waterlogged a reduction of Fe^{III}, to Fe^{2+} takes place accompanied by an increase in Fe solubility. By this process insoluble Fe^{III} compounds become soluble and Fe^{2+} is dissolved in the soil solution, the arabic superscript, *i.e.* Fe^{2+}, indicates that the ion is soluble, the roman superscript, *e.g.* Fe^{III}, that the compound is insoluble. Reduction is brought about by anaerobic bacteria which use Fe oxides as e- acceptors in respiration (Munch and and Ottow 1983). A close contact between the bacteria and the Fe oxides is required for this process. Amorphous Fe is preferred but goethite, haematite and lepidocrocite can be reduced by the action of microbes. This process of Fe reduction is of particular importance in paddy soils where rather high Fe^{2+} concentrations can result. This can often produce toxic effects in rice plants, known as bronzing. In soils subjected to anaerobic conditions the ratio of activities of Fe^{3+}/Fe^{2+} can be an important parameter in relation to crop growth. This ratio can be assessed by measurement of the redox potential according to the equation

$$E = 0.77 + 0.059 \log \frac{a^{Fe\,3+}}{a^{Fe\,2+}}$$

Under anaerobic soil conditions reducing processes are favoured. Hydrous Fe oxides give rise to Fe^{2+} (Ponnamperuma 1972) according to the equation

$$Fe\,(OH)_3 + e^- + 3H^+ \Leftrightarrow Fe^{2+} + 3H_2O$$

From this equation it is evident that the reduction of Fe^{3+} to Fe^{2+} is associated with the consumption of H^+ and thus with an increase in pH. The reverse is the case as soil aeration is increased, a fall in pH being accompanied by the oxidation of Fe^{2+} to Fe^{3+}.

Differences in redox potential can often be observed in the same profile. In the deeper soil layers which are less well aerated, the fraction of Fe^{2+} of the total soluble Fe is frequently higher than in the upper horizons. The redox potential thus generally falls from the upper to the lower horizons.

An important feature of Fe both in soils and plants is the way it readily forms organic complexes or chelates which are called siderophores and are synthesized by bacteria, fungi and plants. They are of of crucial importance for the Fe transport in soils and the Fe supply of plants (Crowley *et al.* 1991, Masalah *et al.* 2000). To date more than 100 different siderophores are known which are produced by bacteria and fungi. The chelator produced by microbes is an organic molecule with a high affinity for Fe, mainly Fe^{III}, and removes the Fe from Fe bearing minerals and contributes to their dissolution. These Fe-chelates (siderophores) are highly soluble and are stable over a wide pH range. This is the reason why Fe is also fairly mobile in soils of high pH including calcareous soils. The structure of these siderophores is different but in all cases Fe is bound by several bonds to the chelator (Neilands and Leong 1986). Numerous siderophores are derived from hydroxamate in which hydroxamic acid (R-CO-NH-OH) binds Fe^{III} with an ionic bond and with a coordinative bond (Figure 13.2) thus forming a ferric mono hydroxamate. Three hydroxamates complex the Fe^{III} to the ferric bi-hydroxamate siderophore, which is common in many fungal species. Such microbial siderophore play an important role in supplying plant roots with Fe (Masalha *et al.* 2000). Siderophores synthesized by plants (phytosiderophores) are derived from nicotianamine (see Figure 13.2), a non-protein amino acid with similarities in structure to the phytosiderophores. In contrast to the siderophores, however, it complexes with Fe^{2+} and in this form plays an essential role in the Fe transport in the symplasm and phloem (Becker *et al.* 1992).

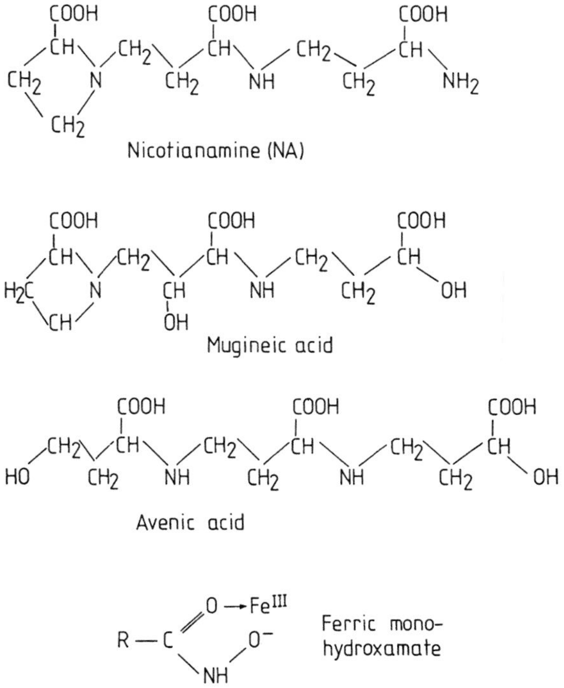

Figure 13.2 Nicotianamine, mugineic acid, avenic acid, and Fe^{III} hydroxamate.

13.2 Iron in Physiology

13.2.1 Uptake and translocation

There is much experimental evidence and it is now generally accepted that the transport of Fe across the plasmamembrane is closely linked to Fe^{III} reduction as shown in Figure 13.3 (Crowley *et al.* 1991). It is Fe^{2+} which is taken up (Fox *et al.* 1996) and which passes through a specific channel of the plasmamembrane (Fox and Guerinot 1998). Fe^{III} siderophores produced in the soil are transported to the roots by diffusion or mass flow and enter the root free space through which they move to the plasmalemma bound Fe^{III} reductase.

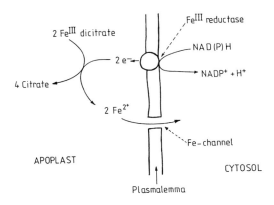

Figure 13.3 Reduction of Fe^{III} dicitrate followed by Fe^{2+} uptake through an Fe channel in the plasmalemma.

As soon as the reductase accepts the e- from the Fe of the siderophore the structure of the siderophore breaks down and Fe^{2+} is released which may then pass through the specific channel of the plasmalemma to enter the cytosol. The reducing equivalents are supplied by NAD(P)H in the cytosol (Rubinstein and Luster 1993, Brüggemann and Moog 1989). This type of Fe uptake occurs in most plant species, but for grasses another mechanism of Fe uptake is operative. As first reported by Takagi (1976) rice and barley roots excrete phytosiderophores, such as mugineic and avenic acid, which are capable of mobilizing Fe^{III} from soil minerals to be taken up by the plant thus exploiting the soil for Fe (Takagi *et al.* 1984). Mugineic acid and its derivatives, avenic acid, 3-hydroxymugineic acid, distichonic acid, are nonprotein amino acids (Figure 13.2) with a unique structure. They have six functional groups by which they bind Fe^{III} in a hexadentate fashion with three carboxyl groups, two N atoms and one OH (Ma and Nomoto 1996). Nicotianamine has only one N atom and presumably for this reason binds Fe^{2+} and not Fe^{3+} (Figure 13.2). Generally the release of mugineic acids takes place during the day with greatest release in the morning. According to Marschner *et al.* (1987) phytosiderophores are secreted by root tips a few mm behind the apex. Siderophore secretion appears to take place as a response to Fe deficiency (Marschner *et al.* 1987, Nishizawa *et al.* 1989) but the regulatory mechanism of synthesis and secretion in relation to the Fe demand is not yet understood. Plant species and cultivars with a high capacity for secreting mugineic acids are able to cope well with the low Fe availability of calcareous soils. The potential of various graminaceous monocots to release phytosiderophores thereby improving the acquisition of Fe follows the sequence:

Barley = wheat > oat = rye > maize = sorghum > rice.

It is supposed that the Fe-phytosiderophore molecule is taken up by a specific transporter located in the plasmalemma. However, neither the transporter nor the genes coding for it have been isolated to date (Ma and Nomoto 1996). In contrast to Fe uptake by dicotyledons Fe uptake in form of phytosiderophores by graminaceous monocots is largely independent of soil pH which is a particular advantage on calcareous soils. Phytosiderophores are quickly decomposed by soil microbes, and siderophores produced by soil bacteria and fungi also contribute to the Fe supply of grasses (Crowley *et al*. 1991). Despite the greater efficiency of the graminaceous species and especially wheat to acquire Fe from the soil, Berg *et al*. (1993) reported the occurrence of Fe chlorosis in various wheat cultivars grown on calcareous soils. Large differences in the degree of Fe chlorosis between cultivars were found with some cultivars showing no chlorosis, results which suggest intercultivar differences in phytosiderophore response to Fe acquisition.

Fe^{III} reduction requires a pH of about 5 at the apoplastic site of the reductase. Although the apoplast is well H^+ buffered at pH 5, a high pH in the apoplast may occur in calcareous soils which may restrict Fe^{III} reduction and thus Fe uptake. The effect of nutrient solution pH on chlorophyll concentration in leaves and Fe concentrations in roots and leaves of sunflower plants is shown in Table 13.1 from the work of Kosegarten *et al*. (1998).

Table 13.1   Effect of pH in the nutrient solution and N source on the Fe concentrations in leaves and roots and chlorophyll concentration in the youngest leaves of *Helianthus annuus* (Kosegarten *et al*. 1998)

Treatment	Chlorophyll, mg/g fr.wt.	Leaf, μg Fe/g dry wt.	Root, μg Fe/g dry wt.
NH_4NO_3, pH4, control	1.50	70.8	156
NH_4NO_3, pH 6.5	0.95***	32.5**	307*
NO_3^-, pH 6.0	0.28***	29.1**	351**

*, **, *** = significantly different from the control at p < 0.05, 0.01 and 0.001, respectively.

At pH 4 in the nutrient solution leaves were green and the leaf Fe concentration was within a normal range whereas at pH 6.5, leaves were chlorotic and and leaf Fe concentration was significantly lower than in the control. For roots the reverse was true; here the high pH in the nutrient solution led to significantly higher Fe concentrations as compared with the control. In the treatment feeding plants exclusively with nitrate the Fe nutritional status of plants was even poorer, with low Fe concentrations in the leaf, highly chlorotic leaves and a high Fe concentration in the roots. The pH in root apoplast differs from that in the nutrient solution because of the high H^+ buffer power of the apoplast. As has been found by various authors high pH in the nutrient solution restricts Fe^{III} reduction (Toulon *et al.* 1992, Susin *et al.* 1996). Manthey *et al.* (1996) reported that the reduction Fe^{III} HEDTA, an artificial siderophore, was highest at pH 5.3 in the outer solution and declined as the pH increased. At pH 7.5 the amount reduced was only a small fraction of that at the maximum at pH 5.3.

As can be seen from the data in Table 13.1 the restricted Fe reduction was associated with an enrichment of Fe in the roots. Presumably this Fe was mainly located in the apoplast contributing to a supply of Fe to the shoot under particular conditions (Bienfait *et al.* 1985, Longnecker and Welch 1990, Becker *et al.* 1992). Such an accumulation of Fe in the root apoplast has not only been found in plants grown in solution culture where Fe is generally supplied to plants as artificial Fe siderophores but was also in plants grown on soils (Masalah *et al.* 2000) and was exceptionally high in the roots of vine and peaches grown in the field on calcareous soils where chlorosis is common (Mengel *et al.* 1994).

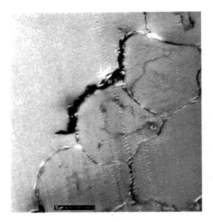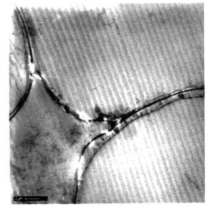

Plate 13.1   Micrograph of epidermal root cells with Al, Si and Fe compounds (dark area) adsorbed to the peripheric cell wall (left hand). Rigt hand: Al, Si and Fe signals from the radial cell walls which means that low molecular Al, Si and Fe compounds have been diffused into the root apoplast. Left hand bar = 5 μm, right hand bar = 2 μm. (EDXA) by courtesy of Schröder.

Form the micrograph shown in Plate 13.1 it is evident that Fe may accumulate at the outer walls of epidermal cells but is also present in the radial cell walls of the peripherical cell layers. This means that Fe has diffused into the root apoplast. From these findings it may be concluded that the chlorosis frequently appearing on calcareous and saline soils with a high pH is not a consequence of low Fe solublity in the soil due to high soil pH (Lindsay 1991) but results from the high pH in the soil solution which permeates the root apoplast to retard Fe^{III} reduction. According to Rubinstein and Luster (1993) Fe^{III} reduction is the limting step in Fe uptake. Thus depression in activity of the reducing enzyme is one reason why many plant species suffer from Fe deficiency under alkaline conditions. Obviously on these soils the microbial production of siderophores is high enough to mobilize sufficient soil Fe to be translocated by diffusion or mass flow to the roots. Iron thus accumulated in the apoplast of roots is very strongly bound (Masalah *et al.* 2000) but can be released by reduction. The nature of such a strong Fe^{III} binding to the root apoplast is not yet understood.

Plants respond to a lack of Fe availablity using different strategies. Plant species which are able to produce siderophores such as grasses respond to Fe deficiency by a strong increase in the release of siderophores (Römheld and Marschner 1986, Marschner *et al.* 1989, Nishizawa *et al.* 1989, Ma and Nomoto 1996) which leads to an enhanced mobilization of soil Fe. Dicotyledons are not able to synthesize siderophores but excrete H^+ into the outer solution which induces the dissolution of insoluble Fe compounds. Protons, released by roots into calcareouas soils, however, are immediately neutralized by the extremely high H^+ buffer power of such soils (Hauter and Mengel 1988) and there is no possibility of dissolving Fe^{III} compounds in the soil. However, H^+ pumped into the root apoplast by the plasmalemma H^+ pump may well depress the pH at microsites of the apoplast so that Fe reduction may proceed. According to Kramer *et al.* (1980) and to Landsberg (1984) both dicotyledons and grasses develop rhizodermal transfer cells which are characterized by invaginations of cell walls and hence by an enlargement of the plasmamembrane to increase the number of plasmalemma H^+ pumps to pump H^+ into the apoplast and thus improve Fe reduction. In addition membranes from deficient plants show a higher Fe reduction power than from plants well supplied with Fe (Holden *et al.* 1991, Susin *et al.* 1996). From this observation one may conclude that the Fe^{III} reduction capacity may also respond to an insufficient Fe supply. Romera *et al.* (1991a) testing various rootstocks of peaches found that roots of chlorosis- resistant cultivars had a higher Fe^{III} reduction power than cultivars sensitive to Fe chlorosis.

Long- distance transport in the xylem occurs as Fe^{III} dicitrate together with an excess of citrate so that all Fe^{III} is complexed by citrate. The Fe^{2+} in the cytosol is presumably oxidized in transport into the xylem (Brown and Jolley 1989). In the phloem and in the symplasm Fe is transported as an nicotianamin Fe^{II} complex (Becker *et al.* 1992, Stephan and Scholz 1993). According to Maas *et al.* (1988) Fe concentration in the phloem reflects the Fe nutritional status of leaves and so may serve as a signal for Fe stress.

13.2.2 Biochemical functions

Iron enters the cytosol as ferrous Fe (Fe^{2+}) which may be toxic since it may easily react with O_2 forming the superoxide radical (van der Mark *et al.* 1981):

$$Fe^{2+} + O_2 \Rightarrow O_2^- + Fe^{3+}$$

The Fe^{2+} concentration must therefore be maintained at a low level and the excess Fe^{2+} is exported into plastids including chloroplasts where it is oxidized by means of O_2. The Fe^{3+} produced reacts with H_2O to form Fe^{III} hydroxide (Laulhere and Briat 1993):

$$Fe^{3+} + 3\ H_2O \Rightarrow Fe^{III}(OH)_3 + 3H^+$$

The Fe hydroxide thus produced is the basic unit of ferritin which is an essential Fe storage form in animals, plants, fungi, and bacteria. Ferritin is a macroprotein Fe complex the core of which consists mainly of Fe hydroxy oxide ($Fe^{III}OOH$) which is surrounded by a protein layer so called apoferritin. The core may comprise several thousand Fe atoms (Theil 1987). Depending on the Fe requirement of the metabolism or the surplus of Fe entering the cell, Fe may become incorporated or released from ferritin. The processes are coupled with redox reactions in which ascorbate plays an essential role as shown in Figure 13.4 (Laulhere and Briat 1993). Owing to ferritin deposition the Fe concentration in chloroplasts may be 5 to 10 times higher than in the total corresponding leaf tissue (Mengel and Bübl 1983).

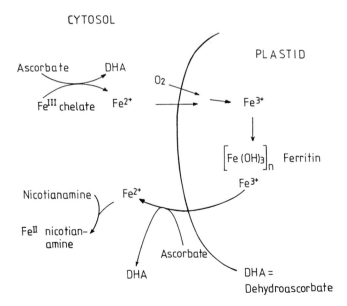

Figure 13.4   Uptake of Fe^{2+} by the plastid, ferritin synthesis in the plastid and release of Fe^{3+} into the cytosol (after Laulhere and Briat 1993).

The Fe provided to the plastid in Figure 13.4 is an Fe^{III} chelate, *e.g.* Fe^{III} citrate, which is reduced by means of ascorbate and the resulting Fe^{2+} enters the plastid where it is oxidized by O_2 and the superoxide thus produced detoxified by a superoxide dimutase. The Fe^{3+} thus obtained reacts with water to produce Fe^{III} hydroxide from which the ferrihydroxy oxide is formed:

$$Fe(OH)_3 - H_2O \Rightarrow FeOOH$$

Release of Fe from ferritin is brought about by the reduction of Fe^{III} by ascorbate and the Fe^{2+} so produced is complexed by nicotianamine which may transport Fe to sites where it is required for synthetic reactions.

The tendency for Fe to form chelate complexes and its ability to undergo a valency change are the two important characteristics which underlie its numerous physiological effects. The most well known function of Fe is in enzyme systems in which haem or haemin function as prosthetic groups (see Figure 1.3). Here Fe plays a somewhat similar role to Mg in the porphyrin structure of chlorophyll. These haem enzyme systems include catalase, peroxidase, cytochrome oxidase as well as the various cytochromes. The Fe centred in the porphyrin ring may accept or release an e^- thus enabling e^- transport as in the photosynthetic e^- - or respiration e^- transport chain. The e^- provided by the Fe porphyrin complex may reduce the different substrates.

Catalase brings about the breakdown of H_2O_2 to water and O_2 as shown below:

$$H_2O_2 + H_2O_2 \Rightarrow 2 H_2O + O_2$$

The enzyme plays an important role in the chloroplasts along with the enzyme superoxide dismutase (see page 150) and in photorespiration and the glycolytic pathway. Peroxidases are widespread and also catalyze a H transfer in which the H acceptor is also H_2O_2:

$$H_2O_2 + RH_2 \Rightarrow 2 H_2O + R$$

There is evidence that cell wall bound peroxidases catalyse the polymerization of phenols to lignin. Peroxidase activity is particularly depressed in Fe deficient roots (Römheld and Marschner 1981b).

As well as haem Fe, iron sulphur proteins play a major role in oxidoreduction. Both binuclear Fe-S clusters (2Fe-2S) and tetranuclear Fe-S clusters (4Fe-4S) occur. Each cluster is surrounded by four cysteine residues associated to a polypeptide chain. These clusters are known as ferredoxins if they act exclusively as electron carriers, are characterized by a very high negative redox potential (see page 142). The significance of ferredoxin as as redox system in photosynthesis as well as in nitrite reduction, sulphate reduction and N_2 assimilation has already been described (see page 167 and Figure 3.20). Nonhaem-iron proteins are widely distributed in photosynthetic and nonphotosynthetic organisms.

Nitrogenase comprises an Fe protein and an FeMo protein the latter being involved in the reduction of N_2 (see page 169). The Fe protein binds the Mg nucleotides (Mg-ATP or Mg-ADP). The most important feature of both these proteins is their extreme lability of conformation to oxygen (Howard and Rees 1994). A further Fe containing enzyme of universal importance is ribonucleotide reductase which brings about the reduction of ribonucleotide diphosphate (ADP; GDP, CDP, TDP) to deoxy-ribonucleotide (Reichard 1993). In this reaction the metabolites for DNA synthesis are produced. Iron is therefore an essential element for the synthesis of genes and without this reaction no nuclear growth can occur. The depressed growth of leaves insufficiently supplied with Fe as reported by Mengel and Malissiovas (1981) and the inhibition of the formation of new leaves (Kosegarten *et al.* 1998) under the condition of insufficient Fe supply may be related to a retarded synthesis of DNA. Plants inadequately supplied with Fe are characterized by small leaves which may be yellow (chlorotic) and some leaves remain green but with restricted growth (Mengel and Malissowas 1981, Kosegarten *et al.* 1998).

Applying radioactive ^{59}Fe to tomato plants suffering from Fe chlorosis Machold and Scholz (1969) observed that the distribution of ^{59}Fe in the leaves corresponded exactly to the areas in which regreening occurred. This is shown in Plate 13.2. It is not surprising from this kind of evidence that the search for a possible function of Fe in the role of chlorophyll formation has received considerable attention. The metabolic pathway involved in chlorophyll formation is shown in Figure 13.5. The same pathway is also operative in the biosynthesis of haem. In Fe deficiency a decrease has been observed in the rate of condensation of glycine and succinyl CoA to form δ-amino-laevulinic acid (ALA) the precursor of porphyrins.

Plate 13.2 Uptake of ^{59}Fe by a chlorotic tomato leaf. The distribution of ^{59}Fe in the autoradiograph (above) corresponds exactly to the area of the leaf in which regreening occurred (below). (Photo Machold and Scholz).

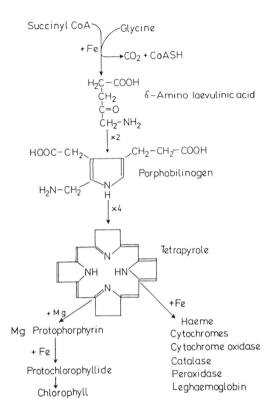

Figure 13.5 Role of Fe in the biosynthesis of chlorophyll and haem containing enzymes.

Results of Miller *et al.* (1982) support the view that the rate of ALA formation is controlled by Fe. In addition Machold and Stephan (1969) have reported that Fe is also necessary in the oxidation step from coproporphyrinogen in chlorophyll synthesis. Pushnik et al. (1984) found that Fe chlorotic leaves were very low in chlorophyll bearing proteins. Terry (1980) has shown that in Fe deficient leaves the rate of photosynthesis decreased per unit area but not per unit chlorophyll indicating that the photosynthetic apparatus remains intact but the number of photosynthetic units was decreased. His results show that as the intensity of Fe deficiency was increased and the chlorophyll per unit leaf area fell, protein content per leaf area, leaf cell volume and chloroplast number were all unaffected, but that chloroplast volume and the amount of protein per chloroplast fell dramatically.

The possible involvement of Fe in protein metabolism has been suspected from the findings of a number of authors who have observed that in Fe deficiency the protein fraction decreased simultaneously with an increase in the level of soluble organic N compounds (Perur *et al.* 1961). If as outlined above DNA synthesis is restricted as a result of Fe deficiency, this will be followed by an inhibition of RNA synthesis including ribosomal RNA which is required for the synthesis of proteins. This is in agreement with earlier results of Price *et al.* (1972) who reported that in Fe deficient algae, the chloroplasts contained less than half the chloroplast RNA and chloroplast ribosomes of the non deficient controls.

Iron deficiency and toxicity

The deficiencies of Fe and Mg are somewhat similar as both are characterized by a failure in chlorophyll production. Iron deficiency, however, unlike Mg deficiency always begins in the younger leaves. In most species chlorosis is interveinal and a fine reticulate pattern can often be observed in the newly formed leaves, the darker green veins contrasting markedly against a lighter green or yellow background. The youngest leaves may often be completely yellow and totally devoid of chlorophyll. Leaf growth is restricted and the formation of new leaves may be completely inhibited (Kosegarten *et al.* 1998). Iron deficency may not always be characterized by chlorotic leaves but simply by a reduction of leaf growth (Kosegarten *et al.* 1998). Besides leaf growth, root growth is also much depressed in iron deficient plants (Bertoni *et al.* 1992).

In the leaves of cereals Fe deficiency is shown by alternate yellow and green stripes along the length of the leaf. As high concentrations of Fe occur in chloroplasts it is not surprising that Fe deficiency causes marked changes in their ultrastructure. This is well shown in the electron micrograph of chloroplasts from pepper leaves (Plate 13.3) from the work of Hecht -Buchholz and Ortmann. This electron micrograph clearly shows that in the chloroplast adequately supplied with Fe, numerous thylakoid grana were formed whereas in the Fe deficient chloroplast, thylakoid grana were lacking. Spiller and Terry (1980) have also reported that Fe deficiency causes a disturbance in the synthesis of thylakoid membranes and a depression of photochemical capacity. Fe deficient plants are characterized by the accumulation of organic anions, amino acids and nitrate. This accumulation is due to the inhibition of growth including protein synthesis as already discussed above. Since growth is inhibited lower amounts of assimilates are needed including NADH and ATP and hence anions of the tricarboxylic acid cycle such as citrate and malate may accumulate.

Iron toxicity is particularly a problem in flooded rice soils. Within a few weeks of flooding, the level of soluble Fe^{2+} in the soil may increase from 0.1 to $50 - 100$ mg/kg Fe and large amounts of Fe can be taken up by plants (Ponnamperuma1978). Iron toxicity in rice is known as *bronzing*. In this disorder the leaves are first covered by tiny brown spots which develop into a uniform brown colour. This frequently occurs in rice leaves containing excessively high Fe concentrations in the range of 300 to 1000 μg Fe/g dry weight (Ottow *et al.* 1983). Iron toxicity is known in various rice growing areas and is especially frequent on heavy soils (Tanaka *et al.* 1973) and is often associated with K^+ deficiency. Trolldenier (1973) reported that when K^+ nutrition is inadequate the capability of rice roots to oxidize Fe^{2+} to Fe^{3+}, is impaired.

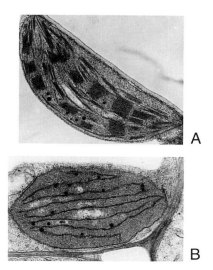

Plate 13.3   Chloroplasts of pepper (*Capsium annuum*). A: sufficient Fe supply, B: Fe deficiency.
Magnification: 29000 x (Photo: Hecht-Buchholz and Ortmann).

13.3   Iron in Crop Nutrition

13.3.1   Iron availability

The Fe concentration of green plant tissues is low in comparison with the macronutrients and generally in the order of 50 to 100 μg Fe/g dry matter. In cereal grains and tubers it is often considerably lower. Total soil Fe is thus always greatly in excess of crop requirements. According to Lindsay (1974) most agricultural crops require less than 0.5 μg/g available Fe in the soil in the plough layer whereas the total Fe level is about 20.000 μg/g Fe in the soil. Any problem of Fe supply to crops from soil is therefore always one of availability.

The solubility of inorganic Fe is highly dependent on soil pH and the redox status of soils. The influence of pH on the solubilities of Fe^{2+}, Fe^{3+} and total soluble Fe in equilibrium with Fe oxide is shown in Figure 13.1. Lindsay (1974) has estimated that to enable mass flow to transport sufficient Fe to the roots, the total solubility must be at least 1 mmol/m^3. As can be seen in the Figure 13.1 this soluble inorganic Fe level is only achieved at pH 3, and by raising the pH to just over 4 only 1% of the Fe demand can be obtained. At normal soil pH levels therefore even allowing for the contribution of diffusion (O'Connor *et al.* 1971) inorganic Fe levels are far below those required by plants. It appears therefore that for plants growing in soil the formation of soluble Fe organic complexes, mainly chelates, must play an important role in Fe supply as already discussed on page 554. According to Raymond (1977), microorganisms suffering from

Fe deficiency are capable of excreting such siderophores which complex with the less available Fe of the medium.

The mechanism by which siderophores (chelates) may operate in the soil is shown in Figure 13.6 (Lindsay 1974). The Fe^{III} complex diffuses to the root cell where the Fe^{III} is reduced and the released Fe^{2+} diffuses through a selective channel into the cytosol (Fe^{2+} transporter). In an analogous way phytosiderophores exploit less soluble Fe sources. The importance of root tips in the Fe nutrition of plants has been confirmed by investigations of Clarkson and Sanderson (1978). This work has revealed that only root tips and not the basal parts of roots are capable of absorbing Fe. The contact zone between roots and soil of relevance to Fe uptake is thus very limited. From these investigations the development of new root tips should also play an important role in determining the Fe uptake potential of plants. More recent research has shown that the selective Fe^{2+} channels (Fe-transporters) are exclusively located in the plasmalemma of the roof epidermis (Guerinot 2000). Not all species are equally susceptible to Fe chlorosis. Iron deficiency is commonly observed in the calcifuge species such as *Azalea*, *Rhododendron*, and blueberry. The most important commercial crops affected are citrus, deciduous fruit trees and vines. Iron chlorosis has also been found in field beans, soya beans, maize, grain sorghum, legumes, rice and tomatoes.

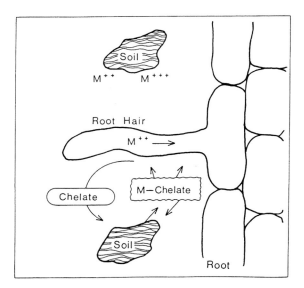

Figure 13.6 Mobilization of soil metal ions by chelates (after Lindsay 1974).

13.3.2 Lime induced chlorosis

Iron chlorosis may result from an absolute Fe deficiency in the soil. Such cases may occur on organic soils and very few mineral and degraded sandy soils, but such cases are rare. However, iron chlorosis occurs frequently on calcareous or saline alkaline soils. This lime induced chlorosis was described precisely for the grape vine at the beginning of the 20. century by Molz (1907, quoted by Mengel 1994) and was a severe problem even at that time. The chlorosis is not caused by absolute Fe deficiency. In contrast to most other plant nutrients where there is an inverse relationship between the intensity of the deficiency and the concentration in the plant tissue, this does not apply to Fe. Here frequently the Fe concentration in the chlorotic leaves can be higher than in the green leaves (Carter 1980, Rashid et la. 1990, Bertoni *et al.* 1992).

Calcareous soils are characterized by high carbonate contents, by high pH and therefore also high HCO_3^- concentrations in the soil solution. Such conditions may depress or even block Fe^{III} reduction in the root apoplast (Toulon *et al.* 1992) as considered above. According to investigations of Kosegarten *et al.* (1998) a high nutrient solution pH associated with a high concentration of HCO_3^- in the nutrient solution did not greatly influence the pH of the xylem sap which was considerably below 7.0. Chlorosis was induced when the plants were supplied exclusively with nitrate as N source at a high pH in the nutrient solution. This is similar to what occurs on calcareous and on saline alkaline soils where the high soil pH is buffered by a high concentration of HCO_3^- and any NH_4-N is either lost by NH_3 volatilization or rapid oxidation to nitrate. As reported by Tagliavini *et al.* (1995) nitrogen nutrition in calcareous soils is predominatly that of nitrate even if N is applied as ammonium. This effect of nitrate nutrition inducing Fe chlorosis has also been reported by others (Aktas and Egmond 1979, Kosegarten and Englisch 1994). Hoffman *et al.* (1992) and Hoffmann and Kosegarten (1995) measuring the pH of the leaf apoplast by means of fluorescence found that it increased with nitrate nutrition whereas it decreased with ammonium nutrition. From this finding it was hypothesized that at high apoplastic leaf pH Fe^{III} reduction in the leaf apoplast is restricted and hence the uptake of Fe from the apoplast into the cytosol impaired (Mengel 1994). In recent experiments Kosegarten *et al.* (1999b) were able to confirm this hypothesis by means of fluorescence ratio imaging. These authors were able to show that at distinct microsites in the apoplast of sunflower leaves the pH was > 6 if the plants were fed exclusively with nitrate. Such microsites with elevated pH levels were not found if the plants were supplied with ammonium-N. In addition the authors showed that Fe^{III} reduction in the leaf apoplast was greatly depressed at high pH. It thus appears that under such conditions the Fe remains in the apoplast where it may even accumulate to high levels when at the same time the Fe concentration in the cytosol is insuffcent. Total leaf Fe concentration may be thus high but metabolically active Fe low.

These findings and interpretations of Kosegarten *et al.* (1999b) have been confirmed by the observation that spraying chlorotic leaves with a diluted acid, as done by Tagliavini *et al.* (1995) with citric acid results in a regreening. Mengel and Geurtzen (1988) reported that spraying chlorotic maize leaves with IAA or fusicoccin induced a regreening of the leaves within two days even though the Fe concentration in leaves did not increase. Fusicoccin as well as IAA stimulate the plasmalemma H^+ pump (see page 118) and thus the apoplastic pH is depressed, promoting the reduction of Fe^{III}. The effects of various treatments for Fe chlorosis in *Pisum sativum* are shown from the work of Sahu *et al.* (1987) in Table 13.2. In agreement with the above explanation it is evident that spraying the leaves with a diluted solution of sulphuric acid cured the chlorosis by increasing the chlorophyll concentration and gave a significant yield increase over the control. This response was greater than that obtained by spraying the leaves with Fe-EDDHA, an artificial siderophore used for treating Fe chlorosis.

Table 13.2 Effect of spraying Fe-EDDHA or diluted H_2SO_4 for curing Fe chlorosis on chlorophyll concentration and pod yield of *Pisum sativum* (Data from Sahu *et al.* 1987).

Treatment	Chlorophyll, mg/g fresh weight	Pod yield, Mg/ha
Control, no spraying	1.37	1.79
H_2SO_4	1.83	3.36
Fe-EDDHA	1.78	3.15
$p < 0.05$	0.676	0.676

Phosphate has often been held to be the main cause of Fe chlorosis. Phosphate concentrations in soil solution, however, are much lower than those normally occurring in solution culture, in which Fe deficiency has been induced by high phosphate levels. One must therefore be very cautious in relating such solution culture experiments to field conditions. Even with extremely high phosphate concentrations in solution culture, Kolesch *et al.* (1982) were unable to induce Fe chlorosis in sunflowers, whereas this was achieved readily by the application of HCO_3^-.

It seems reasonable to assume that on calcareous soils which are generally characterized by low P concentrations in the soil solution, phosphate is not the inducer of Fe chlorosis. This view is supported by experimental data of Kovance *et al.* (1978), Müllner (1979), and Mengel *et al.* (1984). High P concentrations are often found in Fe chlorotic leaves but according to Mengel *et al.* (1984) these occur as the consequence of Fe deficiency and are not the reason for its occurrence.

In calcareous soils HCO_3 may accumulate due to the high pH level and the dissolution of carbonates according to the equation:

$$CaCO_3 + CO_2 + H_2O \Rightarrow Ca^{2+} + 2\ HCO_3^-$$

The dissolution needs CO_2 which is mainly produced by root and microbial respiration. If soil aeration and soil structure are satisfactory CO_2 may escape from the soil medium and HCO_3^- does not accumulate. Under high soil moisture conditions particularly in association with a poor soil structure, however, HCO_3^- may accumulate to concentrations as high as 6 to 8 mol/m^3 HCO_3^-. Chlorosis is then likely to occur. This has been shown for fruit trees by Boxma (1972) and for citrus by Kovance *et al.* (1978). These findings are consistent with the observation that Fe chlorosis is likely to be present under rainy weather conditions (Gärtel 1974), when soil moisture is high and soil aeration poor.

Improving soil structure is therefore one of the most important measures in controlling Fe chlorosis. In vineyards, cultivation of deep rooting crops (*Brassica species*) between the vine rows reduces the hazard of chlorosis. Application of organic manures, however, is a doubtful means of correcting Fe chlorosis for although soil structure may be improved, enhanced CO_2 production in the soil may favour the formation of HCO_3^-. Lime induced Fe chlorosis is mostly controlled by foliar application of Fe chelates. Fe-EDDHA (= Ethylene diamine (di o-hydroxyphenyl) acetate (has proved to be a suitable chelate for this purpose in contrast to Fe-EDTA (= Ethylene diamine tetra acetate) which is not stable enough (see page 7).For successful control more than one application is often necessary. Mixtures of manure with inorganic Fe ($FeSO_4$) have also been recommended for the correction of Fe chlorosis (Chen and Barak 1982).

Plate 13.4   Iron chlorosis in a vineyard. White areas indicate the vines with Fe deficiency. (Photo: Gärtel)

Plate 13.4 shows Fe chlorosis in a vineyard. The light areas and stripes visible on the photograph show the chlorotic leaves. Even in forest trees (*Pinus silvestris*) lime induced chlorosis has been observed (Zech 1970, Carter 1980). Cultivars of a given crop species may also differ considerably in their susceptibility to Fe-chlorosis. This has been shown for soya beans, maize and tomato (Brown *et al.* 1972) as well as for grapes (Saglio1969). Brown (1963) reported that under Fe stress, chlorosis resistant cultivars of soya beans released more H^+ from the roots than did susceptible cultivars. Analogous results have been found by Mengel and Malissiowas (1982) for two vine cultivars. Proton excretion from the roots was greater in the Fe resistant cultivar. The primary effect of this increased H^+ excretion of cultivars resistant to Fe chlorosis is not to dissolve Fe bearing minerals but to reduce the pH in the root apoplast and by this improve the conditions for Fe^{III} reduction as discussed above. The results of Romera (1991a and b) support this interpretation. These authors found that the susceptibility of peach rootstock cultivars to Fe chlorosis decreased as the reduction capacity of the rootstock increased. Highest reduction capacity was found at the lowest pH (pH 4.3 in the outer solution). Similar results have been reported by Treeby and Uren (1993) for citrus rootstocks.

13.3.3   Iron application

In treating Fe chlorosis, the addition of inorganic Fe salts to the soil is mostly without effect for the Fe is rapidly made insoluble as oxides. Foliar treatment with ferrous salts or Fe- chelates is more effective. In soil applications, it is important to consider the stability of the chelate particularly in relation to the soil pH. At higher soil pH levels, soil Ca^{2+} is present in higher concentration and can displace Fe^{3+} from less stable chelates giving rise to a Ca-chelate and precipitated Fe-oxide thus rendering the Fe unavailable. Differences in the stability of chelates are reflected in plant responses. Lindsay *et al.* (1967) in testing various Fe chelates found the best response with Fe EDDHA (ethylenediamine (di o-hydroxyphenyl) acetic acid. This Fe chelate is stable throughout the pH range 4 to 10, whereas the stability of Fe EDTA (ethylene diamine tetraacetic acid) fails above pH 7 so that correspondingly poorer effects were obtained. The question of correcting Fe chlorosis has been discussed by Chen and Barak (1982) and Tagliavini *et al.* (1997) in useful review articles. Chen and Brak (1982) cite cases where mixtures of Fe salts and organic matter such as manure, compost, sewage sludge, and peat were used successfully in controlling Fe chlorosis in various crops. Tagliavini *et al.* (1997) reported results from field trials carried out on calcareous soils in Italy, Spain and Greece. Increasing the chlorophyll concentration of fruit trees by foliar application of of Fe containing solutions had a significant impact on fruit yield of kiwi, peaches and pears as shown in Figure 13.7. Regreening was obtained by spraying Fe chelates, $FeSO_4$ solution alone or in combined with citric acid or sulphuric acid.

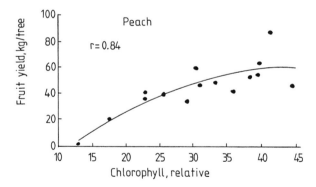

Figure 13.7   Relationship between leaf chlorophyll concentration and fruit yield of peaches (after Tagliavini *et al.* 1997).

Spraying the two acids (sulphuric acid, citric acid) gave only a transitional regreening. Obviously Fe in the apoplast was not sufficient to provide a continuous source of Fe supply. Spraying $FeSO_4$ solutions resulted in a regreening as good as that attained with the Fe chelate (Fe -DTPA). This is of economic importance since the inorganic Fe form is much cheaper than the Fe chelate. Unfortunately the most stable chelates still tend to be too expensive for commercial use. Application of blood meal, $FeSO_4$ + compost and even Fe-EDDHA to calcareous soils have no major influence on the regreening of chlorotic fruit tree leaves as found by Tagliavini *et al.* (1997). Acid fertilizers such as ammonium containing materials or the direct application of acids to soil do not have a long lasting depressive effect on the pH of calcareous soils because of rapid neutralization (Tagliavini *et al.* 1997).

CHAPTER 14

MANGANESE

14.1   Soil Manganese

Manganese occurs in various primary rocks and particularly in ferromagnesian materials. The Mn released from these rocks by weathering forms a number of secondary minerals the most prominent being pyrolusite ($Mn^{IV}O_2$) and manganite [$Mn^{III}O\,(OH)$]. Manganese and Fe oxides often occur together in nodules and iron-pans. Total Mn levels may differ considerably between soils and are in a range from 20 to 3000 μg Mn/g (Gambrell 1996).

The most important Mn soil fractions are Mn^{2+} which may be dissolved in the soil solution or adsorbed to clay and humates and the Mn oxides present in amorphous and crystalline form. According to the prevailing redox potential Mn may be present in di-tri and tetravalent form as shown in Figure 14.1., and even the Mn present in ferromagnesian phyllosilicates may be reduced or oxidized without changing the position in the structure (Allen and Hajek 1989). Under anaerobic conditions Mn^{2+} is produced mainly by microbial respiration functioning as an e$^-$ donor for Mn^{III} and Mn^{IV} compounds and thus Mn oxides are dissolved. Most organisms which reduce Fe^{3+} also reduce Mn^{4+} (Paul and Clark 1996). Drying out soils leads to the reverse effect which also means the formation of Mn oxides.

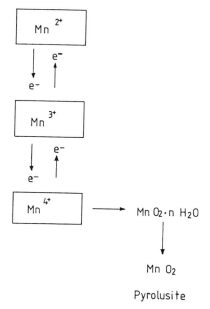

Figure 14.1 Oxidation/reduction processes of manganese in soils.

Besides redox conditions soil pH also has a siginficant influence on the reductive dissolution of Mn oxides and over the range of pH 5 to 6 the reduction rate decreases by a factor of 10 to 100 (Sparks 1988). Frequently organic matter may serve as an e^- donor as is the case of malate and citrate exuded by plant roots (Godo and Reisenauer 1980).

$$Mn^{4+} + HOOC - CH_2 - CHOOH - COOH \rightarrow Mn^{2+} + HOOC - CH_2 - CO - COOH + 2\ H^+$$

Malate   Oxaloacetate

Manganese reduction produces H^+ and according to the equilibrium conditions shown in the equation above the reduction process is depressed with increasing H^+ concentration.

The most important fractions of relevance to plant nutrition are Mn^{2+} and the easily-reducible Mn from Mn oxides, amorphous oxides being more easily reduced than crystalline ones. These combined fractions of Mn^{2+} + easily-reducible Mn are called *active Mn*. As the level of Mn^{2+} in the soil depends on oxidation-reduction reactions and pH, all factors influencing these processes have an impact on Mn availability. In soils rich in organic matter and of low pH (Spodosols) a high proportion of Mn is bound by the organic matter mainly in exchangeable form and therefore represents a pool of available Mn. Raising soil pH generally decreases the organically bound Mn and increases amorphous and crystalline Mn forms (Zhang *et al.* 1997). It is for this reason that Mn deficiency in plants mainly occurs on soils of high pH.

Under waterlogged conditions as for example in paddy soils, reducing processes dominate and thus provide a high level of Mn availability which may even result in Mn toxicity (Tanaka and Yoshida 1970). After submergence and almost parallel with the disappearance of O_2 the level of soluble Mn^{2+} rises. In acid soils high in active Mn the concentration of Mn^{2+} may easily attain toxic levels, whereas in calcareous or sodic soils the Mn level does not rise much after flooding. On these soils Mn deficiency can even occur in rice under submergence conditions (Randhawa *et al.* 1978). The effect of anaerobic soil conditions (flooding) and of liming on Mn availability as reflected in Mn concentration in lucerne grown on the soil is shown in Table 14.1 (Graven *et al.* 1965). The extremely high Mn concentration in the plant material found in the flooding treatment without liming is indicative of Mn toxicity. High Mn concentrations in soils produced by flooding mainly increase the concentration of Mn^{2+} in soils which means the directly plant available Mn (Weil *et al.* 1997). These high Mn concentrations rapidly fall when soils become aerated. This is shown in Figure 14.2 from the work of Weil *et al.* (1997). The investigation comprised three pretreatments lasting 16 days: Air dried soil, flooded soil, and soil at field capacity. The soils were then kept at field capacity. Flooding soils had a major impact on the concentration of exchangeable Mn which was much higher in the soil with the low pH (Jackland, pH 4.8) as compared with the Myersville soil (pH 5.2). These high levels of exchangeable Mn values decreased with time in the aerobic phase (field capacity). Interestingly air drying also increased the level of exchangeable Mn. This observation agrees with the data of Webb *et al.* (1993) who found that the soluble Mn increased during storage of dry soils. The authors suggest that under such conditions the activity of Mn reducing micro-organisms is higher than that of Mn oxidizing microorganisms (Webb *et al.* 1993).

Table 14.1 Effect of liming and a three day period of flooding on dry matter yield and Mn concentrations in lucerne (Graven etal. 1965)

Treatment g CaCO$_3$/pot	Flooding	g, dry matter pot	Mn contencentration μg/dry matter
0	-	3.1	426
0	+	1.2	6067
20	-	5.7	99
20	+	3.0	954

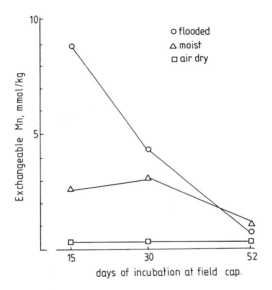

Figure 14.2 Effect of soil pretreatment at different water status followed by an incubation at field capacity. Pretreatment: soil flooded, soil air dry, soil moist.

The microbial activity which brings about Mn oxidation is pH dependent and has an optimum at about pH 7. Killing off these Mn oxidizing microorganisms, as for example by steam sterilization, results in an increase in Mn availability (Roll-Hansen 1952). From the above discussion it is clear that soils of high pH with large organic matter reserves required for intense microbial activity are particularly prone to Mn deficiency. It is understandable that liming depresses Mn availability whereas the application of physiologically acid fertilizers, as for example, $(NH_4)_2SO_4$ has a beneficial effect on Mn uptake by plants (Kühn 1962, Cottenie and Kiekens 1974).

Divalent Mn dissolved in the soil solution is of direct importance in plant nutrition. This dissolved Mn^{2+} is in equilibrium with Mn^{2+} adsorbed to clay minerals and organic matter. According to Geering *et al.* (1969), the Mn^{2+} concentration in the soil solution of acid and neutral soils is in the range of 10^{-3} to 10^{-1} mol/m³. Lindsay (1974) reported that the affinity of Mn^{2+} for synthetic chelates is comparatively low and complexed Mn can easily be replaced by Zn^{2+} and Ca^{2+}. One may assume therefore that Mn in soil organic matter is mainly present as exchangeable Mn^{2+}. The levels of Mn in soil solution are considerably higher than those of Cu and Zn.

Under very dry conditions Mn salts in the soil can be irreversibly dehydrated and thus become less available. Drying, however, may also result in the splitting of Mn double salts. In this process Mn^{2+} is released. Manganese in the divalent form is fairly mobile in the soil and can easily be leached. This happens particularly on acid podzolic soils (Spodosol).

The formation of Mn oxides in soils appears to regulate the levels of Co in soil solution and hence Co availability to plants. According to McKenzie (1975) one of the most important properties of soil Co is its association with Mn oxide minerals.

14.2 Manganese in Physiology

14.2.1 Uptake and translocation

As already observed by Collander (1941) about 60 years ago, the rate of Mn uptake differs considerably between plant species. Generally, however, uptake rates are lower than for other divalent cation species (Ca^{2+}, Mg^{2+}). Presumably phytosiderophores are not involved in exploiting Mn bearing minerals for Mn (Erenoglu *et al.* 1996). According to Ma and Nomoto (1996) the affinity of siderophores for Mn is poor. There is ample evidence that Mn uptake occurs in the form of Mn^{2+} presumably by facilitated diffusion across the plasmalemma (Fox and Guerinot 1998). In a similar way to other divalent cation species, Mn^{2+} participates in cation competition. Magnesium in particular depresses Mn^{2+} uptake (Löhnis 1960, Maas *et al.* 1969). Liming also reduces uptake not only by the direct effect of Ca^+ in the soil solution, but also as a result of the pH increase.

In its chemical behaviour, Mn shows properties of both the alkali earth cations such as Mg^{2+} and Ca^{2+} and the heavy metals (Zn, Fe). It is therefore not surprising that these ion species affect uptake and translocation of Mn in the plant although the mechanism of this effect needs still to be elucidated (Fox and Guerinot 1998). This is particularly true for the depressive effect of Mn on Fe uptake already described many years ago by Somers and Shive (1948). In this context it is of interest that Fe^{2+} is oxidized by Mn^{4+} nonenzymatically (Paul and Clark 1996) which may affect Fe uptake since this is taken up in form of Fe^{2+} (see Figure 13.3). Sideris and Young (1949) found that plants supplied with NH_4^+ took up lower amounts of Mn^{2+} than NO_3^- fed plants. This result, however, can probably be accounted for by the more general effects of different N sources, NH_4^+ bringing about an acidification of the nutrient medium associated with a reflux of H^+ into the cytosol (Yan *et al.* 1992) and thus depressing the plasmalemma polarization which is the driving force for Mn^{2+} uptake.

According to Wittwer and Teubner (1959), Mn is relatively immobile in the plant and it is scarsely translocated in the phloem. This assumption is in line with results of Caballero *et al.* (1996) who found that Mn like to Ca was remobilized from vegetative parts to the seeds only at the beginning of seed formation. At this developmental stage seeds still are supplied by the xylem sap. Tiffin (1972) studied the translocation of a number of heavy metals in tomato plants and in electrophoretic examinations of exudates it was found that Mn migrated towards the cathode. It thus appears that Mn is mainly transported as Mn^{2+} and not as an organic complex. Analogous observations were made in ryegrass extracts by Bremner and Knight (1970). Manganese is preferentially translocated to meristematic tissues.

Young plant organs are thus generally rich in Mn (Amberger 1973). In solution culture experiments Williams and Vlamis (1957) found that the addition of Si enhanced the distribution of Mn in barley plants. This was confirmed by Horst and Marschner (1978b), who showed that at a higher Si supply bean plants could tolerate higher Mn concentrations. In the absence of Si in the nutrient solution the distribution of Mn in the leaf was more heterogenous and characterized by spot-like accumulations.

14.2.2 Biochemical functions

In its biochemical functions Mn^{2+} resembles Mg^{2+}. Both ion species bridge ATP with the enzyme complex (phosphokinases and phosphotransferases). Decarboxylases and dehydrogenases of the TCA cycle are also activated by Mn^{2+}, although it appears that in most cases Mn^{2+} is not specific for these enzymes and can be substituted by Mg^{2+}. *In vivo*, however, the Mn-bridge plays no major role since the cytosolic concentration of Mg^{2+} is normally in the order of 50-100 times greater than that of Mn^{2+} (Clarkson and Hanson (1980).

RNA polymerase activation is brought about by either Mn^{2+} or Mg^{2+} but at low concentrations Mn^{2+} is much more effective (Ness and Woolhouse 1980). Manganese is essential for CO_2 assimilation of C_4 and CAM plants because it activates PEP carboxylase (see page 155). A feature of Mn deficient tissue is the increase which occurs in the activity of peroxidase in contrast to catalase which is not affected (Bar-Akiva and Lavon 1967). This increased activity in peroxidase is probably associated with the high activities of IAA oxidase which are found in Mn deficient plants (Morgan *et al.* 1976). According to Rao *et al.* (1982) peroxidases are the most important constituents of the IAA oxidising system.

There are only a few Mn containing enzymes. A superoxide dismutase (SOD) containing one atom of Mn per enzyme has been isolated from pea leaves and it appears to be widely distributed (Sevilla *et al.* 1980). Mn superoxide dismutase is present in the mitochondria of eukaryots where it detoxifies the superoxide radical (Cadenas 1989). The general process is shown on page 143. Mn SOD enhances resistance to oxygen radicals (oxidative stress tolerance) as was shown by Slooten *et al.* (1995) with transgenic plants overexpressing MnSOD in the chloroplasts.

The most well documented and unique role of Mn in green plants is that of the water splitting and O_2 evolution system in photosynthesis, the so called Hill reaction. Proteins of the photosystem II provide most of the ligands of the Mn cluster which contains 4 Mn atoms in various oxidation states and which is stablilized by Ca^{2+} and chloride (Merchant and Dreyfuss 1998). Light induces oxidation of one Mn^{II} (photoxidation) followed by a coordinated oxidation of the other three Mn atoms so that finally four e^- are released as shown in the simplified scheme of Figure 14.3. The oxidized Mn cluster then oxidizes H_2O by "extracting" four e^- from two H_2O so that $O_2 + 4H^+$ are released (photolysis). This process which occurs in lower and higher plants is of universal importance since it drives photosynthesis and provides oxygen for all aerobic living organisms.

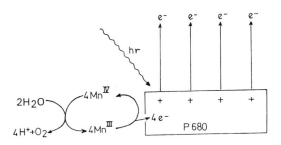

Figure 14.3 The action of Mn in the water splitting protein in which two molecules of water release 1 molecule of O_2 and $4H^+$ with the simultaneous donation of $4e^-$ in turn to each positively charged hole in P 680 created by light excitation.

Because of the importance of Mn for photosynthetic e^- transport it follows that when Mn is deficient the photosynthetic light reaction is seriously disturbed. Other reactions associated with the photosynthetic e^- transport such as photophosphorylation and the reduction of CO_2, nitrite and sulphate are detrimentally affected. When NO_2^- reduction is impaired the accumulated NO_2^- can exert a feedback control on NO_3^- reductase activity so that NO_3^- also accumulates. This is one reason why NO_3^- accumulation is sometimes observed in Mn deficient plants.

14.2.3 Deficiency and toxicity

Chloroplasts are the most sensitive of all cell organelles to Mn deficiency and disorganization of the lamellar system occurs. In whole plants Bussler (1958a) reported that tissues suffering from Mn deficiency have a small cell volume, cell walls dominate and the interepidermal tissue is shrunken. Manganese deficiency resembles Mg deficiency, as in both cases interveinal chlorosis occurs in the leaves. In contrast to Mg deficiency, however, Mn deficiency symptoms are first visible in the younger leaves, whereas in Mg deficiency the older leaves are first affected. Manganese deficiency symptoms in the dicots are often characterized by small yellow spots on the leaves and interveinal chlorosis (Maynard 1979). In this respect this syndrome differs from that of Fe deficiency

Plate 14.1 Upper part, 'grey speck' in oats (Photo: v. Papen) — Lower part, Mn deficiency in sugar beet (Photo: Draycott).

where the whole young leaf becomes chlorotic. A typical example of Mn deficiency is presented in the lower part of Plate 14.1, showing Mn deficiency in sugar beet. The deficiency is in an advanced stage and only the veinal areas are still green. In monocots and particularly in oats, Mn deficiency symptoms appear at the basal part of the leaves (Finck 1956) as greenish grey spots and stripes. Oats in particular are prone to Mn deficiency during the tillering stage. The disease is known as 'grey speck'. The turgor of the affected plants is reduced and at an advanced stage of disease the upper part of the leaf breaks over near the middle (see Plate 14.1).

The Mn nutritional status of plants is reflected in their Mn concentration. The critical deficiency level for most plant species is in the range of 10 to 20 μg Mn/g in the dry matter of mature leaves. Below this level dry matter production, net photosynthesis and chlorophyll are markedly affected (Ohki 1981) whereas rates of respiration and transpiration are largely unaltered. Figure 14.4 shows a clear relationship between the intensity of Mn deficiency and the Mn concentration in the foliage of sugar beet (Farley and Draycott 1973).

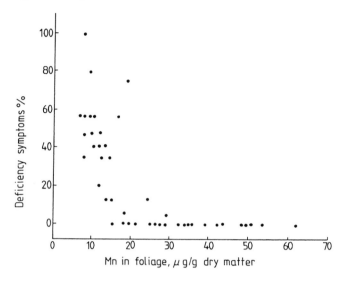

Figure 14.4 Intensity of Mn deficiency and Mn concentration in the foliage of sugar beet (after Farley and Draycott 1973).

Manganese toxicity, unlike the deficiency, is not restricted to a narrow critical concentration range. Edwards and Asher (1982) cite values from 200 $\mu g/g$ in maize to 5300 μg Mn/g dry matter in sunflower associated with a 10% reduction in dry matter yield. These values appear to be particularly temperature dependent with much higher Mn levels associated with higher growing temperatures (Rufty *et al.* 1979). Lower yields obtained with Mn toxicity are related to the depressed CO_2 assimilation while photosynthetic O_2 release was not affected (Kitao *et al.* 1997). Gonzalez and Lynch (1997) found in *Phaseolus vulgaris* that Mn toxicity decreased chlorophyll concentration only in immature leaves with a consequent reduction of leaf CO_2 assimilation while mature leaves even with toxicity symptoms were not affected in CO_2 assimilation. According to Bennett (1974) shoot growth was much more sensitive than root growth to Mn toxicity.

Mn toxicity is mainly found on acid to very acid soils where high concentrations of Mn^{2+} prevail and flooding occurs associated with a high reduction potential (Bergmann 1983). This is particularly the case on acid sulphate soils with a pH even less than 3.0. Here Mn toxicity is frequently associated with with Fe and Al toxicity (Prasad and Goswami 1992). Toxicity symptoms are generally characterized by brown spots of MnO_2 in the older leaves surrounded by, chlorotic areas (Bussler 1958b, Horst and

Marschner 1978a). Mühling *et al.* (1988) reported that Mn toxicity found in spring barley grown on acid soils was characterized by dark brown spots at the leaf tips which were enrolled and had extremely high Mn concentrations. Sometimes Mn excess can induce a deficiency of other mineral nutrients such as Fe, Mg and Ca. The same is true for Mg (Mühling *et al.* 1988). In the case of Mn induced Ca deficiency characterized by *crinkle leaf* which has been reported in cotton (Foy et al. 1981) and beans (Horst and Marschner1978a), it is the transport of Ca to the growing points which is affected. Morgan et al. (1966) considered Mn toxicity to be an expression of auxin deficiency caused by high IAA oxidase activity. This view is consistent with the findings of Horst and Marschner(1978a) that auxin activity was lower in Mn toxic tissue. The lower auxin activity impaired the IAA stimulated proton pump thereby inhibiting cell wall expansion and the formation of new binding sites for the transport of Ca^{2+} to the apical meristems. The authors were able to demonstrate that the CEC of the Mn toxic tissue was lower than that of the normal tissue. Another symptom of Mn toxicity which implicates an interaction between auxin and Mn toxicity is the loss of apical dominance and the proliferation of auxiliary shoots as reported by Kang and Fox (1980) in cowpea suffering from Mn toxicity. Manganese toxicity can be counteracted by Si (Vlamis and Williams (1967). In beans Si induces a more uniform Mn distribution within the leaf (Horst and Marschner (1978b). Manganese toxicity can also be depressed by increased Mg supply whereby Mg can replace Mn from physiologically active sites (Le Bot *et al.* 1990). Excess Mn may also induce Fe deficiency (Foy *et al.* 1978).

14.3   Manganese in Crop Nutrition

Most soils contain adequate levels of available Mn so that Mn applications are unnecessary. The total amount of Mn taken up by arable crops is low and ranges from between 500 to 1000 g Mn/ha. Calcareous peat soils (organic soils, high in pH) are particularly low in available Mn and it is on these soils that Mn deficiency often occurs in crops. In such soils Mn^{2+} oxidation is favoured by high pH and by a high activity of Mn oxidizing microorganisms. Application of Mn salts to the soil, *e.g.* $MnSO_4$ is usually of no use in alleviating deficiency, because the applied Mn^{2+} is rapidly oxidized. When such soils are treated with Mn fertilizers, banded placement rather than broadcast application should be carried out. Randall and Schulte (1971) found that 5.6 kg/ha of banded Mn as sulphate was equivalent to 67.2 kg/ha of broadcast Mn. Generally for soil application $MnSO_4$, is superior to Mn chelates. Foliar application of Mn is frequently recommended on these calcareous organic soils. Draycott and Farley (1973) in comparing soil application of Mn silicate and Mn oxides with Mn foliar sprays found that the soil application did not prevent Mn deficiency in sugar beet whereas Mn foliar application corrected the deficiency and increased sugar yields. Spraying 1 to 5 kg Mn/ha is sufficient to offset deficiency in most crops. According to Ozaki (1955) $MnSO_4$ is considered to be the most effective inorganic carrier of Mn in foliar spray solutions. Of the organic Mn carriers, Mn-EDTA appears to give the best response.

Some podzolic soils (Spodosols) are particularly liable to be low in available Mn. This group of soils however, differs from the organic soils already discussed, in that podzols are inherently low in Mn. This low Mn status mainly results from high leaching. Manganese deficiency on these sites is frequently aggravated by liming (Zhiznevskaya1958) because of the resulting pH increase. Manganese deficiency also occurs on sandy Atlantic Coastal Plain soils (USA) which are inherently low in Mn (Miner *et al.* 1986). Manganese applied in form of $MnSO_4$ to these soils is not so rapidly oxidized as in the calcareous organic soils. Mn soil applications are well able to correct the deficiency. Miner *et al.* (1986) tested 7 different Mn containing materials differing in their water solubiltiy from 1.2 to 99.6%. These materials were band-applied to maize. Manganese uptake and grain yield of maize were higher in the treatments with a higher proportion of water soluble Mn as compared with those low in water-soluble Mn.

The incidence and severity of Mn deficiency appears to depend on seasonal conditions. The deficiency is often worse in cold wet seasons possibly as a result of a reduction in root metabolic activity affecting Mn uptake. Such effects may also account for the increased prevalence of Mn deficiency often observed where good growing conditions follow a cold or dry period (Batey 1971).

Soil analysis is not very reliable in diagnozing available soil Mn status in relation to crop response. According to Browman et al. (1969) who compared a number of the standard methods, NH_4-acetate extraction with a correction for pH gave the most satisfactory results. Similar observations were made by Farley and Draycott (1976). Ammonium acetate extractable Mn correlated best with Mn deficiency, plant Mn, and response to treatment. The most important results of this study are shown in Table 14.2. The Mn concentrations of a large number of crops differing in Mn status is provided in the accumulated data of Labanauskas (1966).

Table 14.2 Sugar beet response to Mn in relation to NH_4-acetate, extractable soil Mn, and Mn in the dried foliage (Farley and Draycott 1976).

Soil Mn, $\mu g/g$	Plants with symptoms, %	Mn in dried foliage, $\mu g/g$	Response
<1.2	100—50	<20	large
1.2—1.8	49—25	20—30	small
>1.8	<25	>30	non

Crop species differ in their susceptibility to Mn deficiency. Two of the most well known deficiency diseases in arable crops are *grey speck* in oats (*Avena sativa*) and *marsh spot* in peas (*Pisum sativum*). Other sensitive crops are: apple, cherry, citrus, raspberry, and sugar beet.

On acid soils high in Mn availability, plants can take up considerable amounts of Mn so that levels in the order of 1000 μg Mn/g dry matter are not uncommon. Löhnis (1960) found Mn concentrations in *Vaccinium myrtillus* higher than 2000 $\mu g/g$ dry matter. There is now considerable interest in screening crop genotypes for resistance to Mn toxicity. This is particularly true for some tropical crop species when growing conditions are commonly acid and available Mn levels high. Foy (1976) recommended that for Mn screening, soils should be high in total Mn within the pH range of 5.0 to 5.5. Here Mn will be soluble in toxic concentration but Al toxicity will be minimal or absent. Manganese tolerant species and cultivars commonly take up Mn at a lower rate (Brown et al. 1972).

CHAPTER 15

ZINC

15.1 Soil Zinc

The average Zn concentration in uncontaminated soils is in the range of 17 to 160 μg Zn/g soil (Reed and Martens 1996). Most of it is present in the lattice structure of primary and secondary minerals (Huang 1989). The radius of Zn^{2+} is very similar to that of Fe^{2+} and Mg^{2+}. To some extent therefore Zn^{2+} may substitute for these ions by isomorphous replacement in mineral structures, in particular in the ferromagnesian minerals, amphiboles (augite, hornblende), and biotite. The occurrence of Zn in these minerals makes up the bulk of Zn in many soils. In addition Zn forms a number of salts including ZnS, sphalerite (ZnFe)S, zincite ZnO, and smithsonite $ZnCO_3$. Apart from ZnS, however, which may be present under reducing conditions, most of these salts are too soluble to persist in soils for any length of time (Lindsay 1972). The two Zn-silicates $ZnSiO_3$, and Zn_2SiO4, (willemite) also occur in some soils.

Zinc is dissolved in the soil solution in ionic or complex form and may be found on exchange sites of clay minerals and organic matter or adsorbed on solid surfaces as Zn^{2+}, $ZnOH^+$ or $ZnCl^+$ and also may become nonextractable possibly by entering holes normally occupied by Al^{3+} in the octahedral layer. The Zn concentration in soil solution is very low and in the range of 3×10^{-5} to 5×10^{-3}mol (Barber 1984). Zn solubility is highly dependent on soil pH and is very low at high soil pH. It is particularly low when $CaCO_3$ is present because of specific adsorption of Zn^{2+} to and occlusion by carbonates. Adsorption affinity to carbonates decreases in the following sequence (Donner and Lynn 1989)

$$\text{magnesite} > \text{dolomite} > \text{calcite}$$

which indicates the higher affinity of Zn to Mg- relative to Ca carbonates. Adsorption and occlusion of Zn by carbonates are the major causes of poor Zn availability and the appearance of Zn deficiency on calcareous soils. Zinc interacts with soil organic matter, and both soluble and insoluble Zn organic complexes are formed. According to Hodgson et al. (1966), on average, 60% of the soluble Zn in soil occurs in soluble Zn organic complexes. Stevenson and Ardakani (1972) in a review of micronutrient organic matter reactions concluded that soluble Zn organic complexes are mainly associated with amino, organic and fulvic acids whilst insoluble organic complexes are derived from humic acids.

The level of Zn in soils is very much related to the parent material. Soils originating from basic igneous rocks are high in Zn. In contrast soils derived from more siliceous parent materials are particularly low. Occasionally very high levels of Zn may occur in soils which have been affected by wastes and sludges (Juste and Mench 1992).

15.2 Zinc in Physiology

15.2.1 Uptake and translocation

Zinc is taken up as Zn^{2+} and currently it is not yet clear whether uptake is as facilitated diffusion through membrane channels specific for Zn^{2+} or whether it is mediated by specific transporters (Fox and Guerinot 1998). Possibly both systems function for transporting Zn^{2+} across the plasmamembrane. In both cases the uptake is driven by the plasmalemma H^+ pump and hence related to the provision of ATP (see page 118). Fox and Guerinot (1998) reported that there are genes for specific Zn transporters mediating a Zn concentration-dependent uptake of Zn^{2+} with K_M values between 10 and 100 (mmol/m^3. These values correspond to the Zn^{2+} concentrations prevailing in the rhizosphere (Norvell and Welch 1993). Interestingly, the proteins found for the specific Zn uptake did not show any activity for the uptake of Fe (Fox and Guerinot 1998).

The supposition that Zn^{2+} is taken up *via* facilitated diffusion or by specific Zn transporters is in line with earlier findings of Schmid *et al.* (1965) with barley roots. These workers observed a steady uptake rate for Zn typical of metabolic uptake. Zn uptake was considerably reduced by low temperature and metabolic inhibitors. The same observation was made in sugar cane leaf (Bowen 1969). In addition both experiments showed that Cu^{2+} strongly inhibited Zn^{2+} uptake. It seems possible that these two ion species compete for the same uptake system. Similar competitive effects of Fe and Mn on Zn uptake have been reported in rice seedlings (Giordano *et al.* 1974). In addition these workers showed that severe retardation of Zn^{2+} absorption is brought about by various metabolic inhibitors. Depressing effects on Zn uptake by the alkaline earths ($Mg^{2+} > Ca^{2+} = Sr^{2+} = Ba^{2+}$) over a wide concentration range were also observed in wheat plants (Chaudhry and Loneragan 1972). Presumably these effects are related to low affinity cation channels which may bring about a non-specific competition between Zn^{2+} and earth alkaline cations (see page 119). More recent data of Gonzales *et al.* (1999) show that cation uptake (Ca^{2+}, Cd^{2+}, Zn^{2+}) across the tonoplast into the vacuole is a H^+/antiport and therefore driven by the pH difference between the vacuole and the cytosol. The authors assume that this tonoplast transport is mediated by specific transporters. If this were the case Zn^{2+} would also compete with other cation species for the driving force (pH difference).

Similarly to Mg^{2+} and Ca^{2+} Zn is also present in the xylem sap as Zn^{2+} and as such is translocated from roots to shoots. Electrophoretic evidence indicates that Zn^{2+} is not bound to stable ligands as is the case with Cu^{2+}, Ni^{2+}, and Fe^{3+}. In tomato exudates Tiffin (1967) observed that Zn is slightly cathodic and concluded that it is not translocated as citrate, as zinc citrate complexes are anodic. Presumably Zn is phloem-mobile according to results of Caballero *et al.* (1996) who found that the redistribution of Zn from the vegetative to the generative plant parts in vetch was as good as that of N and P which are known to be highly mobile in the phloem.

For most plant species Zn concentrations in leaves below 10 to 15 μg Zn/g dry matter are indicative of Zn deficiency and concentrations in the range of 20 to 100 μg Zn/g dry matter are sufficient (Boehle and Lindsay 1969). Figure 15.1 shows the effect of Zn application on the Zn concentration in whole plants, roots, and older leaves of maize from a field experiment (Jarausch-Wehrheim *et al.* 1999). The data relate to trials started in 1974 in which farmyard manure (control) and Zn-contaminated sewage sludge were applied every second year at rates of 10 Mg (Mega gram) and 100 Mg dehydrated sludge/ha. In the control treatment the Zn concentration in whole plants, older leaves, and roots was <100 μg Zn/g dry matter. Highest concentrations were found in the treatment with the heaviest sludge application rate. The Zn concentrations in plants with the low sludge rate were between the other two treatments. Zn concentrations declined in all treatments with time and differences between Zn concentrations in roots and whole plants of the three treatments were lowest at maturation. This contrasts with the Zn concentrations in older leaves which remained at a relatively high level throughout the growth period. In the treatment with the high rate of sludge application Zn concentrations in older leaves even increased with the advance of growth. This clearly shows that the excess Zn is deposited to a large extent in older leaves. Zinc concentrations in the grains were about 50 μg Zn/g dry matter and were not influenced by the contaminated sludge application. From this finding one may conclude that the Zn concentration in grains is hardly influenced by the Zn in the xylem sap.

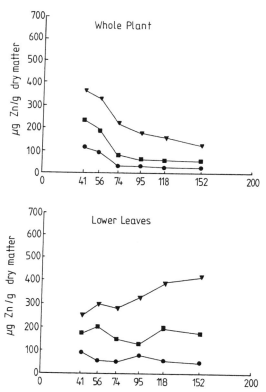

Figure 15.1 Effect of cumulative Zn application from 1974 until 1993 with FYM (●) 29 kg Zn/ha, low rate of sewage sludge (■) 863kg Zn/ha, and high rate of sewage sludge () 5053 kg Zn/ha on Zn concentration in whole plants and older leaves of maize. The figures on the x-axis indicate the days after sowing (after Jarausch-Wehrheim *et al.* 1999).

Interaction between Zn and P has been studied by many workers, and high levels of P supply are well known to induce Zn-deficiency. Leaving aside the soil-plant relationship which is discussed later, it has been shown that excess phosphate results in a metabolic disorder and may lead to Zn deficiency symptoms. Thus Marschner and Schropp (1977) found that high phosphate rates to vine grown in pot experiments using a calcareous soil induced Zn deficieney symptoms in the leaves. In addition growth depression and low Zn concentrations were observed in the younger leaves. In parallel solution culture experiments these workers were not able to induce Zn deficiency, although the Zn concentrations in the leaves of the vine plants were lower than the Zn concentration found in the leaves with Zn deficiency symptoms. It is suggested that phosphate may affect the physiological availability of Zn in plant tissues. The old idea that Zn becomes ineffective in metabolism because it precipitates as $Zn_3(PO_4)_2$ 4 H_2O can now largely be ignored. The solubility of the compound is such that this is too high to cause Zn deficiency.

An interaction between Zn and Fe has been observed by Warnock (1970). Phosphorus induced Zn-deficient maize plants were found to accumulate high levels of Fe, and to a lesser extent Mn. Interference of excess Fe was suggested as a contributory factor in the physiological malnutrition of Zn-deficient plants.

15.2.2   Biochemical functions

According to Fox and Guerinot (1998) Zn is an essential catalytic component of over 300 enzymes. Most of them are Zn-metalloenzymes in which the Zn is bound to three ligands, N or O atoms of the protein and in addition to a solvent water molecule (Coleman 1992). Solvent H_2O means that the water molecule can be replaced mainly by the substrate. Carbonic anhydrase functions in this way as shown in Figure 15.2. In the first step the $Zn-OH_2$ releases a proton so that $Zn-OH^-$ is formed to which a molecule CO_2 is added in the second step. The reaction product is $Zn-HCO_3^-$ the HCO_3^- of which is replaced by H_2O (3. step).

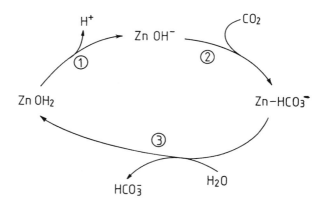

Figure 15.2   Reaction of carbonic anhydrase. 1.Release of H^+. — 2.Carboxylation. — 3. Replacement of HCO_3^- by H_2O (after Coleman 1992).

The reaction cycle may turn in both directions so that either HCO_3^- from CO_2 or CO_2 from HCO_3^- is produced. The overall equation is thus as follows:

$$H_2O + CO_2 \Leftrightarrow H^+ + HCO_3^-$$

The reaction does not need enzymic support but the uncatalyzed interconversion between CO_2 and HCO_3^- is slow and does not match metabolic demand by far. The importance of carbonic anhydrase for photosynthesis is described by Badger and Price (1994) in a useful review article. The enzyme is located in the cytosol and in chloroplasts and according to the metabolic requirement it may provide HCO_3^- or CO_2. Carbonic anhydrase is particularly important for C-4 species in which PEP carboxylase directly uses HCO_3^- as substrate whereas C-3 species have only a low requirement for carbonic anhydrase. In C-4 species carbonic anhydrase is mainly located in the chloroplasts. Cyanobacteria and microalgae require carbonic anhydrase to convert the accumulated pool of HCO_3^- to CO_2 which is the direct substrate for Rubisco (Badger and Price 1994).

Alcohol dehydrogenase is a complex in which the Zn is bound by 3 ligands to the protein and at one site a H_2O molecule is adsorbed. The enzyme catalyses the reaction:

$$CH_3\text{-}CHO + FADH_2 \Rightarrow CH_3\text{-}CH_2OH + FAD$$

In this reaction the acetaldehyde is reduced to ethanol. This reaction plays a role in roots under anaerobic conditions where oxygen is limiting so that the pyruvate originating from the glycolytic pathway cannot be completely used by the TCA cycle owing to a lack of oxygen. Under these conditions the pyruvate is decarboxylated to acetaldehyde which as shown above is then reduced to ethanol. The ethanol then leaks out of the roots into the rhizosphere (Drew 1988). This reaction avoids the accumulation of pyruvate in the root cells and a blockage of the glycolytic pathway.

Cu-Zn superoxide dismutase (SOD) is a complex of about 30 kDa which is required for the detoxification of the superoxide radical (Bowler *et al.* 1992). In green plant tissues the enzyme is mainly located in the stroma of chloroplasts where superoxides may be produced by the e^- uptake of oxygen, mainly supplied by photolysis. Oxygen is a strong e^- scavenger and the superoxide radical thus produced is a toxic radical which may attack and break down various essential molecules (Cadenas 1989). The reaction catalyzed by Cu-Zn SOD is the uptake of $2H^+$ by 2 superoxide radicals and with the production of H_2O_2 as shown in Figure 15.3. The essential step of this reaction is that the single e^-, indicated by a point in Figure 15.3 of the superoxide becomes paired and in this way the toxic radical is neutralized.

Sunscald frequently occurs in countries with hot climates and high light intensities and is characterized by the destruction of photosystems and membranes by oxygen radicals leading to severe loss of fruit and vegetable quality. Plant species which are resistant to sunscald such as tomatoes, pepper, and cucumber possesss high concentrations of SOD enzymes (Bowler *et al.* 1992). Plants well supplied with Zn are protected in this respect (Marschner and Cakmak 1989). Cakmak *et al.* (1997b) reported that cereal species and cultivars differed considerably in efficiency in growing in a Zn deficient soil and that the sensitivity of the cultivars to show the Zn deficiency symptoms were closely related to the Cu/Zn SOD activity in the leaf tissue. In the treatment with no Zn application, rye showed no deficiency symptoms and was not affected in growth, while durum wheat showed severe deficiency symptoms - whitish-brown necrotic patches on leaf blades - and a very much lower shoot dry weight. The sensitivity of bread wheat to Zn deficiency was in between the rye and durum wheat. Interestingly in the no Zn application treatment all three groups of cereals showed low Zn concentration, in the leaves between 6-7 μg Zn/g dry matter. The reason why rye was able to use this low leaf Zn concentration so efficiently as compared with durum wheat is not yet known.

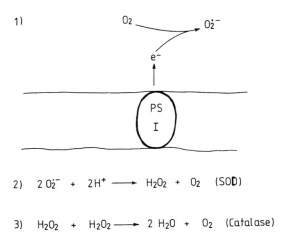

2) $2 O_2^{\cdot-} + 2H^+ \longrightarrow H_2O_2 + O_2$ (SOD)

3) $H_2O_2 + H_2O_2 \longrightarrow 2 H_2O + O_2$ (Catalase)

Figure 15.3 1. Formation of a superoxide radical by O_2 and the uptake of e^- released by photosystem I. — 2. Detoxification of the superoxide radical and formation of superoxide by the Cu-Zn SOD. — 3. Catalase reaction in which 1 H_2O_2 is reduced and 1 H_2O_2 oxidized thus forming water and oxygen.

Zinc containing enzymes are also involved in the carbohydrate metabolism. The Zn containing enzymes aldolase and fructose 1, 6-bisphosphatase have the same substrate, namely 1,6-fructose bisphosphate. Aldolase splits 1,6-fructose bisphosphate into two molecules of triosephosphate and thus "opens" the glycolytic pathway while the 1,6-fructose bisphosphatase dephosphorylates the 1,6-fructose bisphosphate to 6-fructose monophosphate which is the precurser for the synthetic pathway of carbohydrates (sucrose, polysaccharides).

Zinc is very closely involved in the N metabolism of plants. In Zn deficient plants protein synthesis and protein levels are makedly reduced, and amino acids and amides are accumulated. Experiments with *Euglena gracilis* have established that RNA polymerase contains Zn and that when Zn is absent the enzyme is inactivated and RNA synthesis impaired (Falchuk *et al.* 1977). The enzyme catalyses the transcription of the DNA and thus transfers the information of the DNA to RNA. This makes evident the universal importance of Zn for fundamental processes such as the multiplication of genetic information and the transfer of genetic information on protein synthesis. Praske and Plocke (1971) already reported that Zn deficiency affected the structural integrity of the cytoplasmic ribosomes. These organelles contain high amounts of Zn and when Zn is deficient they become extremely unstable.

Alkaline phosphatase is a dimeric Zn-containing enzyme that has its maximum activity at alkaline pH and hydrolyzes phosphate monoesters. The enzyme plays a role in the microbial hydrolyzation of organic phosphates in the rhizosphere (page 458). Zinc storage proteins are of particular importance in mammalia. Here the Zn is bound to the S atoms of cysteine rests. Coleman (1992), who has written a useful review article on Zn proteins emphasizes that Zn is the second most abundant trace element found in eukaryotic organisms, second only to iron.

In Zn deficient tomato plants Tsui (1948) observed low rates of stem elongation, low auxin activities and low tryptophan concentrations. Salami and Kenefick (1970) have confirmed this work growing maize in nutrient solution. These workers found that Zn deficiency symptoms could be eliminated by additions of either Zn or tryptophan to the nutrient medium thus providing indirect evidence of the necessity of Zn for the synthesis of optimum tryptophan levels. Presumably tryptophan synthase requires Zn as was found for *Neurospora*.

15.2.3 Zinc deficiency

Plants suffering from Zn-deficiency often show chlorosis in the interveinal areas of the leaf. These areas are pale green, yellow, or even white. In the monocots chlorotic bands form on either side of the midrib of the leaf appear which later become necrotic. The internodes are short and frquently the grain formation is affected (Brennan 1992). In fruit trees leaf development is adversely affected. Unevenly distributed clusters or rosettes of small stiff leaves are formed at the ends of the young shoots. Frequently the shoots die off and the leaves fall prematurely. In apple trees the disease occurs in the early part of the year and is known as rosette or little-leaf. This deficiency is shown in Plate 15.1. Not only is leaf development restricted. Fewer buds are formed and those that are, many remain closed. Crop yields are consequently drastically reduced. The bark of Zn deficient trees too is characteristically affected and is rough and brittle (Bould *et al.* 1949). Symptoms of Zn deficiency in vegetable crops are more species-related than are deficiency symptoms of other plant nutrients. In most cases, however, Zn deficiency is characterized by short internodes (stunted plants) and chlorotic areas in older leaves. Sometimes chlorosis also appears in younger leaves (Maynard1979). The deficiency prevents the normal development of chloroplast grana and vacuoles are developed in them (Jyung et al. 1975). The level of Zn in Zn-deficient plants is low as shown in Table 15.1 and usually in the range of about 0-15 μg Zn/g dry matter.

Low Zn concentrations in yellow needles of spruce have been reported by Zech and Popp (1983). These low Zn concentrations were associated with low Mg concentrations. Brennan (1992) found that wheat deficient in Zn was less resistant to take all (*Gaeumannomyces graminis*).

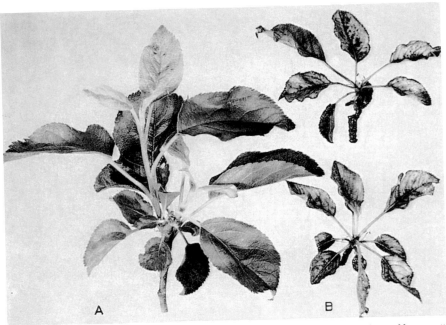

Plate 15.1 Zn deficiency of apple trees (little leaf), A: Normal shoot, B: Zn deficient shoots. Note rosetting, upward curling and waviness of leaf margins, and interveinal chlorosis. (Photo: Bould)

15.2.4   Zinc toxicity and tolerance

Excess Zn supply results in reduction of root growth and leaf expansion which is followed by chlorosis. In soya beans Rauser (1973) observed that a red-brown pigment was formed under conditions of excess Zn. Zinc toxicity may occur in areas particularly in the neighbourhood of Zn ore and spoil heaps. Some plant species, however, are Zn tolerant and are able to grow in soils abnormally high in Zn. Antonovics *et al.* (1971), for example, quote Zn levels of between 600 to 7800 μg Zn/g dry matter of tolerant plant species growing on calamine soils. Generally concentrations in the order of 150 to 200 μg Zn/g dry matter of plant tissue are considered as toxic (Sauerbeck 1982). Bradshaw (1952) working in North Wales, a region that was formerly extensively mined for minerals, showed that populations of the grass *Agrostis tenuis* had evolved which were tolerant to the Zn and Pb contaminated soils of the area. Much valuable information into the mechanism of heavy metal tolerance has stemmed from this observation, for the use of tolerant and nontolerant plant strains has provided a very useful means of investigating Zn and other heavy metal toxicities. Zinc tolerant races of *Agrostis tenuis* take up more Zn into the roots than nontolerant plants (Antonovics *et al.* 1971). Part of the Zn tolerance mechanism is dependent on the ability of the tolerant strains to bind Zn in the cell walls (Turner 1969). In experiments with ^{65}Zn Peterson (1969) observed that Zn is especially associated with the pectate fraction in tolerant ecotypes. There is considerable evidence, however, that cell wall binding in the roots is not the only mechanism by which concentrations of Zn are restricted from active metabolic sites. Godbold *et al.* (1984) found that Zn-tolerant genotypes also accumulated organic anions, especially citrate and malate and therefore it is suggested that these genotypes sequester Zn in form of organic complexes in the vacuole. More recent research of Chardonnens *et al.* (1999) with *Silene vulgaris* provided evidence that sequestering Zn in the vacuole is an important mechanism for detoxifying excess Zn. The authors found that a Zn-tolerant ecotype of *Silene vulgaris* originating from a site at a mine spoil in Belgium had a particular Zn uptake system in the tonoplast which was driven by the hydrolysis of Mg-ATP and even more efficiently by Mg-GTP working like a Zn^{2+} pump. This Zn uptake mechanism in the tonoplast was absent in the Zn sensitive ecotype. In this ecotype root growth was completely inhibited at a concentration of about 700 mmol/m³ in the nutrient solution, whereas for the tolerant ecotype it still grew at a concentration at 4000 mol/m³ Zn. The data of Chardonnens *et al.* (1999) also showed that Zn accumulation in the vacuole was not related to citrate.

Some plants species and ecotypes are able to tolerate very high levels of Zn in the leaves and other upper plant parts. Wainwright and Woolhouse (1975) who found almost equal levels of Zn in Zn tolerant and Zn susceptible *Agrostis tenuis* plants growing in water culture high in Zn. The susceptible plants, however, had lost 50% of their chlorophyll, whereas the tolerant plants were not affected.

There is some biochemical evidence to account for the fact that leaves are able to contain high Zn contents without suffering toxic effects. According to the investigations of Jarausch-Wehrheim *et al.* (1999) excess Zn is mainly deposited in older leaves (Figure 15.1). In a study of a Zn tolerant cultivar of *Phaseolus vulgaris*, Rathore *et al.* (1972) observed no enhanced Zn accumulation in cell walls but very high levels in the cytoplasm. It appears that mitochondria are in some way involved in Zn inactivation. This is strongly suggested from the results of Turner and Marshall (1972) who observed a positive linear relationship between the degree of Zn tolerance of *Agrostis tenuis* and the capacity of mitochondrial preparations from the plants to bind Zn. Wyn-Jones *et al.* (1971) have also observed that oxygen uptake by mitochondria from Zn tolerant *Agrostis tenuis* was inhibited by Zn to a lesser extent than by mitochondria extracted from the susceptible ecotypes. According to Denaeyer-de Smet (1970) some plant species tolerate high Zn levels in the soil due to their low uptake ability for Zn (exclusion mechanism).

15.3   Zinc in Crop Nutrition

15.3.1   Crop requirement and availability

In most soils the total Zn content by far exceeds crop requirement and availability is the important limiting factor. Zinc is present in primary and secondary minerals and may be released by weathering in small amounts. It is adsorbed in form of Zn^{2+}, $ZnOH^+$, and $ZnCl^+$ to clay minerals, carbonates, and organic matter. The adsorption may be nonspecific and in this case the availability is high while the specific adsorption mainly brought about by carbonates is stronger and hence renders Zn less available. Zinc may also be occluded by carbonates and Fe/Al oxides-hydroxides and is in this form not available for plant roots. A major proportion of Zn in soils is bound to organic molecules and this binding mainly includes complexation of Zn due to its affinity to N and O ligands. Also some of these Zn complexes are potentially plant-available since the complexed Zn is equilibrated with the Zn ions in the solution. Iyengar et al. (1981) investigated 19 various soils on their Zn fractions and their availability for maize. The regression equation comprising 6 soil variables accounted for 94% of the Zn uptake of the crop. Zinc uptake was positively influenced by the water soluble + exchangeable Zn, the specifically adsorbed Zn and the organically bound Zn. Uptake was negatively influenced by the Mn oxide bound Zn and by soil pH.

The current method for the determination of available Zn is the soil extraction with diethylene triaminepentaacetic acid (DTPA method) of Lindsay and Norvell (1978). This method is also used for the determination of available Mn and Cu. The DTPA molecules form water-soluble Cu and Zn complexes and therefore decrease the Cu and Zn concentrations in the soil solution. In response Cu^{2+} and Zn^{2+} desorb from soil particle surfaces and are released into the soil solution (Reed and Martens 1996). Dissolved Zn^{2+} in the soil solution and Zn^{2+} bound to organic and inorganic soil particles represent a Zn^{2+} buffer power which according to Nair (1996) is an important component of Zn availability.

Zinc deficiency mainly occurs on soils with high pH, especially on soils rich in carbonates (Cakmak et al. 1996a). Tanaka and Yoshida (1970) in surveying the most important rice growing areas of Asia found that Zn deficiency of rice only occurs on soils with high pH and in particular on calcareous soils of high pH. The availability of Zn is reduced by flooding because of the formation of Zn-sulphides and Zn-carbonates under anaerobic conditions (Yoshida et al. 1971). It appears that Zn deficiency in flooded rice soils results from the combined effect of high pH, high HCO_3^-, levels, sulphide production and impeded internal drainage. The deficiency is often accompanied by visible symptoms of Fe toxicity (Yoshida et al. 1971).

Zinc mobility in soils is important in relation to Zn availability since Zn concentrations in soil solution are so low. The results of Elgawhary (1970) showed that 90.5% of the total Zn required by plants is moved towards the roots by diffusion. Diffusion gradients may therefore occur and root depletion zones similar to those of phosphate have been demonstrated using autoradiographs (Barber et al. 1963). As plant species differ in their Zn requirements, Lindsay (1972) has suggested that this may be one factor explaining differences in sensitivity to Zn deficiency for plants growing in identical environments. Factors which limit the rate of diffusion of Zn to plant roots must also reduce Zn availability. This is probably the most important reason why Zn deficiency often occurs on compacted soils or where root growth is destricted as in container grown plants. Zinc diffusion in soils is also much dependent on soil moisture and this may be the reason why particularly in arid and semi-arid areas Zn deficiency frequently occurs (Cakmak et al. 1996a). According to Lucas and Knezek (1972) organic soils and humic gley soils may contain absolutely low Zn contents and therefore on such soils plants may suffer from Zn deficiency. In practice high available phosphate levels in soil are well known to reduce Zn availability. It was formerly held that this occurred because the formation of zinc phosphate $Zn_3(PO_4)_2 \cdot 4H_2O$ in the soil reduced the Zn concentration in the soil solution to deficiency levels. However, this was not observed in experiments testing the solubility of this compound under various conditions (Jurinak and Inouye 1962).

Plant species and even cultivars of the same species may differ markedly in their potential to exploit soil Zn. This was shown convincingly in field trials in Turkey by Cakmak *et al.* (1997a). Some main results of this investigation are shown in Plate 15.2. From extreme left to right can be seen: Triticale with a normal good growth, Zn inefficient breadwheat cultivar with a poor growth, Zn efficient cultivar breadwheat with a normal growth, durum wheat, highly Zn inefficient, and again Triticale showing vigorous growth.

Plate 15.2 Tolerance and sensitivity of various cereal crops to Zn-deficient soils. From left to right: Triticale, Zn-inefficent bread wheat, Zn-efficient bread wheat, durum wheat and again triticale. In the background same crops with a Zn application of 23 kg Zn/ha. By courtesy of Cakmak.

The reason why some species were able to use the soil Zn more efficiently than others is not yet completely clear. Cakmak *et al.* (1996b) found that the Zn efficient bread wheat cultivars released phytosiderophores at relatively high rates with the onset of visual Zn deficiency symptoms while the release of phytosiderophores by durum wheat remained at a very low level. According to these findings Zn efficient species or cultivars respond to stress by the release of siderophores which can mobilize Zn from soil complexes and thus make it available to roots. Wheat and barley also release phytosiderophores from their roots when the Zn supply is inadequate (Chang *et al.* 1998). The problem needs further investigation since there is not a clear relationship between the release of phytosiderophores and the succeptibility of all breadwheat genotypes to Zn deficiency (Erenoglu *et al.* 1996).

Because of the low mobility of Zn in the soil VA mycorrhizal can substantially increase the uptake of Zn like that of phosphate by the host plant (Kothari *et al.* 1990). This form of acquisition can be of particular importance in ensuring adequate Zn supply in high pH soils. The application of phosphate fertilizers usually decreases mycorrhizal activity so that on soils poorly supplied with Zn this treatment can induce Zn deficiency Martens and Westermann 1991. Zinc concentrations of various plant species differing in Zn supply are shown in Table 15.1 (Boehle and and Lindsay 1969).

Table 15.1   Zinc concentrations of various crops in $\mu g/g$ dry matter (Boehle and Lindsay 1969)

	Deficient	Low	Sufficient	High
Apple Leaves	0–15	16–20	21–50	>51
Citrus Leaves	0–15	16–25	26–80	81–200
Lucerne Tops	0–15	16–20	21–70	>71
Maize Leaves	0–10	11–20	21–70	71–150
Soya Bean Tops	0–10	11–20	21–70	71–150
Tomato leaves	0–10	11–20	21–120	>120

15.3.2   Zinc application

According to FAO investigations on a global scale more than 30% of agricultural soils are deficient in Zn (Sillanpää 1982). This shows that Zn fertilizer application deserves attention. Zn uptake by crops is usually less than 0.5 kg/ha/season. In practice Zn deficiency is easy to correct either by spraying or by soil application with Zn fertilizers. Cakmak *et al.* (1996a) reported that in field trials with rates of 23 kg Zn/ha wheat grain yield in average was increased by 43%. $ZnSO_4$ is the most commonly used fertilizer largely because of its high solubility. On acid sandy soils it may be preferable to spray the crop or use a less readily available Zn source because $ZnSO_4$ is very easily leached. The same applies to alkaline soils which fix Zn very strongly. Under such conditions Zn chelates are often used. With the intensification of agricultural production in tropical and subtropical zones the occurrence of Zn deficiency has increased. De (1974) found enormous grain yield increases of pearl millet by an application of $ZnSO_4$ on sandy soils in India. Zinc application to wheat on two locations in Egypt evidenced that on the alluvial soil the Zn response was only modest and the foliar application was as efficient as the soil application. For the calcareous soil, the soil application of Zn resulted in a marked increase in grain yield (Serry *et al.* 1974). Randhawa et al. (1978) hold the view that many rice growing areas are deficient in Zn and merit Zn application. Rates of 50 to 100 kg Zn/ha in form of $ZnSO_4$ are recommended. Zinc application together with gypsum has often proved to be especially beneficial (Takkar and Singh (1978). It is of particular interest that rice is prone to

Zn deficiency under cold weather conditions. Generally soil Zn application is superior to a foliar spray. A detailed discussion of Zn fertilizers and their use is given by Lindsay (1972). Zinc application may depress the uptake of Cd as was found by McLaughlin *et al.* (1995). Zinc applied with rates of 50 and 100 kg Zn/ha lowered the Cd concentrations in potato tubers significantly.

CHAPTER 16

SOIL COPPER

16.1 Soil Copper

Copper occurs in the soil almost exclusively in divalent form. The largest fraction of Cu is usually present in the crystal lattices of primary and secondary minerals. In addition a high proportion of Cu is bound by the soil organic matter. The Cu ion is adorbed to inorganic and organic negatively charged groups and is dissolved in the soil solution as Cu^{2+} and organic Cu complexes. Copper is specifically adsorbed to carbonates, soil organic matter, phyllosilicates, and hydrous oxides of Al, Fe, and Mn (Reed and Martens 1996). In a fractionation study of Cu in British soils, McLaren and Crawford (1973) found total Cu levels in the range of about 5-50 μg/g soil of which a high proportion was present in occluded or lattice form. The Cu concentration of the soil solution is usually very low being in the range of 1×10^{-5} to 6×10^{-4} mol/m^3. The concentration of Cu^{2+} in the soil solution decreases sharply with an increasing pH whereas the concentration of organic Cu complexes in the soil solution is less dependent on soil pH. Higher pH may even promote the dissolution of organic Cu complexes (Mc Bride 1989). The divalent Cu ion has a strong affinity to soil organic matter compared with other divalent cations as is shown by the following sequence (Schnitzer and Skinner 1965, 1967):

$$Cu > Ni > Pb > Co > Ca > Zn > Mn > Mg$$

Copper may be bound in chelated form to organic matter which means that it is held by covalent and coordinative bonds. It is coordinatively bound to O, N, and S atoms following the preference S > N > O. However, since N is present in soil organic matter in much higher concentrations than S, N atoms play the major role in chelating Cu^{2+} in quantitative terms.

Recent findings indicate that a large proportion of Cu^{2+} is also bound covalently to carboxylic groups as occur in humic and fulvic acids in soil organic matter and as shown below. This covalent monodentate binding of Cu^{2+} to the carboxylic group

$$R\text{-}COO^- + Cu^{2+} \Rightarrow R\text{-}COOCu^+$$

is not as strong as that of the Cu^{2+} chelate. Compared with other divalent cations, however, it is strong as can be seen from Figure 16.1 (Mc Bride 1989). In this work the bonding strengths of various divalent cations (transition elements) were compared with the bonding strength for Ca^{2+} with three different compounds: humic acid, acetic acid and citric acid. For most of the cation species tested citric acid formed by far the strongest bonds because of its capability of chelating the cation. Bonding strengths formed by the cations and carboxylate group of acetic acid and humic acid were much weaker. Of all the cation species tested, however, the bonds formed by Cu^{2+} were by far the strongest.

599

The equilibrium concentration of Cu^+ maintained by sparingly soluble Cu^{2+} salts such as carbonates and oxides is higher than the normal levels of Cu^{2+} in the soil solution. The presence of carbonates or oxides in the soil therefore plays no part in restricting Cu^{2+} solubility. The Cu^{2+} concentration in the soil solution is governed by Cu adsorption to organic and inorganic soil particles. For this reason Cu is very immobile in soils and practically not leached into deeper soil layers. Copper added to the soil as a result of the use of Cu containing sprays or fertilizers is thus largely restricted to the upper soil horizons (Delas 1963). The Cu concentration of many soils therefore decreases down the profile. Copper displacement from soils can be brought about either by strong acids or the use of organic compounds which form Cu complexes. The significance of these complexing reagents can be appreciated by the fact that KCN is capable of extracting more than 50% of the total soil Cu whereas non-complexing reagents extract only very small quantities (Beringer 1963).

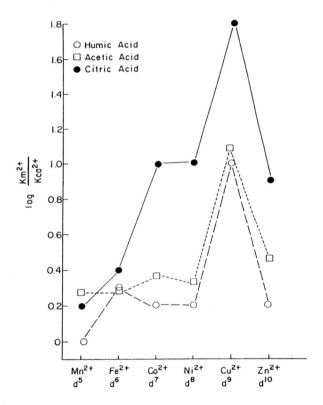

Figure 16.1 Relative bonding strengths of various heavy metals to humic acid, acetic acid and citric acid compared with the bonding strengths of Ca^{2+} (after Mc Bride 1989).

As Cu deficiency occurs primarily on humus rich soils which bind Cu^{2+} strongly, it may be supposed that certain defined organic forms are more readily able to render Cu unavailable. This has been confirmed by neutron activation studies of organically bound Cu in soil solution by Mercer and Richmond (1970). These workers showed that Cu availability in organic soils depended not only on the concentration in soil solution but also on the form in which the Cu occurred. Copper complexes in the soil solution of molecular weight <1000 were much more available to plants than Cu complexes with a molecular weight in excess of 5000.

16.2 Copper in Physiology

16.2.1 Uptake and translocation

Copper is taken up by the plant in only very small quantities since the Cu concentration in most plant species is low and in the range of 5 to 20 $\mu g/g$ dry matter. The mechanism of Cu uptake is still not understood and it is not yet clear whether Cu^{2+} must be reduced before it is transported across the plasmalemma as is the case for yeast (Fox and Guerinot 1998). In *Arabidopsis thaliana* a gene (*COPT1*) was found which encodes for a Cu transporter. This gene is expressed in flowers stems and leaves but not in roots and may be of importance for Cu transport in tissues. Whether it also encodes for a transporter mediating the uptake of Cu from the soil solution across the root plasmalemma remains to be shown. Copper uptake appears to be a metabolically mediated process and there is evidence that Cu strongly inhibits the uptake of Zn and *vice versa* (Bowen1969). This apart, however, the uptake of Cu is largely independent of competitive effects and relates primarily to the levels of available Cu in the soil. Phytosiderophores also bind Cu^{2+} and Römheld (1991) suggests that they may be of importance for exploiting soil Cu. Mugineic acids which are released from the roots of graminaceous species and important for the uptake of Fe^{III} by these plants (see page 555) are less specific for Cu^{2+} and presumably Cu is not taken up in this form (Ma and Nomoto 1996).

Using excised roots from different plant species Keller and Deuel (1958) established that Cu^{2+} is able to displace most other ions from root exchange sites and is very strongly bound in the root apoplast. This observation may well account for the finding that roots are frequently higher in Cu concentration than other plant tissues (Hill 1973).

Copper is not readily mobile in the plant although it can be translocated from older to younger leaves and from vegetative plant parts to seeds as was found for vetch during pod filling by Caballero et al. (1996). Results of Loneragan (1981) show that the movement of Cu is strongly dependent on the Cu status of the plant. In wheat plants well supplied with Cu, movement from the leaves to the grains can take place, but in deficient plants Cu is relatively immobile. Loneragan (1981) suggests that Cu entering leaves is bound by N compounds such as proteins which release little Cu for phloem transport until they are hydrolyzed. This explains why Cu deficiency symptoms develop characteristically in young leaves as found for wheat by Brennan (1994) who emphasizes that the Cu concentration of the youngest leaf blade is a good indicator of the Cu supply to this crop plant. In a number of plant species Tiffin (1972) showed that Cu is present in xylem exudates in anionic Cu complex form. Several anionic forms of Cu have also been observed in ryegrass (Bremner and Knight 1970). Copper has a strong affinity for the N atom of amino groups and it appears quite likely that soluble N compounds such as amino acids act as Cu carriers in xylem and phloem (Loneragan 1981).

16.2.2   Biochemical functions

Copper has a number of properties which to a large extent control its biochemical behaviour. Enzymatically Cu bound participates in redox reactions which are mostly dependent on the valency change.

$$Cu^{2+} + e^- \Leftrightarrow Cu^+$$

In this respect it is similar to iron although Cu^+ is very much less stable than the corresponding Fe^{2+}. In protein complexes, Cu has a particularly high redox potential. One may distiguish between Cu-containing enzymes which react with O_2 the oxidases and those which do not require O_2 This latter group mainly include the "blue proteins" the most well known of which is plastocyanin. As shown in Figure 3.17 plastocyanin is an essential redoxsystem of the photosynthetic e^- transport chain. In most of the blue proteins the Cu atom is coordinated with three ligands, the S of the cysteine and with one N of each of two histidine residues thus forming a trigonal planar structure (Gray and Winkler 1996).

Since plastocyanin is an essential redoxsysten of the photsynthetic e^- transport chain this explains the relatively high concentrations of Cu occurring in chloroplasts more than half of which is present in plastocyanin. Plastocyanin is an acidic protein containing 2 Cu atoms per molecule. It is not surprising therefore that analysis of the chloroplast pigments and activities of photosystem I and II in Cu deficient spinach have revealed that the deficiency has the most marked effect in depressing the content of plastocyanin and the activity of photosystem I (Baszynski et al. 1978).

The most common of the three types of superoxide dismutase isoenzymes contains Cu and Zn (Cu-ZnSOD). This protein has a molecular weight of about 32 000 and contains 2 Zn and 2 Cu atoms. Superoxide dismutases occur in all aerobic organisms and are essential for their survival in oxygen. They protect the organism from damage from superoxide radicals which can be formed when a single electron is transferred to O_2 as may happen particularly at steps where a single e^- is transferred e.g in the e^- transport chains of photosynthesis (see Figure 15.3) and respiration.

$$O_2 + e^- \Rightarrow O_2^- \text{ (superoxide, oxygen radical)}$$
$$O_2^- + O_2^- + 2 H^+ \Rightarrow H_2O_2 + O_2 \text{ (SOD reaction)}$$
$$2 H_2O_2 \Rightarrow 2H_2O + O_2 \text{ (catalase reaction)}$$

The hydrogen peroxide produced is broken down by catalase. The very high proportion of SOD in leaves which appears to be localized in the chloroplast — more than 90% — indicates a role for SOD in protecting the photosynthetic apparatus (Bowler et al. 1992).

Periplasmatic SODs including Cu-ZnSOD also occur in various bacteria species (Fridovich 1995) since the superoxide radical may also be formed on the outer side of the plasmamembrane of living organisms by ultraviolet radiation in surface water.

Cytochrome c oxidase, the terminal oxidase in the mitochondrial transport chain is one of the most well studied of the Cu containing enzymes. The enzyme transfers 4 e$^-$ to O$_2$ of which two are provided by the Fe^{II} in each haem and two by the Cu^I by each of the two Cu atoms (Figure 16.2). This unique step is of paramount importance for all organisms generating ATP by respiration.

The enzyme phenolase which occurs in mitochondria and also in the thylakoid membrane of chloroplasts has two distinct enzyme functions. In its role in the mono-oxygenation (hydroxylation) of monophenols it is sometimes referred to as tyrosinase and as the enzyme bringing about the mono-oxygenation of o-diphenols it is sometimes called polyphenol oxidase. According to Walker and Webb (1981) it is preferable to use the term phenolase to cover the two associated functions. The reaction sequence in phenol oxidation is shown in Figure 16.2.

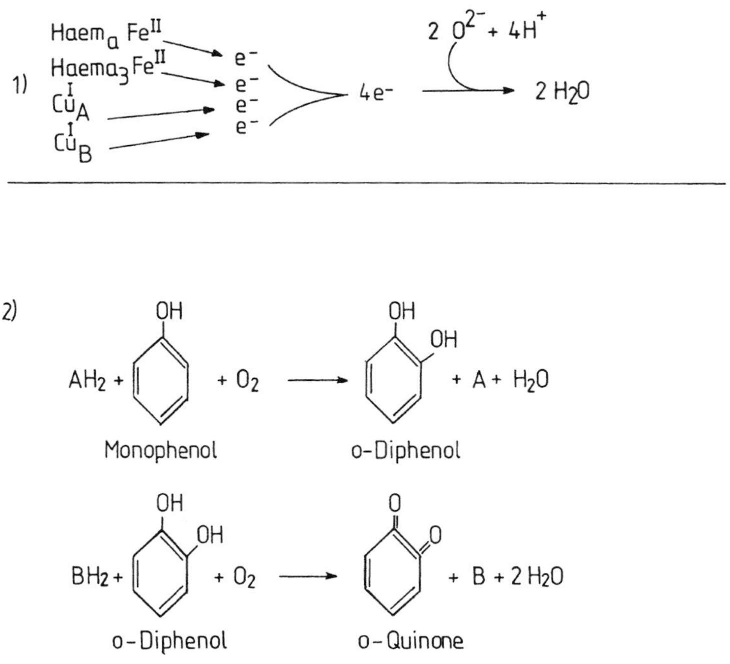

Figure 16.2   1. Oxidation of haem Fe^{II} and of Cu^I leading to the reduction of O$_2$ in the terminal oxido-reduction process of respiration by which 2 H$_2$O are produced. — 2.Reaction of phenolase. In the first step monophenol is oxidized to a diphenol, in the second step diphenol is oxidized to quinone.

Basically this is a hydroxylation reaction by which O$_2$ is split and one O atom is "squeezed" between the C and the H atom of the phenol ring. The other O oxidizes the H donor AH_2 and thus H_2O is formed. In this reaction the monophenol molecule is oxidized to a diphenol which in turn is oxidized to an o-quinone compound. Accumulation of these o-quinones may give rise to polymerisation whereby dark brown melanin compounds are formed. This occurs when plant tissues such as apple or potato are cut open and exposed to the atmosphere. This may also occur in the fermentation of grapes and spoil the wine. It is for this reason that must for fermentation should not contain O_2. For this reason sulphite is added to must at the beginning of fermentation to depress the level of O_2. Phenolase and laccase, another Cu protein which catalyses the oxidation of phenols, are involved in lignin synthesis. Polyphenol oxidase produces phenolic molecules which function as precursors for linin synthesis. A typical symptom of Cu deficiency is thus impairment of lignin synthesis. In Cu deficient tissues phenolase activity is lowered and an accumulation of phenols occurs (Robson *et al.* 1981).

Ascorbic acid oxidase catalyses the oxidation of ascorbic acid to dehydroascorbate, and occurs widely in higher plants. The reaction is analogous to phenol hydroxylation also shown in Figure 16.2. Hydroxyproline is required for the synthesis of collagen and therefore is of particular importance in animal metabolism. The redoxsysem ascorbate/dehydroascorbate represents an essential step in the "Halliwell-Asada reaction pathway" (page 440) which is also of importance for the elimination of toxic O radicals (Bowler *et al.* 1992).

Another group of Cu proteins are the amine oxidases which catalyze oxidative deamination.

$$R\text{-}CH_2\text{-}NH_2 + O_2 + NADH + H^+ \Rightarrow R\text{-}CHO + NH_3 + H_2O + NAD^+$$

These enzymes can also use polyamines such as putrescine and spermidine as substrates. Copper is thus closely related to polyamine metabolism.

Copper influences both carbohydrate and nitrogen metabolism. In the vegetative stage Cu deficiency can induce lower concentration of soluble carbohydrates (Brown and Clark 1977) as might be expected from the role of Cu in photosynthesis. Graham (1980) showed that in Cu deficient wheat plants after anthesis there is an accumulation of soluble carbohydrate in the leaves and roots. This accumulation is attributable to the absence of a sink for carbohydrates resulting from the failure of flower set and consequent lack of the grain filling process because of the sterility of the pollen in the Cu deficient plants. Enough carbohydrates were present in these plants to support excessive tillering. This is an explanation of why excessive tillering is such a common feature in Cu deficient cereals (see Plate 16.1). The pollen sterility found in Cu deficient plants is due to the effect of Cu deficiency in impairing lignification of the anther cell walls.

According to Snowball *et al.* (1980) Cu is specifically required for N fixation, but evidence is still lacking as to the mechanism. It should be borne in mind that Cu can have an indirect effect on N fixation since N fixing nodules have a very high demand for carbohydrate which can be limiting in young plants suffering from Cu deficiency.

16.2.3 Copper deficiency and toxicity

Copper deficiency is well known in a number of different crop plants. In cereal crops the deficiency shows first in the leaf tips at tillering although in severe cases it may appear even earlier. The tips become white and the leaves are narrow and twisted. The growth of the internodes is depressed (Bergmann 1992). As growth progresses the deficiency becomes more severe and in extreme cases ear or panicle formation is absent (see Plate 16.1).

Plate 16.1 Cu deficiency in oats; left normal, right deficient (Photo Mengel).

A typical feature of the deficiency in cereals is the bushy habit of the plants with white twisted tips and a reduction in panicle formation. When the deficiency is less pronounced panicle formation may occur but the ears are not fully developed and may be partially blind (Scharrer and Schaumlöffel 1960). This symptom is associated with the role of Cu in pollen grain viability. It has been shown by Knight *et al.* (1973) that anthers which contain the pollen and ovaries are normally very high in Cu and presumably also have a high Cu demand.

In Cu deficient trees the development of "pendula" forms may occur (Oldenkamp and Smilde (1966). It is believed that this deficiency symptom is indicative of an impairment of lignin synthesis resulting from a lack of the two Cu containing enzymes phenolase and laccase. According to Bussler (1981) inhibition of lignification in Cu deficient tissue is associated with inadequate development in xylem vessels. This function of Cu explains the close relationship in cereals between Cu nutritional status and haulm stability. It also clarifies the interaction between Cu and N fertilizer application in relation to lodging (Vetter and Teichmann 1968). The characteristic behaviour of Cu deficiency affecting newly developing tissues, appears to be dependent on the low mobility of Cu in deficient plants (Loneragan 1981)

For most plant species high concentrations of available Cu in the nutrient medium are toxic to growth. The effect appears to relate in part to the ability of Cu to displace other metal ions and particularly Fe from physiologically important centres. Chlorosis is thus a commonly observed symptom of Cu toxicity, superficially resembling Fe deficiency (Daniels *et al.* 1972). The inhibition of root growth is one of the most rapid responses to toxic Cu levels. Wainwright and Woolhouse (1975) compared the effects of increasing Cu concentrations in a nutrient culture on the plasmalemma of the roots of a non-tolerant and a Cu-tolerant race of *Agrostis tenuis*. Damage to the plasmalemma as measured by K^+ leakage was considerably higher in the non tolerant race. It was concluded that as the effect of excess Cu was to damage membrane structure, part of the Cu tolerance behaviour operates through an exclusion mechanism in the plasmalemma. Calcium plays an essential role in maintaining membrane structure (see page 115). The findings of Wallace *et al.* (1966) that high levels of Ca^{2+} alleviate Cu toxicity thus also supports the view that excess Cu exerts a detrimental influence on membrane structure.

In an investigation of the flora of high Cu bearing soils in Zaire, Duvigneaud and Denaeyer-de-Smet (1959) observed that some plant species were capable of accumulating Cu to levels of the order of 1000 $\mu g/g$ dry matter. Why such plants show no signs of Cu toxicity let alone are able to grow on these soils at all is not yet clear. Some species do accumulate a high proportion of Cu in the roots so that it may be supposed that in part exclusion of uptake is operative. Large amounts of Cu may therefore be bound to the negatively charged sites of the pectic substances (COO$^-$ groups) in the cell walls in the root cortex. In other species, however, there is no doubt that anywhere from 2 to 50 times the normal values of Cu are found in leaves. In these plants the toxic behaviour of excess Cu is in some way prevented. Wu *et al.* (1975) sequestering Cu complexes in the vacuole or as non-toxic complexes in the cytoplasm is an important mechanism of Cu tolerance. In this respect investigations of Nicholson *et al.* (1980) are of interest who found that *Vigna radiata* grown at toxic Cu levels produced a Cu protein with a MG of 7 to 20 kDa which was similar to plastocyanin. Such a molecule could plant the protect against toxic Cu^{2+}.

16.3 Copper in Crop Nutrition

16.3.1 Crop requirement and availability

As the Cu concentration in plant material is normally less than 10 μg Cu/g dry matter the Cu requirement of crop plants is correspondingly small. Thus a cereal crop will take up about 500 g Cu/ha. Most soils contain adequate levels of available Cu to meet this demand. Soils in which Cu deficiency occurs are either inherently low in Cu or more usually are poor in available Cu. The Cu inherently low group includes soils which are excessively leached, such as the sandy podzolic soils and soils developed on parent material poor in copper. Included in the second category where availability limits plant uptake are organic and peaty soils, calcareous soils and some soils high in clay content. Copper deficiency is common on newly reclaimed peats and for this reason the deficiency has been called "reclamation disease".

Crop species differ in their sensitivity to Cu deficiency. In general, the most responsive crops to Cu fertilizers are oats, spinach, wheat, and lucerne. In the medium range are cabbage, cauliflower, sugar beet and maize, whilst beans, grass, potatoes, and soya beans show a low response. The influence of Cu fertilization on oats, one of the most sensitive crops, is shown in Table 16.1 from the results of Scharrer and Schaumlöffel (1960). These findings from a glasshouse experiment clearly show that Cu deficiency decreases grain yield in favour of the formation of vegetative plant material. The study of Cu deficiency in the field is often more complex than under glasshouse conditions. In an extensive survey of Cu deficiency on chalk rendzina soils Davies *et al.* (1971) observed that deficiency symptoms in barley and wheat were aggravated when these crops followed a Brassica crop. Blackening symptoms were observed in wheat and the deficiency symptoms were accentuated by warm wet summer conditions.

Table 16.1 Effect of Cu application on the yield of oats, grown on a Cu deficient soil (Scharrer and Schaumlöffel 1960)

Cu application	Straw, g pot	Grains, g pot
None	72.6	29.6
1.2	57.0	56.7
8.3	58.4	57.7

Fertilizer application can also lead to the onset of Cu deficiency, and particularly where high levels of N are applied. This depressive effect of N fertilier on grain yield of wheat under Cu deficiency conditions is shown in Table 16.2 from the work of Brennan (1994). These field trials were carried out in Western Australia on a gravelly sandy soil low in available Cu.

Table 16.2 Effect of increasing N fertilizer rates on wheat grain yield which has been supplied with different Cu fertilizer rates 20 years before. The soil was a gravelly sandy soil low in available Cu (modified after Brennan 1994)

N rates	Cu rates applied 1967, kg Cu/ha				
kg N/ha	0	0.68	1.39	2.1	2.8
	Grain yield, kg/ha				
18	890	1390	1368	1430	1400
36	750	1480	1590	1530	1520
72	650	1560	1630	1590	1650
110	500	1640	1670	1680	1750

LSD (0.05) for Cu = 185
LSD (0.05) for N = 95

Twenty years previously the plots other than the controls had received increasing rates of Cu application. In the control plots which had not received Cu increasing N fertilizer rates caused a severe depression in grain yield whereas on the plots which had been fertilized with the Cu, N application led to significant increases in grain yield. Brennan's data showed that on this Cu deficient soil N fertilization depressed the Cu concentration of the youngest leaf blade. Presumably the infuence of N fertilization was to stimulate the growth of early developing leaves and thus also the import of Cu into the leaves. Most Cu in wheat leaves is immobile (Loneragan 1981). Application of N under conditions of limited Cu supply is therefore to deprive the younger leaves and generative organs of Cu with consequent detrimental effects in plant growth and development. The prolonged use of phosphatic fertilizers has also been cited as a cause of Cu deficiency in some soils (Bingham 1963). Applications of Zn fertilizers have also been shown to aggravate Cu deficiency in soils with marginal Cu levels (Chaudry and Loneragan 1970).

16.3.2 Copper application

In the assessment of available soil Cu both inorganic and chelating extractants have been used. Extracting with 1000 mol/m^3 HNO$_3$ Henkens (1965) recommended a minimum of

4 μg Cu/g soil Cu as a measure for the adequate growth of cereal crops. Using the chelating reagent DTPA (diethylenetriaminepentacetate) which is more sensitive than EDTA, Follett and Lindsay (1970) suggested a level of 0.2 μg Cu/g soil as being the critical Cu level for this extractant. Extraction of soil Cu by chelates, including DTPA does not always give reliable results. Thus Haq and Miller (1972) found only a poor correlation between the Cu concentration in maize plants and the amount of Cu extracted from the soil by a number of different chelating agents. As shown below in Table 16.4 even when very large amounts of Cu had been applied to soil the correlation between the Cu concentration in leaves and and the Cu extracted from soil by an extractant, may not be high unless additional soil characteristics and especially soil pH are also taken into account (Miner et al. 1997).

Plant analysis may provide a valuable means of testing the Cu nutritional status of crops. According to Brennan (1994) farmers could monitor Cu concentrations in the youngest emerging leaf blade of wheat at the flag leaf stage. This concentration should be in the range of 2.3 to 3.5 μg Cu/g dry matter and is more or less the same as that suggested by Davies et al. (1971) for wheat and barley leaves with values of less than 2 μg Cu/g dry matter indicative of deficiency and concentrations over 3 μg Cu/g dry matter adequate.

When it is considered that a cereal crop of average yield removes only about 30 to 50 g Cu/ha, it is clear that the amount of Cu which is necessary to apply is small. However, as already mentioned, Cu is strongly bound to the soil and for this reason the amount of Cu fertilizer applied must exceed the crop uptake considerably. As shown in Table 16.2, however rates of about 1 kg Cu/ha were still effective 20 years after application.

Both inorganic and organic Cu fertilizers are used to alleviate Cu deficiency (Martens and Westermann 1991). Most frequently $CuSO_4$, particularly in form of $CuSO_4 \cdot 5H_2O$, is applied to the soil by either band or broadcast application. A single application of about $1-10$ kg Cu/ha is usually adequate on mineral soils whereas somewhat higher levels are needed for organic soils (Reuther and Labanauskas 1966). There are, however, some of problems associated with $CuSO_4$. When the salt is applied to the soil a large proportion of the Cu^{2+} is rapidly brought into solution and may be immobilized by strong adsorption to exchange sites. In addition toxic residual effects can result on some soils. Where low or medium responsive vegetable or field crops are being grown, a total application not in excess of 22 kg Cu/ha has been recommended (Murphy and Walsh 1972). Some of the residual effects of soil application have been alleviated either by the use of Cu metal dusts which release Cu at a slower rate (Kühn and Schaumlöffel 1961) or more usually by the use of Cu chelates. Foliar applications of Cu are usually made using $CuSO_4$, Cu oxychloride, Cu oxide or Cu chelates. Again the sulphate form is less satisfactory because of scorching of the foliage. Seed dressings with Cu salts have been tried, although the results have been variable (Murphy and Walsh 1972).

Cu toxicity in plants does not frequently occur in practice because Cu is very strongly bound to soil particles. Toxicity can appear on soils affected by Cu ores or in soils which have been treated over a period of years with Cu salts. In some soils in France on which vines have been cultivated the prolonged use of Bordeaux

Mixture has had such an effect (Delas1963). Copper toxicity appears to occur most severely on acid soils where Cu is not so strongly bound (Drouineau and Mazoyer 1962), and is thus more available for plants. Copper toxicity problems can become acute particularly where Cu containing, wastes are applied to soils regularly. Such soils can be very severely affected because of the low rate at which Cu is leached into the deeper soil layers. This question has been treated by Dam Kofoed (1980). Pig manure and sludges, especially those of industrial origin may be very high in Cu (see Table 16.3). Frequent application of these materials may result in toxic Cu levels in the soil. Crops differ in their susceptibility to copper toxicity. This is shown in Table 16.4. The long-term application of municipial sludge led to a Cu accumulation in the upper soil layer of 18 to 110 kg Cu/ha (Miner *et al.* 1997). Crops grown on these soils had Cu concentrations in their leaves which were 3 to 6 times higher than required for normal growth and which in some cases exceeded the toxic threshold of 50 μg Cu/g dry matter. Copper extracted by the DTPA solution as well as by other extractants gave values which only poorly correlated with the Cu concentration in the leaves. However, also taking soil pH into account in the multiple regression the correlation was much increased. As the pH rose there was a fall in the amount of Cu extracted indicating that the strength of Cu adsorption by the soil matrix had increased considerably. The correlation was further increased by taking into account of soil characteristics which interestingly differed for the three crop species. In the case of swiss chard (*Beta vulgaris*) with increasing humate the amount of Cu extracted decreased, presumably because of strong binding of Cu^{2+} by the carboxylic groups of the humate (see Figure 16.1). For lettuce taking the cation exchange capacity of the soil into account increased the regression coefficient. This is understandable since CEC is a measure of cations held in the soil on exchangeable sites and the binding of Cu to their sites provides an available source of Cu to plant roots while at the same time preventing the leaching of Cu into the deeper layers of the soil and of the root zone.

Table 16.3   Cu concentrations of organic manures and wastes (Dam Kofoed 1980)

	fresh weight, μg g^{-1}	dry weight, μg g^{-1}
Cattle manure	9	34
Pig manure	21	86
Poultry manure	30	69
Slurry, pig	18	265
Slurry, cattle	4	43
Sludge, household	46	113
Sludge, industry	353	1477

Table 16.4 Effect of long-term Cu application with municipal sludge on Cu concentrations in various crops (Data from Miner *et al.* 1997).

Field No.	1	2	3	4	5
Cumulative Cu applied, kg/ha	18	35	52	63	110

Concentration range of Cu in leaves, μg Cu/g dry matter	
Swiss chard (*Beta vulgaris*)	23.5—53.1
Lettuce (*Lactuca sativa*)	11.7—15.1
Tobacco (*Nicotiana tabacum*)	23.4—32.0

Legumes are supposedly especially sensitive to an excess of Cu in soils. Cu enriched plant material can also be a hazard to animals. Ruminants and especially sheep are more suceptible to fodder rich in Cu than pigs, or poultry. A concentration of 50 μg Cu/g dry matter is considered as upper limit for forage. Investigations in USA showed that forage legumes grown under various soil and climatic conditions had Cu concentrations mainly in the range of 6 to 12 μg Cu/ha dry matter (Kubota *et al.* 1987). Too low Cu concentrations in the forage may induce a Cu deficiency which according to Kubota *et al.* (1987) in most cases is caused by excess levels of Mo in the forage in the range of 10 to 20 μg Mo/g dry matter. Cattle and sheep afflicted with this disease (Mo-induced Cu deficiency), called molybdenosis, show symptoms of Cu deficiency and respond to Cu supplements.

CHAPTER 17

MOLYBDENUM

17.1 Soil Molybdenum

The normal range of Mo concentration in agricultural soils is 0.8—3.3 mg/kg (Kubota 1977). Values can vary widely, however, depending on soil parent material. Soils derived from granitic rocks, shells, slates or argillaceous schists are often high in Mo whereas highly weathered acid soils tend to be deficient (Gupta 1997a). Reddy *et al.* (1997) divide Mo into four major fractions in the soil: (1) dissolved Mo (2) Mo occluded with oxides (*e.g.* Al, Fe and Mn oxides), (3) Mo solid phases including molybdenite (MoS_2), powellite ($CaMoO_4$), ferrimolybdite ($Fe_2(MoO_4)_3$) and $PbMo_4$ and (4) Mo associated with organic compounds. In contrast to the heavy metals already discussed Mo forms anonic species in solution, with molybdate (MoO_4^{2-}) the most prominent above a pH of about 4. This property clearly distinguishes Mo from the other heavy metal nutrients and molybdate more resembles phosphate or sulphate in its behaviour in the soil. Molybdate is adsorbed by sequioxides and clay minerals in an analogous way to phosphate. Adsorption in most cases is thus that of ligand exchange and is rather specific. Of all the plant nutrient anions, molybdate ranks second after phosphate in its strength of adsorptive binding (Parfitt 1978). Molybdate adsorption isotherms are similar to those of phosphate and can be described approximately by the Langmuir equation. Typical Mo adsorption curves are shown in Figure 17.1 from the work of Barrow (1970).

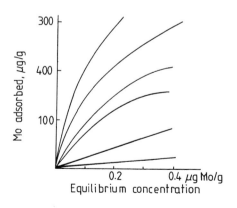

Figure 17.1 Molybdenum adsorption of six soils differing in pH and Mo adsorption capacity (after Barrow 1970).

614

It has been known for many years that the adsorption of Mo in soils, like that for phosphate and sulphate is strongly pH dependent, increasing with decreasing pH. Reisenauer *et al.* (1962) observed maximum retention at about pH 4 and others have reported similar findings (Reddy *et al.* 1997). This effect of pH on adsorption may be explained by competition between OH^- and MoO_4^{2-} ions for adsorption sites and the increased positive charge on adsorbing surfaces at lower pH. This pH dependence of Mo adsorption has practical consequences as Mo deficiency can often be controlled by liming. Adsorbed Mo can be released by other anions than OH^- and phosphate appears to be especially effective (Roy *et al.* 1986).

The Mo concentration in soil solution may thus vary considerably but is usually determined predominantly by soil pH and total Mo content of the soil. As the pH falls the Mo soil solution concentration decreases. Adsorption of molybdate is positively correlated with concentrations of Fe and Al oxides so that highly weathered soils of low pH containing large amounts of amorphous and crystalline Fe oxide minerals are particularly prone to Mo deficiency. Other factors play a part in controlling the Mo concentration soil solution including dissolution and precipitation of Mo containing soil minerals. These include MoS_2 present under reducing conditions and $PbMoO_4$ which appears to control dissolved Mo in alkaline soils (Reddy *et al.* 1997).

There is a lack of any detailed systematic study on the different Mo fractions in soils and their interrelationships (Shuman 1991). In an investigation of the water soluble amorphous Fe oxide and total Mo concentrations of a number of alluvial and calcareous desert soils Elsokkary and Baghdady (1973) reported that from 88—94% of the total Mo was considered unavailable. In the mineral soil horizon 20—50% were present as amorphous and crystalline Fe oxides. In organic samples about half of the Mo present was in the organic plus sulphate fraction.

Soils which are high in Mo and give rise to herbage containing high levels of Mo causing molybdenosis in cattle and sheep. These soils are formed from high Mo-containing parent materials, usually alluvium from igneous rocks and shales (Welch *et al.* 1997). The problem is particularly associated with such poorly drained neutral or alkaline organic soils. Reducing conditions in the soil appear to be important giving rise to ferrous (Fe^{II}) iron which forms soluble ferrous molybdates and molybdites resulting in high concentrations of Mo in the plant.

17.2 Molybdenum in Physiology

Molybdenum is absorbed by plants as molybdate. From water culture experiments it has long been known that uptake can be depressed by SO_4^{2-} (Stout *et al.* 1951). This relationship has been confirmed in both solution culture and soil grown plants (MacLeod *et al.* 1997). The relationship with phosphate uptake is not so clear-cut. Evidence that molybdate is bound and transported across the plasmamembrane by phosphate binding and transport sites has been obtained by Heuwinkel *et al.* (1992). The uptake and transport of $^{99}MoO_4^-$ was found to be markedly enhanced by the roots of P stressed intact tomato plants. Resupply of phosphate depressed the $^{99}MoO_4^-$ uptake. The ionic interactions described above suggest metabolically controlled uptake

(Moore 1974). The form in which Mo is translocated is unknown. Tiffin (1972) has suggested that it may possibly move in the xylem as MoO_4^{2-}, as Mo-S amino acid complex or as a molybdate complex with sugars or other polyhydroxy compounds. Molybdenum is located primarily in the phloem and vascular parenchyma and moderately mobile in the plant.

The Mo concentration of plant material is usually low and plants are adequately supplied with less than 1 mg/kg dry matter (Gupta 1997b). Deficiency is usually under 0.2 mg/kg dry matter. Some typical results for a number of plant species are shown in Table 17.1. Concentrations are low because the MoO_4^{2-} concentration of the soil solution is also normally very low. In marked contrast to other micronutrients Mo can be taken up to much higher concentrations of up to 200 — 1000 mg/kg without adverse effects (Römheld and Marschner 1991). Plant species do vary in sensitivity to Mo toxicity although Mo toxicity in plants in the field is very uncommon.

Table 17.1   Mo concentrations in different plants in μg Mo/g in the dry matter (Johnson 1966)

Lucerne leaves	0.34	Sugar beet tops	0.72
Phaseolus bean tops	0.40	Tomato leaves, healthy	0.68
Spinach leaves	1.60	Tomato leaves, deficient	0.13

The very low concentration of Mo in healthy plants mean that the physiological demand is also low. Molybdenum is an essential component of two major enzymes in plants nitrate reductase and nitrogenase, the effective mechanism of both depending on valency change of Mo (see page 162 and 170). Nitrate reductase the enzyme which brings about the reduction of NO_3^- to NO_2^- occurs in cytosol. It is a dimeric enzyme with three electron transferring prosthetic groups per sub unit Flavin, haem, and Mo. In the reduction electrons are transferred directly from the Mo to NO_3 (Figure 3.34). The leaves of Mo deficient plants are low in nitrate reductase activity which can be induced by infiltration of Mo (Witt and Jungk 1977). Inducible nitrate reductase activity can be used as an indicator of the Mo nutritional status of plants. In plants supplied solely with NH_4-N the Mo requirement is lower and symptoms of deficiency are less severe or absent (Hewitt and McCready 1956).

Nitrogenase which occurs in the nodules of legumes and non legumes consists of two metallo enzyme proteins, a Mo-Fe-S protein and a FeS cluster protein (see Figure 3.38). Electrons from reduced ferredoxin are transferred *via* the Fe protein to the Mo-Fe- protein from which electrons are transferred directly to N_2 to effect the reduction to NH_3. The Mo requirement of nodulated legumes and free living N_2 fixing bacteria is high as observed by Becking (1961) who established the essentiality of Mo for the fixation of N_2 by *Actinomyces alni* occurring in the root nodules of older *Alnus glutinosa*. Some of this important findings are shown in Table 17.2 from results obtained from growing alder seedlings in Mo deficient soil without and with Mo supply. In the absence of Mo the plant became N deficient even though there was a proliferation of very small root nodules. When Mo was supplied nodule number was much decreased but the nodule dry weight per pot increased substantially as did also the weight of roots and shoots and the N concentration in the shoots.

Molybdenum also plays another essential role in the N metabolism of some tropical and subtropical legumes such as soybean and cowpea. In these legumes the dominant long distance N transport from root to shoot are the ureides, allantoin and allantoic acid (see page 210). The precursor of these compounds which is produced in the cytosol of root nodules is uric acid. This compound is an oxidation product of purine (xanthine), which requires the Mo containing metalloprotein xanthine dehydrogenase for its synthesis. In these legumes Mo deficiency can impair growth by either its effect on nitrogenase activity or purine metabolism or by both these causes (Römheld and Marschner 1991)

Molybdenum deficiency frequently begins in the middle and older leaves as Mo is readily retranslocated within the plant. It thus contrasts markedly with other micronutrients. Symptoms of Mo deficiency differ considerably between species but some generalities can be made. Interveinal mottling, marginal chlorosis of the older leaves and upward curling of the leaf margins are all typical. As the deficiency progresses necrotic spots appear at leaf tips and margins which are associated with high NO_3 concentrations in the tissue (Maynard 1979). In the cruciferae in which the deficiency has probably been most often observed the leaf lamella is not properly formed and in extreme cases only the leaf rib is present. This appears rather like a whip and for this reason the deficiency is called "whiptail". Plate 17.1 shows a typical example of "whiptail" in cauliflower.

Table 17.2   Effect of Mo on growth, nodule formation and N uptake by alder plants dependent on N_2 fixation with and without Mo (Becking 1961).

Mo supply	Plant growth gDM/pot	Nodules Nos/pot, gDM/pot		Mo concentr. μg/gDM	N uptake mg/pot
Without	2.76	228-308	7	2.0	53
With	8.82	1-10	132	13.3	245

Curd formation is also distorted. In a number of crop plants Mo deficiency appears to affect the reproductive phase more than vegetative growth. In Mo deficient maize the tasseling stage is delayed a high proportion of flowers fail to open and pollen formation both in the size of grain and viability is greatly reduced (Agarwala *et al.* 1979). Premature sprouting of grains is another problem of Mo deficient maize which can be amended by foliar sprays containing Mo (Cairns and Kritzinger 1992).

Plate 17.1 Mo deficiency (whiptail) in cauliflower. Lower part, beginning of the deficiency in young plants. (Photo Brandenburger)

Molybdenum deficiency in legumes can impair both NO_3 reduction and N_2 fixation so that affected crops appear N-deficient. In clover stands Mo deficiency often occurs very unevenly giving rise to a yellow chequered appearance against a dark green background of normal plants. Molybdenum deficiency symptoms in various crops have been described by Gupta (1997c) and illustrated by (Bergmann 1986, 1992).

17.3 Molybdenum in Crop Nutrition

Most soils contain enough Mo in available form to meet adequately the needs of crop plants. On some acid soils (pH < 5.5), Mo deficiency can arise because of high Mo fixation in the soil and liming can alleviate the problem. In the USA the geographic pattern of Mo deficiency mainly follows the regions of acid sandy soils that are formed from parent material low in molybdenum. Liming these soils to increase the availability of native Mo does not result in yields as high as those obtained by treatment with both lime and molybdenum (Welch *et al.* 1991). Molybdenum deficiency symptoms are commonly observed on soils derived from quartzic material, sandy pebbly alluviums, sandy loams and on soils with high anion exchange capacities (Cheng and Ouellette 1973). Soils with secondary iron oxide accumulations such as the ironstone soils of Australia and Holland are also often Mo deficient as they fix Mo very strongly. Molybdenum deficiency may occasionally appear on peat soils. This is most likely brought about by the retention of Mo by insoluble humic acid from the peat. Humic acid probably reduces the MoO_4^{2-} to Mo^{5+} which becomes fixed in the cationic form (Szalay and Szilagyi 1968). Deficiency has also been reported on calcareous soils as for example those along the Yellow River in China, derived from loess and alluvium lacking in both total and available Mo.

Most frequently liming is enough to prevent Mo deficiency. In some cases — as on Mo deficient soils — it is only by the application of Mo salts that it is posssible to increase yields and plant Mo concentration. Molybdenum application is always preferable to liming when an increase in soil pH is not necessarily desirable. On the other hand great caution must be taken with Mo fertilization as this can result in high Mo levels in fodder which are toxic to animals. Ruminants in particular are susceptible to high Mo levels in the fodder and a Mo concentration of 5 to 10 mg Mo/kg in the dry matter is regarded as suspect for cattle and horses (Miller *et al.* 1991). The disease molybdenosis in ruminants is usually associated with particular soils which are high in Mo. The teart pastures in the UK may range from 20−100 mg/kg Mo and even higher values may be found on poorly drained soils derived from granitic alluvium and black shales ash on highly organic soils (Underwood 1977).

Molybdenosis manifests itself in ruminants by diarrhoea, depigmentation of hair or wool, bone formation and reduction in growth. Essentially it is a secondary Cu deficiency. According to Mackenzie *et al.* (1997) the aetiology of Cu deficiency in cattle has been shown to be due to complex interactions of Cu, Mo, S, and Fe in the rumen which form thiomolybdates. With an insufficient level of Cu, ammonium molybdate is formed which is then absorbed and reacts with Cu in the blood and in the tissues thereby decreasing the activity of the important Cu requiring enzymes. When this occurs, normally black pigmented cattle acquire a ginger tinge on their coats because of impairment of the Cu dependent enzyme tyrosinase which converts tyrosine into the black pigment melanin (see page 603). Slow release Cu frits in the rumen can act to prevent the disorder.

Individual crop plants differ considerably in their requirement for Mo. The Crucuferae and particularly cauliflower and cabbage have a high Mo demand. The same also applies to the legumes because of the requirement of the root nodule bacteria. Differences between the requirements of various legumes may relate to seed size because plants grown from seeds already containing high amounts of Mo, *i.e.* large seeds, may not show deficiency symptoms even when planted on a Mo deficient soil (Adams 1997). Maize appears to have a lower requirement than some of the legumes. Nevertheless, Tanner and Grant (1977) reported widespread Mo deficiency in maize grown on Oxisols derived from ironstone in Zimbabwe. In a field experiment in which the maize seeds were treated with Mo, neither this treatment nor that of a low level of lime increased the yield above the control. When the Mo and lime were applied together, however, a significant increase in yield resulted. This finding highlights an important point. Under the acid soil conditions associated with Mo deficiency, plants can only respond to Mo fertilization when other growth limiting factors associated with the soil acidity are removed, *i.e.* Mn^{2+} and Al^{3+} toxicity. Interactions between nutrients can be of significance in relation to responses of Mo in the field. Thus Rebafka *et al.* (1993) found that the choice of P fertilizer can be important in the Mo nutrition of groundnuts. Replacing the S containing single superphosphate by the non S containing triple superphosphate enhanced Mo uptake and this was the major reason for increased N fixation and hence also crop yield. The Mo/S relationships have been reviewed by MacLeod et. (1997).

Molybdenum deficiency is usually corrected by addition of soluble compounds such as Na^- and NH_4 molybdate or soluble molybdenum trioxide and molybdenized superphosphate. Application rates to the soil of 0.01 to 0.5 kg Mo/ha are usual (Martens and Westermann 1991), but applications can made in a foliar form or directly to seeds before planting. Sedberry *et al.* (1973 quoted by Martens and Westermann 1991) found responses to soyabean by a soil application of 226 g/ha, foliar application of 85 g/ha and a seed treatment of 7 g/ha. Foliar application of Mo is often used for correcting Mo deficiency on crops. Less Mo is used than in soil application and sprays may possibly be more effective under dry conditions. Seed treatment is effective but the rate is dependent on Mo content of the seed. When planted in a Mo deficient soil pea plants with <0.2 mg Mo/kg responded to the treatment but not seeds containing 0.5—0.7 mg Mo/kg. Unfortunately there is no rapid and reliable field method by which farmers can assess the Mo status of the soil so that application rates and forms and levels of application are largely dependent on a knowledge of the demand of the crop, soil type and its pH and Mo status. The same is true for residual effects of Mo fertilization which can vary widely (Martens and Westermann 1991).

CHAPTER 18

BORON

18.1 Soil Boron

The total B concentration in soils is in the range of 20—200 mg/kg dry weight, most of which is inaccessible to plants. The available, hot water soluble fraction in soils adequately supplied with B ranges from 0.5—2.0 mg B/L (Sillanpaä 1982). Boron containing soil minerals include tourmaline (30—40 mg B/kg) which is very insoluble and the very soluble hydrated B minerals. Soluble B consists mainly of boric acid $B(OH)_3$ which under most soil pH conditions (pH 4—8) is undissociated. Boron is thus unlike all other essential plant nutrients which are present in soil solution in ionised form. The fact that boric acid is undissociated in soil solution is the main reason why B can be leached so easily from the soil. Gupta and Cutcliffe (1978) thus reported that more than 60% of applied B was not recovered in the upper layer of a podzolic soil five months after application. However, in soils of arid and semi arid regions, B may accumulate to toxic concentrations in the upper soil layer because of lack of drainage and the reclamation of such soils requires about three times as much water as that of saline soils (Keren and Bingham 1985).

Boron is a very weak monobasic acid which acts as Lewis acid by accepting a hydroxyl ion to form a borate ion.

$$B(OH)_3 + H_2O \Rightarrow B(OH)_4^- + H^+; pK_4 = 9.2$$

The high pK value indicates that the formation of the anion $B(OH)_4^-$ is only of significance towards the upper pH range of soils and therefore the process of surface adsorption of B is closely dependent on soil pH. Adsorption increases in the pH range 5—9, at the expense of the B concentration of the soil solution, then falls off above this level through competition with OH^- as shown in Figure 18.1 (Goldberg 1997). Control of the concentration and amount of B in soil solution through B adsorption reactions is of immediate importance for plant growth since plants respond directly to the concentration of B in soil solution and not to B adsorbed on soil surfaces (Keren et al. 1985).

The occurrence of increasing B adsorption with rising soil pH is in marked contrast to the effect of soil pH on the binding of some other anion species (phosphate, sulphate) where anion release occurs over the same range of increase in soil pH. This difference in effect explains why application of lime to crop plants growing in acid soils marginal in B supply may on the one hand induce deficiency symptoms in the crop, whilst on the other enhance the available P nutritional status of the soil. The lower rate of removal of B via leaching from neutral and alkaline soils is also a consequence of enhanced B adsorption at higher pH.

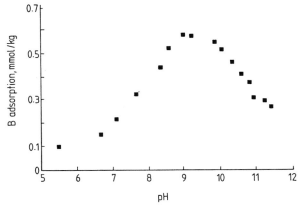

Figure 18.1 Borate adsorption by a top soil as a function of pH (after Goldberg 1997).

Soil constituents capable of adsorbing B include Al and Fe oxides, clay minerals, calcium carbonate, and organic matter. The partition of B between the soil solution and the surfaces of these constituents has been investigated by various workers (see Goldberg 1997). In general, increasing pH up to about pH 8—10 favours adsorption. The mechanism of B adsorption of Al and Fe oxides and clay minerals is considered to be ligand exchange (see page 23) with reactive surface hydroxyl groups. For the clay minerals B adsorption per unit weight of clay follows the order kaolinite < montmorillonite < illite (Keren and Mezuman 1981). On a weight basis, however, clay minerals adsorb significantly less B than do most oxide minerals (Goldberg 1997). Empirical models have been developed to describe adsorption data including the Langmuir adsorption equation developed to quantify gas adsorption on to pure solids (see page 25). Additionally Keren *et al.* (1981) have proposed an adsorption equation to account for competition between $B(OH)_3$, $B(OH)_4^-$ and OH^- for the same adsorption sites, to predict B adsorption under different conditions. Boron adsorption isotherms at different pH's from Keren's data are shown in Figure 18.2.

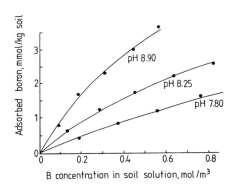

Figure 18.2 Borate adsorption by a soil in relation to soil pH and B soil solution concentration. Dots represent experimentally determined values, solid lines are calculated values (after Keren *et al.* 1981).

The dots represent experimental results of B adsorption dependent on B activity (concentration) of the soil solution and soil pH, whereas the solid lines are calculated values. The very close agreement between these two sets of data confirms the validity of Keren's approach in predicting B adsorption, and the value of the model for adsorption studies of B in soils. From these results it is again clear that increased soil pH from pH 7.8 and 8.9 increases the strength of B adsorption with a corresponding decrease in B concentration in soil solution (*i.e.* available B). This explains why overliming can induce B deficiency in crops.

Soil organic matter is closely associated with the accumulation and availability of B in soils. There is good evidence that native soil B and hot water soluble B in soils are significantly correlated with organic C content (Goldberg 1997). Boron adsorption on both organic soils and on composted organic matter also increased with increasing pH (Yermiyaho *et al.* 1988). These workers suggest that on a weight basis the sorption capacity for B in composed organic matter is about 4 times greater than for soil or clays. Ligand exchange (see page 25) is the most likely mechanism for the B sorption by organic matter in the formation of B-diol complexes as shown in Figure 18.3. Evidence of the formation of these complexes from borate and polyhydroxy compounds has been obtained by the use of ^{31}P nuclear magnetic resonance spectroscopy (Coddington and Taylor 1989).

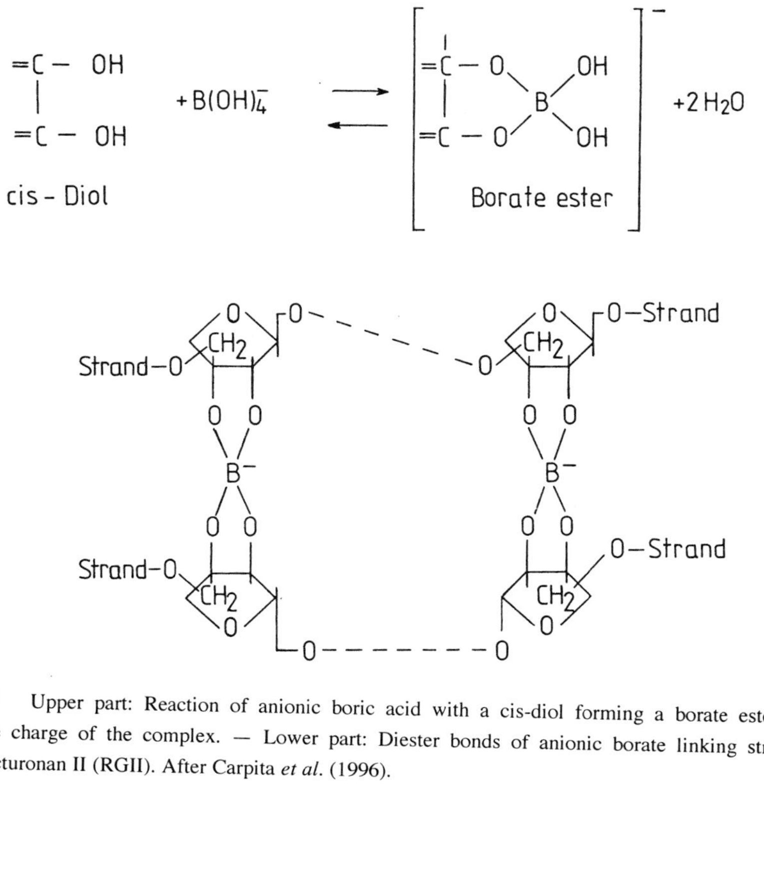

Figure 18.3 Upper part: Reaction of anionic boric acid with a cis-diol forming a borate ester. Note the negative charge of the complex. — Lower part: Diester bonds of anionic borate linking strands of rhamnogalacturonan II (RGII). After Carpita *et al.* (1996).

18.2 Boron in Physiology

18.2.1 Uptake and translocation

Roots take up B from the soil solution mainly as undissociated boric acid and thus in a form potentially permeable to plant cells. From theoretical considerations of membrane permeability coefficient data Raven (1980) concluded that the value is high enough to allow uptake but also pointed out the possibility of active uptake by ester formation with cis diols (see Figure 18.3). A similar conclusion of B uptake as a possible combination of active transport and passive diffusion was suggested in a review by Kochian (1991). The exact nature of boric acid transport across cell membranes is still not totally resolved.

Experimatal evidence in support of a passive process comes from the work of Oertli and Grgurevic (1975) who showed that the amount of B taken up by excised barley roots is dependent on the concentration of free boric acid in the uptake solution. More recently, Brown and Hu (1994) studying B uptake by roots of sunflower and squash as well as tobacco cells over a wide range of B supply ($0.2 - 10$ mol/m^3) reported passive uptake with no indication of saturation kinetics or effects of metabolic inhibition on uptake. However, relationships between B uptake and B concentrations in the uptake medium and rate of transpiration are often not as clear-cut as might be expected from a purely passive process. There are numerous reports in the literature that species can differ significantly in the rate of B accumulation even when grown under identical environmental conditions (see Hu and Brown 1997).

In reviewing the varied and conflicting results on B uptake carried out over the past 30 years, supporting both active and passive transport, Hu and Brown (1997) highlight important difficulties in studying B uptake by plants. These incude:
- the ability of boric acid to form complexes with a variety of compounds including polysaccharides, sugar alcohols etc (see Figure 18.3) both in the cell wall and the cytoplasm;
- the large variations in membrane permeability between species and even between genotypes;
- the fact that uptake studies based on transmembrane electrical gradients, analysis of absorption kinetics and measurements of electrical changes in ion uptake do not readily apply to studies on B.

Hu and Brown (1997) suggest that although the B uptake mechanism has not been completely elucidated, uptake can best be explained by a passive diffusion of free boric acid into the cell followed by a rapid formation of B complexes within the cytoplasm and cell walls. The fall in the concentration of boric acid within the cell associated with the formation of B complexes allows the further absorption of B from the external solution. Uptake is thus seen as a passive process acting in response to external boric acid concentration, membrane permeability, internal complex formation and transpiration rate.

On the other hand the transport of boric acid across plant membranes exclusively in terms of passive transport raises doubts. As pointed out by Kochian (1991) if B uptake is due solely to passive diffusion across the lipid bilayer then it is the only plant nutrient for which this is the case and it is difficult to envisage how plants would regulate B uptake by this transport mechanism. Membrane permeability can differ greatly between species and genotypes in response to environmental factors but it is not clear whether these modifications to the membrane are adequate to regulate uptake. Convincing recent evidence in support of an active component in the uptake process at a low level of B supply (1 mmol/m³) comes from the Dannel et al. (1997). These workers repeated the same kind of experiment as Oertli and Grgurevic (1975) but used an intact root system of sunflower supplied with much lower concentrations of boric acid. As well as measuring B uptake, B translocation was also estimated by analysis of xylem exudate. Both were determined at sufficient and marginal supply of B. At sufficient supply, uptake could be explained by passive diffusion but at marginal supply a concentration mechanism was found to build up a gradient against the nutrient solution to satisfy the B demand of the plant (Dannel et al. 1997). In subsequent experiments (Pfeffer et al. 1999), the uptake at marginal B supply was shown to be turned off by both increased B supply and by metabolic inhibitors. The active uptake together with the formation of B complexes in the root symplasm was shown to be responsible for maintaining the concentration gradient to satisfy the B demand of the plant at low B supply.

In the older literature B is often generally described as having only limited phloem mobility (Oertli and Richardson 1970) and this is still true for the majority of plant species. Boron, however, is now known to be unique among the essential plant nutrients that although restricted in mobility in many plant species is freely mobile in others (Brown and Shelp 1997). This difference can be demonstrated by reference to the distribution of B within a mature leaf of walnut and apple as shown in Figure 18.4.

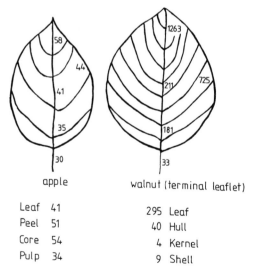

apple	walnut (terminal leaflet)
Leaf 41	295 Leaf
Peel 51	40 Hull
Core 54	4 Kernel
Pulp 34	9 Shell

Figure 18.4 Boron distribution in apple and walnut (mg B/kg dry weight). — Upper part: B distribution in leaves — Lower part: B concentrations (μgB/gDM) in various plant organs — Both species were grown in close proximity and received the same irrgation (after Brown and Shelp 1997).

These leaves were taken from species of similar age and growing under identical management and environmental conditions. Walnut is a striking example of species with restricted B mobility. Within the leaf a very much higher accumulation of B occurred at the leaf tip and leaf margin as a result of transpirational flow of water through the leaf. In such species the rate of transpiration has a decisive influence in determining the upward transport of B and its distribution as observed in tobacco by Michael *et al.* (1969). Redistribution of B from the leaves to other plant organs was much greater in apple than in walnut (lower part of Figure 18.4). Accumulation of B can sometimes lead to toxicity in the margins of older leaves and some plant species are adapted to secret B out of the leaves in guttation droplets (Oertli 1962). The predominant movement of B in the transpiration stream with little phloem movement explains the occurrence of B deficiency symptoms in young growing tissue.

In apple the leaf concentrations of B were significantly lower than in walnut with little evidence of accumulation in leaf tips or margins. The findings are indicative of B mobility in apple trees not determined by transpiration. The same is true for the very even distribution of B between leaf and fruit organs in apple in comparison with the very much higher concentration in the leaf than in kernel or shell of walnut (see bottom part of Figure 184; Brown and Shelp 1997). Further evidence of interspecies differences in phloem mobility of B has been obtained by the study of B concentration gradients along shoots and by foliar ^{10}B labelling (Brown and Hu 1998).

For a broad range of agricultural crops it is now established that B is phloem mobile and is transported as a complex with polyols (borate esters Figure 18.3). Species showing high phloem mobility of B are restricted to those species producing polyols as the primary photosynthetic product. These include celery, carrot, bean, and cauliflower which produce mannitol. For the fruit trees including apple, prune, pear, and apricot, B forms complexes with sorbitol (Brown and Shelp 1997). The isolation and characterization of these B transport molecules from phloem sap by Brown and his colleagues has resolved much of the confusion concerning B transport in plants. Recent exciting findings of Brown *et al.* (1999) have shown that tobacco plants genetically engineered to synthesize sorbitol markedly increased the within-plant B mobility which in turn increased plant growth and yield by overcoming transient deficiencies of B on the soil. The impact on interspecies differences of within-plant boron mobility in the diagnosis of boron status of crops and improved fertilizer strategies is discussed below.

High concentrations of B occur in certain plant organs such as anthers, stigma and ovary, where levels may be twice as high as in stems. These differences are demonstrated in the data of Syworotkin (1958) presented in Table 18.1 showing the B concentrations in various parts of the opium poppy.

Table 18.1 B concentrations of various plant parts of the opium poppy (Syworotkin 1958)

Plant organ	B, mg/kg DM	Plant organ	B, mg/kg DM
Capsule of seeds	69	Culm	17
Upper leaves	45	Roots	20
Medium leaves	34	Seeds	21

18.2.2   General aspects of boron in physiology and metabolism

It is now over 70 years since B was shown to be an essential element for higher plants. The role of B in plant nutrition, however, is still least well understood of all the mineral nutrients. This even though in molar terms it is required at least by the dicotyledons in highest amounts of the micronutrients. It does not appear to be essential for fungi and most of the bacteria and there is no evidence in higher plants that it is either a constituent or an activator of any enzyme. It is relatively easy to induce B deficiency symptoms appear very quickly together with distinct changes in metabolic activity. It is much more difficult though to identify primary roles of the nutrient in relation to these responses. Indeed one of the problems in assessing the role of B in plant is that most of what is known about the role of B in plant nutrition has been determined from what happens when B is withheld or resupplied after deficiency. Suggestion by Parr and Loughman (1983) and others for boron function in plants include sugar transport, cell wall synthesis, lignification, cell wall structure, carbohydrate metabolism, RNA metabolism, respiration, indolic acetic acid metabolism, phenol metabolism, membranes, nitrogen fixation, ascorbate metabolism, and amelioration of Al toxicity. There is increasing evidence that some of these effects do not arise primarily because of a lack of B but by what Marschner (1995) describes as "cascade effect" *i.e.* of secondary origin.

The distribution of plant nutrients within cells is closely related to nutrient function. For B only a limited amount of work has carried out on compartmentation but B has been shown to be present in free space, cytoplasm, vacuole, and cell wall (Thellier *et al.* (1979). Much of the B is present in cell walls associated within pectin but in plants adequately supplied with B as much as 60% of the total can be in a soluble form as boric acid and soluble B complexes (Pfeffer *et al.* 1997). A considerable amount can be in the symplasm as soluble B complexes (see Figure 18.3). This includes the B transport molecules, the polyols in the sieve tubes which mediate in phloem transport of B in some plant species (see page 625). Besides these molecules little is known of the soluble B complexes in cells. They can be present in high concentrations and directly available for possible physiological functions in the cell (Pfeffer *et al.* 1997).

Much is still to be learned of the role of B in plants. In a valuable review on B, Goldbach (1997) distinguishes between facts and hypotheses and in particular draws attention to physiological effects attributed to B where evidence is still lacking. This article together with the excellent review of Blevins and Lukaszewski (1998) dealing with B in plant structure and function, and the Symposium in Boron in Soils and Plants published in Plant and Soil (1997) form the basis of this section.

18.2.3   Cell walls

The chemical composition and ultrastructure of cell walls are quickly affected by a lack of B. Cell walls become thicker in root apical meristems. Primary cell walls lose their smoothness and are characterized by irregular deposits of vesicular aggregates intermixed with membranous material. The timetable of early symptoms has been listed by Hirsch and Torrey (1980). Hemicellular and pectic substances increase and a large proportion of glucose is incorporated into ß 1-3 glucan the main component of callose which blocks sieve cells in B deficient plants.

The dramatic and rapid responses are not surprising since B is now believed to play a key role in the structure and integrity of cell walls. A substantial amount of B is bound in cell walls which under B limiting conditions can be as high as 95—98% of the plant B (see Matoh 1997). The mechanical properties of cell walls can be modified by cross links between major constituents such as hemicellulose and pectic polysaccharides. The role of B in cross linking these pectic polymers was proposed by Loomis and Durst (1992). This binding function of borate between two strands of pectic polysaccharides (rhamnogalacturonan II) is shown in Figure 18.3 (Carpita et al. 1996).

Supporting evidence for such a role comes from the close correlations found in a number of plant species between the uronic fraction, pectin sugars and B concentration (Hu et al. 1996). Isolation of a B containing pectic polysaccharide complex from plants, boron rhamnogalacturonan II (RG II) was achieved by Kobayashi et al. (1996). The B-RG II was found to be composed of boric acid and two chains of pectic polysaccharides cross linked through the borate diester bonding thus forming a network of pectic polysaccharides in cell walls as shown in Figure 18.3 (Carpita et al. (1996). Matoh et al. (1996) suggest that RG II may be the exclusive polysaccharide binding in cell walls and ubiquitous in higher plants.

Boron complexes are acid sensitive. In relation to the process of growth Loomis and Durst (1992) suggest that the effect of auxins in depressing the pH of the cell wall can result in cross linkage breakage and recomplexation to give "creep" thus allowing elongation. The importance of B in maintaining the structure of cell walls through cross linking cell wall macromolecules is also evident from the significant reduction in root cell wall elasticity and increase in cell hydraulic conductivity which occur rapidly under B deficiency (Findeklee et al. 1997). Similarly disturbance of borate diester bonding in cell walls can explain the occurrence of brittle leaves in B deficient plants and the plastic elastic nature of leaves supplied with supraoptimal levels of B (Hu and Brown 1994, Loomis and Durst 1992). Besides pH Ca^{2+} also has an influence on borate binding and stabilizing borate complexes in cell walls (Mühling et al. 1998).

It is important to distinguish between cell wall synthesis and cell wall structure. According to Hu et al. (1996) cell wall synthesis which occurs in the symplasm is not depressed by B deficiency. On the other hand B plays an important role in the ultrastructural arrangement of the cell wall components in the apoplast. The localization of the boron polysaccharide complex (BPC) was determined immuno cytochemically using an antibody towards BPC.

18.2.4   Membrane function

There is increasing evidence for a particular role for B on plasma membrane function. Pollard et al. (1977) reported that B pretreatment of maize root tips being B deficient promoted ATPase activty and the uptake of phosphate and Rb^+. Further evidence of a lower activity of ATPase in B deficient cells and roots which could be restored after resupply with B was obtained by Blaser-Grill et al. (1989) and Schon et al. (1990). Restitution was indicated by hyperpolarization of membrane potential (became more negative) and a net excretion of protons. The effects of B are thus mediated either directly or indirectly by the plasmamembrane bound H^+ pumping ATPase. Obermeyer *et al.* (1996) found that B stimulated the plasmalemma H^+ pump which was associated with a higher growth rate of pollen tubes.

According to Cakmak and Römheld (1997) the effects of B are primarily on plasma membrane itself rather than a direct effect on the ATPase *per se*. Evidence for the role of B in the plasma membrane comes from the dramatic effects of increases in membrane permeability in isolated leaves caused by B deficiency. As compared with B deficient leaves net efflux from severely deficient leaves was 35 fold higher for K^+, 45 fold higher for sucrose and 7 fold higher for phenols and amino acids. Treatment of B deficient leaves with B quickly restored membrane permeability to the level of the B sufficient leaves (Cakmak *et al.* 1995). It is suggested that B stabilizes the structure of the plasma membrane by complexing membrane compounds containing *cis*-diol groups such as glycoproteins and glycolipids to keep channels or enzymes at optimum conformation within the membrane. In accordance with this suggestion lower amounts of these constituents are found in B deficient plants (see Blevins and Lukaszewski 1998). The binding of B to polyhydroxy groups of membranes is seen as maintaining the structure and integrity of membranes, and a major reason for the stimulatory effects of B on membrane-bound ATPase activity and for controlling permeability of plasma membranes (Cakmak *et al.* 1995). Isolation and purification of B complexes within or on the plasma membranes is still needed to confirm unequivocally this role for B.

The close proximity of the plasma membrane and the cell wall and the evidence of the role of B in cell wall structure, phenol metabolism and plasma membrane integrity has lead Römheld and Marschner (1991) to conclude that B exerts its most important influence at the cell wall/plasma membrane interphase. This is also in accordance with experimental evidence of the gradient in distribution of RG II across the cell wall with localization primarily adjacent to the plasma membrane (Matoh 1997). Changes occurring at this interface are regarded as the primary effects of B deficiency which lead on to secondary effects (Römheld and Marschner 1991).

18.2.5   Other aspects of B in biochemistry and nutrition

Accumulation of phenolics can occur rapidly in plants suffering from B deficiency and the formation of *cis*-diol complexes between B and some sugars and phenolics play a vital role in their production. When B is deficient the 6-phosphogluconate borate complex is not formed and 6-phosphogluconate is converted to ribulose-5-phosphate. The substrate flux is thus shifted from glycolysis to the pentose phosphate pathway which leads to the synthesis of phenolic compounds *via* the shikimic pathway.

It has been suggested that the oxidation of these phenols by polyphenol oxidase to form quinones and semiquinones can in turn produce toxic O_2 free radicals to impair plasma membrane function. Cakmak *et al.* (1995) see this reactions as a primary response to B deficiency. Recently, however, Pfeffer *et al.* (1998) in studies on B deficient and sufficient sunflowers grown at different light intensities to provoke changes in phenol metabolism were unable to find any correlations between soluble phenol concentrations, polyphenol oxidase activity and enhanced membrane permeability in response to B deficiency. They were therefore unable to confirm the hypothesis that B stabilizes the structure of the plasma membrane by complexing phenolic compounds thereby preventing damage by O_2 free radicals. In commenting on the part played by phenols, Goldbach (1997) suggests that under B deficiency the use of carbohydrates for the deposition of cell wall material is likely to be inhibited by B deficiency. A principal reason for the accumulation of phenols in B deficient plant tissue could therefore be an intensified flow of C through the pentose phosphate cycle and shikimic pathway in order to mobilize surplus carbohydrates and to deposit them in the apoplast or vacuole. Phenol accumulation would then be a secondary reaction to boron deficiency. Further experimentation is needed to test this hypothesis.

Boron interactions with IAA have also been cited to explain effects of B deficiency. Cessation of root elongation is one of the most rapid responses to B deficiency. In such plants Bohnsack and Albert (1977) observed similarities between the effects of B deficiency in depressing root elongation and increased IAA levels in roots over short term periods. Resupplying B stimulated net root elongation and enhanced IAA oxidase to reduce IAA levels. This close temporal association between B deficiency and increased auxin activity has been observed at the onset of the necrotic disorder in lettuce known as tipburn (Crisp et al. 1976). According to Hirsch and Torrey (1980) however, changes in ultrastructure caused by B deficiency are different from those resulting from enhanced IAA toxicity. Other workers have observed unchanged or reduced IAA levels in apical tissues with the onset of visual symptoms of B deficiency (Fackler et al. 1994). Römheld and Marschner (1991) suggest that IAA accumulation can be a secondary effect of B deficiency caused by accumulation of phenolics which inhibit IAA oxidase.

A common feature in B deficiency is the disturbance in the development of meristematic tissues, whether these are root tips, tips of upper plant parts or tissues of the cambium. Thus Gupta (1979) holds the view that a continuous supply of B is required for maintenance of meristematic activity. The reason for this B requirement is not yet known but it has been shown that B is required for the synthesis of N-bases such as uracil (Albert 1968). Additions of both uracil and orotic acid, an intermediate in uracil biosynthesis were found to alleviate B deficiency symptoms (Birnbaum et al. 1977). This finding strongly suggests that B is involved in uracil synthesis. Uracil is an essential component of RNA and if it is absent, RNA containing assemblies such as ribosomes cannot be formed, thus affecting protein synthesis. Ribonucleic acid synthesis, ribose formation, and the synthesis of proteins are most important processes in meristematic tissues. If they are disturbed by a lack of boron the entire process of meristematic growth is impaired. Krueger et al. (1987) suggest that growth inhibition with B deficiency is caused by decreased concentrations of nucleic acids. A fall in RNA level does precede cessation in cell devision and can be the result not only of inhibited RNA synthesis but also enhanced RNase activity. As pointed out by Römheld and Marschner (1991) however, a decrease in DNA level is a secondary effect since DNA synthesis in B-deficient roots is sustained at least several hours after inhibition of root growth.

Boron exerts diverse influences in plant metabolism. For example results of Blevins' group that B can interfere with Mn dependent enzymic reactions. One such enzyme is allantoate amidohydrolase in soya bean leaves. Boric acid applied as a foliar spray on field grown nodulated soya beans caused a 10 fold increase in allantoate which could be eliminated or greatly reduced in plants resprayed with Mn (Blevins and Lukaszewski 1998).

A most interesting development of studying the role of B in plants has been the observation of Blevins and coworkers that B can ameliorate Al induced root growth inhibition (Blevins and Lukaszewski 1998). Based on the similarity and the characteristic symptoms of Al stressed and B deficient plants mostly involving cell walls, membrane function, and root growth, these workers have hypothesized that Al may exert its toxic influence by inducing B deficiency. This has indeed been found to be the case. In dicotyledenous species which have a higher B requirement than monocots (see page 637) supplementary B in acidic high Al subsoil promoted root growth, cell elongation and cell production. According to these workers root growth inhibition under both B deficient and Al toxic conditions may be a consequence of disrupted ascorbate metabolism.

18.2.6   Boron deficiency

Boron deficiency symptoms relate closely to the mobility of B within plants: In the many plant species in which B is relatively immobile (see page 625) deficiency symptoms first appear as abnormal or retarded growth of the apical growing points. The youngest leaves are misshapen, wrinkled and are often thicker and of a darish blue-green colour. Irregular chlorosis between the intercostal veins may occur. The leaves and stems become brittle because of a disturbance in cell wall growth. As the deficiency progresses the terminal growing point dies and the whole plant is reduced and flower and fruit formation is restricted or inhibited. The typical dying off symptom of the terminal growing point as a result of B deficiency is shown for the tomato plant in Plate 18.1 from the work of Brown (1979). In plant species in which B is phloem mobile relatively higher concentrations of B are found in younger leaves so that "classic" symptoms are not usually observed.

In most plant species the B requirement for reproductive growth is much higher than for vegetative growth (Dell and Huang 1997). Seeds can be lower in B concentration and in viability. These findings reflect the high requirement of pollen tube growth for B. Fruit formation can also be similarly impaired. In some plant species the affected growth of pollen results in parthenogenesis. This is true for grapes and parthenocarpic fruits may result. The fruits developed remain very small and are of poor quality (Gärtel 1974). Plate 18.1 shows a mature bunch of grapes from a B deficient vine. The important role of B on reproductin, pollen tube growth and pollen germination has been reviewed recently by Blevins and Lukaszewski (1998).

The most well known B deficiency symptoms are crown and heart rot in sugar beet. The symptoms begin with anatomical changes at the apical growing points with necrosis of meristematic tissues. The youngest leaves are curled and stunted and turn brown or black. Eventually the inner leaves are affected and the main growing point dies. Older leaves are brittle and chlorotic. The crown of the beet begins to rot and infection then sets in, and the whole plant becomes affected. The healthy part of the beet is low in sugar. In turnips and swedes, B deficiency results in glassy like roots which

Plate 18.1 Boron deficiency. Upper part: Boron deficiency in tomatoes, the plant in the middle showing typical symptoms at the growing point. (Photo Brown). Lower part: Grape bunch with B deficient fruits. (Photo Gärtel)

are hollow and cracked. The appearance of cracked stems is also an indication of B deficiency in celery. These cracked stem symptoms in plants result from an increase in diameters of stems and petioles.

The development of scaly surfaces and the formation of internal and external cork like material is typical of the features asssociated with B deficiency in many plants including tomatoes, cauliflower, citrus and apple. These typical corky like deficiency symptoms possibly relate to the association of B with pectic materials in cell walls. Detailed description of B deficiency symptoms of the major crops can be found in Gupta (1979), Bergmann (1992), Shorrocks (1997). Drop of buds, flowers and fruits is also a common feature of B deficiency.

18.3 Boron in Crop Nutrition

18.3.1 Availability and boron application

Boron is of interest in crop production both from the viewpoint of its effects in deficiency and excess. In reviewing the subject of the occurrence and correction of B deficiency Shorrocks (1997) reports evidence of the worldwide appearance of B deficiency in a great variety of crops with large tracts of land being treated annually with B fertilizers. Boron is thus probably more important than any other micronutrient in obtaining quality high crop yields.

Soils on which B deficiency occurs are from a very wide range of soil types but all have in common a hot water soluble boron concentration below 0.5 mg/L and sometimes much lower. The most important soils carrying B deficient crops are the Spodosols and Podzols and to a lesser extent the Andosols, Lithosols and Luvisols (Shorrocks 1997). Spodosols are strongly weathered coarse textured soils with low base exchange developing under humid conditions. They are widespread throughout the world including China, USA and Brazil, and make up over 50% of the soils in southeast Asia. Podzols which develop under heath and coniferous vegetation are also highly leached course textured soils often formed on parent materials inherently low in B such as sandstones and acid igneous rocks. Sandy soils in particular need regular treatment with B fertilizer for not only may such soils be naturally very low in B but also any B that is present is very prone to leaching.

Boron availability decreases with increasing soil pH. Boron fertilizers should therefore also be applied when liming acid sandy soils with marginal B concentrations. It has to be borne in mind that liming not only increases soil pH but also induces secondary effects including the formation of aluminium hydroxide which can act as a scavenger to immobilize soluble B (Sims and Bingham 1968). Additionally calcium carbonate acts as an important B adsorbing surface in calcareous soils (Goldberg and Forster 1991). A practical example showing the effects of overliming in inducing B deficiency in sugar beet is shown from classic data of Scheffer and Welte (1955) given in Table 18.2.

Table 18.2 Effect of soil pH and carbonate on the proportion of sugar beet infected by crown and hart rot (data of Scheffer and Welte 1955)

pH	Carbonate, %	Percentage of sugar beet		
		Healthy	Infected	Dead
6.7	0.1	100	0	0
7.0	0.1	99	1.0	0
7.5	0.3	46	40	14
8.1	14.4	0	25	75

Boron availability is also related to seasonal behaviour. Deficiency appears to be more prevalent in dry summers. Fleming (1980) has shown that B availabilty generally decreases as soils dry out making deficiency in plants more likely. He suggests that plants encounter lower amounts of available B as they progressively extract moisture from the deeper parts of the profile during dry conditions.

Soil organic matter can act as a reserve of soil B with available B being released by the activity of soil microorganisms on the process of mineralization. Since the rate of mineralization is very much dependent on soil moisture conditions, restriction of mineralization offers another explanation for a decrease of available B under dry conditions. The practical importance of soil organic matter in relation to B availability is suggested from the findings of Mahler *et al.* (1985) from long-term field experiments in which conventional tillage and non tillage treatments are compared. These workers observe that the soil from the "non tillage treatment" was lower in available B and they suggest that this effect is related to the accumulation of organic matter in the upper soil layer of this treatment.

Table 18.3 B-fertilizers, their chemical formulae and their B concentration (after Gupta 1979)

Boron source	Chemical formula	B (%)
Borax	$Na_2B_4O_7 \cdot 10H_2O$	11
Boric acid	H_3BO_3	17
Boron frits (contained in a moderately soluble glass)	$Na_2B_4 \cdot XH_2O$	10-17
Sodium tetraborate Borate-46, Agribor, Tonabor	$Na_2B_4O_7 \cdot 5\,H_2O$	14
Borate-65	$Na_2\,B_4O_7$	20
Sodium pentaborate	$Na_2B_{10}O_{16} \cdot 10\,H_2O$	18
Solubor	$Na_2B_4O_7 \cdot 5\,H_2O + Na_2B_{10}O_{16} \cdot 10\,H_2O$	20-21

The most important straight B fertilizers are given in Table 18.3. Besides those shown other B containing materials are also used *e.g.* borated superphosphate. The most commonly used B fertilizer is borax ($Na_2B_4O_7 \cdot 10H_2O$) but it leaches easily from sandy soils. The other Na borates are also applied directly to the soil except for solubor which is applied both to soil and foliage because of the high solubility. Boric acid (H_3BO_3) is also frequently applied as a leaf spray particularly when the soil is potentially capable of fixing high amounts of boron. One problem of B application not encountered with other micronutrients is the limited range between conditions of deficiency and toxicity. If too much B is applied as for example by uneven application the crop may be damaged. Schumann and Sumner (1999) reported that concentrations of fly ashes ranges from 30 to 700 mg B/kg and high application rates may cause phytotoxicity.

One means of avoided excess B application is the use of boro-silicate glass frits. These are of sintered glass with a large surface area and provide a slow release of

B into the soil solution. The natural ore Colemanite can be used similarly to avoid leaching losses provided that the particle size is small enough. Other materials including sewage sludges, manures and composts can be used as available sources of B. Band or foliar application is often more efficient than broadcast application in correction of B deficiency.

The quantities of B taken up by crops during a season are relatively low. For sugar beet which has a high requirement for B this amounts to 300 g/ha (Shorrocks 1997). Recommended rates of fertilizer B application are several times higher than offtakes (see Martens and Westermann 1991). The fertilizer rate for sugar beet is in the order of 1.0 to 2.0 kg B/ha (about 10 to 20 kg borax/ha). Turnips have a higher B requirement than sugar beet and are frequently treated at higher rates. Because of the risk of B toxicity, however, the rates applied should not be too high. Generally only the crop of rotation, which has the highest B requirement, is treated with B. For arable crops this is usually sugar beet.

Boron fertilization needs to be regulated to crop demand. For most crop species B is relatively immobile within the plant. This means that the plant must be provided with a continuous supply of B to the plants at all phases of growth. Even a short interruption can induce deficiency. The B requirement of reproductive growth may exceed that of vegetative growth (Nyomara et al. 1997). Fruit and seed set may thus be depressed by inadequate B supply even when vegetative growth is optimal. This explains the benefit of foliar sprays of B as a short term measure in particular at flowering. In those crop species in which B is phloem mobile (see page 625), leaf analysis can be used as a guide to foliar B requirement. In apple and some other orchard fruit trees enrichment of floral buds can be ensured by foliar application in early autumn when leaves are still photosynthetically active and the floral buds represent a primary carbohydrate sink (Brown and Shelp 1997). Thus leaf analysis in one year can be used to assist the need for foliar B sprays in the same year in order to increase fruit yield the following year. Effective foliar application rates are 10−50% of that required by broadcast application (Martens and Westermann 1991).

18.3.2   Crop requirement

Crops differ in their sensitivity to B deficiency. Of the dicotyledenous members of the Cruciferae i.e. cabbage, turnips, brussels sprouts, cauliflower and the Chenopodiaceae i.e. sugar beet and swede are all highly sensitive to B deficiency and most responsive to B application. For sugar beet it would appear that evolution of the wild sea beet under saline B rich conditions has resulted in the present agricultural crops of sugar beet and fodder beet with their high B requirements (Shorrocks 1997). Other crops sensitive to B deficiency include celery, groundnut, coffee, oil palm, cotton, sunflower, olive, and pines. Gärtel (1974) claims that B deficiency of grapes is one of the most severe diseases in vine growing. Fruit formation is impaired and yield depressions as high as 80% may occur compaired with plants adequately supplied with B. This is a consequence of the high requirement of B for pollen germination, pollen tube growth and pollen viability (Dugger 1983). Gärtel (1974) reported that the stigmae of vine plants well supplied with B contained 50−60 mg B/kg dry weight in comparison with

8—20 mg B/kg dry weight in the B deficient plants. The importance of adequate B supply at the reproductive stage of growth is also recognized in other seed and fruit forming crops including cotton, sunflower and apples (Shorrocks 1997).

Some legumes also have a high B requirement. Mahler et al. (1985) found that a leguminous rich rotation (lucerne, beans, clover) required more B than did a cereal rotation. The lucerne crop like sugar beet has a particularly high demand for B and responds to B fertilization in 40 states in the USA (Shorrocks 1997). A crop of 7 Mg/ha can remove over 300 g B/ha. This is about 3 times the uptake of most other temperate agricutural crops and about 10 times the uptake of maize. There is some evidence also that a lack of B can depress N_2 fixation. Both the development of N_2 fixation of young soya bean root nodules have been found to be related by a lack of B (see also Blevins and Lukaszewski 1998).

Diagnosis of plant B status is very much dependent on B mobility within the plant. In species in which B is phloem mobile, B does not accumulate in mature leaves (see Figure 18.4). Boron analysis of these tissues can be used to assess B deficiency but not B toxicity. In these species B can accumulate in the fruit. This behaviour provides the basis for the widespread use of hull B concentration as an indicator of B status of almond trees (Nyomara et al. 1997). In most plant species, however, B is reelatively immobile and B analysis of mature leaves largely reflects B accumulation by transpiration and not the current B supply. This can only be achieved by sampling growing tissues. Citing the literature, for a number of dicotyledenous crop species. Bell (1997) reported leaf concentrations of less than 10 mg B/kg dry weight generally associated with deficiency symptoms in sampled young expanding leaves. Critical ranges of B concentration for particular crop species depend on leaf age, plant age, growth stage, and cultivar differences. Values in young expanded leaves in mg B/kg dry weight are for broccoli 9—13, sunflower 31—38, sugar beet 35—40, and wheat 3—7 (Bell 1997). For comparison, sufficiently supplied sugar beet leaves can have B concentrations up to 200 mg B/kg dry weight.

In graminaceous monocots both plant requirement for B and the concentrations of B in plant tissues are much lower than in dicots. Differences in cell wall composition and the lower amount of pectic substances in the graminaceae probably amount for these differences. Boron deficiency in cereals is thus less common. However, the low requirement for vegetative growth can mean that wheat and other small grained cereals can show sterility without exhibiting other symptoms, because of the higher B requirement for reproductive growth (Rerkasem et al. 1993). Syworotkin (1958) distinguishes between three plant groups with regard to B concentration and B requirement: graminaceous species, dicots and dicot species with a latex system such as dandelion, poppy and some Euphorbiaceae (see Table 18.4). The higher B concentrations in dicots with a latex system is presumably because of the greater amounts of B bound to sugars, polyphenols and other constituents of the latex fluid.

Table 18.4   Boron concentrations of various plant groups (Syworotkin 1958). B in $\mu g/g$ DM

Monocots		Dicots		Dicots with a latex system	
Barley	2.3	Peas	22	Dandelion	80
Wheat	3.3	Beets	49	*Euphorbia*	93
Maize	5.0	Lettuce	70	Poppy	94

18.3.3   Boron toxicity

For many species there is only a narrow range in critical tissue concentrations between B deficiency and B toxicity (Blamey *et al.* 1997). B is toxic to many plant species at levels only slightly above that required for normal growth. Toxicity effects may thus occasionally arise by excessive use of B fertilizers or on soils with high B contents such as those derived from marine sediments. The toxicity is, however, more usually associated with arid and semi arid regions where B levels are frequently high in the soil. The B status of irrigation water is particularly important in these regions. Boron is often found in high concentrations in association with saline soils and saline well water. Boron concentrations in irrigation water which can be used safely depend on the B adsorption capacity of the soil since large amounts of B in irrigation water are adsorbed by soil (Keren and Bingham 1985). Plant sensitivity is also of great importance. Nable *et al.* (1997) cite values of up to 0.3—1 mg B/L for sensitive plants, 1—21 mg B/L for semi tolerant plants and 2.1—4 mg B/L for tolerant plants. According to Reisenauer *et al.* (1973) toxicity of crop plants is likely to occur when the level of hot water soluble B exceeds 5 mg B/L.

Boron toxicity may also be caused by industrial pollution. Surface mining can produce waste carbinaceous materials that can be a source of B. Fly ash generated by electrical power plants is another source. This material can be used as an ameliorant to improve physical and chemical properties of the soil but there is a risk of B release into soil solution especially when applied at a high rate to increase soil pH. Amelioration of soils high in B is difficult. Methods used include application of lime and gypsum as well as excessive leaching with low B water (Nable *et al.* 1997). Plants can show very high concentrations of B as a result of industrial pollution. Grasses showing severe toxicity were in the range 270—520 mg B/kg dry matter and polluted needles of *Picea* species were as high as 960 mg B/kg dry weight (Judel 1977).

A valuable approach to B toxicity problems has been the study by Nable *et al.* (1997) of physiological and genotypic variation for tolerance and toxicity to B and the way this variation can be utilized to maximize plant growth on soils high in B. A considerable genetic variation in response to high B has been identified over a wide range of plant species, and it has been possible to breed tolerant genotypes for cultivation on soils high in B. The tolerance mechanisms appear to be a reduced uptake of B by shoots and roots. This is in agreement with findings that cell walls do not contribute effectively to the detoxification of excess B taken up by the plant by providing additional binding sites in response to B toxicity (Dannel *et al.* 1998).

Some of the most sensitive crops to B toxicity are peach, grapes, kidney beans, and figs. Semi-tolerant plants include barley, peas, maize, potato, lucerne, tobacco, and tomato whilst the most tolerant crops are turnips, sugarbeet, and cotton. Toxic effects of B result in leaf tip yellowing followed by progressive necrosis. This begins at the tip and margins and finally spreads between the lateral veins towards the midrib. The leaves take on a scorched appearance and drop prematurely. These effects have been described in detail for a number of plant species (Bergmann 1992).

CHAPTER 19

FURTHER ELEMENTS OF IMPORTANCE

In the previous chapters all elements which have been considered, with the possible exception of Na, are essential plant nutrients. Without any one of these essential elements the plant would be unable to complete its life cycle. All these elements too play a part in plant metabolism although their essential roles are in some cases still incompletely understood.

In addition to these nutrients are a group of elements which can have a beneficial effect on plant growth. A well known example of this type is Si. Under certain conditions this element can stimulate plant growth and may possible be essential for some plant species. The essential element Cl has also been included in this chapter.

19.1 Chlorine

In nature the chlorine ion (Cl⁻) is widely distributed and subject to rapid recycling. Chloride in the soil is not adsorbed by minerals and is one of the most mobile ions, being easily lost by leaching under freely drained conditions. Accumulation can occur mainly under arid condtions where on average the upward transport of water in the soil profile is higher than the leaching of Cl⁻. Soils high in Cl⁻ include those affected by the sea or treated with irrigation water containing Cl⁻, and poorly drained soils receiving run off from other areas.

Most plant species take up Cl⁻ at relatively high rates. The uptake rate depends primarily on the concentration in the nutrient or soil solution and on the individual uptake potential of plant species. According to Gerson and Poole (1972) uptake occurs against an electrochemical gradient presumably mediated by a chloride/H⁺ cotransport across the plasmalemma because low pH promotes Cl⁻ uptake. This means that the plasmalemma H⁺ ATPase activity also drives the Cl⁻ uptake which agrees with earlier findings that Cl⁻ uptake is metabolically controlled. In higher plants specific ion channels for chloride have been found responsible for the plasmalemma and tonoplast transport (Hedrich and Schroeder 1989). Blatt and Thiel (1993) suggest that the plasmamembrane of guard cells has two types of Cl⁻ channels. One is rapidly activated on depolarization of the membrane and is inactivated with time which means that if *e.g.* the plasmalemma is depolarized by high cation uptake, Cl⁻ uptake from the outer solution is favoured. The other Cl⁻ channel is also activated by depolarization but its rate of activation and deactivation is slow. Since the vacuole is positively charged relative to the cytosol, Cl⁻ transport across the tonoplast may be mainly a facilitated diffusion (Hedrich and Schroeder 1989). Usually the Cl⁻ concentration in the cytosol is lower than that in the vacuole. Trebacz *et al.* (1994) investigating liverwort (*Conocephalum conicum*) found a concentration of about 7 mol Cl⁻/m³ in the cytosol and of about 35 mol Cl⁻/m³ in the vacuole. In halophytes, however, the Cl⁻ concentration is much higher and may attain concentrations of about 80 mol Cl^-/m^3 in the cytosol and 300 mol/m^3 in the vacuole. The Cl^- uptake rate is the lower the higher the Cl^- concentration in the cytosol (Glass and Siddiqi 1984). Chloride uptake is inhibited by nitrate (Glass and Siddiqi 1984) and *vice versa* nitrate uptake can be much depressed by high chloride concentrations as found by Wehrmann and Hähndel (1984) for spinach, by Edelbauer (1978) for vine (*Vitis vinifera*) leaves and by James *et al.* (1970) in potato petioles. Presumably this antagonistic behaviour of both anions is due to the facilitated diffusion across the tonoplast through non specific anion channels (Hedrich and Schroeder 1989). In its transport through the root cortex to central cylinder there is evidence that the symplastic pathway represents the main route (Stelzer *et al.* 1975). Chloride is not only taken up by roots it may also be absorbed by aerial plant parts as chloride or chlorine gas (Johnson *et al.* 1957). The quantities of Cl^- in the atmosphere and rain water are considerably influenced by the distance from the sea, falling off rapidly moving inland (see page 352).

Plant tissues usually contain substantial amounts of Cl^- often in the range of 2 to 20 mg Cl/g dry weight but in chlorophilic species Cl^- may amout to 100 mg Cl/g dry weight. The demand for Cl^- for optimum growth, however, is for most species considerably lower, deficiency symptoms occurring in the range of 70 to 700 μg Cl/g dry weight (Clarkson and Hanson 1980), clearly establishing Cl^- as a micronutrient.

According to Merchant and Dreyfuss (1998) Cl^- is required for the fractional assembly of the photosystem II cluster comprising 4 Mn atoms. (see Figure 14.3). This is likely to be an essential function of Cl^- in photosynthetic eukaryots which is in agreement with earlier results of Bove *et al.* (1963) and Kelley and Izawa (1978) who have shown that Cl^- is required in the Hill reaction, the water splitting reaction in Photosystem II. The presence of Cl^- was found to enhance both the evolution of O_2 and photophosphorylation, (Bove *et al.* 1963).

Chloride is an important osmoticum and may be accumulated in the vacuole to higher concentrations. Growth is especially favoured in chlorophilic species like sugar beet and spinach as shown by Terry (1977) and by Wehrmann and Hähndel (1984). The main effect caused by an inadequate supply of sugar beets with Cl^- was a reduction in the rate of cell multiplication in the leaf blades which reduced the leaf surface area and consequently plant growth (Terry 1977). In some plant species chloride may influence photosynthesis indirectly *via* its effect on stomatal regulation of the guard cells. This occurs in onion where the guard cells are poor in starch and are unable to synthesize malate. In stomatal opening the inward flux of K^+ into the guard cells must be accompanied either by the accumulation of malate as a counter-ion or the inward flux of Cl^-. The absence of Cl^- in onion leaves thus inhibits stomatal opening and impairs leaf water potential (Schnabl 1980). The efflux of Cl^- out of guard cells results in a depolarization of the plamamembrane associated by an efflux of K^+ and hence the cell turgor declines (Tyerman 1992). Uptake of K^+ by guard cells has the reverse effect and the polarization of the plasmamembrane thus obtained may favour malate synthesis or the activation of Cl^- channels promoting Cl^- uptake. Plant species which show circadian rhythm of leaflet movement, opening at the morning at the beginning light exposure and closing at dark possess flexor cells and extensor cells. The changes in turgor of the flexor and extensor cells have been shown to result from the movement of K^+

and Cl⁻ through plasmamembrane channels (Kim *et al.* 1993). Blue light initiated the opening of leaflets of *Samanea saman* as well as the uptake of K⁺ and Cl⁻ by the extensor cells. Red light followed by darkness induced the closure of leaflets indicating that the K⁺ and Cl⁻ transport through channels was light-controlled and that light is accepted by phytochromes.

The non-specific roles fulfilled by Cl⁻ largely relate to the high mobility of the ion and the fact that it is tolerated over a wide concentration range. The role of Cl⁻ as a counter-ion to rapid K⁺ fluxes and as a contribution to turgor has already been mentioned. Chloride may also replace nitrate in its turgor function.This has been observed by Wehrmann and Hähndel (1984) who found that Cl⁻ application to spinach in field and solution culture experiments decreased the NO_3^- and increased the Cl⁻ concentration in the plant tops. A growth response also occurred which was probably related to the function of Cl⁻ as an osmoticum favouring water uptake and retention. The same is likely to be true of the favourable influence of Cl⁻ on the growth of coconut and oil palms as found by von Uexküll (1985) and Daniel and Ochs (1975). High concentrations of Cl⁻ in the order of 10 to 20 mg Cl/g dry weight are needed in order to obtain these effects which deserve further detailed investigations. Because of its osmotic functions Cl⁻ is considered an essential plant nutrient by Flowers (1988) who has treated this question in a useful review article. More than 100 chlorine containing natural compounds occur in plants. One of these is 4-chloroindolacetic acid which responded more actively in the *Avena* test than did the nonchlorinated indolacetic acid (Flowers 1988).

Deficiency symptoms have been observed by a number of workers using different crops (Broyer *et al.* 1954, Ulrich and Ohki1956). Wilting of leaves at margins is a typical feature and transpiration is affected. In addition the plants are often chlorotic. In sugar beet leaf growth is slowed down owing to a lower rate of cell multiplication. Leaf area is reduced and partial chlorosis occurs.

Soils considered low in Cl⁻ are below 2 mg water soluble Cl/kg soil which is rare (James et al. 1970). According to these authors water soluble Cl⁻ in soils is a good indicator for Cl⁻ availability. In practice Cl⁻ deficiency very seldom occurs because the ubiquitous presence of Cl⁻ in the atmosphere or in rain and irrigation water is more than enough to meet the 4 to 10 kg Cl/ha/year demand by crops (Reisenauer et al. 1973). Indeed even under laboratory conditions it is difficult to induce Cl⁻ deficiency because of atmospheric contamination (Johnson *et al.* 1957).

The effect of excess Cl⁻ in plants is a more serious problem. Crops growing on salt affected soils often show symptoms of Cl⁻ toxicity. These include burning of leaf tips or margins, bronzing, premature yellowing and abscission of leaves (Eaton 1966). An example of Cl toxicity in maple is shown in Plate 19.1 from the work of Walter et al. (1974). These symptoms arose following salt application to the road and roadside made to prevent snow lying in winter. Plant species differ in their sensitivity to Cl⁻. Sugar beet, barley, maize, and tomatoes are highly tolerant while tobacco, phaseolus beans, citrus, vine, potatoes, lettuce and some leguminous species are very prone to Cl⁻ toxicity. This latter group of chlorophobic crops respond better to sulphate than chloride based fertilizers. In Figure 19.1 an example is shown the effect of K chloride application comparing with that of K sulphate application to vine (*Vitis vinifera*) from the work of Edelbauer (1978). The vine was grown in solution culture with the KCl/K$_2$SO$_4$ ratios of 4/05, 2.5/2, 1/3.5 and trace/4.5. As can be seen from Figure 19.1 grape yield increased considerably with the increase of the sulphate proportion and the yield of the treatments with a chloride/sulphate ratio of 1/3.5 and trace/4.5 was more than twice that of the treatments with high chloride. There was also a clear tendency for the cluster weight and the number of clusters per pot to increase as the chloride supply decreased. It should be emphasized that under field conditions the negative effects of chloride are generally not as spectacular as in a solution culture because under field conditions the chloride and the sulphate applied mix with the chloride and sulphate present in the soil. Additionally chloride leaching must be taken into account. From the KCl applied in autumn a substantial amount of chloride may be leached into deeper soil layers by winter rainfall while the K$^+$ remains adsorbed to soil colloids in the upper soil layer.

Plate 19.1 Chloride toxicity in leaves of maple.

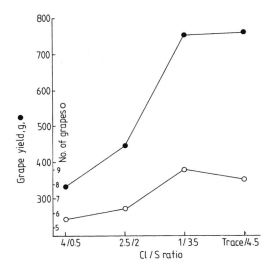

Figure 19.1   Efffect of the Cl/S ratio in the nutrient solution on grape yield and number of grapes (after Edelbauer1978).

Reduction in yield and quality in crops is associated with tissue levels of 5—20 mg Cl/g dry matter for sensitive crops and 40 mg Cl/g dry matter or more of tolerant plant species (Reisenauer *et al.* 1973).

19.2   Silicon

Silicon is the second most abundant element in the lithosphere after oxygen and occurs in almost all minerals. Important Si containing minerals are quartz, opal and cristobalite. Soluble Si in the soil solution is mainly ortho silicic acid ($H_4SiO_4 = Si(OH)_4$) which may form polymers by splitting off H_2O according to the equation:

$$Si\,(OH)_4 + Si(OH)_4 \rightarrow (OH)_3Si\text{-}O\text{-}Si(OH)_3 + H_2O$$

In this manner Si containing chains are formed the solubility of which decrease with increasing chain length. There is a gradual transition from monomeric silicic acid to larger molecules following the sequence (Drees *et al.* 1989):

$$\text{Silicic acid} \Rightarrow \text{Hydrosols} \Rightarrow \text{Hydrogels} \Rightarrow \text{Xerogels}$$

These Si containing sols and gels belong to the fraction of amorphous soil Si which is of high importance for supplying plants with Si because of its lower stability as compared with Si bearing minerals. The accessibility of Si to plants depends largely on how rapidly weathering takes place bringing Si into soil solution. In minerals highly resistant to weathering such as quartz, Si is completely unavailable and in highly weathered soils such as Oxisols and Ultisols crops may suffer from insufficient Si supply (Savant *et al.* 1997, Epstein 1999). Soluble Si is present as monomeric H_4SiO_4 over a wide pH range (2 to 9) and is in equilibrium with amorphous SiO_2 with an equilibrated concentration of about 2 mol/m³. At pH > 9, H_4SiO_4, is deprotonated (Jones and Handreck1965). This relationship is shown in Figure 19.2 (Alexander *et al.* 1954 quoted from Drees *et al.* 1989).

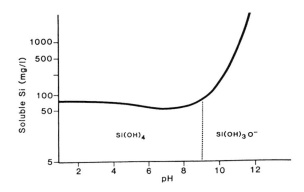

Figure 19.2 Concentration of soluble Si in relation to soil pH (after Drees *et al.* 1989).

The graph shows that the concentration of soluble Si, mainly $Si(OH)_4$, remains at a constant level over a broad pH range until the deprotonation begins at pH 9 with a steep increase of the solubility in the form of $Si(OH)_3O^-$.

$$H_4SiO_4 \Leftrightarrow H_3SiO_4^- + H^+$$

The concentration of deprotonated silicic acid ($H_3SiO_4^-$) is mainly controlled by pH-dependent adsorption reactions on sesquioxides. Adsorption decreases at either side of a maximum at pH 9.5 and is of no major importance at soil pH < 7. Aluminium hydroxyoxides are more effective than iron hydroxyoxides in adsorption although the actual mechanism is still not clear (Jones and Handreck1967). Acid soils tend to contain higher concentrations of Si in the form of non dissociated silicic acid (H_4SiO_4) in the soil solution where its concentration is in the range of 100 to 600 mmol/m³ and hence is generally higher than that of phosphate (Epstein 1999). Liming has been found to decrease Si availablity and uptake of Si by a number of crop plants (Grosse -Brauckmann 1956). An assessment of the available Si in the soil has been obtained from the ratio of easily extractable Si to the free or easily extractable sesquioxides. The higher the ratios of Si/Al or Si/Fe the greater was found to be the uptake of Si by rice (Jones and Handreck 1967).

The form in which Si is taken up by plants is presumably the non dissociated form of silicic acid (H_4SiO_4), although the mechanism of uptake is still not clear. The suggestion that the uptake of nondissociated silicic acid is passive is difficult to justify since other undissociated molecules such as sugars can be taken up by proton cotransport (see page 126) and therefore are dependent on energy. In experiments with oats Jones and Handreck (1965) concluded that uptake was passive observing that values for the uptake by the plant agreed closely with calculated values derived from data of soil solution concentration and water uptake by transpiration. Such an approach is based on the assumption that the Si is passively translocated by the water streaming into the root apoplast. Much of this Si, however must be translocated from the roots to the leaves and the question arises whether this transport is merely an apoplastic one. As shown on page 194 the Casparian strip is a strong barrier for solutes and it is therefore more likely that Si also has at some stage to traverse the plasma membrane to enter the xylem sap. In this context it is of interest that Hildebrand *et al.* (1997, quoted by Epstein 1999) found a cDNA which encodes for a Si transporter in the marine diatom *Cylindrotheca fusiformis*. This observation supports the suggestion that Si transporters may also occur in the membranes of higher plants. This is in accord with numerous observations that the Si concentration differs considerably between species and even cultivars (Epstein 1999). Lafos (1995) found that the Si concentrations in the upper part of *Vitis vinifera* depended much on the type of rootstock. The findings of van der Vorm (1980) showed that Si uptake rates depended much on the concentration of silicic acid in the nutrient solution and differed considerably between species (see Table 19.1). For the three crop species rice, wheat and soya bean growing at three concentrations of Si in solution, a very low one, a medium one and a very high concentration, that uptake rates at all three Si levels in the nutrient solution were highest in rice and lowest in soya bean. Even at the highest Si concentration the Si uptake rate of soya bean was still lower than the uptake rate of rice in the treatment with the lowest Si concentration in the nutrient solution. Transpiration coefficients were similar for rice and wheat; but rice still showed a much higher Si uptake rate than wheat. The last column shows the calculated Si uptake assuming non selective uptake by mass flow. Even at the highest Si concentration in the nutrient medium this did not amount to 50% of the total Si taken up by the roots. These findings indicate a selective uptake of Si by these crop species although the mechanism is not clear. For soya bean at the highest concentration of supply there was an effective exclusion mechanism operative.

Table 19.1 Actual Si uptake and mass-flow supply to the roots of three plant species grown on nutrient solutions varying in Si concentration (Van der Vorm 1980)

Crop species	Si concentration nutr. solution mg Si L^{-1}	Transpiration coefficient L H$_2$O kg^{-1} DM	Actual Si uptake g SiO$_2$	Mass flow supply of Si kg^{-1} DM
Rice	0.75	286	10.9	0.2
	30	248	94.5	7.4
	162	248	124	40.2
Wheat	0.75	295	1.2	0.22
	30	295	18.4	8.9
	162	267	41.0	43.3
Soybean	0.75	197	0.2	0.15
	30	197	1.7	5.9
	162	197	4.0	31.9

contain some Si so that Si at least may function as a micronutrient. In this context it is of particular interest whether Si is essential for lignin synthesis. If this were proved to be the case Si would qualify to be included in the list of classical plant nutrients. Currently though Si is still considered according to Epstein (1999) as "quasi-essential". Generally the Si concentrations in the leaves of plants are in the same order as those for macronutrients and in most cases soils contain substantial amounts of available Si.

For farming, Si may play a role in crop production. The most important crop which responds to Si is rice (Savant *et al.* 1997). As shown in Figure 19.3 from an investigation of Park (1975) a significant relationship occurs between the Si concentration in rice straw and the yield of brown rice. Silicon especially promotes the reproductive organs of rice, as was found by Okuda and Takahashi (1965) in a solution culture experiment with lowland rice.

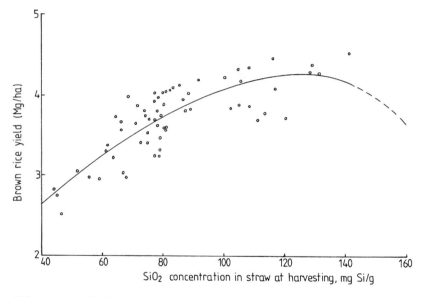

Figure 19.3 Relationship between brown rice yield and SiO_2 concentration in the straw at harvesting stage (after Park 1975).

Table 19.2 The effect of Si supply on the growth and grain yield of lowland rice (Okuda and Takahashi 1965)

	Top length,	Number of panicles/pot	Number of spikelets/panicl.	Percentage of fully ripened grains	Wt. of mature grains (g/pot)
- Si	85	9.5	49.3	55	5.25
+ Si	94.5	11.6	63.2	76	10.83

Data of this investigation shown in Table 19.2, clearly indicate that Si had a particularly beneficial effect on the grain weight. In addition, however, other grain yield components such as number of panicles, number of spikelets per panicle, and the percentage of fully ripened grains were favourably influenced by Si. Continuous rice growing may lead to the depletion of available Si in soils because of the relative high Si uptake rate of rice. Such depletion of available Si causes a decline of rice yield (Epstein 1999). Silicon application to rice may also improve the efficiency of utilization of other plant nutrients as emphasized by Savant *et al.* (1997).

It has been known for many years that the application of silicia containing fertilizers can increase the availability of soil phosphate (Fisher 1929). Experiments on barley at Rothamsted in England quoted by Russell (1973) show that a dressing of 450 kg/ha sodium silicate applied annually was still increasing phosphate availability after a century of use. The mechanism by which this occurs is that the added silicate displaces $H_2PO_4^-$ adsorbed by the Al and Fe oxides. This displacement occurs very readily under acid and neutral conditions. According to Robson and Pitman (1983) silicate can only be adsorbed in the presence of phosphate when it increases the surface negative charge, that is above pH 7. The silicia fertilizers include soluble silicates, sinterphosphates, and Ca silicate slags. Occasionally they are used in soils low in Si to improve crop yields and quality. Ayres (1966) for example claimed increases in yield and sugar concentration of sugar cane by their application in such soils. On paddy rice soils, silicate slags are sometimes used to increase soil pH and soluble silica (Russell 1973). Tokunaga (1991) applied a potassium silicate which is a slow release fertilizer with a favourable effect on rice and cucumber. Savant *et al.* (1997) have published a valuable review article on the importance of Si in rice production.

19.3 Cobalt

In soils the concentration is in a range form 1 to 40 mg/kg soil but may reach concentrations of 100 to 200 mg Co/kg in parent material of volcanic origin (Vanselow 1966). Co occurs in all igneous rocks in concentrations from 1 mg to several hundred mg Co/kg soil, the level of Co closely following the distribution of Mg in the ferromagnesian minerals (Mitchell 1964). In the ultrabasic rocks such as dunite (Mg-rich igneous rock), perioditite and serpentine where the content of Mg rich ferromagnesian minerals is high, levels from 100 to 300 mg Co/kg soil may be present. On the other hand acidic rocks including granites containing high contents of Fe rich ferromagnesian minerals are low in Co with concentrations of 1 to 10 mg Co/kg soil. The distribution in the sedimentary rocks is much dependent on their mode of formation. In the argillaceous rocks such as shales the Co concentration may be relatively high and from 20–40 mg Co/kg soil, whereas sandstones and limestones are usually poor in Co with concentrations below 5 mg Co/kg. There is a clear relationship between the Co concentration and soil texture with an increase in Co with a rise of the clay concentration (Baize 1997). Cobalt occurs primarily in the crystal lattices of ferromagnesian minerals and as such is unavailable to plants (Mitchell 1972). After release from these minerals by weathering, Co^{2+} is held largely in exchangeable form or as organo mineral complexes. Exchangeable Co^{2+} is very firmly bound and like Cu^{2+} the concentration in the soil solution is extremely low. Cobalt is also bound to Fe^{III}/Mn^{IV} oxides and transient waterlogging conditions may lead to a reduction of Fe^{III} and Mn^{IV} associated with a relase of Co^{2+} which may be leached into deeper soil layers (Baize 1997). For this reason the rate of Co release by weathering is more rapid in poorly drained conditions. This is reflected in a higher available Co concentration than in freely drained soils derived from similar parent materials, even though the total soil Co level may be lower in the poorly drained sites (Mitchell 1964). Cobalt like other heavy metals (Cu,Ni, Pb, Zn) can be strongly adsorbed to oxide surfaces, particularly to Mn oxides. According to McKenzie (1989) such adsorption occurs in excess of the surface charge and from the cation only one positive charge is neutralized so that in the case of Co the Co^+ is adsorbed which may render the surface charge positive. This process is known as specific adsorption by which the Co can be rendered unavailable. In some Australian soils Taylor and Mc Kenzie (1966) reported an average of 79% of total soil Co associated with Mn oxide minerals. Similar results have been reported in Ireland from soils on which 'pine', the cobalt deficiency disease of sheep, has been observed (Fleming 1977).

Little is known of the mechanism of Co transport across plant membranes. In experiments with pasture plants using radioactive Co, Handreck and Riceman (1969) found that Co is readily translocated from old leaves to new tissues. By contrast in age sequence studies in lupin Robson *et al.* (1979) observed that Co concentrations in leaves increased as the leaves aged, and that the Co concentrations in the old leaves were much greater than in the young leaves. Even when the root nodules were Co deficient, Co continued to accumulate in the older leaves. These findings confirm some of the reports in the older literature that Co is not readily mobile in the plant (Gustafson and Schlessinger 1956, Langston 1956).

Co behaves like other heavy metals. In a similar way to Fe, Mn, Zn and Cu it is bound by organic molecules as chelate. It can also displace other cations from physiologically important binding sites and can thus affect their uptake and mode of action. Nicholas and Thomas (1954) observed that excess Co supply induced Fe deficiency. Hewitt (1953) reported that the toxic effects of excess Co resembled Mn deficiency. Both these observations indicate that the toxic effects of excess Co relate to the effect of Co in displacing other heavy metals from physiologically important centres. The effects of Co toxicity in plants resulted in leaves which were chlorotic and necrotic, and which frequently wither completely. Cobalt toxicity may be alleviated by Fe (Bollard 1983).

A few plant species are less sensitive to Co toxicity. In some cases Co is accumulated at levels over 100 times that of other plants growing in the same soil. The swamp black gum (*Nyssa sylvatica*) which grows in the south-eastern U.S.A. can have Co concentrations approaching 1000 mg Co/kg dry matter. This species acts as a very good guide to Co availability and values in the leaf of less than 5 mg/kg dry matter are indicative of Co deficiency in ruminants feeding on herbage from the same environment (Vanselow 1966).

Other plant species in which Co accumulates serve to indicate the presence of cobaltiferous ores. One such species found in the Sharba region of Zaire and occurring only in areas rich in Co is *Crotolaria cobalticola* (Fleur du Cobalt). Values from 500 to 800 mg Co/kg have been observed in the dry matter of this species (Duvigneaud and Denaeyer- de Smet 1959).

Cobalt is essential for symbiotic N₂ fixation (Ahmed and Evans 1960) and for rhizobial growth (Cowles *et al.* 1969). The drastic effect of a Co deficient medium on the growth of soya bean in symbiosis with *Rhizobium japonicum* is shown in Plate 19.2. In studies to investigate the effect of Co on symbiotic N₂ fixing bacteria *Rhizobium meliloti* Kliewer and Evans (1963) were able to show that increasing the supply of Co increased rhizobial growth, N₂ fixation and the concentration of vitamin B₁₂ and the formation of leghaemoglobin in the rhizobia. From this work it was concluded that the effects of Co on N₂ fixation and rhizobial growth were mediated by a Co containing coenzyme, cobalamin (= vitamin B₁₂). The structure of the coenzyme is shown in Figure 19.4. Cobalt is at the centre of a corrin ring and is bound by three coordinative bonds to each of the N atoms of three pyroll rings. They form the corrin ring structure which is similar to the porphyrin ring which comprises four pyrrole rings. Cobalt is bound with a further coordinative bond to the ligand below the corrin plane (Figure 19.4).

Plate 19.2   The effect of the addition of 0.1 µg of Co/L in the nutrient solution on the growth of soya bean in symbiosis with *Rhizobium japonicum*. (Photo Ahmed and Evans) By courtesy of Plenum Publishing Co. Ltd.

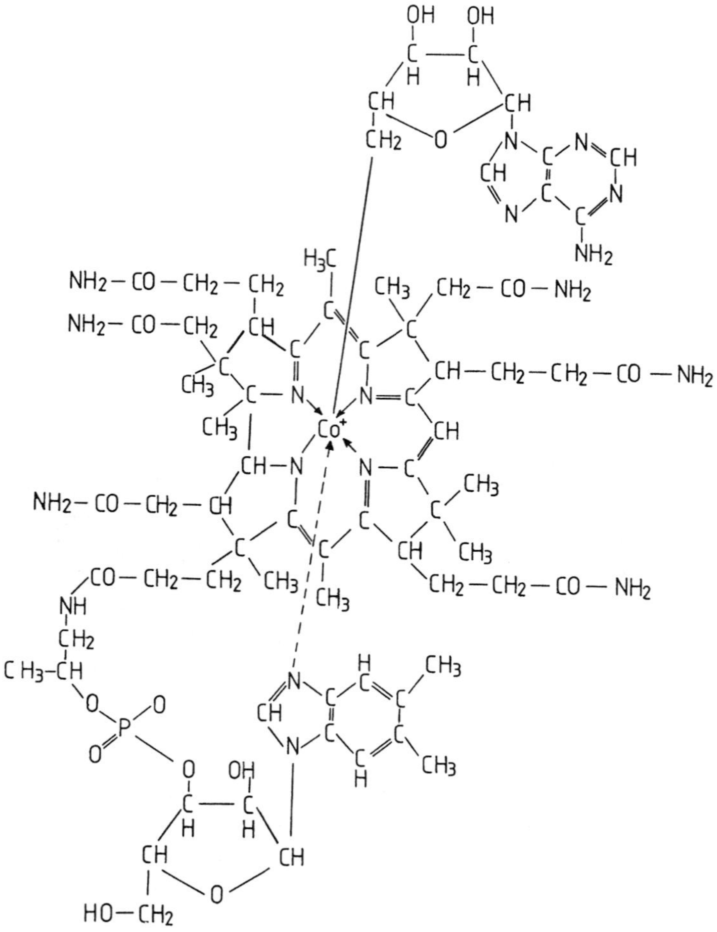

Figure 19.4 Molecular structure of the coenzyme Cobalamine.

In addition there are two covalent bonds so that of Co is centered in a hexadendate structure. One covalent bond attaches Co to the ligand above the corrin plane. This ligand is a adenosyl (Co-carbon bond). The other covalent bond binds Co to a N atom of a pyrrole ring. For the formation of this covalent bond Co provides one e⁻. It is for this reason that the Co has a positive charge (Co^+). The ligand above the plane is adenosyl which is involved in the reduction of nucleotide triphosphate to desoxynucleotide triphosphate (Licht *et al.* 1996).

Cobalamine is the coenzyme of three enzymes required for N_2 fixation and nodule growth (Dilworth *et al.* 1979). These are:

1. Methylmalonyl-coenzyme mutase an enzyme involved in the synthesis of haem in the bacteroids needed for leghaemoglobin production in the nodules. This accounts

for the correlation between Co deficiency and a decrease in leghaemoglobin concentration in the nodules as observed Gladstones *et al.* (1977).

2. Methionine synthase which provides methionine for the protein synthesis in bacteroids.

3. Ribonucleotide reductase which catalyzes the reduction of ribonucleotides to deoxyribonucleotides. The enzyme is therefore essential for DNA synthesis. Cobalt deficiency may thus be expected to result in defective DNA synthesis as well as a reduction in cell division and nodule growth in rhizobia. This is in accordance with the findings of Chatel *et al.* (1978) who observed a smaller number of elongated bacteriods in Co deficient root nodules.

According to Jordan and Reichard (1998) there are three different classes of ribonucleotide reductases. The reductase in the rhizobium belongs to the class II which in contrast to the other two classes of ribonucleotide reductase requires the coenzyme cobalamine and hence Co. The reduction of the ribonucleotide to the de oxyribonucleotide needs a radical which is provided by the cleavage of the carbon- cobalt bond by which the ligand above the corrin plane is bound (Jordan and Reichard 1998). Besides cobalamine further cobamids (= analogs of cobalamin, Co containing coenzymes) are known which differ only slightly in the ligand below the corrin plane. They catalyse a methyl transfer and play a role in methanogenic bacteria (Dimarco *et al.* 1990).

The effects of Co deficiency on nodule formation and function were studied by Dilworth *et al.* (1979) using sweet lupin (*Lupinus angustifolius*) a plant particularly sensitive to Co deficiency. Some of their important results are shown in Table 19.3 in which the *Rhizobium* — inoculated treatments with and without Co are compared. Not only did Co increase the weight of the nodules, the Co concentration in the nodules and the number of bacteriods per nodule, but the cobalamin and leghaemoglobin concentrations were also raised. These workers suggest that the cobalamin concentration determines leghaemoglobin probably *via* its effect on haem synthesis in the bacteriod. In the Co deficient plants the onset of N_2 fixation was delayed by several weeks because of the lower degree of *Rhizobium* infection. Interestingly enough only 12% of the Co present in the nodules occurred as cobalamin which raises the question as to the function, if any, of the remaining Co. Cobalt has been shown to be essential for N_2 fixation not only in legumes but also in *Alnus* (Hewitt and Bond 1966) and *Azolla* (Johnson 1966).

Table 19.3   The influence of addition of Co to inoculated *Lupinus angustifolius* grown on a Co-deficient soil on the weight of the crown nodules and some of their properties, from 6 week old plants (Dilworth *et al.* 1979)

	Nodule weight g/plant fresh weight	Co concentr. ng g-1 dry weight	Bacteroid number x 10^9	Cobalamin concentration ng g^-1 fr. wt.	Leghaemoglobin concentration mg g^-1 fr. wt.
+ Co	0.6	105	27	28.3	1.91
- Co	0.1	45	15	5.9	0.71

It is still in question whether in addition to its requirement in symbiotic N_2 fixation, Co is essential for higher plants. Evidence of beneficial effects of Co for higher plants have been reported for non nodulated subterranean clover Hallsworth *et al.* (1965) and Wilson and Hallsworth (1965). A similar response of wheat to Co was observed by Wilson and Nicholas (1967). In addition these workers reported Co deficiency symptoms in both plant species as shown by chlorosis in the younger leaves. Cobalt complexes of low molecular weight were also detected in plants grown aseptically thus indicating the incorporation of Co into the metabolism. More evidence, however, is still required before Co can be classified as an essential nutrient for higher plants.

Cobalt is also of importance in animal nutrition. It is well established that Co is a metal component of vitamin B_{12} which is essential in N-metabolism in the ruminant. Inadequate levels of Co in the herbage can lead to Co deficiency symptoms in ruminants characterized by a lack of appetite, lack of growth, and poor reproductive ability. The critical Co level of the diet of ruminants is about 0.08 mg Co/kg dry matter of the herbage (Underwood 1971). Ruminants with a diet based on forage grasses and cereal grains usually require Co supplementation as grasses are often low in Co. Considerably higher levels are usually found in legumes. According to Kubota and Allaway (1972) the Co concentration of leguminous forage is in the range of 0.15 to 0.30 and that of grasses in the range of 0.04 to 0.08 μgCo/g dry matter (Kubota and Allaway1972).

Cobalt deficiency occurs in highly leached sandy soils, soils derived from acid igneous rocks or in highly calcareous or peaty soils. It is favoured if the soil pH is neutral to alkaline (Mitchell 1972). Various extractants have been used to determine readily soluble or available Co. These include 2.5% acetic acid (pH 2.5), neutral kmol/m³ ammonium acetate, or 0.05–50 mol/m³ EDTA. Acetic acid extracts of normal agricultural soils give Co levels from 0.05–2 mg Co/kg soil. Where values are less than 0.1 mg Co/kg soil Co deficiency may be suspected. The deficiency can be controlled by administration of a Co salt to the soil at the rate of 1 or 2 kg/ha. If the soil contains large amounts of manganese minerals capable of immobilizing cobalt, higher quantities are required (McKenzie 1975). According to experience of Henkens (1965) in the Netherlands 50 g of Co are required to raise the Co level in the soil by 0.1 mg Co/kg soil in the acetic acid extract. Henkens found that also the application of Co-containig slag could cure Co deficiency.

19.4 Vanadium

Vanadium is widely distributed in biological materials and there are many reports that low concentrations can have a favourable effect on the growth of microorganisms, animals and higher plants (Arnon and Wessel 1953, Pratt 1966). It is claimed to be essential for green algae (Arnon and Wessel, Meisch *et al.* 1975 and 1977) in which it promoted photosynthesis. The precise function of V, however, is still unknown. Some evidence suggests that V may partially substitute for Mo in N_2 fixation in microorganisms but the evidence for the substitution in symbiotic N, fixation is not conclusive (Stewart1967).

In the higher plants there is no evidence as yet that V is an essential element for any plant species. Its concentration in aerial plant parts is usally low and in a range of 0.2—4 mg V/kg dry matter (Bollard 1983). Welch and Huffman (1973) grew lettuce and tomato in nutrient solution much lower in V concentration than than required by *Scenedesmus obliquus*. No evidence of V deficiency was observed and these workers claimed that if V is essential for plants adequate levels in plant tissues are less than 2 mg V/kg dry weight. This concentration is considerably below the normal values found in plant material.

In excess amounts V can be toxic and this has been observed in water culture experiments (Warrington1955). Neither deficiency nor toxicity, however, are of any significance under field conditions.

CHAPTER 20

ELEMENTS WITH MORE TOXIC EFFECTS

There is no clear division between elements which are toxic to plants and those which have a beneficial or even essential effect. The effect of any element on the plant depends not only on its chemical properties but also on its concentration and the presence and concentrations of other elements. The physiological age and species of the plant corncerned as well as other environmental factors are also of importance. Some elements such as Fe, Mn, Cu, B, Zn are essential at low concentrations but are toxic at higher levels. The toxic effects of these elements have already been discussed in the appropriate chapter for each nutrient. In the case of the heavy metals Pb and Cd, toxicity is induced by the mimicking of lighter essential elements in uptake and biochemical behaviour.

20.1 Iodine

Iodine and bromine are typical of most of the elements described above. Neither have been shown to be essential to plants but both are reputed to produce stimulating effects on plant growth at low concentrations. Toxic effects are produced at higher concentrations. For I, the stimulating effect has been observed at I levels in the order of 0.1 $\mu g/g$ in soil and nutrient solution whereas toxic effects occur in excess of about $0.5-1.0$ $\mu g/g$ (Martin 1966). This latter concentration is considerably higher than normal soluble soil I levels so that I toxicity has not been reported on agricultural soils.

The toxic effects of high I begins in the older leaves. In tomato they become chlorotic and absciss, whilst the younger leaves remain a very dark green colour. Growth is severely restricted and the leaves curl back and necrosis occurs at the tip and edges. In severe cases the plant dies. The I concentration at which toxicity appears differs considerably between plan species and was for beans 0.75 μg I/g dry matter in leaves and for leaves of turnips higher than 20 $\mu g/g$ dry matter (Mack and Brasher 1936). According to Whitehead (1973) grass and clover species growing in a nutrient solution with a low I concentration (0.1 mol/m^3) accumulated up to 100 μg I/g dry matter in leaves without growth depressions. Iodine concentrations in roots were much higher than in leaves (Whitehead 1973).

Normal levels in healthy plants range from $0-0.5$ μg I/g dry matter. High levels of chloride can reduce the toxic effects of iodide which suggests that there is a competitive effect between both elements (Lewis and Powers 1941). Iodine is required in animal nutrition, as it is an essential element of the hormone thyroxine.

Higher plants are also capable of incorporating I into tyrosine and thus synthesizing molecules such as 3-iodotyrosine, 3,5-diiodotyrosine, and 3', 3,5-triiodothyronine which are closely related to thyroxine. Iodine 4 derivates may function as inhibitors such as iodoacetate which reacts with suphhydril groups of enzymes (Walton and Dixon 1993).

According to Hsiao (1969) I is essential for the marine brown alga *Patalonia fascia*. Some marine algae may accumulate I at concentrations as high as 10 μg/g dry matter. There is some evidence that in red and brown algae I is accumulated in special cells (quoted after Bollard 1983).

20.2 Bromine

Bromine is taken up by plants as Br⁻. In general the ion is not so toxic as iodide and has been used in many physiological studies in ion uptake. Normally the levels of Br in soils are very low so that Br- toxicity does not occur naturally. In recent years, however, with the use of Br- containing soil fumigants such as methyl bromide, toxic effects have been reported in a number of sensitive plants such as carnations, chrysanthemums, potato, spinach, and sugar beet (Martin 1966). According to Zottini *et al.* (1994) the herbicide Bromoxynil (3,5 dibromo—4-hydroxybenzonitrile) supposedly inhibits the dicarboxylate carrier in the inner mitochndrial membrane and thus the import of malate and succinate. Symptoms of Br⁻ toxicity resemble excess salt effects and leaves often become chlorotic followed by expanding leaf tip necrosis and edge necrosis. Poor seed germination may also result. Some plant species are insensitive to Br⁻ toxicity. These include carrot, tobacco and tomato. These species can accumulate over 2000 μg Br/g dry matter without showing any adverse effects. Normal Br concentrations for plants growing in soil, however, are usually much lower and in the range of 0 to 260 μg Br/g dry matter (Martin 1966), concentrations at the lower end of the scale being more frequent.

To a certain extent Br⁻ can substitute for part of the Cl⁻ requirement of plants (Broyer *et al.* 1954). Ozanne *et al.* (1957) observed that part of this effect resulted from the displacement of Cl⁻ from non effective sites such as in the root to more essential positions. These workers also observed that typical Cl⁻ deficiency symptoms could be alleviated by the addition of Br⁻.

20.3 Fluorine

Fluorine generally occurs in plant material in the range of 2—20 μg F/g dry matter, although some plant species are capable of accumulating much higher amounts. The poisonous South African shrub *Dichapetalum cymosum* for example can accumulate as much as 200 μg/g dry matter. In this species F is present as fluoroacetate. This is toxic to animals for on ingestion it is converted to fluorocitrate. This competitively inhibits the enzyme aconitase responsible for the conversion of citrate to isocitrate in the TCA cycle. Commercial tea has also been shown to have levels of F as high as 400 μg F/g dry matter (Mitchell and Edman 1945) and also to contain fluoroacetate. Very large amounts, however, must be ingested in order to induce toxicity.

High levels of F are usually toxic to plants. Respiration can be both stimulated and inhibited. According to Miller and Miller (1974) fumigation of soya beans with HF at first stimulated respiration and this was followed by respiratory inhibition. It is well known that of the respiratory enzymes enolase in particular is very sensitive even to low F levels. The reason for the primary stimulating effect on respiration, however, is not so clear. According to Lee *et al.* (1965) glucose 6 phosphate dehydrogenase, catalase, peroxidase and cytochrome oxidase activities in soya beans are all increased by fluoride treatment. This may well contribute to the observed primary respiratory stimulation. Fluoride inhibits inorganic pyrophosphatase and hence the oxidation of free fatty acids (Lehninger 1975). It also inhibits an ATP driven H^+ ATPase of the plasmamembrane which is also inhibited by a fluoroaluminate as well as by Al^{3+} (Facanha and de Meis 1995).

Fluorine toxicity is only found under field conditions in industrially polluted regions where hydrofluoric acid occurs. Exposure of plants to even extremely low hydrofluoric acid concentrations over a period of several months gives rise to foliar toxicity effects in many crops. Symptoms differ between plant species but two basic types occur. These are marginal necrosis sometimes called burnt tip, and interveinal chlorosis (Brewer 1966a). Most plant species show marginal chlorosis and damage with interveinal chlorosis as an earlier symptom of a less acute form of toxicity. A few species including maize, however, only show the chlorosis effect. Some crops are more sensitive than others to hydrofluoric acid toxicity and these include grapes and fruit trees. Generally crops which accumulate high levels of F are less sensitive. Toxicity symptoms have been discussed in detail by Brewer (1966a).

The total F content of a soil is usually unrelated to F availability. The form taken up by plants is soluble fluoride, and the predominant factors controlling the level of this ion in soil solution are the soil pH and the amount of soil Ca and P (Hurd-Karrer 1950). When the soil pH is high or when soil Ca or P are present in large amounts, soil fluorine is fixed as calcium fluoride (CaF_2) or aluminium silicofluoride ($Al_2(SiF_6)_3$). Even when soluble fluoride concentrations are high as in acid soil conditions, however, soil F^- is not readily taken up by plant roots. This low uptake potential of F^- was demonstrated by Venkateswarlu *et al.* (1965) who compared the uptake of F^- and Cl^- by barley roots. When the concentration of the two ions at the root surface were identical a 100 fold higher uptake of Cl^- resulted. These effects of poor availability in the soil and low uptake potential account for the normally minute levels of F^- in plants and the infrequency of F toxicity caused by excess uptake from the soil. The subject of fluoride and plant life has been reviewed by Weinstein (1977).

20.4 Aluminium

More than 15% of the earth's crust is made up of Al_2O_3. Aluminium is thus an important constituent of the soil, and along with Si is the major element making up the lattices of primary and secondary clay minerals. The solubility of Al in the soil is too low in neutral and alkaline soils to be toxic to plant growth. Higher plants usually contain in the order of about 200 mg Al/kg dry matter. In tea the levels may be as high as 2000 to 5000 mg/kg dry matter and according to Chenery (1955) Al is necessary for the normal growth of the tea bush. Generally the Al concentration in roots is much higher than in the upper plant parts. Aluminium may be particularly accumulated in the root tips where it is mainly bound in the apoplast. Crossbinding of Al in the pectate moiety of meristematic cell walls may affect the extensibility and water permeability of cell walls as described by Horst (1994) in a useful review article.

As shown on page 55 dissolution of Al hydroxy compounds in soils depends much on soil pH (Lathwell and Peech 1964). Low pH values may result in high levels of soluble Al which are toxic to plants. The dissolution of Al compounds and formation of cationic Al species is complicated. Al^{3+} dissolved in water yields several monomeric soluble Al compounds as shown below (Hsu 1989):

$$Al^{3+} + H_2O \Leftrightarrow Al(OH)^{2+} + H^+ \,(1)$$

$$Al\,(OH)^{2+} + H_2O \Leftrightarrow Al(OH)_2^+ + H^+ \,(2)$$

$$Al(OH)_2^+ + H_2O \Leftrightarrow Al(OH)_3 + H^+ \,(3)$$

From these equations it is evident that with an increasing pH the formation Al-OH cationic species is favoured which promote the arrangement of polyhydroxy Al species. These have a lower solubilty than Al^{3+} and $Al(OH)^{2+}$. Equation (1) has a pK of 5 which means that at pH 5 the Al^{3+} concentration is equal the concentration of $Al(OH)^{2+}$. At this pH 5 the concentrations of both species are already so low that they have no toxic effect (Moore 1974). At lower pH values, however, the solublility of both cationic Al species increases. In a useful review article Kinraide (1991) emphasizes that as yet it is not clear which Al form is actually toxic. There is clear evidence that the solubility of Al bearing minerals increases with a fall in pH. In this low pH range, however, not only mononuclear Al species exist which are molecules with one Al as Al^{3+}, $AlOH^{2+}$, and $Al(OH)_2^+$. Besides these cationic species polynuclear Al species are also formed of which the most toxic is triskaideka Al.

$$Al_{13} = [AlO_4Al_{12}(OH)_{24}(H_2O)_{12}]^{7+}.$$

The formation of such polynuclear forms is not only pH-dependent but also dependent on the concentration of other cationic Al species as well as on the Ca^{2+} and sulphate concentrations. Since these may differ in the rhizosphere and in the root apoplast it is very difficult to test which cationic Al species is especially toxic. The Al_{13} cationic species is toxic at very low concentrations which can hardly be detected. Aluminium is also influenced by proton toxicity, which may be alleviated by Al^{3+}. Presumably the aluminate $[Al(OH)_4^-]$ which is formed at pH > 7.9 is not toxic but it may give rise to Al_{13} in the apoplast because of low pH.

On many acid soils it is not so much the high H^+ concentration in the soil solution as the high Al concentration which is harmful to plants and especially to root growth (Fox 1979). This is also true for the legume-*Rhizobium* symbiosis where Al inhibits the multiplication *Rhizobium* in the soil and thus also the infection and nodulation (Merbach *et al.* 1990). In cultivated naturally acid soils in particular, soil acidity often increases down the profile so that the rooting depth of the plant is restricted and water and nutrients of the subsoil cannot be exploited. The first observable effect of Al on plants is a limitation in root growth (Clarkson and Sanderson1969). Root tips and lateral roots become thickened and turn brown (Mühling *et al.* 1988). According to Wissemeier *et al.* (1987) the restriction of root growth is associated with the formation of callose.

Bennet and Breen (1991) found that Al toxicity causes a loss of the apical dominance. Therefore the growth of the main root is completely inhibited and that of laterals may be stimulated. With Al toxicity the volume of maize root caps was reduced and the root cap cells showed a strong vacuolization; the structure of the nucleus and of the nucleolus was affected as well as structure and function of the Golgi apparatus so that no root slime was produced (Bennet and Breen 1991). These authors conclude from their results that Al toxicity does not originate from Al affecting DNA but is caused by the impact of Al on regulatory mechanisms in the root tips. According to Siegel and Haug (1983) Al binds to calmodulin and may in this way interfere with various enzymatic processes. More recent investigations of Zhang and Rengel (1999) with apical root cells have shown that Al brings about an increase in Ca^{2+} concentration in the cytosol with a detrimental impact on metabolic processes in the cytosol. Recent investigations of Sigavuru et al. (1999) carried out with an Al sensitive cultivar of maize revealed that Al inhibited the formation of microtubles and actin and microfilaments in the transition zone of root maize tips thus affecting the development of the cytoskeleton. This disturbance was associated with the formation of callose and depolarization of the plasmamembrane as reported earlier by Hecht-Buchholz and Foy (1981). Accordingly cationic Al species as well as a fluoroaluminate complex inhibit a plasmamembrane H^+ ATPase (Facanha and de Meis 1995) which is likely to be detrimental to ion uptake. As reported by Grimme (1983) Al especially retards the uptake of Mg^{2+}. Ding et al. (1993) found in membranes isolated from onion bulbs that cationic Al blocked the Ca^{2+} channel and hence the uptake of Ca^{2+}.

Aluminium toxicity is frequently found on acid soils where Al toxicity and H^+ acidity may interact. Both kinds of toxicity depress root growth. Aluminium inhibits cell devision which is an irreversible process in contrast to H^+ inhibition which according to Lazof and Holland (1999) is reversible. Both kinds of toxicity may restrict crop growth considerably. This is particularly true on sulphate acid soils with a pH of about 3 as reported from India by Prasad and Goswami (1992). Frequently under such low pH conditions Mn and Fe toxicities are found when the availabilities of Ca and Mg are low. The application of lime is the most effective means of controlling Al toxicity in acid soils (Haynes 1982).

Plant species and even cultivars of the same species differ considerably in their tolerance of excess soluble or excess exchangeable Al. According to Delhaize et al. (1993) Al tolerant wheat cultivars possess in their genome an Al tolerant locus which encodes for a mechanism inhibiting the uptake of Al. Al tolerant wheat lines have a tolerance which is 5 to 10 fold higher than that of Al sensitive lines. Al tolerant lines responded to an Al addition to the nutrient medium by increasing malate excretion to chelate the Al cation and thus detoxifging it. Such an increased malate release was not found in the Al sensitive lines (Delhaize et al. 1993). The same respponse has also be reported for citrate (Yang et al. 2000) and the secretion of organic anions is regarded as one of the key Al resistance mechanisms in plants (Kochian 1995).

One further reason for cultivar differences in response to Al relates to the varying ability of plants to modify the pH of the soil root interface. Thus nitrate nutrition increases the pH in the rhizosphere as compared with ammonium nutrition (see page 86) and the raised pH lowers the Al solubility in the the soil. Kretschmar et al. (1991)

reported that the incorporation of pearl millet straw increased soil pH from 4.10 to 4.66 associated with a reduction of soluble Al. According to the authors this reduction was not only due to the pH increase but also due to chelation of Al by organic anions produced in the process of microbial straw decomposition. In some Al tolerant wheat cultivars the tolerance mechanism is related to the fact that these cultivars can also take up nitrate at high rates in the presence of ammonium and hence avoid a pH drop in the rhizosphere. Aluminium tolerance can also be dependent on an Al exclusion mechanism. This has been shown by Henning (1975) who experimented with the Al tolerant wheat cultivar `Atlas' and the Al sensitive cultivar 'Brevor'. The tolerant cultivar required about 100—200 times as much Al in the medium as did the Al-sensitive 'Brevor' before Al penetrated the plasmalemma of meristematic root cells. This observation is supported by more recent results of Delhaize *et al.* (1993) who reported that the Al tolerant cultivar has a gene which encodes for a mechanism inhibiting Al uptake. Once inside the cell Al was equally harmful to Al tolerant and sensitive cultivar. This example shows that the plasmalemma may exclude Al, and that in this respect marked differences between cultivars do exist. From experiments with cowpeas (*Vigna unguiculata*) Horst *et al.* (1982) suggest that the mucilage layer around the root tip acts to protect the young root tissue from Al toxicity by adsorbing Al.

Aluminium tolerance may also be brought about by organic acids and polyphenols which detoxify Al by chelation. A typical example of this kind of tolerance is the tea crop, which takes up a large amount of Al, which is stored in the older leaves where it is detoxified by organic compounds (Sivasubramaniam and Talibudeen 1972). A very useful review paper on metal toxicity and especially on Al toxicity has been published by Foy et al. (1978).

20.5 Nickel

Nickel is closely related to Co both in its chemical and physiological properties. It readily forms chelate compounds and can replace other heavy metals from physiologically important centres. High Ni concentrations have a toxic effect on plants. In oats Vergnano and Hunter (1952) observed that Ni toxicity closely resembled Fe deficiency, a finding which may well relate to the displacement of Fe by Ni. High Ni concentrations in the nutrient medium reduce the uptake of most other nutrients (Crooke and Inkson 1955). According to Knight and Crooke (1956) this reduction in uptake results from the damaging effects of high Ni concentrations on the roots. The phytotoxicity of Ni has been reviewed by Mishra and Kar (1974).

Acute Ni toxicity gives rise to chlorosis. In the cereals this shows as pale yellow stripes running the length of the leaves. Eventually the whole leaf may turn white and in extreme cases necrosis occurs at leaf margins. In the dicots Ni toxicity appears as chlorotic markings between the leaf veins, the symptoms being similar to those of Mn deficiency (Hewitt 1953).

Most soils contain only very small quantities of Ni, usually less than 100 μg Ni/g soil, and well below the level at which Ni toxicity occurs. Long term heavy applications of contaminted sewage sludge, however, may lead to consideralbe rise of the Ni level in the top soil (Juste and Mench 1992). Soils derived from ultrabasic igneous rocks and particularly serpentine, however, may contain 20 to 40 times this concentration and Ni toxicity in plants is common. Serpentine soils occur in various regions throughout the world as far apart as the mountain ranges of the Pacific Coast of the U.S.A., the North of Scotland, parts of the Balkans, the South of Russia and Zimbabwe. On weathering, the mineral serpentine [$Mg_3Si_2O_5$ (OH_4)) produces a characteristic soil with a distinctive associated flora and a comparatively sparse vegetation. The soils are rich in Mg and Fe and poor in Ca. In addition they contain relatively high levels of Ni, Co and Cr. Krause (1962) reported 250 μg exchangeable Ni/g soil in the surface layer of red earth soil derived from serpentine as compared with about 1 μg usually occurring in agricultural soils. In the serpentine soil the ratio of exchangeable Ca to Mg (Ca/Mg) was extremely low and mostly less than 0.4. The cereals, with the exception of oats, are the agricultural crops best able to cope with these conditions.

Pioneer species on serpentine soils are *Cerastium holostoides* and *Silene vulgaris*, which according to Rodenkirchen and Roberts (1993) have different strategies to cope with the conditions on serpentine soils. *Cerastium* can tolerate high Mg and heavy metal concentrations in the leaves while in *Silene* the tolerance against high Mg and heavy metal concentrations in soils is based on an excluion of these elements. This behaviour is shown in Table 20.1. in which the N and mineral concentration of both species are listed.

Table 20.1 Mean mineral concentrations in *Cerastium holosteoides* and *Silene vulgaris* (Rodenkirchen and Roberts 1993)

	Cerastium	Silene
N, mg/g TM	11.1	17.2
P, mg/g TM	0.95	1.25
K, mg/g TM	13.1	26.6
Ca, mg/g TM	15.4	10.8
Mg, mg/g TM	34.1	17.3
Mg/Ca	2.2	1.6
Fe, μg/g TM	4099	280
Mn, μg/g TM	405	114
Zn, μg/g TM	29	21
Cu, μg/g TM	37	3
Ni	888	18
Cr	296	6

It is evident that particularly the heavy metal concentrations in the upper plant part of *Cerastium* are much higher than those of *Silene*. The reason why *Cerastium* can tolerate such high concentrations of Fe, Ni, and Cr is not yet understood. Presumably these elements are chelated and thus detoxified.

Nickel toxicity can frequently be greatly alleviated by liming. Liming not only decreases the availability of Ni and Cr, it also increases the low ratio of exchangeable Ca/Mg. Potassium application also reduces the appearance of Ni toxicity but phosphate fertilizers have the reverse effect (Crooke and Inkson 1955). Normally the Ni concentration of plant material is about 0.1 to 5 μg Ni/g dry matter. On Serpentine soils, however, these Ni concentrations may be much higher as shown in Table 20.1. Such levels are toxic to plants not adapted to these soils. In a sand culture experiment supplied with increasing levels of Ni, Vergano and Hunter (1952) found Ni concentrations in the dry plant material varying from 1 to 1000 μg Ni/g dry matter. Toxic symptoms in oats, a Ni sensitive crop, were observed in plants with Ni contents in excess of 100 μg Ni/g dry matter (Crooke 1956). The uptake by less sensitive crops is lower. Nickel appears to be particularly mobile in the phloem and is present in the xylem sap as free Ni^{2+} and chelated to citrate (Gerendas et al. 1999). After uptake considerable amounts of Ni are thus transferred to seeds and fruits (Cataldo et al. 1978, Mitchell et al. 1978).

The biological significance of Ni as a possible micronutrient has been reviewed by Welch (1981) and Gerendas et al. (1999). It is now established that Ni is essential for animals. Brown et al. (1987) reported that Ni is essential for the germination of barley seeds and plants grown for three generations in a nutrient solution deprived of Ni produced seeds with extremely low Ni concentrations. The germination of seeds was dramatically depressed below concentrations of 100 ng/g in the seed. The authors were able to obtain deficiency symptoms including intraveinal chlorosis in wheat, barley, and oats and suggest that Ni is a micronutrient for these cereal crops (Brown et al. 1987). Nickel has been shown to be an integral part of the enzyme urease isolated from jack bean seeds (Dixon et al. 1975). Therefore plants grown solely on urea N have a Ni requirement (Gordon et al. 1978, Gerendas and Sattelmacher 1997), and in the absence of Ni may even accumulate urea to toxic levels. The question of Ni essentiality has been treated in a useful review article by Gerendas et al. (1999). Further Ni containing enzymes have been found in bacteria. The methylcoenzyme-M reductase, which catalyzes the last step in methane production of methanogene bacteria is an enzyme with a tetrapyrol ring with a Ni located at the centre (Thauer 1986). Some hydrogenases contain also Ni and in this respect hydrogenases in N_2 fixing bacteria and contributing to the efficiency of the nitrogenase are of interest (Cammack 1995). Tropical N_2 fixing legume species (e.g. soya beans) accumulate large amounts of ureides. The enzymic breakdown of these metabolites, however, does not require urease and thus there is also no requirement for Ni for this purpose (Gerendas et al. 1999).

20.6 Chromium

There has been much interest in Cr since the discovery that it participates in mammalian glucose metabolism and appears to be essential to man and animals. As yet, however, there is no evidence of an essential role in plant metabolism (Huffman and Allaway 1973).

Total Cr levels in igneous and sedimentary rocks are usually in the range of up to about 100 μg/g. For most soils the Cr concentration is in the range of 15 to 100 μg/g soil and increases with the proportion in clay (Baize 1997). Those, however, derived from serpentine may contain several per cent Cr. Soil Cr is largely unavailable to plants because it occurs in relatively insoluble compounds such as chromite $FeCr_2O_4$ in mixed oxides of Cr, Al and Fe, or in silicate lattices. In addition Cr^{3+} binds tenaciously to negatively charged sites on clays and organic matter. For this reason the translocation of Cr from soils into plants is generally insignificant (Juste and Mench 1997). Chromates (hexavalent Cr) in soils are relatively rare and are only stable in alkaline oxidizing conditions (Allaway 1968). It is supposed that Cr^{3+} and CrO_4^{2-} are taken up by two different mechanisms (Bollard 1983). Uptake of CrO_4^{2-} is depressed by SO_4^{2-}.

Frank *et al.* (1976) in investigating 296 agricultural soils in Ontario found no accumulation of chromium in soils which had been dressed with NPK fertilizers. Similar observations have been reported by Watanabe (1984) who investigated plots from various field experiments to which different rates of mineral fertilizers and compost had been applied. The levels found in the soil were in the range of 50 to 100 $\mu g/g$ soil. There was a tendency for the application of fused phosphate to increase the level of soil Cr.

Toxicity effects have been observed by Hunter and Vergnano (1953) in oats. Plants suffering from severe Cr toxicity had small roots and narrow brownish red leaves, covered in small necrotic spots. Acute infertility in some serpentine soils may result from Cr toxicity although other factors are usually involved (see page 663). The literature of Cr toxicity in relation to serpentine soils has been discussed by Pratt (1966b).

20.7 Selenium

Selenium resembles sulphur in its chemical properties. In uptake a competitive effect occurs between selenate and sulphate indicating that both ions have an affinity for the same uptake system (Leggett and Epstein 1956). Selenium is present in soils in organic form as well as in inorganic anion form such as selenide (Se^{-2}), selenite (SeO_3^{2-}), and selenate (SeO_4^{2-}). Selenite and selenate may be adsorbed to clay minerals, the adsorption strength increasing with a fall in soil pH (Gissel-Nielsen *et al.* 1984). In most soils Se occurs in very low concentrations and often less than 0.2 $\mu g/g$ soil. In acid and neutral soils, availability is low and Se is often present which may be fixed as ferric selenite. The element may occur in soils in various redox states depending on soil redox potential, soil pH and microbiological activity (Allaway 1968). Recent investigations of Wright (1999) have shown that nitrate may oxidize Se of lower oxidation status (Se^{2-}, SeO_3^{2-}) to SeO_4^{2-} and by this mobilize soil Se. The process plays a particular role in hydrogeologic settings where irrigation occurs on Cretaceous shale. Oxidation-reduction processes are important for the mobilization of Se into the environment and nitrate can contribute to this mobilization (Wright 1999). Laborartory and field experiments have shown that selenate reduction in wetland sediments is a process mediated by microorganisms (Zhang and Moore 1997). Reduction products are mainly elemental Se or organically bound Se with a low solubility.

Selenate, which occurs only in soils under well aerated alkaline conditions is taken up by plants at much higher rates than selenite (Mikkelsen *et al.* 1989) and therefore is the most important direct source for plant uptake. Such Se rich soils with high selenate levels are found primarily under arid climatic conditions in soils with seleniferous formations. Mining of uranium, coal or betonite frequently resulted in severe soil pollution with excess Se (Boon 1989). Selenium toxicity increases after oxidation to selenate which occurs with mining when deeper soil layers come in contact with air. Selenium is the most highly enriched element in coal and carbonaceous materials. Increased solubility and mobility of Se leads to a contamination of soils, surface and ground water with Se (Boon 1989).

Various methods for the determination of available Se in soils are discussed by Huang and Fujii (1996) and according to Jump and Sabey (1989) Se concentrations in extracts of water-saturated soil pastes correlated highest with Se concentration in plants. Selenium concentrations in the upper plant parts of most crop species are in the range of 1 to 10 μg Se/g dry matter. Selenium toxicity symptoms may differ for various plant species (Mikkelsen *et al.* 1989). In grasses snow-white chlorosis of leaves and pink root tissues have been observed. Excess of Se gives rise to stunting and yellow chlorosis in plants; growth is retarded and the Se concentrations in plants are increased.

On Se-rich sites, ecotypes have been developed which tolerate high Se concentration in the plant. Such Se accumulator species belong to the genera *Astragalus*, *Stanleya*, and *Haplopappus* (Gissel-Nielsen *et al.* 1984). Seleniferous indicator plant species such as the milk vetches (*Astragalus*) are indigenous to these soils and contain high Se concentrations while native grasses growing on the same soil are low in Se concentrations. According to Shrift (1969) of the about 500 *Astragalus* species in North America about 475 are non-Se accumulators. They have a normal Se concentration of about 5 to 30 μg Se/g dry matter in contrast to the Se accumulators with several 1000 μg Se/g dry matter. Interestingly Se stimulates the growth of the Se accumulator species while in non-accumulators the growth is depressed. This is shown for two *Astragalus* species grown in sand culture with and without Se from the work of Trelease and Trelease (1939 quoted by Shrift 1969) in Table 20.2. In *Astragalus crassicarpus*, the non-accumulator, Se caused a severe growth depression while in the accumulator species *Astragalus racemosus* growth was considerably enhanced. According to these results Shrift (1969) suggests that "there is considerably evidence that selenium serves as a micronutrient for plants other than those in the genus *Astragalus*".

Table 20.2 Effect of Se on the growth of a Se-accumulator *Astragalus* species (*A. racemosus*) and of a non Se accumulator species (*A. crassicarpus*) grown in sand culture (Trelease 1939, quoted after Shrift 1969)

Se in medium	Astragalus racemosus	Astragalus crassicarpus
	g dry weight/10 plants	
0	9.4	8.44
9	26.4	1.13

Selenium toxicity relates to the incorporation of Se-cysteine into proteins. In Se-cysteine the Se stands for the S in the cysteine molecule. Shrift (1969) reported that Se tolerant plant species accumulate Se in the soluble Se-methyl cyteine form which obviously is not toxic in contrast to Se-methionine which is found in non Se accumulators. Selenium-methionine may be incorporated into proteins which may have toxic effects in plants (Gissel-Nielsen et al. 1984).

The differences between plants in their ability to accumulate and to tolerate Se have not been fully explained. It has been suggested that in non accumulator plants, Se is mostly found in the proteins whereas the accumulator plants have the ability to synthesize Se containing non-protein amino acids and this prevents toxicity (Shrift 1969). The later findings of Nigam and McConnell (1976), however, using radioactive selenate found a considerable percentage of radioactivity in the proteins of the Se accumulator Astragalus bisucatus. These workers suggested that differences in the toxicity of Se towards plant species are difficult to explain in terms of interspecies differences in protein incorporation.

The effect of Se in animal nutrition is particularly important. Selenium is an essential element for animals and is required in very low concentrations. Deficiency has been shown to give rise to a muscular dystrophy in livestock known as 'white muscle disease' as well as hair and feather loss. In bacteria and animals Se cysteine is an essential amino acid which has its specific t-RNA species and which recognize the UGA codon on the mRNA. Thus by means of ribosome, Se-cysteine is incorporated into polypeptides which are essential for the synthesis of some enzymes such as hydrogenases in various bacteria (Stadtman 1996) and glutathione peroxidase in mammals. This enzyme decomposes H_2O_2 (see page 440). For the reaction it is important that the SeH group is deprotonated and since the pK of the SeH group is lower than that of the SH group the former is more efficient in the detoxification of H_2O_2 (Stadtman 1990). Therefore Se is also an essential element for human beings.

Gissel-Nielsen et al. (1984) have treated the problem of Se in soils and plants and its importance in lifestock and human nutrition in a uesful review article. They state that Se malnutrition in humans is more widespread than is generally assumed. It is found frequently in children in the South East of China and in Europe particularly in the Scandinavian countries. For this reason the fertilzation of crops with Se may be recommended in order to increase the Se concentration in animal and human food (Gissel-Nielsen et al. 1984). According to Mikkelsen et al. (1989) selenate as Se fertilizer is more efficient than selenite, and elemental Se is hardly taken up by crops.

Allaway (1968) has suggested that it is desirable to control the Se levels in food and feed crops to a range between 0.1 to1 Se $\mu g/g$, and cites a survey in United States which showed that one-third of forage and grain crops were below this optimum level. At higher concentrations in excess of about 5 $\mu g/g$ in the diet there is danger of Se toxicity. This is well known as "alkali disease " in farm animals and occurs on Se rich alkali soils. Selenium toxicity has been reported since 1856 for many areas of western USA especially of the Great Plains (Boon 1989). A useful review article on "Selenium in Agriculture and the Environment" which considers Se in soils, in crops, its poisining in lifestock and its impact on the environment has been edited by Jacobs (1989).

20.8 Lead

Lead is a major chemical pollutant of the environment, and is highly toxic to man. No other pollutant has accumulated in man to average levels so close to those which are potentially clinically poisonous. The major source of Pb pollution arises from petrol combustion. This source accounts for about 80% of the total Pb in the atmosphere. Lead is added to petrol as tetra ethyl lead, and is emitted in exhaust fumes largely as minute particles of inorganic Pb compounds. About 50% of this falls somewhere within the region of 100 m from the road. The remainder is distributed widely in the biosphere. This is evident from the results of Murozumi et al. (1969) which shows the dramatic increase in the Pb content of snow over the past decades in core samples taken from the North Greenland Icecap. This rise must be almost entirely attributable to increased petrol consumption. Industrial regions are particularly polluted by airborne Pb. In Manchester, for example, a large industrial city in England, levels in the region of 1000 μg/g air have been observed in street dust (Day et al. 1975). The average Pb level in soils in comparison is about 15 μg/g. Blood Pb levels have also been found to rise considerably in humans living near motorways. Lead toxicity in man has also been caused by water contamination from Pb piping. This can also be an important source of Pb pollution.

Lead is toxic because it mimics many aspects of the metabolic behaviour of Ca, and inhibits many enzyme systems. In man, one of the chief concerns of Pb toxicity is its effect in causing brain damage particularly to the young. There is evidence that Pb pollution can induce aggressive behaviour in animals. It is believed that this can also occur in humans and Bryce-Smith and Waldron (1974) have presented a very strong case for implicating Pb as one of the causal factors for the increased rate of delinquency in large industrial cities. From the viewpoint of plant nutrition it is important to remember that Pb pollution mainly arises from airborne sources.

The total Pb concentration of agricultural soils lies between 2 to 200 μg/g soil. Agricultural soils in France have an average Pb concentration of 30 μg/g soil with a minium of 7.5 μg/g and a maximum of 1560 μg/g soil. This latter value is an exceptionally high concentration occurring in clay rich soils in the area of Sinemurien near Morvan, Burgundy (Baize 1997). Soils with such high levels are limited to a relatively few regions, frequently where Pb mineral deposits occur. In such soils Pb in the upper horizons may reach over 3000 μg/g soil. Toxic effects of Pb can result in a reduction in plant growth but this is not generally seen in the field and almost all detailed observations of Pb toxicity in plants are restricted to water culture experiments (Brewer 1966b). Gobold et al. (1988) cultivating spruce seedlings in nutrient solution found that in the Pb treatment root was severely retarded even at a low concentration. Lead was mainly found in the cell walls, in the Hartig net of mycorrhiza but was also present in vacuolar deposits of root samples taken from forest soils.

Page et al. (1971) carried out an extensive study on the Santa Ana Freeway in Southern California, a highway with a very high traffic density of 70 000 cars per 24 hrs. These workers analysed the Pb fall out and Pb concentrations of 27 crops at various distances from the road. It was concluded that the concentration of Pb contaminate was dependent on several factors including distance from the highway, nature of the collecting surface of the plant, duration of the exposure, the traffic density and the direction of the prevailing winds. Many workers have shown that Pb contamination very clearly follows the motorway areas. Vegetation at the side of the road may have levels of 50 μg Pb/g dry matter but a distance of only 150 m away from the motorway the level is normally about 2 to 3 μg/g dry matter. Contamination occurs only on the outer part of plant seed or leaves and stem, and a high proportion can be removed by washing. Levels of Pb in grain, tubers and roots are very little affected and do not deviate very much from normal levels for such tissues of about 0.5 μg Pb/g dry matter (Foy *et al.* 1978). The same observation was made in field trials in which Pb was enriched in the soil by long term sludge application (Juste and Mench 1992, Logan *et al.* 1997).

This is true if the Pb is provided in inorganic form such as Pb^{2+}. Organic Pb, however, such as Pb tetraethyl, Pb triethyl, and Pb diethyl as well as other alkyl derivates of Pb are extremely mobile in the soil and are taken up by plants much more rapidly than Pb^{2+}. Such organic Pb forms may be released in the petrol combustion fumes if combustion is incomplete. Diehl *et al.* (1983) found that concentrations of 100 μg Pb^{2+}/g soil had no effect on growth and yield of spring wheat while the same concentrations of Pb in organic form (Pb tetraalkyl) almost completely inhibited plant growth. Even a rate as low as l0 μg Pb/g soil as organic Pb resulted in a drastic depression in growth. These authors, who cultivated spring wheat in pot experiments on a normal soil, found that Pb provided in organic form was translocated to the grains in relatively large amounts. Thus Pb levels of about 20 μg/g dry matter were found in the grains. This level is in the range of that associated with severe growth reductions in other species (Judel and Stelte 1977). According to Bryce-Smith (1975) Pb trialkyl is a powerful mutagenic agent and is known to derange the spindle of the fibre mechanism of cell division in both plant and animal cells. In maize plants, Malone *et al.* (1974) have shown that Pb is first concentrated in dictyosome vesicles which fuse together to encase Pb deposits. These are then removed from the cytoplasm to outside the plasmalemma to fuse with the cell wall where much Pb can be accumulated.

Lead contamination by the inorganic form in soils is usually restricted to the top few cm of the soil profile (Heilenz 1970). This retention in the upper part of the profile relates to its high affinity to organic matter (Baize 1997) with which it forms insoluble Pb chelates (Lagerwerff 1972). The availability of soil Pb is usually low and can be decreased even further by liming. A high soil pH may precipitate Pb as hydroxide, phosphate or carbonate, as well as possibly promoting the formation of Pb organic matter complexes.

In animals, Pb toxicity interferes with Fe metabolism and the formation of haem. Lead inhibits two steps in the conversion of delta-amino-laevulnic acid to haem (see page 563). It inhibits the enzyme ALA dehydrase in the conversion of delta-amino-laevulnic acid (ALA) to porphobilinogen (PBG). It also blocks the formation of haem from coproporphyrinogen III. In the blood and urine of Pb toxic patients there is therefore a marked increase in the levels of ALA and coproporphyrin III (the oxidation product of coproporphyrinogen III). It is not known whether Pb has the same effect in haem synthesis in plant cells.

20.9   Cadmium and Further Heavy Metals

There is considerable current interest in Cd in plant nutrition. Cadmium and Zn are chemically very similar. Cadmium is thus able to mimic the behaviour of the essential element Zn in its uptake and metabolic functions. Unlike Zn, however, Cd is toxic both to plants and animals. The basic cause of the toxicity probably lies in the much higher affinity of Cd for thiol groupings (SH) in enzymes and other proteins. The presence of Cd therefore disturbs enzyme activity. In plants excess Cd may also disturb Fe metabolism and cause chlorosis.

In animal nutrition Cd is a cumulative poison. It is mainly stored in the kidneys and to some extent also in the liver and spleen. Excess Cd results in damage to the kidney tubules, rhinitis (inflammation of the mucous membrane of the nose), emphysema (a chronic disease of the lungs in which the alveoli become excessively distended) as well as other chronic disorders. In marked contrast to the effects of Pb, however, neurological disorders are not induced by Cd.

A condition of chronic Cd poisoning which has been observed in the Toyama city region of Japan is known as Itai-Itai disease. Excess Cd in the diet has been found to impair kidney function and hence disturb the metabolism of Ca and P and cause bone disease. The disease which is very painful causes excessive demineralization and embrittlement of the skeleton. It has been observed particularly in middle-aged women whose stores of Ca have been depleted by frequent childbirth. The cause of this disease has been traced back to the diet of rice grown on paddy soils polluted with Cd from a mine source.

Cadmium concentration in non-contaminated soils is generally lower than 1 $\mu g/g$ soil. Baize (1997) reported that the average concentration (median) of 460 arable soils from France was 0.22 mg Cd/kg soil and varied between 0.08 to 0.75 mg Cd/kg. Generally the Cd concentration in soils increases with the clay concentration. According to investigations of Wopereis et al. (1988), who studied the spatial variability of heavy metals in a forest acid brown earth soil on granite, of the heavy metals studied (Ni, Cu, Zn, Cd, Pb), Cd showed the greatest variability with a coefficient of variation of 55% while that of Pb, Ni, Zn, and Cu was 9, 13, 13, and 20%, respectively. In a highly informative review paper Sauerbeck and Lübben (1991) emphasize that heavy metals were not evenly distributed in soils and are bound mainly to organic matter, carbonates and Mn- and Fe oxides but are hardly present in the interlayers of 2:1 phyllosilicates. Plant availability of Cd, Zn and Ni is better reflected by the soil extraction with neutral solutions such as ammonium nitrate than with strong acids or chelates. Generally the concentrations of Cu, Pb, Ni, and Cr are higher in roots than in tops. For Zn and Cd this is reverse showning that Zn and Cd are easily translocated from roots to tops. Cadmium retention by the solid soil phase increases exponentially pattern with an increase of soil pH (Baize 1997). Acid sandy soils are hence particularly prove to the possible hazard of Cd leaching into deeper soil layers and into the ground water. In a useful review article McBride (1989) treating the solubility of heavy metals in soils, reported that the affinity of Cd^{2+} to amorphous Fe hydroxides, Al hydroxide, and the silanol groups of silica is low as compared with other heavy metals (Pb, Cu, Zn, Ni). This is in agreement with results of Sloan et al. (1997) who investigated the availability of heavy metals in soils by a sequential extraction techniqe. It was shown that after 15 years biosolid application (sludge, filter cake and other organic residues) the relative bioavailabilty of heavy metals followed the sequence: Cd > Zn > Ni > Cu >> Cr > Pb. Helal *et al.* (1999) reported that the amendment of sandy soils with clay of canal sediments enriched the soils with Cd and Zn the bioavailability of which was increased by irrigation with saline water. The solubilty of Cd and Zn in soils may be increased by root exudates as was found by Zhu *et al.* (1999) in column experiments. The dissolved Cd and Zn thus may be leached into the subsoil where it is fixed again.

There is not much known about the uptake mechanism for Cd^{2+} by plants. Presumably it is a facilitated diffusion across the plasmalemma. Recent results of Gonzalez *et al.* (1999) suggest, that Cd^{2+} like Ca^{2+} and Zn^{2+} is translocated across the tonoplast by a proton antiport. Accumulation of Cd in the vacuole is related to the Cd tolerance of ecotypes as reported by Chardonnes *et al.* (1998). Uptake rates depend on the Cd concentration in the nutrient medium as found by Logan *et al.* (1997) in field studies in which for several years heavy metal contaminated sludge was applied. As shown in Figure 20.1 Cd concentration in maize biomass increased with the total Cd concentration in the soil. The curve fitted to a Mitscherlich type although the points are much scattered indicating that also other factors such as the Cd concentration in soil influenced the Cd concentration in the crop.

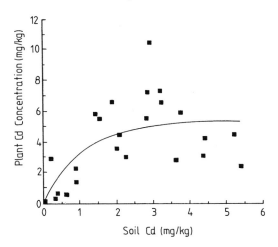

Figure 20.1 Cd concentration in maize biomass *versus* Cd concentartion in soil (Logan *et al.* 1997).

Cadmium can be transported readily from the soil *via* the plant root to the upper plant parts. Its availability depends much on soil pH and the presence of other cation species. Calcium and Zn^{+} in particular depress the Cd^{2+} uptake. Homma and Hirata (1984) found that Cd^{2+} uptake rates by rice seedlings were as high as the uptake rates for Zn^{2+} if the concentration of both ion species were lower than 1.0 $mmol/m^3$. At higher concentrations, Zn^{2+} uptake rates were more than twice as those of Cd^{2+} when both ion species were present in equal concentrations in the nutrient solution. Translocation from the leaves into seeds is low presumably because Cd^{2+} is like Ca^{2+} not phloem-mobile. For this reason the Cd concentrations of cereal grains from highly contaminated soils only exceeded by 1 $\mu g/g$ (Sommer 1979). According to Chaney and Hornick (1978) this is the source of 50% of Cd in the average US diet. There is much evidence that crop species and genotypes vary markedly in Cd uptake, thus offering a means of retaining Cd at a low level in plant products. In some species such lettuce and celery the uptake rate of Cd is particularly high and approaching the non tolerant level (Sauerbeck and Lübben 1991).

The main source of Cd pollution in the environment are metal smelters. Zinc smelters are particularly notorius in this respect as the Cd: Zn ratio in Zn ores is usually in the order of 1:350. In addition, soils along roadside are polluted with Cd from tyres and lubricant oils (Lagerwerff 1972). Cadmium in soils may also result from the application of sewage sludges. As shown in Figure 20.1 repeated sludge application may lead to Cd concentrations in the soil which exceeed the critical level of 3 mg Cd/kg soil. The Cd concentration of sludges may differ considerably. Sludges from industrial areas generally have higher heavy metal concentrations including Cd than sludges from residential areas. Vlamis et al. (1985) reported a Cd concentration of 37 mg/kg in the sludge from an industrial urban area while that from the sludge from a residential area was only 8 mg/kg. These authors tested both types of sludges in field experiments. Over the seven year experimental period during which barley was grown on the plots the application of the industrial sludge led to a marked increase of Cd in the grains to a concentration of 0.6 $\mu g/g$. By contrast the barley grains from the plots which had been treated with residential sludge the increase in Cd concentration was only slight with a final value of 0.1 μg Cd/g. The problem of long-term sewage sludge application and ist effect on heavy metal uptake by crops has been considered by Juste and Mench (1992) in an informative review article.

Cadmium is also added to soils in small amounts in phosphate fertilizers (Baize 1997). Mortvedt and Osborn (1982) found that increasing rates of phosphate fertilizer application resulted in higher Cd concentrations in the grains of winter wheat and maize. Uptake and solubility of Cd phosphates decreased according to the sequence: $Cd(H_2PO_4)_2 > CdHPO_4 > Cd_3(PO_4)_2$. Cadmium concentrations in rock phosphates can vary widely depending on their source (Baechle and Wolstein 1984).

According to Sauerbeck and Lübben (1991) decontamination of soils by the extracting of heavy metals by plants and repeated cropping has hardly any chance of success. Even for Cd, which is taken up at relatively high rates, about 200 years would be required to depress the soil level from 3 to 1 mg Cd/kg soil. This decontamination of soils by means of plant uptake of organic and/or inorganic contaminants is called phytoremediation and has recently been reviewed by Salt et al. (1998). Presently, about 45 plant families are known to contain metal-accumulating species, so called hyperaccumulators. The problem with these hyperaccumulators as yet is that most of them have a poor biomass production and have low rates of growth whereas rapidly growing crop species are sensitive to metals and accumulate only low concentrations in the shoots.

The particular hazard with Cd is that the plant does not necessarily act as an indicator of levels toxic to humans and animals since plants tolerate higher levels of Cd than do animals. The same is true for mercury an tellurium (Hg and Tl) as can be seen in Table 20.3, although these elements are of less practical significance. Plants can thus appear quite healthy but may contain high concentrations of these three elements (Cd, Hg, Tl) completely unacceptable in an animal and human diet.

Table 20.3 Critical concentrations of various heavy metals in plants and in the diet of animals. Higher levels are toxic (after Sauerbeck 1982)

	Plants	Animals
	μg g^{-1} in the dry matter	
Cd	5—10	0.5—1
Hg	2—5	1
Tl	20—30	5
Co	10—20	10—50
Cr	1—2	50—3000
Cu	15—20	30—100
Ni	20—30	50—60
Pb	10—20	10—30
Zn	150—200	500

Sauerbeck (1982) cites experiments to show that Tl uptake may differ between plant species considerably, Tl uptake of rape being 40 times greater than that of grass or spinach.

As already mentioned above uptake, of heavy metals depends much on soil conditions. Generally availability decreases as the pH level rises and the concentrations of clay and humus increase. For this reason it is very difficult to establish critical soil levels for heavy metals. Nevertheless as a first approach to control heavy metal concentrations in agricultural and horticultural soils, critical concentrations for a number of heavy metals have been proposed for soils in the Federal Republic of Germany as shown in Table 20.4.

Table 20.4 Critical concentratios of heavy metals in soils (μg g^{-1} air dry soil)

Pb	100	Ni	50
Cd	3	Hg	2
Cr	100	Zn	300
Cu	100		

GENERAL READING

General Reading, Chapter 1

Asher, C. J. Beneficial elements, functional nutrients and possible new essential elements, p. 703-723 In: Micronutrients in Agriculture, 2nd ed. Edited by Mortvedt, J. J., Cox, F. R., Shuman, L. M. and Welch, R. M., Soil Sci. Amer. Book Series, Madison WI, USA 1991.

Clarkson, D. T. and Hanson, J. B. The mineral nutrition of higher plants. Annu. Rev. Plant Physiol. 31, 239-298, 1980.

Epstein, E. Silicon. Annu. Rev. Plant Physiol. Plant Mol. Biol. 50, 641-664, 1999.

Fageria, N. K., Baliger, V. C., Jones, C. A. Growth and Mineral Nutrition of Field Crops. (Books in Soils, Plants, and the Environment Series), Marcel Dekker, New York 1991.

Flowers, T. J. Chloride as a nutrient and as an osmoticum, p. 55 -78. In: Adv. in Plant Nutrition, edited by Tinker, B. and Läuchli, A., Praeger, New York 1988.

Loué, A. (F) Microelements in Agriculture. Agri-Nathan International, Paris 1986.

Marschner, H. Mineral Nutrition of higher Plants. 2nd. edition, Academic Press, San Diego, London. 1995.

Mengel, K. (G) Nutrition and Metabolism of Plants, 7. edition. Gustav Fischer Verlag, Jena, 1991.

Schilling, G. (G) Plant Nutrition and Fertilizer Application. Verlag Eugen Ulmer, Stuttgart 2000.

General Reading, Chapter 2

Appel, T. and Mengel, K. Prediction of mineralizable nitrogen in soils on the basis of an analysis of extractable organic N. Z. Pflanzenernähr. Bodenk. 161, 433-452, 1998.

Barber, S. A. Soil Nutrient Bioavailability - A Mechanism Approach, 2nd ed. Wiley, New York, 1995.

Barea, J. M. and Azcon-Aguilar, C. Mycorrhizas and their significance in nodulating nitrogen -fixing plants. Adv. Agron. 36, 1-54, 1983.

Bergmann, W. (G) Coloured Atlas. Nutritional Disorders of Crop Plants. VEB Gustav Fischer Verlag, Jena 1986

Bergmann, W. Nutritional Disorder of Plants, Development Visual and Analytical Diagnosis, Gustav Fischer, Jena 1992.

Bollons, H. M. and Barraclough, P. B. Inorganic orthophosphate for diagnosing the phosphorus status of wheat plants. J. Plant Nutr. 20, 641-655, 1997.

Brady, N. C. and Weil, R. R. The Nature and Properties of Soils. 12. edition. Prentice-Hall, London 1999

Drew, M. C. Effects of flooding and oxygen deficiency on plant mineral nutrition. Adv. Plant Nutrition 3, 115-159, 1988.

Flowers, T. J., Troke, P. F. and Yeo, A. R. The mechanism of salt tolerance in halophytes. Annu. Rev. Plant Physiol. 28, 89-121, 1977.

Jones, J. B., Wolf, B. and Mills, H. A. Plant Analysis Handbook, Athens, Georgia, USA:Micro-Macro Publishing Inc., 1991.

Hall, J. R. A review of VA mycorrhizal growth responses in pastures. Angew. Botanik 61, 127-134, 1987.

Haynes, R. J. Lime and phosphate in soil-plant system. Adv. Agron. 37, 249-315, 1984.

Hinsinger, P. How do plant roots acquire mineral nutrients? Chemical processes involved in the rhizosphere. Adv. Agron. 64, 225-241, 1998.

Jungk, A. and Claassen, N. Ion diffusion in the soil-root system. Adv. Agron. 61, 53-110, 1997.

Kinraide, T. B. Identity of the rhizotoxic aluminium species. Plant and Soil 134, 167-178, 1991.

Loll, M. J. and Bollag, J. M. Protein transformation in soil. Adv. Agron. 36, 351-382, 1983.

Marschner, H. Plant-soil relationships: acquisition of mineral nutrients by roots from soils, p. 125 -155. In: Plant Growth: Interactions with Nutrition and Environment, edited by Porter, J. R. and Lawlor, D. W. Cambridge University Press, Cambridge 1991.

Marschner, H. and Römheld, V. Root-induced changes in the availability of micronutrients in the rhizosphere, p. 557 579. In: Plant Roots, the *Hidden Half*, edited by Waisel, Y., Eshel, A. and Kafkafi, U. Marcel Dekker New York 1996.

Molina, J, A. E., Clapp, C. E., Shaffer, M. J., Chichester, F. W., and Larson W. E. . NCSOIL, a model of nitrogen and carbon transformation in soils: description, calibration, and behavior. Soil. Sci. Soc. A. J. 47. 85-93, 1983

Nair, K. P. P. The buffering power of plant nutrients and effects on availability. Adv. Agron. 57, 237-287, 1996.

Nemeth, K. The availability of nutrients in the soil as determined by electro-ultrafiltration (EUF). Adv. Agron. 31, 155-188, 1979.

Neyra, C. A. and Döbereiner J. Nitrogen fixation in grasses. Adv. in Agron. 29, 1-38, 1977.

Nye, P. H. Diffusion of ions in uncharged soils and clays. Adv. Agron. 31, 225-272, 1979.

Nye, P. and Tinker, P. B. Solute Movement in the Root Soil System. Blackwell, Oxford, 1977.

Parfitt, R. L. Anion adsorption by soil and soil materials. Adv. Agron. 30, 1-50, 1978.

Paul, E. A. and Clark, F. E. Soil Microbiology and Biochemistry. London, Academic Press 1996.

Pinton, R., Varanini, Z. and Nannipieri, P. editors. The Rhizosphere. Biochemistry and Organic Substances at the Soil Plant Interface. Marcel Dekker, New York, 2000

Read, D. J., Lewis, D. H. and Alexander, I. J. editors. Mycorrhizas in Ecosystems. CAB International, 1992

Sahrawat, K. L. Nitrogen availability indexes for submerged rice soils. Adv. Agron. 36, 415-451, 1983.

Scheffer, F. and Schachtschabel, P. (G) Textbook of Soil Science, 11th edition, F. Enke-Verlag, Stuttgart, 1982.

Schulten, H. R. and Leinweber, P. Dithionite-citrate-bicarbonate-extractable organic matter in particle size fractions of a Haplaquoll. Soil Sci. Soc. Am. J. 59, 1019-1027, 1995.

Shainberg, J., Sumner, M. E., Miller, W. P., Farina, M. P. W., Pavan, M. A. and Fey, M. V. Use of gypsum on soils: A review. Adv. Soil Sci. 9, 1-111, 1989.

Smith, S. E. and Read, D. J. Mycorrhizal Symbiosis, 2nd edition, Academic Press 1997

Sposito, G. The Chemistry of Soils. Oxford University Press, Oxford 1989.

Uren, N. C. and Reisenauer, H. M. The role of root exudates in nutrient acquisition. Adv. Plant Nutrition 3, 79-114, 1988.

Van Wambeke, A. Soils of the Tropics, McGraw-Hill, New York 1991.

Westermann, R. L. Baird, J. V., Christensen, N. W., Fixen, P. E. and Whitney, D. A. editors. Soil Testing and Plant Analysis. 3rd edition, No3, SSSA Book Series, Madison USA 1990.

Walburg, G., Bauer, M. E., Daughtry, C. S. T. and Housley, T. L. Effects of nitrogen nutrition on the growth, yield and reflectance characteristics of corn. Agron. J. 74, 677-683, 1982.

Walworth, J. L. and Sumner, M. E. Foliar diagnosis: A review. Adv. Plant Nutrition 3, 193-241, 1988.

Watt, M. and Evans, J. R. Proteoid roots. Physiology and development. Plant Physiol. 121, 317-321, 1999.

General Reading, Chapter 3

Badger, M. R. and Price, G. D. The role of carbonic anhydrase in photosynthesis. Annu. Rev. Plant Physiol. Plant Mol. Biol. 45, 369-392, 1994.

Barkla, B. J. and Pantoja, O. Physiology of ion transport across the tonoplast of higher plants. Annu. Rev. Plant Physiol. Plant Mol. Biol. 47, 159-184, 1996.

Bowler, C., Van Montague, M. and Inze, D. Superoxide dismutase and stress tolerance. Annu. Rev. Plant Physiol. Plant Mol. Biol. 43, 83-116, 1992.

Brunhold, C. Reduction of sulfate to sulfide, p. 13-31. In: Sulfur Nutrition and Sulfur Assimilation in Higher Plants, edited by Rennenberg, H., Brunhold, C., DeKok, L. J. and Stulen, I., SPB Acadernic Publishing, The Hague 1990,

Bush, D. R. Proton-coupled sugar and amino acid transporters in plants. Annu. Rev. Plant Physiol. Plant Mol. Biol. 44, 513-542, 1993.

Cadenas, E. Biochemistry of oxygen toxicity. Annu. Rev. Biochem. 58, 79-110, 1989.

Campbell, W. H. Nitrate reductase biochemistry comes of age. Plant Physiol. 111, 355-361, 1996.

Chrispeels, M-J. and Maurel, C. Aquaporins: The molecular basis of facilitated water movement through fiving plant cells. Plant Physiel. 105, 9-13, 1994.

Fox, T. C. and Guerinot, M. L. Molecular biology of cation transport in plants. Annu. Rev. Plant Physiol. Plant Mol. Biol. 49, 669-696, 1998.

Glass, A. D. M. and Siddiqi, M. Y. The control of nutrient uptake rates in relation to the inorganic composition of plants, p. 103-147. In: Adv. in Plant Nutrition, edited by Tinker, P. B. and Läuchli, A., Praeger, New YorK 1984.

Hedrich, R. and Schroeder, J. I. The physiology of ion channels and electrogenic pumps in higher plants. Annu. Rev. Plant Physiol. 40, 539-569, 1989.

Kaiser, M. W., Weinert, H. and Huber, S. C. Nitrate reductase in higher plants: a case study for transduction of environmental stimuli into control of catalytic activity. Physiol. Plant. 105, 385-390, 1999.

Lea, P. J., Blackwell, R. D. and Joy, K. W. Ammonia assimilation in higher plants, p. 153-186. In: Nitrogen Metabolism of Plants, edited by Mengel, K. and Pilbeam, D. J.: Clarendon Press, Oxford 1992.

Leustek, T. and Saito, K. Sulphate transport and the assimilation in plants. Plant Physiol. 120, 637-643, 1999.

McIntyre, G. I. The role of nitrate in the osmotic and nutritional control of plant development. Aust. J. Plant Physiol. 24, 103-118, 1997.

Morgan, M. A., Jackson, W. A. Pan, W. L. and Volk, R. J. Partitioning of reduced nitrogen derived from exogeneous nitrate in maize roots: initial priority of protein synthesis. Plant and Soil 91, 343-347, 1986.

Stitt, M. Nitrate regulation of metabolism and growth. Current Opinion in Plant Biology 2, 178-186, 1999

General Reading, Chapter 4

Addiscott, T. M. Potassium in relation to transport of carbohydrates and ions in plants, p. 205 -220. In: Potassium Research and Agricultural Production, 10th Congr. Intern. Potash Institute, Bern 1974.

Boyer, J. S. Biochemical and biophysical aspects of water deficits and the predisposition to disease. Annu. Rev. Phytopathol. 33, 251-274, 1995.

Boyer, J. S. Water transport. Annu. Rev. Plant Physiol. 36, 473-516, 1985.

Canny, M. J. Transporting water in plants. Amer. Sci. 86, 152-159, 1998.

Canny, M. J. Apoplastic water and solute movement: new rules for an old space. Annu. Rev. Plant Physiol. Plant Mol. Biol. 46, 215-236, 1995.

Chrispeels, M. J., Crawford, N. M. and Aschroeder, J. I. Proteins for transport of water and mineral nutrients across the membranes of plant cells. The Plant Cell 11, 661-675, 1999.

Clarkson, D. T. Roots and the delivery of solutes and water to the xylem. Phil. Trans. Roy. Soc. London Ser. B 341, 5-17, 1993.

Dainty, J. Water relations of plant cells, p. 12-35. In: Transport in Plants, edited by Lüttge, U. and Pitman, M. Encyclopedia of Plant Physiology, New series, Vol. 2, Springer Verlag Berlin 1976.

De Boer, A. H. Potassium translocation into the root xylem. Plant Biol. 1, 36-45, 1999.

Ehlers, W. (G) Transpiration coefficients of crops under field conditions. Pflanzenbauwissenschaften 1, 97-108, 1997.

Flowers, T. J. Physiology of halophytes. Plant and Soil 89, 41-56, 1985.

Flowers, T. J. and Yeo, A. R. Ion relations in plants under drought and salinity. Austr. J. Plant Physiol. 13, 75-91, 1986.

Flowers, T. J. and Läuchli, A. Sodium *versus* potassium substitution and compartmentation, p. 651-681. In: Encyclopedia of Plant Physiology New Series. Vol 15B., edited by Läuchli, A. and Bieleski, R. L., Springer Verlag, Berlin 1983.

Greenway, H. and Munns, R. Mechanism of salt tolerance in nonhalophytes. Annu. Rev. Plant Physiol. 31, 149-190, 1980.

Hanks, R. J. and Rasmussen, V. P. Predicting crop production as related to plant water stress. Adv. Agron. 35, 193-215, 1982.

Hanson, A. D. and Hitz, W. D. Metabolic responses of mesophyte to plant water deficits. Annu. Rev. Plant Physiol. 33, 163-203, 1982.

Hsiao, T. C. Plant responses to water stress. Annu. Rev. Plant Physiol. 24, 519-570, 1973.

Hsiao, T. C. and Acevedo, E. Plant responses to water deficits, water use efficiency, and drought resistance. Agricult. Meteorol. 14, 59-84, 1974.

Jungk, A. and Claassen, N. Ion diffusion in the soil-root system. Adv. Agron. 61, 53-110, 1997.

Kinzel, H. (G) Plant Ecology and Mineral Metabolism. Verlag Eugen Ulmer, Stuttgart 1982.

Kramer, P. J. and Boyer, J. S. Water relations of plants and soils. Acad. Press, San Diego, 1995.

Kuntze, H., Roeschmann, G., Schwerdtfeger, G. (G) Soil Science. 5th Edition, Verlag Eugen Ulmer Stuttgart, 1994.

Koyro H. -W. and Huchzermeyer, B. Salt and drought stress effects on metabolic regulation in maize, p. 843-878. In: Handbook of Plant and Crop Stress, edited by Pessarakli, M., Marcel Dekker, New York 1999.

Läuchli, A. Potassium interactions in crop plants. In: Frontiers in Potassium Nutrition. New Perspectives on the Effect of Potassium on Physiology of Plants, p. 71-76. edited by Oosterhuis, D. M. and Berkowitz, G. A. Potash and Phosphate Institute, Norcross, Georgia USA, 1999

680

Lieth, H., Moschenko, M., Lohmann, M., Koyro, H. -W. and Hamdy, A. Halophyte uses in different climates I. Progress in Biometerology, Vol. 13. Backhuys Publ., Leiden, 1999.

Maathuis, F. J. M. and Amtmann, A. K^+ nutrition and Na^+ toxicity: The basis of cellular K^+/Na^+ ratios. Ann. Bot. 84, 123-133, 1999.

Marschner, H. Mineral Nutrition of Higher Plants. 2nd Edition Acad. Press, London 1995.

Marschner, H., Kirkby, E. A. and Engels, C. Importance of cycling and recycling of mineral nutrients within plants for growth and development. Bot. Acta 110, 265-273, 1997.

McIntyre, G. I. The role of nitrate in the osmotic and nutritional control of plant development. Aust. J. Plant Physiol. 24, 103-118, 1997.

Morgan, J. M. Osmoregulation and water stress in higher plants. Annu. Rev. Plant Physiol. 35, 299-319, 1984.

Milburn. J. A. Water flow in plants. Longman, London and New York, 1979.

Munns, R. Physiological processes limiting plant growth in saline soils. Some dogmas and hypotheses. Plant Cell. Environ. 16, 15-24, 1993.

Neumann, E. J. Root and soil water relations, p. 362-440. In: The Plant Root and its Environment, edited by Carson, E. W. University Press of Virginia, Charlottesville 1974.

Nobel, P. Physicochemical and environmental plant physiology. Acad. Press, San Diego, 1991.

Passioura, J. B. and Fry, S. C. Turgor and cell expansion: beyond the Lokhardt equation. Aust. J. Plant Physiol. 19, 565-576, 1992.

Pate, J. S. Transport and partitioning of nitrogenous solutes. Annu. Rev. Plant Physiol. 31, 313-340, 1980.

Schroeder, J. I., Ward, J. M. and Gassmann, W. Perspectives on the physiology and structure of inward-rectifying K^+ channels in higher plants: Biophysical implications for K^+ uptake. Annu. Rev. Biophys. Biomol. Struct. 23, 441-471, 1994.

Schurr, U. Dynamics of nutrient transport from the root to the shoot. Progress in Botany 60, 234-253, 1999.

Serrano, R., Mulet, J. M., Rios, G., Marquez, J. A., de Larrinoa, I. F., Leube, M. P., Mendizabal, I., Pascual-Ahuir, A., Proft, M., Ros, R. and Montesinos, C. A glimpse of the mechanisms of ion homeostasis during salt stress. J. Exp. Bot. 50, 1023-1036, 1999.

Slatyer, R. O. Plant-Water Relationships. Academic Press, London, New York 1967.

Smith, J. A. C. and Griffiths, H. Water deficits. Plant responses from cell to community. BIOS Scientific, Oxford, 1993.

Smith, J. A. C. Ion transport and the transpiration stream. Bot. Acta 104, 416-421, 1991.

Somero, G. N., Osmond, C. B. and Bolis, C. L. Water and Life. Springer Verlag Berlin, Heidelberg, 1992.

Stanhill, G. Water use efficiency. Adv. Agron. 39, 53-85, 1986

Steudle, E. and Peterson, C. A. How does water get through roots? J. Exp. Bot. 49, 775-788, 1998.

Talbott, L. D. and Zeiger, E. The role of sucrose in guard cell osmoregulation. J. Exp. Bot. 49, 329-337, 1998.

Tyree, M. T. and Sperry, J. S. Vulnerability of xylem to cavitation and embolism. Annu. Rev. Plant Physiol. 40, 19-38, 1989.

Yeo, A. Molecular biology of salt tolerance in the context of whole-plant physiology. J. Exp. Bot. 49, 915-929, 1998.

Zeiger, E., Farqhuar, G. and Gowan, I. Stomatal Function. Stanford University Press, Stanford, CA, 1987.

General Reading, Chapter 5

Blatt R. M., Thiel G. Hormonal control of ion channel gating. Annu. Rev. Plant Physiol. Plant Mol. Biol. 44, 543-567, 1993.

Bowler, C., Van Montague, M. and Inze, D. Superoxide dismutase and stress tolerance. Annu. Rev. Plant Physiol. Plant Mol. Biol. 43, 83-116, 1992.

Cadenas, E. Biochemistry of oxygen toxicity. Annu. Rev. Biochem. 58, 79-110 1989.

Chapin, F. S., Schulze, E. D. and Mooney, H. A. The ecology and economics of storage in plants. Annu. Rev. Ecol. Syst. 21, 423-447, 1990.

Clouse S. D. and Sasse J. M. Brassinosteroids: esssential regulators of plant growth and development. Annu. Rev. Plant Physiol. Plant Mol. Biol. 49, 427 -451, 1998.

Creelman R. A, Mullet J. E. Oligosaccharins, brassinolides, and jasmonates: Nontraditional regulators of plant growth, development and gene expression. The Plant Cell 9, 1211-1223, 1997.

Davies W. J., Zhang J. Root signals and the regulation of growth and development of plants in drying soils. Annu. Rev. Plant Physiol. Plant Mol. Biol. 42, 55-76, 1991.

Evans L. T. Feeding the Ten Billion - Plants and Population Growth. Cambridge University Press, Cambridge 1998.

Evans L. T. Gibberellins and flowering in long day plants, with special reference to *Lolium temulentum*. Austr. J. Plant Physiol. 26, 1-8, 1999.

Geiger, D. R. Phloem loading, p. 396-431. In: Transport in Plants, I. Phloem Transport, edited by Zimmermann, M. H. and Milburn, J. A. Springer-Verlag Berlin, Heidelberg, New York, Tokyo, 1975.

Geiger, D. R., Servaites, J. C. and Shieh, W. J. Balance among parts of source sink system: a factor in crop productivity, p. 155- 192. In: Crop Photosynthesis: Spatial and Temporal Determinants. Elsevier, Amsterdam 1992.

George B. J. Design and interpretation of nitrogen response experiments. The nitrogen requirement of cereals, p. 139-149. In: Ministry of Agriculture and Fisheries, ed. Ministry of Agriculture Fisheries and Food Fish Reference Book 385. London 1983.

Huglin, P. (F) Biology and Ecology of Vine. Editions Payot, Lausanne 1986.

Lambers, H. The physiological significance of cyanide-resistant respiration in higher plants p. 113-128. In: Energy Metabolism in Higher Plants in Different Environments, edited by Lambers, H., Ph. D. Thesis of the Rijks-Universiteit Groningen, Netherlands, 1979.

Kende, H. and Zeevaart, H. D. The five „Classical" plant hormones. The Plant Cell 9, 1197-1210, 1997.

Kosegarten, H. and Mengel, K. Starch deposition in storage organs and the importance of nutrients and external factors. Z. Pflanzenernähr. Bodenk. 161, 273-287, 1998.

Millard, P. Ecophysiology of the internal cycling of nitrogen for tree growth. Z. Pflanzenernähr. Bodenk. 159, 1-10, 1996.

Morell, M. K. S., Rahman, S., Abrahams, S. L. and Appels, R. The biochemistry and molecular biology of starch synthesis in cereals. Austr. J. Plant Physiol. 22, 647-660, 1995.

Nelson, O. E. and Pan, D. Starch synthesis in maize endosperms. Annu. Rev. Plant Physiol. Plant Mol. Biol. 46, 475-496, 1995.

Oparka, K. J. and Van Bel, A. J. E. Pathways of phloem loading and unloading: a plea for a uniform terminology, p. 249-254 . In: Carbon Partitioning within and between Organisms, edited by Pollock, C. J., Farrar, J. and Gordon, A. J., BIOS Scientific Publ. Limited 1992.

Patrick, J. W. Phloem unloading: sieve element unloading and post-sieve element transport. Annu. Rev. Plant Physiol. Plant Mol. Biol. 48, 191-222, 1997.

682

Patrick, J. W. and Offler, C. E. Post-sieve element transport of sucrose in developing seeds. Austr. J. Plant Physiol. 22, 681-702, 1995.

Payne, P. J. Genetics of wheat storing proteins and the effect of allelic variation on bread making quality. Annu. Rev. Plant Physiol. 38, 141-153, 1987.

Pollock, C. J. and Cairns, A. J. Fructan metabolism in grasses and cereals. Annu. Rev. Plant Physiol. Plant Mol. Biol. 42, 77-101, 1991.

Porter J. R. and Gawith M. Temperatures and growth and the development of wheat: a review. European J. Agron. 10, 23-26, 1999.

Preiss, J. Biology and molecular biology of starch synthesis and its regulation, p. 59-114. In: Oxford Surveys of Plant Molecular and Cell Biology, edited by Miflin, B. J. Vol. 7, Oxford 1991.

Schapendonk., A. H. C. M., Stol, W. Van Kraalingen. D. G. and Bouman, B. M. LINGRA, a sink/source model to simulate grassland productivity in Europe. European J. Agron. 9, 87-100, 1998.

Shewry, P. R., Napier, J. A. and Tatham, A. S. Seed storage proteins: structures and biosynthesis. The Plant Cell 7, 945-956, 1995.

Sembdner G. and Parthier, B. The biochemistry and the physiological and molecular actions of jasmonates. Annu. Rev. Plant Physiol. Plant Mol. Biol. 44, 569-589, 1993.

Smith, A. M., Denyer, K. and Martin, C. The synthesis of the starch granule. Annu. Rev. Plant Physiol. Plant Mol. Biol. 48, 67-87, 1997.

Solomos, T. Cyanide-resistant respiration in higher plants. Annu. Rev. Plant Physiol. 28, 279-297, 1977.

Ten Berge, H. F. M., Aggarwal, P. K. and Kropff, M. J. Application of Rice Modelling. Amsterdam: Elsevier, 1997.

Van Ittersum, M. K. and Van de Geijn, S. C. Perspectives for Agronomy- Adopting Ecological Principles and Managing Resoure Use. Amsterdam: Elsevier, 1997.

Warren-Wilson, J. Maximum yield potential, p. 34-56. In: Transition from Extensive to Intensive Agriculture with Fertilizers. Proc. 7th Colloq. Int. Potash Inst., Bern, 1969.

General Reading, Chapter 6

Aulakh, M. S., Doran, J. W. and Moisier, A. R. Soil denitrification-significance, measurement, and effects on management. Adv. Soil Sci. 18, 17-57, 1992.

Bacon, P. E. editor. Nitrogen Fertilization in the Environment. Marcel Dekker, New York, 1995

Bar-Yosef, B. Advances in fertigation. Adv. Agron. 65, 2-77. 1999.

Barrow, N. J. Plant Nutrition from Genetic Engineering to Field Practice, editor. Kluwer Academic Publishers, Dordrecht, 1993.

Batts, G. R., Morison, J. I. L., Ellis, R. H., Hadley, P. and Wheeler, T. R. Effects of CO_2 and temperature on growth and yield of crops of winter wheat over four seasons p. 67-76. In: Proceedings. ESA Congress, Veldhoven 1996, edited by Ittersum, van M. K. and Geijn, van de S. C. Elsevier, Amsterdam 1997.

Bowman, R. A., Reeder, J. D. and Lober, R. W. Changes in soil properties in a central plane rangeland soil after 3, 20, and 60 years of cultivation. Soil Sci. 150, 851-857, 1990.

Braun, H. Fertilizers and Food Production, Rome:FAO, 1989.

Bresler, E. Trickle-drip irrigation: Principles and application to soil water management. Adv. Agron. 29, 343-393, 1977.

Chapin, F. S. Ecological aspects of plant mineral nutrition. Adv. Plant Nutrition. 3:161-191, 1988.

Fageria, N. K., Baligar, V. C. and Jones, C. A. Growth and Mineral Nutrition of Field Crops, Marcel Dekker, New York 1991.

Fragoso, M. A. C. and Van Beusichem, M. L. Optimization of Plant Nutrition, Dordrecht, Kluwer Academic Publishers, 1993.

Fuchs, A. Potentials for non-food utilization of fructose and inulin. Starch/stärke 10, 335-343, 1987.

Gales, K. Yield variation of wheat and barley in Britain in relation to crop growth and soil conditions - a review. J. Sci. Food Agric. 34, 1085-1104, 1983.

Hanks, R. J. and Rasmussen, V. P. Predicting crop production as related to plant water stress. Adv. Agron. 35, 193-215, 1982.

Haygarth, P. M. and Jarvis, S. C. Transfer of phosphorus from agricultural soils Adv. Agron. 66, 196-249, 1999.

Johnston, A. E. The Rothamsted classical experiments, p. 9-37. In: Long-term Experiments in Agricultural and Ecological Sciences, edited by Leigh, R. A. and Johnston, A. E. CAB International, Wallingford, UK 1994

Juste, C. and Mench, M. Long-term application of sewage sludge and its effect on metal uptake by crops, p. 159-193. In: Biogeochemistry of Trace Metals, edited by Adriano, D. C. Lewis Publishers Ann Arbor, London, Tokyo 1992.

Kafkafi, U. Combined irrigation and fertilization in arid zones. Israel J. Plant Sci. 42, 301-320, 1994.

Mengel, K. Turnover of organic nitrogen in soils and its availability to crops. Plant and Soil 181, 83-93, 1996.

Moffat, A. S. Higher yielding perennials point the way to new crops. Science. 274:1469-1470, 1996.

Morard, P. (F) Soilless Cultivation of Vegetables. S. A. R. L. Publications Agricoles ISBN 2-9509297, Agen, France 1995.

Neumann, R. Chemical crop protection research and development in Europe, p. 49-55. In: Proceedings of the 4. th ESA Congress, Veldhoven 1996, edited by Ittersum, van M. and Geijn, van S. C., Elsevier, Amsterdam 1997.

Oertli, J. J. Controlled-release fertilizers. Fertil. Res. 1, 103-123, 1980.

Otter-Nacke, S. and Kuhlmann, H. A. A comparison of the performance of N simulation models in the prediction of N-min on farmers' fields in the spring. Fert. Res. 27, 341-347, 1991.

Owen, T. R. and Jürgens-Gschwind, S. Nitrates in drinking water: a review . Fert. Res. 10, 3-35, 1986.

Paramasivam, S. and Alva, A. K. Nitrogen recovery from controlled-release fertilizers under intermittent leaching and dry cycles. Soil Sci. 162, 447-453, 1997.

Parker, D. R. and Norvell, W. A. Advances in solution culture methods for plant mineral nutrient research. Adv. Agron. 65, 151-313, 1999.

Picard, D. New approaches for cropping system studies in the tropics, p. 30-36. In: 3. Congress of the European Society of Agronomy, edited by Borin, M. and Sattin, M. Colmar, France: European Society of Agronomy, Colmar 1994.

Sanchez, P. A. and Leakey, R. R. B. Land use transformation in Africa: three determinants for balancing food security with natural resources. Eur. J. Agron. 7, 15-23, 1997.

Smith, S. J., Schepers, J. S. and Porter, L. K. Assessing and managing agricultural nitrogen losses to the environment. Adv. Soil Sci. 14, 1-43, 1990.

Tunney, H., Carton, O. P., Brookes, P. C. and Johnston, A. E. editors. Phosphorus Loss from Soil Water. CAB International, Wallingford, UK 1998

Van Ittersum, M. K. and Van de Geijn, S. C. editors, Perspectives for Agronomy. Adopting Ecological Principles and Managing Resource Use. Elsevier, Amsterdam 1997.

Vlek, P. L. G. editor. Nutrient Cycling in Agroecosystems. Vol. 46, Kluwer Academic Publishers, Dordrecht, 1996.

Wild, A., Jones, L. H. P. and Macduff, J. H. Uptake of mineral nutrients and crop growth: the use of flowing nutrient solutions. Adv. Agron. 41, 171-219 1987.

Woolhouse, H. W. Crop physiology in relation to agricultural production: the genetic link, p. 1-21. In: Physiological Processes Limiting Plant Productivity, edited by Johnson, C. B. London: Butterworth, 1981.

General Reading, Chapter 7

Appel, T. and Mengel, K. Prediction of mineralizable nitrogen in soils on the basis of an analysis of extractable organic N. Z. Pflanzenernähr. Bodenk. 161, 433-452, 1998.

Blevins, D. G. An overview of nitrogen metabolism in higher plants, p. 1-41. In: Plant Nitrogen Metabolism, edited by Poulton, J. E., Romeo, J. T. and Conn, E. C. Plenum, New York 1989.

Bradbury, N. J., Whitmore, A. P., Hart, P. B. S. and Jenkinson, D. S. Modelling the fate of nitrogen in crop and soil in the years following application of 15-N labelled fertilizer to wheat. J. Agric. Sci. 121, 363-379, 1993.

Boddey, R. M., Urquiaga, S., Reis, V. and Döbereiner, J. Biological nitrogen fixation associated with sugar cane. Plant and Soil 137, 11-117, 1991.

Craswell, E. T. and Godwin, D. C. The efficiency of nitrogen fertilizers applied to cereals in different climates. Adv. Plant Nutrition 1, 1-55, 1984.

Döbereiner, L. Dinitrogen fixation in rhizosphere and phyllosphere associations, p. 330-350. In: Inorganic Plant Nutrition, Encycl. Plant Physiol. New Series, Vol. 15A, edited by Läuchli, A and Bieleski, R. L., Springer Verlag Berlin 1983.

Döbereiner, J. Biological nitrogen fixation in the tropics: social and economic contributions. Soil Biol Biochem. 29, 771-174, 1997.

Engels, C. and Marschner, H. Plant uptake and utilization of nitrogen, p. 41-81. In: Nitrogen Fertilization in the Environment, edited by Bacon, P. E. Marcel Dekker, New York 1995.

Glass, A. D. M. et al. Nitrogen transport in plants, with an emphasis on the regulation of fluxes to match plant demand. Journal of Plant Nutrition and Soil Science 164, 199-207, 2001.

Greenwood, D. J. Modelling N-response of field vegetable crops grown under diverse conditions with N_ABLE: A Review. J. Plant Nutr. 24, in press 2001

Haynes, R. J. editor. Mineral Nitrogen in the Soil Plant System, Academic Press, London 1996

Hecht-Buchholz, C. The apoplast-habitat of endophytic dinitrogen-fixing bacteria and their significance for the nutrition of nonleguminous plants. Z. Pflanzenernähr. Bodenk. 161, 509-520, 1998.

Jenkinson, D. S. The turnover of organic carbon and nitrogen in soil. Phil. Trans Royal Soc. B 329, 361-368, 1990.

Layzell, D. B. Oxygen and the control of nodule metabolism and N_2 fixation p. 435-440. In: Biological Nitrogen Fixation for the 21st Century, edited by Elmerich, C., Kondorosi, A. and Newton, W. E. Kluwer Academic Publishers, Dordrecht 1998.

Martins-Loucao, R. amd Lips, S. H. editors. Nitrogen in a Sustainable Ecosystem from Cell to the Plant. Backhuys Publishers, Leiden, The Netherlands 2000.

Mengel, K. Turnover of organic nitrogen in soils and its availability to crops. Plant and Soil 181, 83-93, 1996.

Mengel, K. and Pilbeam, D. J. Nitrogen Metabolism of Plants. Clarendon Press, Oxford, 1992.

Millard, P. Ecophysiology of the internal cycling of nitrogen for tree growth. Z. Pflanzenernähr. Bodenk. 159, 1-10, 1996.

Mylona, P., Pawlowski, K. and Bisseling, T. Symbiotic nitrogen fixation. The Plant Cell 7, 869-885, 1996.

Neeteson, J. J. and Hassink, J. editors. Nitrogen Mineralization in Agricultural soils. DLO Res. Institute for Agrobiology and Soil Fertility, Wageningen/Haren 1993.

Nicolardot, B., Molina, J. A. E. and Allard, M. R. C and N fluxes between pools of soil organic matter: model calibration with long-term incubation data. Soil. Biol. Biochem. 26, 235-243, 1994.

Salsac, L. Chaillou, S. Morot-Gaudry, J. F. Lesaint, C. and Jolivet, E. Nitrate and ammonium nutrition in plants. Plant Physiol. Biochem. 25, 805-812, 1987.

686

Scherer, H. W. Dynamics and availability of the non-exchangeable NH_4-N - a review. Eur. J. Agron. 2, 149-160, 1993.

Schulten, H. R. and Schnitzer, M. The chemistry of soil. Organic nitrogen: a review. Biol. Fert. Soils 26, 1-15, 1998.

Ullrich, W. R. Transport of nitrate and ammonium through plant membranes, p. 121-137 . In: Nitrogen Metabolism in Plants, edited by Mengel, K. and Pilbeam, D. J. Clarendon Press, Oxford 1992.

Van Cleemput, O., Hofman, G. and Vermoesen, A, editors. Progress in Nitrogen Cycling Studies. Kluwer Academic Publishers, Dordrecht 1996.

Wiren, N. von, Gazzarrini, S., Goyon, A. and Frommer, W. The molecular physiology of ammonium uptake and retrival. Current Opinion in Plant Biology 3, 254-261, 2000

General Reading, Chapter 8

Bell, C. I., Cram, W. J. and Clarkson, D. T. Turnover of sulfate in leaf vacuoles limits retranslocation under sulfur stress, p. 163 165 . In: Sulfur Nutrition and Sulfur Assimilation in Higher Plants, edited by Rennenberg, H., DeKok, L. J. and Stulen, I. SPB Academic Publishing bv, The Hague 1990.

Clarkson, D. T., Hawkesford, M. J. and Davidian, J. C. Membrane long-distance transport of sulfate p. 3 -19. In: Sulfur Nutrition and Assimilation, edited by De Kok, L. J., Stulen, I., Renenberg, H. Brunhold, C. Rauser, W. E. SPB Academic Publishing bv, The Hague 1993.

Cram, W. J., Dekok, L., Stule, I., Brunhold, C, and Rennenberg, H. editors. Sulphur Metabolism in Higher Plants, Molecular, Ecophysiological and Nutritional Aspects. Backhuys Publishers, Leiden, The Netherlands, 1997

De Kok, Stulen, I., Rennenberg, H., Brunhold, C. and Rauser, W. E. editors. Sulfur Nutrition and Assimialtion in Higher Plants. SPB Academic Publishing bv, The Hague, 1993.

Jolivet, P. Elemental sulfur in agriculture, p. 193-206. In: Sulfur Nutrition and Assimilation in Higher Plants, editors De Kok, L. J., Stulen, I, Rennneberg, H., Brunhold, C. and Rauser, W. E. SPB Academic Publishing bv, The Hague 1993.

Leustek, T. and Saito, K. Sulphate transport and the assimilation in plants. Plant Physiol. 120, 637-643, 1999.

Jürgens-Gschwind, S. and Jung, J. Results of lysimeter trials at the Limburger Hof facility: The most important findings from 50 years of experiments. Soil Sci. 127, 146-160, 1979.

Marsh, K. B., Tillman, R. W. and Syers, J. K. Charge relationships of sulphate sorption by soils. Soil Sci. Soc. Am. J. 51, 318-323, 1987.

Maynard, D. G. editor. Sulfur in the Environment. Marcel Dekker, New York, 1998

McGrath, S. P., Zhao, F. J. and Withers, P. J. A. Development of sulphur deficiency in crops and ist treatment, p. 47. Proc. No 379. The Fertilizer Society, Peterborough UK 1996.

Pasricha, N. S. and Fox, R. L. Plant nutrient sulfur in the tropics and subtropics. Adv. Agron. 50, 210-269, 1993.

Randall, P. J. and Wrigley, C. W. Effects of sulphur supply on the yield, composition and quality of grain from cereals oilseeds and legumes. Adv. Cereal Sci. Techn. 8, 171-206, 1986.

Rennenberg, H. The fate of excess sulfur in higher plants. Annu. Rev. Plant Physiol. 25, 121-153, 1984.

Schiff, J. A., Stern, A. I., Saidha, T. and Li, J. Some molecular aspects of sulfate metabolisn in photosynthetic organisms p. 21-35 . In: Sulfur Nutrition and Assimilation of Higher Plants, edited by DeKok, L. J., Stulen, I., Rennenberg, H., Brunold, C. and Rauser, W. E. SPB Academic Publishing bv, The Hague, 1993.

Schnug, E. Physiological functions and environmental relevance of sulfur -containing secondary metabolites p. 179-190. In: Sulfur Nutrition and Assimilation of Higher Plants, edited by DeKok, L. J., Stulen, I., Rennenberg, H., Brunhold, C. and Rauser, W. E. SPB Academic Publishing bv, The Hague 1993.

Scott, N. M. Sulphur in soils and plants p. 379-401. In: Organic Matter and Biological Activity, edited by Vaugham, D. and Malcolm, R. E. Martinus Nijhoff, The Hague 1985.

Syers, J. K., Skinner, R. J. and Curtin, D. Soil and Fertilizer Sulphur in UK Agriculture, p. 43. Proc. No 264, The Fertilizer Soc., London 1986.

Wainwright, M. Sulfur oxidation in soils. Adv. Agron. 37, 449-396, 1984.

Wu, J., O'Donnell, A. G. and Syers, J. K. Microbial growth and sulphur immobilization following incorporation of plant residues in soil. Soil Biol. Biochem. 25, 1567-1573, 1993.

Zhao, F. J., Wu, J. and Mc Grath, S. P. Soil organic sulphur and its turnover p. 467- 505. In: Humic Substances in Terrestrial Ecosystems, edited by Piccolo, A. Elsevier, Amsterdam 1996.

General Reading, Chapter 9

Bucher, M., Rausch, C. and Daram, P. Molecular and biochemical mechanisms of phosphorus uptake into plants. Journal of Plant Nutrition and Soil Science 164, 209-218, 2001.

Bolland, M. D. A. and Gilkes, R. J. Rock phosphates are not effective fertilizers in Western Australia soils: a review. Fert. Res. 22, 79-85, 1990.

Bolland, M. D. A., Gilkes, R. J. and Allen, D. G. The residual value of superphosphate and rock phosphate for lateric soils and its evaluation using three soil phosphate tests. Fert. Res. 15, 253-280, 1988.

Dalal, R. C. Soil organic phosphorus. Adv. Agron. 29, 83-117, 1977.

Gerke, J. Kinetics of soil phosphate desorption as affected by citric acid . Z. Pflanzenernähr. Bodenk. 157, 17-22, 1994.

Haynes, R. J. Lime and phosphate in soil-plant system. Adv. Agron. 37, 249 -315, 1984.

Jungk, A. Phosphorus supply of plants- how is it accomplished ? Proceedings of the National Science Council, ROC Part B: Life Sciences 18, 187-197, 1994

Jungk, A. and Claassen, N. Ion diffusion in the soil-root system. Adv . Agron. 61, 53-110, 1997.

Khasawneh, F. E. and Doll, E. C. The use of phosphate rock for direct applications to soils. Adv. Agron. 30, 159-206, 1978.

Mengel, K. Agronomic measures for better utilization of soil and fertilizer phosphates. Eur. J. Agron. 7, 221-233, 1997.

Sanyal, S. K. and De Datta, S. K. Chemistry of phosphorus transformation in soil. Adv. Soil Sci. 16, 1-120, 1991.

Ratcliffe, R. G. *In vivo* NMR studies of higher plants and algae. Adv. Bot. Res. 20, 43-123, 1994.

Schachtman, D. P., Reid, R. J. and Ayling, S. M. Phosphorus uptake by plants: from soil to cell. Plant Physiol. 116, 447-453, 1998.

Shainberg, J., Sumner, M. E., Miller, W. P., Farina, M. P. W., Pavan, M. A. and Fey, M. V. Use of gypsum on soils: A review. Adv. Soil Sci. 9, 1-111, 1989.

Steffens, D. Phosphorus release kinetics and extractable phosphorus after long-term fertilization. Soil Sci. Soc. Am. J. 58, 1702-1708, 1994.

Stevenson, F. J. and Cole, M. A. The phosphorus cycle, p. 279-329 . In Cycles of Soil Carbon, Nitrogen, Phosphorus Sulfur, Micronutrients 2nd ed. John Wiley, Weinheim 1999

General Reading, Chapter 10

Aktar, M. S. Potassium availability as affected by soil mineralogy, p. 33-48. In: Potasssium in Agriculture, edited by Mengel, K. International Potash Institute, Basel 1994.

Chen, S. Z., Low, P. F. and Roth, C. B. Relation between potassium fixation and the oxidative state of octahedral iron. Soil Sci. Soc. Am. J. 51, 82-86, 1990.

Evangelou, V. P., Wang, J. and Phillips, R. E. New developments and perspectives on soil potassium quantity/intensity relationships. Adv. Agron. 52, 173-227, 1994.

Fernando, M., Mehroke, I. and Glass, A. D. M. *De novo* synthesis of plasma membrane and tonoplast polypeptides of barley roots during short term K deprivation. Plant Physiol. 100, 1269-1276, 1992.

Fox, T. C. and Guerinot, M. L. Molecular biology of cation transport in plants. Annu. Rev. Plant Physiol. Plant Mol. Biol. 49, 669-696, 1998.

Greenwood, D. J. and Karpinets, T. V. Dynamic model for the effects of K -fertilizer on crop growth, K-uptake and soil-K in arable cropping. 1. Description of the model. Soil Use and Management. 13, 178-183, 1997.

Hinsinger, P. and Jaillard, B. Root-induced release of interlayer potassium and vermiculization of phlogopite as related to potassium depletion in the rhizosphere of ryegrass. J. Soil Sci. 44, 525-534, 1993.

Johnston, A. E., editor, Food Security in the WANA region, the essential need for balanced fertilization. Intern. Potash Institute, Basel, 1997.

Kochian, L. V. and Lucas, W. J. Potassium transport in roots. Adv. in Bot. Res. 15, 93-178, 1988.

Leigh, R. A. Potassium homeostasis and membrane transport. Journal of Plant Nutrition and Soil Science 164, 193-198, 2001.

Maathuis, F. J. M. and Sanders, D. Mechanism of potassium absorption by higher plants. Physiol. Plant. 96, 158-168, 1996.

Mengel, K. Integration and functions and involvement of potassium metabolism at the whole plant level, p. 1-11 In: Frontiers in Potassium Nutrition: New Perspectives on the Effects of Potassium on Physiology of Plants, edited by Oosterhuis, D. M. and Berkowitz, G. A. The Potash and Phosphate Institute, Norcross, Georgia USA 1999.

Munson, R. D.: Potassium in Agriculture. Proc. Intern. Symposium, Atlanta 1985. Am. Soc. Agron. Madison USA 1985.

Oosterhuis, D. M. and Berkowitz, G. A., editors. Frontiers in Potassium Nutrition: New Perspectives on the Effects of Potassium on Physiology of Plants. Potash and Phosphate Institute, Norcross, Georgia USA 1999.

Rea, F. A. Poole, R. J. Vacuolar H^+ -translocating pyrophosphatase. Annu . Rev. Plant Physiol. Plant Mol. Biol. 44, 157-180, 1993.

Sharpley, A. N. Reaction of fertilizer potassium in soils of differtent mineralogy. Soil Sci. 149, 44-51, 1990.

Sparks, D. L. Potassium dynamics in soils. Adv. Soil Sci. 6, 1-63, 1987.

Toney, M. D., Hohenester, E., Cowan, S. W. and Jansonius, I. N. Dialkylglycine decarboxylase structure: bifunctional active site and alkali metal sites. Science. 261, 756-759, 1993.

Walker, N. A., Sanders, D. and Maathuis, F. J. M. High-affinity potassium uptake in plants. Science. 273, 977-979, 1996.

White, P. J. The regulation of K^+ influx into roots of rye *(secale cereale* L.) seedlings by negative feedback via the K^+ flux from shoot to root in the phloem. J. Expt. Bot. 48, 2063-2073, 1997.

690

General Reading, Chapters 11 and 12

Bangerth, F. Calcium-related physiological disorders of plants. Ann. Rev. Phytopathol. 17, 97-122, 1979.

Barkla, B. J. and Pantoja, O. Physiology of ion transport across the tonoplast of higher plants. Annu. Rev. Plant Physiol. Plant Mol. Biol. 47, 159-184 1996.

Bush, D. S. Calcium regulation in plant cells in its role in protein signaling. Annu. Rev. Plant Physiol. Plant Mol. Biol. 46, 95-122, 1995.

Cohen, P. The structure and regulation of protein phosphatases. Annu. Rev. Biochem. 58, 453-508, 1989.

Evans, D. E., Briars, S. -A. and Williams, L. E. Active calcium transport by plant cell membranes. J. Exp. Bot. 42, 285-303, 1991.

Foy, C. D. Effects of soil calcium availability on plant growth, p. 565-600. In: The Plant Root and its Environment, edited by Carson, E. W. Univ. Press of Virginia, Charlottesville 1974.

Hanson, J. B. The functions of calcium in plant nutrition. Adv. Plant Nutr. 1, 149-208, 1984.

Haynes, R. J. Effects of liming on phosphate availability in acid soils. Plant and Soil 68, 289-308, 1982.

Hepler, P. K. and Wayne, R. O. Calcium and plant development. Annu. Rev. Plant Physiol. 36, 397-439, 1985.

Huguet, C. and Coppenet, M., editors. (F) Magnesium in Agriculture, INRA editions, Paris 1992

Kerschberger, M. (G) Effect of pH on the lactate-soluble P level in soils (DL-method). Arch. Acker-Pflanzenbau Bodenkd. 31, 313-319, 1987.

Kirkby, E. A. and Mengel, K. The role of magnesium in plant nutrition. Z. Pflanzenernähr.
 Bodenk. 2, 209-222, 1976.

Kirkby, E. A. and Pilbeam, D. J. Calcium as a plant nutrient. Plant Cell and Environment 7, 397-405, 1984

Marschner, H. Calcium nutrition of higher plants. Neth. J. agric. Sci. 22, 275-282, 1974

McLaughlin, S. B. and Wimmer, R. Calcium physiology and terrestrial ecosystem processes. New Phytol. 142, 373-417, 1999.

Shainberg, J., Sumner, M. E., Miller, W. P., Farina, M. P. W., Pavan, M. A. and Fey, M. V. Use of gypsum on soils: A review. Adv. Soil Sci. 9, 1-111, 1989.

Jaillard, B., Guyon, A. and Maurin, A. F. Structure and composition of calcified roots, and their identification in calcareous soils. Geoderm. 50, 197-210, 1991.

Sumner, M. E. Gypsum and acid soils: the world scene. Adv. Agron. 51, 1-33, 1993.

Zhang, W. H. and Rengel, Z. Cytosolic Ca^{2+} activities in intact wheat root apical cells subjected to aluminium toxicity, p. 353-358. In: Plant Nutrition - Molecular Biology and Genetics, edited by Gissel-Nielsen, G. and Jensen, A. Kluwer Academic Publishers, Dordrecht 1999.

Zielinski, R. E. Calmodulin and calmodulin binding proteins in plants. Annu. Rev. Plant Physiol. Plant Mol. Biol. 49, 697-725, 1998.

General Reading, Chapter 13

Abadia, J., editor. Iron Nutrition in Soils and Plants. Kluwer Academic Publishers, Dordrecht, 1995.

Brown, J. C. and Jolley, V. D. Plant metabolic responses to iron-deficiency stress. Biological Sciences. 39, 546-551, 1989.

Chen, Y. and Barak, Y. Iron nutrition of plants in calcareous soils. Adv . Agron. 35, 217-240, 1982.

Chen, Y. and Hadar, Y. editors. Iron Nutrition and Iron Interactions in Plants. Kluwer Academic Publishers, Dordrecht 1991.

Crowley, D. E., Wang, Y. C., Reid, C. P. P. and Szaniszlo, P. J. Mechanisms of iron acquisition from siderophores by microorganisms and plants, p. 213-232. In: Iron Nutrition and Interactions in Plants, edited by Y. Chen, and Y. Hadar, Kluwer Academic Publishers, Dordrecht 1991.

Kosegarten, H., Hoffmann, B. and Mengel, K. Apoplastic Fe^{3+} reduction in intact sunflower leaves. Plant Physiol. 121, 1069-1079, 1999.

Lindsay, W. Iron oxide solubilization by organic matter and its effect on iron availability, p. 29-36. In: Iron Nutrtion and Interactions in Plants, edited by Y . Chen and Y. Hadar, Kluwer Academic Publishers, Dordrecht 1991.

Mengel, K. Iron availability in plant tissues - iron chlorosis on calcareous soils. Plant and Soil 165, 275-283, 1994.

Neilands, J. B. and Leong, S. A. Siderophores in relation to plant growth and disease. Annu. Rev. Plant Physiol. 37, 187-208, 1986.

Römheld, V., editor Iron Nutrition and Interactions in Plants. Proceed. 9th Symposium, J. Plant Nutrition 23, 1549-2102, 2000.

Römheld, V. and Marschner, H. Mobilization of iron in the rhizosphere of different plant species. Adv. Plant Nutrition. 2, 155-204, 1986.

Schwertmann, U. Solubility and dissolution of iron oxides p. 3-27 . In: Iron Nutrition and Interactions in Plants, edited by Chen, Y. and Hadar, Y., Kluwer Academic Publishers, Dordrecht 1991.

Stephan, U. W. and Scholz, G. Nicotianamine: mediator of transport of iron and heavy metals in the phloem? Physiol. Plant. 88, 522-529, 1993.

Tagliavini, M., Abadia, J., Rombola, A. D., Abadia, A., Tsipouridis, C. and Marangoni, B. Agronomic means for the control of iron chlorosis in deciduous fruit plants. In: Proceedings 9th Iron Symposium, Iron Nutrition and Interaction in Plants, edited by Römheld. V., J. Plant Nutrition 23, 2007-2027, 2000.

Theil, E. C. Ferritin: structure, gene regulation, and cellular function in animals, plants and microorganisms. Annu. Rev. Biochem. 56, 289-315, 1987.

General Reading, Chapters 14 to 20

Baize, D. (F) Total Contents of Metallic Trace Elements in Soils. Paris:INRA, 1997.

Bergmann, W. Nutritional Disorder of Plants, Development Visual and Analytical Diagnosis. Gustav Fischer, Jena 1992.

Blevins, D. G. and Lukaszewski, K. M. Boron in plant structure and function Annu. Rev. Plant Physiol. Mol. Biol. 49, 481-500, 1998.

Boon, D. Y. Potential selenium problems in Great Plains soils, p. 107-121. In: Jacobs, L. W. editor., ASA, SSSA, Madison USA 1989.

Bowler, C., Van Montague, M. and Inze, D. Superoxide dismutase and stress tolerance. Annu. Rev. Plant Physiol. Plant Mol. Biol. 43, 83-116, 1992.

Bromfield, S. M., Cumming, R. W., David, D. J. and Williams, C. H. Change in soil pH, manganese and aluminium under subterranean clover pasture. Austr. Exp. Agric. Anim. Husb. 23, 181-191, 1983.

Cakmak, I. Possible roles of zinc in protecting plant cells from damage by reactive oxygen species. New Phytologist 146, 185 -205, 2000

Cocker, K. M., Evans, D. M. and Hodson, M. J. The amelioration of aluminium toxicity by silicon in higher plants: Solution chemistry or an in planta mechanism? Physiol. Plant. 104, 608-614, 1998.

Coleman, J. E. Zinc proteins: Enzymes, storage proteins, transcription factors, and replication proteins. Annu. Rev. Biochem. 61, 897-946, 1992.

De la Fuente, J. M. and Herrera-Estrella, L. Advances in the understanding of aluminium toxicity and the development of aluminium-tolerant transgenic plants. Adv. Agron. 66, 103-120, 1999.

Dell, B., Brown, P. H. and Bell, R. W. editors Boron in Soils and Plants: Reviews, Plant and Soil 193, 1-209, 1997

Epstein, E. Silicon. Annu. Rev. Plant Physiol. Plant Mol. Biol. 50, 641-664, 1999.

Flowers, T. J. Chloride as a nutrient as an osmoticum. Adv. Plant Nutrition 3, 55-78, 1988

Foy, C. D. General principles of screening plants for aluminium and manganese tolerance, 255-267. In: Plant Adaption to Mineral Stress in Problem Soils. Workshop Proceedings, Agricultural Library Beltsville, Maryland USA 1976.

Foy, C. D., Chaney, R. L. and White, M. C. The physiology of metal toxicity in plants. Annu. Rev. Plant Physiol. 29, 511-566, 1978.

Graham, R. D., Hannan, R. J. and Uren, N. C. editors. Manganese in Soils and Plants, Kluwer Academic Publishers, Dordrecht, 1988.

Gerendas, J., Polacco, J. C., Freyermuth, S. K. and Sattelmacher, B. Significance of nickel for plant growth and metabolism. J. Plant Nutr. Soil Sci. 162, 241-256, 1999.

Gissel-Nielsen, G., Gupta, U. C., Lamand, M. and Westermarck, T. Selenium in soils and plants and its importance in livestock and human nutrition. Adv . Agron. 37, 397-460, 1984.

Goldbach, H. E. A critical review on current hypothesis concerning the role of boron in higher plants: suggestions for further research and methodological requirements. J. Trace Microprobe Technique 15, 51-91, 1997.

Graham, R. D., Hannan, R. J. and Uren, N. C., editors. Manganese in Soils and Plants. Kluwer Academic Publishers, Dordrecht 1988.

Gupta, U. C. Soil and plant factors affecting molybdenum uptake by plants, p. 721 -91. In: Molybdenum in Agriculture, edited. by U. C. Gupta, Cambridge University Press, Cambridge 1997a.

Gupta, U. C. Deficient, sufficient, and toxic concentrations of molydenum in crops, p. 150-159. In: Molybdenum in Agriculture, edited by U. C. Gupta, Cambridge University Press, Cambridge 1997b.

Gupta, U. C. Symptoms of molybdenun deficiency and toxicity in crops, p. 160 -170 . In: Molybdenum in Agriculture, edited by U. C. Gupta, Cambridge University Press, Cambridge 1997c.

Horst, W. J. The role of apoplast in aluminium toxicity and resistance of higher plants: a review. Z. Pflanzenernähr. Bodenk. 158, 419-428, 1994.

Jacobs, L. Selenium in Agriculture and the Environment, ASA, Madison USA 1989.

Juste, C. and Mench, M. Long-term application of sewage sludge and its effect on metal uptake by crops p. 159-193. In: Biogeochemistry of Trace Metals, edited by Adriano, D. C. Ann Arbor, London, Tokyo: Lewis Publishers 1992.

Kinraide, T. B. Identity of the rhizotoxic aluminium species. Plant and Soil 134, 167-178, 1991.

Kochian, L. V. Cellular mechanisms of aluminium toxicity and resistance in Plants. Annu. Rev. Plant Physiol. Plant Mol. Biol. 46, 237-260, 1995.

Loneragan, J. F., Robson, A. D. and Graham, R. D. editors. Copper in Soils and Plants. Academic Press, London 1981.

Marschner, H. and Römheld, V. Root-induced changes in the availability of micronutrients in the rhizosphere, p. 557-579. In: Plant Roots, the Hidden Half, edited by Waisel, Y., Eshel, A. and Kafkafi, U. Marcel Dekker, New York 1996.

McBride, M. B. Reactions controlling heavy metal solubility in soils. Adv. Soil Sci. 10, 1-56, 1989.

McGrath, S. P., Sanders, J. R. and Shalaby, M. H. The effect of soil organic matter levels on soil solution concentrations and extractabilities of manganese, zinc and copper. Geoderma. 42, 177-188, 1988.

McLaughlin, M. J., Maier, N. A., Freeman, K., Tiller, K. G., Williams, C. M. J. and Smart, M. K. Effect of potassic and phosphatic fertilizer type, phosphate fertilizer Cd content and additions of zinc on cadmium uptake by commercial potato crops. Fert. Res. 40, 63-70, 1995.

Mikkelsen, R. L., Page, A. L. and Bingham, F. T. Factors affecting selenium accumulation by agricultural crops, p. 65-94. In: Selenium in Agriculture and the Evironment, edited by Jacobs, L. W. ASA. SSSA, Madison USA 1989.

Miner, G. S., Gutierrez, R. and King, L. D. Soil factors affecting plant concentrations of cadmium, copper, and zinc on sludge-amended soils. J. Environ. Qual. 26, 989-994, 1997.

Mordvedt, J. J. Cox, R. F. Shuman, L. M. and Welch, R. M. editors. Micronutrients in Agriculture. 2nd edition. SSSA, Book Series, Madison, USA, 1991

Nable, R. O., Banuelos, G. S. and Paull, J. Boron toxicity. Plant and Soil 193, 181-198, 1997.

Prasad, R. and Goswami, N. N. Soil fertility restoration and management for sustainable agriculture in South Asia. Adv. Soil Sci. 17, 37-77, 1992.

Römheld, V. and Marschner, H. Functions of micronutrients in plants, p. 297-328. In: Micronutrients in Agriculture, edited by Mortvedt, J. J., Cox, F. R., Shuman, L. M., Welch, R. M. SSSA, Madison, USA 1991.

Salt, D. E., Smith, R. D. and Raskin, I. Phytoremediation. Annu. Rev. Plant Physiol. Plant Mol. Biol. 49, 643-668, 1998.

Sauerbeck, D. and Lübben, S. (G) Effects of municipal disposals on soils, soil organisms and plants, p. 1-32. In: Berichte aus der ökologischen Forschung, Vol. 6, edited by Forschungszentrum Jülich, Zentralbibliotek, Jülich, Germany 1991.

Shorrocks, V. M. The occurrence and correction of boron deficiency. Plant and Soil 193, 121-148, 1997.

Weinstein, L. H. Fluoride and plant life. J. of Occupational Medicine 19, 49-78, 1977.

Terry, N, Zayed, A. M., de Souza, M. P. and Tarun, A. S. Selenium in higher plants Annu. Rev. Plant Physiol. Plant Mol. Biol. 51, 401-432, 2000.

Xu, G., Magen, H., Tarchitsky, J. and Kafkafi, U. Advances in chloride nutrion of plants. Adv. Agron. 68, 98 -150, 2000

REFERENCES

(F) = French text
(G) = German text
(R) = Russian text

Acevedo, E., Hsiao, T. C. and Henderson, D. W. Immediate and subsequent growth responses of maize leaves to changes in water status. Plant Physiol. 48, 631-636, 1971.

Ackerson, R. C., Krieg, D. R. and Sung, F. J. M. Leaf conductance and activity. J. Plant Nutr. 7, 609-621, 1980.

Adams, F. Soil solution. In: The Plant Root and its Environment, edited by Carson, E. W. University Press of Virginia p. 441-481, Charlottesville, USA 1974.

Adams, J. F. Yield response to molybdenum by field and horticultural crops. In: Molybdenum in Agriculture, ed. by U. C. Gupta, Cambridge University Press, p. 182-201, Cambridge 1997.

Adatia, M. H. and Besford, R. T. The effects of silicon on cucumber plants grown in recirculating nutrient solution. Annals of Bot. 58, 343-351, 1986.

Addiscott, T. M. Potassium and the distribution of calcium and magnesium in potato plants. J. Sci. Fd Agric. 25, 1173-1183, 1974.

Adepetu, J. A. and Akapa, L. K. Root growth and nutrient uptake characteristics of some cowpea varieties. Agron J. 69, 940-943, 1977.

Adjei-Twum, D. C. and Splittstoesser, W. E. The effect of soil water regimes on leaf water potential, growth and development of soybeans. Physiol. Plant. 38, 131-137, 1976.

Ae, N., Arihara, J., Okada, K., Yoshihara, T. and Johansen, C. Uptake mechanism of iron-associated phosphorus in pigeon pea growing on Indian Alfisols and its significance to phosphorus availability in cropping systems. In: Transactions 14. International Congr. Soil Science, ed. by Masayoshi Koshino, Tokyo: International Soil Science Society., p. 164-169, 1990.

Agarwala, S. C., Chatterjee, C., Sharma, P. N. and Nautiyal, N. Pollen development in maize plants subjected to molybdenum deficiency. Can. J. Bot. 57, 1946-1950, 1979.

Agble, W. K. Agronomic practices under favourable rain-fed conditions. In: Proc. 1. FAO/SIDA Seminar for Plant Scientists from Africa and Near East, Cairo, 1974.

Aguilars, A. and van Diest, A. Rock phosphate mobilization induced by the alkaline uptake pattern of legumes utilizing symbiotically fixed nitrogen. Plant and Soil 61, 27-42, 1981.

Ahmad, N. and Wilson, H. W. Acid sulfate soils of the Carribean region - their occurrence, reclamation, and use. Soil Sci. 153, 154-164, 1992.

Ahmed, S. and Evans, H. J. Cobalt: a micronutrient element for the growth of soybean plants under symbiotic conditions. Soil Sci. 90, 205-210, 1960.

Ahsen, von U. and Noller, H. F. Identification of bases in the 16S rRNA essential for tRNA binding at the 30S ribosomal P site. Science. 267, 234-237, 1995.

Ahti, E. Correcting stem girth measures for variations induced by soil moisture changes. Communicationes Instituti Forestalis Fenniae 78. 4, Helsinki, 1973.

Aktar, M. S. Potassium availability as affected by soil mineralogy. In: Potasssium in Agriculture, ed. by

Mengel, K. Basel: International Potash Institute, p. 1994.

Aktas, M. and van Egmond, F. Effect of nitrate nutrition on iron utilization by an Fe-efficient and an Fe-inefficient soybean cultivar. Plant and Soil 51, 257-274, 1979.

Alazard, D. and Duhoux, E. Nitrogen fixing stem nodules on *Aeschynomene afraspera*. Biol. Fertil. Soils 4, 61-66, 1987.

Alberda, T. Crop photosynthesis: methods and compilation of data obtained with a mobile field equipment. 3. Perennial ryegrass. Agr. Res. Rep. 865, Centre for Agricultural Publishing and Documentation, Wageningen p. 4-11, 1977.

Albert, L. S. Induction and antagonism of boron-like deficiency symptoms of tomato plants by selected nitrogen-bases. Plant Physiol. 43, S-51, 1968.

Aldhous, P. Tropical deforestation: Not just a problem in Amazonia. Science. 259, 1390, 1993.

Allaway, W. G. and Ashford, A. E. Structure of hair roots in *Lysinema ciliatum* R. Br. and its implications for their water relations. Ann. Bot. 77, 383-388, 1996.

Allaway. W. H.: Trace element cycling. Adv. Agron. 20, 235-274 1968.

Allen, B. L. and Hajek, B. F. Mineral occurrence in soil environments. In: Minerals in Soil Environments. 2. ed, ed. by Dixon, J. B. and Weed, S. B., SSSA America, p. 199-278, Madison, USA 1989.

Allen, G. J., Wyn Jones, R. L. and Leigh, R. A. Sodium transport measured in plasma membrane particles isolated from wheat genotypes with differing K^+/Na^+ discrimination. Plant Cell Environ. 18, 105-115, 1995.

Allen, S. E., Terman, G. L. and Hunt, C. M. Soluble and slow-release nitrogen fertilizer effects on grass forage, as influenced by rate and placement. J. agric. Sci. 77, 397-404, 1971.

Altherr, L.: Organic farming on trial. Natural History 81, 16-24, 1972.

Altherr, E. and Evers, F. H. (G) Effects of magnesia when fertilizing a spruce stand on bunter sandstone in the Odenwald. Allg. Forst- u. Jagdzeitung 146, 217-224, 1975.

Amberger, A. (G) The role of manganese in the metabolism of plants. Agrochimica 17 69-83, 1973.

Amberger, A. and Aigner, H. (G) Experimental results of a straw application trial lasting eight years. Z. Acker- u. Pflanzenbau 130, 291-303, 1969.

Amberger, A. and Schweiger, P. (G) Effect of straw incorporation combined with an application of Ca cyanamide in long-term field trials. Z. Acker- u. Pflanzenb. 134, 323-334, 1971.

Amberger, A., Vilsmeier, K. and Gutser, R. (G) Nitrogen fractions in various types of slurries and their effects in plant trials. Z. Pflanzenernähr. Bodenk. 145, 325-336, 1982.

Ambus, P. and Christensen, S. Spatial and seasonal nitrous oxide and methane fluxes in Danish forest-grassland- and agroecosystems. J. Environ. Qual. 24, 993-1001, 1995.

Amin, J. V. and Joham, H. E. The cations of the cotton plant in sodium substituted potassium deficiency. Soil Sci. 105, 248-254, 1968.

Anac, D. and Colakoglu, H. Response of some major crops to K fertilization. In: K Availability of Soils in West Asia and North Africa -Status and Perspectives, ed. by Mengel, K. and Krauss, A., International Potash Institute, p. 235-247, Basel 1995.

Anac, D., Okur, B., Kilic, C., Aksoy, U., Can, Z., Hepaksoy, S., Anac, S., Ul, M. A. and Dorsan, F. Potassium fertilization to control salinization effects. In: Food Security in the WANA region, the essential need for balanced fertilization, ed. by Johnston, A. E., International Potash Institute, p. 370-377, Basel 1997.

Anderson, D. L., Kussow, W. R. and Corey, R. B. Phosphate rock dissolution in soil: Indications from plant growth studies. Soil Sci. Soc. Am. J. 49, 918-925, 1985.

Anderson, G. D. Potassium response of various crops in East Africa, in Potassium in Tropical Crops and Soils, p. 287-309, Proc. 10th Coll. Int. Potash Inst., Bern, 1973.

Andrews, T. J. and Hatch, M. D. Activities and properties or ribulose diphosphate carboxylase from plants with the C_4 dicarboxylic pathway of photosynthesis. Phytochernistry 10, 9-15, 1971.

Anghioni, I. and Barber, S. A. Phosphorus influx and growth charateristics of corn roots as influenced by phosphorus supply. Agron. J. 72, 685-688, 1980.

Anonymus, Y. Asiatic people and alcohol. Naturw. Rundschau 27, 280-281, 1974.

Antonovics, J., Bradshaw, A. D. and Turner, R. G. Heavy metal tolerance in plants. Adv. in Ecol. Res. 7, 1-85, 1971.

Appel, T. Non-biomass soil organic N - the substrate for N mineralization flushes following soil drying -rewetting and for organic N rendered $CaCl_2$- extractable upon soil drying. Soil. Biol. Biochem. 30, 1445-1456, 1998.

Appel, T. (G) The function of extractable organic N fractions in soils and their relation to a computersimulation of nitrogen turnover. Habilitation Thesis, Justus-Liebig-University. Giessen, 1998.

Appel, T. and Mengel, K. Importance of organic nitrogen fractions in sandy soils, obtained by electro-ultrafiltration or $CaCl_2$ extraction, for nitrogen mineralization and nitrogen uptake of rape. Biol. Fertil. Soils 10, 97-101, 1990.

Appel, T. and Mengel, K. Nitrogen uptake of cereals on sandy soils as related to nitrogen fertilizer application and soil nitrogen fractions obtained by Electro-Ultrafiltration (EUF) and $CaCl_2$ extraction. Eur. J. Agron. 1, 1-9, 1992.

Appel, T. and Mengel, K. Prediction of mineralizable nitrogen in soils on the basis of an analysis of extractable organic N. Z. Pflanzenernähr. Bodenk. 161, 433-452, 1998.

Appel, T. and Mengel, K. Nitrogen fractions in sandy soils in relation to plant nitrogen uptake and organic matter incorporation. Soil Biol. Biochem 25, 685-691, 1993.

Appel, T. and Steffens, S. (G) Comparison of electroultafiltration (EUF) and extraction with 0,01 molar $CaCl_2$ solution for the determination of plant availble nitrogen in soils. Z. Pflanzenernaehr. Bodenk. 151, 127-130, 1988.

Appelquist, L. A. Lipids in Cruciferae. II. Fatty acid composition of *Brassica napus* seed as affected by nitrogen, phosphorus, potassium and sulfur nutrition of the plants. Physiol. Plant. 21, 455-465, 1968.

Apse, M. P., Aharon, G. S., Snedden, W. A. and Blumwald, E. Salt tolerance conferred by overexpression of a vacuolar Na^+/H^+ antiport in *Arabidopsis*. Science 285, 1256-1258, 1999.

Archibold, O. W. Ecology of World Vegetation, Chapman and Hall, London 1995.

Arissian, M., Perrissin-Fabert, D., Blouet, A., Morel, J. L. and Guckert, A. Effect of imazaquin on the absorption, translocation, and pattern of distribution of chloromequat chloride in winter wheat. J. Plant Growth Regul. 10, 1-4, 1991.

Arisz, W. H. Significance of the symplsm theory for transport across the roots. Protoplasma 46, 1-62, 1956.

Armstrong, M. J. and Kirkby, E. A. The influence of humidity on the mineral composition of tomato plants with special reference to calcium distribution Plant and Soil 52, 427-435, 1979.

Arnon, D. I. Photosynthesis 1950-1975: Changing concepts and perspectives. In: Photosynthesis I, Plant Physiol. New Series, Vol. 5, ed. by Trebst, A. and Avron, M. p. 7-56, Springer-Verlag, Berlin, New York 1977.

Arnon, D. I. and Stout, P. R. The essentiality of certain elements in minute quantity for plants with special reference to copper. Plant Physiol. 14, 371-375, 1939.

Arnon, D. I. and Wessel, G. Vanadium as an essential element for green plants. Nature 172, 1039-1040, 1953.

Asada, K. Radical production and scavenging in the chloroplasts. In: Photosynthesis and the environment, ed. by Baker, N. R. Kluwer Academic Publisher, p 123-150, Dordrecht1996.

Asche, N. (G) Deposition, interception and leaching in the canopy of an oak/hornbeam stand. Z. Pflanzenernähr.

Bodenk. 151, 103-107, 1988.

Asher, C. J. Beneficial elements, functional nutrients, and possible new essential elements. In: Micronutrients in Agriculture, ed. by Mordtvedt, J. J., Cox, F. R., Shuman, L. M. and Welch, R. M. SSSA, p. 703-723, Madison, USA 1991.

Asher, C. J. and Edwards, D. G. Modern solution culture techniques. In: Inorganic Plant Nutrition, Encycl. of Plant Physiology, New Series Vol. 15. A.,ed. by A. Läuchli and R. L. Bieleski, p. 94 - 119. Springer-Verlag, Berlin, New York 1983.

Ashworth, E. N., Christiansen, M. N., John, J. B. S, and Patterson, G. W. Effect of temperature and BASF 13 338 on the lipid composition and respiration of wheat roots. Plant Physiol. 67, 711-715, 1981.

Aslam, M. and Huffaker, R. C. Dependence of nitrate reduction on soluble carbohydrates in primary leaves of barley under aerobic conditions. Plant Physiol. 75, 623-628, 1984.

Aslam, M., Travis, R. L. and Huffaker, R. C. Comparative kinetics and reciprocal inhibition of nitrate and nitrite uptake in roots of uninduced and induced barley (*Hordeum vulgare* L.) seedlings. Plant Physiol. 99, 1124-1133, 1992.

Asmus, F., Kittelmann, G. and Görlitz, H. (G) effect of long-term manure application on physical properties of a loamy soil. Arch. Acker-Pflanzenbau Bodenkd. 31, 41-46, 1987.

Assmann, S. M., Simoncini, L. and Schroeder, J. L. Blue light activates electrogenic ion pumping in guard cell protoplasts of *Vicia faba*. Nature 318: 285-287, 1985.

Auerswald, H., Schwarz, D., Kornelson, C., Krumbein, A. and Brückner, B. Sensory analysis, sugar and acid content of tomato at different EC values of the nutrient solution. Sci. Hortic. 82: 227-242, 1999.

Aufhammer, W., Kübler, E. and Bury, M. (G) Nitrogen uptake and nitrogen residuals of a winter oil-seed rape and fallout rapE. J. Agron. and Crop Sci. 172, 255-264, 1994.

Aufhammer, W. and Solansky, S. (G) Influence of kinetin application on photosynthate storage in the ears of spring barley by kinetin application. Z. Pflanzenern. Bodenk., 4, 503-515, 1976.

Aulakh, M. S., Doran, J. W. and Moisier, A. R. Soil denitrification - significance, measurement, and effects on management. Adv. Soil Sci. 18, 17-57, 1992.

Austin, R. B., Ford, M. A., Morgan, C. L. and Yeoman, D. Old and modern wheat cultivars compared on the Broadbalk wheat experiment. Eur. J. Agron. 2, 141-147, 1993.

Ayala, F., O'Leary, J. W. and Schumaker, K. S. Increased vacuolar and plasma membrane H^+-ATPase activities in *Salicornia bigelovii* Torr in response to NaCl. J. Exp. Bot. 47, 25-32, 1996.

Ayres, A. S. Calcium silicate slag as a growth stimulant for sugarcane on low-silicon soils. Soil Sci. 101, 216-227, 1966.

Bache, D. T. Soil acidification and aluminium mobility. Soil Use and Management 1, 10-14, 1985.

Badger, M. R. and Price, G. D. The role of carbonic anhydrase in photosynthesis. Annu. Rev. Plant Physiol. Plant Mol. Biol. 45, 369-392, 1994.

Baechle, H. T. and Wolstein, F. Cadmium compounds in mineral fertilizers. The Fertilizer Society, Proc. No 226, London 1984.

Baize, D. (F) Total contents of heavy metals in soils. Paris, INRA 1997.

Baker, D. A. and Weatherley, P. E. Water and solute transport by exuding root systems of *Ricinus communis*. J. exp. Bot. 20, 485-496, 1969.

Balcar, J., Brezinova-Doskarova, A. and Eder, J. Dependence of radiostrontium uptake by pea and lupin on the content of calcium in the nutrient solution. Biol. Plant. 11, 34-40, 1969.

Baldocchi, D. D. Deposition of gaseous sulfur compounds to vegetation. In: Sulfur Nutrition and Assimilation in Higher Plants, ed. by De Kok, L-J., Stulen, I., Rennenberg, H., Brunold, C. and Rauser, W. E. SPB Academic Publishing bv. The Hague, 1993.

Balfour, E. The Living Soil and the Haughley Experiment. Faber and Faber. London 1975

Ballif, J. L. (F) Loss of magnesium by leaching from chalk soil: eleven years of lysimeter measurements, 1974-1985. In: Le magnesium en agriculture, ed. by Huguet, C. and Coppenet, M. INRA editions, p. 86-92, Paris 1992.

Ballif, J. L. and Muller, J. C. (F) Contribution to investigation of the sulphate dynamics in a chalk soil: ten years of lysimeter measurememts -1974-1984. C. R. Acad. Agric. 71, 1385-1397, 1985.

Bange, G. G. J. On the quantitative explanation of stomatal transpiration. Acta Bot. Neerl. 2, 255-296, 1953.

Bangerth, F. Calcium-related physiological disorders of plants. Ann. Rev. Phytopathol. 17, 97-122, 1979.

Bangerth, F., Dilley, D. R. and Dewey, D. H. Effect of postharvest calcium treatment on internal breakdown and respiration of apple fruit J. Ame. Soc. Hort. Sci. 97, 679-682, 1972.

Bar-Akiva, A. and Lavon, R. Visible symptoms of some metabolic patterns in micronutrient-deficient Eureka lemon leaves. Israel J. Agr. Res. 17, 7-16, 1967.

Bar-Yosef, B. Advances in fertigation. Adv. Agron. 65, 1-77, 1999.

Barber, D. A. and Martin, J. K. The release of organic substances by cereal roots in soil. New Phytol. 76, 569-578, 1966.

Barber, D. A. and Shone, M. G. T. The absorption of silica from aqueous solutions by plants. J. exp. Bot. 17, 569-578, 1966.

Barber, J. Influence of surface charges on thylakoid structure and function. Annu. Rev. Plant Physiol. 33, 261-295, 1982.

Barber, J. Light induced uptake of potassium and chloride by *Chlorella pyrenoidosa*. Nature 217, 876-878, 1968.

Barber, S. A. A diffusion and mass flow concept of soil nutrient availability. Soil Sci. 93, 39-49, 1962.

Barber, S. A. Mechanism of potassium absorption by plants. In: The Role of Potassium in Agriculture, ed. by Kilmer et al. p. 293 -310, Madison, USA1968.

Barber, S. A. Influence of the plant root on ion movement in the soil. In: The Plant Root and its Environment, ed. by Carson, E. W., University Press of Virginia p. 525-564, Charlottesville 1974.

Barber, S. A. Growth requirements of nutrients in relation to demand at the root surface. In: The Soil-Root Interface, ed. by Harley, J. L. and Scott Russell, R. Academic Press p. 5-20, London, New York 1979.

Barber, S. A. Soil Nutrient Bioavailability. A Mechanistic Approach. John Wiley, New York. 1984.

Barber, S. A. Soil Nutrient Bioavailability. A Mechanistic Approach, 2. Ed. John Wiley, New York, 1995.

Barber, S. A., Walker, J. M. and Vasey, E. H. Mechanism for the movement of plant nutrients from the soil and fertilizer to the plant root. J. Agr. Food Chem. 11, 204-207, 1963.

Barea, J. M. and Azcon-Aguilar, C. Mycorrhizas and their significance in nodulating nitrogen-fixing plants. Adv. Agron. 36, 1-54, 1983.

Barekzai, A. (G) Problems with slurry application in agriculture. Naturwiss. 79, 457-461, 1992.

Barekzai, A., Becker, A. and Braschkat, J. (G) Effects of an increased slurry application on the N uptake and the N balance of maize and on N fractions in the soil (NO_3, NH_4 und N_{org}-N). Arch. Acker-Pflanzenbau Bodenkd. 37,. 341-353, 1993.

Barekzai, A., Becker, A. and Mengel, K. (G) Effect of slurry application on the EUF-N-fractions in an arable and a grassland soils. VDLUFA-Schriftenr. 28, Congr. Vol. Part II, p. 27-42, 1988.

Barekzai, A. and Mengel, K. (G) Aging of phosphate fertilizers in different soil types. Z. Pflanzenernaehr. Bodenkd. 148, 365-378, 1985.

Barekzai, A. and Mengel, K. Effect of microbial decomposition of mature leaves on soil pH. Z. Pflanzenernaehr. Bodenk. 156, 93-94, 1993.

Barekzai, A., Steffens, D., Bohring, J. and Engels, T. (G) Principle and evaluation of the „Giessener-Model"

700

for N fertilizer recommendations to winter cereals by means of the EUF method. Agribiol. Res. 45, 65-76, 1992.

Baringa, M. A new face for the glutamate receptor. Science. 267, 177-178, 1995.

Barker, A. V., Maynard, D. N., Mioduchowska, B. and Buch, A. Ammonium and salt inhibition of some physiological processes associated with seed germination. Physiol. Plant. 23, 898-907, 1970.

Barkla, B. J., Charuk, J. H. M., Cragoe, E. J. and Blumwald, E. Photolabeling of tonoplast from sugar beet cell suspensions by 3-(H)-5-(N-methyl-N-isobutyl)-amiloride, an inhibitor of the vacuolar Na^+/H^+ antiport. Plant Physiol. 93: 924-930, 1990.

Barkla, B. J. and Pantoja, O. Physiology of ion transport across the tonoplast of higher plants. Annu. Rev. Plant Physiol. Plant Mol. Biol. 47, 159-184 1996.

Barnhisel, R. I. and Bertsch, P. M. Chlorites and hydroxy- interlayered vermiculite and smectites. In: Minerals in Soil Environments, ed. by Dixon, J. B. and Weed, S. B. SSSA, p. 729-788, Madison, USA 1989.

Barraclough, P. B. The growth and activity of winter wheat roots in the field: Nutrient inflow of high yielding crops. J. Agric. Sci. 106, 53-59, 1986.

Barraclough, P. B. and Leigh, R. A. Critical plant K concentrations for growth and problems in the diagnosis of nutrient deficiencies by plant analysis. Plant and Soil 155/156, 219-222, 1993.

Barrow, N. J. Comparison of the adsorption of molybdate, sulfate and phosphate by soils. Soil Sci. 109, 282-288, 1970.

Barrows, H. L., Taylor, A. W. and Simpson, E. C. Interaction of limestone particle size and phosphorus on the control of soil acidity. Proc. Soil Sci. Soc. Amer. 32, 64-68, 1968.

Bartlett, R. J. and Simpson, T. J. Interaction of ammonium and potassium in a potassium-fixing soil. Proc. Soil Sci. Soc. Amer. 31, 219-222, 1967.

Bassham, J. A. and Calvin, M. The path of carbon in photosynthesis. Prentice Hall Inc., Englewood Cliffs, NX. p. 104, 1957.

Basu, P. S., Sharma, A., Garg, I. D. and Sukumaran, N. P. Tuber sink modifies photosynthetic response in potato under water stress. Environ. Expt. Bot. in press, 1999.

Baszynski, T., Warcholowa, M., Krupa, Z., Tukendorf, A., Krol, M. and Wolinska, D. The effect of magnesium deficiency on photochemical activities of rape and buckwheat chloroplasts. Z. Pflanzenphysiol. 99, 295-303, 1978.

Batey, T.: Manganese and boron deficiency. In: Trace Elements in Soils and Crops. Techn. Bull. 21, Ministry of Agric., Fisheries and Food, UK,, p. 137-149, 1971.

Batts, G. R., Morison, J. I. L., Ellis, R. H., Hadley, P. and Wheeler, T. R. Effects of CO_2 and temperature on growth and yield of crops of winter wheat over four seasons. In: Proceedings. ESA Congress, Veldhoven 1996, edited by van Ittersum, M. K. and van de Geijn, S. C., Elsevier, p. 67-76, Amsterdam 1997.

Baule, H. (G) Relationships between the nutrient content and diseases in forest trees. Landw. Forsch. 2311. Sonderh., 92-104, 1969.

Baule, H. and Fricker, C. The fertiliser treatment of forest trees. BLV-Verlagsges. München 1970.

Baumgärtel, G., Engels, Th. and Kuhlmann, H. (G) How can the appropriate N fertilizer application be verified? DLG-Mitteil. 9, 472-474, 1989.

Bayzer, H. and Mayr, H. H. (G) Amino acid composition of rye grain proteins influenced by nitrogen application and chlorine-choline-chloride. Z. Lebensmittel-Untersuchung u.-Forsch. 133, 215-217, 1967.

Becana, M. and Rodriguez-Barrueco, C. Protective mechanisms of nitrogenase against oxygen excess and partially-reduced oxygen intermediates. Physiol. Plant. 75, 429-438, 1989.

Beck, E. The degradation of transitory starch granules in chloroplasts. In: Regulation of carbon partitioning in photosynthetic tissue. ed. by Heath, R. and Preiss, J., American Society Plant Physiologists, Rockville,

p. 27-44, 1993.

Beck, E. and Ziegler, P. Biosynthesis and degradation of starch in higher plants. Annu. Rev. Plant Physiol. Plant Mol. Biol. 40: 95-117, 1989.

Beck, Th. (G) Mineralization of soil nitrogen in laboratory incubation experiments. Z. Pflanzenernaehr. Bodenkd. 146, 243-252, 1983.

Becker, M., Ali, M., Ladha, J. K. and Ottow, J. C. G. Agronomic and economic evaluation of *Sesbania rostrata* manure estabishment in irrigated rice. Field Crop Res. 40, 135-141, 1995.

Becker, R., Grün, M. and Scholz, G. Nicotianamine and the distribution of iron in the apoplasm and symplasm of tomato (*Lycopersicon esculentum* Mill.). Planta 187, 48-52, 1992.

Becking, J. H. A requirement of molybdenum for the symbiotic nitrogen fixation in alder. Plant and Soil 15, 217-227, 1961.

Beevers, L. and Hageman, R. H. Uptake and reduction of nitrate: bacteria and higher plants. In: Inorganic Plant Nutrition Encycl. Plant Physiol. New Series Vol. 15A, ed. by Läuchli, A. and Bieleski, R. L. Springer Verlag p. 351-375, Berlin, NewYork 1983.

Bell, A. W. Diagnosis and prediction of boron deficiency for plant production. Plant and Soil 193, 149-168, 1997.

Bell, C. I., Clarkson, D. T. and Cram, W. J. Sulphate supply and its regulation of transport in roots of tropical legume *Macroptilium artopurpureum* cv. Siratro. J. Exp. Bot. 46, 65-71, 1995.

Bell, O. T., Koeppe, D. E. and Miller, R. J. The effects of drought stress on respiration of isolated corn mitochondria. Plant Physiol. 48, 413-415, 1971.

Ben-Zioni, A., Vaadia, Y. and Lips, S. H. Nitrate uptake by roots as regulated by nitrate reduction products of the shoot. Physiol. Plant. 24: 288-290, 1971.

Benckiser, G., Haider, K. and Sauerbeck, D. Field measurements of gaseous nitrogen losses from an Alfisol planted with sugar-beets. Z. Pflanzenernähr. Bodenk. 149, 249-261, 1986.

Bender, J., Herstein, U. and Black, C. Growth and yield responses of spring wheat to increasing carbon dioxide, ozone and physiological stresses: a statistical analysis of IIESPACE-wheat program results. European J. Agron. 10, 185-195, 1999.

Bennett, A. C.: Toxic effects of aqueous ammonia, copper, zinc, lead, boron, and manganese on root growth. In: The Plant Root and its Environment, ed. by Carson, E. W. University of Virginia, Charlottesville p. 669-683,1974.

Bennett, A. C. and Adams, F. Concentration of NH_3 (aq) required for incipient NH_3 toxicity to seedlings. Soil Sci. Soc. Amer. Proc. 34, 259-263, 1970.

Bennet, R. J. and Breen, C. M. The recovery of root cells of *Zea mays* L. from various aluminium treatments: Towards elucidating the regulatory processes that underly root growth control. Environ. Experiment. Bot. 31, 153 -163, 1991.

Bennie, A. T. P. Growth and mechanical impedance. In: Plant Roots, the Hidden Half, edited by Waisel, Y., Eshel, A. and Kafkafi, U. New York: Marcel Dekker, p. 313-414, 1996.

Berczi, A., Larsson, C., Widell, S. and Moller, I. M. On the presence of inside-out plasma membrane vesicles and vanadate-inhibited K^+,Mg^{2+}- ATPase in microsomal fractions from wheat and maize roots. Physiol. Plant. 77, 12-19, 1989.

Berendes, R., Voges, D., Demange, P., Huber, R. and Burger, A. Structure -function analysis of the ion channel selectivity filter in human annexin V. Science. 262, 427-430, 1993.

Berg, P. and Rosswall, T. Ammonium oxidizer numbers, potential and actual oxidation rates in two Swedish arable soils. Biol. Fertil. Soils 1, 131 -140, 1985.

Berg, W. A., Hodges, M. E. and Krenzer, E. G. Iron deficiency in wheat grown on the southern plains.

J. Plant Nutr. 16, 1241-1248, 1993.

Bergmann, W.: (G) Recommendations for establishing fertilizer application schemes for large areas of industrialized plant production according to soil tests. VEB Chemiehandel, 113 Berlin 1969.

Bergmann, W. (G) Nutritional Problems with Crops, Development and Diagnosis. VEB Fischer Verlag, Jena, 1983

Bergmann, W. Nutritional Disorder of Plants, Development Visual and Analytical Diagnosis, Jena:Gustav Fischer, 1992.

Bergmann, W. (G) Coloured Atlas of Nutritional Disorders of Crop Plants — Visual and Analytical Diagnosis. VEB G. Fisher Verlag Jena 1986.

Bergmann, W. (G) Nutrient Deficiencies in Crops, Jena:Gustav Fischer Verlag, 1993.

Beringer, H. (G) Uptake and effect of the micronutrient copper applied in ionic and chelated form to barley. Z. Pflanzenernähr. Düng. Bodenk. 100, 22-34, 1963.

Beringer, H. Influence of temperature and seed ripening on the in $vivo$ incorporation of $^{14}CO_2$ into the lipids of oat grains ($Avena$ $sativa$ L.). Plant Physiol. 48, 433-436, 1971.

Beringer, H., Haeder, H. E. and Lindauer, M. Water relationships and incorporation of C-14 assimilates in tubers of potato plants differing in potassium nutrition. Plant Physiol. 73, 956-960, 1983.

Beringer, H. and Saxena, N. P.: Effect of temperature on the content of tocopherol in seed oils. Z. Pflanzenernähr. Bodenk. 120, 71-78, 1968.

Berkowitz, G. A. and Peters, J. S. Chloroplast inner-envelope ATPase acts as a primary H^+ pump. Plant Physiol. 102:261-267, 1993.

Bernstein, L. Salt tolerance of plants. Agric. Inform. Bull. No. 283, 1970.

Bernstein, L. and Hayward, A. E. Physiology of salt tolerance. Annu. Rev. Plant Physiol. 9, 25-46, 1958.

Berridge, M. J. The AM and FM of calcium signaling. Nature 386, 759-760, 1997.

Bertholdsson, N. O. Characterization of malting barley cultivars with more or less stable grain protein content under varying environmental conditions. European J. Agron. 10, 1-8, 1999.

Bertoni, G. M., Pissaloux, A., Morard, P. and Sayag, D. R. Bicarbonate - pH relationship with iron chlorosis in white lupins. J. Plant Nutr. 15, 1509-1518, 1992.

Bhadoria, P. B. S., Kaselowsky, J., Claasen, N. and Jungk, A. Phosphate diffusion coefficients in soil as affected by bulk density and water content. Z. Pflanzenernaehr. Düng. Bodenk. 154, 53-57, 1991

Bhat, K. K. S. and Nye, P. H. Diffusion of phosphate to plant roots in soil. II. Uptake along the roots at different times and the effect of different levels of phosphorus. Plant and Soil 41, 365-382, 1974.

Bhuija, Z. H. and Walker, N. Autotrophic nitrifying bacteria in acid tea soils from Bangladesh and Sri LankA. J. appl. Bact. 42, 253-257, 1977.

Biddulph, O., Nakayama, F. S. and Cory, R. Transpiration stream and ascension of calcium. Plant Physiol. 36, 429-436, 1961.

Biederbeck, V. O., Campbell, C. A., Bowren, K. E. and McIver, R. N. Effect of burning cereal straw on soil properties and grain yields in Saskatchewan. Soil Sci. Soc. Am. J. 44, 103-111, 1980.

Bienfait, H. F., Briel, W. van den and Mesland-Mul, N. T. Free space iron pools in roots, generation and mobilization. Plant Physiol. 78, 596-600, 1985.

Bingham, F. T. Relation between phosphorus and micronutrients in plants. Soil Sci. Soc. Amer. Proc. 27, 389-391, 1963.

Binns, A. N. Cytokinin accumulation and action: biochemical, genetic and molecular appraoches. Annu. Rev. Plant Mol. Biol. 45, 173-196, 1994.

Birnbaum, E. H., Dugger, W. M. and Beasley, B. C. A. Interaction of boron with components of nucleic acid metabolism in cotton ovules cultured in $vitro$. Plant Physiol. 59, 1034-1038, 1977.

Bisson, L. F. Influence of Nitrogen on Yeast Fermentation of Grapes. In: International Symposium on nitrogen in grapes and wine, edited by Rantz, J. M. and Lewis, K. L. Davis,USA. The Am. Soc. for Enology and Viticulture., p. 78-89, 1991.

Blackmer, A. M., Pottker, D., Cerrato, M. E. and Webb, J. Correlations between soil nitrate concentrations in late spring and corn yields in Iowa. J. Prod. Agric. 2, 103-109, 1989.

Blaise, D., Amberger, A. and Tucher, v. S. Influence of iron pyrites and dicyandiamide on nitrification and ammonia volatilization from urea applied to loess brown earths (Luvisol). Biol. Fertil. Soils 24, 179-182, 1997.

Blamey, F. P. C., Asher, C. J. and Edwards, D. G. Boron deficiency in sunflower. In: Boron in Soils and Plants, edited by Bell, R. W. and Rerkasem, B. Dordrecht, Kluwer Academic Publisher, p. 145-149, 1997.

Blaser-Grill, D., Amberger, A. and Goldbach H. E. Influence of boron on the membrane potential in *Elodea densa* and *Helianthus annuus* roots and H+ extrusion of suspension cultured *Daucus carota* cells. Plant Physiol. 90, 280-284, 1989.

Blatt, M. R. and Armstrong, F. Potassium channels of stomatal guard cells: Abscisic acid-evoked control of the outward rectifier mediated by cytoplasmic pH. Planta 191, 330-341, 1993.

Blatt, M. R. and Thiel, G. Hormonal control of ion channel gating. Annu. Rev. Plant Physiol. Plant Mol. Biol. 44, 453-467, 1993.

Blevins, D. G. An overview of nitrogen metabolism in higher plants In: Plant Nitrogen Metabolism, edited by Poulton, J. E., Romeo, J. T. and Conn, E. C. New York, Plenum, p. 1-41, 1989.

Blevins, D. G. and Lukaszewski, K. M. Boron in plant structure and function Annu. Rev. Plant Physiol. Mol. Biol. 49, 481-500, 1998.

Blevins, R. L., Frye, W. W., Baldwin, P. L. and Robertson, S. D. Tillage effects on sediment and soluble nutrient losses. J. Environ. Qual. 19, 683-686, 1990.

Blizzard, W. E. and Boyer, J. S. Comparative resistance of the soil and the plant to water transport. Plant Physiol. 66, 809-814, 1980.

Bloem, E. M. (G) Sulphur balance of agroecosystems with particular regard to hydrolic and soil -physical properties of sites. Wiss. Mitteil. FAL Braunschweig:ISBN 3-933140-13-7, 1998.

Blondel, A. and Blanc, D. (F) Influence of ammonium ion uptake and reduction in young wheat plants. C. R. Acad. Sci. (Paris) Ser. D, 277, 1325-1327, 1973.

Bloodworth, J. E., Page, J. B. and Cowley, W. R. Some applications of the thermoelectric method for measuring water flow in plants. Agron. J. 48, 222-228, 1956.

Blumwald, E. and Poole, R. J. Salt tolerance in suspension cultures of sugar beet. Induction of Na+/H+ antiport activity at the tonoplast by growth in salt. Plant Physiol. 87, 104-108, 1987.

Boardman, N. K. Comparative photosynthesis of sun and shade plants. Annu. Rev. Plant Physiol. 28, 355-377, 1977.

Boddey, R. M., Urquiaga, S., Reis, V. and Döbereiner, J. Biological nitrogen fixation associated with sugar cane. Plant and Soil 137, 11-117, 1991.

Boeckx, P. and van Cleemput, O. Methane oxidation in a neutral landfill cover soil:influence of moisture content temperature and nitrogen-turnover. J. Environ. Qual. 25, 178-183, 1996.

Boehle, J, and Lindsay, W. L. Micronutrients. The Fertilizer Shoe-Nails. Pt. 6. In the Limelight-Zinc, Fertilizer Solutions 13 (1), 6-12, 1969.

Bogoslavsky, L. and Neumann, P. M. Rapid regulation by acid pH of cell wall adjustment and leaf growth in maize plants responding to reversal of water stress. Plant Physiol. 118, 701-709, 1998.

Boguslawski, E. von. (G) Law of yield formation. In: W. Ruhland: Encycl. of Plant Physiol. Vol 4, Springer

Verlag Berlin, Göttingen, Heidelberg, p. 943-976, 1958.

Boguslawski, E. von and Lach G. (G) The K release of soils measured by plant uptake in comparison with the exchangeable potassium. Z. Acker- u. Pflanzenbau 134, 135-164, 1971.

Boguslawski, E. von and Schildbach, R. (G) Effects of sites, years, fertilizer application and irrigation on quality and yield level of sugar beets. Zucker 22, 123-132, 1969.

Boguslawski, E. von and Schneider, B. (G) Third approximation of the yield law. 3. Communication. Z. Acker-u. Pflanzenb. 119, 1-28, 1964.

Böhm, W. (G) The assessment of the root system under field conditions. Kali-Briefe (Büntehof) 14 (2), 91-101, 1978.

Böhm, W. (G) The fixation of dinitrogen by root nodules of legumes. In memory of Hermann Hellriegel's epoch-making discovery in 1886. Angew Botanik 60, 1-5, 1986.

Bohnsack, C. W. and Albert, L. S. Early effects of boron deficiency on indoleacetic acid oxidase levels of squash root tips. Plant Physiol. 59, 1047 -1050, 1977.

Bolan, J. S., Syers, J. K., Tillman, W. R. and Scotter, D. R. Effect of liming and phosphate additions on sulphate leaching in soils. J. Soil Sci. 39, 493-504, 1988.

Bole, J. B. Influence of root hairs in supplying soil phosphorus to wheat. Can. J. Soil Sci. 53, 196-175, 1973

Bolland, M. D. A. Effectiveness of Ecophos compared with single and coastal superphosphate. Fert. Res. 45, 37-49, 1996.

Bolland, M. D. A., Gilkes, R. J. and Allen, D. G. The residual value of superphosphate and rock phosphate for lateric soils and its evaluation using three soil phosphate tests. Fert. Res. 15, 253-280, 1988.

Bolland, M. D. A. and Gilkes, R. J. Rock phosphates are not effective fertilizers in Western Australia soils: a review. Fert. Res. 22, 79-85, 1990.

Bollard, E. G. Transport in the xylem. Annu. Rev. Plant Physiol. 11, 141-166, 1960.

Bollard, E. G. Involvement of unusual elements in plant growth and nutrition. In: Inorganic Plant Nutrition, Encycl. Plant Physiol. New Series Vol 15B, edited by Läuchli, A. and Bieleski, R. L. Springer Verlag Berlin, Heidelberg, New York, Tokyo, p. 695-744, 1983.

Bollons, H. M. and Barraclough, P. B. Inorganic orthophosphate for diagnosing the phosphorus status of wheat plants. J. Plant. Nutr. 20, 641-655, 1997.

Bolt, G. H. Ion adsorption by clays. Soil Sci. 79, 267-276, 1955.

Bonifacio, E. and Barberis, E. Phosphorus dynamics during pedogenesis on serpentinite. Soil Sci. 164, 960-968, 1999.

Boon, D. Y. Potential selenium problems in Great Plains soils. In: Selenium in Agriculture and the Environment, edited by Jacobs, L. W., Madison WS. Soil Sci. Soc. America, p. 107-121, 1989.

Borchardt, G. Smectites. In: Minerals in Soil Environments, edited by Dixon, J. B. and Weed, S. B. Madison, USA. Soil Science Soc. of America, p. 675-727, 1989.

Borin, M., Giupponi, C. and Morari, S. Effects for cultivation systems for maize on nitrogen leaching. 1. Field experiments. Eur. J. Agron. 6, 101-112, 1997.

Borst, N. P. and Mulder, C. (N) Nitrogen contents, nitrogen fertilizer rates and yield of winter barley on sandy, clay, and silty soils in North Holland. Bedryfsontwikkeling 2, 31-36, 1971.

Bosabalidis, A. M. and Tsaftaris, A. Development and senescence of infected root nodule cells of *Vicia faba* L. Biol. Fertil. Soils 1991.

Bouche-Pillon, S., Fleurat-Lessard, P., Fromont, J. C., Serrano, R. and Bonnemain, J. L. Immunolocalization of the plasma membrane H^+ -ATPase in minor veins of *Vicia faba* in relation to phloem loading. Plant Physiol. 105, 691-697 1994.

Boulaine, J. (F) History of Pedology and Soil Science. INRA, Paris, 1989.

Bould, C., Nicholas, D. J. D., Tolhurst, J. A. H. and Wallace, T. Zinc deficiencies of fruit trees in Britain. Nature 164, 801-882, 1949.

Bouma, D. Nutrient uptake and distribution in subterranean clover during recovery frorn nutritional stresses. 1. Experiments with phosphorus. Aust. J. biol. Sci. 20, 601-612, 1967.

Bouma, D. Nutrient uptake and distribution in subterranean clover during recovery from nutritional stresses. II. Experiments with sulphur. Aust. J. Biol. Sci. 20 613-621, 1967.

Bouteau, F., Bousquet, U., Pennarum, M., Convert, M., Dellis, O., Cornel, D. and Rona, J. P. Time dependent K^+ currents through plasmalemma of lactifer protoplasts from *Hevea brasiliensis*. Physiol. Plant. 98, 97-104, 1996.

Bove, J. M., Bove, C., Whatley, F. R. and Arnon, D. I. Chloride requirement for oxygen evolution in photosynthesis. Z. Naturforsch. 18b, 683-688, 1963.

Bowden, J. W., Posner, A. M. and Quirk, J. P. Ionic adsorption on variable charge mineral surfaces. Theoretical-charge development and titration curves. Austr. J. Soil Res. 15, 121-136, 1977.

Bowen, G. D. and Rovira, A. D. The rhizosphere, the hidden half of the hidden half. In: The Plant Roots, the Hidden Half, edited by Waisel, Y., Eshel, A. and Kafkafi, U. New York: Marcel Dekker, p. 641-669, 1991.

Bowen, J. E. Adsorption of copper, zinc and manganese by sugar cane tissue. Plant Physiol. 44, 255-261, 1969.

Bowler, C., Montague, M. van and Inze, D. Superoxide dismutase and stress tolerance. Annu. Rev. Plant Physiol. Plant Mol. Biol. 43, 83-116, 1992.

Boxma, R. Bicarbonate as the most important soil factor in lime-induced chlorosis in the Netherlands. Plant and Soil 37, 233-243, 1972.

Boyd, D. A. Some recent ideas on fertilizer response curves. In: Role of Fertilization in the Intensification of the Agricultural Production. Proc. 9. Congr. International Potash Institute, edited by International Potash Institute, Bern: International Potash Institute, p. 461-473, 1970.

Boyer, J. S. Relationship of water potential to growth of leaves. Plant Physiol. 43, 1056-1062, 1968.

Boyer, J. S. Leaf enlargement and metabolie rates in corn, soybean and sunflower at various leaf water potentials. Plant Physiol. 46, 233-235, 1970.

Boyer, J. S. Plant productivity and environment. Science 218, 443-448, 1982.

Boyer, J. S. Hydraulics, wall extensibility and wall proteins. Proc. of II. Annual Penn. State Symposium on Plant Physiology, American Society of Plant Physiologists, Pennsylvania State University, University Park, PA, 109-121, 1987.

Boyer, J. S. Cell enlargement and growth-induced water potentials. Physiol. Plant. 73, 311-316, 1988.

Boyer, J. S. Biochemical and biophysiccal aspects of water deficits and the predisposition to disease. Annu. Rev. Phytopath. 33, 251-274, 1995.

Bradbury, N. J., Whithmore, A. P., Hart, P. B. S. and Jenkinson, D. S. Modellimg the fate of nitrogen in crop and soil in the years following the application of ^{15}N-labelled fertilizer to winter wheat. J. Agric. Sci. 121, 363-379, 1993.

Bradshaw, A. D. Populations of *Agrostis tenuis* resistant to lead and zinc poisoning. Nature 169, 1098, 1952.

Brady, N. C. The Nature and Properties of Soils. 9. ed., Collier Macmillan Publishers, London, p. 143-147, 1984.

Brady, N. C. and Weil, R. R. The Nature and Properties of Soils. 12th ed. Prentice Hall Inc. International (UK) London 1999.

Brändlein, W. Fertilizers, consumption, production, world trade. In: Fertilizers, edited by Ullman's Encyclopedia of Industrial Chemistry. VHC Verlagsgesellschaft, Weinheim, p. 414-421, 1987.

Brag, H. The influence of potassium on the transpiration rate and stomatal opening in *Triticum aestivum* and

Pisum sativum. Physiol. Plant. 26, 250-257, 1972.

Braschkat, J. (G) Ammonia losses after the application of cow slurrry on permanent grassland: Verlag Ulrich E. Grauw, Stuttgart 1996.

Braschkat, J., Mannheim, T. and Marschner, H. Estimation of ammonia losses after application of liquid cattle manure on grassland. Z. Pflanzenernähr. Bodenk. 160, 117-123, 1997.

Braun, H. Fertilizers and Food Production, p. 43-44 Rome: PAO, 1989.

Braunschweig, L. C. von: Results of longterm field trials testing the optimum soil potassium status. Landw. Forsch. Sonderh. 35, 219-231, 1978.

Bravo, L. A., Zuniga, G. E., Alberdi, M. and Corcuera, L. J. The role of ABA in freezing tolerance and cold acclimation in barley. Physiol. Plant. 103, 17-23, 1998.

Bray, R. H. and Kurtz, L. T. Determination of total, organic, and available phosphorus in soils. Soil Sci. 59, 39-45, 1945.

Breisch, H., Guckert, A. and Reisinger, O. (F) Electromicroscopic studies on the apical zone of maize roots. Soc. bot. Tr. Coll. Rhizosphere 122, 55-60, 1975.

Breitenbeck, G. A. and Bremner, J. M. Effects of various nitrogen fertilizers on emission of nitrous oxide from soils. Biol. Fertil. Soils 2, 195 -199, 1986.

Breitenbeck, G. A. and Paramasivam, S. Availability of ^{15}N - labelled nonexchangeable ammonium to soil microorganisms. Soil Sci. 159, 301-310, 1995.

Bremner, J. and Knight, A. H. The complexes of zinc, copper, and manganese present in ryegrass. Brit. J. Nutr. 24, 279-290, 1970

Brennan, R. F. The effect of zinc fertilizer on take-all and the grain yield of wheat on zinc deficient soils of the Esperance region, Western Australia. Fert. Res. 31, 215-219, 1992.

Brennan, R. F. The residual effectiveness of previously applied copper fertilizer for grain yield of wheat grown on soils of south-west Australia. Fertil Res. 39, 11-18, 1994.

Bresler, E. Trickle-drip irrigation: Principles and application to soil water management. Adv. Agron. 29, 343-393, 1977.

Breteler, H. Nitrogen fertilization, yield and protein quality of a normal and a high lysine maize variety. J. Sci. Fd Agric. 27, 978-982, 1976.

Breteler, H. and Smit, A. L. Effect of ammonium nutrition on uptake and metabolism of nitrate in wheat. Neth. J. agric. Sci. 22, 73-81, 1974.

Brevedan, E. R. and Hodges, H. F. Effects of moisture deficits on 14C translocation in corn (*Zea mays* L.). Plant Physiol. 52, 436-439, 1973.

Brewer, R. F. Fluorine In: Diagnostic Criteria for Plants and Soils, edited by Chapman, H. D. Univ. of California. Div. of Agric. Sciences, p. 180-196, 1966.

Brewer, R. F. Lead. In: Diagnostics Criteria for Plants and Soils, edited by Chapman, H. D. Univ. of California. Div. of Agric. Sciences, p. 213-217, 1966.

Brewster, J. L., Bhat, K. K. S. and Nye, P. H. The possibility of predicting solute uptake and plant growth response from independently measured soil and plant characteristics. V. The growth and phosphorus uptake of rape in soil at a range of phosphorus concentrations and a comparison of results with the prediction of a simulation model. Plant and Soil 44, 295-328, 1976.

Briskin, D. P. and Poole, R. J. Plasma membrane ATPase of red beet forms as phosphorylated intermediate. Plant Physiol. 71, 507-512, 1983.

Briskin, D. P. and Poole, R. J. Characterization of the solubilized plasma membrane ATPase of red beet. Plant Physiol. 76, 26-30, 1984.

Briskin, D. P., Thornley, W. R. and Wyse, R. E. Membrane trransport in isolated vesicles from sugar beet tap root. Plant Physiol. 78, 871-875, 1985.

Bristow, A. W. and Garwood, E. A. Deposition of sulphur from the atmosphere and sulphur balance in four soils under grass. J. Agric. Sci. 103, 463-468, 1984.

Brockman, J. S. Quality and timing of fertiliser N for grass and grass/clover swards. Proc. Fert. Soc. 142, 5-13, 1974

Bromfield, S. M., Cumming, R. W., David, D and Williams, C. H. Change in soil pH, manganese ans aluminium under subterranean clover pasture. Aust. J. Exp. Agric. Anim. Husb. 23, 181-191, 1983.

Bronner, H. (G) Relation between the easily soluble nitrogen in soils and the development of beets. Landw. Forsch. 30/II Sonderh. 39-44, 1974.

Brouwer, R., Kleinendorst, A. and Locher, J. T. Growth response of maize plants to temperature. Proc. Uppsala Symp. 1970. Plant response to climatic factors, Unesco 1973, p. 169-174

Browman M. G., Chesters, G. and Pionke, H. B. Evaluation of tests for predicting the availability of soil manganese to plants. J. Agric. Sci. 72, 335-340, 1969.

Brown, J. C. Iron chlorosis in soybeans as related to the genotype of rootstock. Soil Sci. 96, 387-394, 1963.

Brown, J. C. Mechanism of iron uptake by plants. Plant, Cell and Environment 1, 249-257, 1978.

Brown, J. C. Effects of boron stress on copper enzyme activity in tomato. J. of Plant Nutrition 1, 39-53, 1979.

Brown, J. C., Ambler, J. E., Chaney, R. L. and Foy, C. D. Differential responses of plant genotypes to micronutrients. In: Micronutrients in Agriculture, edited by Soil Sci. Soc. Amer., p. 389-418, 1972.

Brown, J. C. and Clark, R. B. Copper as essential to wheat production. Plant and Soil 48, 509-523, 1977.

Brown, J. C. and Jolley, V. D. Plant metabolic responses to iron -deficiency stress. Biological Sciences. 39, 546-551, 1989.

Brown, P. H., Bellaloui, N., Hu, H. and Dandakar, A. Transgenically enhanced sorbitol synthesis facilitates phloem boron transport and increases tolerance of tobacco to boron deficiency. Plant Physiol. 119, 17-20, 1999.

Brown, P. H. and Hu, H. Boron uptake by sunflower, squash and tobacco cells. Physiol. Plant. 91, 435-441, 1994.

Brown, P. H. and Hu, H. Phloem boron mobility in diverse plant species. Bot. Acta. 11, 331-335, 1998.

Brown, P. H. and Shelp, B. T. Boron mobility in plants. Plant a. Soil 193, 85 -101, 1997.

Brown, P. H., Welch, R. M. and Carey, E. E. Nickel: an essential micronutrient for higher plants. Plant Physiol. 85, 801-803, 1987.

Brownell, P. F. and Bielig, L. M. The role of sodium in the conversion of pyruvate to phosphoenolpyruvate in mesophyll chloroplasts of C_4 plants. Aust. J. Plant Physiol. 23, 171-177, 1996.

Brownell, P. F. and Crossland, C. J. The requirement for sodium as a micronutrient by species having the C_4 dicarboxylic photosynthetic pathway. Plant Physiol. 49, 794-797, 1972.

Broyer, T. C., Carlton, A. B., Johnson, C. M. and Stout, P. R. Chlorine: A micronutrient element for higher plants. Plant Physiol. 29, 526-532, 1954.

Broyer, T. C. and Stout, P. R. The macronutrient elements. Annu. Rev. Plant Physiol. 10, 277-300, 1959.

Brüggemann, W. and Moog, P. R. NADH-dependent Fe-3⁺ EDTA and oxygen reduction by plasma membrane vesicles from barley roots. Physiol. Plant. 75, 245-254, 1989.

Bruin, P. and Grootenhuis, J. A. Interrelation of nitrogen, organic matter, soil structure and yield, Stikstof, Dutch Nitrogenous Fertilizer Review 12, 157-163, 1968.

Brumagen, D. M. and Hiatt, A. J. The relationship of oxalic acid to the translocation and utilization of calcium in *Nicotiana tabacum*. Plant and Soil 24, 239-249, 1966.

Brummel, D. and MacLachlan, Calcium antagonist interferes with auxin-regulated xyloglucan glycosyltransferase levels in pea membranes. Biochemistry and Biophysics 1014, 298-304,

Brun, W. A. and Cooper, R. L. Effects of light intensity and carbon dioxide concentration on photosynthetic rate of soybean. Crop Sci. 7, 451-454, 1967.

Brune, H., Thier, E. and Borchert, E. (G) Variabillty of biological protein quality of different cereals (variations in fertilizer application and sites). 3ʳᵈ Comm.: Experimental results of investigations into the metabolism of pigs and rats with regard to amino acid indices. Z. Tierphysiol., Tierernähr. u. Futtermittelkd. 24, 89-107, 1968.

Brunhold, C. Reduction of sulfate to sulfide. In: Sulfur Nutrition and Sulfur Assimilation in Higher Plants, edited by Rennenberg, H., Brunhold, C., DeKok, L. J. and Stulen, I. The Hague: SPB Academic Publishing, p. 13-31, 1990.

Bryce-Smith, D. and Waldron, H. A.: Lead, behaviour and criminality. The Ecologist 4, 367-377, 1974.

Bryce-Smith, D. Heavy metals as contaminants of the human environment. The Educational Techniques Subject Group. Chemistry Cassette. The Chem. Soc. London 1975.

Buchner, A. and Sturm, H. (G) Fertilizer application in intensive agriculture, 3ʳᵈ ed. DLGVerlag Frankfurt/Main, p. 156, 1971.

Budig, M. (G) Fertilizer placement to overcorne potassium deficiency of grapes on loess soils. Ph. D. Thesis, Justus-Liebig-Universität, Giessen 1970.

Bünemann, G. and Lüdders, P. (G) Effect of seasonal nitrogen supply on the growth of apple trees. II. Bitter pit in 'Cox' related to N-timing and Ca-supply. Die Gartenbauwiss. 34, (16), 287-302, 1969.

Bunescu, S., Tomoroga, P. and Iancu, C. (Ru) The influence of some phytotechnical factors on the phytosanitary state of wheat under irrigation conditions. Probleme Agricole No. 5, 45-52, 1972.

Burgstaller, G. and Huber, A. (G) Potato protein - its nutritional value and application in pig feeding. Landw. Forsch. Sonderh. 37, 383-391, 1980.

Burkart, N. and Amberger, A. (G) Effect of a potassium fertilizer application on mineral nutrient uptake, yield and quality of agricultural crops grown on K fixing soils in Southern Bavaria. Bayer. Landw. Jahrb. 54, 615-626, 1977.

Burkart, N. and Amberger, A. (G) Effect of potassium fertilizer application on the availability of potassium in K fixing soils during the growth period. Z. Pflanzenernaehr. Bodenk. 141, 167-179, 1978.

Burnham, C. P. and Lopez-Hernandez, D. Phosphate retention in different soil taxonomic classes. Soil Sci. 134, 376-380, 1982.

Burns, I. G., Walker, R. L. and Moorby, J. How do nutrients drive growth? Plant and Soil 196, 321-325, 1997.

Burström, H. G. Calcium and plant growth. Biol. Rev. 43, 287-316, 1968.

Burt, R. L. Carbohydrate utilization as a factor in plant growth. Aust. J. Biol. Sci. 17, 867-877, 1964.

Bush, D. R. Proton-coupled sugar and amino acid transporters in plants. Annu. Rev. Plant Physiol. Plant Hol. Biol. 44, 513-542, 1993.

Bush, D. S. Calcium regulation in plant cells and its role in signaling. Annu. Rev. Plant Physiol. Plant Mol. Biol. 46, 95-122, 1995.

Bush, D. S. and Wang, T. Diversity of calcium-efflux transporters in wheat aleurone cells. Planta 197, 19-30, 1995.

Bussler, W. (G) Manganese deficiency symptoms in higher plants. Z. Pflanzenernähr. Düng. Bodenk. 81 (126) 225-241, 1958.

Bussler, W. (G) Manganese toxicity in higher plants. Z. Pflanzenernähr. Düng. Bodenk. 81, 256-265, 1958.

Bussler, W. (G) The development of calcium deficiency symptoms. Z. Pflanzenernähr. Düng. Bodenk. 100, 53-58, 1963.

Bussler, W. Physiological functions and utilization of copper. In: Copper in Soils and Plants, edited by Loneragan, J. F., Rorson, A. D. and Graham, R. D., Academic Press, London, p. 213-234, 1981.

Byers, M., Franklin, J. and Smith, S. J. The nitrogen and sulphur nutrition of wheat and its effect on baking quality of grain. Cereal Quality 15, 337-344, 1987.

Caballero, R., Arauzo, M. and Hernaiz, P. J. Accumulation and redistribution of mineral elements in common vetch during pod filling. Agron J. 88, 801-805, 1996.

Cadenas, E. Biochemistry of oxygen toxicity. Annu. Rev. Biochem. 58, 79-110 1989.

Cadish, G. and Giller, K. E. Estimating the contribution of legumes to soil organic matter build up in mixed communities of C_3/C_4 plants. Soil Biol. Biochem. 28, 823-825, 1996.

Cairns, A. L. P. and Kritzinger, J. H. The effect of molybdenum in seed dormancy in wheat. Plant and Soil 145, 295-297, 1992.

Cakmak, I., Alti, M., Kaya, R., Evliya, H. and Marschner, H. Differential response of rye, triticale, bread and durum wheat to zinc deficiency in calcareous soils. Plant and Soil 188, 1-10, 1997a.

Cakmac, I., Hengeler, C. and Marschner, H. Changes in phloem export of sucrose in leaves in response to phosphorus, potassium, and magnesium deficiency in bean plants. J. Exp. Bot. 45, 1251-1257, 1994.

Cakmak, I., Kurz, H. and Marschner, H. Short-term effects of boron, geranium and high light intensity on membrane permeability in boron deficient leaves of sunflower. Physiol. Plant. 95, 11-18, 1995.

Cakmak, I. and Marschner, H. Magnesium deficiency and high light intensity enhance activities of superoxide dismutase, ascorbate peroxidase and gluthatione reductase in bean plants. Plant Physiol. 98, 1222-1227, 1992.

Cakmak, I., Öztürk, L., Eker, S., Torun, B., Kalfa, H. I. and Yilmaz, A. Concentration of zinc and activity of copper/zinc- superoxide dismutase in leaves of rye and wheat cultivars differing in sensitivity to zinc deficiency. J. Plant Physiol. 151, 91-95, 1997b.

Cakmak, I. and Römheld, V. Boron deficiency induced impairment of cellular functions in plants. Plant and Soil 193, 71-83, 1997.

Cakmak, I., Sari, N., Marschner, H., Ekiz, H., Kalayci, M., Yilmaz, A. and Braun, H. J. Phytosiderophore release in bread wheat and durum wheat genotypes differing in zinc effciency. Plant and Soil 180, 183-189, 1996.

Cakmak, I., Yilmaz, A., Kalayci, M., Torun, B., Erenoglu, B. and Braun, H. J. Zinc deficiency as a critical problem in wheat production in Central Anatolia. Plant and Soil 180, 165-172, 1996.

Caldwell, C. R. and Haug, A. Divalent cation inhibition of barley root plasma membrane-bound Ca^{2+}-ATPase activity and its reversal by monovalent cations. Physiol. Plant. 54, 112-118, 1982.

Caley, C. Y., Duffus, C. M. and Jeffcoat, B. Effects of elevated temperature and reduced water uptake on enzymes of starch synthesis in developing wheat grains. Austr. J. Plant Physiol. 18, 431-439, 1990.

Calvacante, V. A. and Döbereiner, J. A new acid-tolerant nitrogen-fixing bacterium associated with sugar cane. Plant and Soil 108, 23-31, 1988.

Calvin, M. (G) The photosynthetic cycle. Angew. Chemie 68, 253-264, 1956.

Calvin, M. Hydrocarbons from plants: Analytical methods and observations Naturwiss. 67, 525-533, 1980.

Cammack, R. Splitting molecular hydrogen. Nature 373, 556-557, 1995.

Campbell, C. A. Soil organic carbon, nitrogen and fertility. In: Development in Soil Science 8, Soil Organic Matter, edited by Schnitzer, M. and Khan, S. U., Elsevier, Amsterdam, p. 174-271. 1978.

Campbell, W. H. Nitrate reductase biochemistry comes of age. Plant Physiol. 111, 355-361, 1996.

Canny, M. J. Apoplastic water and solute movement: new rules for an old space. Annu. Rev. Plant Physiol. Plant Mol. Biol. 46, 215-236, 1995.

Canny, M. J. Vessel contents during transpiration-embolisms and refilling. Amer. J. Bot. 85, 1223-1230, 1997.

Canny, M. J. Transporting water in plants. Amer. Sci. 86, 152-159, 1998.

Carlile, M. J. and Watkinson S. C. The Fungi. Academic Press, London, New York, 1994.

Carpenter, J. F. and Crowe, J. H. An infrared spectroscopic study of the interactions of carbohydrates with dried proteins. Biochem. 28, 3916-3922, 1989.

Carpita, N., McCann, M. and Griffing, L. R. The plant extracellular matrix: News from the cell's frontier. The Plant Cell 8, 1451-1463, 1996.

Carter, M. R. Association of cation and organic anion accumulation with iron chlorosis of Scots pine on prairie soils. Plant and Soil 56, 293-300, 1980.

Casella, S. Positive and negative aspects of denitrification in Rhizobium-legume associations. In: Physiological Limitations and the Genetic Improvement of Symbiotic Nitrogen Fixation, edited by O'Gara, F., Manian, S. and Drevon, J. J. Dordrecht ISBN 90-247 3692-7: Kluwer Academic Publishers, p. 117-125, 1988.

Castillo, L. D., Hunt, S. and Layzell, D. B. The role of oxygen in the regulation of nitrogenase activity in drought-stressed soybean nodules. Plant Physiol. 106, 949-955, 1994.

Cataldo, D. A., Garland, T. R., Wildung, R. E. and Drucker, H. Nickel in plants. II Distribution and chemical form in soybean plants. Plant. Physiol. 62, 566-570, 1978.

Catroux, G. and Schnitzer, M. Chemical, spectroscopic, and biological charateristics of the organic matter in particle size fractions separated from an Aquoll. Soil Sci. Soc. Am. J. 51, 1200-1207, 1987.

Catterall, W. A. Structure and function of voltage-gated ion channels. Annu. Rev. Biochem. 64, 493-531, 1995.

Cavalieri, A. J. Proline and glycine betaine accumulation by *Spartina alterniflora* Loisel. In response to NaCl and nitrogen in a controlled environment. Oecologia 57, 20-24, 1983.

Cecotti, S. P. A. A global review on nutrient sulphur balance, fertilizers and the environment. Agro-Food-Industry Hi-Tech. 7, 18-22, 1996.

Chan, M. K., Kim, J. and Rees, D. C. The nitrogenase FeMo-cofactor and P-Cluster pair: 2. 2 A° resolution structures. Science. 260, 792-794, 1993.

Chandler, R. F. Overcoming physiological barriers to higher yields through plant breeding. In: Role of Fertilization in the Intensification of Agricultural Production. Int. Potash Inst. Bern p. 421-434, 1970.

Chaney, R. L. and Hornick, S. B. Accumulation and effects of cadmium on crops. In: Proc. 1st Int. Cadmium Conf. p. 136-150 Bulletin London 1978.

Chang-Chi Chu. Carbon dioxide in the open atmosphere and in a field of sugarcane at Tainan, Taiwan. Taiwan Sugar Experiment Station, Tainan, Research Report No. 1, 1-18, 1968.

Chang, Y. C., Ma, J. F. and Matsumoto, H. Mechanism of Al-induced iron chlorosis in wheat (*Triticum aestivum*). Al-inhibited biosynthesis and secretion of phytosiderophore. Physiol. Plant. 102, 9-15, 1998.

Chapin, F. S., Schulze, E. D. and Mooney, H. A. The ecology and economics of storage in plants. Annu. Rev. Ecol. Syst. 21, 423-447, 1990.

Chapman, D. L. (G) Quoted from G. Kortüm: Textbook of Electrochemistry, Verl. Chemie, p. 345. Weinheim 1957.

Chapman, H. D. Diagnostic Criteria for Plants and Soils. University of California, Div. of Agric. Sciences, 1966.

Chapman, M. A. and Keay, J. The effect of age on the response of wheat to nutrient stress. Aust. J. Exp. Agric. and Animal Husbandry 11, 223-228, 1971.

Chardonnens, A. N., Bookum, W. M., Kuijper, L. D. J., Verkleij, J. A. C. and Ernst, W. H. O. Distribution of cadmium in leaves of cadmium tolerant and sensitive ecotypes of *Silene vulgaris*. Physiol. Plant. 104, 75-80, 1998.

Chardonnens, A. N., Koefoets, P. L. M., Zanten, A., Schat, H. and Verkleij, J. A. C. Properties of enhanced tonoplast zinc transport in naturally selected zinc-tolerant *Silene vulgaris*. Plant Physiol. 120, 779-785, 1999.

Charley, J. L. and McGarity, J. L. High soil nitrate-levels in patterned saltbush communities. Nature 201,

1351-1352, 1964.

Chartzoulakis, K. S. and Loupassaki, M. H. Effects of NaCL salinity on germination, growth, gas exchange and yield of greenhouse eggplant. Agricultural Water Management 32, 215-225, 1997.

Chatel, D. L., Robson, A. D., Gartrell, J. W. and Dilworth, M. J. The effect of inoculation and cobalt application on the growth and nitrogen fixation by sweet lupins. Aust. J. Agric. Res. 29, 1191-1202, 1978.

Chatterton, N. J. and Silvius, J. E. Photosynthate partitioning into starch in soybean leaves. I. Effects of photoperiod vs. photosynthetic period duration. Plant Physiol. 64, 749-753, 1979.

Chaudhry, F. M. and Loneragan, J. F. Effects of nitrogen, copper and zinc nutrition of wheat plants. Aust. J. Agric. Res. 21, 865-879, 1970.

Chaudhry, F. M. and Loneragan, J. F. Zinc adsorption in wheat seedlings: Inhibition by macronutrient ions in short term experiments and its relevance to long term zinc nutrition. Soil Sci. Soc. Amer. Proc. 36, 323-327, 1972.

Chazen, O. and Neumann, M. Hydraulic signals from the roots and rapid cell-wall hardening in growing maize (*Zea mays* L.) leaves are primary response to polyethylene glycol-induced water deficits. Plant Physiol. 104, 1385-1392, 1994.

Chen, S. Z., Low, P. F. and Roth, C. B. Relation between potassium fixation and the oxidative state of octahedral iron. Soil Sci. Soc. Am. J. 51, 82-86, 1990.

Chen, Y. and Barak, P. Iron nutrition of plants in calcareous soils. Adv. Agron. 35, 217-240, 1982.

Chenery, M. A preliminary study of aluminium and the tea bush. Plant and Soil 6, 174-200, 1955.

Cheng, B. T. and Oullette, G. J. Molybdenum as a plant nutrient. Soil and Fertilizers 36, 207-215, 1973

Chevalier, S. and Huguet, C. (F) Magnesium deficiency effects on apple trees. Ultrastructural evolution in deficient leaves of apple trees. Ann. agron. 26, 351-362, 1975.

Chien, S. H., Sale, P. W. G. and Friesen, D. K. A discussion of the methods for comparing the relative effectiveness of phosphate fertilizers varying in solubility. Fertil Res. 24, 149-157, 1990.

Chimiklis, P. E. and Karlander, E. P. Light and calcium interactions in Chlorella inhibited by sodium chloride. Plant Physiol. 51, 48-56, 1973.

Chiu, T. F. and Bould, C. Effect of shortage of calcium and other cations on Ca-45 mobility, growth and nutritional disorders of tomato plants (*Lycopersicon esculentum*) J. Sci. Fd. Agric. 27, 969-977, 1976.

Chrispeels, M. J., Crawford, N. M. and Schroeder, J. I. Proteins for transport of water and mineral nutrients across the membranes of plant cells. The Plant Cell 11, 661-675, 1999.

Chrispeels, M. J. and Maurel, C. Aquaporins: the molecular basis of facilitated water movement through living plant cells. Plant Physiol. 105, 9 -15, 1994.

Christen, O., Sieling, K., Richter-Harder, H. and Hanus, H. Effects of temporary water-stress before anthesis on growth, development and grain yield of spring wheat. Eur. J. Agron. 4, 27-36, 1995.

Christenson, D. R. and Doll, E. C. Release of magnesium from soil clay and silt fractions during cropping. Soil Sci. 116, 59-63, 1973.

Claassen, N. and Barber, S. A. Simulation model for nutrient uptake from soil by growing plant root system. Agron. J. 68, 961-964, 1976.

Claassen, N., Hendriks, K. and Jungk, A. (G) Rubidium depletion on the soil-root interface by maize plants. Z. Pflanzenernähr. Bodenkd. 144, 553-545, 1981.

Claassen, N. and Jungk, A. (G) Potassium dynamics at the soil-root interface in relation to the uptake of potassium by maize plants. Z. Pflanzenernähr. Bodenkd. 145, 513-525, 1982.

Claassen, N. and Jungk, A. (G) Effect of K uptake rate, root growth and root hairs on potassium uptake efficiency of several plant species. Z. Pflanzenernähr. Bodenkd. 147, 276-289, 1984.

Clarkson, D. T. Calcium uptake by calciole and calcifuge species in the genus Agrostis L., J. Ecol. (Oxford)

712

53, 427-435, 1965.

Clarkson, D. T. Roots and the delivery of solutes and water to the xylem. Phil. Trans. Roy. Soc. London Ser. B 341, 5-17, 1993.

Clarkson, D. T. and Hanson, J. B. The mineral nutrition of higher plants. Annu. Rev. Plant Physiol. 31, 239-298, 1980.

Clarkson, D. T., Hawkesford, M. J. and Davidian, J.-C. Membrane and long-distance transport of sulfate. In: Sulfur Nutrition and Assimilation in Higher Plants, edited by L. J. De Kok, I. Stulen, H. Rennenberg, C. Brunhold and W. E. Rauser SPB Academic Publishing bv, The Hague, p. 3-19, 1993.

Clarkson, D. T. and Sanderson, J. Sites of absorption and translocation of iron in barley roots. Tracer and microautoradiographic studies. Plant Physiol. 61, 731-736, 1978.

Clarkson, D. T. and Sanderson, J. The uptake of a polyvalent cation and its distribution in the root apices of *Allium cepa*. Tracer and autoradiographic Studies. Planta 89, 136-154, 1969.

Clarkson, D. T. and Scattergood, C. B. Growth and phosphate transport in barley and tomato plants during the development of and recovery from phosphate stress. J. Exp. Bot. 33, 865-875, 1982.

Cleland, R. E. A dual role of turgor pressure in auxin-induced cell elongation in Avena coleeoptiles. Planta 77, 182-191, 1967.

Cleland, R. E., Virk, S. S., Taylor, D. Björkmann, T. Calcium, cell walls and growth. In: Calcium in Plant Growth and Development, edited by Leonard, R. T. and Hepler, P. K. The American Society of Plant Physiology, Symposium Series, Vol. 4., p. 9-16, 1990.

Clouse, S. D. and Sasse, J. M. Brassinosteroids: esssential regulators of plant growth and development. Annu. Rev. Plant Physiol. Plant Mol. Biol. 49, 427-451, 1998.

Cocker, K. M., Evans, D. M. and Hodson, M. J. The amelioration of aluminium toxicity by silicon in higher plants: Solution chemistry or an in planta mechanism? Physiol. Plant. 104, 608-614, 1998.

Cocucci, M. C. and Dalla Rosa, S. Effects of canavanine on IAA- and fusicoccin-stimulated cell enlargement, proton extrusion and potassium uptake in maize coleoptiles. Physiol. Plant. 48, 239-242, 1980.

Coddington, J. M. and Taylor, M. J. High field ^{11}B and ^{13}C NMR investigations of aqueous borate solutions and borate diol complexes. J. Coord. Chem. 20, 27-38, 1989.

Cohen, P. The structure and regulation of protein phosphatases. Annu. Rev. Biochem. 58, 453-508, 1989.

Coic, Y., Fauconneau, G., Pion, R., Busson, F., Lesaint, C. and Labonne, F. (F) Effect of the mineral nutrition on the composition of grain proteins in cereals (wheat and barley). Ann. Physiol. vég. 5 (4), 281-292, 1963.

Coic, Y., Lesaint, C. and Le Roux, F. (F) Effects of ammonium and nitrate nutrition and a change of ammonium and nitrate supply on the metabolism of anions and cations in tomatoes. Ann. Physiol. vég. 4, 117-125, 1962.

Coleman, J. E. Zinc proteins: Enzymes, storage proteins, transcription factors, and replication proteins. Annu. Rev. Biochem. 61, 897-946, 1992.

Collander, R. Selective absorption of cations by higher plants. Plant Physiol. 16, 691-720, 1941.

Collier, G. F. and Tibbits, T. W. Tipburn of lettuce. Hort. Rev. 4, 49-65, 1982.

Colombo, R. and Cerana, R. Inward rectifying K^+ channels in the plasma membrane of *Arabidopsis thaliana*. Plant Physiol. 97, 1130-1135, 1991.

Comin, J. J., Barloy, J., Bourrie, G. and Trolard, F. Differential effects of monomeric and polymeric aluminium on the root growth and the biomass production of root and shoot of corn in solution culture. European J. Agron. 11, 115-122, 1999.

Conry, M. J. and MacNaeidhe, F. Comparative nutrient status of a peaty gleyed podzol and its plaggen counterpart on the Dingle penninsula in the south-west of Ireland. European J. Agron. 11, 85-90, 1999.

Cook, H. and Oparka, K. J. Movement of fluorescein into isolated caryopses of wheat and barley. Plant Cell Environ. 6, 239-242, 1983.

Cooke, G. W. Phosphorus and Potassium Fertilizers: their Forms and their Places in Agriculture. The Fertilizer Soc. Proc. 92, 1966.

Cooke, G. W. Fertilizing for Maximum Yield. Crosby Lockwood and Son Ltd. London 1972.

Cooke, G. W. Change in the amounts of fertilizers used and the forms in which they are produced, together with comments on current problems in valuing fertilizers and using them efficiently. Cento Seminar on Fertilizer Analytical Methods, Sampling and Quality Control, Pakistan 1974.

Cooke, G. W. Sources of protein for people and livestock; the amounts now available and future prospects. In: Fertilizer Use and Protein Production, edited by IPI, Berne: International Potash Institute, p. 29-51, 1975.

Cooke, G. W. The role of organic manures and organic matter in managing soils for higher crop yields - a review of the experimental evidence. Proc. Int. Seminar on Soil Environnement and Fertility Management in Intensive Agriculture, p. 53-64. Tokyo 1977.

Cooke, G. W. and Williams, R. J. B. Losses of nitrogen and phosphorus from agricultural land. Water Treatm. Exam. 19, 253-276, 1970.

Coombs, J. Enzymes of C_4 metabolism In: Photosynthesis II, Encycl. Plant Physiol. New Series Vol. 6, edited by Gibbs, M. and Latzko, E., Springer-Verlag Berlin, Heidelberg New York p. 251-262, 1979.

Cooper, A. J. Crop production with nutrient film- technique. Proc. 4th International Soilles Culture Las Palmas, 1976.

Coors, J. G. Resistance to the European corn borer, *Ostrinia nubialis* (Hubrier) in maize, *Zea mays* L., as affected by soil silicia, plant silicia, structural carbohydrates, and lignin. Plant and Soil 1987.

Coppenet, M. (F) Results from observations of a twelve years lasting lysimeter experiment in Quimper (1954-1965). Ann. agron. 20, 111-143, 1969.

Corbeels, M., Hofman, G. and Cleemput, O. Residual effect of nitrogen fertilization in a wheat sunflower cropping sequence on a Vertisol under semi -arid mediterranian conditions. European J. Agron. 9, 109-116, 1998.

Cornish, K. and Zeevart, J. A. D. Movement of abscisic acid into the apoplast in response to water stress in *Xanthium strumarium* L. Plant Physiol. 78 623-626, 1985.

Cosgrove, D. J. Plant cell enlargement and the action of expansins. Bioessays 18, 533-540, 1996.

Cosgrove, D. J. Relaxation in a high-stress environment: The molecular bases of extensible cell walls and cell enlargement. Plant Cell 9, 1031-1041, 1997.

Coté G. G. and Crain, R. C. Biochemistry of phosphoinositides. Annu. Rev. Plant Physiol. Plant Mol. Biol, 44, 333-356, 1993.

Cottenie, A. and Kiekens, L. Quantitative und qualitative plant response to extreme nutritional conditions. In: Plant Analysis and Fertilizer Problems, Wehrmann, J., Vol. 2, German Soc. Plant Nutrition, p. 543-556, Hannover 1974.

Cowles, J. R., Evans, H. J. and Russell, S. B_{12} co-enzyme-dependent ribonucleotide reductase in Rhizobium species and the effect of cobalt deficiency on the activity of the enzymE. J. Bacteriol. 97, 1460, 1969.

Cox, G., Moran, K. J. Sanders, F., Nockolds, C. and Tinker, P. B. Translocation and transfer of nutrients in vesicular-arbuscular mycorrhizas. III. Polyphosphate granules and phosphorus translocation. New Phytol. 84, 649-659, 1980.

Cox, W. J. and Reisenauer, H. M. Growth and ion uptake by wheat supplied nitrogen as nitrate, or ammonium, or both. Plant and Soil 38, 363-380, 1973.

Cramer, G. R. Kinetics of maize leaf elongation. II. Response of a Na-excluding cultivar and a Na-including cultivar to varying Na/Ca salinities. J. Exp. Bot. 43, 857-864, 1992.

Cramer, G. R. and Bowman, D. C. Short-term leaf elongation kinetics of maize in response to salinity are dependent of the root. Plant Physiol. 95, 965-967, 1991.

Cramer, G. R., Läuchli, A. and Polito, V. S. Displacement of Ca^{2+} by Na^+ from the plasmalemma of root cells. A primary response to salt stress? Plant Physiol. 79, 207-211, 1985.

Craswell, E. T. and Godwin, D. C. The efficiency of nitrogen fertilizers applied to cereals in different climates. Adv. Plant Nutrition 1, 1-55, 1984.

Crawford, N. W., Campbell, W. H. and Davis, R. W. Nitrate reductase from squash:cDNA clonoing and nitrate regulation. Proc. Natl. Acad. Sci. 83, 8073-8076, 1986.

Creelman, R. A. and Mullet, J. E. Oligosaccharins, brassinolides, and jasmonates: Nontraditional regulators of plant growth, development and gene expression. The Plant Cell 9, 1211-1223, 1997.

Crisp, P., Collier, G. F. and Thomas, T. H. The effect of boron on tipburn and auxin activity in lettuce. Sci. Hortic. 5, 215-226, 1976.

Cronshaw, J. Phloem structure and function. Annu. Rev. Plant Physiol. 32, 465-484, 1981.

Crooke, W. M. Effect of soil reaction on uptake of nickel from a serpentine soil. Soil Sci. 81, 269-2762, 1956.

Crooke, W. M. and Inkson, H. E. The relationship between nickel toxicity and major nutrient supply. Plant and Soil 6, 1-15, 1955.

Crowe, J. H. and Crowe, L. M. Membrane integrity in anhydrobiotic organisms: toward a mechanism for stabilizing dry cells. In: Water and Life, edited by Somero, G. N., Osmond, C. B. and Bolis, C. L. Springer Verlag Berlin, Heidelberg, 87-103, 1992.

Crowe, J. H., Crowe, L. M., Leslie, S. B. and Fisk, E. Mechanism of stabilization of dry biomolecules in anhydrobiotic organisms. In: Responses to Cellular Dehydration during Environmental Stress, edited by Close, T. J. and Bray, E. A., MD: Am. Soc. Plant Physiol., p. 11-20, 1993.

Crowley, D. E., Wang, Y. C., Reid, C. P. P. and Szaniszlo, P. J. Mechanisms of iron acquisition from siderophores by microorganisms and plants. In: Iron Nutrition and Interactions in Plants, edited by Y. Chen, and Y. Hadar, Dordrecht: Kluwer Academic Publishers, ISBN 0-7923-1095-0, p. 213-232, 1991.

Croy, L. I. and Hageman, R. H. Relationsbip of nitrate reductase activity to grain protein production in wheat. Crop Sci. 10, 280-285, 1970.

Crutzen, P. J. Atmospheric chemical processes of the oxides of nitrogen, including nitrous oxide. In: Denitrification, Nitrification, Nitrous Oxides, edited by Delwiche, John Wiley, New York, p. 17-44, 1981.

Cruz, R. T., Jordan, W. R. and Drew, M. C. Structural changes and associated reduction of hydraulic conductance in roots of Sorghum bicolor L. following exposure to water deficits. Plant Physiol. 99, 203-212, 1992.

Cuevas, E. and Medina, E. Nutrient dynamics within amazonian forests. II. Fine root growth, nutrient availability and leaf litter decomposition. Oecologia 76, 222-235, 1988.

Curtin, D. and Smillie, G. W. Effects of liming on soil chemical characteristics and grass growth in laboratory and long term field amended soils. Plant and Soil 95, 23-31, 1986.

Daie, J., Wyse, R., Hein, M. and Brenner, M. L. Abscisic acid metabolism by source and sink tissues of sugar beet. Plant Physiol. 74, 810-814, 1984.

Dainty, J. Ion transport and electrical potentials in plant cells. Annu. Rev. Plant Physiol. 13, 379-402, 1962

Dakora, F. J. and Keya, S. O. Contribution of legume nitrogen fixation to sustainable agriculture in Sub-Saharan Africa. Soil Biol. Biochem. 29, 809-817, 1997.

Dalal, R. C. Soil organic phosphorus. Adv. Agron. 29, 83-117, 1977.

Dam Kofoed, A. Copper and its utilization in Danish agriculture. Fertilizer Research 1, 63-71, 1980.

Dam Kofoed, A. and Hojmark, J. V. (D) Field experiments with magnesium fertilizers. Tidsskrift tot Planteavl 75, 349-376, 1971.

Dam Kofoed, A., Lindhard, J. and Sondergard-Klausen, P. (D) Experiments with anhydrous ammonia as a nitrogenous fertilizer. Tidsskrift for Planteavl 71, 145-225, 1967.

Dam Kofoed, A. and Lindhard, J. (D) Removal of plant nutrients from grass-covered soils in lysimeters. Tidsskrift for Planteavl 72, 417-437, 1968.

Dam Kofoed, A. and Sondergard-Klausen, P. (D) Field application of fertilizer nitrogen to grass and to clover-grass mixtures. Tidsskrift for Planteavl 73, 203-246, 1969.

Daniel, C. and Ochs, R. (F) Improvernent of production of young oil palms in Peru by chloride fertilizer application. Oléagineux 30, 295-298, 1975.

Danielli, J. F. and Davson, H. A. A contribution to the theory of the permeability of thin films. J. Cellular comp. Physiol. 5, 495-508, 1935.

Daniels, R. R., Stuckmeyer, B. E. and Peterson, L. A. Copper toxicity in *Phaseolus vulgaris* L. as influenced by iron nutrition. I. An anatomical study. J. Amer. Soc. Hort. Sci. 9, 249-254, 1972.

Dannel, F., Pfeffer, H. and Römheld, V. Effect of pH and boron concentration in the nutrient solution on translocation of boron in the xylem of sunflower. In: Boron in Soils and Plants, edited by Bell, R. W. and Rerkasem, B., Kluwer Academic Publishers, Dordrecht, p. 183-186, 1997.

Dannel, F., Pfeffer, H. and Römheld, V. Compartmentation of boron in roots and leaves of sunflower as affected by boron supply. J. Plant Physiol. 153, 615-622, 1998.

Darré, J. and Gilleron, G. (F) Magnesium fertilizers. In: Le magnesium en agriculture, edited by Huguet, C. and Coppenet, M. Paris, INRA editions, p. 259-270, 1992.

Dasberg, S., Erner, J. and Bielorai, H. Nitrogen balance in a citrus orchard. J. Environ. Qual. 13, 352-356, 1984.

Dauzat, J. and Eroy, M. N. Simulating light regime and intercropping yields in coconut based farming systems. In: Proceed. ESA Congress, edited by Ittersum, M. K. van and Geijn, S. C. van de Amsterdam: Elsevier, p. 87 - 98, 1997.

Davenport, R. J., Reid, R. J. and Smith, F. A. Sodium-calcium interactions in two wheat species differeing in salinity tolerance. Physiol. Plant. 99, 323-327, 1997.

Davidson, E. A. Fluxes of nitrous oxide and nitric oxide from terrestical ecosystems. In: Microbial Production and Consumption of Greenhouse Gases:Methane, Nitogen Oxides and Halomethanes, edited by Rogers, J. E. and Whitman, W. B. Washington: American Soc. Microbiology, p. 219-235, 1991.

Davies, D. B., Hoopere, L. J. and Charlesworth, R. R. Copper deficiency in crops: III Copper disorders in cereals grown in chalk soils in South Eastern and Central Southern England in 'Trace Elements in Soils and Crops', Tech. Bulletin, Min. of Agric., Fisheries and Food 21, 88-118, 1971.

Davies, D. D. The fine control of cytosolic pH. Physiol. Plant. 67, 702-706, 1986.

Davies, E. Plant responses to wounding. In: The biochemistry of plants, vol 12. Academic Press, 1987.

Davies, W. J. and Zhang, J. Root signals and the regulation of growth and development of plants in drying soil. Annu. Rev. Plant Physiol. Plant Mol. Biol. 42 55-76, 1991.

Davis, R. F. and Higinbotham, N. Electrochemical gradients and K^+ and Cl^- fluxes in excised corn roots. Plant Physiol. 57, 129-136, 1976.

Davison, A. and Jefferies, B. J. Some experiments on the nutrition of plants growing on coal mine waste heaps. Nature 210, 649-650, 1966.

Day, A. D. and Intalap, S. Some effects of soil moisture stress on the growth of wheat (*Triticum aestivum* L. em Thell). Agron. J. 62, 27-29, 1970.

Day, J. P., Hart, H. and Robinson, M. S. Lead in urban street dust. Nature 253, 343-345, 1975.

716

Dayly, G. C. Restoring value to the world's degraded lands. Science. 269, 350-354, 1995.

De, R. Cultural practices for maize. sorghum and millets. 1. FAO/SIDA Seminar for plant scientists from Africa and Near East, FAO Rome, p. 440-451, 1974.

De Boer, A. H. Potassium translocation into the root xylem. Plant Biol. 1, 36-45, 1999.

De Boer, A. H. and Wegner, L. H. Regulatory mechanisms of ion channels in xylem parenchyma cells. J. Exp. Bot. 48, 441-449, 1997.

De Boodt, M. and De Leenheer, L. Investigations into pore distribution in soils. Medel. Landbouwhogeschool, 98-130, 1955.

De Datta, S. K. and Patrick, W. H. Nitrogen Economy of Flooded Rice Soils. Martinus Nijhoff Publishers, Dordrecht, Boston, 1986.

De Swart, P. H. and van Diest, A. The rock phosphate solubility capacity of *Puraria javanica* as affected by soil pH, superphosphate priming effect and symbiotic N fixation. Plant and Soil 100, 135-147, 1987.

De Witt, C. T. Transpiration and Crop Yield. No 64. 6 Verslag van Lanbouwk, Douderzock 1958.

De Witt, N. D. and Sussman, M. R. Immunocytological localization of an epitope-tagged plasma membrane proton pump (H^+-ATPase) in phloem companion cells. Plant Cell 7, 2053-2067, 1995.

Decau, J. and Pollacsek, M. Improving plant protein by nuclear techniques. Intern. Atomic Energy Agency, 132/17, 357-366, 1970.

Decau, J. and PujoL, B. (F) Comparative effects of irrigation and nitrogen fertilizer on the qualitative and quantitative production of different maize cultivars. Ann agron. 24, 359-373, 1973.

Degenhardt, J., Larsen, B. P., Howell, S. H. and Kochian, L. V. Aluminium resistance in the Arabidopsis mutant alr-104 is caused by an aluminium induced increase in rhizosphere pH. Plant Physiol. 117, 19-27, 1999.

Dejou, J. (F) Review of minerals and rocks containing magnesium. In: Le magnesium en agriculture, edited by Huguet, C. and Coppenet, M. Paris, INRA, p. 35-49, 1992.

DeKock, P. C., Dyson, P. W., Hall, A. and Grabowska, F.: Metabolic changes associated with calcium deficiency in potato sprouts. Potato Res. 18, 573-581, 1975.

Dela Guardia, M. D. and Benlloch, M. Effects of potassium and gibberellic acid on stem growth of whole sunflower plants. Physiol. Plant. 49, 443-448, 1980.

Delas, L (F) The toxicity of copper accumulated in soils. Agrochimica 7, 258-288, 1963

Delas, J. (F) Fertilizer appliction to vine and quality, Toulouse:COMIFER, 1985.

Delas, J. (F) Magnesium in vine culture. In: Le magnesium en agriculture, edited by Huguet, C. and Coppenet, M., Paris, INRA editions, p. 158-173, 1992.

Delauney, A. J. and Verma, D. P. S. Proline biosynthesis and osmoregulation in plants. Plant J. 4, 215-223, 1993.

Delhaize, E., Ryan, P. R. and Randall, P. J. Aluminium tolerance in wheat (*Triticum aestivum* L.). II. Aluminium stimulated excretion of malic acid from root apices. Plant Physiol. 103, 695-702, 1993.

Delhon, P., Gojon, A., Tillard, P. and Passama L. Diurnal regulation of NO_3- uptake in soybean plants. I. Changes in NO_3- influx, efflux, and N utililisation in the plant during the day/night cyclE. J. Exp. Bot. 46, 1585-1602, 1995.

Dell, B. and Huang, L. Physiological response of plants to low boron. Plant and Soil 193, 103-120, 1997.

Deloch, W. H. (G) Analytical determination of sulphur in biochemical materials and the uptake of sulphur by crops in relation to fertilizer application. Ph. D. Thesis Justus Liebig University Giessen 1960.

Delwiche, C. C. Cycling of elements in the biosphere. In: Inorganie Plant Nutrition, Encycl. Plant Physiol. New Series Vol 15, edited by Läuchli, A. and Bieleski, R. L., Springer Verlag Berlin, Heidelberg, New York, Tokyo, p. 212-238, 1983.

Demmig, B. and Gimmler, H. Properties of the isolated intact chloroplast at cytoplasmic K⁺ concentrations. I. Lightinduced cation uptake into intact chloroplasts is driven by an electrical potential difference. Plant Physiol. 73, 169-174, 1983.

Denaeyer-de Smet, S. (F) Aspects of the accumulation of zinc in plants growing on calamine soils. Bull. Inst. RES. Sci. Nat. Belg. 46, 1-13, 1970.

Devel, J. (F) Technical aspects of the use of biofuels: the case of plant oils. Aspects techniques de llutilisation des biocarburants: le cas particulier des huiles vegetales. C. R. Acad. Agric. 78(3), 3-18, 1992.

Dhindsa, R. S. and Cleland, R. E. Water stress and protein synthesis. 1. Differential inhibition of protein synthesis. Plant Physiol. 55, 778-781, 1975.

Dickinson, D. B. Permeability and respiratory properties of germinating pollen. Physiol. Plant. 20, 118-127, 1967.

Diederichs, C. Influence of different P sources on the efficiency of several tropical endomycorrhizal fungi in promoting the growth of *Zea mays* L. Fert. Res. 30, 39-46, 1991.

Diehl, K. H., Rosopulo, A., Kreuzer, W. and Judel, G. K. (G) Behaviour of lead tetraethyl in soil and its uptake by plants. Z. Pflanzenernähr. Bodenk. 146, 551-559, 1983.

Dijkshoorn, W. and Ismunadji, M. Nitrogen nutrition of rice plants measured by growth and nutrient content in pot experiments. 2. Uptake of ammonium and nitrate from a waterlogged soil. J. agric. Sci. 20, 44-57, 1972.

Dijkshoorn, W. and Ismunadji, M. Nitrogen nutrition of rice plants measured by growth and nutrient content in pot experiments. 3 Changes during growth. Neth. J. agric. Sci. 20, 133-144, 1972.

Dijkshoorn, W. and van Wijk, A. L. The sulphur requirements of plants as evidenced by the sulphur-nitrogen ratio in the organic matter, a review of published data. Plant and Soil 26, 129-157, 1967.

Dilworth, M. J., Robson, A. D. and Chatel, D. L. Cobalt and nitrogen fixation in *Lupinus angustifolius* L. II. Nodule formation and function. New Phytol. 83, 63-79, 1979.

Dimarco, A. A., Bobik, T. A. and Wolfe, R. S. Unusual coenzymes of methanogenesis. Annu. Rev. Plant Physiol. Plant Mol. Biol. 59, 335-394, 1990.

Ding, J. P., Badot, P. M. and Pickard, B. G. Aluminium and hydrogen ions inhibit a mechanosensory Calcium-selective cation channel. Aust. J. Plant Physiol. 20, 771-778, 1993.

Ding, J. P. and Pickard, B. G. Activation of a Ca²⁺ channel by ABA and resultant inhibition of a K⁺ channel in excised plasmalemma of onion epidermis. Plant Physiol. 96, Suppl.:137-No. 911, 1991.

Dinkelacker, B., Römheld, V. and Marschner, H. Citric acid excretion and precipitation of calcium citrate in the rhizosphere of white lupin (*Lupinus albus* L.). Plant,Cell and Environm. 12, 285-292, 1989.

Dixon, J. B. Kaolin and serpentine group minerals. In: Minerals in Soil Environments, edited by Kissel, D. E., Madison, Wisconsin, USA: Soil Sci. Soc. America, p. 467-525, 1989.

Dixon, N. E., Gazzola, C., Blakeley, R. L. and Zerner, B. Jack bean urease (EC 3. 5. 1. 5.) a metalloenzyme. A simple biological role for nickel? J. Am. Chem. Soc. 97, 4131-4132, 1975.

Döbereiner, L. Dinitrogen fixation in rhizosphere and phyllosphere associations. In: Inorganic Plant Nutrition, Encycl. Plant Physiol. New Series, Vol. 15A, edited by Läuchli, A and Bieleski, R. L., Springer Verlag Berlin, Heidelberg, New York, Tokyo, p 330 -350, 1983.

Döbereiner, J. Biological nitrogen fixation in the tropics: social and economic contributions. Soil Biol. Biochem. 29, 771-774, 1997.

Doll, E. C. and Lucas, R. E. Testing soils for potassium, calcium, and magnesium. In: Soil Testing and Plant Analysis, edited by Wailh, L. M. and Beaton, J. D. Soil Sci. Soc. America, Madison/USA, p. 133-151, 1973.

Donner, H. E. and Lynn, W. C. Carbonate, halide, sulfate, and sulfide minerals. In: Minerals in Soil Environments, edited by Kissel, D. E., Madison, Wisconsin: Soil Sci. Soc. of America, p. 279-330, 1989.

718

Dou, H. and Steffens, D. (G) Mobility and plant-availability of phosphorus in organic and inorganic forms in the rhizophere of *Lolium perenne*. Z. Pflanzenernaehr. Bodenk. 156, 279-285, 1993.

Dou, H. and Steffens, D. Recovery of ^{15}N labelled urea as affected by fixation of ammonium by clay minerals. Z. Pflanzenernähr. Bodenk. 158, 351-354, 1995.

Douglas, L. A. Vermicilites. In: Minerals in Soil Environments edited by Kissel, D. E. SSSA, Madison, Ws USA p. 635-674.

Downes, R. W. Differences and transpiration rates between tropical and temperate grasses under controlled conditions. Planta 88, 261-273, 1969.

Draycott, A. P. and Durrant, M. J. Plant and soil magnesium in relation to response of sugar beet to magnesium applications. J. of the Int. Inst. for Sugar Beet Research 5, 129-135, 1971.

Draycott, A. P. and Farley, R. F. Response by sugar beet to soil dressings and foliar sprays of manganesE. J. Sci. Fd Agric. 24, 675-683, 1973.

Draycott, A. P., Marsh, J. A. P. and Tinker, P. B. H. Sodium and potassium relationships in sugar beet. J. agric. Sci. 74, 567-573, 1970.

Drees, L. R., Wilding, L. P., Smeck, N. E. and Senkayi, A. L. Silica in soils: quartz and disordered silica polymorphs. In: Minerals in Soil Environments, edited by Kissel, D. E., Wadison,WS: Soil Sci. Soc. America, p. 913-974, 1989.

Drew, M. C. Effects of flooding and oxygen deficiency on plant mineral nutrition. Adv. Plant Nutrition. 3, 115-159, 1988.

Drew, M. C. and Goss, M. J. Effect of soil physical factors on root growth. Chem. and Ind. No. 14, 679-684, 1973.

Drew, M. C. and Nye, H. P. The supply of nutrient ions by diffusion to plant roots in soil. II. The effect of root hairs on the uptake of potassium by roots of rye grass (*Lolium multiflorum*). Plant and Soil 31, 407-424, 1969.

Drew, M. C., Nye, P. H. and Vaidyanathan, L. V. The supply of nutrient ions by diffusion to plant roots in soil. I. Absorption of potassium by cylindrical roots of onion and leek. Plant and Soil 30, 252-270, 1969.

Drew, M. C. and Saker, L. R. Uptake and long-distance transport of phosphate, potassium and chloride in relation to internal ion concentrations in barley: evidence of non-allosteric regulation. Planta 160, 500-507, 1984.

Drews, G. (G) Two hundred years photosynthesis research. Naturw. Rundschau 51, 417-424, 1998.

Drouineau. G. and Mazoyer, R. (F) Contribution to the study of copper toxicity in soils. Ann. agronom. 13, 31-53, 1962.

Dubcovsky, J., Maria, G. S., Epstein, E., Luo, M. C. and Dvorak, J. Mapping of the K^+/Na^+ discrimination locus kna1 in wheat. Theor. Appl. Genet. 92, 448-454, 1996.

Düring, H. (G) Abscisic acid in fruits of *Vitis vinifera*. Naturw. 60, 301-302, 1973.

Duffus, D. and Binnie, J. Sucrose relationships during endosperm and embryo development in wheat. Plant Physiol. Biochem. 28, 161-165, 1990.

Dugger, W. M.: Boron in plant metabolism. In: Inorganic Plant Nutrition, Encycl. Plant Physiol. New Series Vol. 15B, edited by Läuchli, A. and Bieleski, R. L., Springer Verlag Berlin, Heidelberg, New York, Tokyo, p. 626-650, 1983.

Duisberg, P. C. and Buehrer, T. F. Effect of ammonia and its oxidation products on rate of nitrification and plant growth. Soil Sci. 78, 37-49, 1954.

Dujardin, P. Molecular physiology of starch synthesis in the potato (*Solanum tuberosum* L.). Acta Bot. Gall. 142, 361-370, 1995.

Durand, J. L., Gastal, F., Etchebest, S., Bonnet, A. C. and Ghesquiere, M. Interspecific variability of plant water status and leaf morphogenesis in temperate forage grasses under summer water deficit. In: Proceedings of the 4th ESA Congress, Veldhoven, edited by Van Ittersum, M and Van Geijn, S. C., Elsevier, Amsterdam, p. 135-143, 1997.

Duru, M., Balent, G. and Langlet, A. Mineral nutritional status of pastures. I. Effects on herbage accumulation. Eur. J. Agron. 3, 43-51, 1994.

Duthion, M. (F) Potassium in soils. Revue Agricol. France - Fertilisation No. 2, 1966.

Duvigneaud, P. and Denaeyer-De Smet, S. (F) Effect of some heavy metals in the soil (copper, cobalt, manganese, uranium) on the vegetation in Upper Katanga. Ier Colloq. Soc. Bot. France 121, 1959.

Duxbury, J. M. and McConnaughei, P. K. Effect of fertilizer source on denitrification and nitrous oxide emissions in a maize-field. Soil Sci. Soc. Am. J. 50, 644-648, 1986.

Eakin, J. H. Food and fertilizers. In: The Fertilizer Handbook, edited by The Fertilizer Institute, Washington, p. 1-21, 1972.

Eaton, F. M. Chlorine. In: H. D. Chapman: Diagnostic Criteria for Plants and Soils. Univ. of California, Div. of Agric. Sciences, p. 98-135,1966.

Edelbauer, A. (G) Investigations on the effect of various KCl/K_2SO_4 ratios on grape yield, juice quality and amino acid pattern in the juice of Vitis vinifera grown in solution culture. In: 4th Int. Coll. on the Control of Plant Nutrition, Vol. I, edited by Cottenie, A. p. 293-303. Gent 1976.

Edelbauer, A. (G) Composition of juice of Vitis vinifera at different chloride/sulphate ratios in the nutrient solution. Mitt. Klosterneuburg, Rebe u. Wein, Obstbau u. Früchteverwertung, Jahrg. 27, 217-222, 1977.

Edelbauer, A. (G) Grape yield, mineral concentration in leaves and first-year shoots and frost sensitivity of buds of Vitis vinifera L. at different Cl-/SO_4^- ratios. Z. Pflanzenernähr. Bodenk. 141, 83-94, 1978.

Edelmann, J. and Jefford, T. G. The mechanism of fructosan metabolism in higher plants as exemplified in Helianthus tuberosus. New Phytol. 76 517-531, 1968.

Edwards, D. G. and Asher, C. J. Tolerance of crop and pasture species to manganese toxicity. Proc. of the 9th Inter. Plant Nutrition Colloq. Commonwealth Agricultural Bureau p. 145-150, 1982.

Edwards, G. E. Carbon fixation and partitioning in the leaf. In: Regulation of Carbon and Nitrogen Reduction and Utilization in Maize, edited by Shannon, J. C., Knievel, D. P. and Boyer, C. D., Rockville,Md. Am. Soc. Plant Physiologists, p. 51-65, 1986.

Effmert, E. (G) The effect of fertilizer application on the phosphate content of potato starch (1). Effect of fertilizer application on the ratio of amylose/amylopectin in potato starch. ThaerArchiv 11, 745-753 and 755-759, 1967.

Egli, R. Climatic effects of air traffic. Environmental Conserv. (Genf) 22, 196-198, 1995.

Egmond, F. van, and Breteler, H. Nitrate reductase activity and oxalate content of sugarbeet leaves. Neth. J. agric. Sci. 20, 193-198, 1972.

Ehlers, W. (G) Transpiration coefficients of crops under field conditions. Pflanzenbauwissenschaften 1, 97-108, 1997.

Ehlers, W., Gebhardt, H. and Meyer, B. (G) Investigations into the position specific bonds of potassium to illite, kaolinite, montmorillonite, and humus. Z. Pflanzenernähr. Bodenk. 119, 173-186, 1968.

Ehrler, W. L. Transpiration of alfalfa as affected by low root temperature and other factors of a controlled environment. Plant Physiol. 37, Supplm. 843, 1962.

El-Hassanin, A. S. and Lynd, J. Q.: Soil fertility effects with tripartite symbiosis for growth, nodulation and nitrogenase activity of Vicia faba L. J. Plant Nutrition 8 (6) 491-504, 1985.

El-Sheikh, A. M. and Ulrich, A. Interactions of rubidium, sodium and potassium on the nutrition of sugar beet plants. Plant Physiol. 46, 645-649, 1970.

Elgawhary, S. M., Lindsay, W. L. and Kemper, W. D. Effect of complexing agent and acids on the diffusion of zinc to a simulated root. Soil Sci. Soc. Amer. Proc. 34 211-214, 1970.

Ellenberg, H. (G) Nitrogen as a factor of ecological sites. Ber. Dtsch. Bot. Ges. 77, 82-92, 1964.

Elsokkary, I. H. and Baghdady, N. H. Studies of molybdenum in some soils of Egypt. Alexandria J. Agric. Res. 21, 451-460, 1973.

Embleton, T. W., Matsumura, M., Stolzy, M. H., Devitt, D. A., Jones, W. W. and El-Motaium, R. Citrus nitrogen fertilizer management, ground water pollution, soil salinity and nitrogen balance. Appl. Agric. Res. 1, 57-64, 1986.

Engel, T., Mangstl, A., Reiner, L. and Graff, M. Prognois of the Nmin-store in the soil layer of 60-90 cm depth. VDLUFA-Schriftenr. 30, 313-318, 1989.

Engels, K. A., Becker, M., Ottow, J. C. G. and Ladha, J. K. Influence of phosphorus or phosphorus-potassium fertilization on biomass and dinitrogen fixation of the stem-nodulating green manure legume *Sesbania rostrata* in different marginally productive wetland rice soils. Biol. Fertil. Soils 20, 107-112, 1995.

Engels, C. and Marschner, H. Allocation of photosynthate to individual tubers of *Solanum tuberosum* L. II. Relationship between growth rate, carbohydrate concentration and ^{14}C partitioning within tubers. J. Exp. Bot. 37, 1804-1812, 1986.

Engels, C. and Marschner, H. Effects of reducing leaf area and tuber number on the growth rates of tubers on individual potato plants. Potato Res. 30, 177-186, 1987.

Engels, C. and Marschner, H. Adaptation of potassium translocation into the shoot of maize (*Zea mays*) to shoot demand: Evidence for xylem loading as a regulating step. Physiol. Plant. 86, 263-268, 1992.

Eppendorfer, W. The effect of nitrogen and sulphur on changes in nitrogen fractions of barley plants at various early stages of growth and on yield and amino acid composition of grain. Plant and Soil 29, 424-438, 1968.

Epstein, E. Ion absorption by roots: The role of micro-organism. New Phytol. 71, 873-874, 1972a.

Epstein, E. Mineral Nutrition of Plant: Principles and Perspectives. John Wiley and Sons, Inc., New York, London, Sydney, Toronto 1972b.

Epstein, E. The anomaly of silicon in plant biology. Proc. Natl. Acad. Sci. 91, 11-17, 1994.

Epstein, E. Silicon. Annu. Rev. Plant Physiol. Plant Mol. Biol. 50, 641-664, 1999.

Erenoglu, B., Cakmak, I., Marschner, H., Römheld, V. Eker, S., Daghan, H., Kalayci, M. and Ekiz, H. Phytosiderophore release does not relate well with zinc efficiency in different bread wheat genotypes. J. Plant Nutr. 19, 1596-1580, 1996.

Esch, A. and Mengel, K. Combined effects of acid mist and frost drought on the water status of young spruce trees (*Picea abies*). Environ. Expt. Bot. 39, 57-65, 1998.

Eschrich, W. Free space invertase, its possible role in phloem unloading. Ber. Dtsch. Bot. Ges. 93, 363-378, 1980.

Eschrich, W. Phloem unloading of photoassimilates. In: Transport of Photoassimilates, edited by Baker, D. A. and Milburn, J. A. New York: John Wiley & sons, p 206-263, 1989.

Estruch, J. J., Chriqui, D., Grossman, K., Schell, J. and Spena, A. The plant oncogene role is responsible for the release of cytokinins from glucoside conjugates. European MBO Journal 10, 2889-2895, 1991.

Etherton, B. Relationship of cell transmembrane electropotential to potassium and sodium acculmulation ratios in oat and pea seedlings. Plant Physiol. 38, 581-585, 1963.

Etherton, B. and Higinbotham, N. Transmembrane potential measurement of cells of higher plants as related to salt uptake. Science 131, 409-410, 1961.

Evans, D. E., Briars, S.-A. and Williams, L. E. Active calcium transport by plant cell membranes. J. Exp. Bot. 42, 285-303, 1991.

Evans, H. J. and Barber, L. E. Biological nitrogen fixation for food and fiber production. Science 197, 332-339, 1977.

Evans, H. J. and Russell, S. A. Physiological chemistry of symbiotic nitrogen fixation by legumes. In: The Chemistry and Biochemistry of Nitrogen Fixation, Plenum Publishing Co. edited by Postgate, J. R., p. 191-244, 1971.

Evans, H. J. and Sorger, G. J. Role of mineral elements with emphasis on the univalent cations. Annu. Rev. Plant Physiol. 17, 47-77, 1966.

Evans, L. T. The physiological basis of crop yield. In: Crop Physiology, edited by Evans, L. T. Cambridge: Cambridge University Press, p. 327-355, 1975.

Evans, L. T. Feeding the Ten Billion - Plants and Population Growth, Cambridge:Cambridge University Press, 1998.

Evans, L. T. Gibberellins and flowering in long day plants, with special reference to *Lolium temulentum*. Austr. J. Plant Physiol. 26, 1-8, 1999.

Evans, L. T. and Rawson, H. M. Photosynthesis and respiration by the flag leaf and components of the ear during grain development in wheat. Aust. J. Biol. Sci. 23, 245-254, 1970.

Evans, L. T., Wardlaw, I. F. and Fischer, R. A. Wheat. In: Crop Physiology, Cambridge University Press, edited by Evans, L. T., p. 101-149, Cambridge 1975.

Everaarts, A. P. and De Moel, C. P. The effect of nitrogen and the method of application on yield and quality of whit cabbage. European J. Agron. 9, 203 -211, 1998.

Evert, R. F. Vascular anatomy of angiospermous leaves, with special consideration of the maize leaf. Ber. Deutsch. Bot. Ges. 93, 43-55, 1980.

Evert, R. F. and Russin, W. A. Structurally, phloem unloading in the maize leaf cannot be symplastic. Am. J. Bot. 90, 1310-1317, 1993.

Ewart, J. A. D. Glutenin and dough tenacity. J. Sci. Fd Agric. 29, 551-556, 1978.

Ewert, F. and Honermeier, B. Spikelet initiation of winter triticale and winter wheat in response to nitrogen fertilization. European J. Agron. 11, 107-113, 1999.

Fabian, P. (G) Air contamination and greenhouse effect: anthropogenic actions and their impact on climate. In: Klima und Mensch, edited by Schubert, V. and Quenzel, H. EOS-Verlag, St. Ottilien, p. 57-97, 1997.

Facanha, A. R. and de Meis, L. Inhibition of maize root H^+-ATPase by fluoride and fluoroaluminate complexes. Plant Physiol. 108, 241-246, 1995.

Fackler, U., Goldbach, H. E., Weiler, E. W. and Amberger, A. Influence of boron deficiency on indole-3-acetic acid and abscisic acid levels in roots and shoot tips. J. Plant Physiol. 119, 295-299, 1994.

Fageria, N. K., Baligar, V. C. and Jones, C. A. Growth and Mineral Nutrition of Field Crops, New York, Basel, Hong Kong: Marcel Dekker,Inc., 1991.

Fairley-Grenot, K. and Assmann, S. M. Whole-cell K^+ current across the plasma membrane of guard cells from a grass: *Zea mays*. Planta 186, 282-292, 1992.

Faiz, S. M. A. and Weatherley, P. E. Further investigations into the location and magnitude of the hydraulic resistance in the soil-plant system. New Phytol. 81 19-28, 1978.

Falchuk, K. H., Ulpino, L, Mazus, B. and Vallee, B. L. *E. gracilis* RNA polymerase I: A zinc metalloenzyme. Biochem. Biophys. Res. Commun 74, 1206-1212, 1977.

Fallon, K. M., Shacklock, P. S. and Trewavas, A. J. Detection *in vivo* of very rapid red light-induced calcium-sensitive protein phosphorylation in etiolated wheat (*Triticum aestivum*) leaf protoplasts. Plant Physiol. 101, 1039-1045, 1993.

Fanning, D. S., Keramidas, V. Z. and El-Desoky, M. A. Micas. In: Minerals in Soil Environments, edited by Dixon, J. B. and Weed, S. B. Madison, USA: Soil Science Soc. of America, p. 551-634, 1989.

Farley, R. F. and Draycott, A. P. Manganese deficiency of sugar beet in organic soil. Plant and Soil 38, 235-244, 1973.

Farley, R. F. and Draycott, A. P. Diagnosis of manganese deficieney in sugar beet and response to manganese applications. J. Sci. Fd. Agric. 27, 991-998, 1976.

Farmer, V. C. and Wilson, M. J. Experimental conversion of biotite to hydrobiotite. Nature 226, 841-842, 1970.

Farquhar, G. D., Firth, P. M., Wetselar, R. and Weir, B. On the gaseous exchange of ammonia between leaves and the environment: determination of the ammonia compensation point. Plant Physiol. 66, 710-714, 1980.

Farquhar, D. G., Wetselaar, R. and Weir, B. Gaseous nitrogen losses from plants. In: Gaseous Loss of Nitrogen from Plant-Soil-Systems, edited by Freney, J. R. and Simpson, J. R. The Hague: Martinus Nijhoff / Dr. W. Junk Publishers, p. 159-180, 1983.

Faust, M. and Shear, C. B. Biochemical changes during the development of cork spot of apples. Qual. Plant Mater. Veg. 19, 255-265, 1969.

Feigenbaum, S., Edelstein, R. E. and Shainberg, J. Release rate of potassium and structural cations from mica to ion exchangers in dilute solutions. Soil Sci. Soc. Am. J. 45, 501-506, 1981.

Felle, H. The apoplastic pH of Zea $mays$ root cortex as measured with pH sensitive microelectrodes: aspects of regulation. J. Expt. Bot. 49, 987-995, 1998.

Felle, H., Peters, W. and Palme, K. The electrical response of maize to auxins. Biochim. Biophys. Acta 1064, 199-204, 1991.

Felle, H. H. The H^+/Cl^- symporter in root-hair cells of $Sinapis$ $alba$. Plant Physiol. 106, 1131-1136, 1994.

Feng, J., Volk, R. J. and Jackson, W. A. Source and magnitude of ammonium generation in maize roots. Plant Physiol. 118, 835-841, 1999.

Feng, K. (G) Dynamics of interlayer NH_4^+ in flooded chinese rice soils. Ph. D. Thesis, Justus Liebig University, Giessen, 1995.

Fenn, L. B. and Hossner, L. R. Ammonia volatilization from ammonium or ammonium-forming fertilizers. Adv. Soil Sci. 1, 123-169, 1985.

Fernando, M., Mehroke, J. and Glass, A. D. M. De $novo$ synthesis of plasma membrane and tonoplast polypeptides of barley roots during short term K deprivation. Plant Physiol. 100, 1269-1276, 1992.

Feuerle, R. (G) Effect of shading and reduction of stem by treatment with CCC on the transiett storage of sugars in stems and the inulin yield of tubers of $Helianthus$ $tuberosus$ L. Diploma work, Faculty Agriculture, Justus Liebig University Gießen, 1992.

Finck, A. (G) Manganese requirement of oats at various growth stages. Plant and Soil 7, 389-396, 1956.

Findeklee, P., Wimmer, M. and Goldbach, H. E. Effects of boron deficiency on physical cell wall parameters, hydraulic conductivity and plasma bound reductase activities in young C. pepo and V. faba roots. In: Boron in Soils and Plants, edited by Bell, R. and Rerkasem, B., Kluwer Academic Publish. Dordrecht, p. 221-227, 1997.

Findenegg, G. R. A comparative study of ammonium toxicity at different pH of the nutrient solution. Plant and Soil 103, 239-243, 1987.

Finger, H. and Werk, 0. (G) Increase of the sodium and magnesium content in pasture herbage and the influence of Magnesia-Kainite application on the forage uptake by cows. Landw. Forsch. 28/II. Sonderh., 190-196, 1973.

Firestone, M. K. Biological denitrification. In: Nitrogen in Agricultural Soils, edited by Stevenson, F. J. Am. Soc. Agronomy, Madison USA, p. 289-326, 1982.

Fischer, E. S., Lohaus, G., Heineke, D. and Heldt, H. W. Magnesium deficiency results in accumulation of carbohydrates and amino acids in source and sink leaves of spinach. Physiol. Plant. 102, 16-20, 1998.

Fischer, R. A. Stomatal opening: role of potassium uptake by guard cells. Science 160, 784-785, 1968.

Fischer, R. A. and Hsiao, T. C. Stomatal opening in isolated epidermal strips of *Vicia faba*. II. Responses to KCl concentration and the role of potassium adsorption. Plant Physiol. 43, 1953-1958, 1968.

Fisher, D. B. Structure of functional soybean sieve elements. Plant Physiol. 56, 555-569, 1975.

Fisher, D. B. An evaluation of the Münch hypothesis for phloem transport in soybean. Planta 139, 25-28, 1978.

Fisher, D. B. and Wang, N. Sucrose concentration gradients along the postphloem transport pathway in the maternal tissues of developing wheat grains. Plant Physiol. 109, 587-592, 1995.

Fisher, R. A. A preliminary note on the effect of sodium silicate in increasing the yield of barley. J. Agric. Sci. 19, 132-139, 1929.

Fismes, J., Vong, P. C., Guckert, A. and Frossard, E. Influence of sulphur on apparent N-use efficiency, yield and quality of oilseed rape (*Brassica napus* L.) grown on a calcareous soil. European J. Agron. 12, 127-141, 2000.

Fleming, G. Mineral disorders associated with grassland farming. In Proc. Intern. Meeting on Animal Production from Temperate Grassland. An Foras Taluntais, Dublin, p. 88-95, 1977.

Fleming, G. A. Essential micronutrients. In: Applied Soil Trace Elements, edited by Davies, B. E., John Whiley, New York, p. 155-176, 1980.

Flowers, T. J. Cloride as a nutrient and as an osmoticum. In: Advances in Plant Nutrition, edited by Tinker, B. and Läuchli, A. New york: Praeger, p. 55-78, 1988.

Flowers, T. J., Troke, P. F. and Yeo, A. R. The mechanism of salt tolerance in halophytes. Annu. Rev. Plant Physiol. 28, 89-121, 1977.

Flügge, U.-I. Phosphte translocators in plastids. Annu. Rev. Plant Physiol. Plant Mol. Biol. 50, 27-45, 1999.

Föhse, D., Claassen, N. and Jungk, A. Phosphorus efficiency of plants. II Significance of root hairs and cation-anion balance for phosphorus influx in seven plant species. Plant and Soil 132, 261-272, 1991.

Föhse, D. and Jungk, A. Influence of phosphate and nitrate supply on root hair formation of rape, spinach and tomato plants. Plant and Soil 74, 359-386, 1983.

Follett, R. H. and Lindsay, W. L. Profile distribution of zinc, iron, manganese and copper in Colorado soils. Colorado Exp. Station Techn. Bull. 110, 1970.

Fondy, B. R. and Geiger, D. R. Diurnal pattern of translocation and carbohydrate metabolism in source leaves of *Beta vulgaris* L. Plant Physiol. 70 671-676, 1982.

Forster, H. (G) Effect of some interruptions in the nutrient supply on the development of yield and quality characteristics in sugar beets. Landw. Forsch. 25/II. Sonderh., 99-105, 1970.

Forster, H. (G) Effect of the potassium and nitrogen to plants on yield components and yield formation of cereals. Landw. Forsch. 26, 221-227, 1973a.

Forster, H. Relationship between the nutrition and the appearance of'greenback' and 'blossomend rot' in tomato fruits. Acta Hort. 29, 319-326, 1973b.

Forster, H. and Mengel, K. (G) The effect of a short term interruption in the K supply during the early stage on yield formation, mineral content and soluble amino acid content. Z. Acker-u. Pflanzenbau 130, 203-213, 1969.

Forster, H. and Venter, F. (G) The effect of the K nutrition on 'green back' in tomato fruits. Gartenbauwiss. 40, 75-78, 1975.

Fortmeier, H. (G) Na+/H+ antiport in maize roots? Investigations as to the mechanism of the active Na+ transport in the plasmalemma of maize root cells (*Zea mays* L.) Giessen, Ph. D. Thesis, 2000.

Fortmeier, R. and Schubert, S. Salt tolerance of maize (*Zea mays* L.): the role of sodium exclusion. Plant, Cell Environ. 18, 1041-1047, 1995.

Fox, R. H. Soil pH, aluminium saturation, and corn grain yield. Soil Sci. 127, 330-334, 1979.

724

Fox, T. C. and Guerinot, M. L. Molecular biology of cation transport in plants. Annu. Rev. Plant Physiol. Plant Mol. Biol. 49, 669-696, 1998.

Fox, T. C., Shaff, J. E., Grusak, M. A., Norvell, W. A., Chen, Y., Chaney, R. L. and Kochian, L. V. Direct measurement of ^{59}Fe labeled Fe^{2+} influx in roots of pea using a chelator buffer system to control free Fe in solution. Plant Physiol. 111, 93-100, 1996.

Foy, C. D.: Effects of soil calcium availability on plant growth. In: The Plant Root and its Environment, edited by Carson, E. W. Univ. Press of Virginia, Charlottesville, p. 565-600, 1974.

Foy,C. D. General principles of sceening plants for aluminium and manganese tolerance. In: Plant Adaption to Mineral Stress in Problem Soils. Workshop Proceedings, Agricultural Library Beltsville, Maryland USA, p. 255 -267, 1976.

Foy, C. D., Chaney, R. L. and White, M. C. The physiology of metal toxicity in plants. Annu. Rev. Plant Physiol. 29, 511-566, 1978.

Foy, C. D., Webb, H. W. and Jones, J. E. Adaptation of cotton genotypes to an acid manganese toxic soil. Agron. J. 73, 107-111, 1981.

Foy, R. H. and Withers, P. J. A. The contribution of agricultural phosphorus to eutrophication, London: The Fertilizer Soc., Proc. No. p. 365, 1995.

Foyer, C. H., Lelandais, M., Edwards, E. A. and Mullineaux, P. M. The role of ascorbate in plants, interactions with photosynthesis, and regulatory significance. In: The Active Oxygen/Oxidative Stress and Plant Metabolism, edited by Pell, E. and Steffens, K. Am. Soc. Plant Physiologists, Rockville. p. 131-144, 1991.

Franco, A. A. and Faria, de S. M. The contribution of N_2-fixing tree legumes to land reclamation and sustainability in the tropics. Soil Biol. Biochem. 29, 897-903, 1997.

Francois, A. (F) The role of alcoholic liquors in nutrition. C. R. Acad. Agric. 76, 73-86, 1990.

Frank, R., Ishida, K. and Suda, P. Metals in agricultural soils of Ontario. Can. J. Soil Sci. 56, 191-196, 1976.

Franke, W. Mechanisms of foliar penetration of solutions. Annu. Rev. Plant Physiol. 18, 281-300, 1967.

Fredeen, A. L., Rao, I. M. and Terry, N. Influence of phosphorus nutrition on growth and carbon partitioning of *Glycine max*. Plant Physiol. 89, 225-230, 1989.

Frehner, M., Keller, F., Matille, P. and Wiemken, A. Sugar transport across the tonoplast of vacuoles isolated from protoplasts of Jerusalem artichoke tubers (*Helianthus tuberosus* L.). In: Plant vacuoles, edited by D. Marin, Plenum Press, New York, pp. 281-286, 1987.

Frémond, Y. and Ouvrier, M. (F) Importance of an adequate mineral nutrition for the establishment of a coco plantation on sandy soils. Oléagnieux, 26e année, No. 10, 609-616, 1971.

Freney, J. R., Denmead, O. T., Wood, A. W., Saffigna, P. G., Chapman, L. S., Ham, G. J., Hurney, A. P. and Stewart, R. T. Factors controlling ammonia loss from trash covered sugar cane fields fertilized with urea. Fert. Res. 31, 341 -349, 1992.

Frensch, J. and Hsiao, T. C. Hydraulic propagation of pressure along immature and mature xylem vessels of roots of *Zea mays* measured by pressure-probe techniques. Planta 190, 263-270, 1993.

Frensch, J. and Steudle, E. Axial and radial hydraulic resistance to roots of maize (*Zea mays* L.). Plant Physiol. 91, 719-726, 1989.

Fridovich, I. Superoxide radical and superoxide dismutases. Annu. Rev. Biochem. 64, 97-112, 1995.

Fried, M. 'E', 'L', and 'A' values. 8th Intern. Congress of Soil Science, Bucharest, Romania IV, 29-39, 1964.

Fried, M. and Broeshart, H. The Soil-Plant System, Acadernic Press, New York, London p. 183-206, 1969.

Fried, M. and Shapiro, R. E. Soil-plant relationships in ion uptake. Annu. Rev. Plant Physiol. 12, 91-112, 1961.

Fries-Nielsen, B. An approach towards interpreting and controlling the nutrient status of growing plants by means of chemical plant analyses. Plant and Soil 24, 63-80, 1966.

Fritz, D., Habben, J., Reuff, B. and Venter, F. (G) The variability of quality-relevant molecules in tomatoes. Gartenbauwiss. 41, 104 -109, 1976.

Frota, J. N. E. and Tucker, T. C. Salt and water stress influences nitrogen metabolism in red kidney beans. Soil Sci. Soc. Am. J. 42, 743-746, 1978.

Früchtenicht, K., Hoffmann, G. and Vetter, H. (G) Is fertilizer application correct with regard to soil fertility, yield, and profit? In: Düngung, Umwelt, Nahrungsqualität, VDLUFA, Darmstadt, p. 152-168, 1978.

Fry, S. C. Cellulase, hemicelluloses and auxin-stimulated growth: a possible relationship. Physiol. Plant. 75, 532-536, 1989.

Fuchs, W. (G) Investigations on the effect of nitrogen fertilization on the setting and the development of the yield character 'number of spikelets per ear' in case of winter rye, winter wheat, and two row spring barley. Arch. Acker- u. Pflanzenbau and Bodenk. 19, (4) 277-286, 1975.

Fung, L. E., Wang, S. S., Altman, A. and Hüttermann, A. Effect of NaCl on growth, photosynthesis, ion and water relations of four popular genotypes. Forest, Ecology and Management 107, 135-146, 1998.

Funk, R., Maidl, F. X., Wagner, B. and Fischbeck, G. (G) Vertical water ant nitrate transport into deeper soil layers of arable land in Southern Germany. Z. Pflanzenernähr. Bodenk. 158, 399-406, 1995.

Fürstenfeld, F., Deller, B. and Schultz, R. (G) Nitrogern fertilizer application systems in comparison. VDLUFA-Schriftenr. 38, 215-218, 1994.

Gabathuler, R. and Cleland, R. E. Auxin regulation of a proton translocating ATPase in pea root plasma membrane vesicles. Plant Physiol. 79, 1080-1085, 1985.

Gaffney, T., Friedrich, L., Vernooij, B., Negrotto, D., Nye, G., Uknes, C., Ward, E., Kessmann, H. and Ryals, J. Requirement of salicylic acid for the induction of systemic aquired resistance. Science. 261, 754-756, 1993.

Gahoonia T. S., Claassen, N. and Jungk, A. Mobilization of phosphate in different soils by ryegrass supplied with ammonium or nitrate, Plant a. Soil. 140, 241-248, 1992.

Gale, J. The combined effect of environmental factors and salinity on plant growth. In: Plants in saline environments, edited by Poljakoff-Mayber, H. & Gale, J. New York: Springer Verlag, 186-192, 1975.

Gale, J. and Hagen, R. M. Plant antitranspirants. Annu. Rev. Plant Physiol. 17, 269-282, 1966.

Gales, K. Yield variation of wheat and barley in Britain in relation to crop growth and soil conditions - a review. J. Sci. Food Agric. 34, 1085-1104, 1983.

Gamalei, Y. Structure and function of leaf minor veins in trees and herbs. A taxonomic review. Trees 3, 96-110, 1989.

Gambrell, R. P. Manganese. In: Methods of Soil Analysis, edited by Bigham, J. M. Madison, Wisconsin: Soil Sci. Soc. of America, p. 665-682, 1996.

Ganmore-Neumann, R. and Kafkafi, U. The effect of root temperature and NO_3^-/NH_4^+ ratio on strawberry plants. I. Growth, flowering and root development. Agron. J. 75, 941-947, 1983.

Ganßmann, W. (G) β-glucan in oats and its physiological effect. Getreide, Mehl u. Brot. 47, 47-52, 1993.

Gapon, E. N. On the theory of exchange adsorption in soils. J. Gen. Chem. (USSR)3,144-163, 1933.

Garbarino, J. and DuPont, F. M. Rapid induction of Na^+/H^+ exchange activity in barley root tonoplast. Plant Physiol. 89, 1-4, 1989.

Gardner, D. J. C. and Peel, A. J. ATP in sieve tube sap from willow. Nature 222, 774, 1969.

Gardner, W. K., Barber, D. A. and Parberry, D. G. The acquisition of phosphorus by *Lupinus albus* L. Plant and Soil 70, 107-124, 1983.

Gardner, W. K. and Parbery, D. G. The acquisition of phosphorus by *Lupinus albus* L. I. Some characteristics of the soil root interface. Plant and Soil 68, 19-32, 1982.

Gardner, W. R. Dynamic aspects of soil-water availability to plants. Annu. Rev. Plant Physiol. 16, 323-342,

726

1965.

Gärtel, W. (G) The micronutrients - their importance for the nutrition of grapes with particular regard to deficiency and toxicity symptoms. Weinberg u. Keller 21, 435-507, 1974.

Gartner, J. A. Effect of fertilizer nitrogen on a dense sward of Kikuyu. Paspalum and carpet grass. 2. Interactions with phopshorus and potassium. Queensl. J. of Agric. and Anim. Sci. 26, 365-372, 1969.

Garz, J. and Chaanin, A. (G) Relationships between nitrogen fertilizer application and the turnover of soil organic matter. Tag. Ber. Akad. Landwirtsch. Berlin 289, 193-200, 1990.

Gasser, J. K. R. The efficiency of energy used in the production of carbohydrates and lipids. In: Fertilizer Use and Production of Carbohydrates and Lipids. Int. Potash Inst. Bern, p. 341-352, 1977.

Gate, P., Crosson, P. and Lehe, D. (G) Can one assess the risk of cereal lodging ? Perspectives Agric. 209, 84-88, 1996.

Gavalas, N. A. and Manetas, Y. Calcium inhibition of phosphoenolpyruvate carboxylase. Possible physiological consequences for 4-carbon-photosynthesis. Z. Pflanzenphysiol. 100, 179-184, 1980.

Gazzarrini, S., Lejay, J., Gojon, A., Ninnemann, O., Frommer, W. B. and von Wiren, N. Three functional transporters for constitutive, diurnally regulated and starvation-induced uptake of ammonium into Arabidopsis roots. The Plant Cell 11, 937-947, 1999.

Geering, H. R., Hodgson, J. F. and Sdano, C. Micronutrient cation complexes in soil solution: IV. The chemical state of manganese in soil solution. Soil. Sci. Soc. Amer. Proc. 33, 81-85, 1969.

Geiger, D. R. Phloem loading. In: Transport in Plants, I. Phloem Transport, edited by Zimmermann, M. H. and Milburn, J. A. Springer-Verlag Berlin, Heidelberg, New York, Tokyo p. 396-431, 1975.

Geiger, D. R. Control of partitioning and export of carbon in leaves of higher plants. Bot. Gaz. 140, 241-248, 1979.

Geiger, D. R. and Cataldo, D. A. Leaf structure and translocation in sugar beet. Plant Physiol. 44, 45-54, 1969.

Geiger, D. R., Koch, K. E. and Shieh, W.-J. Effect of environmental factors on whole plant assimilate partitioning and associated gene expression. J. Exp. Bot. 47 (special issue): 1229-1238, 1996.

Geiger, D. R., Servaites, J. C. and Shieh, W. J. Balance among parts of source sink system: a factor in crop productivity. In: Crop Photosynthesis: Spatial and Temporal Determinants, Elsevier, Amsterdam, p. 155-192, 1992.

Geiger, D. R. and Servaites, J. C. Diurnal regulation of photosynthetic carbon metabolism in C_3 plants. Annu. Rev. Plant Physiol. Plant Mol. Biol. 45, 235-256, 1994.

Geiger, D. R., Sovonick, S. A., Shock, T. L. and Fellows, R. J. Role of free space in translocation in sugar beet. Plant Physiol. 54, 892-898, 1974.

George, B. J. Design and interpretation of nitrogen response experiments. In: The nitrogen requirement of cereals, edited by MAFF Reference Books, p. 135-148, 1983.

George, J. R., Pinheiro, M. E. and Bailey, T. B. Long-term potassium requirements of nitrogen fertilized smooth brome-grass. Agron. J. 71, 586-591, 1979.

Gepstein, S. and Thimann, K. V. The role of ethylene in the senescence of oat leaves. Plant Physiol. 68: 349-354, 1981.

Gerdemann, LW.: Mycorrhizae. In: The Plant Root and its Environment, edited by E. W. Carson, University Press of Virginia, p. 205 - 217, Charlottesville 1974.

Gerendas, J., Polacco, J. C., Freyermuth, S. K. and Sattelmacher, B. Significance of nickel for plant growth and metabolism. J. Plant Nutr. Soil Sci. 162, 241-256, 1999.

Gerendas, J., Ratcliffe, R. G. and Sattelmacher, B. [31]P nuclear magneteic resonance evidence for differences in intercellular pH in the roots of maize seedlings grown with nitrate or ammonium. J. Plant Physiol.

137, 125-128, 1990.

Gerendas, J. and Sattelmacher, B. Significance of Ni supply for growth, urease activity and the contents of urea, amino acids, and mineral nutrients of urea -grown plants-. Plant and Soil 190, 153 162, 1997.

Gerke, J. Orthophosphate and organic phosphate in the soil solution of four sandy soils in relation to pH - evidence for humic-Fe (Al) phoshate complexes. Commun. Soil Sci. Plant Anal. 23, 601-612, 1992.

Gerke, J. Kinetics of soil phosphate desorption as affected by citric acid. Z. Pflanzenernähr. Bodenk. 157, 17-22, 1994.

Gerke, J. Beissner, L. and Römer,W. The quantitative effect of chemical phosphate mobilization by carboxylate anions on P uptakke by a single root. I. The basic concept and determination of soil parameters.: Plant Nutr. Soil Sci. 163, 207 - 212, 2000.

Gerke, J. and Hermann, R. Adsorption of orthophosphate to humic-Fe -complexes and to amorphous Fe-oxide. Z. Pflanzenernähr. Bodenk. 155, 233-236, 1992.

Gerke, J., Meyer, U. and Römer, W. Phosphate, Fe and Mn uptake of N_2 fixing red clover and ryegrass from an Oxisol as affected by P and model humic substances application. 1. Plant parameters and soil solution composition. Z. Pflanzenernähr. Bodenk. 158, 261-268, 1995.

Gerke, J., Römer, W. and Jungk, A. The excretion of citric and malic acid by proteoid roots of *Lupinus albus* L.; effects on soil solution concentrations of phosphate, iron and aluminium in the proteoid rhizosphere in samples of an oxisol and a luvisol. Z. Pflanzenernähr. Bodenk. 157, 289-294, 1994.

Gerson, D. F. and Poole, R. J.: Chloride accumulation by mung bean root tips. A low affinity active transport system at the plasmalemma. Plant Physiol. 50, 603-607,1972.

Getz, H. P., Thom, M. and Maretzki, A. Proton and sucrose transport in isolated tonoplast vesicles from sugarcane stalk tissue. Physiol. Plant. 83: 404-410, 1991.

Giaquinta, R. T. and Geiger, D. R.: Mechanism of inhibition of translocation by localized chilling. Plant Physiol. 51, 372-377, 1973.

Gibson, T. S., Speirs, J. and Brady, C. J. Salt tolerance in plants. II. *In vitro* translation of m-RNA from salt-tolerant and salt-sensitive plants on wheat germ ribosomes. Responses to ions and compatible organic solutes. Plant, Cell Environ. 7: 579-587, 1984.

Gifford, R. M. and Evans, L. T. Photosynthesis, carbon partitioning, and yield. Annu. Rev. Plant Physiol. 32, 485-509, 1981.

Gifford, R. M., Thorne, J. H., Hitz, W. D. and Giaquinta, R. T. Crop productivity and photoassimilate partitioning. Sci. 225, 801-808, 1984.

Giordano, P. M. and Lindsay, W. L. Soil Sci. Soc. America, p. 505-524, 1972.

Giordano, P. M. and Mortvedt, J. J. Agronomic effectiveness of micronutrients in macronutrient fertilizers. In: Micronutrients in Agriculture, edited by Mortvedt, J. J.

Giordano, P. M., Noggle, J. C. and Mortvedt, J. J. Zinc uptake by rice as affected by metabolic inhibitors and competing cations. Plant and Soil 41, 637-646, 1974.

Gisiger, L. and Hasler, A. (G) Causes of grey speck in oats. Plant and Soil 1, 19-30, 1949.

Giskin, M. and Majdan, A. Problems of plant nutrition and fertilizer use in Huleh muck soils. In: Transition from Extensive to Intensive Agriculture with Fertilizers. Proc. 7th Colloq. Intern. Potash Institute Berne, p. 249-252, 1969.

Gissel-Nielsen, G., Gupta, U. C., Lamand, M. and Westermarck, T. Selenium in soils and plants and its importance in livestock and human nutrition. Adv. Agron. 37, 397-460, 1984.

Gladstones, J. S., Loneragan, J. F. and Goodchild, N. A. Field responses to cobalt and molybdenum by different legurne species with inferences on the role of cobalt in legume growth. Austr. J. of Agric, Res. 28, 619-628, 1977.

728

Glass, A. D. M. Regulation of ion transport. Annu. Rev. Plant Physiol. 34, 311-326, 1983.

Glass, A. D. M. Nitrogen transport in plants, with an emphasis on the regulation of fluxes to meet plant demand. Presentation at the Annual Meeting of the German Society of Plant Nutrition. Sept. 2000.

Glass, A. D. M. and Siddiqi, M. Y. The control of nutrient uptake rates in relation to the inorganic composition of plants. In: Advances in Plant Nutrition, edited by Tinker, P. B. and Läuchli, A. New York: Praeger, p. 103-147, 1984.

Glendening, T. M. and Poulton, J. E. Glucosinolate biosynthesis. Plant Physiol. 86, 319-321, 1988.

Glenn, E. P., Brown, J. J. and Blumwald, E. Salt tolerance and crop potential of halophyrtes. Crit. Rev. Plant Sci., 18, 227-255, 1999.

Glüsenkamp, K.-H.,. Kosegarten, H., Mengel, K., Grolig, F., Esch, A. and Goldbach, H. E. A fluorescein boronic acid conjugate as a marker for borate binding sites in the apoplast of growing roots of *Zea mays* L. and *Helianthus annuus* L. In: Boron in Soils and Plants, edited by Bell, R. W. and Rerkasem, B. Dordrecht, Netherlands: Kluwer Academic Publishers, p. 229-235, 1997.

Glynne, M. D. Effect of potash on powdery mildew in wheat. Plant Path. 8, 15-16, 1959.

Gobold, D. L., Fritz, E. and Hüttermann, A. Lead contents and distribution in spruce roots. In: Air Pollution and Ecosystems, edited by Mathy, P. Dordrecht: D. Reidel Publishing Company, p. 864-869, 1988.

Godbold, D. L., Horst, W. J., Collins, J. C., Thurman, D. A. and Marschner, H. Accumulation of zinc and organic acids in roots of zinc-tolerant and non-tolerant ecotypes of *Deschampsia caespitosa*. J. Plant Physiol. 116, 59-69, 1984.

Godo, G. H. and Reisenauer, H. M. Plant effects on soil manganese availability. Soil Sci. Soc. Am. J. 44, 993-995, 1980.

Goh, K. M. Status and distribution of soil sulphur fractions, total nitrogen and organic carbon in camp and non-camp soils of grazed pastures receiving long-term superphosphate applications. Biol. Fertil. Soils in press:1992.

Gök, M. and Ottow, J. C. G. Effect of cellulose and straw incorporation in soil on total denitrification and nitrogen immobilization at intially aerobic and permanently anaerobbic conditions. Biol. Fertil. Soils 5, 317-322, 1988.

Goldbach, H. E. A critical review on current hypothesis concerning the role of boron in higher plants: sugestions for further research and methodological requirements. J. Trace and Microprobe,Techniques 15, 51-91,1997.

Goldberg, S. Reactions of boron with soils. Plant and Soil 193, 35-48, 1997.

Goldberg, S. and Forster, H. S. Boron sorption on calcareous soils and reference calcites. Soil Sci. 152, 304-310, 1991.

Gollmick, F., Neubert, P. and Vielemeyer, H. P. (G) Possibilities and limitations of plant analysis in estimating the nutrient requirement of crops. Fortschrittsberichte f. d. Landw. u. Nahrungsgüterwirtschaft 8, H. 4 Berlin, 1970.

Gomez-Cadenas, A., Tadeo, F. R., Primo-Millo, E. and Talon, M. Involvement of abscisic acid and ethylene in the response of citrus seedlings to salt shock. Physiol. Plant. 103, 475-484, 1998.

Gong, Z. Origin, evolution and classification of paddy soils in China. Adv. Soil Sci. 5, 179-200, 1986.

Gonzalez, A., Korenkov, V. and Wagner, G. J. A comparison of Zn, Mn, and Ca transport mechanisms in oat root tonoplast vesicles. Physiol. Plant. 106, 203-209, 1999.

Gonzalez, A. and Lynch, J. P. Effects of mangenese toxicity on leaf CO_2 assimilation of contrasting common bean genotypes. Physiol. Plant. 101, 872-880, 1997.

Good, N. E., Winget, G. D., Winter, W., Connoly, T. N., Izawa, S. and Singh, R. M. M. Hydrogen ion buffers for biological research. Biochemistry 5, 467 -477, 1966.

729

Gordon, W. R., Schwemmer, S. S. and Hillman, W. S. Nickel and the metabolism of urea by *Lemna pancicostata*. Hegelm. 6746. Planta 140, 265-268, 1978.

Gorham, J., Wyn Jones, R. G. and McDonnell, E. Some mechanisms of salt tolerance in crop plants. Plant Soil 89, 15-40, 1985.

Görlitz, H. (G) Effect of fertilizer application on properties of potato starch. In: Mineralstoffversorgung von Pflanze und Tier, Tagungsberichte Nr. 85, Dt. Akad. d. Landw. Wiss. Berlin, p. 93-100, 1966.

Goss, R. L. The effects of potassium on disease resistance. In: The Role of Potassium in Agriculture, Madison, USA, p. 221-241, 1968.

Goswami, A. K. and Willcox, J. S. Effect of applying increasing levels of nitrogen to ryegrass. 1: Composition of various nitrogenous fractions and free amino acids. J. Sci. Food Agric. 20, 592-595, 1969.

Goulding, K. W. T., Hütsch, B. W., Webster, C. P., Willison, T. W. and Powlson, D. S. The effect of agriculture on methane oxidation in soil. Phil. Trans. R. Soc. Lond. 351, 313-325, 1995.

Gouy, G. (G) quoted from D. Kortüm: Textbook of Electrochemistry. Verl. Chemie, Weinheim, p. 345, 1957.

Govers, F. and Bisseling, T. Nodulins in root nodule development: function and gene regulation. In: Nitrogen Metabolism of Plants, edited by Mengel, K. and Pilbeam, D. J., Oxford, Clarendon Press, p. 31-37, 1992.

Graebe, J. E. Gibberellin biosynthesis and control. Annu. Rev. Plant Physiol. 38, 419-465, 1987.

Graham, E. R. and Fox, R. L. Tropical soil potassium as related to labile pool and calcium exchange equilibria. Soil Sci. 111, 318-322, 1971.

Graham, E. R. and Kampbell, D. H. Soil potassium availability and reserve as related to the isotopic pool and calcium exchange equilibria. Soil Sci. 106, 101-106, 1968.

Graham, E. R. and Lopez, P. L. Freezing and thawing as a factor in the release and fixation of soil potassium as demostrated by isotopic exchange and calcium equilibria. Soil Sci. 108, 143-147, 1969.

Graham, J. H., Leonard, R. T. and Menge, J. A. Membrane mediated decrease in root exudation responsible for phosphorus inhibition of vesicular-arbuscular mycorrhiza formation. Plant Physiol. 68, 548-552, 1981.

Graham, R. D. The distribution of copper and soluble carbohydrates in wheat plants grown at high and low levels of copper supply. Z. Pflanzenernähr. Bodenk. 143, 161-169, 1980.

Graham, R. D. Absorption of copper by plant roots. In: Copper in Soils and Plants, edited by Loneragan, J. F., Robson, A. D. and Graham, R. D., Academic Press, p. 141-160, 1981.

Granli, T. and Bockman, O. C. Nitrous oxide from agriculture. Norwegian J. Agric. Sci. Suplement No. 12, 7-128, 1994.

Graven, E. H., Attoe, O. J. and Smith, D. Effect of liming and flooding on manganese toxicity in alfalfa. Soil Sci. Soc. Amer. Proc. 29, 702-706, 1965.

Gray, B. G. and Winkler, J. R. Electron transfer in proteins. Annu. Rev. Biochem. 65, 537-561, 1996.

Green, J. The effect of potassium and calcium on cotyledon expansion and ethylene evolution induced by cytokinins. Physiol. Plant. 57, 57-61, 1983.

Green, J. F. and Muir, R. M. The effect of potassium on cotyledon expansion induced by cytokinins. Plant Physiol. 43, 213-218, 1978.

Greenway, H. and Munns, R. Mechansim of salt tolerance in nonhalophytes. Annu. Rev. Plant Physiol. 31, 149-190, 1980.

Greenway, H. and Pitman, M. G. Potassium retranslocation in seedlings of *Hordeum vulgare*. Aust J. biol. Sci. 18, 235-247, 1965.

Greenwood, D. J. Studies on the distribution of oxygen around the roots of mustard seedlings (*Sinapis alba* L.). New Phytol. 70, 97-101, 1971.

Greenwood, D. J. Fertilizer food production: world scene. Fertilizer Research 2, 31-51, 1981.

Greenwood, D. J. Modelling N-response of field vegetable crops grown under diverse conditions with N_ABLE: A Review. J. Plant Nutrition 24, In Press 2001

Greenwood, D. J., Cleaver, I. J. and Turner, M. K. Fertilizer requirements of vegetable crops. The Fertilizer Soc., London, p. 4-30, 1974.

Greenwood, D. J., Cleaver, T. J., Turner, M. K., Hunt, J., Niendorf, K. B. and Loquens, S. M. H. Comparison of the effects of nitrogen fertilizer on the yield, nitrogen content and quality of 21 different vegetable and agriculture crops. J. agric. Sci. 95, 471-485, 1980.

Greenwood, D. J. and Karpinets, T. V. Dynamic model for the effects of K fertilizer on crop growth, K uptake and soil-K in arable cropping. 1. Description of the model. Soil Use and Management 13, 178-183, 1997a.

Greenwood, D. J. and Karpinets, T. V. Dynamic model for the effects of K fertilizer on crop growth, K uptake and soil-K in arable cropping. 2. Field test of the method. Soil Use and Management. 13, 184-189, 1997b.

Greenwood, D. J., Rahn, C., Draycott, A., Vaidyananthan, L. V. and Peterson, C. Modelling and measurement of the effects of fertilizer-N and crop residue incorporation on N-dynamics in vegetable cropping. Soil Use and Management 12, 13-24, 1996.

Greenwood, D. J. and Stone, D. A. Prediction and measurement of the decline in the critical-K and total cation plant concentration during the growth of field vegetable crops. Annals of Bot. 82, 871-881, 1998.

Greenwood, D. J. and Walker, A. Modelling soil productivity and pollution. Phil. Trans. R. Soc. London B, 329, 309-320, 1990.

Grignon, C. and Sentenac, H. pH and ionic conditions in the apoplast. Annu. Rev. Plant Physiol. Plant Mol. Biol. 42, 103-128, 1991.

Grill, E., Winnacker, C. L. and Zenk, M. H. Phytochelatins, a class of heavy-metal-binding peptides from plants, are functionally analogous to metallothionins. Proc. Natl. Acad. Sci. 84, 439-443, 1987.

Grimme, H. Aluminium induced magnesium deficiency in oats. Z. Pflanzenernähr. Bodenk. 146, 666-676, 1983.

Grimme, H., Braunschweig, L. C. von, and Nemeth, K. (G) Potassium, calcium and magnesium interactions as related to cation uptake and yield. Landw. Forsch. 30/II. Sonderh., 93-100, 1974.

Groot, J. J. R. and Houba, V. J. G. A comparison of different indices for nitrogen mineralization. Biol. Fertil. Soils 19, 1-9, 1995.

Grosse-Brauckmann, E. (G) Influence of N, CaO and P_2O_5 on SiO_2 uptake by cereals. Landw. Forsch. 9, 196-203, 1956.

Grove, M., Spencer, G., Rohwedder, P., Mandava, N., Worley, J., Warten, J., Steffens, G., Flippen-Anderson, J. and Cook, J. Brassinolide a plant growth-promoting steroid isolated from *Brassica napus* pollen. Nature 281, 216 -217, 1979.

Gruters, U., Fangmeier, A. and Läger, H. J. Modelling stomatal responses of spring wheat (*Triticum aestivum* L. cv Turbo) to ozone and different levels of water supply. Environ. Poll. 87, 141-149, 1995.

Guckert, A., Breisch, H. and Reisinger, O. (F) Interface soil-roots. L Electron microscopic study on the relationship between mucilage, clay minerals and microorganisms. Soil Biol. Biochem. 7, 241-250, 1975.

Guerin, V., Trinchant, J.-C. and Rigaud, J. Nitrogen fixation (C_2H_2 reduction) by broad bean (*Vicia faba* L.) nodules and bacteroids under water restricted conditions. Plant Physiol. 92, 595-601, 1990.

Guerinot, M. L. Molecular characterization of iron uptake in *Arabidopsis*. Abstract. 10th Intern. Symposium on Iron Nutrition and Interactions in Plants 33, 2000.

Guerrero, M. G., Vega, J. M. and Losada, M. The assimilatory nitrate reducing system and its regulation. Annu. Rev. Plant Physiol. 32, 169-204, 1981.

Guo, P. C., Bohring, J. and Scherer, H. W. (G) Turnover of fertilizer NH_4^+ in soils of different clay mineral composition. Z. Pflanzenernähr. Bodenk. 146, 752-759, 1983.

Gupta, U. C. Boron nutrition of crops. Adv. Agron. 31, 273-307, 1979.

Gupta, U. C. Soil and plant factors affecting molybdenum uptake by plants. In: Molybdenum in Agriculture, ed. by U. C. Gupta, Cambridge University Press, Cambridge, p. 71-91, 1997a.

Gupta,U. C. Deficient, sufficient, and toxic concentrations of molydenum in crops. In: Molybdenum in Agriculture, ed. by U. C. Gupta, Cambridge University Press, p. 150 -159, Cambridge p. 150 - 159, 1997b.

Gupta, U. C. Symptoms of molybdenun deficiency and toxicity in crops. In: Molybdenum in Agriculture, ed. by U. C. Gupta, Cambridge University Press, p. 160 -170, Cambridge 1997c

Gupta, U. C. and Cutcliffe, J. A. Effects of methods of boron application on leaf tissue concentration of boron and control of brown-heart in rutabaga. Can. J. Plant Sci. 58, 63-68, 1978.

Gustafson, F. G. and Schlessinger, M. J. Absorption of ^{60}Co by bean plants in the dark. Plant Physiol. 31, 316-318, 1956.

Gutierrez-Boem, F. H. and Thomas, G. W. Phosphorus nutrition affects wheat response to water deficit. Agron. J. 90, 166-171, 1998.

Gutser, R. and Teicher, K. (G) Changes of soluble nitrogen in an arable brown earth under winter wheat during a growth period. Bayr. Landw. Jahrb. 53, 215 -226, 1976.

Hadas, A. and Kafkafi, U. Kinetics of the mineralization of ureaform as influenced by temperature. Soil Sci. 118, 16-21, 1974.

Haeder, H. E. (G) The influence of chloride nutrition in comparison with sulphate nutrition on assimilation and translocation of assimilates in potato plants. Landw. Forsch. 3211. SH, 121-131, 1975.

Haeder, H. E. and Beringer, H. Influence of potassium nutrition and water stress on the content of abscisic acid in grains and flag leaves of wheat during grain development. J. Sci. Food Agric. 32, 552-556, 1981.

Haeder, H. E, and Mengel, K. (G) The absorption of potassium and sodium in dependence on the nitrogen nutrition level of the plant. Landw. Forsch. 23/I. Sonderh., 53-60, 1969.

Haeder, H. E., Mengel, K. and Forster, H. The effect of potassium on translocation of photosynthates and yield pattern of potato plants. J. Sci. Fd Agric. 24, 1479-1487, 1973.

Hager, A., Menzel, H. and Krauss, A. (G) Experiments and hypothesis of the primary effect of auxins on expansion growth. Planta 100, 47-75, 1971.

Hajibagheri, M. A. and Flowers, T. J. X-ray microanalysis of ion distribution within root cortical cells of the halophyte Suadea maritima (L.) Dum. Planta 177, 131-134, 1989.

Hajibagheri, M. A., Yeo, A. R. and Flowers, T. J. Quantitative ion distribution within maize root cells in salt-sensitive and salt-tolerant varieties. New Phytol. 105, 367-379, 1987.

Hak, T. A. Diseases of wheat, barley and rice and their control. 1. FAO/SIDA Seminar for plant scientists from Africa and Near East, Cairo 1973, FAO Rome, p. 542-549, 1974.

Halim, R. A., Buxton, D. R., Hattendorf, M. J. and Carlson, R. E. Water-stress effects on alfalfa forage quality after adjustment for maturity differences. Agron. J. 81, 189-194, 1989.

Hall, D. A. The influence of varied calcium nutrition on the growth and ionic compostion of plants. Ph. D. Thesis. University of Leeds 1971.

Hall, D. O. Solar energy and biology for fuel food and fibre. TIBS 2, 99-101, 1977.

Hall, D. O., Cammack, R. and Rao, K. K.: Role for ferredoxins in the origin of life and biological evolution. Nature 233, 136-138, 1971.

Hall, J. R. A review of VA mycorrhizal growth responses in pastures. Angew. Botanik 61, 127-134, 1987.

Hall, S. M. and Baker, D. A. The chemical composition of Ricinus phloem exudate. Planta 106, 131-140, 1972.

Hallbäcken, L. Long term changes of base cation pools in soil and biomass in a beech and in a spruce forest

732

of Southern Sweden. Z. Pflanzenernähr. Bodenk. 155, 51-60, 1992.

Hallsworth, E. G., Wilson, S. B. and Adams, W. A. Effect of cobalt on the non nodulated legume. Nature 205, 307, 1965.

Handreck, K. A. and Riceman, D. S. Cobalt distribution in several pasture species grown in culture solutions. Aust. J. Agric. Res. 20, 213-226, 1969.

Handyanto, E., Cadisch, C. and Giller, K. E. Nitrogen release from prunings of legume hedge row tree in relation to quality of the prumings and incubation method. Plant and Soil 160, 237-248, 1994.

Haneklaus, S., Knudsen, L. and Schnug, E. Relationship between potassium and sodium in sugar beet. Commun. Soil Science and Plant Analysis 29, 1793-1798, 1998.

Hanks, R. J. and Rasmussen, V. P. Predicting crop production as related to plant water stress. Adv. Agron. 35, 193-215, 1982.

Hanotiaux, G. (F) Soil sampling for chermcal analysis. Bull. Rech. Agron. de Gembloux, N. S. 1, Nr. 3, 1966.

Hanson, A. D. and Hitz, W. D. Metabolic response of mesophytes to plant water deficits. Annu. Rev. Plant Physiol. 33, 163-203, 1982.

Hanson, J. B. The functions of calcium in plant nutrition. Adv. Plant Nutrition. 1, 149-208, 1984.

Hanson, J. B. and Bonner, J. The relationship between salt and water uptake in Jerusalem artichoke tuber tissue. Ann. J. Bot. 41, 702-710, 1954.

Hantschel, R., Kaupenjohann, M., Horn, R., Gradl, J. and Zech, W. Ecologically important differences between equilibrium and percolation soil extracts. Geoderma 43, 213-227, 1988.

Haq, A. U. and Miller, M. H. Prediction of available soil Zn, Cu and Mn using chemical extractants. Agron. J. 64, 779-782, 1972.

Haque, I., Lupwayi, NA. and Ssali, H. Arognomic evaluation of unacidulated and partially acidulated Minjingu amd Chilembwe phosphate rocks for clover production in Ethiopia. European J. Agron. 10, 37-47, 1999.

Harley, J. L. Mycorrhiza, Oxford University Press, Oxford 1971.

Harris, P. and James, A. T. Effect of low temperature on fatty acid biosynthesis in seeds. Biochern. Biophys. Acta 187, 13-18, 1969.

Harrison, M. A. and Walton, D. C. Abscisic acid in water stressed bean leaves. Plant Physiol. 56, 250-254, 1975.

Hartt, C. E. Effect of potassium deficiency upon translocation of ^{14}C in detached blades of sugarcane. Plant Physiol. 45, 183-187, 1970.

Hartung, W., Radin, J. W. and Hendrix, D. L. Abscisic acid movement into the apoplastic solution of water-stressed cotton leaves. Plant Physiol. 86, 908-913, 1988.

Hassidim, M., Braun, Y., Lerner, H. R. and Reinhold, L. Na^+/H^+ and K^+/H^+ antiport in root membrane vesicles isolated from the halophyte Atriplex and the glycophyte cotton. Plant Physiol. 94, 1795-1801, 1990.

Hauck, R. D. Quantitative estimates of nitrogen-cycle-processes: Concepts and review. In: Nitrogen-15 in Soil Plant Studies, IAEA,Vienna, p. 65-80, 1971.

Hauter, R. and Mengel, K. Measurement of pH at the root surface of red clover (*Trifolium pratense*) grown in soils differing in proton buffer capacity. Biol. Fertil. Soils 5, 295-298, 1988.

Hauter, R. and Steffens, D. (G) Influence of mineral and symbiotic nitrogen nutrition on proton release of roots, phosphorus uptake and root development of red clover. Z. Pflanzenernaehr. Bodenk. 148, 633-646, 1985.

Havlin, J. L., Westfall, D. G. and Olsen, S. R. Mathematical models for potassium release kinetics in calcareous soils. Soil Sci. Soc. Am. J. 49, 371-376 1985.

Hawkesford, M. J., Schneider, A., Belcher, A. R. and Clarkson, D. T. Regulation of enzymes in the sulphur-assimilatory pathway. Z. Pflanzenernähr. Bodenk. 158, 55-57, 1995.

Haygarth, P. M., Hepworth, L. and Jarvis, S. C. Forms of phosphorus transfer in hydrological pathways from soil under grazed grassland. European J. Soil Sci. 1997.

Hayman, D. S. Mycorrhiza and crop production. Nature 287, 487-488, 1980.

Haynes, R. J. Ion exchange properties of roots and ionic interactions within the root apoplasm: Their role in ion accumulation by plants. The Botanical Review 46, 74-99, 1980a.

Haynes, R. J. Effects of liming on phosphate availability in acid soils. Plant and Soil 68, 289-308, 1982.

Haynes, R. L. Competitive aspects of the grass-legume association. Adv. Agron. 33, 227-261, 1980b.

Haynes, R. J. Effect of lime, silicate, and phosphate applications on the concentrations of extractable aluminium and phosphate in a spodosol. Soil Sci. 138, 8-14, 1984.

Haynes, R. J. and Goh, K. M. Ammonium and nitrate nutrition of plants. Biol. Rev. 53, 465-510, 1978.

Heathcote, R. C. (G) Fertilization with potassium in the Savanna zone of Nigeria. Potash Review, Subject 16, 57th suite, 1972.

Heatherly, L. G. and Russell, W. J. Effect of soil water potential of two soils on soybean emergence. Agron. J. 71, 980-982, 1979.

Heber, U. and Heldt, H. W. The chloroplast envelope: Structure, function, and role in leaf metabolism. Annu. Rev. Plant Physiol. 32, 139-168, 1981.

Heber, U. and Purczeld, P. Substrate and product fluxes across the chloroplast envelope during bicarbonate and nitrite reduction. Proc. 4th Int. Congr. on Photosynthesis, p. 107-118, 1977.

Hecht-Buchholz, C. The apoplast-habitat of endophytic dinitrogen-fixing bacteria and their significance for the nutrition of nonleguminous plants. Z. Pflanzenernähr. Bodenk. 161, 509-520, 1998.

Hecht-Buchholz, C. and Foy, C. D. Effect of aluminium toxicity on root morphology of barley. Plant and Soil 63, 93-95, 1981.

Heckrath, G., Brooks, P. C., Poulton, P. R. and Goulding, K. W. T. Phosphorus leaching from soils containing different phosphorus concentrations in the Broadbalk experiment. J. Environ. Qual. 24, 904-919, 1995.

Hedrich, R. and Schroeder, J. I. The physiology of ion channels and electrogenic pumps in higher plants. Annu. Rev. Plant Physiol. 40, 539-569, 1989.

Hehl, G. and Mengel, K. (G) The effect of varied applications of potassium and nitrogen on the carbohydrate content of several forage crops. Landw. Forsch. 27/II. Sonderh., 117-129, 1973.

Heilenz, S. (G) Investigations into the lead content of plants on sites with heavy traffic. Landw. Forsch. 25/I. Sonderh., 73-78, 1970.

Heineke, D., Sonnewald, U., Bussis, D., Günther, G. and Leidreiter, K. Apoplastic expression of yeast-derived invertase in potato. Effects on photosynthesis, leaf solute composition, water relations, and tuber composition. Plant Physiol. 100, 301-308, 1992.

Helal, M. and Dressler, A. Mobilization and turnover of soil phosphorus in the rhizosphere. Z. Pflanzener. Bodenk. 152, 175-180, 1989.

Helal, H. M. and Mengel, K. Nitrogen metabolism of young barley plants as affected by NaCl salinity and potassium. Plant and Soil 51, 457-462, 1979.

Helal, H. M. and Mengel, K. Interaction between light intensity and NaCl salinity and their effects on growth, N assimilation, and photosynthetic conversion in young broad beans. Plant Physiol. 67, 999-1002, 1981.

Helal, H. and Sauerbeck, D. Influence of roots on C and P metabolism in soil. Plant and Soil 76, 175-182, 1984.

Helal, H. M., Upenov, A. and Issa, J. Growth and uptake of Cd and Zn by Leucaena leucocephala in reclaimed soils as affected by Na Cl salinity. J. Plant Nutr. Soil Sci. 162, 589-592, 1999.

Heldt, H. W., Ja Chong, C., Maronde, D., Herold, A., Stankovic, Z. S., Walker, D. A., Kraminer, A., Kirk, M. R, and Heber, U. Role of orthophosphate and other factors in the regulation of starch formation in leaves and isolated chloroplasts. Plant Physiol. 59, 1146-1155, 1977.

Helyar, K. R. Nitrogen cycling and soil acidigication. J. Austral. Institute Agric. Sci. 1976:217-221, 1976.

Hendricks, L., Claassen, N. and Jungk, A. (G) Phosphate depletion at the soil-root interface and the phosphate uptake of maize and rape. Z. Pflanzenernähr. Bodenk. 144, 486-499, 1981.

Henkens, C. H. (G) General lines for the application of trace elements in Holland. Landw. Forsch. 18, 108-116, 1965.

Henning, S. J. Aluminium toxicity in the primary meristem of wheat roots. Ph. D. Thesis. Oregon State Univ. Corvallis, Oregon, USA 1975.

Hentschel, G. The uptake of ^{15}N-Iabelled urea by bush beans. In: Nitrogen Nutrition of the Plant, edited by E. A. Kirkby. University of Leeds, Agricultural Chemistry Symposium p. 30-34, 1970.

Hepler, P. K. and Wayne, R. O. Calcium and plant development. Annu. Rev. Plant Physiol. 36, 397-439, 1985.

Herppich, W. B., Flach, B. M. T., von Willert, D. J. and Herppich, M. Field investigations of photosynthetic activity, gas exchange and water potential at different leaf ages in *Welwitschia mirabilis* during a severe drought. Flora 191, 59-66, 1996.

Herppich, W. B. and Peckmann, K. Responses of gas exchange, photosynthesis, nocturnal acid accumulation and water relations of *Aptenia cordifolia* to short-term drought and rewaterinG. J. Plant Physiol. 150, 467-474, 1997.

Herrmann, B. (G) Fatty acid composition of the crude fat fraction of winter rape seed as influenced by nitrogen fertilization. Arch. Acker- und Pflanzenbau und Bodenk. 21, 141-148, 1977.

Herschbach, C. and Rennenberg, H. Influence of glutathione (GSH) on net uptake of sulphate and sulphate transport in tobacco plants. J. Exp. Bot. 45, 1069-1076, 1994.

Herzog, H. and Geisler, G. (G) Effect of cytokinin application on assimilate storage and endogenous cytokinin activity in the caryopsis of two spring wheat cultivars. Z. Acker- und Pflanzenbau 144, 230-242, 1977.

Hesse, P. R. A Textbook of Soil Chemical Anaylsis, John Murry London 1971.

Heuwinkel, H., Kirkby, E. A., Le Bot, J. and Marschner, H. Phosphorus deficiency enhances molybdenum uptake by tomato plants. J. Plant Nutr. 15, 549 -568, 1992.

Hewitt, E. J. Metal interrelationship in plant nutrition. J. Exper. Bot. 4, 59-64, 1953.

Hewitt, E. J. Sand and Water Culture Methods used in the Study of Plant Nutrition. 2nd Commonwealth Agricultural Bureaux, Farnham Royal, England 1966.

Hewitt, E. J. Physiological and biochemical factors which control the assimilation of inorganic nitrogen supplies by plants, p. 78-103. In: Nitrogen Nutrition of the Plant, edited by E. A. Kirkby. The University Leeds 1970

Hewitt, E. J. Assimilatory nitrate-nitrite reduction. Annu. Rev. Plant Physiol. 26, 73-100, 1975.

Hewitt, E. J. and Bond, G. The cobalt requirement of non-legume root nodule plants. J. Exp. Bot. 17, 480-491, 1966.

Hewitt, E. J. and McCready, C. C. Molybdenum as a plant nutrient. VII. The effect of different molybdenum and nitrogen supplies on yields and composition of tomato plants grown in sand culturE. J. Horticult. Sci. 31, 284-290, 1956.

Hewitt, E. J. and Smith, T. A. Plant Mineral Nutrition. English Univ. Press London 1975.

Heyn, J. and Brüne, H. (G) A comparison N fertilizer recommendations to sugar beet according to Nmin and EUF soil analysis in field trials in Hessia. VDLUFA Schriftenreihe 30, 195-200, 1990.

Higinbotham, N. The mineral absorption process in plants. The Botanical Review 99, 15-69, 1973.

Hignett, T. P. Liquid fertilizer production and distribution. UNO, Second Interregional Fertilizer Symposium,

Kiev, Ukraine, 1971; New Delhi, India, 1971.

Hildebrandt, E. A. (G) Uptake and degradation of nitrosamines in sunflower seedlings. Landw. Forsch. Sonderh. 36, 187-195, 1979.

Hill, J. M. The changes with age in the distribution of copper and some copper containing oxidase in red clover (*Trifolium pratense* L. cv Dorset Marlgrass). J. Exp. Bot. 24, 525-536, 1973.

Hille, B. A K$^+$ channel worthy of attention. Science. 273, 1677, 1996.

Hingston, F. J., Posner, A. M. and Quirk, J. P. Anion adsorption by goethite and gibbsite. I. Desorption of anions by hydrous oxide surfaces. J. Soil Sci. 25, 16-26, 1974.

Hinsinger, P. How do plant roots require mineral nutrients? Chemical processes involved in the rhizosphere. Adv. Agron. 64, 225-252, 1998.

Hinsinger, P. and Jaillard, B. Root-induced release of interlayer potassium and vermiculization of phlogopite as related to potassium depletion in the rhizosphere of ryegrass. J. Soil Sci. 44, 525-534, 1993.

Hipp, B. W. and Thomas, G. W. Method for predicting potassium uptake by grain sorghum. Agron. J. 60, 467-469, 1968.

Hirsch, A. M. and Torrey, I. G. Utrastructural changes in sunflower root cells in relation to boron deficiency and added auxin. Can. J. Bot. 58, 856-866, 1980.

Hoagland, D. R. Lectures on the inorganic nutrition of plants. Chronica Botanica Company, Waltharn, Mass. USA, p. 48-71, 1948.

Hoagland, R. R. and Arnon, D. I. The water culture methods for growing plants without soil. Cal. Agric. Exp. ST. Circ. 347, 1-32, 1950.

Hodgson, J. F., Lindsay, W. L. and Trierweiler, J. F. Micronutrient cation complexing in soil solution. II. Complexing of zinc and copper in displacing solution from calcareous soils. Soil Sci. Soc. Amer. Proc. 30, 723-726, 1966.

Hoffland, E., Findenegg, G. R. and Nelemans, J,A. Solubilization of rock phosphate by rape. II. Local root exudation of organic acids as a response to P- starvation. Plant and Soil 113, 161-165, 1989.

Hoffland, E., Findenegg, G. R., Leffelar, P. A. and Nelemans, J. A. Use of simulation model to quantify the amount of phosphate released from rock phosphate by rape. In: Transactions 14. Intern. Congr. Soil Science, edited by Masayoshi Koshino, Tokyo: International Society of Soil Science, p. 170-175, 1990.

Hoffmann, B. and Kosegarten, H. FITC - dextran for measuring apoplast pH gradients between various cell types in sunflower leaves. Physiol. Plant. 95, 327-335, 1995.

Hoffmann, B., Plänker, R. and Mengel, K. Measurements of pH in the apoplast of sunflower leaves by means of fluorescence. Physiol. Plant. 84, 146 -153, 1992.

Hoffmann, G. (G) The Investigations of Soils. VDLUFA-Verlag, Darmstadt, 1991

Höfner, W., Feucht, D. and Schmitz, M. (G) Modification of morphological and physiological yield formation factors in wheat by N-fertilization and growth regulator application. Ber. Deutsch. Bot. Ges. 97, 139-150, 1984.

Holbrook, N. M., Burns, M. J. and Field, C. B. Negative xylem pressures in plants: a test of the balancing pressure technique. Sci. 270, 1193-1194, 1995.

Holden, M. J., Luster, D. G., Chaney, R. L., Buckhout, T. J. and Robinson, C. Fe - chelate reductase activity of plasma membranes isolated from tomato (*Lycopersicon esculentum* Mill.) roots. Plant Physiol. 97, 537-544, 1991.

Holder, C. B. and Brown, K. W. The relationship between oxygen and water uptake by roots of intact bean plants. Soil Sci. Soc. Am. J. 44, 21-25, 1980.

Holford, I. C. R. and Mattingly, G. E. G. Phosphate sorption by jurassic eolitic limestones. Geoderma 13, 257-264, 1975.

736

Holford, I. C. R. Effects of phosphate buffer capacity of soil on the phosphate requirements of plants. Plant and Soil 45, 433-444, 1976.

Holliday, R. The efficiency of solar energy conversion by the whole crop. In: Food Production and Consumption, edited by Duckham, A, N., Jones, J. G. W. and Roberts, E. H. North Holland Publishing Company, Amsterdam, Oxford, p. 127-146, 1976.

Holmgren, P., Jarvis, P. G. and Jarvis, M. S. Resistance to carbon dioxide and water vapour transfer in leaves of different plant species. Physiol. Plant. 18, 557-573, 1965.

Homma, Y. and Hirata, H. Kinetics of cadmium and zinc absorption by rice seedling roots. Soil Sci. Plant Nutr. 30, 527-532, 1984.

Honeycutt, C. W., Potaro, L. J. and Halteman, W. A. Predicting nitrate formation from soil, fertilizer, crop residue, and sludge with thermal units. J. Environ. Qual. 20, 850-856, 1991.

Hoogerkamp, M. Ley, periodically reseeded grassland or permanent grassland. Agric. Res. Rep. 812, 5-35, 1974.

Hooker, M. L., Sander, D. H., Peterson, G. A. and Daigger, L. A. Gaseous N losses from winter wheat. Agron. J. 72, 789-792, 1980.

Horst, W. J. The role of apoplast in aluminium toxicity and resistance of higher plants: a review. Z. Pflanzenernähr. Bodenk. 158, 419-428, 1994.

Horst, W. J. and Marschner, H. Effect of excessive manganese supply on uptake and translocation of calcium in bean plants (*Phaseolus vulgaris* L.). Z. Pflanzenphysiol. 87, 137-148, 1978a.

Horst, W. J. and Marschner, H. Effect of silicon in manganese tolerance of bean plants (*Phaseolus vulgaris* L). Plant and Soil 50, 287-303, 1978b.

Horst, W. J. and Marschner, H. Effect of silicon on manganese tolerance of bean plants (*Phaseolus vulgaris* L.). Plant and Soil 50, 287-303, 1987.

Horst, W. J., Wagner, A. and Marschner, H. Mucilage protects root meristems from aluminium injury. Z. Pflanzenphysiol. 105, 435-444, 1982.

Hossner, L. R., Freeouf, J. A, and Folsom, B. L. Solution phosphorus concentration and growth of rice (*Oryza sativa* L.) in flooded soils. Proc Soil Sci. Soc. Amer. 37, 405-408, 1973.

Houba, V. J. G., Novozamsky, I., Huybregts, A. W. M. and van der Lee, J. J. Comparison of soil extractions by 0. 01 M $CaCl_2$,by EUF and by some conventional extraction procedures. Plant and Soil 96, 433-437, 1986.

Houseley, T. L. and Daughtry, C. S. T. Fructan content and fructosyltransferase activity during wheat seed growth. Plant Physiol. 83, 4-7, 1987.

Howard, J. B. and Rees, D. C. Nitrogenase: a nucleotide-dependent molecular switch. Annu. Rev. Biochem. 63, 235-264, 1994.

Hsiao, S. J. C. Life history and iodine nutrition of the marine brown alga *Petalonia fascia* (O. F. Müll.) Kuntze. Can. J. Bot. 47, 1611-1616, 1969.

Hsiao, T. C. and Acevedo, E. Plant responses to water deficits, water use efficiency, and drought resistance. Agricult. Meteorol. 14, 59-84, 1974.

Hsiao, T. C., Acevedo, E., Fereres, E. and Henderson, D. W. Water stress, growth and osmotic adjustment. Phil. Trans. Royal Soc. London 273, 479-500, 1976.

Hsiao, T. C. and Läuchli, A. A role of potassium in plant-water relations. Adv. Plant Nutr. 2, 281-312, 1986.

Hsiao, T. C. Plant responses to water stress. Annu. Rev. Plant Physiol. 24, 519-570, 1973.

Hsu, P. H. Aluminium hydroxides and oxyhydroxides. In: Minerals in Soil Environments, edited by Hissel, D. E. Madison,USA: Soil Sci. Soc. of America, p. 331-378, 1989.

Hu, H. and Brown, P. H. Localization of boron in cell walls of squash and tobacco and its association of pectin. Plant Physiol. 105, 681-689, 1994.

Hu, H., Brown, P. H. and Labavitch, J. M. J. Species variability in boron requirement is correlated with cell wall pectin. J. Exp. Bot. 47, 227-232, 1996.

Hu, H. and Brown, P. H. Absorption of boron by plants. Plant and Soil 193, 49 -58, 1997.

Huang, A. H. C. Oil bodies and oleosins in seeds. Annu. Rev. Plant Physiol. Plant Mol. Biol. 43, 177-200, 1993.

Huang, C. X. and Van Steveninck, R. F. M. Maintenance of low Cl- concentrations in mesophyll cells of leaf-blades of barley seedlings exposed to salt stress. Plant Physiol. 90, 1440-1443, 1989.

Huang, P. M. Feldspars, olivines, pyroxenes, and amphibioles. In: Minerals in Soil Environments, edited by Dixon, J. B. and Weed, S. B., Madison, Soil Sci. Soc. America, p. 975-1050, 1989.

Huang, P. M. and Fujii, R. Selenium and arsenic. In: Methods of Soil Analysis, edited by Bingham, J. M. Madison, Soil Sci. Soc. America, p. 793-831, 1996.

Huber, D. M., Warren, H. L., Nelson, D. W. and Tsai, C. Y. Nitrification inhibitors - new tools for food production. Bio Science 27, 523-529, 1977.

Huber, D. M., Warren, H. L., Nelson, D. W., Tsai, C. Y. and Shaner, G. E. Response of winter wheat to inhibiting nitrification of fall-applied nitrogen. Agron. J. 72, 632-637, 1980.

Huber, S. C., Huber, J. L and Mc Michael, R. W. Control of plant enzyme activity by reversible protein phosphorylation. Int. Rev. of Cytology 149, 47-98, 1994.

Huettl, R. F.-J. Decline of Norway Spruce (*Picea abies* Karst.) and Silver Fir (*Abies alba* Mill.) stands in the Southwest of West Germany from the viewpoint of forest nutrition. IUFRO, Intern Symposium «Human Impacts on Forests», Strasbourg, France, p. 16-22, 1984.

Huffman, E. W. D. and Allaway, W. H. Growth of plants in solution culture containing low levels of chromium. Plant Physiol. 52, 72-75, 1973.

Huglin, P. (F) Biology and Ecology of vine. Editions Payot, ISBN-2-601-0301, Lausanne 1986.

Huguet, C. and Coppenet, M. (F) Magnesium in Agriculture. Paris, INRA, 1994.

Humble, G. D. and Hsiao, T. C. Specific requirement of potassium for light-activated opening of stomata in epidermal strips. Plant Physiol. 44, 230-234, 1969.

Humble, G. D. and Hsiao, T. C. Light-dependent influx and efflux of potassium of guard cells during stomatal opening and closing. Plant Physiol. 46, 483-487, 1970.

Humble, G. D. and Raschke, K. Stomatal opening quantitatively related to potassium transport. Plant Physiol. 48, 447-453, 1971.

Hunter, T. G. and Vergnano, O.: Trace element toxicities in oats. Ann. App. Biol. 40,761-777, 1953.

Huppert, V. and Buchner, A. (G) Recent experimental results on the effect of several N forms with particular regard to environmental conditions. Z. Pflanzenernähr. Düng. Bodenk. 60, 62-92, 1953.

Hurd-Karrer, A. M. Comparative fluorine uptake by plants in limed and unlimed soil. Soil Sci. 70, 153-159, 1950.

Husted, S. and Schjoerring, J. K Apoplastic pH and ammonium concentration in leaves of *Brassica napus* L. Plant Physiol. 109, 1453-1460, 1995.

Hutchin, M. E. and Vaughan, B. E. Relation between simultaneous Ca and Sr transport rates in isolated segments of vetch, barley and pine roots. Plant Physiol. 43, 1913-1918, 1968.

Hutchinson, T. C. Lime-chlorosis as a factor in seedling establishment on calcareous soils. I. A comparative study of species from acidic and calcareous soils in their susceptibility to limechlorosis. New Phytol. 66, 697-705, 1967.

Hütsch, B. W. Methane oxidation in arabie soil as inhibited by ammonium, nitrite, and organic manure with respect to soil pH. Biol. Fert. Soils 28, 27-35, 1998.

Hütsch, B. W. Sources and sinks of methane in German agrosystems in context of the global methane budget. Agribiol. Res. 51, 75-87, 1998.

Hütsch, B. W. Tillage and land use effects on methane oxidation rates and their vertical profile in soil. Biol. Fert. Soils 27, 284-292, 1998.

Hütsch, B. W. and Steffens, D. (G) Determination of soluble aluminium at ist impact on yield formation of spring barley. VDLUFA-Schriftenr. 20 Kongr. Bd., 317-332, 1986.

Hütsch, B. W. and Steffens, D. (G) Effect of different soil cultivation on the vertical distribution of available phosphate and potassium in the profile of four different soil types. Agribiol. Res. 54, 352-358, 1992.

Hütsch, B. W., Webster, C. P. D and Powlson, D. S. Long-term effects of nitrogen fertilization on methane oxidation in soil of the Broadbalk wheat experiment. Soil Biol. Biochem. 25, 1307-1315, 1993.

Hütsch, B. W., Webster, C. P. and Powlson, D. S. Methane oxidation in soil as affected by land use, soil pH and N fertilization. Soil Biol. Biochem. 26, 1613-1622, 1994.

Hylton, L. O., Ulrich, A. and Cornelius, D. R. Potassium and sodium interrelations in growth and mineral content of Italian ryegrass. Agron. J. 59, 311-314, 1967.

Imsande, J. and Touraine, B. N demand and the regulation of nitrate uptake. Plant Physiol. 105, 3-7, 1994.

Incoll, N. D., Long, S. P. and Ashmore, M. R. SJ units in Plant Science. Current Adv. Plant Sci. 28, 331-343 (1977).

Ingestad, T. and Lund, A. B. Nitrogen stress in birch seedlings. I. Growth technique and growth. Physiol. Plant. 59, 179-183, 1979.

Inoue, H. and Tanaka, A. Comparision of source and sink potentials between wild and cultivated potatoes. J. Sci. Soil & Manure Japan 49, 321. 327, 1978.

Irving, H. R., Gehring, C. A. and Parish, R. W. Changes in cytosolic pH and calcium of guard cells precede stomatal movements. Proc. Natl. Acad. Sci. USA 89, 1790-1794, 1992.

Isermann, K. (G) The effect of adsorption processes in the xylem on the calcium distribution in higher plants. Z. Planzenernähr. Bodenk. 126, 191-203, 1970.

Isermann, K. Environmental aspects of fertilizer application. In: Ullmann's Encyclopedia of Industrial Chemistry, Weinheim: VCH Verlagsgesellschaft, p. 400-409, 1987.

Isermann, K. Share of agriculture in nitrogen and phosphorus emmissions into the surface waters of Western Europe against the background of their eutrophication. Fert. Res. 26, 253-269, 1990.

Israel, D. W. and Jackson, W. A. The influence of nitrogen nutrition on ion uptake and transportation by leguminous plants. In: Mineral Nutrition of Legumes in Tropical and Subtropical Soils, edited by Andrew, C. S. and Kamprath, E. J. CSIRO, Australia, p. 113-128 1978.

Itoh, S. and Barber, S. A. Phosphorus uptake by six plant species as related to root hairs. Agron. J. 75, 457-461, 1983.

Ivanov, S. (G) The climatic zones of the earth and the chemical activities of plants. In: Fortschritte der naturwiss. Forschung, edited by Abderhalden, H., Heft 5, p., Berlin 1929.

Iyengar, S. S., Martens, D. C. and Miller, W. P. Distributions and plant availability of soil Zn fractions. Soil Sci. Soc. Am. J. 45, 735-739, 1981.

Jacob, A. and von Uexküll, H. R. Fertilizer Use, Nutrition, and Manuring of Crops. Verlagsgesellschaft für Ackerbau, Hannover, 1963.

Jacobs, L. Selenium in Agriculture and the Environment. Madison WS. Amer. Soc. Agron. 1989.

Jahn-Deesbach, W. and May, H. (G) The effect of variety and additional late nitrogen application on the thiamin (vitamin B_1) content of the total wheat grain, various flour types, and secondary milling products. Z. Acker- u. Pflanzenbau 135, 1-18, 1972.

Jaillard,. B., Guyon, A. and Maurin, A. F. Structure and composition of calcified roots, and their identification in calcareous soils. Geoderm. 50, 197-210, 1991.

Jama, B., Buresh, R. J. and Place, F. M. Sesbania tree fallows on phosphorus-deficent sites: Maize yield and financial benefit. Agron. J. 90, 717 -726, 1998.

Jama, B., Swinkels, R. A. and Buresh, R. J. Agronomic and economic evaluation of organic and inorganic sources of phosphorus in Western Kenya. Agron. J. 89, 597-604, 1997.

James, D. W, Weaver, W. H. and Reeder, R. L. Chloride uptake by potatoes and the effects of potassium, chloride, nitrogen and phosphorus fertilization. Soil Sci. 109, 48-52, 1970.

James, D. W., Weaver, H. W. and Reeder, R. L. Chloride uptake by potatoes and the effects of potassium, chloride, nitrogen and phosphorus application. Soil Sci. 100, 48-52, 1970.

Janzen, H. H. Deposition of nitrogen into the rhizosphere by wheat roots. Soil Biol. Biochem. 22, 1155-1160, 1990.

Jarausch- Wehrheim, B., Moquot, B. and Mench, M. Absorption and translocation of sludge-borne zinc in field-grown maize (*Zea mays* L.). European J. Agron. 11, 23-33, 1999.

Jardine, P. M. and Sparks, D. L. Potassium-calcium exchange in a multireactive system: II thermodynamics. Soil Sci. Soc. Am. J. 48, 45-50, 1984.

Jenkinson, D. S. The turnover of organic carbon and nitrogen in soil. Phil. Trans Royal Soc. B 329, 361-368, 1990.

Jenkinson, D. S., Bradbury, N. J. and Coleman, K. How the Rothamsted classical experiments have been used to develop and to test models for the turnover of carbon and nitrogen in soil. In: Long-term Experiments in Agricultural and Ecological Sciences, edited by Leigh, R. A. and Johnston, A. E. Wallingford: CAB International, p. 117-138, 1994.

Jenkinson, D. S. and Ladd, J. N. Microbial biomass in soil: measurement and turnover. In: Soil Biochemistry, Vol. 5, edited by Paul, E. A. and Ladd, J. N. New York, Basel: Marcel Dekker, p. 415-471, 1981.

Jenkinson, D. S. and Rayner, J. H. The turnover of soil organic matter in some of the Rothamsted classical experiments. Soil Sci. 123, 298-305, 1977.

Jenner, C. F. An investigation of the association between hydrolysis of sucrose and its absorption by grains of wheat. Austr. J. Plant Physiol. 1, 319-329, 1974.

Jenner, C. F. and Rathjen, A. J. Limitations to the accumulation of starch in the developing wheat grain. Ann. Bot. 36, 743-754, 1972.

Jenner, C. F. and Rathjen, A. J. Factors regulating the accumulation of starch in ripening wheat grain. Aust. J. Plant Physiol. 2, 311-322, 1975.

Jennings, D. H. The effects of sodium chloride on higher plants. Biol. Rev. 51, 453-486, 1976.

Jenny, H. and Overstreet, R. Contact effects between plant roots and soil colloids. Proc. Nat. Sci. 24, 384-392, 1938.

Jensen, E. S. Nitrogen immobilization and mineralization during initial decomposition of pea and barley residues. Biol. Fertil. Soils 23, 459 - 464, 1996.

Jensen, L. S. and Sorensen, J. Microscale fumigation-extraction and substrate induced respiration methods for measuring microbial biomass in barley rhizosphere. Plant and Soil 162, 151-161, 1994.

Jensen, P. J., Hangarter, R. P. and Estelle, M. Auxin transport is required for hypocoyl elongation in leght-grown but not in dark-grown Arabidopsis. Plant Physiol. 116, 455-462, 1998.

Jeschke, W. D. K$^+$/Na$^+$ exchange and selectivity in barley roots. Effect of Na$^+$ fluxes. J. Exp. Bot. 28, 1289-1305, 1977.

Jeschke, W. D., Kirkby, E. A., Peuke, A. D., Pate, J. S. and Hartung, W. Effects of P deficiency on accumulation and transport of nitrate and phosphate in intact plants of castor bean(*Ricinus communis*). J. Exp. Bot. 48, 75-91, 1997.

Jeschke, W. D., Klagges, S. and Bhatti, A. S. Collection and composition of xylem sap and root structure in two halophytic species. Plant Soil 172, 97-106, 1995.

Jeschke, W. D., Pate, J. S. and Atkins, C. A. Partitioning of K+, Na+, Mg^{2+}, and Ca+ through xylem and phloem to component organs of nodulated white lupin under mild salinity. J. Plant Physiol. 128, 77-93, 1987.

Jeschke, W. D. and Pate, J. S. Modelling of the partitioning, assimilation and storage of nitrate within root and shoot organs of castor bean (*Ricinus communis* L.). J. Exp. Bot. 4, 1091-1103, 1991a.

Jeschke, W. D. and Pate, J. S. Cation and chloride partitioning through xylem and phloem within the whole plant of *Ricinus communis* L. under conditions of salt stress. J. Exp. Bot. 42, 1105-1116, 1991 b.

Jeschke, W. D., Wolf, O. and Pate, J. S. Solute exchanges from xylem to phloem in the leaf and from phloem to xylem in the root. In: Recent Advances in Phloem Transport and Assimilate Compartmentation, edited by Bonnemain J. L., Delroit, S., Lucas, W. J. and Dainty, J. Nantes, Cedex, France: ouest Editions, Presses Academiques, pp. 196-203, 1991.

Jia, Z. P., McCullough, N., Martel, R. Hemmingsen, S. and Young, P. G. Gene amplification at a locus encoding a putative Na+/H+ antiporter confers sodium and lithium tolerance in fission yeast. EMBO J. 11, 1631-1640, 1992.

Joergensen, R. G., Meyer, B. and Mueller, T. Time-course of the soil microbial biomass under wheat: a one year field study. Soil Biol. Biochem. 26, 987-994, 1994.

Johannes, E., Brosnan, J. M. and Sanders, D. Parallel pathways for interacellular Ca^{2+} release from the vacuole of higher plants. Plant J. 2, 97-102, 1992.

Johnson, C. M. Molybdenum In: Diagnostic Criteria for Plants and Soils, edited by Chapman, p. 286-301, 1966.

Johnson, C. M., Stout, P. R., Broyer, T. C. and Carlton, A. B. Comparative chlorine requirements of different plant species. Plant and Soil 8, 337-353, 1957.

Johnston, A. E. The Rothamsted classical experiments. In: Long-term Experiments in Agricultural and Ecological Sciences, edited by Leigh, R. A. and Johnston, A. E. Wallingford, UK: CAB International, p. 9-37, 1994.

Johnston, A. E., Goulding, K. W. T. and Poulton, P. R. Soil acidification during more than 100 years under permanent grassland and woodland at Rothamsted. Soil Use Manage. 2, 3-10, 1986.

Jolivet, P. Elemental sulfur in agriculture. In Sulfur Nutrition and Assimilation in Higher Plants, edited by L. J. De Kok, I. Stulen, H. Rennenberg, C. Brunold, and W. E. Rauser. SPB Academic Publishing bv, The Hague p. 193 - 206, 1993.

Jones, A. M. Auxin-binding proteins. Annu. Rev. Plant Physiol. Plant Mol. Biol 45, 393-420, 1994.

Jones, C. A. C-4 Grasses and Cereals. Growth, Development and Stress Response. Wiley, New York, 1985.

Jones, J. B., Wolf, B. and Mills, H. A. Plant Analysis Handbook, Athens, Georgia, USA:Micro-Macro Publishing Inc., 1991.

Jones, L. H. P. and Handreck, K, A. Studies of silica in the oat plant. III. Uptake of silica from soils by the plant. Plant and Soil 23, 79-96, 1965.

Jones, L. H. P. and Handreck, K. A, Silica in soils, plants and animals. Adv. in Agronomy 19, 107-149, 1967.

Jones, L. H. P., Hartley, R. D. and Jarvis, S. C. Mineral content of forage plants in relation to nutritional quality-silicon. Ann. Rep. of Grassland Res. Inst. p. 25-26, 1978.

Jordan, A. and Reichard, P. Ribonucleotide reductases. Annu. Rev. Biochem 67, 71-98, 1998.

Jorden, W. R. and Sullivan, C. Y. Reaction and resistance of grain sorghum to heat and drought. Internat. Crops Research Institute for the semi-arid tropics. Sorghum in the eightees. Proc. of the Internat. Symp. on Sorghum, Patansheru. A. P. India, Icrisat, pp 131-143, 1982.

Jordi, W., Dehuijzen, H. M., Stoopen, G. M. and Overbeek, J. H. M. Role of other plant organs in gibberellic acid-induced delay of leaf senescence in *Alstromeria* cut flowering stems. Physiol. Plant.

87, 426-432, 1993.

Judel, G. K. (G) Fixation and mobilization of boron in soils with high B contents toxic to crops. Landw. Forsch. Sonderh. 34/II, 103-108, 1977.

Judel, K. G., Gebauer, W. G. and Mengel, K. Yield response and availability of various phosphate fertilizer types as estimated by EUF. Plant and Soil 83, 107-115, 1985.

Judel, G. K. and Kühn, H. (G) The effect of sodium application to sugar beet at sufficient supply of potasium in pot trials. Zuck. 28, 68-71,1975.

Judel, G. K. and Mengel, K. Effect of shading on nonstructural carbohydrates and their turnover in culms and leaves during the grain filling period of spring wheat. Crop. Sci. 22, 958-962, 1982.

Judel, G. K. and Stelte, W. (G) Pot experiments with vegetables studying lead uptake from soil. Z. Planzenernähr. Bodenk. 140, 421-429, 1977.

Jump, R. K. and Sabey, B. R. Soil test extractants for prediciing selenium in plants. In: Selenium in Agriculture and the Environment, edited by Jacobs, L. W. Madison W. S., USA: Am. Soc. Agron., Soil Sci. Soc. Am., p. 95-105, 1989.

Jung, J. and Dressel, J. (G) Behaviour of magnesium in soil and plants studied in a lysimeter experiment lasting 10 years. Z. Acker- u. Pflanzenbau 130, 122-135, 1969.

Jungk, A. (G) Content of minerals and water in dependence on the development of plants. Z. Planzenernähr. Bodenk. 125, 119-129, 1970.

Jungk, A., Claassen, N., Schulz, V. and Wendt, J. (G) Plant- availability of phosphate stores in arable soils. Z. Pflanzenernähr. Bodenk. 156, 397-406, 1993.

Jungk, A. and Claasen, N. Ion diffusion in the soil-root system. Adv. Agron. 61, 53-110, 1997.

Jürgens-Gschwind, S. and Jung, J. Results of lysimeter trials at the Limburger Hof facility: The most important findings from 50 years of experiments. Soil Sci. 127, 146-160, 1979.

Jurinak, J. J. and Inouye, T. S. Some aspects of zinc and copper phosphate formation in aqueous systems. Soil Sci. Amer. Proc. 26, 144-147, 1962.

Juste, C. and Mench, M. Long-term application of sewage sludge and ist effect on metal uptake by crops. In: Biogeochemistry of Trace Metals, edited by Adriano, D. C. Ann Arbor, London, Tokyo, Lewis Publishers, p. 159-193, 1992.

Jyung, W. H., Ermann, A., Schlender, K. K. and Scala, J. Zinc nutrition and starch metabolism in *Phaseolus vulgaris* L. Plant Physiol. 55, 414-420, 1975.

Kafkafi, U. Combined irrigation and fertilization in arid zones. Israel J. Plant Sci. 42, 301-320, 1994.

Kaiser, G. and Heber, U. Sucrose transport into vacuoles isolated from barley mesophyll protoplasts. Planta 161, 562-568, 1984.

Kaiser, M. W., Weinert, H. and Huber, S. C. Nitrate reductase in higher plants: a case study for transduction of environmental stimuli into control of catalytic activity. Physiol. Plant. 105, 385-390, 1999.

Kallsen, C. E., Sammis, T. W. and Gregory, E. J. Nitrogen and yield as related to water use of spring barley. Agron. J. 76, 59-64, 1984.

Kamachi, K., Yamaya, T., Hayakawa, T., Mae, T. and Ojima, K. Vascular bundle-specific localization of cytosolic glutamine synthetase in rice leaves. Plant Physiol. 99, 1481-1486, 1992.

Kane, Y. (G) Comparative investigations of nitrogen fertilizer recommendation at differnt soil cultivation. Ph. D. Thesis Justus Liebig University Giessen, 2000.

Kang, B. T. and Fox, R. L. A methodology for evaluating the manganese tolerance of cowpea (*Vigna unguiculata*) and some preliminary results of field trials. Field Crops Res. 3, 199-210, 1980.

Kappers, I. F., Jordi, W., Maas, F. M., Stoopen, G. M. and Van der Plas, L. H. W. Gibberellin and phytochrome control senescence in *Alstromeria* leaves independently. Physiol. Plant. 103, 91-98, 1998.

Karbachsch, M. (G) Potassium nutrition of tobacco on a K⁺ fixing northwest Iranian soil. Z. Pflanzenernähr. Bodenk. 141, 513-522, 1978.

Karlen, D. L., Arny, D. C. and Walsh, L. M. Incidence of chocolate spot (*Pseudomonassyringae*), Northern corn leaf blight (*Helminthosporium tureicum*) and lodging of corn as influenced by soil fertility. Comm. in: Soil Science and Plant Analysis 4, 359-368, 1973.

Katznelson, H. The rhizosphere effect of mangels on certain groups of microorganisms. Soil Sci. 62, 343-354, 1946.

Kaufmann, M. R. Water relations of pine seedlings in relation to root and shoot growth. Plant Physiol. 43, 281-288, 1968.

Kauss, H. Some aspects of calcium-dependent regulation in plant metabolism. Annu. Rev. Plant Physiol. 38, 47-72, 1987.

Kawai, K. The relationship of phosphorus adsorption to amorphous aluminium for characterizing andosols. Soil Sci. 129, 186-190, 1980.

Keerthisinghe, D. G., Freney, J. R. and Moisier, A. R. Effect of wax-coated calcium carbide and Nitrapyrin on nitrogen loss and methane emission from dry-seeded flooded rice. Biol. Fertil. Soils 16, 71-78, 1993.

Keerthisinghe, G., Mengel, K. and DeDatta, S. K. The release of nonexchangeable ammonium (¹⁵N labelled) in wetland rice soils. Soil Sci. Soc. Am. J. 48, 291-294, 1984.

Keita, S. and Steffens, D. (G) Effect of soil structure on root growth and phosphate uptake of spring wheat. Z. Pflanzenernähr. Bodenk. 152, 345-351,1989.

Keller, P. and Deuel, H. (G) Cation exchange equilibrium with dead plant roots. Trans, Comm. II and IV. Int. Soc. Sci., Vol. II, Hamburg 1958, p. 164-168, Verlag Chemie, Weinheim/Bergstr. 1958.

Kelley, P. M. and Izawa, S. The role of chloride ion in photosystem II. I. Effects of chloride on photosystem II electron transport and hydroxylamine inhibition. Biochim. Biophys. Acta 502, 198-210, 1978.

Kelly, G. J., Latzko, E. and Gibbs, M. Regulatory aspects of photosynthetic carbon metabolism. Annu. Rev. Plant Physiol. 27, 181-205, 1976.

Kemmler, G. (G) Fertilizer application to modern rice- and wheat cultivars in developing countries. In: Proc. VIIth Fertilizer World Congress, Vienna, p. 545-563, 1972.

Kemp, A., Deijs, W. B., Hemkes, O. J. and van Es, A. J. H. Hypomagnesaemia in milking cows: intake and utilization of magnesium from herbage by lactating cows. Neth. J. agric. Sci. 9, 134-149, 1961.

Kende, H. Ethylene biosynthesis. Annu. Rev. Plant Physiol. Plant Mol. Biol. 44, 283-307, 1993.

Kende, H. and Zeevaart, H. D. The five "Classical" plant hormones. The Plant Cell 9, 1197-1210, 1997.

Kerby, T. A., Hake, K. and Keeley, M. Cotton fruiting modification with mepiquat chloride. Agron. J. 78, 907-912, 1986.

Keren, R. and Bingham, F. T. Boron in water, soils and plants. Adv. Soil Sci. 1, 229-276, 1985.

Keren, R., Bingham, F. T. and Rhoades, J. D. Plant uptake of boron as affected by boron distribution between liquid and solid phases in soil. Soil Sci. Soc. Am. J. 48, 297-302, 1985.

Keren, R., Gast, R. G. and Bar Yosef, B. pH-dependent boron adsorption by Na-montmorillonite. Soil Sci. Soc. Am. J. 45, 45-48, 1981.

Keren, R. and Mezuman, U. Boron adsorption by clay minerals using a phenomenological equation. Clay Miner. 29, 198-203, 1981.

Kerschberger, M. (G) Effect of pH on the lactate-soluble P level in soils (DL-method). Arch. Acker-Pflanzenbau Bodenkd. 31, 313-319, 1987.

Kersebaum, K. C. and Richter, J. Modelling nitrogen dynamics in a plant-soil system with a simple model for advisory purposes. Fert. Res. 27, 273-281,1991.

Keys, A. J., Bird, I. F., Cornelius, M. J., Lea, P. J., Wallsgrove, R. M. and Miflin, B. J. Photorespiratory nitrogen cycle. Nature 275, 741-743, 1978.

Khamis, S., Chaillou, S. and Lamaze, T. CO$_2$ assimilation and partitioning of carbon in maize plants derived of orthophosphat E. J. Exp. Bot, 41, 1619-1625, 1990.

Khan, A. A. and Sagar, G. R. Translocation in tomato: the distribution of the products of photosynthesis of the leaves of tomato plant during the phase of food production. Hort. Res. 7, 60-69, 1967.

Khasawneh, F. E. and Doll, E. C. The use of phosphate rock for direct applications to soils. Adv. Agron. 30, 159-206, 1978.

Kim, H. J., Cote, G. G. and Crain, R. C. Potassium channels in *Samanea sama* protoplasts controlled by phytochrome and the biological clock. Science. 260, 960-962, 1993.

Kimball, B. A. Carbon dioxide and agricultural yield: An assemblage and analysis of 430 prior observations. Agron. J. 75, 779-788, 1983.

King, G. M. and Schnell, S. Effect of increasing atmospheric methane concentration on ammonium inhibition of soil methane consumption. Nature 370, 282-284, 1994.

Kinraide, T. B. Identity of the rhizotoxic aluminium species. Plant and Soil 134, 167-178, 1991.

Kinzel, H. Pflanzenökologie und Mineralstoffwechsel. Stuttgart, Verlag Eugen Ulmer, 1982.

Kirchmann, H. and Thorvaldsson, G. Challenging targets for future agriculture. E. J. Agron. 12, 145-161, 2000.

Kirkby, E. A. Influence of ammonium and nitrate nutrition on the cation-anion balance and nitrogen and carbohydrate metabolism of white mustard plants grown in dilute nutrient solutions. Soil Sci. 105, 133-141, 1968.

Kirkby, E. A. Maximizing calcium uptake. Comm. Soil Sci Plant Anal. 10, 89-113, 1979.

Kirkby, E. A.: Plant growth in relation to nitrogen supply. In: Terrestrial Nitrogen Cycles, edited by Processes, Ecosystem Strategies and Management Impacts, edited by Clarke, F. E. and Rosswall, T. Ecol Bull Stockholm 33, 249-267, 1981.

Kirkby, E. A. and Armstrong, M. J. Nitrate uptake by roots as regulated by nitrate assimilation in the shoot of castor oil plants. Plant Physiol. 65, 286-290, 1980.

Kirkby, E. A., Armstrong, M. J. and Leggett, J. E. Potassium recirculation in tomato plants in relation to potassium supply. J. Plant Nutr. 3, 955-966, 1981.

Kirkby, E. A. and Hughes, A. D. Some aspects of ammonium and nitrate nutrition in plant metabolism. In: Nitrogen Nutrition of the Plant, edited by Kirkby, E. A., Univ. of Leeds, p. 69-77, 1970.

Kirkby, E. A. and Knight, A. H. The influence of the level of nitrate nutrition on ion uptake and assimilation, organic acid accumulation and cation-anion balance in whole tomato plants. Plant Physiol. 60, 349-353, 1977.

Kirkby, E. A. and Mengel, K. Ionic balance in different tissues of the tomato plant in relation to nitrate, urea or ammonium nutrition. Plant Physiol. 42, 6-14, 1967.

Kirkby, E. A. and Mengel, K. Preliminary observations on the effect of urea nutrition on the growth and nitrogen metabolism of sunflower plants. In: Nitrogen Nutrition of the Plant, edited by Kirkby, E. A. The University of Leeds, p. 35-38, 1970.

Kirkby, E. A. and Mengel, K. The role of magnesium in plant nutrition. Z. Planzenernähr. Bodenk. 2, 209-222, 1976.

Kitao, K., Lei, T. T. and Koike, T. Effect of mangenese toxicity on photosynthesis of white birch (*Betula platyphylla* var. *japonica*) seedlings. Physiol. Plant. 101, 249-256, 1997.

Kjellerup, V. and Dam Kofoed, A. Nitrogen fertilization in leaching of plant nutrients from soil. Lysimeter experiments with [15]N. Tidsskr. Planteavl. 87, 1-22, 1983.

Klapheck, S., Chrost, B., Starke, J. and Zimmermann, H. γ-glutamylcysteinylserine - A new homologue of glutathione in plants of the family Poaceae. Bot. Acta 105, 174-179, 1992.

Klapp, E. (G) Textbook of Husbandry and Crop Science. 3rd. ed., P. Parey-Verlag Berlin p. 63, 1951.

Kleczkowski, L., Villand, A. P., Lüthi, E., Olsen, A. O. and Preiss, J. Insensitivity of barley endosperm ADPglucosepyrophosphorylase to 3-phosphoglycerate and orthophosphate regulation. Plant Physiol. 101, 179-186, 1993.

Klepper, L. and Hagemen, R. H. The occurence of nitrate reductase in apple leaves. Plant Physiol. 44, 110-114, 1969.

Kliewer, M. and Evans, H. J. Cobamide coenzyme contents of soybean nodules and nitrogen fixing bacteria in relation to physiological conditions. Plant Physiol. 38, 99-104, 1963.

Kliewer, W. M. Influence of environment on metabolism of organic acids and carbohydrates in *Vitis vinifera*. I. Temperature. Plant Physiol. 39, 869-880, 1964.

Kliewer, W. M. and Cook, J. A. Arginine and total free amino acids as indicators of the nitrogen status of grapevines. J. Am. Soc. Hort. Sci. 96, 581-587, 1971.

Kluge, H. (G.) Nutrient uptake of various crops. Ministry of Agriculture, Fed. Rep. Germany 1992.

Kluge, M. The flow of carbon in Crassulacean Acid Metabolism (CAM). In: Photosynthesis II, Encycl. Plant Physiol., New Series, Vol. 6, edited by Gibbs, M. and Latzko, E., Springer, Berlin, Heidelberg, New York. p. 112-123, 1979.

Knight, A. H. and Crooke, W. M. Interaction between nickel and calcium in plants. Nature 178, 220, 1956.

Knight, A. H., Crooke, W. M. and Burridge, T. C. Cation exchange capacity, chemical composition and the balance of carboxylic acids in the floral parts of various plant species. Ann. Bot. 37, 159-166, 1973.

Knight, H., Trewavas, A. J. and Knight, M. R. Calcium signalling in *Arabidopsis thaliana* responding to drought and salinity. Plant Journal 12, 1067-1078, 1997.

Knight, W. E., Hagedorn, C., Watson, V. H. and Friesner, D. L. Subterranean clover in the United States. In: Advances in Agronomy, edited by Brady, N. C. New York, London: Academic Press, p. 165-191, 1982.

Knoblauch, M. and van Bel, A. J. E. Sieve tubes in action. The Plant Cell 10, 35-50, 1998.

Kobayashi, M., Matoh, T. and Azuma, J. Two chains of rhamnogalacturonan II are crossed-linked by borate-diol ester bonds in higher plant cell walls. Plant Physiol. 110, 1017-1020, 1996.

Kobrehel, K., Wong, J. H., Balogh, A., Kiss, F., Yee, B. C. and Buchanan, B. B. Specific reduction of wheat storing proteins by thioredoxin h. Plant Physiol. 99, 919-924, 1992.

Koch, K. and Mengel, K. Effect of varied potassium nutrition on the uptake and incorporation of labelled nitrate by young tobacco plants (*Nicotiana tabacum* L.). J. Sci. Food Agric. 23, 1107-1112, 1972.

Koch, K. and Mengel, K. The influence of the level of potassium supply to young tobacco plants (*Nicotiana tabacum* L.) on short-term uptake and utilization of nitrate nitrogen (^{15}N). J. Sci. Food Agric. 25, 465-471, 1974.

Koch, K. and Mengel, K. The influence of potassium nutritional status on the absorption and incorporation of nitrate nitrogen. In: Plant Analysis and Fertilizer Problems, Vol. I, Proc. 7th Intern. Colloq. Hanover, p. 209-218,1974.

Koch, K. and Mengel, K. The effect of K on N utilization by spring wheat during grain formation. Agron. J. 69, 477-480, 1977.

Koch, K. E. Carbohydrate-modulated gene expression in plants. Annu. Rev. Plant Physiol. Plant Mol. Biol. 47, 509-540, 1996.

Kochian, L. V. Mechansism of micronutrient uptake and translocation in plants. In: Micronutrients in Agriculture, 2nd ed. edited by J. J. Mortuedt, F. R. Cox, L. H. Shuman and R. M. Welch SSSA Book series, Madison, USA p. 229-298, 1991.

Kochian, L. V. Cellular mechanisms of aluminium toxicity and resistance in plants. Annu. Rev. Plant Physiol.

Plant Mol. Biol. 46, 237- 260, 1995.

Kochian, L. V. and Lucas, W. J. Potassium transport in roots. Adv. in Bot. Res. 15, 93-178, 1988.

Koda, Y. The role of jasmonic acid and related compounds of plant development. Int. Rev. Cytol. 135, 155-199, 1992.

Kohl, A. and Werner, W. (G) Investigations concerning the seasonal change of EUF-fractions for the characterization of easily mobilizable soil nitrogen by electro-utlrafiltration (EUF). VDLUFA Schriftenreihe 20, Kongr. Band, 333-341, 1982.

Köhler, B. and Raschke, K. The delivery of salts to the xylem. Three types of anion conductance in the plasmalemma of the xylem parenchyma of roots of barley. Plant Physiol., 122, 243-254, 2000.

Köhn, W. (G) Effect of long term tillage-fertilization and rotation measurements on chemical and physical properties and on the yield level of a loamy sand soil. Part 2. Long term changes in yield and investigations on yield components of cereals. Bayerisch. Landw. Jahrbuch Heft 4, 419-442, 1976.

Kolesch, H., Oktay, M. and Höfner, W. Effect of iron chlorosis-inducing factors on the pH of the cytoplasm of sunflower (*Helianthus annuus*). Plant and Soil 82, 215-221, 1982.

Kong, T. and Steffens, S. (G) Importance of the potassium depletion in the rhizosphere and clay minerals for the release of non-exchangeable potassium and its determination with HCl. Z. Pflanzenernaehr. Bodenk. 152, 337-343, 1989.

Koontz, H. V. and Foote, R. E. Transpiration and calcium deposition by unifoliate leaves of *Phaseolus vulgaris* differing in maturity. Physiol. Plant. 19, 313-321, 1966.

Korensky, F. and Neuberg, J. Autumn application of anhydrous ammonia for spring cultures in Czechoslovakia. Rostlinná Výroba 14, 803-814, 1968.

Kosegarten, H. and Englisch, G. Effect of various nitrogen forms on the pH in leaf apoplast and on iron chlorosis of *Glycine max* L. Z. Pflanzenernähr. Bodenk. 157, 401-405, 1994.

Kosegarten, H., Grolig, F., Wieneke, J., Wilson, G. and Hoffmann, B. Differential ammonia-elicited changes of cytosolic pH in root hair cells of rice and maize as monitored by 2',7'-bis-(2-carboxyethyl)-5 (and-6) -carboxyfluorescein-fluorescence ratio. Plant Physiol. 113, 451-461, 1997.

Kosegarten, H., Grolig, F., Esch, A., Glüsenkamp, K. H. and Mengel, K. Effects of NH_4^+, NO_3^- and HCO_3^- on apoplast pH in the outer cortex of root zones of maize, as measured by the fluorescence ratio of fluorescein boronic acid. Planta 209, 444-452, 1999a.

Kosegarten, H., Hoffmann, B. and Mengel, K. Apoplastic pH and Fe^{3+} reduction in intact sunflower leaves. Plant Physiol. 121, 1069-1079, 1999b.

Kosegarten, H. and Mengel, K. Evidence for a glucose 1-phosphate translocator in storage tissue amyloplasts of potato (*Solanum tuberosum*) suspension-cultured cells. Physiol. Plant. 91, 111-120, 1994.

Kosegarten, H. and Mengel, K. Starch deposition in storage organs and the importance of nutrients and external factors. Z. Pflanzenernähr. Bodenk. 161, 273-287, 1998.

Kosegarten, H., Wilson, G. H. and Esch, A. The effect of nitrate nutrition on iron chlorosis and leaf growth in sunflower (*Helianthus annuus* L.). Eur. J. Agron. 8, 4283-293, 1998.

Köttgen, P. (G) Determination of easily soluble nutrients released by electrical current, a means of estimation of the fertility status of agricultural soils. Z. Pflanzenernähr. Düng. Bodenk. 29A, 275-290, 1933.

Kothari, S. K., Marschner, H. and Römheld, V. Contribution of the VA myccorhizal hyphae in acquisition of phosphorus and zinc by maize grown in a calcareous soil. Plant and Soil 131, 177-185, 1990.

Kovacevic, V. and Vukadinovic, V. The potassium requirement of maize and soyabean on a high K-fixing soil. South. Afr. J. Plant and Soil 9, 10-13, 1992.

Kovanci, I., Hakerlerler, H. and Höfner, W. (G) Cause of iron chlorosis in mandarins (*Citrus reticulata blanco*) in the Aegean area. Plant and Soil 50, 193-205, 1978.

Kowalenko, C. G. and Cameron, D. R. Nitrogen transformations in soil-plant systems in three years of field experiments using tracer and non-tracer methods on an ammonium-fixing soil. Can. J. Soil Sci. 58, 195-208, 1977.

Koyro, H.-W. Ultrastructural and physiological changes in root cells of sorghum plants (*Sorghum bicolor* x *S. sudanensis* cv. Sweet Sioux) induced by NaCL. J. Exp. Bot. 48, 693-706, 1997.

Koyro, H.-W. and Huchzermeyer, B. Influence of high NaCl salinity on growth, water and osmotic relations of the halophyte *Beta vulgaris* ssp. maritima – Development of a quick check. In: Halophyte uses in different climates, edited by Lieth, H., Moschenko, M., Lohmann,

Koyro, H.-W. and Hamdy, A. I. Progress in Biometerology, Vol. 13. Backhuis Publ., Leiden, pp. 89-103, 1999.

Koyro, H.-W. and Stelzer, R. Ion concentrations in the cytoplasm and vacuoles of rhizodermal cells from NaCl treated Sorghum, Spartina and Puccinellia plants. J. Plant Physiol. 133, 441-446, 1988.

Koyro, H.-W., Wiegmann, L., Lehmann, H. and Lieth, H. Physiological mechanisms and morphological adaptation of *Languncularia racemosa* to high NaCl salinity. In: Water management, salinity and pollution control towards sustainable irrigation in the mediterranean region. Bari: Tecnomack, edited by Lieth, H., Hamdy, A. &. Koyro, H.-W. B. ari: Tecnomack, 51-78, 1997.

Kramer, D., Römheld, V., Landsberg, E. and Marschner, H. Induction of transfer-cell formation by iron deficiency in the root epiderrnis of *Helianthus annuus* L. Planta, 147, 335-339, 1980.

Kramer, P. J. Water relations of plant cells and tissues. Annu. Rev. Plant Physiol. 6, 253-272, 1955.

Kramer, P. J. and Boyer, J. S. Water relations of plants and soils. Acad. Press, San Diego, 1995.

Krapp, A. and Stitt, M. An evaluation of direct and indirect mechanisms for the "sink-regulation" of photosynthesis in spinach: changes in gas exchange, carbohydrates, metabolites, enzyme activities, and steady-state transcript levels after cold-girdling source leaves. Planta 195, 313-323, 1995.

Krause, W. (G) Soils and plant communities. In: Encylcopedia of Plant Physiology. W. Ruhland ed. Springer-Verlag, Berlin, Vol. 4, 807-850, 1958.

Krause, W. (G) Flora and vegetation on Serpentine sites of the Balkans. Z. Pflanzenernähr. Düng. Bodenk. 99, 97-107, 1962.

Krauss, A. Influence of nitrogen nutrition on tuberinitiation of potatoes. In: Physiological Aspects of Crop Productivity, p. 175-184 Int. Potash Inst., Bern, 1980.

Krauss, A. and Marschner, H. (G) Influence of nitrogen nutrition and application of growth regulators on tuber initiation in potato plants. Z. Pflanzeüernähr. Bodenk. Heft 2, 143-155, 1976.

Krauss, A. and Marschner, H. (G) Influence of nitrogen nutrition of potatoes on tuber induction and tuber growth rate. Z. Pflanzenernähr. Bodenk. 128, 153-168, 1971.

Kreis, W. and Hölz, H. (G) Cellular transport and storgar of natural molecules. Naturw. Rundschau. 44, 463-470, 1992.

Krenzer, E. G., Moss, D. N. and Crookston, R. K. Carbon dioxide compensation points of flowering plants. Plant Physiol. 56, 194-206, 1975.

Kretzschmar, R. M., Hafner, H., Bationo, A. and Marschner, H. Long- and short-term effects of crop residues on aluminium toxicity, phosphorus availability and growth of pearl millet in a sandy acid soil. Plant and Soil 136, 215-223, 1991.

Krohne-Ehrich, G. (G) Alcoholism and alcohol dehydrogenase. Naturw. Rundschau 31, 382, 1978.

Kronzucker, H. J., Siddiqi, M. Y. and Glass, D. M. Kinetics of NH_4^+ influx in spruce trees. Plant Physiol. 110, 773-779, 1996.

Kronzucker, H. J., Siddiqi, M. Y., Glass, A. D. M. and Kirk, G. J. D. Nitrate-ammonium synergism in rice. A subcellular flux analysis. Plant Physiol. 119, 1041-1045, 1999.

Krueger, R. W., Lovatt, C. J. and Albert, L. S. Metabolic requirements of *Cucurbita pepo* for boron. Plant Physiol. 83, 254-258, 1987.

Krüger, W. The influence of fertilizers on fungal disease of maize. In: Fertilizer Use and Plant Health, p. 145-156. Int. Potash Inst., Bern, 1976.

Kubota, J. Molybdenum status of United States soils and plants. In: Molybdenum in the Environment, edited by Chappell, W. R. and Peterson, K. K. New York: Marcel Dekker, p. 558-581, 1977.

Kubota, J. and Allaway, W. H. Geographic distribution of trace element problems. In: Micronutrients in Agriculture, edited by Mortvedt, J. J., Giordano, P. M. and Lindsay, W. L. Soil Sci. Soc. America, Madison/USA, p. 525-554, 1972.

Kubota, J., Welch, R. M. and van Campen, D. R. Soil related nutritional problem areas for grazing animals. Adv. Soil Sci. 6, 189-215, 1987.

Kucey, R. M. N., Janzen, H. H. and Leggett, M. E. Microbiocally mediated increases in plant available soil phosphorus. Adv. Agron. 42, 199-228, 1989.

Kuchenbuch, R. and Jungk, A. (G) Effect of potassium fertilizer application on the potassium availability in the rhizosphere of rape. Z. Pflanzenernähr. Bodenk. 147, 435-448, 1984.

Kücke, M. and Kleeberg, P. Nitrogen balance and soil nitrogen dynamics in two areas with different soil, climatic and cropping conditions. Eur. J. Agron. 6, 89-100, 1997.

Kühbauch, W. (G) Intensity of land use in historical times. Geowiss. 11, 121-129, 1993.

Kuhlmann, K. P., Stripf, R., Wätjen, U., Richter, F. W., Gloystein, F. and Werner, D. Mineral composition of effective and ineffective nodules of *Glycine max* in comparison to roots: Characterization of developmental stages by differences in nitrogen, hydrogen sulfur, molybdenum, potassium and calcium content. Angew. Botanik 56, 315-323, 1982.

Kuhlmann, H. and Wehrmann, L (G) Testing different Methods of soil analysis for their applicability for the determination of K fertilizer requirement of loess soils. Z. Planzenernähr. Bodenk. 147, 334-348, 1984.

Kuhn, A. J., Bauch, J. and Schröder, W. H. Monitoring uptake and contents of Mg, Ca and K in Norway spruce as influenced by pH and Al, using microprobe analysis and stable isotope labelling. Plant and Soil 168, 135-150, 1995.

Kuhn, C., Franceschi, V. R., Schulz, A., Lemoine, R. and Frommer, W. B. Macromolecular trafficking indicated by localization and turnover of sucrose transporters in enucleate sieve elements. Science 275, 1298-1300, 1997.

Kühn, H. (G) Possibilities for the enrichment of vegetables with micronutrients by fertilizer application. Landw. Forsch., 16 Sonderh., 112-120, 1962.

Kühn, H. and Schaumlöffel, E. (G) The effect of high copper application on the growth of cereals. Landw. Forsch. 14, 82-98, 1961.

Kukaszewski, K. M. and Blevins, D. G. Root growth inhibition in boron deficient or aluminium-stressed squash plants may be a result of impaired ascorbate metabolism. Plant Physiol. 112, 1-6, 1996.

Kuntz, J. E. and Riker, A. J. The use of radioactive isotopes to ascertain the role of root grafting in the translocation of water, nutrients, and disease-inducing organisms. Proc. Int. Conf. Peaceful Uses At. Energy 12, 144-148, 1955.

Kuntze, H. and Bartels, R. (G) Nutrient status and yield production on peat grassland. Landw. Forsch. Sonderh. 31/I, 208-219, 1975.

Kuntze, H., Roeschmann, G., Schwerdtfeger, G. Bodenkunde. 5 neubearb. und erweit. Aufl. Verlag Eugen Ulmer Stuttgart, 1994.

Kursanov, A. L. Transport of assirnilates and sugar storage in sugar beet. Z. Zuckerind. 24, 478-487, 1974.

748

Kursanov, A. L. and Vyskrebentzewa, E. (F) The role of potassium in plant metabolism and the biosynthesis of compounds important for the quality of agricultural products. In: Potassium and the Quality of Agricultural Products. Proc. 8th Congr. Intern. Potash Institute, Bern p. 401-420, 1966.

Kutschera, U. Fusicoccin-induced growth and dark respiration in rye coleoptiles. J. Plant Physiol. 154, 554-556, 1999.

Kutschera, U. and Schopfer, P. Evidence against the acid-growth theory of auxin action. Planta 163, 483-493, 1985.

Kutschera, U. and Schopfer, P. *In vivo* measurement of cell wall extensibility in maize coleoptiles: Effect of auxin and abscisic acid. Planta 169, 437-447, 1986.

Labanauskas, C. K.: Manganese. In: Diagnostic criteria for plant and soils, edited by Chapman, H. D., University of California, p. 264-285, 1966.

Lacan, D. and Durand, M. Na^+-K^+ exchange at the xylem / symplast boundary. Plant Physiol. 110, 705-711, 1996.

Ladha, J. K., Pareek, R. P. and Becker, M. Stem -nodulating legume - Rhizobium symbiosis and its agronomic use in lowland rice. Adv. Soil Sci. 20, 147-192, 1992.

Lafos, K. (G) Uptake and distribution of silicon in vine (*Vitis*). Ph. D. Thesis, Justus Liebig University, Giessen, 1995.

Lag, J. Relationships between the chemical composition of the precipitation and the contents of exchangeable ions in the humus layer of natural soils. Acta Agric. Scand. 18, 148-152, 1968.

Lagerwerff, J. V. Lead, mercury and cadmium as environmental contaminants. In: Micronutrients in Agriculture, edited by Mortvedt, J. J. Giordano, P. M. and Lindsay, W. L. Soil Sci. Soc. America, Madison/USA, p. 593-636, 1972.

Lagerwerff, J. V. and Bolt, G. H. Theoretical and experimental analysis of Gapon's equation for ion exchange. Soil Sci. 87, 217-222, 1959.

Lambers, H. The physiological significance of cyanide-resistant respiration in higher plants. In: Energy Metabolism in Higher Plants in Different Environments, edited by Lambers, H., PH. D. Thesis of the Rijks-Universiteit Groningen, Netherlands, p. 113-128, 1979.

Lambert, R. G. and Linck, A. J. Comparison of the uptake of P-32 and K-42 intact alfalfa and oat roots. Plant Physiol. 39, 920-924, 1964.

Landsberg, E. C. Regulation of iron-stress-response by whole plant activity. J. Plant Nutr. 7, 609-621, 1984.

Lang, A. Turgor-regulated translocation. Plant, Cell and Environment 6, 683-689, 1983.

Lang, A., Thorpe, M. R. and Edwards, W. R. N. plant water potential and translocation. In: Phloem Transport, edited by Cronshaw, J., Lucas, W. J. and Giaquinta, R. T. A. R. Liss, Inc., New York, pp. 193-194, 1986.

Langston, R. Studies on marginal movement of cobalt-60 in cabbage. Proc. Amer. Soc. hort. Sci. 68, 366-369, 1956.

Lapushner, D., Frankel, R. and Fuchs, Y. Tomato cultivar response to water and salt stress. Acta Horticult. 190, 247-252, 1986.

Larsen, S. The use of ^{32}P in studies on the uptake of phosphorus by plants. Plant and Soil 4, 1-10, 1952.

Larsen, S. Isoionic exchange of phosphate in paddy soils. Plant and Soil 27, 401-407, 1967a.

Larsen, S. Soil phosphorus. Adv. in Agron. 19, 131-206, 1967b.

Larsen, S. and Widdowson, A. E. Chemical composition of soil solution. J. Sci. Fd Agric. 19, 693-695, 1968.

Lathwell, D. J. and Peech, M. Interpretation of chemical soil tests. Cornell Univ. Agric. Exp. Stat., New York State College of Agriculture, Ithaca, New York, Bulletin 995, October 1964.

Laties, G. G. Active transport of salt into plant tissue. Annu. Rev. Plant Physiol. 10, 87-112, 1959.

Latimore, M., Giddens, J. and Ashley, D. A. Effect of ammonium and nitrate nitrogen upon photosynthate supply and nitrogen fixation by soybeans. Crop Sci. 17, 399-404, 1977.

Latiri-Souki, K., Nortcliffe, S. and Lawlor, D. W. Nitrogen fertilizer can increase dry matter, grain production and radiation and water use efficiency for durum wheat under sem-arid conditions. Eur. J. Agron. 9, 21-34, 1998.

Läuchli, A. Function of the root in relation to the structural aspects and localization of ions. XII, Intern. Botanical Congr. Leningrad 1975.

Läuchli, A. Salinity-potassium interactions in crop plants. In: Frontiers in Potassium Nutrition: New Perspectives on the Effects of Potassium on Physiology of Plants, edited by Oosterhuis, D. M. & Berkowitz, G. A. Potash & Phosphate Institute, Georgia USA /Potash & Phosphate Institute of Canada, Saskatoon, Canada, 71-76, 1999.

Läuchli, A., Kramer, D., Pitman, M. G. and Lüttge, U. Ultrastructure of xylem parenchyma cells of barley roots in relation to ion transport to the xylem. Planta 119, 85-99, 1974.

Läuchli, A. and Pflüger, R. Potassium transport through plant cell membranes and metabolic role of potassium in plants. In: Potassium Research - Review and Trends. Potash Inst. Bern, p. 111-163, 1978.

Läuchli, A. and Wieneke, J. Studies on growth and distribution of Na, K, and Cl in soybean varieties differing in salt tolerance. Z. Pflanzenernähr. Bodenk. 142, 3-13, 1979.

Lauer, M. J., Blevins, D. G. and Sierzputowska-Gracs, H. P. ^{31}P nuclear magnetic resonance determination of phosphate compartmentation in leaves of reproductive soybeans (Glycine max) as affected hy phosphate nutrition. Plant Physiol. 89, 1331-1336, 1989.

Laulhere, J.-P. and Briat, J.-F. Iron release and uptake by plant ferritin: effects of pH, reduction and chelation. Biochem. J. 290, 693-699, 1993.

Laves, D. (G) Potassium transformation in soil. Arch. Acker- u. Pflanzenbau u. Bodenk. 22 (8), 521-528, 1978.

Layzell, D. B. Oxygen and the control of nodule metabolism and N_2 fixation. In: Biological Nitrogen Fixation for the 21st Century, edited by Elmerich, C., Kondorosi, A. and Newton, W. E. Dordrecht: Kluwer Academic Publishers, p. 435-440, 1998.

Layzell, D. B., Hunt, S. and Palmer, G. R. Mechanism of nitrogenase inhibition in soybean nodules. Plant Physiol. 92, 1101-1107, 1990.

Lazof, D. B. and Holland, M. J. Evaluation of the aluminium induced root growth inhibition in isolation from low pH effect in Glycine max, Pisum sativum and Phaseolus vulgaris. Austr. J. Plant Physiol. 26, 147-157, 1999.

Lea, P. J., Blackwell, R. D. and Joy, K. W. Ammonia assimilation in higher plants. In: Nitrogen Metabolism of Plants, edited by Mengel, K. and Pilbeam, D. J. Oxford: Clarendon Press, p. 153-186, 1992.

Leach, G. Energy and Food Production. IPC Science and Technology Press, Guilford 1976.

Le Bot, J. and Kirkby, E. A. Diurnal uptake of nitrate and potassium during the vegetative growth of tomato plants. J. Plant Nutr. 15, 247-264, 1992.

Le Bot, J., Kirkby,E. A. and van Beusichem, M. L. Manganese toxicity in tomato plants: effects on cation uptake and distribution. J. Plant Nutr. 13, 513-525, 1990

Le Bot, J., Gos, M. J., Carvalho, M. J. G., van Beusichem, M. L. and Kirkby, E. A. The significance of the magnesium to manganese ratio in plant tissues for growth and alleviation of manganese toxicity in tomato (Lycopersicon esculentum) and wheat (Triticum aestivum) plants. Plant and Soil.... 1990

Lee, C., Miller, G. W. and Welkie, G. W. The effects of hydrogen fluoride and wounding on respiratory enzymes in soybean leaves. Air Water Pollut. Int. J. 10, 169-181, 1965.

Lee, R. B. and Ratcliffe, R. B. Subcellular distribution of inorganic phosphate, and levels of nucleoside triphosphate, in mature maize roots at low external phosphate concentrations: Measurements with ^{31}P

NMR. J. Expt. Bot. 44, 587-598, 1993.

Leggett, J. E. and Epstein, E. Kinetics of sulfate absorption by barley roots. Plant Physiol. 31, 222-226, 1956.

Leggett, J. E. and Gilbert, W. A. Magnesium uptake by soybeans. Plant Physiol. 44, 1182-1186, 1969.

Leggewie, G., Wilmitzer, L. and Riesmeier, J. G. Two cDNAs from potato are ableto complement a phoshphate uptake-deficient yeast mutant: identification of phosphate transporters from higher plants. Plant Cell 9, 381-392, 1997.

Le Gouis, J., Delebarre, O., Beghin, D., Heumez, E. and Pluchard, P. Nitrogen uptake and utilization efficiency of two-row and six- row winter barley cultivars grown at two N levels. European J. Agron. 10, 73-79, 1999.

Lehman, W. F., Rutger, J. N., Rownson, F. E. and Kaddah, M. Value of rice characteristics in selection for resistance to salinity in an arid environment. Agron. J. 76, 366-370, 1984.

Lehninger, A. L. Biochemistry, the Molecular Basis of Cell Structure and Function. Worth Publishers, Inc., New York 1975.

Leigh, R. A. and Johnston, A. E. Concentrations of potassium in the dry matter and tissue water of field grown barley and their relationship to grain yield. J. of Agric. Sci. Camb. 161, 675-685, 1983.

Leigh, R. A. and Wyn Jones, R. G. Cellular compartmentation in plant nutrition:The selective cytoplasm and the promiscuous vacuole. Adv. Plant Nutrition. 2, 249-279, 1986.

Leinweber, P. Phosphorus fractions in soils from an area with high density of livestock population. Z. Pflanzenernähr. Bodenk. 159, 251-256, 1996.

Leinweber, P., Geyer-Wedell, K. and Jordan,E. (G) Phosphorus concentrations in soils in an area with high livestock density. Z. Pflanzener. Bodenk. 157, 383 385, 1994.

Leinweber, P. and Schulten, H. R. Nonhydrolyzable organic nitrogen in soil zize separates from long-term agricultural experiments. Soil Sci. Soc. Am. J. 62, 383-393, 1998.

Lemon, E. and van Houtte, R. Ammonia exchange at the land surface. Agron. J. 72, 876-883, 1980.

Lenzi, M. A. and Di Luzio, M. Surface runoff, soil erosion and water quality modelling in the Alpone watershed using AGNPS integrated with a geographic information system. Eur. J. Agron. 6, 1-14, 1997

Lerchl, D., Hillmer, S., Grotha, R. and Robinson, D. G. Ultrastructural observations on CTC-induced callose formation in *Riella helicophylla*. Bot. Acta 102, 62-70, 1989.

Lesaint, C. and Coic, Y. (F) Hydroponic Cultures, Paris: Maison Rustique, 1983

Leskosek, M. (G) Field trials comparing different forms of phosphate fertilizer on meadows in Slowenia. Landw. Forsch. 30/11 Sonderheft:112-122, 1973.

Leskosek, M. (G) Relationships between soil K and plant K and the K impact on grass yield. Das wirtschaftseigene Futter 24, 65-74, 1978.

Leskosek, M., Sestic, S. and Derkacev, E. (G) Phosphate fertilizer application to wheat/maize rotation on Tschernosem. Die Phosphors. 29, 163-176, 1972.

Lessani, H. and Marschner, H. Relation between salt tolerance and long distance transport of sodium and chloride in various crop species. Aust. J. Plant Physiol. 5, 27-37, 1978.

Lessard, R., Rochette, P., Gregorich, E. G., Pattey, E. and Desjardins, R. L. Nitrous oxide fluxes from manure-amended soil under maize. J. Environ. Qual 25, 1371-1377, 1996

Letham, D. S. and Palni, L. M. S. The biosynthesis and metabolism of cytokinins. Annu. Rev. Plant Physiol. 34, 163-197, 1983.

Leustek, T. and Saito, K. Sulphate transport and the assimilation in plants. Plant Physiol. 120, 637-643, 1999.

Lewis, D. A. and Tatchell, J. A. Energy in UK agriculturE. J. Sci. Food Agric. 30, 449-457, 1979.

Lewis, D. G. and Quirk, J. P. Phosphate diffusion in soil and uptake by plants. III. ^{31}P- movement and uptake by plants as indicated by ^{32}P autoradiography. Plant and Soil 26, 445-453, 1967.

Lewis, J. C. and Powers, W. L. Antagonistic action of chlorides on the toxicity of iodides to corn. Plant Physiol. 16, 393-398, 1941.

Lhuillier-Soundele, A., Munier-Jolain, N. G. and Ney, B. Dependence of seed nitrogen on plant nitrogeri availability during the seed filling in pea. European J. Agron. 11, 157-166, 1999.

Li, C., Fan, X. and Mengel, K. Turnover of interlayer ammonium in loess-derived soil grown with winter wheat in the Shaanxi Province of China. Biol. Fertil. Soils 9, 211-214, 1990.

Licht, S., Gerfen, G. J. and Stubbe, J. Thiyl radicals in ribonucleotide reductase. Science. 271, 477-481, 1996.

Lickfett, T., Matthäus, B., Velasco, L. and Möllers, C. Yield and quality parameters of two oilsed rape cultivars as affected by different phosphorus supply. Eur. J. Agron. 11, 293-299, 1999.

Liebig, J. (G) Organic Chemistry and its Application to Agriculture and Physiology. Verlag Viehweg, Braunschweig 1841.

Liebig, J. Organic Chemistry in its Application to Agriculture and Physiology. Vieweg Verlag, Braunschweig, 8th Ed. 1865.

Lieth, H., Hamdy, A. and Koyro, H.-W. Water management, salinity and pollution control towards sustainable irrigation in the mediterranean region. Valenzano, Bari, Instituto Agronomico Mediterraneo, pp 1-209, 1997.

Likens, G. E., Driscoll, C. T. and Buso, D. C. Long-term effects of acid rain: Response and recovery of a forest ecosystem. Science 272, 244-246, 1996.

Lin, C., Motto, H. L., Douglas, L. A. and Busscher, W. J. Multifactor kinetics of phospate reactions with minerals in acid soils: 11 Experimental curve fitting. Soil Sci. Soc. Am. J. 47, 1103-1109, 1983.

Lin, W. Inhibition of anion transport in corn root protoplasts. Plant Physiol. 68, 435-438, 1981.

Lind, A. M. and Pedersen, M. B. Nitrate reduction in subsoil. II. General description of boring profiles, and chemical investigations on the profile cores. Tidsskr. Planteavl. 80, 82-99, 1976.

Lindhauer, M. G. The role of K^+ in cell extension, growth and storage of assimilates. In: Methods of K-Research in Plants, edited by International Potash Institute, Bern, Switzerland: International Potash Institute, p 161-187, 1989.

Lindsay, W. L. Zinc in soils and plant nutrition. Adv. in Agron. 24, 147-186, 1972.

Lindsay, W. L.: Role of chelation in micronutrient availability. In: The Plant Root and Its Environment, edited by E. W. Carson, University Press of Virginia, p. 507-524, 1974.

Lindsay, W. Iron oxide solubilization by organic matter and its effect on iron availability. In: Iron Nutrition and Interactions in Plants, edited by Y. Chen and Y. Hadar, Dordrecht: Kluwer Academic Publishers, ISBN 0-7923-1095-0, p. 29-36, 1991.

Lindsay, W, L., Hodgson, J. F. and Norvell, W. A. The physiochemical equilibrium of metal chelates in soils and their influence on the availability of metal cations. Trans. Comm. II and IV. Int. Soc. Soil Sci. (Aberdeen 1966), p. 305-316, 1967.

Lindsay, W. L. and Norvell, W. A. Development of a DTPA test for zinc, iron, amnganese, and copper. Soil Sci. Soc. Am. J. 42, 421-428, 1978.

Lindsay, W. L. and Schwab, A. P. The chemistry of iron in soils and its availability to plants. J. Plant Nutrition 5, 821-840, 1982.

Lindsay, W. L., Vlek, P. L. and Chien, S. H. Phosphate minerals. In: Minerals in Soil Environments, edited by Dixon, J. B. and Weed, S. B., Madison USA: Soil Science Soc. Am., p. 1089, 1989.

Linser, H. and Herwig, K. (G) Investigations into the relationship between nutrient uptake and the osmotic pressure of the outer solution. Protoplasma LVII, 588-600, 1963.

Linser, H. and Herwig, K. (G) Relationships between wind, transpiration and nutrient translocation in flax with particular regard to a varied water and potash application. Kali-Briefe, Fachgeb. 2, 2. Folge, 1968.

752

Linser, H., Kühn, H. and Schlögl, G. (G) A field technique for distinguishing between sulphur and nitrogen deficiency. V. Simposio Internazionale di Agrochimica su 'Lo zolfo in agricoltura', Palermo, p. 90-103, 1964.

Linser, H., Mayr, H. and Bodo, G. (G) Effect of chlorocholine chloride on spring wheat. Bodenkultur 12, 279-280, 1961.

Lioi, L. and Giovanneti, M. Variable effectivity of three vesicular -arbuscular mycorrhizal endophytes in *Hedysarum coronarum* and *Medicago sativa*. Biol. Fertil. Soils 4, 193-197, 1987.

Lippert, E. (G) Textbook of Plant Physiology.: Fischer-Verlag, Jena, 1993.

Liu Zhi-Yu and Qin Sheng-Wu The study of nitrogen distribution around rice rhizosphere. Proc. Sympos. of Paddy Soil. Ed. Institute of Soil Science. Academia Sinica, Science Press, Beijing. Springer Verlag Berlin, Heidelberg, New York, Tokyo, p. 511-546, 1981.

Liu, Z., Shi, W. and Fan, X. The rhizosphere effects of phosphorus and iron in soils. In: Transactions, 14. International Congr. Soil Science, edited by Masayoshi Koshino, Tokyo: International Society of Soil Science., p. 147-152, 1990.

Locher, J. T. and Brouwer, R. Preliminary data on the transport of water, potassium and nitrate in intact and bleeding maize plants. Mededeling 238 van het 1. B. S. (Wageningen), p. 41-49, 1964.

Logan, T. J., Linday, B. J., Goins, L. E. and Ryan, J. A. Field assessment of sludge metal bioavailability to crops: sludge rate response. J. Environ. Qual. 26, 534-550, 1997.

Lohaus, G., Winter, H., Riens, B. and Heldt, H. W. Further studies of the phloem loading process in leaves of barley and spinach. The comparison of metabolite concentrations in the apoplastic compartment with those in the cytosolic compartment and in the sieve tubes. Bot. Acta 108, 270-275, 1995.

Löhnertz, O., Schaller, K. and Mengel, K. (G) Nutrient dynamics in vine. I. Communication: nitrate in wood and vegetative parts of vine. Wein-Wiss. 44, 20 -27, 1989a.

Löhnertz, O., Schaller, K. and Mengel, K. (G) Nutrient dynamics in vine: nitrate in vegetative and reproductive organs. Die Weinwiss. 44, 77-86, 1989b.

Löhnis, M. P. Effect of magnesium and calcium supply on the uptake of manganese by various crop plants. Plant and Soil 12, 339-376, 1960

Loll, M. J. and Bollag, J. M. Protein transformation in soil. Adv. Agron. 36, 351-382, 1983.

Loneragan, J. F. Distribution and moverment of copper in plants. In: Copper in Soils and Plants, edited by Loneragan, J. F., Robson, A. D. and Graham, R. D., Academic Press, London p. 165-188, 1981.

Loneragan, J. F., Snowball, K. and Robson, D. A. Response of plants to calcium concentration in solution culture. Austr. J. Agric. Res. 19, 845-897, 1968.

Loneragan, J. F. and Snowball, K. Calcium requirements of plants. Aust. J. agric. Res. 20, 465-478, 1969.

Long, J. M. and Widders, I. E. Quantification of apoplastic potassium content by elution analsysis of leaf lamina tissue from pea (*Pisum sativum* L. cv Argenteum). Plant Physiol. 94, 1040-1047, 1990.

Long, S. R. and Ehrhardt, D. W. New route to a sticky subject. Nature 338, 545-546, 1989.

Longnecker, N. and Welch, R. M. Accumulation of apoplastic iron in plant roots. Plant Physiol. 92, 17-22, 1990.

Loomis, R. S., Williams, W. A. and Hall, A. E. Agricultural productivity. Annu. Rev. Plant Physiol. 22, 431-463, 1971.

Loomis, W. D. and Durst, R. W. Chemistry and biolgy of boron. Biofactors. 3, 229-239, 1992.

Lorenz, H. Nitrate ammonium and amino acids in bleeding sap of tomato plants in relation to the form and concentration of nitrogen in the medium. Plant and Soil 45, 169-176, 1976.

Lorenzen, J. H. and Ewing, E. E. Starch accumulation in leaves of potato (*Solanum tuberosum,* L.) during the first 18 days of photopeeriod treatment. Ann. Bot. 69, 481-485, 1992.

Lorenzini, G., Guidi, L., Nali, C., Ciompi, S. and Soldatini, G. F. Photosynthetic response of tomato plants to vascular wilt diseases. Plant Sci. 124, 143-152, 1997.

Loreto, F. and Sharkey, T. D. A gas-exchange study of photosynthesis and isoprene emission in *Qercus rubra* L. Planta 182, 523-531, 1990.

Loué, A.: (F) Average effect of potassium fertilization to arable crops in long term field trials. Potash Review (Berne) Subj. 16, Suite 79th No. 4, 1979.

Loué, A. Experimental evidence of NxK interactions. In: Potassium and Fertilizer Use Efficiency, edited by National Fertilizer Development Centre, Islamabad: National Fertilizer Development Centre, p. 123-149, 1989.

Low, A. J. and Armitage, E. R. The composition of the leachate through cropped and uncropped soils in lysimeters compared with that of the rain. Plant and Soil 33, 393-411, 1970.

Lu, J., Ertl., J. R. and Chen, C. Transcriptional regulation of nitrate reductase mRNA levels by cytokinin-abscisic acid interactions. Plant Physiol. 98, 1255-1260, 1992.

Lu, Z. and Neumann, P. M. Water stress inhibits hydraulic conductance and leaf growth in rice seedlings but not the transport of water *via* mercury-sensitive water channels in the root. Plant Physiol. 120, 143-151, 1999.

Lucas, R. E. and Davis, J. F. Relationships between pH values of organic soils and availabilities of 12 plant nutrients. Soil Sci. 92, 177-182, 1961.

Lucas, R. E. and Knezek, B. D. Climatic and soil conditions promoting micronutrient deficiencies in plants. In: Micronutrients in Agriculture, edited by Mortvedt, J. J. Giordano, P. M. and Lindsay, W. L. Soil Sci. Soc. America, p. 265-288, Madison 1972.

Ludewig, M., Dörffling, K. and Seifert, H. Abscisic acid and water transport in sunflowers. Planta 175, 325-333, 1988.

Luetzelschwab, M., Asard, H., Ingold, U. and Hertel, R. Heterogenity of auxin-accumulating nembrane vesicles from *Cucurbita* and *Zea*: a possible reflection of cell polarity. Planta 177, 304-311, 1989.

Lundegardh, H. (G) Leaf analysis. Verlag G. Fischer, Jena 1945

Lynch, J.,Cramer, G. R. and Läuchli, A. Salinity reduces membrane associated calcium in corn root protoplasts. Plant Physiol. 83, 390-394, 1987.

Lynch, J. and Läuchli, A. Salinity affects intracellular calcium in corn root protoplasts. Plant Physiol. 87, 351-356, 1988.

M. A. A. F. Ministry of Agriculture, Fisheries and Food, Fertilizer Recommendations. GFJ ADAS HMSO London 1979.

Ma, B. L., Morrison, M. J. and Dwyer, L. M. Canopy light reflectance and field greenness to assess nitrogen fertilization and yield of maize. Agron. J 88, 915-920, 1996.

Ma, J. F. and Nomoto, K. Effective regulation of iron acquisition in graminaceous plants. The role of mugineic acids as phytosiderophores. Physiol. Plant. 97, 609-617, 1996.

Maas, E. V., Moore, D. P. and Mason, B. J. Influence of calcium and magnesium on manganese absorption. Plant Physiol. 44, 796-800, 1969.

Maas, F. M., van de Weterig, D. A. M., van Beusichem, M. L. and Bienfait, H. F. Characterization of phloem iron and its possible role in the Fe -efficiency reactions. Plant Physiol. 87, 167-171, 1988.

Maathuis, F. J. M. and Amtmann, A. K^+ nutrition and Na^+ toxicity: The basis of cellular K^+/Na^+ ratios. Ann. Bot. 84, 123-133, 1999.

Matthuis, F. J. M. and Sanders, D. Regulation of K^+ absorption in plant root cells by external K^+: interplay of different plasma membrane K^+ transporters. J. Expt. Bot. 48, 451-458, 1997.

Maathuis, F. J. M. and Sanders, D. Mechanism of potassium absorption by higher plants. Physiol. Plant.

96, 158-168, 1996.

Macduff, J. H., Bakken, A. K. and Dhanoa, M. S. An analysis of the physiological basis of commonality between diurnal patterns of NH$_4^+$, NO$_3^-$ and K$^+$ uptake by *Phleum pratense* and *Festuca pratensis*. J. Exp. Bot. 48, 1691-1701, 1997.

Macduff, J. H. and Dhanoa, M. S. Diurnal and ultradian rhythms in K$^+$ uptake by *Trifolium repens* under natural light patterns: Evidence for segmentation at different root patterns. Physiol. Plant. 98, 298-308, 1996.

MacLeod, J. E., Gupta, U. C. and Stanfield, B. Molybdenum and sulphur relationship in plants. In: Molybdenum in Agriculture, ed. by Gupta, U. C., Cambridge University Press, p. 229-249, 1997.

Machet, J. M. and Hebert, J. (F) Results of six years field trials on recommendation of nitrogen fertilizer application to sugar beet. In: Symposium "Nitrogen and Sugar Beet", edited by International Institute for Sugar Beet Research, Bruxelles: International Institute for Sugar Beet Research, p. 493-507, 1983.

Machold, O. and Scholz, G. (G) Iron status and chlorophyll synthesis in higher plants. Naturwiss. 56, 447-452, 1969.

Machold, O. and Stephan, U. W. The function of iron in porphyrin and chlorophyll biosynthesis. Phytochemistry 8, 2189-2192, 1969.

Mack, W. B. and Brasher, E. P. The influence of commercial fertilizers, potassium iodine, and soil acidity on the iodine content of certain vegetables. J. Agric. Res. 53, 789-800, 1936.

Mackay, A. D., Kladivko, E. J., Barber, S. A. and Griffith, D. R. Phosphorus and potassium uptake by corn in conservation tillage. Soil Sci. Soc. Am. J. 51, 970-974, 1987.

Mackenzie, A. M., Illingworth, D. V., Jackson, D. W. and Telfer, S. B. A comparison of methods of assessing copper status in cattle. In: Trace Elements in Man and Animals, edited by Fischer, P-W. F., L'Abbe, M. R., Cockell, K. A. and Gibson, R. S. Ottawa: NRC Research Press, p. 301-302, 1997.

Macklon, A. E. S. and De Kock, P. C. Physiological gradients in the potato tuber. Physiol. Plant 20, 421-429, 1967.

Maertens, M. C. (F) Experirnental investigation into the nutrition of maize with minerals and water. Comparison between the requirement of the plant and the uptake potential of the roots for nitrogen, phosphorus, and potassium. C. R. Acad. Sc. (Paris) 273, Serie D, 682-684, 1971.

Mahapatra, J. C., Prasad, R. and Leelavathi, C. R. Complex fertilizers based on nitrophosphate process, FAO Symposium. Fertilizer Ass. of India, New Dehli, 1973.

Mahler, R. L., Hammel J. E. and Harder, R. W. The influence of crop rotation and tillage methods on the distribution of extractable boron in Northern Idaho soils. Soil Sci. 139, 67-73, 1985.

Mahler, R. L. and McDole, R. E. Effect of soil pH on crop yield in Northern Idaho. Agron. J. 79, 751-755, 1987.

Maier-Maercker, U. The role of peristomatal transpiration in the mechanism of stomatal movement. Plant Cell Environ. 6, 369-380, 1983.

Major, D. J. and Charnetski, W. A. Distribution of ^{14}C labelled assimilates in rape plants. Crop Sci. 16, 530-532, 1976.

Malavolta, E., Dantas, J. P., Morias, R. S. and Nogueira, F. D. Calcium problems in Latin America. Comm. in Soil Sci. and Plant Anal. 10, 29-40, 1979.

Maldiney, R., Pelese, F., Pilate, G., Sotta, B., Sossountzov, L. and Miginiac, E. Endogenous levels of abscisic acid, indole-3-acetic acid,zeatin and zeatin-riboside during the course of adventious root formation in cuttings of Craigella and Craigella lateral supressor tomatoes. Physiol. Plant 68, 426-430, 1986.

Mallarino, A. P. Evaluation of excess soil phosphorus supply for corn by ear-leaf test. Agron. J. 87, 687-691, 1995.

Malone, C., Koeppe, D. E. and Miller, R. J. Localization of lead accumulated in corn plants. Plant Physiol. 53, 388-394, 1974.

Maloth, S. and Prasad, R. Relative efficiency of rock phosphate and superphosphate for cowpea (*Vigna sinensis* Savi) fodder. Plant and Soil 45, 295-300, 1976.

Mandal, S. C. Phosphorus management of our soils. Need for a more rational approach. 40th Sess. Indian Soc. of Soil Science, Bhubaneswar 1975.

Manguiat, I. J., Mascarina, G. B., Ladha, J. K., Buresh, R. J. and Tallada, J. Prediction of nitrogen availability and rice yield in soils: Nitrogen mineralization parameters. Plant and Soil 160, 131-137, 1994.

Mannheim, T., Braschkat, J., Dörr, J. and Marschner, H. (G) Computer supported assessing scheme for the determination and reduction of ammonia emmission after slurry application Z. Pflanzenernähr. Bodenk. 160, 133-140, 1997.

Manschadi, A. M., Sauerborn, J., Stützel, H., Göbel, W. and Saxena, M. C. Simulation of faba bean (*Vicia faba*) root system development under Mediterranean conditions. Eur. J. Agron. 9, 259-272, 1998.

Manthey, J. A., Tisserat, B. and Crowley, D. E. Root responses of sterile -grown onion plans to iron deficiency. J. Plant Nutr. 19, 145-161, 1996.

Marcelle, R. and Bodson, M. Greenback disease and mineral content of tomato fruit. J. Plant Nutrition 1, 207-217, 1979.

Marinos, N. C. Studies on submicroscopic aspects of mineral deficiencies. 1. Calcium deficiency in the shoot apex of barley. Am. J. Bot. 49, 834-849, 1962.

Marion, G. M., van Cleve, K., Dyrness, C. T. and Black, C. H. The soil chemical environment along a forest primary successional sequence on the Tanana River floodplain, interior Alaska. Can. J. Forest Res. 23, 914-922, 1993.

Marmé, D. Calcium transport and function. In: Inorganic Plant Nutrition, edited by Läuchli, A. and Bieleski, R. L. Encycl. Plant Physiol. New Series Vol. 15B, Springer Verlag Berlin, Heidelberg, New York, Tokyo, p. 599-625, 1983.

Marquard, R., Kühn, H. and Linser, H. (G) The effect of the sulphur nutrition on the synthesis of mustard oils. Z. Pflanzenernähr. Bodenk. 121, 221-230, 1968.

Marre, E. Fusicoccin: a tool in plant physiology. Annu. Rev. Plant Physiol. 30, 273-288, 1979.

Marschner, H. Why can sodium replace potassium in plants? In: Potassium in Biochemistry and Physiology, Proc. 8th Colloq. Int. Potash Inst., Bern, p. 50-63, 1971.

Marschner, H. (G) Effect of O_2 supply of roots on mineral uptake and plant growth. In: Pseudogley and Gley, Trans. Comm. V and VI of the Int. Soc. Soil Sci., 541-555, 1972.

Marschner, H. Mechanisms of regulation of mineral nutrition in higher plants. In: Mechanisms of Regulation of Plant Growth, edited by Bieleski, R. L., Ferguson, A. R. and Cresswell, M. M. Bulletin 12, The Royal Society of New Zealand, p. 99-109, 1974.

Marschner, H. (G) Nutritional and yield physiological aspects of plant nutrition. Angew. Botanik 52, 71-87, 1978.

Marschner, H. Mineral Nutrition of Higher Plants, London:Academic Press, 1995a.

Marschner, H. and Rhizosphere pH effects on phosphorus nutrition. In: Genetic Manipulation of Crop Plants to Enhance Integrated Nutrient Mangement in Cropping Systems. 1 Phosphorus edited by C. Johansen et al. JRISAT, Asia Center India, p. 107-115, 1995b.

Marschner, H. and Cakmak, I. Mechanisms of phosphorus induced zinc deficiency in cotton. II Evidence of impaired shoot control of phosphorus uptake and translocation under zinc deficiency. Physiol. Plant. 68, 491-496,1986.

Marschner, H. and Cakmak, I. High light intensity enhances chlorosis and necrosis in leaves of zinc, potassium and magnesium deficient bean (*Phaseolus vulgaris*) plants. J. Plant Physiol. 134, 308-315, 1989.

756

Marschner, H., Häussling, M. and George, E. Ammonium and nitrate uptake rates and rhizosphere pH in non-mycorrizal roots of Norway spruce (*Picea abies* L. Karst.). Trees 5, 14-21, 1991.

Marschner, H., Kirkby, E. A. and Cakmak, I. Effect of mineral nutritional status on shoot-root partitioning of photoassimilates and cycling of mineral nutrients. J. Exp. Bot. 47, 1255-1263, 1996.

Marschner, H., Kirkby, E. A. and Engels, C. Importance of cycling and recycling of mineral nutrients within plants for growth and development. Bot. Acta 110, 265-273, 1997.

Marschner, H., Kuiper, P. J. C. and Kylin, A. Genotypic differences in the response of sugar beet plants to replacement of potassium by sodium. Physiol. Plant. 51, 239-244, 1981.

Marschner, H., Kylin, A. and Kuiper, P. J. C. Differences in salt tolerance of three sugar beet genotypes. Physiol. Plant. 51, 234-238, 1981.

Marschner, H. and Ossenberg-Neuhaus, H. (G) Effect of 2, 3, 5 tri iodobenzoic acid (TIBA) on calcium transport and cation exchange capacity in sunflowers. Z. Pflanzenphysiologie 85, 29-44, 1977.

Marschner, H. and Richter, C. (G) Calcium translocation in roots of maize and bean seedlings. Plant and Soil 40, 193-210, 1974.

Marschner, H., Römheld, V. and Kissel, M. Localization of phytosiderophore release and iron uptake along intact barley roots. Physiol. Plant. 71, 157-162, 1987.

Marschner, H. and Römheld, V. Root-induced changes in the availability of micronutrients in the rhizosphere. In: Plant Roots, the Hidden Half, edited by Waisel, Y., Eshel, A. and Kafkafi, U. New York: Marcel Dekker, p. 557-579, 1996.

Marschner, H., Sattelmacher, B. and Bangerth, F. Growth rate of potato tubers and endogenous contents of indolylacetic acid and abscisic acid. Physiol. Plant. 60, 16-20, 1984.

Marschner, H. and Schropp, A. (G) Comparative studies on the sensitivity of six rootstock varieties of grapevine to phosphate-induced Zn deficiency. Vitis 16, 79-88, 1977.

Marschner, H., Treeby, M. and Römheld, V. Role of root induced changes in the rhizosphere for iron acquisition in higher plants. Z. Pflanzenernähr. Bodenk. 152, 197-104, 1989.

Marsh, K. B., Tillman, R. W. and Syers, J. K. Charge relationships of sulphate sorption by soils. Soil Sci. Soc. Am. J. 51, 318-323, 1987.

Marshall, C. J. Protein prenylation: a mediator of protein-protein interactions. Science. 259, 1865-1866, 1993.

Martens, D. A., Johanson, J. B. and Frankenberger, W. T. Production and persistence of soil enzymes with repeated additions of organic residues. Soil Sci. 153, 53-61, 1992.

Martens, D. C. and Westermann, D. T. Fertilizer application for correcting micronutrient deficiencies. In: Micronutrients in Agriculture, 2th edition, edited by J. J. Mortvedt. F. R. Cox, L. M. Shuman, R. M. Welch. SSSA Book Series, Madison, USA, p. 549-592, 1991.

Martin, H. W. and Sparks, D. L. Kinetics of nonexchangeable potassium release from two coastal plain soils. Soil Sci. Soc. Am. J. 47, 883-887, 1983.

Martin, H. W. and Sparks, D. L. On the behaviour of nonexchangeable potassium in soils. Commun. Soil Sci. Plant Anal. 16(2), 133-162, 1985.

Martin, J. P.: Bromine, p. 62-64. In: Diagnostic Criteria for Plants and Soils, edited by Chapman, H. D., Univ. of California, Div. of Agric. Sciences 1966.

Martini, J. A. and Mutters, R. G. Effect of liming and fertilization on sulfur availability, mobility, and uptake in cultivated soils of South Carolina. Soil Sci. 138, 403-410, 1984.

Martins, D. C. and Shelp, B. J. Fertilizer applications for correcting micronutrient deficiency. In: Micronutrients in Agriculture, edited by Soil Sci. Soc. America, p. 549-592, 1991.

Mary, B., Fresneau, C., Morel, J. L. and Mariotti, A. C and N cycling during decomposition of root mucilage, roots and glucose in soil. Soil Biol. Biochem. 25, 1005-1014, 1993.

Marzadori, C., Antisari, L. V. and Gioachini, P. Turnover of interlayer ammonium in soil cropped with sugar beet. Biol. Fertil. Soils in press:18, 27 - 31, 1994.

Masalha, J., Kosegarten, H., Elmaci, Ö. and Mengel, K. The central role of microbial activity for iron acquisition in maize and sunflower. Biol. Fertil. Soils 30, 433-439, 2000.

Mathur, B. N., Agrawal, N. K. and Singh, V. S. Effect of soil versus foliar application of urea on the yield of American cotton variety '320'. Indian J. agric. Sci. 38, 811-815, 1968.

Matoh, T. Boron in plant cell walls. Plant and Soil 193, 59-70, 1997.

Matsubayashi, M., Iro, R., Nomoto, T., Takase, and Yamada, N. Some properties of paddy field soils. In: Theory and practice of fertilizer application, p. 183-227, 1963.

Matsuda, K. and Riazi, A. Stress-induced osmotic adjustment in growing regions of barley leaves. Plant Physiol. 68, 571-576, 1981.

Matthews, M. A., Van Volkenburgh, E. and Boyer, J. S. Acclimation of leaf growth to low water potentials in sunflower. Plant Cell Environ. 7, 199-206, 1984.

Matzke, H. and Mengel, K. Importance of the plasmalemma ATPase activity on the retention and exclusion of inorganic ions. Z. Pflanzenernähr. Bodenk. 156, 515-519, 1993.

Maynard, D. N. Nutritional disorders of vegetable crops. A review. J. Plant Nutrition 1, 1-23, 1979.

Maynard, D. N., Barker, A. V., Minotti, P. L. and Peck, N. H. Nitrate accumulation in vegetables. Adv. Agron. 28, 71-118, 1976.

Maynard, D. G., Stewart, W. J. and Bettany, J. R. The effects of plants on soil sulfur transformations. Soil Biol. Biochem. 17, 127-134, 1985.

McAinsh, M. R, Brownlee, C. and Hetherington, A. M. Calcium ions as second messengers in guard cell signal transduction. Physiol. Plant. 100, 16-29, 1997.

McAinsh, M. R. and Hetherington, A. M. Encoding specificity in Ca^{2+} signalling systems. Perspectives 3, 22-26, 1998.

McAuliffe C. F., Hall, N. S., Dean, L. A. and Hendricks, S. B. Exchange reactions between phosphates and soils. Hydroxylic surfaces of soil minerals. Soil Sci. Soc. Amer. Proc. 12, 119-123, 1947.

McBride, M. B. Reactions controlling heavy metal solubility in soils. Adv. Soil Sci. 10, 1-56, 1989.

McCree, K. J. and Richardson, S. G. Stomatal closure vs. osmotic adjustment: a comparison of stress responses. Crop Sci. 27, 539-543, 1987.

McDonald, R., Wang, H. L., Patrick, J. W. and Offler, C. E. The cellular pathway of sucrose transport in developing cotyledons of Vicia faba L. and Phaseolus vulgaris L.: a physiological assessment. Planta 196, 659-667, 1995.

McEwen, J. and Johnston, A. E. Factors affecting the production and composition of mixed grass/clover swards containing modern high-yielding clovers. In: Nutrient Balances and Fertilizer Needs in Temperate Agriculture. Intern. Potash Institute Bern, p. 41-55, 1984.

McGrath, S. P., Sanders, J. R. and Shalaby, M. H. The effect of soil organic matter levels on soil solution concentrations and extractabilities of manganese, zinc and copper. Geoderma 42, 177-188, 1988.

McGrath, S. P., Zhao, F. J. and Withers, P. J. A. Development of sulphur deficiency in crops and its treatment. London: Proc. No. 379, pp. 47, 1994.

McGrath, S. P. and Zhao, F. J. A risk assessment of sulphur deficiency in cereals using soil and atmospheric deposition data. Soil Use and Management, 11, 110-114, 1995.

McIntyre, G. I. The role of nitrate in the osmotic and nutritional control of plant development. Aust. J. Plant Physiol. 24, 103-118, 1997.

McKenzie, R. M. Soil cobalt. In: Trace Elements in Soil-Plant-Animal Systems, edited by Nicholas, D. J. D. and Egan, A. R. Academic Press London, p. 83-93, 1975.

758

McKenzie, R. M. Manganese oxides and hydroxides. In: Minerals in Soil Environments, edited by Kissel, D. E., Madison WS, USA. Soil Sci. Soc. of America, p. 439-465, 1989.

McLaren, R. G. and Crawford, D. V. Studies on soil copper. I. The fractionation of Cu in soils. J. Soil Sci. 24, 172-181, 1973.

McLauglin, M. J., Maier, N. A., Freeman, K., Tiller, K. G., Williams, C. M. J. and Smart, M. K. Effect of potassic and phosphatic fertilizer type, phosphate fertilizer Cd content and additions of zinc on cadmium uptake by commercial potato crops. Fert. Res. 40, 63-70, 1995.

McLaughlin, S. B. and Wimmer, R. Calcium physiology and terrestrial ecosystem processes. New Phytol. 142, 373-417, 1999.

Mc Lean, E. O. and Watson, M. E. Soil measurements of plant-available potassium. In: Potassium in Agriculture, edited by Munson, R. D. Madison,Ws. USA: Am. Soc. Agronomy, p. 277-308, 1985.

McNeal, F. H., Watson, C. A. and Kittmas, H. A. Effects of dates and rates of nitrogen fertilisation on the quality and field performance of five hard red spring wheat varieties. Agron. J. 55, 470-472, 1963.

McNeil, S. D., Nuccio, M. L. and Hanson, A. D. Betaines and related osmoprotectants. Targets for metabolic engineering of stress resistance. Plant Physiol. 120, 945-949, 1999.

McQueen-Mason, S., Durachko, D. M. and Cosgrove, D. J. Two endogenous proteins that induce cell wall expansion in plants. Plant Cell 4, 1425-1433, 1992.

Mehl, M. (G) Practicable prognosis model for the assessment of Nmin nitrogen in spring. Several- years investigations on representative arable soils in various climatic areas of Hessia. Fachverlag Köhler, Giessen. ISBN 3-922-IS306-81-0, 1999.

Mehrhoff, R. and Kühbauch, W. (G) Yield structure of old and modern winter wheat cultivars related to transient storage and remobilization of fructans in the wheat culm. J. Agron. Crop. Sci. 165 47-53, 1990.

Meidner, H. Vapour loss through stomatal pores with the mesophyll tissue excluded. J. Exp. Bot. 27, 172-174, 1976.

Meidner, H. Three hundred years of research into stomata. In: Stomatal function, edited by Zeiger, E., Farquhar, G. D. & Cowan, I. R. Stanford University Press, Stanford, CA, pp. 7-27, 1987.

Meinke, H., Hammer, G. L., Keulen, v. H., Rabbinge, R. and Keating, B. A. Improving wheat simulation capabilities in Australia from a cropping system perspective. I. Water and nitrogen effects on spring wheat in a semi-arid environment. In: Perspectives for Agronomy, Development in Crop Science 25, edited by Ittersum, van M. and Geijn, van S. C. Amsterdam: Elsevier, p. 99-112, 1997.

Meisch, H.-U., Benzschawel, H. and Bielig, H.-J. The role of vanadium in green plants. II. Vanadium in green algae - two sites of action. Arch. Microbiol. 105, 77-82, 1975.

Meisch, H.-U., Benzschawel, H. and Bielig, H. J. The role of vanadium in green plants. II. Vanadium in green plants - two sites of action. Arch. Microbiol. 114, 67-70, 1977.

Mengel, D. B. and Barber, S. A. Rate of nutrient uptake per unit of corn root under field conditions. Agron. J. 66, 399-402, 1974.

Mengel, K. (G) Uptake and reduction of nitrate and nitrate concentration in plants. Landwirtsch. Forsch. 37. Congr. Vol., 146-157, 1984.

Mengel, K. Potassium movement within plants and its importance in assimilate transport. In: Potassium in Agriculture, edited by Munson, R., Am. Soc. Agronomy, Madison USA p. 397-411, Madison 1985.

Mengel, K. Ernährung und Stoffwechsel der Pflanze, 7th ed. Gustav Fischer Verlag, Jena, Stuttgart 1991.

Mengel, K. Nitrogen: agricultural productivity and environmental problems. In: Nitrogen Metabolism of Plants, edited by Mengel, K. and Pilbeam, D. J. Oxford: Oxford University Press, p. 1-15, 1992.

Mengel, K. Iron availability in plant tissues - iron chlorosis on caicareous soils. Plant and Soil 165,

275-283, 1994.

Mengel, K. Symbiotic dinitrogen fixation-its dependence on plant nutrition and its ecophysiological impact. Z. Pflanzenernähr. Bodenk. 157, 233 -241, 1994b.

Mengel, K. Turnover of organic nitrogen in soils and its availability to crops. Plant and Soil 181, 83-93, 1996.

Mengel, K. Agronomic measures for better utilization of soil and fertilizer phosphates. Eur. J. Agron. 7, 221-233, 1997.

Mengel, K. Integration of functions and involvement of potassium metabolism at the whole plant level. In: Frontiers in Potassium nutrition: New Perspectives on the Effects of Potassium on Physiology of Plants, edited by Oosterhuis, D. M. & Berkowitz, G. A. Potash & Phosphate Institute, Georgia USA /Potash & Phosphate Institute of Canada, Saskatoon, Canada, 1-11, 1999.

Mengel, K. and Arneke, W. W. Effect of potassium on the water potential, the pressure potential, the osmotic potential and cell elongation in leaves of *Phaseolus vulgaris*. Physiol. Plant. 54, 402-408, 1982.

Mengel, K., Breininger, M. T. and Bübl, W. B. icarbonate, the most important factor inducing iron chlorosis in vine grapes on calcareous soils. Plant and Soil 81, 333-344, 1984.

Mengel, K. and Bübl, W. (G) Distribution of iron in vine leaves with HCO_3^- induced chlorosis. Z. Pflanzenernähr. Bodenk. 146, 650-571, 1983.

Mengel, K. and Busch, R. The importance of the potassium buffer power on the critical potassium level in soil. Soil Sci. 133, 27-32, 1982.

Mengel, K. and Forster, H. (G) The effect of the potassium concentration in the soil solution on the yield and the water consumption of sugar beets. Z. Pflanzenernähr. Bodenk. 134, 148-157, 1973.

Mengel, K. and Forster, H. (G) Yield performance of some spring rape cultivars (*Brassica napus* L., ssp. oleifera) with different potassium supply. Kali-Briefe Fachgeb. 3, 1. Folge, 1976.

Mengel, K., Friedrich, B. and Judel, G. K. Effect of light intensity on the concentrations of phytohormones in developing wheat grains. J. Plant Physiol. 120, 255-266, 1985.

Mengel, K. and Geurtzen, G. Relationship between iron chlorosis and alkalinity in *Zea mays*. Physiol. Plant. 72, 460-465, 1988.

Mengel, K. and Haeder, H. E. Effect of potassium supply on the rate of phloem sap exudation and the composition of phloem sap of *Ricinus communis*. Plant Physiol. 59, 282-284, 1977.

Mengel, K. and Helal, M. (G) The influence of the exchangeable Ca^{2+} of young barley roots on the fluxes of K, and phosphate - an interpretation of the Viets effect. Z. Pflanzenphysiol. 57, 223-234, 1967.

Mengel, K., Hogrebe, A. M. R. and Esch, A. Effect of acidic fog on needle surface and water relations of *Picea abies*. Physiol. Plant. 75, 201-207, 1989.

Mengel, K., Lutz, H. J. and Breininger, M. T. (G) Leaching of nutrients out of young intact spruce needles. Z. Pflanzenernähr. Bodenk. 150, 61-68, 1987.

Mengel, K. and Malissiovas, N. (G) Bicarbonate as inducing factor of iron chlorosis in vine (Vitis vinifera). Vitis, 20, 235-243, 1981.

Mengel, K. and Malissiovas, N. Light dependent proton excretion by roots of entire vine plants (*Vitis vinifera*). Z. Pflanzenernähr. Bodenk. 145, 261-267, 1982.

Mengel, K. and Pflüger, R. (G) The influence of several salts and several inhibitors on the root pressure of *Zea mays*. Physiol. Plant. 22, 840-849, 1969.

Mengel, K., Plänker, R. and Hoffmann, B. Relationship between leaf apoplast pH and iron chlorosis of sunflower (*Helianthus annuus* L.) J. Plant Nutr. 17, 1053-1065, 1994.

Mengel, K. and Rahmatullah, X. Exploitation of potassium by various crop species from primary minerals in soils rich in micas. Biol. Fertil. Soils 17, 75-79, 1994.

Mengel, K., Rahmatullah, X. and Dou, H. Release of potassium from the silt and sand fraction of loess derived soils. Soil Sci. 163, 805-813, 1998.

Mengel, K., Robin, P. and Salsac, L. Nitrate reductase in shoots and roots of maize seedlings as affected by the form of nitrogen nutrition and the pH of the nutrient solution. Plant Physiol. 71, 618-622, 1983.

Mengel, K. and Scherer, H. W. Release of non exchangeable (fixed) soil NH_4 under field conditions during the growing season. Soil Sci., 131, 226-232, 1981.

Mengel, K., Schneider, B. and Kosegarten, H. Nitrogen compounds extracted by electroultrafiltration (EUF) or $CaCl_2$ solution and their relationships to nitrogen mineralization in soils. J. Plant Nutr. Soil Sci. 162, 139-148, 1999.

Mengel, K. and Schubert, S. Active extrusion of protons into deionized water by roots of intact maize plants. Plant Physiol. 79, 344-348, 1985.

Mengel, K., Secer, M. and Koch, K. Potassium on protein formation and amino acid turnover in developing wheat grain. Agron. J. 73, 74-78, 1981.

Mengel, K. and Steffens, D. (G) Relationship between the cation/anion uptake and the release of protons by roots of red clover. Z. Pflanzenernähr. Bodenk. 145, 229-236, 1982.

Mengel, K. and Steffens, D. Potassium uptake of rye-grass (*Lolium perenne*) and red clover (*Trifolium pratense*) as related to root parameters. Biol. Fert. Soils 1, 53-58, 1985.

Mengel, K. and Uhlenbecker, K. Determination of available interlayer potassium and its uptake by ryegrass. Soil Sci. Soc. Am. J. 57, 761-766, 1993.

Mengel, K. and Viro, M. Effect of potassium supply on the transport of photosynthates to the fruits of tomatoes (*Lycopersicon esculentum*). Physiol. Plant 30, 295-300, 1974.

Mengel, K. and Viro, M. The significance of plant energy status for the uptake and incorporation of NH_4-nitrogen by young rice plants. Soil Sci. Plant Nutr. 24, (3) 407-416, 1978.

Merbach, W., Augustin, J. and Mirus, E. (G) Effect of aluminium on the legume -rhyzobium- symbiosis. Zentralbl. Mikrobiol. 145, 521-527, 1990.

Merbach, W., Mirus, E., Knof, G., Remus, R., Ruppel, S., Russow, R., Gransee, A. and Schulze, J. Release of carbon and nitrogen compounds by plant roots and their possible ecological importancE. J. Plant Nutr. Soil Sci. 162, 373-383, 1999.

Mercer, E. R. and Richmond, J. L. Fate of nutrients in soil: Copper. In: Letcombe Laboratory Annual Report, p. 9, 1970

Merchant, S. and Dreyfuss, B. W. Posttranslational assembly of photosynthetic metalloproteins. Annu. Rev. Plant Physiol. Plant Mol. Biol. 49, 25 -51, 1998.

Mertz, E. T., Bates, L. S. and Nelson, O. E. Mutant gene that changes protein composition and increases lysine content of maize endosperm. Science 145, 279-280, 1964.

Michael, G. (G) Phosphate fractions in oat grains and spinach related to a varied application of phosphorus. Bodenk. u. Pflanzenernähr. 14, 148-171, 1939.

Michael, G. and Beringer, H. The role of hormones in yield formation. In: Physiological Aspects of Crop Productivity, 15th Colloq. Int. Potash Inst. Bern, p. 85-116, 1980.

Michael, G. and Blume, B. (G) The influence of a nitrogen application on the protein composition of barley grains. Z. Pflanzenernähr. Düng. Bodenk. 88, 237-250, 1960.

Michael, G. and Djurabi, M. (G) Effect or 30 years mineral and farm yard manure application on the properties of a loamy soil. Z. Pflanzenernähr. Düng. Bodenk. 107, 40-50, 1964.

Michael, G., Martin, P. and Owassia, I. The uptake of ammonium and nitrate from labelled ammonium nitrate in relation to the carbohydrate supply of the roots. In: Nitrogen Nutrition of the Plant, edited by Kirkby, E. A. Univ. of Leeds, p. 22-29 1970.

Michael, G., Wilberg, E. and Kouhsiahi-Tork, K. (G) Boron deficiency induced by high air humidity. Z.

Pflanzenernähr. Bodenk. 122, 1-3, 1969.

Mikkelsen, R. L., Page, AQ. L. and Bingham, F. T. Factors affecting selenium accumulation by agricultural crops. In: Selenium in Agriculture and the Environment, ed. by L. W. Jacobs, SSSA p. 65-94, Madison, USA 1989.

Millard, P. Ecophysiology of the internal cycling of nitrogen for tree growth. Z. Pflanzenernähr. Bodenk. 159, 1-10, 1996.

Miller, C. Potassium selectivity in proteins: oxygen cage or in the face. Science. 261, 1692-1693, 1993.

Miller, E. R., Lei, X. and Ullrey, D. E. Trace Elements in Animal Nutrition. In:Micronutrients in Agriculture, ed. J. J. Mortvedt, F. R. Fox, L. M. Shuman, R. M. Welch, SSSA p. 593-662, Madison, USA, 1991.

Miller, G. W., Denney, A., Pushnik, J. and Ming-Ho Yu. The formation of δ-aminolevulinate a precursor of chlorophyll in barley and the role of iron. J. Plant Nutr. 5, 289-300, 1982.

Miller, J. E. and Miller, G. W. Effects of fluoride on mitochondrial activity in higher plants. Physiol Plant. 32, 115-121, 1974.

Millet, E., Avivi, Y. and Feldman, M. Yield response of various wheat genotypes to inoculation with Azospirillum brasilense. Plant and Soil 80, 261-266, 1984.

Miner, G. S., Gutierrez, R. and King, L. D. Soil factors affecting plant concentrations of cadmium, copper and zinc on sludge amended soils. J. Environ. Qual. 26, 989-994, 1997.

Miner, G. S., Traore, S. and Tucker, M. R. Corn response to starter fertilizer acidity and manganese materials varying in water solublity. Agron. J. 78, 291-295, 1986.

Miner, G. S., Ully, J. P. and Terry, D. L. Nitrogen release characteristics of isobutylidene diurea and its effectiveness as a source of N for flue-cured tobacco. Agron. J. 70, 434-438, 1978.

Minorsky, P. V. An heuristic hypothesis of chilling injury in plants: a role for calcium as the primary phyiological transducer of injury. Plant, Cell and Environment 8, 75-94, 1985.

Minotti, P. L., Williams, D. Craig, and Jackson, W. A.: Nitrate uptake by wheat as influenced by ammonium and other cations. Crop Sci. 9, 9-14, 1969.

Mir, N. A., Salon, C. and Canvin, D. T. Inorganic carbon-stimulated O_2 photoreduction is suppressed by NO_2 assimilation in air-grown cells of Synechococcus UTEX 625. Plant Physiol. 109, 1295-1300, 1998.

Mishra, D. and Kar, M. Nickel in plant growth and metabolism. Bot. Rev. 40, 395-452, 1974.

Mitchell, G. A., Bingham, E. T. and Page, A. L. Yield and metal composition of lettuce and wheat grown on soils amended by sewage sludge enriched with cadmium, copper, nickel, and zinc. J. Environ. Qual. 7, 165-171, 1978.

Mitchell, H. H. and Edman, M. Fluorine in soils, plants and animals. Soil Sci. 60, 81-90, 1945.

Mitchell, H. H., Hamilton, T. S. and Beadles, J. R. The relationship between the protein content of corn and the nutritional value of the protein. J. Nutr. 48, 461-476, 1952.

Mitchell, P. Coupling of phosphorylation to electron and hydrogen transfer by a chemiosmotic type of mechanism. Nature 191, 144-148, 1961.

Mitchell, P. Promotive chemiosmotic mechanism in oxidative and photosynthetic phosphorylation. Trends in Biochechnical Sciences 3, N58-N61, 1978.

Mitchell, R. A. C., Black, C. R., Burkart, S., Burke, J. I., Donnelly, A., Temmerman, L., Fangmeier, A., Mulholland, B. J., Theobald, J. C. and Oijen, M. Photosynthetic responses in spring wheat grown undet elevated CO_2 concentrations and stress conditions in European, multiple-site experiment IIESPACE-wheat. European J. Agron. 10, 205-214, 1999.

Mitchell, R. L. Trace elements in Soil. In: Chemistry of the Soil, edited by Bear, E. E. New York, Reinhold, p. 320-368, 1964.

Mitchell, R. L. Cobalt in soil and its uptake by plants. Agrochimica 16 521-532, 1972.

Mitscherlich, E. A. (G) Soil Science for Farmers, Foresters and Gardeners. 6. ed. Max Niemeyer Verlag, Halle 1950.

Mitscherlich, E. A. (G) Soil Science for Farmers, Foresters and Gardeners. 7th ed., Verlag P. Parey, Berlin, Hamburg 1954.

Mitsui, S. and Takatoh, H. Nutritional study of silicon in graminaceous crops. Part I. Soil Sci. Plant Nutr. 9, 49-53, 1963.

Mitsui, T., Christeller, J. T., Hara-Niseumura, J. and Akazawa, T. Possible roles of calcium and calmodulin in the biosynthesis and secretion of α-amylase in rice seed scutellar epithelium. Plant Physiol. 75, 21-25, 1984.

Miyake, Y. and Takahashi, E. Silicon deficiency of tomato plant. Soil Sci. Plant Nutr. 24, 175-189, 1978.

Mohr, H. D. (G) Soil penetration by roots in relation to important soil characteristics. Kali-Briefe (Büntehof) 14, (2) 103-113, 1978.

Moiser, A. R., Guenzi, W. D. and Schweizer, E. E. Soil losses of dinitrogen and nitrous oxide from irrigated crops in Northeastern Colorado. Soil Sci. Soc. Am. J. 50, 344-348, 1986.

Moitra, R. Tillage and much management of rainfed rapeseed, Sriniketan,India:Ph. D. Thesis. Institute of Agriculture., 1996.

Moldrup, P., Yamaguchi, T., Hansen, J. A. and Rolston, D. E. An accurate and numerical stable model for one dimensional solute transport in soils. Soil Sci. 92, 261-273, 1992.

Molina, J. A. E., Clapp, C. E., Shaffer, M. J., Chichester, F. W. and Larson, W. B. NCSOIL, a model of nitrogen and carbon transformation in soils: description, calibration and behavior. Soil Sci. Soc. Am. J. 47, 85-91, 1983.

Molle, K. G. and Jessen, T. (Danish) Increasing amounts of nitrogen to spring cereals grown on low areas 1960-67. T. Planteavl 72, 489-502, 1968.

Moloney, A. H., Guy, R. D. and Layzell, D. B. A model of the regulation of nitrogenase electron allocation in legume nodules. Plant Physiol. 104, 541-550, 1994.

Momoshima, N. and Bondietti, E. A. Cation binding in wood: Applications to understanding historical changes in divalent cation availability to red spruce. Can. J. Forest Res. 20, 1840-1849, 1990.

Moorby, J. The influence of carbohydrate and mineral nutrient supply on the growth of potato tubers. Ann. Bot. 32, 57-68, 1968.

Moore, D. P. Physiological effects of pH on roots. In: The Plant Root and Its Environment, edited by E. W. Carson., University Press of Virginia, Charlottesville, p. 135-151, 1974.

Moore, P. H. and Cosgrove, D. J. Developmental changes in cell and tissue water relations parameters in storage parenchyma of sugarcane. Plant Physiol. 96, 794-801, 1991.

Morard, P. (F) Contribution to the study of the potassium nutrition of sorghum. Ph. D. Thesis, University Toulouse 1973.

Morard, P. (F) Soilless Cultivation of Vegetables, Agen, Freance:S. A. R. L. Publications Agricoles ISBN 2-9509297, 1995.

Morel, J. L. Mench, M. and Guckert, A. Measurement of Pb^{2+}, Cu^{2+}, and Cd^{2+} binding with the mucilage exudates from maize (Zea mays L.) roots. Biol. Fert. Soils 2, 29-34, 1986.

Morell, M. K. S., Rahman, S., Abrahams, S. L. and Appels, R. The biochemistry and molecular biology of starch synthesis in cereals. Austr. J. Plant Physiol. 22, 647-660, 1995.

Morgan, J. M. Osmoregulation and water stress in higher plants. Annu. Rev. Plant Physiol. 34, 299-319, 1984.

Morgan, P. W., Joham, H. E. and Amin, J. V. Effect of manganese toxicity on the indoleacetic acid oxidase system in cotton. Plant Physiol. 41, 718-724, 1966.

Morgan, P. W., Taylor, D. M. and Joham, H. E. Manipulation of IAA oxidase activity and auxin deficiency symptoms in intact cotton plants with manganese nutrition. Physiol. Plant. 37 149-156, 1976.

Morré, D. J. Membrane biogenesis. Annu. Rev. Plant Physiol. 26, 441-481, 1975.

Mortvedt, J. J. and Osborn, G. Studies on the chemical form of cadmium contaminants in phosphate fertilizers. Soil. Sci. 134, 185-192, 1982.

Mosse, B. Advances in the study of vesicular-arbuscular mycorrhizas. Annu. Rev. Phytopathol. 11, 171-176, 1973.

Mosse, B., Powell, C. W. and Hayman, D. W. Plant growth responses to vesicular arbuscular mycorrhiza. Interactions between VA. mycorrhyiza, rock phosphate and symbiotic nitrogen fixation. New Phytol 76, 331-342, 1976.

Mostafa, M. A. E. and Ulrich, A. Absorption, ditribution, and form of Ca in relation to Ca deficiency (tip burn) of sugar beets. Crop Sci. 16, 27 - 39, 1976

Mozaffari, M. and Sims, J. T. Phosphorus transformations in poultry litter-amended soils of the Atlantic Coastal Plain. J. Environ. Qual. 25, 1357 -1365, 1996.

Mühling, K. H. Characerizationof ion relationships in leaf apoplasts by means of fluorescence ratio imaging microscopy. Shaker Verlag, Aachen 1998.

Mühling, K. H., Breininger, M. T. and Barekzai, A. (G) Dynamics of different N fractions in permanent grassland after slurry application, investigated with the Nmin and $CaCl_2$ method. Landwirtsch. Forsch. 42, 61-71, 1989.

Mühling, K. H. and Läuchli, A. Effect of K^+ nutrition, leaf age and light intensity on apoplastic pH in leaves of Vicia fabA. J. Plant Nutr. Soil Sci. 162, 571-576, 1999.

Mühling, K. H., Schubert, S. and Mengel, K. Role of plasmalemma H^+ ATPase in sugar retention by roots of intact maize and field bean plants. Z. Pflanzenernähr. Bodenk. 156, 155-161, 1993.

Mühling, K. H., Steffens, D. and Mengel, K. Determination of phytotoxic soil aluminium by electroultrafiltration. Z. Pflanzenern. Bodenk. 151:267 -271, 1988.

Mühling, K. H., Wimmer, M. and Goldbach, H. E. Apoplastic and membrane-associated Ca in leaves and roots as affected by boron deficiency. Physiol. Plant. 102, 179-184, 1998.

Müller, H., Deigele, C. and Ziegler, H. Hormonal interactions in the rhizosphere of maize (Zea mays L.) and their effects on plant development. Z. Pflanzenernähr. Bodenk. 152, 247-254, 1989.

Müller, W., Gärtel, W. and Zakosek, H. (G) Leaching of plant nutrients from vineyard soils. Z. Pflanzenernähr. Bodenk. 148, 417-428, 1985.

Müller-Röber, B. and Koßmann, J. Approaches to influence starch quantity and starch quality in transgenic plants. Plant, Cell Environ. 17, 601-613, 1994.

Müllner, L. (G) Results of research project concerning chlorosis. Mitt. Klosterneuburg 29, 141-150, 1979.

Münch, E. (G) Translocation of materials in plants. Fischer Verlag, Jena 1930.

Munch, J. C. and Ottow, J. C. G. (F) Bacterial reductions of amorphous and crystalline iron oxides. Science du Sol. 3/4, 205-125, 1983.

Mundel, G. and Krell, W. (G) Changes in chemical criteria of a grassland soil due to long term application of high nitrogen rates. Arch. Acker- u. Pflanzenbau u. Bodenkd. 22, 643-651, 1978.

Munk, H. (G) The nitrifieation of ammonium salts in acid soils. Landw. Forsch. 11, 150-156, 1958.

Munk, H. (G) Vertical migration of inorganic phosphate under conditions of a high phosphate application. Landw. Forsch. 27/I. Sonderh., 192-199, 1972.

Munk, H. and Rex, M. (G) Calibration of phosphate soil tests. Agribiol. Res. 43, 164-174, 1990.

Munns, R. Why measure osmotic adjustment? Aust. J. Plant Physiol. 15, 717-726, 1988.

Munns, R. Physiological processes limiting plant growth in saline soils. Some dogmas and hypotheses. Plant Cell Environ. 16, 15-24, 1993.

Munson, R. D.: Potassium in Agriculture. Proc. Intern. Symposium, Atlanta 1985. Am. Soc. Agron. Madison

USA 1985.

Müntz, K., Jung, R. and Saalbach, G. Synthesis, processing, and targeting of legume seed proteins. In: Seed Storage Compounds, edited by Shewry, P. R. & Stobart, K. Clarendon Press, Oxford, p 129-146, 1993.

Murata, N., Ishizaki-Nishizawa, O., Higashi, S., Hayashi, H., Taska, Y. and Nishida, I. Genetically engineered alteration in the chilling sensitivity of plants. Nature 356, 710-713, 1994.

Murata, Y. and Matsushima, S.: Rice In: Crop Physiology wedited by L. T. Evans, p. 73-99. Cambridge University Press 1975.

Murozumi, M., Chow, T. J. and Patterson, C.: Chemical concentrations of pollutant lead aerosols, terrestrial dust and sea salts in Greenland and Antarctic snow strata. Geochim. Cosmochim. Acta 33, 1247-1294, 1969.

Murphy, D. J. Designer of oil crops. VCH Verlagsgesellschaft, Weinheim, 1994.

Murphy, L. S. and Walsh, L. M.: Correction of micronutrient deficiencies with fertilizers. In: Micronutrients in Agriculture, edited by J. J. Mortvedt, P. M. Giordano, and W. L. Lindsay Soil Sci. Soc. America, Madison, p. 347-387, 1972.

Mylona, P., Pawlowski, K. and Bisseling, T. Symbiotic nitrogen fixation. The Plant Cell 7, 869-885, 1996.

Mythili, J. B. and Nair, T. V. R. Relationship between photosynthetic carbon exchange rate, specific leaf mass and other leaf characteristics in chickpea genotypes. Aust. J. Plant Physiol. 23, 617-622, 1996.

Myttenaere, C. (F) Effect of the strontium-calcium ratio on the localisation of strontium and calcium in *Pisum sativum*. Physiol. Plant 17, 814-827, 1964.

Nable, R. O., Banuelos, G. S. and Paull, J. Boron toxicity. Plant and Soil 193, 181-198, 1997.

Naeem, M., Tetlow, I. J. and Emes, M. J. Starch synthesis in amyloplasts purified from developing potato tubers. Plant J. 11, 1095-1103, 1997.

Nair, K. P. P. and Mengel, K. Importance of phosphate buffer power for phosphate uptake by rye. Soil Sci. Soc. Am. J. 48, 92-95, 1984.

Nair, K. P. P. The buffering power of plant nutrients and effects on availability. Adv. Agron. 57, 237-287, 1996.

Nanzyo, M. and Watanabe, Y. Diffuse reflectance infrared spectra and ion adsorption properties of the phosphate surface complex on goethite. Soil Sci. Plant Nutr. 28, 359-368, 1982.

Neeteson, J. J. and Rassink,J. Nitrogen Mineralization in Agricultiral Soils, Wageningen/Haren:DLO Res. Institute for Agrobiology and Soil Fertility 1993.

Neilands, J. B. and Leong, S. A. Siderophores in relation to plant growth and disease. Annu. Rev. Plant Physiol. 37, 187-208, 1986.

Nelson, O. E. and Pan, D. Starch synthesis in maize endosperms. Annu. Rev. Plant Physiol. Plant Mol. Biol. 46, 475-496, 1995.

Nelson, W. L. Plant factors affecting potassium availability and uptake. In: The Role of Potassium in Agriculture, ewdited by V. J. Kilmer, S. E. Younts and N. C. Brady. Madison/USA, p. 355-380, 1968.

Nemeth, K. (G) The characterization of the K status of soils by desorption curves. Geoderm. 5, 99-109, 1971.

Nemeth, K. The availability of nutrients in the soil as determined by electro-ultrafiltration (EUF). Adv. Agron. 31, 155-188, 1979.

Nemeth, K., Irion, H. and Maier, J. (G) Influence of EUF-K, EUF-Na and EUF- Ca fraction on the K uptake and the yield of sugar beet. Kali-Briefe 18, 777-790, 1987.

Nemeth, K., Mengel, K. and Grimme, H. The concentration of K, Ca and Mg in the saturation extract in relation to exchangeable K, Ca and Mg. Soil Sci. 109, 179-185, 1970.

Nemeth, K., Recke, H. and Heuer, C. (G) Assessment of the nitrogen fertilizer demand of sugar beets by means of EUF. Zuckerindustrie 116, 991-996, 1991.

Ness, P. J. and Woolhouse, H. W. RNA synthesis in *Phaseolus vulgaris* cultivar Canadien Wonder. RNA synthesis in chloroplast preparations from *Phaseolus vulgaris* leaves and solubilization of the RNA polymerasE. J. Exp. Bot. 31, 223-234, 1980.

Neubauer, H. and Schneider, W. (G) The nutrient uptake of seedlings and its application for the estimation of the nutrient content in soils. Z. Pflanzenernähr. Düng. Bodenk. A 2, 329-362, 1923.

Neubert, P., Wrazidlo, W., Vielemeyer, H. P. Hundt, I., Gollmick, F. and Bergmann, W. (G) Tables of plant analysis. Inst. of Plant Nutrition, Jena 1970.

Neumann, R. Chemical crop protection research and development in Europe. In: Proceedings of the 4. th ESA Congress, Veldhoven, edited by van Ittersum, M. and van Geijn, S. C. Amsterdam: Elsevier, p. 49 - 55, 1997.

Newman, E. I.: Root and soil water relations. In: The Plant Root and Its Environment, edited by E. W. Carson, University Press of Virginia, Charlottesville, p. 362-440, 1974

Newman, E. I. and Andrews, R. E. Uptake of phosphorus and potassium in relation to root growth and root density. Plant and Soil 38, 49-69, 1973.

Newport, J. W. and Forbes, D. J. The nucleus:structure, function, and dynamics. Annu. Rev. Biochem. 56, 535-565, 1987.

Neyra, C. A. and Döbereiner J. Nitrogen fixation in grasses. Adv. in Agron. 29, 1-38, 1977.

Nguyen, M. L. and Goh, K. M. Sulphur mineralization and release of soluble organic sulphur from camp and non-camp soils of grazed pastures receiving long-term superphosphate applications. Biol. Fertil. Soils 14, 272-279, 1992.

Nicholas, D. J. D. Determination of minor element levels in soils with the *Aspergilles niger* method. Trans. Intern. Congr. Soc. Soil Sci. Vol. III, Madison, Wisc., p. 168-182, 1960.

Nicholas, D. J. D. and Thomas, W. D. E. Some effects of heavy metals on plants grown in soil culture. Plant and Soil 5, 67-80, 1954.

Nicholson, C., Stein, J. and Wilson, K. A. Identification of the low weight copper protein from copper-intoxicated mung bean plants. Plant Physiol. 66, 272-275, 1980.

Nicolardot, B., Molina, J. A. E. and Allard, M. R. C and N fluxes between pools of soil organic matter: model calibration with long-term incubation data. Soil. Biol. Biochem. 26, 235-243, 1994.

Nigam, S. N. and McConnell, W. B. Metabolism of Na_2SeO_4 in *Astragalus bisulcatus* lima bean and wheat: a comparative study. J. Exp. Bot. 27, 565-571, 1976.

Niinemets, U., Sober, A., Kull, O., Hartung, W. and Tenhunen, J. D. Apparent controls of leaf conductance by soil water availability and via light-acclimation of foliage structural and physiological properties in a mixed deciduous, temperate forest. Intern. J Plant Sci. 160, 707-721, 1999.

Nishizawa, N. K., Mori, S., Takahashi, S. and Uedai, T. Mugineic acid secretion by cultured barley cells derived from anther. Protoplasma 148, 164 -166, 1989.

Nitsos, R. E. and Evans, H. J. Effect of univalent cations on the activity of particulate starch synthetase. Plant Physiol. 44, 1260-1266, 1969.

Nobel, P. Physicochemical and environmental plant physiology. Acad. Press, San Diego, 1991.

Nodvin, S. C., Driscoll, C. T. and Likens, G. E. The effect of pH on sulfate adsorption by forest soil. Soil Sci. 142, 9-75, 1986.

Noller, H. F. Ribosomal RNA and translation. Annu. Rev. Biochem. 60, 191-227, 1991.

Nolte, C. and Werner, W. Investigations on the nutrient cycle and its components of a biodynamically-managed farm. Biol. Agricult. Horticult. 10, 235 -254, 1994.

Norman, R. J. and Gilmour, J. T. Utilization of anhydrous ammonia fixed by clay minerals and soil organic matter. Soil Sci. Soc. Am. J. 51, 959-962, 1987.

Normanly, J., Slovi, J. P. and Cohen, J. D. Rethinking auxin biosynthesis and metabolism. Plant Physiol. 107, 323-329, 1995

North, G. B. and Nobel, P. S. Changes in hydraulic conductivity and anatomy caused by drying and rewetting roots of *Agave deserti* (Agavaceae). Am. J. Bot. 78, 906-915, 1991.

Norvell, W. A. and Welch, R. M. Growth and nutrient uptake by barley (*Hordeum vulgare* L. cv Herta): studies using and N-(hydroxyethyl)ethylenedinitrotriacetic acid-buffered nutrient solution. Plant Physiol. 101, 619-625, 1993.

Nowakowski, T. Z. Effects of nitrogen fertilizers on total nitrogen soluble nitrogen and soluble carbohydrate contents of grass. J. Agric. Sci. 59, 387-392, 1962.

Nuccio, M. L., Russell, B. L., Nolte, K. D., Rathinasabapathi B., Gage D. A. and Hanson A. D. The endogenous choline supply limits glycine betaine synthesis in transgenic tobacco expressing choline monooxygenase. Plant J. 16, 487-498, 1998.

Nye, P. H. Processes in the root environment. J. Soil Sci. 19, 205-215, 1968.

Nye, P. H.: Soil properties controlling the supply of nutrients to the root surface. In: The Soil - Root Interface, edited by Harley, J. L. and Scott Russell, R., Academic Press, p. 39-49, 1979.

Nye, P. H. and Tinker, P. B. Solute Movement in the Soil Root System. Blackwell Scientific Publications, Oxford, London, Edinburgh, Melbourne 1977.

Nyomara, A. M. S., Brown, P. H. and Freeman, M. Foliar applied boron increases tissue boron concentration and nut set of almond. J. Amer. Hort. Sci. 122, 405-410, 1997.

O'Connell, A. M. and Grove, T. S. Acid phosphatase activity in karri (*Eucalyptus diversicolor* F Muell). in relation to soil phosphate and nitrogen supply. J. Exp. Bot. 36, 1359-1372, 1985.

O'Connor, G. A., Lindsay, W. L. and Olsen, S. R. Diffusion of iron and iron chelates in soil. Soil Sci. Soc. Am. Proc. 35, 407-410, 1971.

O'Toole, J. C., Ozbun, J. L. and Wallace, D. H. Photosynthetic response to water stress in *Phaseolus vulgaris*. Physiol. Plant. 40, 111-114, 1977.

Oaks, A. Efficiency of nitrogen utilization in C 3 and C 4 cereals. Plant Physiol. 106, 407-414, 1994.

Oaks, A. Primary nitrogen assimilation in higher plants and its regulation. Can. J. Bot. 72, 739-750, 1994.

Oaks, A. and Hirel, B. Nitrogen metabolism in roots. Annu. Rev. Plant Physiol. 36, 345-365, 1985.

Oaks, A., Wallace, W. and Stevens, D. Synthesis and turnover of nitrate reductase in corn roots. Plant Physiol. 50, 649-654, 1972.

Obermeyer, G., Kriechbaumer, R., Strasser, D., Maschessni, A. and Bentrup, F. W. B. oric acid stimulates the plasma membrane H -ATPase of ungerminated lily pollen grains. Physiol. Plant. 98, 281-290, 1996.

Oberson, A., Besson, J. M., Maire, N. and Sticher, H. Microbiological processes in soil organic phosphorus transformations in conventional and biological cropping systems. Biol. Fertil. Soils 21, 138-148, 1996.

Obigbesan, G. O. The influence of potassium nutrition on the yield and chemical composition of some tropical root and tuber crops. In: Potassium in Tropical Crops and Soils, 10th Colloq. Intern. Potash Institute, Bern, p. 311-322, 1973.

Obigbesan, G. O. and Mengel, K. Use of electroultrafiltration (EUF) method for investigating the behaviour of phosphate fertilizers in tropical soils. Fert. Res. 2, 169-176, 1981.

Odell, R. T., Melsted, S. W. and Walker, W. M. Changes in organic carbon and nitrogen of Morrow plot soils under different treatments, 1904-1973. Soil Sci. 137, 160-171, 1984.

Oenema, O. Calculated rates of soil acidification of intensively used grassland in the Netherlands. Fert. Res. 26, 217-228, 1990.

Oenema, O., Velthof, G. E. and Bussink, D. W. Emission of ammonia, nitrous oxide and methane from cattle slurry. In: Biogeochemistry of Global Change: radioactively active trace gases, edited by Oremland, R. S., p. 419-433, 1993.

Oertli, J. J. Loss of boron from plants through guttation. Soil Sic. 94, 214-219, 1962.

Oertli, J. J. Controlled-release fertilizers. Fertil. Res. 1, 103-123, 1980.

Oertli, J. J. and Grigurevic, E. Effect of pH on the absorption of boron by excised barley roots. Agron. J. 67, 78-280, 1975.

Oertli, J. J. and Richardson, W. F. The mechanism of boron immobility in plants. Physiol. Plant. 23, 108-116, 1970.

Ogren, W. L. Photorespiration: pathways, regulation and modification. Annu. Rev. Plant Physiol. 35, 415-452, 1984.

Ohinshi, J., Flügge, U. I., Heldt, H. W. and Kanai, R. Involvement of Na^+ in active uptake of pyruvate in mesophyll chloroplasts of some C-4 plants. Plant Physiol. 94, 950-959, 1990.

Ohki, K. Manganese critical levels for soybean growth and physiological processes. J. Plant Nutr. 3, 271-284, 1981.

Ohwaki, Y. and Hirata, H. Differences in carboxylic acid exudation among P starved leguminous crops in relation to carboxylic acid contents in plant tissues and phospholipid levels in roots. Soil Sci. Plant Nutr. 38, 235-243, 1992.

Oijen, M. and Ewert, F. The effects of climatic variations in Europe on the yield response of spring wheat cv. Minaret to elevated CO_2 and O_3; an analysis of open-top chamber experiments by means of two crop growth simulation models. European J. Agron. 10, 249-264, 1999.

Ojima, D. S., Valentine, D. W., Moiser, A. R., Parton, W. J. and Schimmel, D. S. Effect of land use change on methane oxidation in temperate forest and grassland soils. Chemosphere. 26, 675-685, 1993.

Okuda, A. and Takahashi, E. The role of silicon. In: The Mineral Nutrition of the Rice Plant, Proc. Symp. Intern. Rice Res. Inst., John Hopkins Press, Baltimore/USA, p. 123-146, 1965.

Okusanya, O. T. and Ungar, L. A. The effect of time of seed production on the germination response of Spergularia marina. Physiol. Plant. 59, 335-342, 1983.

Oldenkamp, L. and Smilde, K. W. Copper deficiency in douglas fir Pseudotsuga menziesii Mirb. Franco. Plant and Soil 25, 150-152, 1966.

Ollagnier, M. and Ochs, R. (F) The chlorine nutrition of oil palm and coconut. Olbgineux, 26e annee, No. 6, 367-372, 1971.

Olsen, S. R., Cole, C. V., Watanabe, F. S. and Dean, C. A. Estimation of available phosphorus in soils by extraction with sodium bicarbonate. US. Dep. Agric. Cir. No. 939, 19, 1954.

Olsen, S. R. and Watanabe, F. S. Diffusive supply of phosphorus in relation to soil texture variations. Soil Sci. 110, 318-327, 1970.

Oparka, K. J. Phloem unloading in the potato tuber. Pathways and sites of ATPase. Protopl. 131, 201-210, 1986.

Oparka, K. J. and van Bel, A. J. E. Pathways of phloem loading and unloading: a plea for a uniform terminology. In: Carbon Partitioning within and between Organisms, edited by Pollock, C. J., Farrar, J. and Gordon, A. J., BIOS Scientific Publ. Limited, p. 249-254, 1992.

Oparka, K. J., Prior, D. A. M. and Wright, K. M. Symplastic communication between primary and developing lateral roots of Arabidopsis thalianA. J. Exp. Bot. 46, 187-197, 1995.

Oparka, K. J., Viola, R., Wright, K. M. and Prior, D. A. M. Sugar transport and metabolism in the potato tuber. In: Carbon Partitioning within and between organisms, edited by Pollock, C. J., Farrar, J. F. and Gordon, A. J. BIOS Scientific Publishers Limited, Oxford, GB, p. 255-260, 1992.

Oparka, K. J. and Wright, K. M. Influence of cell turgor on sucrose partitioning in potato tuber storage tissues. Planta 175, 520-526, 1988.

Oscarson, P. and Larsson, C. M. Relationship between uptake and utilization of NO_3^- in Pisum growing exponentially under nitrogen limitation. Physiol. Plant. 67, 109-117, 1986.

Osmond, C. B. Crassulacean acid metabolism: A curiosity in context. Annu. Rev. Plant Physiol. 29, 379-414, 1978.

Otter-Nacke, S. and Kuhlmann, H. A comparison of the performance of N simulation models in the prediction of N-min on farmers' fields in the spring. Fert. Res. 27, 341-347, 1991.

Ottow, J. C. G., Benckiser, G., Watanabe, I. and Santiago, S. Multiple nutritional soil stress as the prerequisite for iron toxicity of wetland rice (*Oryza sativa* L.). Trop. Agric. (Trinidad) 60, 102-106, 1983.

Ottow, J. C. G., Prade, K., Bertenbreiter, W. and Jacq, V. A. Strategies to alleviate iron toxicity of wetland rice on acid sulphate soils. In: Rice Production on Acid Soils in the Tropics, edited by De Turck, P. and Ponnamperuma, F. N. Kandy,Sri Lanka: Institute of Fundamental Studies, p. 205-211, 1991.

Outlaw, W. H., Jr. Current concepts on the role of potassium in stomatal movements. Physiol. Plant. 59, 302-311, 1983.

Overnell, L. Potassium and photosynthesis in the marine diatom *Phaeodactylum tricornutum* as related to washes with sodium chloride. Physiol. Plant. 35, 217-224, 1975.

Oweis, T., Pala, M. and Ryan, J. Management alternatives for improvement durum wheat production under supplemental irrigation in Syria. European J. Agron. 11, 255-266, 1999.

Owen, T. R. and Jürgens-Gschwind, S. Nitrates in drinking water: a review. Fert. Res. 10, 3-35, 1986.

Oyonarte, C., Perez-Pujalte, A., Delgado, G., Delgado, R. and Almendros, G. Factors affecting soil organic matter turnover. in a Mediterranean ecosystem from Sierra de Gador (Spain): An analytical approach. Communications in Soil Science and Plant Analysis 25, 1929-1945

Ozaki, L. G. Effectiveness of foliar manganese sprays on peas and beans. Amer. Soc. Hort. Proc. 66, 313-316, 1955.

Ozanne, P. G. Phosphate nutrition of plants - a general treatise. In: The Role of Phosphorous in Agriculture, edited by Khasawneh, F. E., Sample, E. C. and Kamprath, E. J., Amer. Soc. Agron., Madison, USA, p. 559-590, 1980.

Ozanne, P. G., Wooley, J. T. and Broyer, T. C. Chlorine and bromine in the nutrition of higher plants. Austr. J. Biol. Sci. 10, 66-79, 1957.

Pace, G. M., Volk, R. J. and Jackson, W. A. Nitrate reduction in response to CO_2-limited photosynthesis. Plant Physiol. 92, 286-292, 1990.

Padurariu, A., Horovitz, C. T., Paltineanu, R. and Negomireanu. On the relationship between soil moisture and osmotic potential in maize and sugar beet plants. Physiol. Plant, 22, 850-860, 1969.

Page, A. L., Ganse, T. J. and Joshi, M. S. Lead quantities in plants, soils and air near some major highways in Southern California. Hilgardia 41, 1-31, 1971.

Page, E. R. Studies in soil and plant manganese. II. The relationship of soil pH to manganese availability. Plant and Soil 16, 247-257, 1962.

Page, M. B. and Talibudeen, O. Nitrate concentrations under winter wheat and in fallow soil during summer at Rothamsted. Plant and Soil 47, 527-540, 1977.

Pagel, H. and Van Huay, H. (G) Important parameters of phosphate adsorption curves from tropical and subtropical soils and their changes resulting from P application. Arch. Acker-Pflanzenbau Bodenk. 20, 765-778, 1976.

Parfitt, R. L. and Smart, R. S. C. The mechanisrn of sulfate adsorption on iron oxides. Soil Sci. Soc. Am. J. 42, 48-50, 1978.

Park, C. S. The micronutrient problem of Korean agriculture. In: Symposium Commemorating the 30th Anniversary of Korean Liberation, edited by Nat. Acad. Sci. Rep. Korea, Seoul, p. 847-862, 1975.

Parker, D. R. and Norvell, W. A. Advances in solution culture methods for plant mineral nutrient research. Adv. Agron. 65, 151-313, 1999.

Parker, J. H. How fertilizer moves and reacts in the soil. Crops and Soils Magazine, Nov. 1972.

Parr, A. and Loughman, B. C. Boron membrane functions in plants. In: Metals and Micronutrients: Uptake and Utilization by Plants, edited by Robb, D. A. and Pierpoint, W. S. London: Academic Press, p. 87-107, 1983.

Pasricha, N. S. and Fox, R. L. Plant nutrient sulfur in the tropics and subtropics. Adv. Agron. 50, 209-269, 1993.

Passioura, J. B. and Fry, S. C. Turgor and cell expansion: beyond the Lokhardt equation. Aust. J. Plant Physiol. 19, 565-576, 1992.

Pate, J. S. Movement of nitrogenous solutes in plants, IAEA-PI-341/13. In: Nitrogen-15 in Soil-Plant Studies. International Atomic Energy Agency, Vienna, p. 165-187, 1971.

Pate, J. S. Uptake, assimilation and transport of nitrogen compounds by plants. Soil Biol. Biochem. 5, 109-119, 1973.

Pate, J. S. Transport and partitioning of nitrogenous solutes. Annu. Rev. Plant Physiol. 31, 313-340, 1980.

Pate, J. S. Exchange of solutes between the phloem and the xylem and circulation in the whole plant. In: Transport in Plants I - Phloem Transport. Encycl. Plant Physiology, Vol. I, edited by Zimmermann, M. H. and Milburn, J. A. Springer Verlag Berlin, Heidelberg, New York, p. 451-473, 1975.

Pate, J. S., Atkins, C. A., Hamel, K., McNeil, D. L. and Layzell, D. B. Transport of organic solutes in phloem and xylem of a nodulated legume. Plant Physiol. 63, 1082-1088, 1979.

Patrick, J. W. Assimilate partitioning in relation to crop productivity. Hort. Sci. 23, 33-40, 1988.

Patrick, J. W. Sieve element unloading: cellular pathway, mechanism and control. Physiol. Plant. 78, 298-308, 1990.

Patrick, J. W. Phloem unloading: sieve element unloading and post-sieve element transport. Annu. Rev. Plant Physiol. Plant Mol. Biol. 48, 191-222, 1997.

Patrick, J. W. and Offler, C. E. Post-sieve element transport of sucrose in developing seeds. Austr. J. Plant Physiol. 22, 681-702, 1995.

Patrick, W. H, jr. and Reddy, K. R. Fertilizer nitrogen reactions in flooded soils. Proc. Intern. Seminar on Soil Environment and Fertility Management in Intensive Agriculture, p. 275-281, Tokyo 1977.

Pätzold, C. and Dambroth, M. (G) Sensitivity to injury. Der Kartoffelbau 15, 291-292, 1964.

Paul, E. A. and Clark, F. E. Soil Microbiology and Biochemistry, London: Academic Press, 1996.

Paul, R. (G) Field trials for the assessment nitrogen release fom crop harvest residues by means of electro-ultrafiltration (EUF). Ph. D. Thesis Justus Liebig University 1994.

Paul, R. E., Johnson, C. M. and Jones, R. L. Studies on the secretion of maize root cap slime. I. Some properties of the secreted polymer. Plant Physiol. 56, 300-306, 1975.

Paul, R. E. and Jones, R. L. Studies on the secretion of maize root cap slime. II. Localization of slime production. Plant Physiol. 56, 307-312, 1975.

Paul, R. E. and Jones, R. L. Studies on the secretion of maize root cap slime. IV Evidence for the involvement of dictyosomes. Plant Physiol. 57, 249-256, 1976.

Payne, P. J. Genetics of wheat storing proteins and the effect of allelic variation on bread making quality. Annu. Rev. Plant Physiol. 38, 141-153, 1987.

Pedrazzini, F., Tarsitano, R. and Nannipieri, P. The effect of phenyl phosphorodiamidate on urease activity and ammonia volatilization in flooded rice. Biol. Fertil. Soils 3, 183-188, 1987.

Pedro, G. (F) Pedogenesis in the humid tropics and the dynamics of potassium. In: Potassium in Tropical Crops and Soils. Proc. 10th Colloq. Intern. Potash Institute, Berne, p. 23-49, 1973.

Peech, M. Lime requirements vs. soil pH curves for soils of New York State. Ithaca, N. Y. Agronomy, Cornell University 1961.

Pei, Z.-M., Kuchitsu, K., Ward, J. M., Schwarz, M. and Schroeder, J. I. Differential abscisic acid regulation of guard cell slow anion channels in Arabidopsis wild-type and abi1 and abi2 mutants. Plant Cell 9, 409-423, 1997.

Pei, Z. M., Ward, J. M. and Schroeder, J. I. Magnesium sensitizes slow vacuolar channels to physiological cytosolic calcium and inhibits fast vacuolar channels in faba bean guird cell vacuoles. Plant Physiol. 121, 977-986, 1999.

Pelton, W. L. Influence of low rates on wheat yield in southwestern Saskatchewan. Canad. J. Plant Sci. 49, 607-614, 1969.

Penningsfeld, F. and Forchthammer, L. (G) Response of the most important vegetables on a varied nutrient ratio in fertilizer application. Die Gartenbauwiss. 8, 347-372, 1961.

Peoples, T. R. and Koch, D. W. Role of potassium in carbon dioxide assimilation in *Medicago sativa* L. Plant Physiol. 63, 878-881, 1979.

Perur, N. G., Smith, R. L. and Wiebe, H. H. Effect of iron chlorosis on protein fractions of corn leaf tissues. Plant Physiol. 36, 736-739, 1961.

Perry, C. C., Williams, R. J. P. and Fry, S. C. Cell wall biosynthesis during silicification of grass hairs. J. Plant Physiol. 126, 437-448, 1987.

Peters, J. M. and Basta, N. T. Reduction of excessive bioavailable phosphorus in soils by using municipal and industrial wastes. J. Environ. Qual 25, 1236-1241, 1996.

Peterson, C. A. Exodermal Casparian bands: their significance for ion uptake by roots. Physiol. Plant. 72, 204-208, 1988.

Peterson, C. A., Murrmann, M. and Steudle, E. Location of major barriers to water and ion movement in young roots of *Zea mays* L. Planta 190, 127-136, 1993.

Peterson, C. A. and Steudle, E. Lateral hydraulic conductivity of early metaxylem vessels in *Zea mays* L. roots. Planta 189, 288-297, 1993.

Peterson, P. J.: The distribution of Zn-65 in *Agrostis tenuis* and *A. stolonifera* tissues. J. Exp. Bot. 20, 863-875, 1969.

Peterson, R. B. and Zelitch, I. Relationship between net CO_2 assmilation and dry weight accumulation in field-grown tobacco. Plant Physiol. 70, 677-685, 1982.

Peuke, A. D., Glaab, J., Kaiser, W. M. and Jeschke, W. D. The uptake and flow of C, N and ions between roots and shoots in *Ricinus communis* L. IV. Flow and metabolism of inorganic nitrogen and malate depending on nitrogen nutrition and salt treatment. J. Exp. Bot. 47, 377-385, 1996.

Peuke, A. D. and Jeschke, W. D. The uptake and flow of C, N and ions between roots and shoots in *Ricinus communis* L. I. Growth with ammonium or nitrate as nitrogen sourcE. J. Exp. Bot. 44, 1167-1176, 1993.

Peynaud, E. (F) Knowledge and Work of Wine. ISBN 2-04-011417-3, Bordas, Paris 1981.

Pfeffer, H., Dannel, F. and Römheld. V. Compartmentation of boron in roots and its translocation to the shoot of sunflower as affected by short term changes in boron supply. In: Boron in Soils and Plants, edited by Bell, R. W. and Rerkasem, B., Kluwer Academic Publishers, Dordrecht, p. 203-207, 1997.

Pfeffer, H., Dannel, F. and Römheld, V. Are there connections between phenol metabolism, ascorbate metabolism and membrane integrity in leaves of boron-deficient sunflower plant? Physiol. Plant. 104, 479-485, 1998.

Pfeffer, H., Dannel, H. and Römheld, V. Are there two mechanisms for boron uptake in sunflower. J. Plant Physiol. 155, 34-40, 1999.

Pfeiffenschneider, Y. and Beringer, H. Measurement of turgor potential in carrots of different K nutrition by using the cell pressure probe. In: Methods of K-Research in Plants, edited by International Potash Institute, Bern, Switzerland: International Potash Institute, p. 203-217, 1989.

Pflüger, R. and Mengel, K. (G) The photochemical activity of chloroplasts obtained from plants with a different potassium nutrition. Plant and Soil 36, 417-425, 1972.

Phelong, P. C. and Siddique, K. H. M. Contribution of stem dry matter to grain yield in wheat cultivars. Aust. J. Plant Physiol. 18, 53-64, 1991.

Picard, D. New approaches for cropping system studies in the tropics. In: 3. Congress of the European Society of Agronomy, edited by Borin, M. and Sattin, M. Colmar, France. European Society of Agronomy, p. 30-36, 1994.

Pickard, P. G. Voltage transients elicited by brief chilling: short communications. Plant, Cell and Environment 7, 679-682, 1984.

Piekielek, W. P. and Fox, R. H. Use of a chlorophyll meter to predict sidedress nitrogen requirements for maize. Agron. J. 84, 59-65, 1992.

Pier, P. A. and Berkowitz, G A. Modulation of water stress effects on photosynthesis by altered leaf K. Plant Physiol. 85, 655-661, 1987.

Pimentel, D., Hurd, L. E., Bellotti, A. C., Forster, M. J., Oka, I. N., Sholfs, O. D. and Whitman, R. J. Food production and the energy crisis. Science 182, 443-449, 1973.

Pimpini, F., Venter, F. and Wünsch, A. (G) Investigations on the nitrate concentration in cauliflower. Landwirtsch. Forsch. 23, 363-370, 1970.

Pirela, H. J. and Tabatabai, M. A. Sulfur mineralization rates and potentials of soils. Biol. Fertil. Soils 6, 26-32, 1988.

Pissarek, H. P. (G) Influence of intensity and performance of Mg deficiency on the grain yield of oats. Z. Acker- u. Pflanzenbau 148, 62-71, 1979.

Pissarek, H. P. (G) The development of potassium deficiency symptoms in spring rape. Z. Pflanzenernähr. Bodenk. 136, 1-96, 1973.

Pitman, M. G. Ion transport into the xylem. Annu. Rev. Plant Physiol. 26, 71 -88, 1977.

Plies-Balzer, E., Kong, T., Schubert, S. and Mengel, K. Effect of water stress on plant growth, nitrogenase activity and nitrogen economy of four different cultivars of Vicia faba L. Eur. J. Agron. 4, 167-173, 1995.

Pockman, W. T., Sperry, J. S. and O'Leary, J. W. Sustained and significant negative pressure in xylem. Nature 378, 715-716, 1995.

Pohlhil, R. M. The Papillionideae. In: Advances in Legume Systematics, edited by Pohlhil, R. M. and Raven, P. H., Royal Botanic Gardens Kew, p. 191-204, London 1981.

Poljakoff-Mayber, A. and Gale, J. Plants in Saline Environments. Ecological Studies, Vol. 15. Springer-Verlag, Berlin, Heidelberg, New York, 1975.

Pollard, A. S., Parr, A. J. and Loughman, B. C. Boron in relation to membrane function in higher plants. J. Exp. Bot. 28, 831-841, 1977.

Pollock, C. J. and Cairns, A. J. Fructan metabolism in grasses and cereals. Annu. Rev. Plant Physiol. Plant Mol. Biol. 42 77-101, 1991.

Pollock, C. J., Farrar, J. F. and Gordon, A. J. BIOS Scientific Publishers Limited, Oxford, GB, p. 249-254, 1992.

Ponnamperuma, F. N. Dynamic aspects of flooded soils and the nutrient of the rice plant. In: The Mineral Nutrition of the Rice Plant, Proc. of a Symposium at The Intern. Rice Res. Inst., Febr. 1964, p. 295-328. The Johns Hopkins Press, Baltimore, Maryland 1965.

Ponnamperuma, F. N. The chemistry of submerged soils. Adv. Agron. 24, 29-96, 1972.

Ponnamperuma, F. N. Electrochemical changes in submerged soils and the growth of rice. In: Soils and Rice. (The Intern. Rice Research Institute, ed.) Los Banos, Philippines, p. 421-441, 1978.

Poole, R. J. Energy coupling for membrane transport. Annu. Rev. Plant Physiol. 29, 437-460, 1978.

Poovaiah, B. W. Role of calcium in ripening and senescence. Comm. Soil. Sci. Plant Anal. 10, 83-88, 1979.

Poovaiah, B. W. and Leopold, A. C. Inhibition of abscission by calcium. Plant Physiol. 51, 848-851, 1973.

Pop, V. V., Postolache, T., Vasu, A. and Craciun, C. Calcophilous earthworm activity in soil; an experimental approach. Soil Biol. Biochem. 24, 1483-1490, 1992.

Portela, E. A. C. Potassium supplying capacity of northeastern Portuguese soils. Plant and Soil 154, 13-20, 1993.

Porter, G. A., Knievel, D. P. and Shannon, J. C. Assimilate unloading from maize (*Zea mays* L.) pedical tissue. II. Effects of chemical agents on sugar, amino acid, and ^{14}C unloading. Plant Physiol. 85, 558-565, 1987.

Porter, J-R. and Gawith, M. Temperatures and growth and the development of wheat: a review. European J. Agron. 10, 23-26, 1999.

Poskuta, J. and Kochanska, K. The effect of potassium glycidate on the rates of CO_2 exchange and photosynthetic products of bean leaves. Z. Pflanzenphysiol. 89, 393-400, 1978.

Post-Beittenmiller, D. Biochemistry and molecular biology of wax production in plants. Annu. Rev. Plant Physiol. Plant Mol. Biol. 47, 405-430, 1996.

Postgate, J. Nitrogenase. Biologist 32, 43-48, 1985.

Pottosin, I. I. and Andjus, P. R. Depolarization activated K^+ channel. In: Chara droplets. Plant Physiol. 106, 313-319, 1994.

Poulton, P. R., Tunney, H. and Johnston, A. E. Comparison of fertilizer P recommended in Ireland and England and Wales. In: Phosphorous Loss from Soil to Water, edited by Tunney, H., Carton, O. C., Brookes, P. C. and Johnston, A. E., CAB International, p., 1997.

Pozuelo, J. M., Espelie, K. E. and Kollatukudi, P. E. Magnesium deficiency results in increased suberisation in endodermis and hypodermis of corn roots. Plant Physiol. 74, 256-260, 1984.

Prasad, R. and Goswami, N. N. Soil fertility restoration and management for sustainable agriculture in South Asia. Adv. Soil Sci. 17, 37-77, 1992.

Praske, J. A. and Plocke, D. J. A role for zinc in the structural integrity of the cytoplasmic ribosomes of *Euglena gracilis*. Plant Physiol. 48, 150-155, 1971.

Pratt, P. F. Potassium. In: Methods of Soil Analysis, part 2, edited by Black, C. A. et al., Am. Soc. of Agron. Madison, Wisc. Agronomy 9, 1023-1031, 1965.

Pratt, P. F.: Chromium. In: Diagnostic Criteria for Plants and Soils, edited by Chapman, H. D., University of California, Div. of Agric. Sciences, p. 136-141, 1966.

Pratt, P. F.: Vanadium. In: Diagnostic Criteria for Plants and Soils, edited by Chapmen H. D. University of California, Riverside, p. 480-483, 1966.

Prausse, A. (G) Results of a three years trial with phosphorus following the application of various phosphate forms. Thaer-Archiv 12, 97-114, 1968.

Preiss, J. Regulation of the biosynthesis and degradation of starch. Annu. Rev. Plant Physiol. 33, 431-454, 1982.

Preiss, J. Biology and molecular biology of starch synthesis and its regulation. In: Oxford Surveys of Plant Molecular and Cell Biology, edited by Miflin, B. J. Vol. 7, p 59-114, 1991.

Preiss, J. and Levi, C. Metabolism of starch in leaves. In: Photosynthesis II, New Series, Vol. 6, edited by Gibbs, M. and Latzko, E., Springer-Verlag Berlin, Heidelberg, New York p. 282-312, 1979.

Price, C. A., Clark, H. E. and Funkhouser, H. E. Functions of micronutrients in plants. In: Micronutrients in Agriculture. edited by J. J. Mortvedt, P. M. Giordano and W. L. Lindsay. Soil Sci. Soc. of America, Madison/Wisconsin, p. 731-742, 1972.

Priebe, A., Klein, H. and Jagger, H. Role of polyamines in SO_2 polluted pea plants. J. Exptl. Bot. 29, 1045-1050, 1978.

Primost, E. (G) The influence of fertilizer application on the quality of wheat. Landw. Forsch., 22. Sonderh., 149-157, 1968.

Prince, A. L., Zimmerman, M. and Bear, F. E. The magnesium supplying powers of 20 New Jersey soils. Soil Sci. 63, 69-78, 1947.

Prins, W. H., Dilz, K. and Neeteson, J. J. Current recommendation for nitrogen fertilization within the EEC in relation to nitrate leaching. In: -, edited byFertilizer Society, U. K. London, p. 27, 1988.

Prior, B. (G) Effect of nitrogen supply on the soluble amino acids in the organs of *Vitis vinifera* (c. v. Riesling) and on the quality of must and wine. Ph. D. Thesis, Justus Liebig University, 1997.

Prummel, J. Fertilizer placement experiments. Plant and Soil 8, 231-253, 1957.

Pulss, G. and Hagemeister, H. (G) Hypomagnesaemie after the feeding of wilted silage of pasture herbage during the stable period. Z. Tierphysiol., Tierernähr. u. Futtermittel 25, 32-42, 1969.

Pushnik, J., Miller, G. and Giannini, J. Re-establishment of photochemical activities in iron chlorotic leaves by foliar iron application. VIth International Colloq. for the Optimization of Plant Nutrition, edited by P. M. Prével Vol. 4, p. 1229-1238, 1984.

Pyle, A. M. Ribozymes: a distinct class of metalloenzymes. Science 261, 709-714, 1993.

Quick, W. P. and Schaffer, A. A. Sucrose metabolism in sources and sinks. In: Photoassimilate distribution in plants and crops. Source-Sink relationships, edited by Zamski, E. & Schaffer, A. Marcel Dekker, New York, pp. 115-156, 1996.

Quispel, A. Dinitrogen - fixing symbioses with legumes, non-legumes angiosperms and associative symbioses. In: Inorganic Plant Nutrition. Encycl. Plant Physiol. New Series Vol. 15A, edited by Läuchli, A. and Bieleski, R. L. Springer Verlag Berlin, Heidelberg, New York, Tokyo, p. 286-329, 1983.

Radin, J. W. Responses of transpiration and hydraulic conductance in nitrogen and phosphorus cotton seedlings. Plant Physiol. 92, 855-857, 1990.

Raghavendra, A. S. and Das, V. S. R. Photochernical activities of chloroplasts isolated from plants with the C-4 pathway of photosynthesis and from plants with the Calvin cycle. Z. Pflanzenphysiol. 88, 1-11, 1978.

Rahmatullah, and Mengel, K. Potassium release from mineral structures by H$^+$ ion resin. Geoderma 96, 291-305, 2000.

Raikov, L. Reclamation of solonetz soils in Bulgaria, p. 35-47. In: European Solonetz Soils and Their Reclamation, edited by Szabolcs, I. Akademiai Kiadö, Budapest 1971.

Raman, K. V. and Jackson, M. L. Vermiculite surface morphology. In: Clays and Clay Minerals. 12th Natl. Conf. Pergamon Press, New York, p. 423-429, 1964.

Randall, G. W. and Schulte, E. E. Manganese fertilization of soybeans in Wisconsin. Proc. Wis. Fert. and Aglime Conf. 10, 4-10, 1971.

Randall, P. J. and Wrigley, C. W. Effects of sulphur supply on the yield, composition and quality of grain from cereals oilseeds and legumes. Adv. Cereal Sci. Techn. 8, 171-206, 1986.

Randhawa, N. S., Sinha, M. K. and Takkar, P. N. Micronutrients, In: Soils and Rice, edited by the Intern. Rice Research Institute, Los Banös, Philippines, p. 581-603, 1978.

Rao, I. M., Sharp, R. E. and Boyer, J. S. Leaf magnesium alters photosynthetic response to low water potentials in sunflower. Plant Physiol. 84, 1214-1219, 1987.

Rao, I. M. and Terry, N. Leaf phosphate status, photosynthesis and carbon partitioning in sugar beeet. I Changes in growth, gas exchange and Calvin cycle enzymes. Plant Physiol. 90, 14-819, 1989.

Rao, N. R., Naithaini, S. C., Jasdanwala, R. T and Singh, Y. D. Changes in indolacetic acid oxidase and peroxidase activities during cotton fibre development. Z. Pflanzenphysiol. 106, 157-165, 1982.

Rapp, A. (G) Aroma ingredients of wine. Weinw.,Technik 7, 17-27, 1989.

Rashid, A., Couvillon, G. A. and Benton Jones, J. Assessment of Fe status of peach rootstocks by techniques used to distinguish chlorotic and non -chlorotic leaves. J. Plant Nutr. 13, 285-307, 1990.

Ratcliffe, R. G. *In vivo* NMR studies of higher plants and algae. Adv. Bot. Res. 20, 43-123, 1994.

Rathore, V. S., Bajaj, Y. P. S. and Wittwer, S. H.: Sub cellular localization of zinc and calcium in bean (*Phaseolus vulgaris* L.) tissues. Plant Physiol. 49, 207-211, 1972.

Rathsack, K. (G) The nitrificide effect of dicyandiamide. Landwirtsch. Forsch. 31, 347-358, 1978.

Rathsack, R. (G) The nitrification inhibiting effect of dicyan diamide. Landw. Forsch. 31, 347-358, 1978.

Rauhut, D. (G) Quality reducing sulfur containing compounds in wine. Ph. D. Thesis, Justus Liebig University, Giessen 1996.

Rauser, W. E. Zinc toxicity in hydroponic culture. Can. J. Bot. 51, 301-304, 1973.

Rauser, W. E. Metal-binding peptides in plants. In: Sulfur Nutrition and Assimilation in Higher Plants, edited by De Kok, L. J., Stulen, I., Rennenberg, H., Brunold, C. and Rauser, W. E. The Hague: SPB Academic Publishing bv, p. 239-251, 1993.

Raven, J. A. Short- and long distance transport of boric acid in plants. New Phytol. 84 231-249, 1980.

Raven, J. A. The transport and function of silicon in plants. Biol. Rev. 58, 179-207. 1983.

Raven, J. A. and Smith, F. A. Nitrogen assimilation and transport in vascular land plants relation, to intracellular pH regulation. New Phytol 76, 415-431, 1976.

Ray, T. B. and Black, C. C. The C4 pathway and its regulation. In: Photosynthesis II, Encycl. Plant Physiol. New Series Vol 6, edited by Gibbs, M. and Latzko, E., Springer Verlag Berlin, Heidelberg, New York, p. 77-101,1979.

Rayle, D. L. and Cleland, R. E. The acid growth theory of auxin-induced cell elongation is alive and well. Plant Physiol. 99, 1271-1274, 1992.

Raymond, K. N. Kinetically inert complexes of the siderophores in studies of microbial iron transport. In: Bioinorganie chemistry II, Adv. Chemistry, edited by Raymond, K. N., Amer. Chem. Soc. Washington DC., p. 33-54, 1977.

Rayssiguier, Y. (F) Metabolism of magnesium in animals and pathology related to a deficit in magnesium. In: Le magensium en agriculture, edited by Huguet, C. and Coppenet, M., INRA editions, p. 21-31, Paris 1992.

Rea, F. A. and Poole, R. J. Vacuolar H $^+$translocating pyrophosphatase. Annu. Rev. Plant Physiol. Plant Mol. Biol. 44, 157-180, 1993.

Rease, J. T. and Drake, S. R. Yield increased and fruit disorders decreased with repeated annual calcium sprays on „Anjour" pears. J. Tree Fruit Production 1, 51-59, 1996.

Rebafka, F. P., Ndunguru, B. J. and Marschner, H. Single superphosphate depresses molybdenun uptake and limits yield response to phosphorus in groundnut (*Arachus hypogaea* L.) grown on an acid soil in Niger West Africa. Fertil. Res. 34, 233-242, 1993.

Reddy, K. J., Munn, L. C. and Wang, L. Chemistry of Mineralogy of Molybdenum in Soils. In: Molybdenum in Agriculture, edited by Gupta, U. C., Cambridge University Press, p. 4 - 22 Cambridge 1997.

Reddy, K. R. and Rao, P. S. C. Nitrogen and phosphorus fluxes from a flooded organic soil. Soil Sci. 136, 300-307, 1983.

Reddy, V. R., Baker, D. N. and Hodges, H. F. Temperature and mepiquat chloride effects on cotton canopy architecture. Agron. J. 82, 190-195, 1990.

Reed, S. T. and Martens, D. C. Copper and zinc. In: Methods of Soil Analysis, edited by Bigham, J. M., Madison: Soil Sci Soc. Am., p. 703-722, 1996.

Rees, R. M., Yan, L. and Ferguson, M. The release and plant uptake of nitrogen from some plant and animal manures. Biol. Fertil. Soils 15, 285-293, 1993.

Reeve, N. G. and Sumner, M. E. Amelioration of subsoil acidity in Natal Oxisols by leaching of surface-applied amendments. Agrochemophysica 4, 1-6, 1972.

Reich, P. B., Ellsworth, D. S. and Uhl, C. Leaf carbon and nutrient assimilation and conservation in species of differing successsional status in an oligotrophic Amazonian forest. Functional Ecology 9, 65-76, 1995.

Reichard, P. From RNA to DNA, why so many ribonucleotide reductases? Science. 260, 1773-1777, 1993.

Reicosky, D. C. and Ritchie, J. T. Relative importance of soil resistance and plant resistance in root water absorption. Soil Sci Soc. Am. J. 40, 293-297, 1976.

Reisenauer, H. M., Tabikh, A. A. and Stout, P. R. Molybdenum reactions with soils and the hydrous oxides of iron, aluminium and titanium. Soil Sci. Soc. Amer. Proc. 26, 23-27, 1962.

Reisenauer, H. M., Walsh, L. M. and Hoeft, R. G. Testing soils for sulphur, boron, molybdenum and chlorine. In: Soil Testing and Plant Analysis, edited by Walsh, L. M. and Beaton, J. D. Soil Sci. Soc. of America, Madison/Wisconsin, p. 173-200, 1973.

Reissig, H. (G) The influence of liming on the Sr-90 uptake by crops under field conditions. Kernenergie 5, 678-684, 1962.

Reith, W. J. S. Soil properties limiting the efficiency of fertilizers. In: 7th Fertilizer World Congress, Zürich, C. I. E. C., p. 275-278, 1972.

Rendig, V. V., Oputa, C. and McComb, E. A. Effects of sulphur deficiency on non-protein nitrogen, soluble sugars and N/S ratios in young corn plants. Plant and Soil 44, 423-437, 1976.

Renelt, D. (G) Consideration of precrop and weather for N fertilizer application to wheat on soils derived from loess in the eastern Harz area. Ph. D. Thesis, University Halle, Germany, 1993.

Rengel, Z. Role of calcium in aluminium toxicity. New Phytol. 121, 499-514, 1992.

Rennenberg, H. Role of O-acetylserine in hydrogen sulfide emission from pumpkin leaves in response to sulfate. Plant Physiol. 73, 560-565, 1983.

Rennenberg, H. The fate of excess sulfur in higher plants. Annu. Rev. Plant Physiol. 25, 121-153, 1984.

Rennenberg, H., Pol, A., Martini, N. and Thoene, B. Interaction of sulphate and glutathione transport in cultured tobacco cells. Planta 176, 68 -74, 1988.

Rennenberg, H., Schmitz, K. and Bergmann, L. Long distance transport of sulphur in *Nicotiana tabacum*. Planta 147, 57-62, 1979.

Rerkasem, B., Netsangtip, S. and Cheng, C. Grain set failure in boron-deficient wheat. Plant and Soil 155/156, 309-312, 1993.

Resseler, H. and Werner, W. Properties of untreated rock residues in partially acidulated phophate rocks affecting their reactivity. Fert. Res. 20, 135-142, 1989.

Reuther, W. and Labasnauskas, C. K.: Copper. In: Diagnostic Criteria for Plants, edited by Chapman, H. C., Univ. of California, Agric. Pub. Berkley U. S., p. 157-179, 1966.

Reuveni, M. R., Colombo, H. R., Lerner, H. R., Pradet, A. and Poljakoff-Mayber, A. Osmotically proton extrusion from carrot cells in suspension culture. Plant Physiol. 85, 383-388, 1987.

Rex, M. (G) The influence of soil-rooting depth on the yield and nutrient uptake of cereals. Ph. D. Thesis, Justus-Liebig-University, Giessen, 1984.

Rheinbaben, v. W. (G) Impacts of optimum and suboptimum supplied plants on the denitrification intensity of plants grown in solution culture, pot experiments and field. Habilitation Thesis, Justus-Liebig-University, Giessen 1990.

Ribailler, D. and d'Auzac, J. (F) New perspectives in hormonale stimulation of the production of *Hevea brasiliensis*. R. G. C. P. 47, 433-439, 1970.

Rich, C. J. Mineralogy of soil potassium. In: The Role of Potassium in Agriculture. edited by V. J. Kilmer,

S. E. Younts and N. C. Brady, Amer. Soc. Agron., Madison/USA, p. 79-96, 1968.

Rich, C. J. and Black, W. R. Potassium exchange as affected by cation size, pH and mineral structure. Soil Sci. 97, 384-390, 1964.

Richards, L. A. and Wadleigh, C. H. Soil water and plant growth. In: Soil Physical Conditions and Plant Growth, edited by Shaw, B. T. Academic Press, New York, pp. 73-251, 1952.

Riehm, H. and Quellmalz, E. (G) The determination of plant nutrients in rain water and in the air and their importance for the agriculture. In: 100 Jahre Staatl. Landw. Versuchs- und Forschungsanstalt Augustenberg, edited by Riehm, H., p. 171-183, 1959.

Riesmeier, J. W., Flügge, U.-I., Schulz, B., Heineke, D., Heldt, H.-W., Willmitzer, L. and Frommer, W. B. Antisense repression of the chloroplast triose phosphate translocator affects carbon partitioning in transgenic potato plants. Proc. Natl. Acad. Sci. USA 90, 6160-6164, 1993.

Riesmeier, J. W., Willmitzer, L. and Frommer, W. B. Evidence for an essential role of the sucrose transporter in phloem loading and assimilate partitioning. EMBO 13, 1-7, 1994.

Riga, A., Fischer, V. and van Praag, H. J. Fate of fertilizer nitrogen applied to winter wheat as Na $^{15}NO_3$ and ($^{15}NH_4$)$_2$ SO$_4$ studied in microplots through a four-course rotation. 1. Influence of fertilizer splitting on soil and fertilizer nitrogen. Soil Sci. 130, 88-99, 1980.

Rinne, R. W. and Langston, R. G. Effect of growth on redistribution of some mineral elements in peppermint. Plant Physiol. 35, 210-215, 1960.

Rios, M. A. and Pearson, R. W. The effect of some chemical environmental factors on cotton behavior. Soil Sci. Soc. Amer. Proc. 28, 232-235, 1964.

Röbbelen, G. The state of new crops development and their future prospects in Northern Europe. In: New Crops for Temperate Regions, edited by Anthony, K. R. M., Meadley, J. and Röbbelen, G. London: Chapman and Hall, p. 22-34, 1993.

Roberts, D. M. and Harmon, A. C. Calcium-modulated proteins: targets of intracellular calcium signals in higher plants. Annu. Rev. Plant Physiol. Plant Mol. Biol. 43, 375-414, 1992.

Roberts, S. and Tester, M. Permeation of Ca^{2+} and monovalent cations through an outwardly rectifying channel in maize root stelar cells. J. Exp. Bot. 48, 839-846, 1997a.

Roberts, S. and Tester, M. A patch clamp study of Na$^+$ transport in maize roots. J. Exp. Bot. 48, 431-440, 1997b.

Robson, A. D., Dilworth, M. J. and Chatel, D. L. Cobalt and nitrogen fixation in Lupinus angustifolius L. I. Growth, nitrogen concentrations and cobalt distribution. New Phytol. 83, 53-62, 1979.

Robson, A. D., Hartley, R. D. and Jarvis, S. C. Effect of copper deficiency on phenolic and other constituents of wheat (Triticum aestivum cultivar Sappo) cell walls. New Phytol 89, 361-372, 1981.

Robson, A. D. and Pitman, M. G. Interactions between nutrients in higher plants. In: Inorganic Plant Nutrition, edited by Läuchli, A. and Bieleski, R. L. Encycl., Plant Physiol. New Series Vol. ISA, Springer Verlag Berlin, Heidelberg, New York, Tokyo, p. 147-173, 1983.

Rocznik, K. (G) The meteorological year 1997 in Germany. Naturw. Rundschau 51, 191-19-, 1998.

Rodenkirchen, H. and Roberts, B. A. Soils and plant nutrition on a serpentinized ridge in South Germany. II. Foliage macro-nutrient and heavy metal concentrations. Z. Pflanzenernähr. Bodenk. 156, 407-410, 1993.

Roeb, G. and Führ, F. (G) Short term isotopes in plant physiology shown with ^{11}C as example. Reprint. Reinisch-Westfälische Akademie der Wissenschaften, Westdeutscher Verlag, p. 55-80, 1990.

Roelofs, J. G. M., Boxma, A. W. and Van Dijk, H. F. G. Effects of airborne ammonium on natural vegetation and forests. In: Air Pollution and Ecosystems, edited by Mathy, P. Dordrecht: D. Reidel Publishing Company, p. 876-880, 1988.

Roelofs, J. G. M., Kempers, A. J., Houdijk, A. L. F. M. and Jansen, J. The effect of air -borne ammonium sulphate on *Pinus nigra* var. maritima in the Netherlands. Plant and Soil 84, 45-56, 1985.

Roland, J. C. and Bessoles, M. (F) Evidence for calcium in cells of the collenchyma. C. R. Acad. Agric. 267, 589-592, 1968.

Roll-Hansen, J. Steaming of soil for tomatoes. State Experiment Station Kvitzmar, Stjördal, Norway, Report No. 10, 1952.

Rolston, D. E., Fried, M. and Goldhamer, D. A. Denitrification measured directly from nitrogen and nitrous oxide gas fluxes. Soil Sci. Soc. Am. J. 40, 259-266, 1976.

Romera, F. J., Alcantara, E. and de la Guardia, M. D. Characterization of the tolerance to iron chlorosis in different peach rootstocks grown in nutrient solution. II. Iron-stress response mechanisms. In: Iron Nutrition and Interactions in Plants, edited by Chen, Y. and Hadar, Y. Kluwer Academic Publishers, Dordrecht, p. 151-155, 1991a.

Romera, F. J., Alcantara, E. and de la Guardia, M. D. Characterization of the tolerance to iron chlorosis in different peach rootstocks grown in nutrient solution. I. Effect of bicarbonate and phosphate. In: Iron Nutrition and Interactions in Plants, edited by Chen, Y. and Hadar, Y. Kluwer Academic Publishers, Dordrecht, ISBN 0-7923-1095-0, p. 145-149, 1991b.

Römheld, V. The role of phytosiderophores in acquisition of iron and other micronutrients in graminaceous species: an ecological approach. In: Iron Nutrition and Interactions in Plants, edited by Chen, Y. and Hadar, Y. Dordrecht: Kluwer Academic Publishers ISBN 0-7923-1095-0, p. 159-166, 1991.

Römheld, V. and Marschner, H. Rhythmic iron stress reactions in sunflower at suboptimal iron supply. Physiol. Plant 53, 347-353, 1981a.

Römheld, V. and Marschner, H. Iron deficiency stress induced morphological changes and physiological changes in the root tips of sunflower (*Helianthus annuus*, cv Sobrid). Physiol. Plant. 53, 354-360, 1981b.

Römheld, V. and Marschner, H. Mobilization of iron in the rhizosphere of different plant species. Adv. Plant Nutr. 2, 155-204, 1986.

Römheld, V. and Marschner, H. Functions of micronutrients in plants. In: Micronutrients in Agriculture, ed. by J. J. Mortvedt, F. R. Cox, L. M. Shuman, R. M. Welch SSSA, p. 297-328, Madison, USA 1991.

Ron Vaz, M. D., Edwards, A. C., Shand, C. A. and Cresser, M. S. Phosphorus fractions in soil solution: influence of soil acidity and fertilizer addition. Plant and Soil 148, 175-183, 1993.

Rorison, I. H. Some experimental aspects of the calcicole-calcifuge problem. I. The effects of competition and mineral nutrition upon seedling growth in the field. J. Ecol. 48, 585-599, 1960a.

Rorison, I. H. The calcicole-calcifuge problem. II. The effects on mineral nutrition on seedling growth in solution culturE. J. Ecol. 48, 679-688, 1960b.

Rorison, I. H. Ecological Aspects of the Mineral Nutrition of Plants. Blackwell Scientific Publ. Oxford and Edinburgh 1969.

Roscoe, B. The distribution and condition of soil phosphate under old permanent pasture. Plant and Soil 12, 17-29, 1960.

Roth, D., Günther, R. and Roth, R. (G) Transpiration coefficients and water use efficiency of crops. 1. Comm. Transpiration coefficients and water use of cereals, sugar beet, and silage maize under field conditions and in pot experiments. Arch. Acker-Pflanzenbau Bodenkd. 32, 397-403, 1988a.

Roth, D., Roth, R., Günther, R. and Spengler, R. (G) Transpiration coefficients and water use efficiency of crops. 2. Commun. Effect of varied water supply on transpiration coefficient and water use efficiency. Arch. Acker-Pflanzenbau Bodenkd. 32, 405-410, 1988b.

Roth-Bejerano, N. and Neyidat, A. Phytochrome effects onK fluxes in guard cells of *Commelina communis*. Physiol. Plant. 71, 345-351, 1987.

Roussel, N., van Stallen, R. and Vlassak, K. (F) Results of experimentation over two years with unhydrous ammoniA. J. Intern. Inst. Sugar Beet Res., 2, 35-52, Tienen 1966.

Roux, L. (F) Condensed phosphates in mineral nutrition. Application of a highly condensed potassium polyphosphate to barley grown in solution culture. Ann. Physiol. vég. 10, 83-98, 1968.

Rovira, A. D., Bowden, G. D. and Foster, R. C. The significance of rhizosphere microflora and mycorrhizas in plant nutrition. In: Inorganic Plant Nutrition, New Series, Vol. 15A, edited by Läuchli, A. and Bieleski, R. L., Springer Verlag Berlin, Heidelberg, New York, Tokyo, p. 61-93, 1983.

Rovira, A. D. and Davey, C. B. Biology of the rhizosphere. In: The Plant Root and its Environment, edited by E. W. Carson, University Press of Virginia, Charlottesville, p. 153-204, 1974.

Roy, W. R., Hasselt, J. J. and Griffin, R. A. Competitive coefficients for the adsorption of arsenate, molybdate, and phosphate mixtures by soils. Soil Sci. Soc. Am. J. 50, 1176-1182, 1986.

Rroco, E. and Mengel, K. Nitrogen losses from entire plants of spring wheat (*Triticum aestivum*) from tillering to maturation. Eur. J. Agron., in press, 2000.

Rubinstein, B. and Luster, D. G. Plasma membrane redox activity: Components and role in plant processes. Annu. Rev. Plant Physiol. Plant Mol. Biol. 44, 131-155, 1993.

Rubio, F., Gassmann, W. and Schroeder, J. I. Sodium- driven potassium uptake by the plant potassium transporter HKTI and mutations conferring salt tolerance. Science. 270, 1660-1663, 1995.

Ruck, A., Palme, K., Venis, M. A., Napier, R. M. and Felle, H. H. Patch-clamp analysis establishes a role for an auxin binding protein in the auxin stimulation of plasma membrane current in *Zea mays* protoplasts. Plant J. 4, 41-46, 1993.

Rufty, T. W., Miner, G. S. and Raper, C. D. jr. Temperature effects on growth and manganese tolerance in tobacco. Agron. J. 71, 638-644, 1979.

Rufty, T. W. jr., Mackown, C. T. and Volk, R. J. Effects of altered carbohydrate availability on whole plant assimilation $^{15}NO_3^-$. Plant Physiol 89, 457-463, 1989.

Rush, D. W. and Epstein, E. Differences between salt-sensitive and salt-tolerant genotypes of the tomato. Plant Physiol. 57, 162-166, 1976.

Russell, E. W. Conditions and Plant Growth, 10th Edition, Longman 1973.

Russell, R. S. and Clarkson, D. T. Ion transport in root systems. In: Perspectives in Experimental Biology. Vol. 2. Botany, edited by Sunderland, N., Pergamon Press, Oxford and New York, p. 401-411, 1976.

Russell, R. S., Rickson, J. B. and Adams, S. N. Isotopic equilibria between phosphates in soil and their significance in the assessment of fertility by tracer methods. J. Soil Sci. 5, 85-105, 1954.

Ryan, J., Abdel Monem, M., Shroyer, J. P., El Bouhssini, M. and Nachit, N. N. Potential of nitrogen fertilization and Hessian fly-resistance to improve Morrocco's dryland wheat yields. Eur. J. Agron. 8, 153-159, 1998.

Ryan, P. F. Fertilizer placement for kale. Irish J. Agric. Res. 1, 231-236, 1962.

Ryden, J. C., Syers, J. K. and Harris, R. F. Phosphorus in runoff and streams. Adv. in Agron. 25, 1-45, 1973.

Saab, I. N., Ho, T. H. D. and Sharp, R. E. Translatable RNA populations associated with maintainance of primary root elongation and inhibition of mesocotyl elongation by abscisic acid in maize seedlings at low water potentials. Plant Physiol. 109, 593-601, 1995.

Saalbach, E. (G) The significance of atmospheric sulphur compounds for the supply of agricultural crops. AngeW. B. ot. 58, 147-156, 1984.

Sadeghian, E. and Kühn, H. (G) Effect of ancymidol and ethrel on the gibberellic acid content (GA3) of cereal species. Z. Pflanzenernähr. Bodenkd. 135, 309-314, 1976.

Sadiq, M. Potash status of soils in Pakistan. In: Proc. 12 th International Forum on Soil Taxonomy and Agrotechnology Transfer, edited by US-AID US Dep. Agric., Soil Survey Pakistan, p. 113-118, 1986.

Saftner, R. A., Daie, J. and Wyse, R. E. Sucrose uptake and compartmentation in sugar beet taproot tissue.

Plant Physiol. 72, 1-6, 1983.

Saftner, R. A. and Wyse, R. E. Alkali cation/sucrose co-transport in the root sink of sugar beet. Plant Physiol. 66, 884-889, 1980.

Saglio, P. (F) Iron nutrition of grapes. Ann. Physiol. vég. 11, 27-35, 1969.

Sah, R. N. and Mikkelsen, D. S. Effects of anaerobic decomposition of organic matter on sorption and transformations of phosphate in drained soils: Effects on amorphous iron content and phosphate transformation. Soil Sci. 142, 346-551, 1986.

Sahu, M. P., Sharma, D. D., Jain, G. L. and Singh, H. G. Effects of growth substances, sesquestrene 138-Fe and sulphuric acid on iron chlorosis of garden peas (*Pisum sativum* L.). J. Horticult Sci. 62, 391-394, 1987.

Sahu, M. P. and Singh, H. G. Effect of sulphur on prevention of iron chlorosis and plant composition of ground nut on alkaline calcareous soils. J. agric. Sci. Camb. 1109, 73-77, 1987.

Salami, U. A. and Kenefick, D. G. Stimulation of growth in zinc deficient corn seedlings by the addition of tryptophan. Crop Sci. 10, 291-294, 1970.

Salsac, L., Chaillou,, S., Morot-Gaudry, J. F., Lesaint, C. and Jolivet, E. Nitrate and ammonium nutrition in plants. Plant Physiol. Biochem. 25, 805-812, 1987.

Salt, D. E., Smith, R. D. and Raskin, I. Phytoremediation. Annu. Rev. Plant Physiol. Plant Mol. Biol. 49, 643-668, 1998.

San Valentin, G. O., Robertson, W. K., Johnson, J. T. and Weeks, W. W. Effect of slow-release fertilizer in fertilizer residues and on yield and composition on flue-cured tobacco. Agron. J. 70, 345-348, 1978.

Sanchez, P. A. and Leakey, R. R. B. Land use transformation in Africa: three determinants for balancing food security with natural resources. Eur. J. Agron. 7, 15-23, 1997.

Sanchez-Blanco, M. J., Morales, M. A., Torrecillas, A. and Alarcon, J. J. Diurnal and seasonal changes in *Lotus creticus* plants grown under saline stress. Plant Sci. 136, 1-10, 1998.

Sanders, F. E. and Tinker, P. B. Phosphate flow into mycorrhizal roots. Pestic. Sci. 4, 385-395, 1973.

Sanders, J. L. and Brown, D. A. A new fiber optic technique for measuring root growth of soybeans under field conditions. Agron. J. 70, 1073-1076, 1978.

Sanderson, G. W. and Cocking, E. C. Enzymic assimilation of nitrate in tomato plants. 1. Reduction of nitrate to nitrite. Plant Physiol. 39, 416-422, 1964.

Sangster, A. G., Hodson, M. J. and Wynn Parry, D. Silicon deposition and anatomical studies in the inflorescence tracts of four *Phalaris* species with their possible relevance to carcenogenesis. New Phytol. 93, 105-122, 1983.

Santakumari, M. and Berkowitz, G. A. Protoplast volume: water potential relationships and bound water fraction in spinach leaves. Plant Physiol. 91, 13-18, 1989.

Sanvicente, P., Lazarevitch, S., Blouet, B. and Guckert, A. Morphological and anatomical modifications in winter barley culm after late plant growth reulator treatment. European J. Agron. 11, 45 -51, 1999.

Sauerbeck, D. (G) Which heavy metal concentrations in plants should not be exceeded in order to avoid detrimental effects on their growth. Landw. Forsch. Sonderh. 39, 108-129, 1982.

Sauerbeck, D. and Lübben, S. (G) Effects of municipal disposals on soils, soil organisms and plants. In: Berichte aus der ökologischen Forschung, Vol. 6, edited by Forschungszentrum Jülich, Jülich: Zentralbibliotek, p. 1-32, 1991.

Sauter, J. J. Analysis of the amino acids and amides in the xylem sap of *Salix caprea* L. in early spring. Pflanzenphysiol. 79, 276-280, 1976.

Sautter, U. (F) Fermentation at low temperature. Alles über Wein. 1, 149-14-, 1999.

Savant, N. K. and DeDatta, S. K. Nitrogen transformations in wetland rice soils. Adv. Agron. 35, 241-302, 1982.

Savant, N. K., Snyder, G. H. and Datnoff, L. E. Silicon management and sustainable rice production. Adv. Agron. 58, 151-199, 1997.

Savin, R. and Nicolas, M. E. Effects of short periods of drought and high temperature on grain growth and starch accumulation of two malting barley cultivars. Aust. J. Plant Physiol. 23, 201-210, 1996.

Scaife, M. A. and Clarkson, D. T. Calcium related disorders in plants - a possible explanation for the effect of weather. Plant and Soil. 50, 723-725, 1978.

Schacherer, A. and Beringer, H. (G) Number and size distribution of endosperm cells in developing cereal grains as an index for their sink capacity. Ber. Deutsch. Bot. Ges. 97, 183-195, 1984.

Schachtman, D. P., Reid, R. J. and Ayling, S. M. Phosphorus uptake by plants: from soil to cell. Plant Physiol. 116, 447-453, 1998.

Schachtman, D. P. and Schroeder, J. I. Structure and transport mechanism of a high-affinity potassium uptake transporter from higher plants. Nature 370, 655-658, 1994.

Schachtman, D. P., Tyerman, S. T. and Terry, B. R. The K^+/Na^+ selectivity of a cation channel in the plasma membrane of root cells does not differ in salt tolerant and salt sensitive wheat species. Plant Physiol. 97, 598-605, 1991.

Schachtschabel, P. (G) Investigations into the sorption of clay minerals and organic soil colloids and the determination of the proportion of these colloids on the total sorption of soils. Kolloid Beiheft 51, 199-276, 1940.

Schachtschabel, P. (G) The available magnesium in soils and its determination. Z. Pflanzenernähr. Düng. Bodenk. 67, 9-23, 1954.

Schachtschabel, P. (G) Fixation and release of potassium and ammonium ions. Assessment and determination of the potassium availability in soils. Landwirtsch. Forsch. 15, 29-47, 1961.

Schäfer, P. and Siebold, M. (G) Influence of increasing potash application rates on yield and quality of the spring wheat 'Kolibri'. Results from a potash fixing location. Bayer. Landw. Jahrb. 49, 19-39, 1972.

Schäfer, W., Klank, I. and Kretschmer, H. (G) Dayly rhythm of net photosynthesis of winter wheat and sugar beet under field conditions. Arch. Acker-Pflanzenbau Bodenkd. 28, 441-450, 1984.

Schäffer, H. J., Forsthöfel, N. R. and Cushman, J. C. Identification of enhancer and silencer regions involved in salt-responsive expression of Crassulacean acid metabolism (CAM) genes in the facultative halophyte Mesembryanthemum crystallinum. Plant Mol. Biol. 28, 205-218, 1995.

Schapendonk, A. H. C. M., Dijkstra, P., Groenwold, J., Pot, C. S. and Van de Gejn, S. C. Carbon balance and water use efficiency of frequently cut Lolium perenne L. swards at elevated carbon dioxide. Global Change Biology 3, 207-216, 1997.

Schapendonk, A. H. C. M., Stol, W., Van Kraalingen, D. W. G. and Bouman, B. A. M. LINGRA, a sink/source model to simulate grassland productivity in Europe. European J. Agron. 9, 87-100, 1998.

Scharpf, H. C. and Wehrmann. J. (G) Importance of mineral nitrogen quantity in the soil profile at the beginning of the growth period for the N-application rate for winter wheat. Landw. Forsch. 3211, Sonderheft, 100-114, 1975.

Scharrer, K. and Jung, J. (G) The influence of the nutrition on the ratio of cations in plants. Z. Pflanzenernähr. Düng. Bodenk. 71, 76-94, 1955.

Scharrer, K. and Mengel, K. (G) On the transient occurrence of visible magnesium deficiency in oats. Agrochimica 4, 3-24, 1960.

Scharrer, K. and Schaumlöffel, E. (G) The uptake of copper by spring cereals grown on copper deficient soils. Z, Pflanzenernähr. Düng. Bodenk. 89, 1-17, 1960.

Scheffer, F. and Schachtschabel, P. (G) Textbook of Soil Science, 11th edition, F. Enke-Verlag, Stuttgart, 1982.

Scheffer, F. and Welte, E.: (G) Plant Nutrition, 3rd ed., P. -Enke-Verlag, Stuttgart, p. 163, 1955.

Scherer, H. W. Dynamics and availability of the non-exchangeable NH$_4$-N - a review. Eur. J. Agron. 2, 149-160, 1993.

Scherer, H. W. and Ahrens, G. R. Depletion of non-exchangeable NH$_4$-N in the soil root interface in relation to clay mineral composition and plant species. Eur. J. Agron. 5, 1-7, 1996.

Scherer, H. W. and Danzeisen, L. (G) Effect of increasing nitrogen rates on the development of root nodules, the symbiotic nitrogen assimilation and growth and yield of faba beans (*Vicia faba* L.). Z. Pflanzenernähr. Bodenk. 143, 464-470, 1980.

Scherer, H. W. and Lange, A. N fixation and growth of legumes as affected by sulphur fertilization. Biol. Fertil. Soils 23, 449-453, 1996.

Scherer, H. W., MacKown, C. T. and Leggett, J. E. Potassium-ammonium uptake interactions in tobaccco seedlings. J. Exptl. Bot. 35, 1060-1070, 1984.

Scherer, H. W. and Mengel, K. (G) Effect of soil moisture on the release of non exchangeable NH$_4$, and its uptake by plants. Mitt. Deutsch. Bodenk. Ges. 32, 429-438, 1981.

Scherer, H. W. and Mengel, K. (G) Turnover of ^{15}N labelled nitrate nitrogen in soil as related to straw application and soil moisture. Z. Pflanzenernähr. Bodenk. 146, 109-117, 1983.

Scherer, H. W., Schubert, S. and Mengel, K. (G) The effect of potassium nutrition on growth rate, carbohydrate content, and water retention in young wheat plants. Z. Pflanzenernähr. Bodenk. 145, 237-245, 1982.

Scherer, H. W. and Weimar, S. (G) Importance of the potassium levels in soils and the proportion of expansible clay minerals of total clay for the dynamics of the nonexchangeable NH$_4$-N after slurry application. Agribiol. Res. 47, 123-139, 1994.

Scherer, H. W. and Werner, W. Significance of soil microorganisms for the mobilization of nonexchangeable ammonium. Biol. Fertil. Soils 22, 248-251, 1996.

Schiff, J. A. Reduction and other metabolic reactions of sulfate. In: Inorganic Plant Nutrition, Encycl. Plant Physiol. New Series Vol 15A, edited by Läuchli, A. and Bieleski, R. L., Springer Verlag Berlin, Heidelberg, New York, Tokyo, p. 401-421, 1983.

Schiff, J. A., Stern, A. I., Saidha, T. and Li, J. Some molecular aspects of sulfate metabolism in photosynthetic organisms. In: Sulfur Nutrition and Assimilation in Higher Plants, edited by DeKok, L. J., Stulen, I., Rennenberg, H., Brunold, C. and Rauser, W. E. the Hague: SPB Academic Publishing bv., p. 21-35, 1993.

Schildbach, R. (G) Relationships between fertilizer application to brewing barley and the beer quality. Z. Acker- u. Pflanzenbau 136, 219-237, 1972.

Schimanski, C. (G) Investigations into the translocation of magnesium (Mg-28) in sun flowers. Z. Pflanzenernähr. Bodenk. 136, 68-81, 1973.

Schimanski, C. (G) The influence of certain experirnental pararnenters on the flux characteristics of Mg-28 in the case of barley seedlings grown in hydroculture. Landw. Forsch. 34,154-165, 1981.

Schirmer, T., Keller, T. A., Wang, Y. F. and Rosenbusch, J. P. Structural basis for sugar translocation through maltoporin channels at 3. 1 A resolution. Science. 267, 512-514, 1995.

Schlatter, C. (G) Toxicological evaluation of extragenous compounds in food. In: Wie sicher sind unsere Lebensmittel. Bund. f. Lebensmittelrecht und Lebensmittelkunde e. V. (Ed.). B. Behr's Verlag, Hamburg, p. 169, 1983.

Schlechte, G. (G) Nutrient uptake of plants and mycorrhiza. 1 Ectotrophic rnycorrhiza. Kali Briefe, (Büntehof) Fachgeb. 2, 6. Folge, 1976.

Schmalfuss, K.: (G) Plant Nutrition and Soil Science. 9th ed. S. -Hirzel-Verlag, Stuttgart, p. 160, 1963.

Schmalfuß, K. and Kolbe, G. (G) The fertilizer farmyard manure. Albr.-Thaer-Archiv 7, 199-213, 1963.

Schmeer, H. and Mengel, K. (G) The influence of straw incorporation on the nitrate concentration in the soil throughout the winter period. Landwirtsch. Forsch. 37. Congr. Vol. 214-229, 1984.

Schmid, R. (G) Nitrogen fixing plants. Naturw. Rdsch. 21, 384-386, 1968.

Schmid, W. E., Haag, H. P. and Epstein, E. Absorption of zinc by excised barley roots. Physiol. Plant. 18, 860-869, 1965.

Schmidt, A.: Photosynthetic assirnilation of sulfur compounds. In: Photosynthesis II, Encycl. Plant Physiol. New Series Vol. 6, edited by Gibbs, M. and Latzko, E., Springer Verlag Berlin, Heidelberg, New York, p. 481-486, 1979.

Schmitt, L. (G) Benefit of a Correct Fertilizer Application, Frankfurt: DLG Verlag, 1958.

Schmitt, L. and Brauer, A. (G) Compound fertilizers and straight fertilizers in trials lasting 10 years. Landw. Forsch. 22, 244-261, 1969.

Schmitt, L. and Brauer, A. (G) High nitrogen rates to sugar beets and malting barley quality. Landwirtsch. Forsch. 26, 326-331, 1975.

Schmitt, L. and Brauer, A. (G) Seventy-five years Fertilizer Application Experiments on Meadows of the Agricultural Experimental Station of Darmstadt - Results of the Oldest Exact Experiments of the European Continent. J. D. Sauerländer's Verlag Frankfurt/Main 1979.

Schnabl, H. Anion metabolism as correlated with volume changes in guard cell protoplasts. Z. Naturforsch. Sect. C Biosci. 35, 621-626, 1980.

Schneider, U. and Haider, K. Denitrification- and nitrate leaching losses in intensively cropped water shed. Z. Pflanzenernähr. Bodenk. 155:135 -141, 1992.

Schneiders, M. (G) Investigations into the specifically bound ammonium in cultivation of flooded rice with particular consideration of the redox potential in the rhizosphere. Ph. D. Thesis, University Bonn, 1998.

Schneiders, M. and Scherer, H. W. Fixation and release of ammonium in flooded rice soils as affected by redox potential. Eur. J. Agron. 8, 181-189, 1998.

Schnier, H. F., De Datta, S. K. and Mengel, K. Dynamics of ^{15}N-labeled ammoniumsulfate in various inorganic and organic fractions of wetland rice soils. Biol. Fertil. Soils 4, 171-177, 1987.

Schnier, H. F., DeDatta, S. K., Mengel, K., Marqueses, E. P. and Faronilo, J. E. Nitrogen use efficiency, floodwater properties, and nitrogen-15 balance in transplanted lowlwand rice as affected by liquid urea band placememt. Fert. Res. 16, 241-255, 1988.

Schnier, H. F., Dingkuhn, M., De Datta, S. K., Mengel, K., Wijangco, E. and Javellana, C. Nitrogen economy and canopy carbon dioxide assimilation of tropical lowland rice. Agron. J. 82, 451-459, 1990a.

Schnier, H. F., Dingkuhn, M., DeDatta, S. K., Marqueses, E. P. and Faronilo, J. E. Nitrogen-15 balance in transplanted and direct-seeded rice as affected by different methods of urea application. Biol. Fertil. Soils 10, 89-96, 1990b.

Schnitzer, M. and Skinner, S. I. M. Organo-metallic interactions in soils. IV. Carboxyl and hydroxyl groups in organic matter and metal retention. Soil Sci. 99, 278-284, 1965.

Schnitzer, M. and Skinner, S. I. M. Organo-metallic interactions in soils: Stability constants of Pb-2, Ni-2, Mn-2, Co-2, Ca-2, and Md-2 fulvic acid complex. Soil Sci. 103, 247-252, 1967.

Schnug, E. (G) Quantitative und qualitative Aspects of diagnosis and therapy of rape (*Brassica napus* L.) related to glucosinolate-low cultivars. Habilitationsschrift Thesis, University Kiel 1989.

Schnug, E. Sulphur nutritional status of European crops and consequences for agriculture. Sulphur in Agric. 15, 7-12, 1991.

Schnug, E. Physiological functions and environmental relevance of sulfur-containing secondary metabolites. In: Sulfur Nutrition and Assimilation of Higher Plants, edited by De Kok, L. J., Stulen, I., Rennenberg, H., Brunhold, C. and Rauser, W. E. The Hague: SPB Academic Publishing bv, p. 179-190, 1993.

Schnyder, H., Nelson, C. J. and Spollen, W. G. Diurnal growth of tall fescue leaf blades. II. Dry matter partitioning and carbohydrate metabolism in the elongation zone and adjacent expanding tissue. Plant Physiol. 86, 1077-1083, 1988.

Schoen, R. and Rye, R. O. Sulfur isotope distribution in solfataras, Yellowstone National Park, Science 170, 1082-1084, 1971.

Schofield, R. K. Can a precise meaning be given to 'available'soil phosphorus? Soils and Fertilizers 28, 373-375, 1955.

Schön, H. G., Mengel, K. and DeDatta, S. K. The importance of initial exchangeable ammonium in the nitrogen nutrition of lowland rice soils. Plant and Soil 86, 403-413, 1985.

Schön, M., Niederbudde, E. A. and Markorn, A. (G) Results of a 20 years lasting experirnent with mineral fertilization and farm yard manure application in the loess area near Landsberg (Lech). Z. Acker- u. Pflanzenbau 143, 27-37, 1976.

Schönwiese, C. D. (G) The impact of man on the climate. Naturw. Rundschau 41, 387-390, 1988.

Schon, M. K., Novacky, A. and Blevins, D. J. Boron induced hyperpolarization of sunflower root cell membranes and increases in membrane permeability to K^+. Plant Physiol. 93, 556-573, 1990.

Schouwenburg, J. Ch. and Schuffelen, A. C. Potassium exchange behaviour of an illite. Neth. J. Agric. Sci 11, 13-22, 1963.

Schrader, L. E., Ritenour, G. L., Eilricri, G. L. and Hageman, R. H. Some characteristics of nitrate reductase from higher plants. Plant Physiol. 43, 930-940, 1968.

Schreiber L. Chemical composition of Casparian strips isolated from *Clivia miniata* reg. roots: evidence for lignin. Planta 199, 596-601, 1996.

Schroeder, D. (G) Potassium fixation and potassium release of loess soils. Landw. Forsch. 8, 1-7, 1955

Schroeder, D. Structure and weathering of potassium containing minerals. In: Potassium Research-Review and Trends. Proc. 11th Congr. Int. Potash Inst., Bern, p. 43-63, 1978

Schroeder, D. Soils - Facts and Concepts, Bern/Switzerland:Intrenat. Potash Institue. 4th p ed. 32-33, 1984.

Schroeder, J. I. and Hagiwara, S. Cytosolic calcium regulates ion channels in the plasma membrane of *Vicia faba* guard cells. Nature 338, 427-430, 1989.

Schroeder, J. I. and Hagiwara, S. Repetitive increases in cytosolic Ca2+ of guard cells by abscisic acid activation of nonselective Ca2+ permeable channels. Proc. Natl. Acad. Sci. USA 87, 9305-9309, 1990.

Schroeder, J. I., Ward, J. M. and Gassmann, W. Perspectives on the physiology and structure of inward-rectifying K^+ channels in higher plants: Biophysical implications for K^+ uptake. Annu. Rev. Biophys. Biomol. Struct. 23, 441-471, 1994.

Schubert, E., Mengel, K. and Schubert, S. Soil pH and calcium effect on nitrogen fixation and growth of broad bean. Agron. J. 82, 969-972, 1990.

Schubert, K. R., Jennings, N. T. and Evans, H. J. Hydrogen reactions of nodulated leguminous plants. Plant Physiol. 61, 398-401, 1978.

Schubert, S. and Läuchli, A. Na^+ exclusion, H^+ release, and growth of two different maize cultivars under NaCl salinity. J. Plant Physiol. 126, 145-154, 1986.

Schubert, S. and Läuchli, A. Metabolic dependance of Na^+ efflux from roots of intact maize seedlings. J. Plant. Physiol. 193, 193-198, 1988.

Schubert, S. and Matzke, H. Influence of phytohormones and other effectors on proton extrusion by isolated protoplasts from rape leaves. Physiol. Plant. 64, 285-289, 1985.

Schubert, S., Serray, R., Plies-Balzer, E. and Mengel. K., Effect of draught stress on growth, sugar concentrations and amino acid accumulation in N_2 fixing Alfalfa (*Medicago sativa*). J. Plant. Physiol. 146, 541-546, 1995.

784

Schuffelen, A. C. (G) Nutrient content and nutrient release in soils. Vortragstagung Chemie und Landwirtschaftliche Produktion. 100 Jahre Landwirtschaftlich-chemische Bundesversuchsanstalt Wien, 27-42, 1971.

Schüller, H. (G) The CAL-method, a new method for the determination of available phosphate in soil. Z. Pflanzenernähr. Bodenk. 123, 48-63, 1969.

Schulte-Altedorneburg, M. (G) the quatitaive determination of the phytohormone abscisic acid and ist synthesis in beets (*Beta vulgaris* L. ssp altissima) under drought stress. Ph. D. Thesis, University Köln, 1990.

Schulten, H. R. and Leinweber, P. Dithionite-citrate-bicarbonate-extractable organic matter in particle size fractions of a Haplaquoll. Soil Sci. Soc. Am. J. 59, 1019-1027, 1995.

Schulten, H. R. and Leinweber, P. New insights into organic-mineral particles: composition, properties and models of molecular structure. Biol. Fert. Soils 30, 399-432, 2000.

Schulten, H. R., Monreal, C. M. and Schnitzer, M. Effect of long term cultivation on the chemical structure of soil organic matter. Naturwiss. 82, 42-44, 1995.

Schulten, H. R. and Schnitzer, M. A state of the art structural concept for humic substances. Naturw. 80, 4-30, 1993.

Schulten, H. R. and Schnitzer, M. Three-dimensional models for humic acids and soil organic matter. Naturwiss. 82, 478-498, 1995.

Schulten, H. R. and Schnitzer, M. The chemistry of soil organic nitrogen: a revieW. B. iol. Fert. Soils 26, 1-15, 1998.

Schulz, A. Phloem transport and differential unloading in pea seedlings after source and sink manipulations. Planta 192, 239-248, 1994.

Schulze, E.-D. and Bloom, A. J. Relationship between mineral nitrogen influx and transpiration in radish and tomato. Plant Physiol. 76, 827-828, 1984.

Schumacher, R. and Frankenhauser, F. (G) Fight against bitter pit. Schweiz. Z. f. Obst- u. Weinbau 104, No 16424, 1968.

Schumann, A. W. and Sumner, M. E. Plant nutrient availability from mixtures of fly ashes and biosolids. J. Environ. Qual. 28, 1651-1657, 1999.

Schürmann, P. Plant thioredoxins. In: Sulfur Nutrition and Assimilation in Higher Plants, edited by De Kok, L. J., Stulen, I., Rennenberg, H. Brunold, C. and Rauser, W. E. SPB. Academic Publishing bv, The Hague, p. 153-162, 1993.

Schurr, U. Dynamics of nutrient transport from the root to the shoot. Progress in Botany 60, 234-253, 1999.

Schurr, U., Gollan, T. and Schulze, E. D. Stomatal response to drying soil in relation to the changes in the xylem sap composition of *Helianthus annuus*. II. Stomatal sensitivity to abscisic acid imported from the xylem sap. Plant Cell Environ. 15, 561-567, 1992.

Schurr, U. and Schulze, E.-D. The concentration of xylem sap constituents in root exudate, and in sap from intact, transpiring castor bean plants (*Ricinus communis* L.). Plant, Cell Environ. 18, 409-420, 1995.

Schwarz, K. A bound form of silicon in glycosaminoglycans and polyuronides. Proc. Nat. Acad. Sci. 70, 1608-1612, 1973.

Schwertmann, U. (G) The selective cation adsorption of the clay fraction of some soils developed from sediments. Z. Pflanzenernähr. Düng. Bodenk. 97, 9-25, 1962.

Schwertmann, U. Solubility and dissolution of iron oxides. In: Iron Nutrition and Interactions in Plants, edited by Chen, Y. and Hadar, Y Dordrecht: Kluwer Academic Publishers ISBN 0-7923-1095-0, p. 3-27, 1991.

Schwertmann, U., Süsser, P. and Nätscher, L. (G) Proton buffer systems in soils. Z. Pflanzenernähr. Bodenk. 150, 174-178, 1987.

Scott, N. M. Sulphur in soils and plants. In: Organic Matter and Biological Activity, edited by Vaugham, D. and Malcolm, R. E. Martinus Nijhoff, p. 379-401, 1985.

Secer, M. (G) Effect of potassium on nitrogen metabolization and grain protein formation in spring wheat. Kali-Briefe (Büntehof) 14 (6), 393-402, 1978.

Secer, M. and Unal, A. (G) Nutrient concentrations in leaf lamina and petioles of sugar melons and their relationship to yield and quality. Gartenbauwiss. 55, 37-41, 1990.

Sembdner, G. and Parthier, B. The biochemistry and the physiological and molecular actions of jasmonates. Annu. Rev. Plant Physiol. Plant Mol. Biol. 44, 569-589, 1993.

Senn, A. P. S. and Goldsmith, M. H. M. Regulation of electrogenic proton pumping by auxin and fusicoccin as related to the growth of *Avena* coleoptiles. Plant Physiol. 88, 131-138, 1988.

Sentenac, H. and Grignon, C. Effect of pH on orthophosphate uptake by corn roots. Plant Physiol. 77, 136-141, 1985.

Sentenac, H., Mousin, D. and Salsac, L. (F) Measurement of phosphatase activity in cellulose cell wall material obtained with the help of a non ionic detergent. C. R. Acad. Sci. Paris, 290, Serie D-21, 1980.

Serrano, R. Structure and function of plasmamembrane ATPase. Annu. Rev. Plant Physiol. Plant Mol. Biol. 40, 61-94, 1989.

Serrano, R., Mulet, J. M., Rios, G., Marquez, J. A., de Larrinoa, I. F., Leube, M. P., Mendizabal, I., Pascual-Ahuir, A., Proft, M., Ros, R. and Montesinos, C. A glimpse of the mechanisms of ion homeostasis during salt stress. J. Exp. Bot. 50, 1023-1036, 1999.

Serry, A., Mawardi, A., Awad, S. and Aziz, I. A. Effect of zinc and manganese on wheat production. 1. FAO/SIDA Seminar for Plant Scientists from Africa and Near East, FAO Rome, p. 404-409, 1974.

Servaites, J. C., Schrader, L. E. and Jung, D. M. Energy dependent loading of amino acids and sucrose into the phloem of soybean. Plant Physiol. 64, 546-550, 1979.

Sevilla, F., Lopez-George, J., Gomez, M. and Del Rio, L. A. Manganese superoxide dismutase (EC 1, 15, 1, 1) from a higher plant purification of a new manganese enzyme. Planta 150, 153-157, 1980.

Sewell, P. L. and Ozanne, P. G. The effect of modifying root profiles and fertilizer solubility on nutrient uptake. In: Proc. Australian Plant Nutr. Conf. Mt. Gambier, edited by Miller, T. C., CSIRO Australia, p. 6-9, 1970.

Shainberg, I., Sumner, M. E., Miller, W. P., Farina, M. P. W., Pavan, M. A. and Fey, M. V. Use of gypsum on soils: A review. Adv. Soil Sci. 9, 1-111, 1989.

Sharma, G. C. Controlled-release fertilizers and horticultural applications. Scientia Hortic. 11, 107-129, 1979.

Sharples, R. O. The structure and composition of apples in relation to storage quality. Rep. E. Malling Res. Stn for 1967, 185-189, 1968.

Sharpley, A. N. Reaction of fertilizer potassium in soils of different mineralogy. Soil Sci. 149, 44-51, 1990.

Sharpley, A. N. Assessing phosphorus bioavailibility in. agricultural soils and runoff. Fert. Res. 36, 259-272, 1993.

Sharpley, A. N. and Smith, S. J. Phosphorus transport in agricultural runoff: The role of soil erosion. In: Soil Erosion and Agricultural Land, edited by Boardman, J, Foster, L. D. D. and Dearing, J. A., John Wiley, New York, p. 351-366, 1990.

Shear, C. B. Calcium-related disorders of fruits and vegetables. Hort. Sci. 10, 361-365, 1975.

Sheriff, D. W. Epidermal transpiration and stomatal response to humidity: some hypotheses explored. Plant Cell. Ebnviron. 7, 669-677, 1984.

Shewry, P. R. Barley seed storage proteins - structure, synthesis, and deposition. In: Nitrogen Metabolism of Plants, edited by Mengel, K. and Pilbeam, D. Clarendon Press, Oxford p. 201-227, 1992.

Shewry, P. R., Napier, J. A. and Tatham, A. S. Seed storage proteins: structures and biosynthesis. The

Plant Cell 7, 945-956, 1995.

Shimazaki, K., Iino, M. and Zeiger, E. Blue light-dependent proton extrusion by guard cell protoplasts of *Vicia faba*. Nature 319, 324-326, 1986.

Shimshi, D. Interaction between irrigation and plant nutrition. In: Transition from Extensive to Intensive Agriculture with Fertilizers. Proc. 7th Colloq. Int. Potash Inst., Bern, p. 111-120 1969.

Shingles, R. and McCarty, R. E. Direct measurement of ATP-dependent proton concentration changes and characterization of a K^+-stimulated ATPase in pea chloroplast inner envelope vesicles. Plant Physiol. 106, 731-737, 1994.

Shingles, R., Roh, M. H. and McCarty, R. E. Nitrite transport in chloroplast inner envelop vesicles. I. Direct measuremnet of proton-linked transport. Plant Physiol. 112, 1375-1381, 1996.

Shorrocks, V. Micronutrient News, Hertforshire UK:Micronutrient Bureau, 1994.

Shorrocks, V. M. The occurrence and correction of boron deficiency. Plant and Soil 193, 121-148, 1997.

Shrift, A. Aspects of selenium metabolism in higher plants. Annu. Rev. Plant Physiol. 20, 475-494, 1969.

Shuman, L. M. Chemical forms of micronutrients in soils. In: Micronutrients in Agriculture 2nd ed. edited by J. J. Mordredt, F. R. Cox, L. M. Shuman and R. M. Welch, SSSA Book Series, Madison USA, p. 113-144, 1991.

Sibbesen, E. An investigation of the anion exchange resin method for soil phosphate extraction. Plant and Soil 50, 305-321, 1978.

Siddiqi, M. Y., Glass, A. D. M., Ruth, T. J. and Fernando, M. Studies of the regulation of the nitrate influx by barley seedlings using $^{13}NO_3$. Plant Physiol. 90, 806-813, 1989.

Sideres, C. P. and Young, H. J. Growth and chemical composition of *Ananas comosus* in solution cultures with different iron-manganese ratios. Plant Physiol. 24, 416-440, 1949.

Siegel, N. and Haug, A. Calmodulin-dependent formation of membrane potential in barley root plasma membrane vesicles: A biochemical model of aluminium toxicity in plants. Physiol. Plant. 59, 285-291, 1983.

Siegl, V. (G) Grüner Veltliner, a successful Austrian wine. Alles über Wein. 16, 98-101, 1998.

Siemes, J. (G) Maintenance of human nutrition by mineral fertilizer application. Pflug u. Spaten No. 6, p. 2, 1979.

Sigavuru, M., Baluska, F., Volkmann, D., Felle, H. H. and Horst W. J. Impacts of aluminium on the cytoskleton of the maize root apex. Short term effects on the distal part of transition zone. Plant Physiol. 119, 1073-1082, 1999.

Sigg, L., Stumm, W., Zobrist, J. and Zürcher, F. The chemistry of fog: factors regulating ist composition. Chimica 41, 159-164, 1987.

Silberbusch, M. and Barber, S. A. Sensitivity analysis of parameters used in simulating potassium uptake with a mechanistic-mathematical model. Agron. J. 75, 851-854, 1983a.

Silberbusch, M. and Barber, S. A. Sensitivity of simulated phosphorus uptake to parameters used by a mechanistic-mathematical model. Plant and Soil 74, 93-100, 1983b.

Sillanpaä, M. Micronutruiemts and the nutrient status of soils. A global study. FAO Soils Bul. No. 48, 1982.

Silvius, J. E., Ingle, M. and Baer, C. H. Sulphur dioxide inhibition of photosynthesis in isolated spinach chloroplasts. Plant Physiol. 56, 434-437, 1975.

Simard, R. R., Tran, T. S. and Zizka, J. Evaluation of the electro -ultrafiltration technique (EUF) as a measure of the K supplying power of Quebec soils. Plant and Soil 132, 91-101, 1991.

Simpson, R. J., Lambers, H. and Dalling, M. J. Translocation of nitrogen in a vegetative wheat plant (*Triticum aestivum*). Physiol. Plant. 56, 11-17, 1982.

Sims, J. R. and Bingham, F. T. Retention of boron by layer silicates, sesquioxides and soil materials: II. Sesquioxides. Soil Sci. Soc. Amer. Proc. 32, 364-369, 1968.

Sims, J. T. Molybdenum and cobalt. In: Methods of Soil Analysis, edited by Bigham, J. M., Soil Sci. Soc.

America, p. 723-737, Madison, USA 1996.

Sims, J. T. and Ellis, B. G. Adsorption and availability of phosphorus following the application of limestone to an acid, aluminous soil. Soil Sci. Soc. Am. J. 47, 888-893, 1983.

Sindhu, R. K. and Walton, D. C. Conversion of xanthoxin to abscisic acid by cell-free preparations from bean leaves. Plant Physiol. 85, 916-921, 1987.

Sinden, S. and Deahl, K. L. Effect of glycoalcaloids and phenolics on potato flavor. J. Food Sci. 41, 520-523, 1976.

Singh, B. and Brar, S. P. S. Dynamics of native and applied potassium in maize-wheat rotation. Potash Review, Subj. 9, 35th suite, No 6, 1977.

Singh, R., Goyal, R. K., Bhullar, S. and Goyal, R. Factors, including enzymes, controlling the import and transformation of sucrose to starch in the developing sorghum caryopsis. Plant Physiol. Biochem. 29, 177-183, 1991.

Sionit, N., Teare, I. D. and Kramer, P. J. Effects of repeated application of water stress on water status and growth of wheat. Physiol. Plant. 50, 11-15, 1980.

Sivaguru, M., Baluska, S., Volkmann, D., Felle, H. H. and Horst, W. J. Impacts of aluminium on the cytoskeleton of the maize root apex. Short-term effect on the distal patt of the transition zone. Plant Physiol. 119, 1073 -1982, 1999.

Sivasubramanian, S. and Talibudeen, O. Effects of aluminium on the growth of tea (*Camellia sinensis*) and its uptake of potassium and phosphorus. Tea 43, 4-13, 1972.

Skelton, B. J. and Shear, G. M. Calcium translocation in the peanut (*Arachis hypogea* L.). Agron. J. 63, 409-412, 1971.

Skole, D. and Tucker, C. Tropical deforestation and habitat fragmentation in the Amazon: Satellite data from 1978 to 1988. Science. 260, 1905-1910, 1993.

Slack, A. V. Chemistry and Technology of Fertilizers. John Wiley and Sons, New York, London, Sidney 1967.

Slack, C. R. and Hatch, M. D. Comparative studies on the activity of carboxylases and other enzymes in relation to the new pathway of photosynthetic carbon dioxide fixation in tropical grasses. Biochem. J. 103, 660-665, 1967.

Slangen, J. H. G. and Kerkhoff, P. Nitrification inhibitors in agriculture and horticulture: A literature review. Fertilizer Research 5, 1-76, 1984.

Slatyer, R. O. The significance of the permanent wilting percentage in studies of plant and soil water relations. Bot. Rev. 23, 585-636, 1957.

Slatyer, R. O. Climatic control of plant water relations. In: Environmental Control of Plant Growth, edited by Evans, L. T., New York, Academic, p. 34-54, 1963.

Slatyer, R. O. Plant-Water Relationships. Academic Press, London, New York 1967.

Slatyer, R. O. and Bierhausen, J. F. The influence of several transpiration suppressants on transpiration, photosynthesis, and water-use efficiency of cotton leaves. Aust. J. Biol. Sci. 17, 131-146, 1964.

Sloan, J. J., Dowdy, R. H., Dolan, M. S. and Linden, D. R. Long-term effects of biosolids application in agricultural soils. J. Environ. Qual. 26, 966-974, 1997.

Slooten, L., Capiau, K., Van Camp, W., Van Montagu, M., Sybesma, C. and Inze, D. Factors affecting the enhancement of oxidative stress tolerance in transgenic tobacco overexpressing manganese superoxide dismutase in the chloroplasts. Plant Physiol. 107, 737-750, 1995.

Sluijsmans, C. M. J. and Kolenbrander, G. J. The significance of animal manure as a source of nitrogen in soils. In: Proc. Intern. Seminar on Soil Environment and Fertility Management in Intensive Agriculture. Tokyo, p. 403-411, 1977.

Smiley, R. W. Rhizosphere pH as influenced by plants,soils and nitrogen fertilizers. Soil Sci. Soc. Am. J. 38, 795-799, 1974.

Smith, A. M., Denyer, K. and Martin, C. The synthesis of the starch granule. Annu. Rev. Plant Physiol. Plant Mol. Biol. 48, 67-87, 1997.

Smith, F. A. and Raven, J. A. Intercellular pH and its regulation. Annu. Rev. Plant Physiol. 30, 289-311, 1979.

Smith, J. A. C. Ion transport and the transpiration stream. Bot. Acta 104, 416-421, 1991.

Smith, J. A. C. and Milburn, J. A. Phloem turgor and the regulation of sucrose loading in *Ricinus communis* L. Planta 148, 42-48, 1980.

Smith, J. U., Bradbury, N. J. and Addiscott, T. M. SUNDIAL: A PC based system for simulating nitrogen dynamics in arable land. Agron. J. 88, 38-43, 1996.

Smith, P. F. Mineral analysis of plant tissues. Ann. Rev. Plant Physiol. 13, 81-108, 1962

Smith, S. E. and Read, D. J. Mycorrhizal Symbiosis, London San Diego:Academic Press, 1997.

Smith, S. J., Power, J. F. and Kemper, W. D. Fixed ammonium and nitrogen availability indexes. Soil Sci. 158, 132-140, 1994.

Smith, S. J., Schepers, J. S. and Porter, L. K. Assessing and managing agricultural nitrogen losses to the environment. Adv. Soil Sci. 14, 1-43, 1990.

Smith, T. A. and Sinclair, C. The effect of acid feeding on amine formation in barley. Ann. Bot. 31, 103-111, 1967.

Smyth, T. J. and Chevalier, P. Increases in phosphatase and β-glucosidase activities in wheat seedlings in response to phosphorus- deficient growth. J. Plant Nutr. 7, 1221-1231, 1984.

Snedden, W. A., Fromm, Calmodulin, calmodulin-related proteins and plant responses to the environment. Trends in Plant Science 3, 299-304, 1998.

Snitwongse, P., Satrusajang, A. and Buresh, R. J. Fate of fertilizer applied to lowland rice on a Sulfic Tropaquet. Fert. Res. 16, 227-240, 1988.

Snowball, K., Robson, A. D. and Loneragan, J. F. The effect of copper on nitrogen fixation in subterranean clover (*Trifolium subterraneum*). New Phytol. 85, 63-72, 1980.

Solomonson, L. P. and Barber, M. J. Assimilatory nitrate reductase:functional properties and regulation. Annu. Rev. Plant Physiol. Plant Mol. Biol. 41, 225-253, 1990.

Solomos, T. Cyanide-resistant respiration in higher plants. Annu. Rev. Plant Physiol. 28, 279-297, 1977.

Somero, G. N., Osmond, C. B. and Bolis, C. L. Water and Life. Springer Verlag Berlin, Heidelberg, 1992.

Somers, J. J. and Shive, J. W. The iron-manganese relation in plant metabolism. Plant Physiol. 17, 582-602, 1942.

Sommer, G. (G) Pot experiments to establish the danger levels of cadmium copper, lead and zinc in relation is the application of refuse materials in agriculture. Landw. Forschung. Sonderheft 35, 350-364, 1979.

Sommer, S. G., Sherlok, R. R. and Khan, R. Z. Nitrous oxide and methane emissions from slurry amended soils. Soil. Biol. Biochem. 28, 1541-1544, 1996.

Sonntag, C. and Michael, G. (G) Influence of a late nitrogen application on the protein content and protein composition of grains obtained from conventional and lysine rich varieties of maize and barley. Z. Acker- u. Pflanzenbau 138, 116-128, 1973.

Soper, R. J. and Huang, P. M. The effect of nitrate nitrogen in the soil profile on the response of barley to fertilizer nitrogen. Can. J. Soil Sci. 43, 350-358, 1962.

Sovonick, S. A., Geiger, D. R. and Fellows, R. J. Evidence for active phloem loading in the minor veins of sugar beet. Plant Physiol. 54, 886-891, 1974.

Spanswick, R. M. and Williams, E. J. Electrical potentials and Na, K and Cl concentrations in the vacuole and cytoplasm of *Nitella translucens*. J. Exp. Bot. 15, 193-200, 1964.

Sparks, D. L. Potassium release from sandy soils. In: Nutrient Balances and the Need for Potassium, edited by International Potash Institute, Bern: International Potash Institute, p. 93-107, 1986.

Sparks, D. L. Potassium dynamics in soils. Adv. Soil Sci. 6, 1-63, 1987.

Sparks, D. L. Kinetics of Soil Chemical Processes, London:Academic Press, p. 166-167, 1988.

Sparks, D. L., Carski, T. H. and Fendorf, S. E. Kinetic methods and measurements. In: Methods of Soil Analysis, edited by Bigham, J. M. Madison: Soil Sci. Soc. America, p. 1275-1307, 1996.

Spiller, S. and Terry, N. Limiting factors in photosynthesis II. Iron stress dimishes photochemical capacity by reducing the number of photosynthetic units. Plant Physiol. 65, 121-125, 1980.

Sposito, G. The Chemistry of Soils. Oxford University Press, Oxford 1989.

Srivastava, A. and Zeiger, E. Guard cell zeaxanthin tracks photosynthetic active radiation and stomatal apertures in *Vicia faba* leaves. Plant Cell Environ. 18, 813-817, 1995.

Stadtman, T. C. Selenium biochemistry. Annu. Rev. Biochem. 59, 111-127, 1990.

Stadtman, T. C. Selenocysteine. Annu. Rev. Biochem. 65, 83-100, 1996.

Stahli, D., Perrissin-Fabert, A., Blouet, A. and Guckert, A. Contribution of the wheat (*Triticum aestivum*) flag leaf to grain yield in response to plant growth regulators. Plant Growth Reg. 16, 293-297, 1995.

Stählin, A. (G) Response of grassland and forage crops on 'Floranid'. Z. Acker- u. Pflanzenbau 126, 301-316, 1967.

Stanford, G. and Smith, S. J. Nitrogen mineralization potentials of soils. Soil Sci. Soc. Am. Proc. 36, 465-472, 1972.

Stanhill, G. Water use efficiency. Adv. Agron. 39, 53-85, 1986.

Stanier, R. Y., Ingrham, J. L., Wheelis, M. L. and Painter, P. R. General Microbiology, Mac Millan Press, London, 5th Ed., 1995.

Stapp, C. and Wetter, C. (G) Contributions to the quantitative microbiological determination of magnesium, zinc, iron, molybdenum and copper in soils. Landw. Forsch. 5, 167-180, 1953.

Staswick, P. E. Storage proteins of vegetative plant tissues. Annu. Rev. Plant Physiol. Plant Mol. Biol. 45, 303-322, 1994.

Steffens, D. (G) The effect of long-term application of different phospahate fertilizers on the phosphate availability in the rhizosphere of rape Z. Pflanzenernähr. Bodenk. 150, 75-80, 1987.

Steffens, D. Phosphorus release kinetics and extractable phosphorus after long-term fertilization. Soil Sci. Soc. Am. J. 58, 1702-1708, 1994.

Steffens, D., Barekzai, A. and Boguslawski, v. E. (G) Effect of sewage sludge of different preparation on the hydrolyzable and EUF extractable nitrogen in arable land and and grassland. VDLUFA-Schriftenr. 23. Congr. Vol. 405-423, 1987.

Steffens, D. and Mengel, K. (G) The uptake potential of *Lolium perenne* and *Trifolium pratense* for interlayer K^+ of clay minerals. Landw. Forsch. Sonderh. 36, 120-127, 1979.

Steffens, D. and Sparks, D. L. Kinetics of nonexchangeable ammonium release from soils. Soil Sci. Soc. Am. J. 61, 455-462, 1997.

Steffens, D. and Sparks, D. L. Effect of residence time on the kinetics of nonexchangeable ammonium release from illite and vermiculitE. J. Plant Nutr. Soil Sci. 162, 599-605, 1999.

Steingröver, E., Oosterhuis, R. and Wieringa, F. Effect of light treatment and nutrition on nitrate accumulation in spinach (*Spinacia oleraceae*) Z. Pflanzenphysiol. 107, 97-102, 1982.

Stelzer, R., Läuchli, A. and Kramer, D. (G) Intereellular pathways of chloride in roots of intact barley plants. Cytobiologie 10, 449-457, 1975.

Stephan, U. W. and Scholz, G. Nicotianamine: mediator of transport of iron and heavy metals in the phloem? Physiol. Plant. 88, 522-529, 1993.

Stephens, O. Changes in yields and fertilizer response with continuous cropping in Uganda. Experimental Agriculture 5, 263-269, 1969.

Steucek, C. G. and Koontz, H. V. Phloem mobility of magnesium. Plant Physiol. 46 50-52, 1970.

Steudle, E. Water flow in plants and its coupling to other processes: An overview. In: Methods in Enzymology edited by Fleischer, S. & B. Vol. 174. Academic Press, New York, pp. 183-225, 1989.

Steudle, E. Water transport across roots. Plant and Soil 167, 79-90, 1994.

Steudle, E. and Frensch, J. Osmotic responses of maize roots: Water and solute relations. Planta 177, 281-295, 1989.

Steudle, E. and Peterson, C. A. How does water get through roots? J. Exp. Bot. 49, 775-788, 1998.

Stevenson, F. J. and Ardakani, M. S. Organic matter reactions envoling micronutrients in soils. In: Micronutrients in Agriculture, edited by J. J. Mortvedt, P. M. Giordano and W. L. Lindsay, Soil Sci. Soc. America, Madison/Wisonsin, p. 79-114, 1972.

Stewart, W. D. P. Nitrogen-fixing plants. Science 158, 1426-1432, 1967.

Stitt, M. Fructose-2,6 bisphosphate as a regulatory molecule in plants. Annu. Rev. Plant Physiol. Plant Mol. Biol. 41, 153-158, 1990.

Stockdale, E. A., Gaunt, J. L. and Vos, J. Soil plant nitrogen dynamics -what concepts are required? In: Proceedings ESA Congr. Veldhoven 1996, edited by van Ittersum, M. K. and Van Geijn, S. C. Amsterdam: Elevier, p. 201 - 215, 1997.

Stockley, C. Conference report: Wolf Blass foundation; International wine and health conference: „mecically, is wine just another alcoholic beverage?" J. Wine Res, 8:55-59, 1997.

Stone, D. A. The effect of starter fertilizer injection on the growth and yield of drilled vegetable crops in relation to soil nutrient status J. Hort. Sci. and Biotechnology 73, 441-445, 1998.

Stoop, J. M. H., Williamson, J. D. and Pharr, D. M. Mannitol metabolism in plants: a method for coping with stress. Trends Plant Sci 1, 139-144, 1996.

Storey, H. H. and Leach, R. Sulphur deficiency disease in the tea bush. Ann. Appl. Biol. 20, 23-56, 1933.

Storey, R. and Wyn Jones, R. G. Salt stress and comparative physiology in the gramineae. III. Effect of salinity upon ion relations and glycine betaine and proline levels in Spartina x townsendii. Aust. J. Plant Physiol. 5, 831-838, 1978.

Stout, P. R., Meagher, W. R., Pearson, G. A. and Johnson, C. M. Molybdenum nutrition of crop plants. 1. The influence of phosphate and sulfate on the absorption of molybdenum from soils and solution cultures. Plant and Soil 3, 51-87, 1951.

Stoy, V. (G) Assimilate synthesis and distribution as components for the yield formation of cereals. Symp. German Ass. Appl. Botany, Hanover 1972.

Stoy, V. (G) Photosynthetic performance and the distribution of assimilates as yileld-limiting factors in cereal cultivation. Vorträg für Pflanzenzüchter, p. 34-51, 1973.

Strasser, O., Köhl, K. and Römheld, V. Overestimation of apoplastic Fe in roots of soil grown plants. Plant and Soil 210, 179-187, 1999.

Strebel, O., Böttcher, J. and Duynisveld, W. H. M. (G) Effect of site conditions and soil use on nitrate leaching and nitrate concentration in the groundwater. Landwirtsch. Forsch. Congr. Vol. 34-44, 1985.

Strebel, O., Duynisveld, W. H. M., Grimme, H., Renger, M. and Fleige, H. (G) Water uptake of root and nitrate supply (mass flow, diffusion) as a function of soil profile depth and time in a sugar beet stand. Mitteil. Deutsch. Bodenkundl. Ges. 38, 153-158, 1983.

Streeter, J. G. Nitrate inhibition of legume nodule growth and activity. II. Short term studies with high nitrate supply. Plant Physiol. 77, 325-328, 1985.

Streeter, J. G. and Salminen, S. O. Distribution of the two types of polysaccharide formed by *Bradyrhizobium japonicum* bacteroids in nodules of field-grown soybean plants (*Glycine max* L. Merr.). Soil Biol. Biochem. 25, 1027 -1032, 1993.

Sturm, H. (G) Good argicultural practice - aspects and concepts of the agrar chemistry. In: Zukunftsorientierter Pflanzenbau im Spannungsfeld zwischen Reglementierung und Eigenverantwortung, edited by BASF AG, Limburgerhof: Uhlenbecker, K., p. 61-91, 1992.

Sturm, H. and Isermann, K. (G) The long-term utilization of mineral fertilizer phosphate on agricultural soils. Landw. Forsch. Sonderh. 35, 180-192, 1978.

Stutte, C. A., Weiland, R. T. and Blem, A. R. Gaseous nitrogen loss from soybean foliage. Agron. J. 71, 95-97, 1979.

Subra Rao, N. S. Prospects of bacterial fertilization in India. Fertil. News 19, 32-36, 1974.

Sumner, M. E. Gypsum and acid soils: the world scene. Adv. Agron. 51, 1-32, 1993.

Susin, S., Abadia, A., Gonzalez-Reyes, J. A., Lucena, J. J. and Abadia, J. The pH requirement for *in vivo* activity of the iron-deficiency-induced "Turbo Ferric chelate reductase. Plant Physiol. 110, 111-123, 1996.

Sutcliffe, J. Plants and Water, 2nd edition (Arnold ed.) Studies in Biology 14, 1979.

Suzuki, A., Oaks, A., Jacquot, J. -P., Vidala, A. and Gadal, P. A natural electron donor from maize roots for glutamate synthase and nitrite reductase. Plant Physiol. Supplement to Vol 75, 850, 1984.

Suzuki, I., Sugiyama, T. and Omata, T. Regulation of nitrate reductase activity under CO_2 limitation in the cyanobacterium *Synechococcus* sp. PCC7942. Plant Physiol. 107, 791-796, 1995.

Syed-Omar, S. R. and Sumner, M. E. Effect of gypsum on soil potassium and magnesium status and growth of alfalfa. Commun. Soil Sci. Plant Anal. 22, 2017-2028, 1991.

Syers, J. K., Skinner, R. J. and Curtin, D. Soil and Fertilizer Sulphur in UK Agriculture. Proc. No 264, The Fertilizer Soc., London, pp. 43, 1987.

Syworotkin, G. S. (G) The boron content of plants with a latex systern. Spurenelemente in der Landwirtschaft. Akademie-Verlag Berlin, 283-288, 1958.

Szabolcs, I. Solonetz soils in Europe, their formation and properties with particular regard to utilization. In: European Solonetz Soils and their Reclamation, edited by Szabolcs, I., Akadémiai Kiadó Budapest, p. 9-33, 1971.

Szalay, A. and Szilagyi, M. Laboratory experiments on the retention of micronutrients by peat humic acids. Plant and Soil 29, 219-224, 1968.

Tadano, T. and Yoshida, S. Chemical changes in submerged soils and their effect on rice growth. In: Soils and Rice, edited by International Rice Research Institute, Los Bagnos, Phil., p. 399-420, 1978.

Tagliavini, M., Abadia, J., Rombola, A. D., Abadia, A., Tsipouridis, C. and Marangoni, B. Agronomic means for the control of iron chlorosis in deciduous fruit plants. J. Plant Nutrition 23, in print, 2000.

Tagliavini, M., Millard, P., Quartieri, M. and Marangoni, B. Timing of nitrogen uptake affects winter storage and spring remobilization of nitrogen in nectarine (*Prunus persica* var. nectarina trees) Plant and Soil, 211 149 - 153, 1999.

Tagliavini, M., Scudellari, D., Marangoni, B. and Toselli, M. Acid-spray regreening of kiwifruit leaves affected by lime-induced iron chlorosis. In: Iron Nutrition in Soils and Plants, edited by Abadia, J., Kluwer Academic Publishers, Dordrecht, p. 191-195, 1995.

Taiz. L. Plant cell expansion: regulation of cell wall mechanical properties. Annu. Rev. Plant Physiol. 35, 585-657, 1984.

Takagi, S. Naturally occurring iron-chelating compounds in oat- and rice-root washings. Soil Sci. Plant Nutr. 22, 423-433, 1976.

Takagi, S., Nomoto, K. and Takemoto, T. Physiological aspect of mugeneic acid, a possible phytosiderophore of graminaceous plants. J. Plant Nutr. 7, 469-477, 1984.

Takai, Y., Koyama, T. and Kamura, T. Microbial metabolism of paddy soils. J. Agr. Chem. Soc. 31, 211-220, 1957.

Takkar, P. N. and Singh, T. Zn nutrition of rice as influenced by rates of gypsum and Zn fertilization of alkali soils. Agron. J. 70, 447-450, 1978.

Talbott, L. D. and Zeiger, E. The role of sucrose in guard cell osmoregulation. J. Exp. Bot. 49, 329-337, 1998.

Talbott, L. D., Srivastava, A. and Zeiger, E. Stomata from growth-chamber-grown *Vicia faba* have an enhanced sensitivity to CO_2. Plant Cell Environm. 19, 1184-1194, 1996.

Talbott, L. D. and Zeiger, E. General role of potassium and sucrose in guard-cell osmoregulation. Plant Physiol. 111, 1051-1057, 1996.

Tanaka, A. The relative importance of the source and the sink as the yield-lirniting factors of rice. ASPAC, Technical Bulletin No. 6, p. 1-18, 1972.

Tanaka, A. Influence of special ecological conditions on growth, metabolism and potassium nutrition of tropical crops as exemplified by the case of rice. In: Potassium in Tropical Crops and Soils. l0th Colloq. Int. Potash Inst., Bern, p. 97-116, 1973.

Tanaka, A., Yamaguchi, J. and Kawaguchi, K. A note on the nutritional status of the rice plant in Italy, Portugal, and Spain. Soil Sci. Plant Nutr. 19, (3) 161-171, 1973.

Tanaka, A. and Yoshida, S. Nutritional disorders of the rice plant in Asia. Intern. Rice Res. Inst., Technical Bulletin 10, 1970.

Tanner, C. B. Transpiration efficiency of potato. Agron. J. 73, 59-64, 1981.

Tanner, P. D. and Grant, P. M. Response of maize (*Zea mays* L.) to lime and molybdenum on acid red and yellow-brown clays and clay loams. Rhod. J. agric. Res. 15, 143-150, 1977.

Tanner, W. and Beevers, H. Does transpiration have an essential function in long-distance ion transport in plants? Plant, Cell Environm. 13, 745-750, 1990.

Tanner, W. and Caspari, T. Membrane transport carriers. Annu. Rev. Plant Physiol. Plant Mol. Biol. 47, 595-626, 1996.

Tarafdar, J. C. and Claassen, N. Organic phosphorus compounds as a source for P for higher plants through phosphatases produced by plant roots and microorganisms. Biol. Fertil. Soils 5, 308-312, 1988.

Tarafdar, J. C. and Jungk, A. Phosphatase activity in the rhizosphere and its relation to the depletion of soil organic phosphorus. Biol. Fertil. Soils 3, 199-204, 1987.

Tashiro, T. and Wardlaw, I. F. The effect of high temperature on the accumulation of dry matter, carbon and nitrogen in the kernel of rice. Aust. J. Plant Physiol. 18, 259-265, 1991.

Taylor, C. B. Plant vegetative development: from seed and embryo to shoot and root. The Plant Cell 9, 981-988, 1997.

Taylor, H. M. and Klepper, B. Water uptake by cotton root systems: An examination of assumptions in the single root model. Soil Sci. 120, 57-67, 1975.

Taylor, H. M. and Klepper, B. The role of rooting characteristics in the supply of water to plants. Adv. Agron. 30, 99-128, 1978.

Taylor, R. M. and McKenzie, R. M. The association of trace elements with manganese minerals in Australian soils. Aust. J. Soil Res. 4, 29-39, 1966.

Teermaat, A. and Munns, R. Use of concentrated macronutrient solutions to seperate osmotic from NaCl-specific effects on plant growth. Austr. J. Plant Physiol. 13, 509-522, 1986.

Temperli, A., Künsch, U., Schärer, P. and Konrad, P. (G) Effect of two different cultivation systemes on the nitrate concentration in lettuce. Schweiz. Landw. Forsch. 21, 167 -195, 1982.

Ten Berge, H. F. M., Aggarwal, P. K. and Kropff, M. J. Application of Rice Modelling. Elsevier, Amsterdam, 1999.

Terman, G. J. Effect of rate and source of potash on yield and starch content of potatoes. Maine Agric. Expt. Sta. Bull. 581, 1-24, 1950.

Terman, G. L. Volatilization losses of nitrogen as ammonia from surface-applied fertilizers, organic amendernents, and crop residues. Adv. Agron. 31, 189-223, 1979.

Terman, G. L. and Allen, S. E. Leaching of soluble and slow-release N and K fertilizers from lakeland sand under glass and fallow. Soil and Crop Science Society of Florida, Proceed. 30, 130-140, 1970.

Terry, N. Photosynthesis growth and the role of chloride. Plant Physiol. 60, 69-75, 1977.

Terry, N. Limiting factors in photosynthesis 1. Use of iron stress to control phytochemical capacity in *vivo*. Plant Physiol 65, 114-120, 1980.

Teske, W. and Matzel, W. (G) Nitrogen leaching and nitrogen utilization by the plants as established in field lysirneters using ^{15}N urea. Arch. Acker- u. Pflanzenbau u. Bodenk. 20, 489-502, 1976.

Tetlow, I. J., Blissett, K. J. and Emes, M. J. Starch synthesis and carbohydrate oxidation in amyloplasts from developing wheat endosperm. Planta 194, 454-460, 1994.

Thauer, R. K. (G) Nickel enzymes in the metabolism of methanogeneous bacteria. Naturw. Rundschau. 39, 426-431, 1986.

Theil, E. C. Ferritin: structure, gene regulation, and cellular function in animals, plants and microorganisms. Annu. Rev. Biochem. 56, 289-315, 1987.

Thellier, M., Duval, Y. and Demarty, M. Borate exchanges of *Lemna minor* L. with the help of enriched stable isotopes and of (n,a) nuclear reaction. Plant Physiol. 63, 283-288, 1979.

Thicke, F. E., Russelle, M. P., Hesterman, O. P. and Sheaffer, C. C. Soil nitrogen mineralization indexes and corn response in crop rotations. Soil Sci. 156, 322-335, 1993.

Thies, W., Becker, K. W. and Meyer, B. (G) Balance of labelled fertilizer N ($^{15}NH_4$ and $^{15}NO_3$) in *in situ* sand lysimeters and time dependent fertilizer - and soil N leaching - comparison between soils with vegetation and fallow. Landw. Forsch. Congr. Vol. II, 55-62, 1977.

Thiman, K. V. Senescence in Plants, Boca Rato:CRC Press Inc., p. 85-115, 1980.

Thiyagarajah, M., Fry, S. C. and Yeo, A. R. *In vitro* salt tolerance of cell wall enzymes from halophytes and glycophytes. J. Exp. Bot. 47, 1717-1724, 1996.

Thomas, G. W., Haszler, G. R. and Blevins, R. L. The effects of organic matter and tillage on maximum compactabilty of soils using the Proctor test. Soil Sci. 161, 502-508, 1996.

Thompson, C. A. and Whitney, D. A. Long-term tillage and nitrogen fertilization in a West Central Great Plains wheat-sorghum-fallow rotation. J. Prod. Agric. 11, 353-359, 1998.

Thomsen, I. K. Turnover of ^{15}N-straw and NH_4 NO_3 in a sandy loam soil: effects of straw disposal and N fertilization. Soil Biol. Biochem. 25, 1561 -1566, 1993.

Thomson, W. W. and Weier, T. E. The fine structure of chloroplasts from mineral-deficient leaves of *Phaseolus vulgaris*. Am. J. Bot. 49, 1047-1055, 1962.

Thorup-Kristensen, K. The effect of nitrogen catch crop species on the nitrogen nutrition of succeeding crops. Fert. Res. 37, 227-234,1994

Tiffin, L. O. Translocation of manganese, iron, cobalt and zinc in tomato. Plant Physiol. 42, 1427-1432, 1967.

Tiffin, L. O. Translocation of micronutrients in plants. In: Micronutrients in Agriculture, edited by J. J. Mortvedt, P. M. Giordano and W. L. Lindsay, Soil Sci. Soc. America, Madison, p. 199-229, 1972.

Tinker, P., B. The effects of magnesium sulphate on sugar beet yield and its interactions with other fertilizers. J. Agric. Sci. 68, 205-212, 1967.

Tinker, P. B. The role of microorganisms in mediating and facilitating the uptake of plant nutrients from soil. Plant and Soil 76, 77-91, 1984.

Tokunaga, Y. Potassium silicate: as slow-release fertilizer. Fert. Res. 30, 55-59, 1991.

Tolbert, N. E. (2-chloroethyl) trimethylammonium chloride and related compounds as plant growth substances. II. Effect of growth on wheat. Plant Physiol. 35, 380-385, 1960.

Tolbert, N. E. Glycolate rnetabolism by higher plants and algae. In: Photosynthesis II, Encycl. Plant Physiol. New Series, Vol. 6, edited by Gibbs, M. and Latzko, E., Springer-Verlag Berlin, Heidelberg, New York, p. 338-352, 1979.

Tomar, J. S. and Soper, R. J. Fate of tagged urea N in the field with different methods of N and organic matter placement. Agron. J. 73, 991-995, 1981.

Tomm, G. O., Kessel, C. and Slinkard, A. E. Bi-birectional transfer of nitrogen between alfalfa and bromegrass: Short and long term evidence. Plant and Soil 164, 77-86, 1994.

Toney, M. D., Hohenester, E., Cowan, S. W. and Jansonius, J. N. Dialkylglycine decarboxylase structure: bifunctional active site and alkali metal sites. Science. 261, 756-759, 1993.

Toriyama, K. Development of agronomic practices for production of field crops under irrigated conditions-rice. In: 1. FAO/SIDA Seminar for Plant Scientists from Africa and the Near East, Cairo 1973, FAO, Rome, p. 452-456, 1974.

Touchton, J. T., Hoeft, R. G. and Weich, L, F. Nitrapyrin degradation and movement in soil. Agron. J. 70, 811-816, 1978.

Toulon, V., Sentenac, H., Thibaud, J. B., Davidian, C., Moulineau, C. and Grignon, C. Role of apoplast acidification by the H^+ pump. Effect on the sensitivity to pH and CO_2 of iron reduction. Planta 186, 212-218, 1992.

Touraine, B., Muller, B. and Grignon, C. Effect of phloem-translocated malate on NO_3^- uptake by roots of intact soybean plants. Plant Physiol. 93, 1118-1123, 1992.

Traore, A. and Maranville, J. W. Nitrate reductase activity of diverse grain sorghum genotypes and ist relationship to nitrogen use efficiency. Agron. J. 91, 863-869, 1999.

Travers, A. A. DNA conformation and protein binding. Annu. Rev. Biochem. 58, 427-452, 1989.

Trebacz, K., Simonis, W. and Schönknecht, G. Cytoplasmatic Ca^{2+}, K^+, Cl^-, and NO_3^- activities in the liverwort *Conocephalum conicum* L. at rest and during action potentials. Plant Physiol. 106, 1073-1084, 1994.

Treeby, M. and Uren, N. Iron deficiency stress responses amongst citrus rootstocks. Z. Pflanzenernähr. Bodenk. 156, 75-81, 1993.

Trepp, G. P., Plank, D. W., Gantt, J. S. and Vance, C. P. NADH-glutamate synthase in alfalfa root nodules. Immunocytochemical localization. Plant Physiol. 119, 829-837, 1999.

Trethewey, R. N. Ap Rees, T. A mutant of *Arabidopsis thaliana* lacking the ability to transport glucose across the chloroplast membrane. Biochem. J. 301, 449-454, 1994.

Trewavas, A. J. and Malho, R. Signal perception and transduction: The origin of the phenotype. The Plant Cell 9, 1181-1195, 1997.

Triboi, E., Abad, A., Michelena, A., Lloveras, J., Ollier, L. and Daniel, C. Environmental effects on the quality of two wheat genotypes: 1. Quantitative and qualitative variation of storage proteins. Eur. J. Agron. 13, 47-64, 2000.

Tributh, H., Boguslawski, v. E., Liers, v. A., Steffens, S. and Mengel, K. Effect of potassium removal by crops on transformation of illitic clay minerals. Soil Sci. 143, 404-409, 1987.

Trip, P. Sugar transport in conducting elements of sugar beet leaves. Plant Physiol. 44, 717-725, 1969.

Trolldenier, G. (G) Cereal diseases and plant nutrition. Potash Review Subj. 23, suite 24, 1969.

Trolldenier, G. (G) Soil Biology, the Soil Organisms in the Economy of Nature. Franckh'sche Verlagshandlung, Stuttgart, p. 116, 1971a.

Trolldenier, G. Recent aspects of the influence of potassium on stomatal opening and closing. In: Potassium in Biochemistry and Physiology. Proc. 8th Colloq. Int. Potash Inst., Bern, p. 130-133, 1971b.

Trolldenier, G. Secondary effects of potassium and nitrogen nutrition of rice: Change in microbial activity and iron reduction in the rhizosphere. Plant and Soil 38, 267-279, 1973.

Trolldenier, G. Visualization of oxidizing power of rice roots and possible participation of bacteria in iron deposition. Z. Pflanzenernähr. Bodenk. 151, 117-121, 1988.

Trolldenier, G. and Zehler, E. Relationship between plant nutrition and rice diseases. In: Fertilizer Use and Plant Health. Proc. 12th Colloq. Int. Potash Inst., Bern, p. 85-93, 1976.

Truog,E. Determination of readily available phosphorus in soils. J. Am. Soc. Agron. 22, 874 - 882, 1930

Tsui, C. The role of zinc in auxin synthesis in the tomato plant. Amer. J. Bot. 35, 172-179 1948.

Tuberosa, R., Sanguinetti, M. C., Stefanelli, S. and Quarrie, S. A. Number of endosperm cells and endosperm abscisic acid content in relation to kernel weight in four barley genotypes. Eur. J. Agron. 1, 125-132, 1992.

Tubiello, F. N., Donatelli, M., Rosenzweig, C. and Stockle, C. O. Effects of climate change and elevated CO_2 on cropping systems: Model prAdictions at two Italien locations. European J. Agron. in press:-, 2000.

Tukey, H. B., Wittwer, S. H. and Bukovac, M. J. The uptake and loss of materials by leaves and other above-ground plant parts with special reference to plant nutrition. Nutrient Uptake of Plants, 4. Intern. Symposium, Agrochimica Pisa, Florenz, p. 384-413, 1962.

Tunney, H., Carton, O. T., Brookes, P. C. and Johnston, A. E. (Editors) Phosphorus Loss from Soil to Water. CAB International, 1997.

Turgeon, R. and Webb, J. A. Leaf development and phloem transport in *Cucurbita pepo*: Carbon economy. Planta 123, 53-62, 1975.

Turner, D. W. and Barkus, B. The effect of season, stage of plant growth and leaf position on nutrient concentrations in the banana leaf on a Kraznozem in New South Wales. Aust. J. Exp. Agric. and Animal Husb. 14, 112-117, 1974.

Turner, L. J. and Kramer, J. R. Sulfate ion binding on goethite and hematite. Soil Sci. 152, 226-230, 1991.

Turner, R. G. The subcellular distribution of zinc and copper within the roots of metal-tolerant clones of *Agrostis tenuis* Sibth. New Phytol. 69, 725-731, 1969.

Turner, R. G. and Marshall, C. The accumulation of Zn-65 by root homogenates of Zn tolerant and non tolerant clones of *Agrostis tenuis* Sibth. New Phytol. 70, 539-545, 1972.

Tyerman, S. D. Anion channels in plants. Annu. Rev. Plant Physiol. Plant Mol. Biol. 43, 351-373, 1992.

Tyerman, S. D., Bohnert, H. J., Maurel, C., Steudle, E. and Smith, J. A. C. Plant aquaporins: their molecular biology, biophysics and significance for plant water relations. J. Exp. Bot. 50, 1055-1071, 1999.

Tyree, M. T. The cohesion-tension theory of sap ascent. Current controversities. J. Exp. Bot. 48, 1753-1765, 1997.

Tyree, M. T., Salleo, S., Nardini, A., Lo Gullo, M. A. and Mosca, R. Refilling of embolized vessels in young stems of Laurel. Do we need a new paradigm? Plant Physiol. 120, 11-21, 1999.

Tyree, M. T. and Sperry, J. S. Vulnerability of xylem to cavitation and embolism. Annu. Rev. Plant Physiol. 40, 19-38, 1989.

Tyson, R. H. and Ap Rees, T. Starch synthesis by isolated amyloplasts from wheat endosperm. Planta 175, 33-38, 1988.

Uebel, E. (G) Results of long-term potassium/magnesium fertilizer application. Allgemeine Forst Zeitschr. 9, 1-4, 1996.

Uexküll, H. R. von. Response of coconuts to potassium chloride in the Philippines. Oléagineux 27, 31-91, 1972.

Uexküll, H. R. von. Potassium nutrition in some tropical plantation crops. In: Potassium in Agriculture, edited by Munson, R. D., Am. Soc. Agronomy, Madison USA, p. 929-954, 1985.

Ullrich, W. R. Transport of nitrate and ammonium through plant membranes. In: Nitrogen Metabolism in Plants, edited by Mengel, K. and Pilbeam, D. J. Oxford Universiry Press, Oxford p. 121-137, 1992.

Ullrich, W. R. and Novacky, A. Nitrate-dependent membrane potential changes and their induction in *Lemna gibba*. G1. Plant Sci. Lett. 22, 211-217, 1981.

Ullrich-Eberius, C. I., Novacky, A., Fischer, E. and Lüttge, U. Relationship between energy-dependent phosphate uptake and the electrical membrane potential in *Lemna gibba* Gl. Plant Physiol. 67, 797-801, 1981.

Uloro, Y. The effect of nitrogen, phosphorus, potassium, and sulphur on the yield and yield components of (*Ensete ventricosum* W.) in southwest Ethiopia. Wissenschaftlicher Fachverlag, Giessen 1994.

Uloro, Y. and Mengel, K. *Ensete ventricosum*, an important food plant in Ethiopia. In: Plant Research and Development, edited by Bittner, A., Institut f. wissenschaftliche Zusammenarbeit, ISSN 0340-2843, p. 75-80, 1996.

Ulrich, A. and Hills, F. L. Plant analysis as an aid in fertilizing sugar crops. Part 1.: Sugar beets. In: Soil Testing and Plant Analysis, edited by Walsh, L. M. and Beaton, J. D., S. S. S. A., p. 271-288, 1973.

Ulrich, A. and Ohki, K. Chlorine, bromine and sodium as nutrients for sugar beet plants. Plant Physiol. 31, 171-181, 1956.

Ulrich, A. and Ohki, K. Potassium. In: Diagnostic Criteria for Plants and Soils, edited by Chapman, H. D., University of California, Riverside, Dif. of Agric. Sciences, p. 362-393, 1966.

Ulrich, A., Ripie, D., Hills, F. J., George, A. G. and Morse, M. D. Principles and practices of plant analysis. In: Soil Testing and Plant Analysis. II. Plant Analysis. Soil Sci. Soc. America, Madison, Wisc., p. 11-24, 1967.

Ulrich, B. Effects of air pollution on forest ecosystems and waters - the principles demonstrated at a case study in Central Europe. Atmospheric Environment 18, 621-628, 1984.

Underwood, E. J. Trace Elements in Human and Animal Nutrition. Academic Press, New York 1971.

Underwood, E. J. Trace Elements in Human and Animal Nutrition. Academic Press, New York, 1977.

Unsworth, M. H., Crawford, D. V., Gregson, S. K. and Rowlatt, S. M. Pathways for sulfur from the atmosphere to plants and soil. In: Sulfur Dioxide and Vegetation, edited by Winner, W. E., Mooney, H. A. and Goldstein, R. A. Stanford: Stanford University Press, California, p. 375-388, 1985.

Uren, N. C. and Reisenauer, H. M. The role of root exudates in nutrient acquisition. Adv. Plant Nutrition. 3, 79-114, 1988.

Vaadia, Y., Raney, F. C. and Hagan, R. M. Plant water deficits and physiological processes. Annu. Rev. Plant Physiol. 12, 265-292, 1961.

Van Bavel, C. H. M., Fritschen, L. J. and Lewis, W. E. Transpiration by sudangrass as an externally controlled process. Science 141, 269-270, 1963.

Van Bel, A. J. E. Different phloem-loading machineries correlated with the climate. Acta Bot. Neerl. 41, 121-141, 1992.

Van Bel, A. J. E. and Van Erven, A. J. A model for proton and potassium co-transport during thew uptake of glutamine and sucrose by tomato discs. Planta 145, 77-82, 1979.

Van Beusichem, M. L., Kirkby, E. A. and Baas, R. Influence of nitrate and ammonium nutrition on the uptake, assimilation, and distribution of nutrients in *Ricinus communis*. Plant Physiol. 86, 914-921, 1988.

Van Bremen, N. and Moorman, F. R. Iron-toxic soils. In: Soils and Rice, edited by Intern Rice Research Inst., Los Banos, Phillip.: International Rice Res. Institute, p. 781-800, 1978.

Van den Geijn, S. C. and Petit, C. M. Transport of divalent cations. Plant Physiol. 64, 954-958, 1979.

Van den Honert, T. Water transport in plants as a catenary process. Disc. Faraday Soc. 3, 146-153, 1948.

Van der Mark, F., Lange, de T. and Bienfait, H. F. The role of ferritin in developing primary bean leaves under various light conditions. Planta 153, 338-342, 1981.

Van der Paauw, F. Relations between the potash requirements of crops and meteorological conditions. Plant and Soil 3, 254-268, 1958.

Van der Paauw, F. Fertilization with phosphorus. Intern. superphosphate manufacturers' association. Extr. Bull. Docum. No. 32, Paris, 1962.

Van der Paauw, F. Effect of winter rainfall on the amount of nitrogen available to crops. Plant and Soil 16, 361-380, 1962.

Van der Paauw, F. Factors controlling the efficiency of rock phosphate for potatoes and rye on humic sandy soils. Plant and Soil 22, 81-98, 1965.

Van der Paauw, F. (G) Development and evaluation of a new water extraction technique for the determination of available phosphate. Landw. Forsch. 23/II. Sonderh., 102-109, 1969.

Van der Vorm, P. D. J. Uptake of Si by five plant species as influenced by variation in Si supply. Plant and Soil 56, 153-156, 1980.

Van der Vorm, P. D. J. and van Diest, A. Aspects of the Fe and Mn nutrition of rice plants. II. Iron and manganese uptake by rice plants grown on aerobic water cultures. Plant and Soil 52, 12-29, 1979.

Van Diest, A. Soil structural problems associated with intensive farming in the Netherlands. In: Proc. of Intern. Seminar on Soil Environmental and Fertility Management in Intensive Agriculture, Tokyo p. 145-153, 1977.

Van Genuchten, M. A comparison of numerical simulations of the one-dimensionnal unsaturated-saturated flow and mass transport equations. Adv. Mater Resour. 5, 47-55, 1982.

Van Noordwijk, N., de Willigen, P., Ehlert, P. A. I. and Chardon, W. J. A simple model of P uptake by crops as a possible basis for P fertilizer. Netherlands J. Agric. Sci. 38, 317-332, 1990.

Van Praag, H. J., Fischer, V. and Riga, A. Fate of fertilizer nitrogen applied to winter wheat as Na $^{15}NO_3$ and ($^{15}NH_4$)$_2$ SO$_4$ studied in microplots through a four-course rotation: 2. Fixed ammonium turnover and nitrogen reversion. Soil Sci. 130, 100-105, 1980.

Van Riemstdijk, W. H., Lexmond, T. M., Enfield, C. G. and Van der Zee, S. E. Phosphorus and heavy metals. Accumulation and consequences. In: Animal Manure on Grassland and Fodder Crops. Fertilizer or Waste? edited by Van der Meer, H. G., Martinus nijhoff, Dordrecht, p. 213-227, 1987.

Van Steveninck, R. F. M. The significance of calcium on the apparent permeability of cell membranes and the effects of substitution with other divalent ions. Physiol. Plant. 18, 54-69, 1965.

Van Steveninck, R. F. M. and Van Steveninck, M. E. Abscisic acid and membrane transport. In: Abscisic acid, edited by F. T. Addicott, Praeger, New York, p 171-235, 1983.

Van Wambeke, A. Soils of the Tropics, New York: McGraw-Hill, 1991.

Vance, C. P. and Heichel, G. H. Carbon in N$_2$ fixation: limitation or exquisite adaption. Annu. Rev. Plant Physiol. Mol. Biol. 42, 373-392, 1991.

Vanselow, A. P. Cobalt. In: Criteria for Plants and Soils, edited by Chapman, H. D., University of California, Div. of Agric. Sciences, p. 142-156, 1966.

Veihmeyer, F. J. and Hendrickson, A. H. The moisture equivalent as a measure of the field apacity of soils. Soil Sci. 32, 181-193, 1931.

Veit-Köhler, U., Krumbein, A. and Kosegarten, H. Effect of different water supply on plant growth and fruit quality of Lycopersicon esculentum. J. Plant Nutr. Soil Sci. 162, 583-588, 1999.

Veleuthambi, K. and Poohvaiah, B. W. Calcium and calmodulin-regulated phosphorylation of soluble and mernbrane proteins from corn coleoptiles. Plant Physiol. 76, 359-365, 1984.

Venkateswarlu, P., Armstrong, W. D. and Singer,L. Absoprtion of fluoride and chloride by barley roots. Plant Physiol. 40, 255-261, 1965.

Vergnano, O. and Hunter, J. G. Nickel and cobalt toxicities in oat plants. Ann. Bot. NS 17, 317-328, 1952.

Vertregt, N. Relation between black spot and composition of the potato tuber. Eur. Potato J. 11, 34-44, 1968.

Vessey, J. K. and Waterer, J. In search of the mechanism of nitrate inhibition of nitrogenase activity in legume nodules: Recent developments. Physiol. Plant. 84, 171-176, 1992.

Vetter, H. and Klasink, A. (G) Nutrient contents on slurries and faeces. In: Wieviel düngen? DLG-Verlag Frankfurt, p. 189-194, 1977.

Vetter, H. and Teichmann, W. (G) Field trials with varied copper and nitrogen treatrnents in Weser-Ems. Z. Pflanzenernähr. Bodenk. 121, 97-111, 1968.

Viets, F. G. Calcium and other polyvalent cations as accelerators of ion accumulation by excised barley roots. Plant Physiol. 19, 466-480, 1944.

Vincent, J., Leggett, J. E. and Egli, D. B. Cation accumulation by *Glycine max.* (L) Merr. as related to maturity stages. In: The Soil Root Interface, edited by J. L. Harley and R. S. Russell, Academic Press, London, New York, San Francisco, p. 440, 1979.

Viro, M. (G) The effect of a varied nutrition with potassium on the translocation of assirnilates and minerals in *Lycopersicon esculentum*. Ph. D. Thesis, Justus-Liebig-University Giessen 1973.

Vlamis, J. and Williams, D. E. Manganese and silicon interactions in the gramineae. Plant and Soil 27, 131-140, 1967.

Vlamis, J., Williams, D. E., Corey, J. E., Page, A. L. and Ganje, T. J. Zinc and cadmium uptake by barley in field plots fertilized seven years with urban and suburban siudge. Soil Sci. 139, 81-87, 1985.

Vömel, A. (G) Nutrient balance in various lysimeter soils. 1. Water leaching and nutrient balance. Z. Acker- u. Pflanzenbau 123, 155-188, 1965/66.

Vorm, P. D. J. Uptake of Si by five plant species, as influenced by variation in Si-supply. Plant and Soil 56, 153-156, 1980.

Wada, K. Allophane and imogolite. In: Minerals in Soil Environments, edited by Dixon, J. B. and Weed, S. B. Madison,USA: Soil Science Soc. of America, p. 1051-1087, 1989.

Wagner, G. H. and Broder, M. W. Microbial progression in the decomposition of corn stalk residue in soil. Soil Sci. 155, 48-52, 1993.

Wainwright, M. Sulfur oxidation in soils. Adv. Agron. 37, 349-396, 1984.

Wainwright, S. J. and Woolhouse, H. W. Physiological mechanisms of heavy metal tolerance in plants. In: The Ecology of Resource Degradation and Renewal, edited by Chadwick, M. J. and Goodman, G. T., Blackwell, Oxford, p. 231-257, 1975.

Walburg, G., Bauer, M. E., Daughtry, C. S. T. and Housley, T. L. Effects of nitrogen nutrition on the growth, yield and reflectance characteristics of corn. Agron. J. 74, 677-683, 1982.

Walker, C. D. and Webb, J. Copper in plants. Form and behaviour In: Copper in Soils and Plants, edited by Loneragan, J. F., Robson, A. D. and Graham, R. D., Academic Press, p. 189-212, 1981.

Walker, D. A. Chloroplast and cell - The movement of certain key substances across the chloroplast envelope In: Int. Review of Science, Plant Biochemistry Series 1, Vol. II, edited by Northcote, D. H., Butterworths, p. 1-49, 1974.

Walker, D. A. Regulation of starch synthesis in leaves - the role of orthophosphate. In: Physiological Aspects of Crop Productivity. Proc. 15th Colloq. Int. Potash Inst., Bern, p. 195-207, 1980.

Walker, N. Report of the Rothamsted Experimental Station, Part 1. p. 283, 1976.

Walker, T. W. and Adams, A. F. R. Competition for sulphur in a grass-clover association. Plant and Soil 9, 353-366, 1958.

Wallace, A., Frolich, E. and Lunt, O. R. Calcium requirements of higher plants. Nature 209, 634, 1966.

Wallace, A., Patel, P. M., Berry, W. L. and Lunt, O. R. Reclaimed sewage water: a hydroponic growth medium for plants. Res. Rec. Conserv. 3, 191-199, 1978.

Wallace, T. The Diagnosis of Mineral Deficiencies in Plants by Visual Symptoms. A colour Atlas and Guide. Her Majesty's Stationery Office, London 1961.

Wallace, W. and Pate, J. S. Nitrate assimilation in higher plants with special reference to cocklebur (*Xanthium pennsylvatieum* Walir). Ann. of Bot. 31, 213-228, 1967.

Walter, B., Bamen, D., Patenburg, H. and Koch, W. (G) Effect of gas and salt on soil and plant. Das Gartenamt 10, 578-581, 1974.

Walter, B., Koch, W. and Bastgen, D. (G) Experiences and results of urea foliar applications to grapes. Weinberg und Keller 20, 265-274, 1973.

Walton, K. M. and Dixon, J. E. Protein tyrosine phosphatases. Annu. Rev. Biochem. 62, 101-120, 1993.

Walworth, J. L., Lezsch, W. S. and Sumner, M. E. Use of boundary lines in establishing diagnostic norms. Soil Sci. Soc. Am. J. 50, 123-128, 1986.

Walworth, J. L. and Sumner, M. E. Foliar diagnosis: A review. Adv. Plant Nutrition. 3, 193-241, 1988.

Wang, M. Y., Glass, A. D. M., Shaff, J. E. and Kochian, L. V. Ammonium uptake by rice. Plant Physiol. 104, 899-906, 1994.

Ward, J. M., Pei, Z.-M. and Schroeder, J. I. Roles of ion channels in initiation of signal transduction in higher plants. Plant Cell 7, 833-844, 1995.

Waring, S. A. and Bremner, J. M. Ammonium production in soil under waterlogged conditions as an index of nitrogen availability. Nature 201, 951-952, 1964.

Warner, R. L. and Kleinhofs, A. Nitrate utilization by nitrate reductase-deficient barley mutants. Plant Physiol. 67, 740-743, 1981.

Warnock, R. E. Micronutrient uptake and mobility within corn plants (*Zea mays* L.) in relation to P induced zinc deficiency. Soil Sci. Soc. Amer. Proc. 34, 765-769, 1970.

Warren Wilson, J. Maximum yield potential. In: Transition from Extensive to Intensive Agriculture with Fertilizers. Proc. 7th Colloq. Int. Potash Inst., Bern, p. 34-56, 1969.

Warren Wilson, J. Control of crop processes. In: Crop Processes in Controlled Environments, edited by A. R. Rees, K. E. Cockshill, D. W. Hand, & R. G. Hurd, Academic Press, London / New York, p. 7-30, 1972.

Warrington, K. The influence of iron supply on toxic effects of manganese molybdenum and vanadium on soybeans, peas and flax. Ann. Appl. Biol. 41, 1-22, 1955.

Watanabe, H. Accumulation of chromium from fertilizers in cultivated soils. Soil Sci. Plant Nutr. 30, 543-554, 1984.

Watanabe, I., Berja, N. S. arnd Del Rosario, D. C. Growth of *Azolla* in paddy fields as affected by phosphorus fertilizer. Soil Sci. Plant Nutr. 26, 301-307, 1980.

Watanabe, I., Espinas, C. R., Berja, N. S. and Alimagno, B. V. Utilization of the *Azolla-Anabaena* complex as a nitrogen fertilizer tot rice. IRRI Research Paper Series No. 11, 3.-4. November, 1977.

Watson, C. J., Stevens, R. J. and Laughlin, R. J. Effectiviness of the urease inhibitor NBPT (N-(n-butyl)thiophosphoric triamide) for improving the efficiency of urea for ryegrass production. Fert. Res. 24, 11-15, 1990.

Watson, D. J. The physiological basis of variation in yield. Adv. Agron. 4, 101-144, 1952.

Watson, J. D. Molecular Biology of the Gene. Benjamin, New York, p. 494, 1965.

Watt, M. and Evans, J. R. Proteoid roots. Physiology and development. Plant Physiol. 121, 317-321, 1999.

Webb, M. J., Dinkelaker, B. E. and Graham, R. D. The dynamic nature of Mn availability during storage of calcareous soil: Its importance in plant growth experiments. Biol. Fertil. Soils 15, 9-15, 1993.

Webster, C. W., Poulton, P. R. and Goulding, K. W. T. Nitrogen leaching from winter cereals grown as apart of a 5-year ley-arable rotation. European J. Agron. 10, 99-109, 1999.

Wedin, D. A. and Tilman, D. Influence of nitrogen loading and species composition on the carbon balance of grasslands. Science. 274, 1720-1723, 1996.

Wegner, L. H. and Raschke, K. Ion channels in the xylem parenchyma of barley roots. Plant Physiol. 105, 799-813, 1994.

Wehrmann, J. and Hähndel, R. Relationship between N- and Cl-nutrition and NO_3 -content of vegetables. VIth Intern. Colloquium for the Optimization of Plant Nutrition, Montpellier. Proceed. Vol. 2, 679-685, 1984.

Wehrmann, J. and Scharpf, H. C. (G) The mineral content of the soil as a measure of the nitrogen fertilizer requirement (N_{min} method). Plant and Soil 52, 109-126, 1979.

Wehrmann, J. and Scharpf, H. C. The Nmin-method - an aid to integrating various objectives of nitrogen fertilization. Z. Pflanzenernähr. Bodenk. 149, 428-440, 1986.

Wei, C., Tyree, M. T. and Steudle, E. Direct measurement of xylem pressure in leaves of intact maize plants. A test of the cohesion-tension theory taking hydraulic architecture into consideration. Plant Physiol. 121, 1191-1205, 1999.

Weig, A., Deswarte, C. and Chrispeels, M. J. The major intrinsic protein family of *Arabidopsis* has 23 members that form three distinct groups with functional aquaporins in each group. Plant Physiol. 114, 1347-1357, 1997.

Weil, R. R., Foy, C. D. and Coradetti, C. A. Influence of soil moisture regimes on subsequent soil manganese availability and toxicity in two cotton genotypes. Agron. J. 89, 1-8, 1997.

Weimberg, R., Lerner, H. R. and Poljakoff-Mayber, A. A relationship between potassium and proline accumulation in salt-stressed *Sorghum bicolor*. Physiol. Plant. 55, 5-10, 1982.

Weinstein, L. H. Fluoride and plant lifE. J. of Occupational Medicine 19, 49-78, 1977.

Weiss, A. and Herzog, A. Isolation and characterization of a silicon-organic complex from plants. In: Biochemistry of Silicon and Related Problems, edited by G. Bendz and I. Lindquist., Plenum New York, 109-127, 1978.

Welch, L. F., Johnson, P. E., McKibben, G. E., Boone, L. V. and Pendleton, I. W. Relative efficiency of broadcast *versus* banded potassium for corn. Agron. J. 58, 618-621, 1966.

Welch, R. M. The biological significance of nickel. J. Plant Nutrition 3, 345-356, 1981.

Welch, R. M., Allaway, W. H., House, W. A. and Kubota, J. Geographic distribution of trace element problems. In Micronutrients in Agriculture, ed. by Mortvedt, J. J., Cox, F. R., Shuman, L. M. and Welch, R. M. Madison USA. SSSA Book Series No. 4, Madison USA p. 31-57, 1991.

Welch, R. M. and Huffman, W. D. Vanadium and plant nutrition. Plant Physiol. 52,183-185, 1973.

Welch, R. W. Genotypic variation in oil and protein in barley grain. J. Sci. Fd Agric 29, 953-958, 1978.

Weller, F. A method for studying the distribution of absorbing roots of fruit trees. Expl. Agric. 7, 351-361, 1971.

Weller, H. and Höfner, W. (G) Photosynthetic O_2 production of *Chlorella pyrenoidosa* in relation to potassium nutrition. Kali-Briefe (Hannover) Fachgeb. 2, 4. Folge, 1974.

Werner, D. (G) Dinitrogen fixation and primary production. AngeW. B. otanik 54, 67-75, 1980.

Werner, D. (G) Plant and Microbial Symbiosis. G. Thieme Verlag, Stuttgart, 1987.

Werner, D. and Roth, R. Silicia metabolism. In: Inorganic Plant Nutrition, Encycl. Plant Physiol. New Series, Vol. 15B, edited by Läuchli, A. and Bieleski, R. L., Springer Verlag Berlin, Heidelberg, New York, Tokyo, p. 682-694, 1983.

Werner, W. (G) Characterization of the available phosphate after an application of different phosphate forms for some years. Z. Pflanzenernähr. Bodenk. 122, 19-32, 1969.

Werner, W., Fritsch, F. and Scherer, H. W. (G) Effect of long-term slurry application on the nutrient status of soils. 2. Commun. Sorption and solubility criteria of soil phosphates. Z. Pflanzenernähr. Bodenk. 151, 63-68, 1988.

Westermark, U., Hardell, H.-L. and Iversen, T. The content of protein and pectin in the lignified middle lamella/primary wall from spruce fibers. Holzforschung 40, 65-68, 1986.

Westgate, M. E. and Boyer, J. S. Transpiration and growth - induced water potentials in maize. Plant Physiol. 74, 882-889, 1984.

Whalen, S. C., Reedburgh, W. S. and Sandbeck, K. A. Rapid methane oxidation in a land fill cover soil. Appl. Environ. Microbiol. 56, 3405-3411, 1990.

Wheeler, G. L., Jones, M. A. and Smirnoff, N. The biosynthetic pathway of vitamin C in higher plants. Nature 393, 365-369, 1998.

Whipps, J. M. and Lynch, J. M. The influence of the rhizosphere on the crop productivity. Adv. Microbial Ecol. 6, 187-244, 1986.

White, D. J. Energy use in agriculture. In: Aspects of Energy Conversion, edited by Blair, Jones and van Horn. Pergamon Press, Oxford and New York 1976.

White, G. C. and Greenham, D. W. P. Seasonal trends in mineral nitrogen content of the soil in a long-term NPK trial on dessert apples. J. horticult. Sci. 42, 419-428, 1967.

White, P. J. The regulation of K$^+$ influx into roots of rye (Secale cereale L.) seedlings by negative feedback via the K$^+$ flux from shoot to root in the phloem. J. Expt. Bot. 48, 2063-2073, 1997.

Whitehead, D. C. Uptake and distribution of iodine in grasses and clover plants grown in solution. J. Sci. Food Agric. 24, 43-50, 1973.

Widdowson, F. V., Penny, A. and Williams, R. J. B. Experiments measuring effects of ammonium and nitrate fertilizers, with and without sodium and potassium, on spring barley. J. agric. Sci. 69, 197-207, 1967.

Wiebe, H. H. and Al-Saadi, H. A. Matric bound water of water tissue from succulents. Plant Physiol. 36, 47-51, 1976.

Wiersum, I. K. Calcium content of the phloem sap in relation to the Ca status of the plant. Acta bot. neerl. 28, 221-224, 1979.

Wiklander, L.: The soil. In: Encyclopedia of Plant Physiology, Vol. 4, Springer-Verlag, Berlin, Göttingen, Heidelberg, p. 118-164, 1958.

Wild, A., Jones, L. H. P. and Macduff, J. H. Uptake of mineral nutrients and crop growth: the use of flowing nutrient solutions. Adv. Agron. 41, 171-219 1987.

Wildhagen, H., Lacher, P. and Meyer, B. (G) Model-experiment „Göttinger Komposttonne": N-, P-, and K-fertilizer effect of the biodisposals on cereals in field trials. Mitl. Deutsch. Bodenkundl. Ges. 55/II, 667-692, 1987.

Wildhagen, H., Styperek, P. and Meyer, B. (G) P balances and P fractionation in soils of long-term phosphate fertilizer trials on loess soils. Mitt. Dtsch. Bodenkundl. Ges. 38, 423-428, 1983.

Wilkinson, B. G. Mineral composition of apples. IX. Uptake of calcium by the fruit. J. Sci. Fd. Agric. 19, 446-447, 1968.

Wilkinson, S., Corlett, J. E., Oger, L. and Davies, W. J. Effects of xylem sap pH on transpiration from wild-type and flacca tomato leaves. Plant Physiol. 117, 103-709, 1998.

Wilkinson, S. and Davies, W. J. Xylem sap pH increase: A drought signal received at the apoplastic face of the guard cell that involves the suppression of saturable abscisic acid uptake of the epidermal symplast. Plant Physiol. 113, 559-573, 1997.

Willenbrink, J. (G) The plant vacuole as store. Naturwissenschaften 74, 22-29, 1987.

Williams, C. H. Soil acidification under clover pasture. Aust. J. Exp. Agric. Anim. Husb. 20, 561-567, 1980.

Williams, D. E. and Vlamis, J. The effect of silicon on yield and manganese 54 uptake and distribution in the leaves of barley plants grown in culture solutions. Plant Physiol. 32, 404-409, 1957.

Williams, E. G. Factors affecting the availability of soil phosphate and efficiency of phosphate fertilizers. Anglo-Soviet Symposium on Agrochemical Research on the Use of Mineral Fertilizers, Moscow, 1970.

Williams, E. G. and Knight, A. H. Evaluations of soil phosphate status by pot experiments, conventional extraction methods and labile phosphate values estimated with the aid of phosphorus-32. J. Sci. Fd Agirc. 14, 555-563, 1963.

Williams, J., Phillips, A. L., Gaskin, P. and Hedden, P. Function and substrate specifity of the gibberellin 3-beta hydroxylase encoded by the Arabidopsis GA4 gene. Plant Physiol, 117, 559-563, 1998.

Williams, W. A., Morse, M. D. and Ruckman, J. R. Burning vs incorporation of rice crop residues. Agron. J. 64, 467-468, 1972.

Williamson, R. E. and Ashley, C. C. Free Ca^{2+} and cytoplasmatic streaming in the Alga chara. Nature 296, 647-650, 1982.

Wilson, C. and Shannon, M. C. Salt-induced Na^+/ H^+ antiport in root plasma membrane of a glycophytic and halophytic species of tomato. Plant Sci. 107, 147-157, 1995.

Wilson, G. H. The regulation of intercellular pH and ammonium in intact rice (Oryza sativa) and maize (Zea mays) roots: an investigation into the mechanism of NH_3 toxicity. Ph. D. Thesis, School of Chemical and Life Sciences, University of Greenwich, London 1998.

Wilson, G. H., Grolig, F. and Kosegarten, H. Differential pH restoration after ammonia-elicited vacuolar alkalisation in rice and maize root hairs as measured by fluorescence ratio. Planta 206, 154-161, 1998.

Wilson, L. G., Bressan, R. A. and Filner, P. Light-dependent emission of hydrogen sulfide from plants. Plant Physiol. 61, 184-189, 1978.

Wilson, S. B. and Hallsworth, E. G. Studies of the nutrition of the forage legumes. IV. The effect of cobalt on the growth of nodulated and non nodulated Trifolium subterraneum L. Plant and Soil 22, 260, 1965.

Wilson, S. B. and Nicholas, D. J. D. A cobalt requirement for non-nodulated legumes and for wheat. Phytochemistry 6, 1057-1060, 1967.

Winner, C., Feyerabend, L and Müller, A. von (G) Investigation in the content of nitrate nitrogen in a soil profile and its uptake by sugar beet. Zucker 29, 477-484, 1976.

Winsor, G. W.: Potassium and the quality of glasshouse crops. In: Potassium and the Quality of Agricultural Products. Proc. 8th Congn Int. Potash Inst., Bern, p. 303-312, 1966.

Winteringham, F. P. W. B. iogeochemical cycling of phosphorus. In: Phosphorus Life and Environment - from Research to Application, edited by Cottenie, A. Casablanca, Marocco: World Phosphate Institute, p. 325-336, 1992.

Wirén, N. von, Lauter, F.-R., Ninnemann, O., Gillisen, B., Walch-Liu, P., Engels, C., Jost, W. and Frommer, W. B. Differential regulation of three functional ammonium transporter genes by nitrogen in root hairs and by light in leaves of tomato. The Plant Journal 21, 167-176, 2000.

Wissemeier, A. H., Klotz, F. and Horst, W. J. Aluminium induced callose synthesis in roots of soybean (Glycine max L.). J. Plant. Physiol. 129, 487-492, 1987.

Witt, H. H. and Jungk, A. (G) Assessment of molybdenum supply of plants by means of Mo-induced nitrate reductase activity. Z. Pflanzenern. Bodenk. 140, 209-222, 1977.

Wittwer, S. H. and Teubner, F. G. Foliar absorption of mineral nutrients. Annu. Rev. Plant Physiol. 10, 13-32, 1959.

Wolf, O. and Jeschke, W. D. Sodium fluxes, xylem transport of sodium, and K/Na selectivity in roots of seedlings of Hordeum vulgare, cv. VillA. J. Plant Physiol. 125, 243-256, 1986.

Wolf, O., Munns, R., Tonnet, M. L. and Jeschke, W. D. Concentrations and transport of solutes in xylem and phloem along the leaf axis of NaCl-treated *Hordeum vulgare*. J. Exp. Bot. 41, 1131-1141, 1991.

Wollny, E. (G) Investigations into the capillary movement of water in soils. Forsch. -Gebiete Agr. Phys. 8, 206-220, 1885.

Wolt, J. D. Soil Solution Chemistry. John Wiley & Sons, New York, 1994.

Wooley, J. T. Sodium and silicon as nutrients for the tomato plant. Plant Physiol. 32, 317-321, 1957.

Woolhouse, H. W. Light gathering and carbon assimilation processes in photosynthesis; their adaptive modifications and significance for agriculture. Endeavour, New Series 2, 35-46, 1978.

Woolhouse, H. W. Crop physiology in relation to agricultural production: The genetic link. In: Physiological Processes Limiting Plant Productivity, edited by Johnson, C. B., Butterworth, London, p. 1-21, 1981.

Wopereis, M. C., Gascuel-Odoux, C., Bourrie, G. and Soignet, G. Spatial variability of heavy metals in soil on a one hectare scale. Soil Sci. 146, 113 -123, 1988.

Wright, J. P. and Fisher, D. B. Direct measurement of sieve tube turgor pressure using severed aphid stylets. Plant Physiol. 65, 1133-1135, 1980.

Wright, W. G. Oxidation and mobilization of selenium by nitrate in irrigation drainage. J. Environ. Qual. 28, 1182-1187, 1999.

Wu, J., O'Donnell, A. G. and Syers, J. K. Microbial growth and sulphur immobilization following incorporation of plant residues in soil. Soil Biol. Biochem. 25, 1567-1573, 1993.

Wu, J. and Seliskar, D. M. Salinity adaptation of plasma membrane H^+-AtPase in the salt marsh plant *Spartina patens*: ATP hydrolysis and enzyme kinetics. J. Exp. Bot. 49, 1005-1013, 1998.

Wu, L., Thurman, D. A. and Bradshaw, A. D. The uptake of copper and its effect upon respiratory processes of roots of copper-tolerant and non -tolerant clones of *Agrostis stolonifera*. New Phytol. 75, 225-229, 1975.

Wu, W. and Assmann, S. M. Is ATP required for K^+ channel activation in Vicia guard cells? Plant Physiol. 107, 101-109, 1995.

Wu, W. and Berkowitz, G. A. Stromal pH and photosynthesis are affected by electroneutral K^+ and H^+ exchange through chloroplast envelope ion channels. Plant Physiol 98, 666-672, 1992.

Wulff, F., Schulz, V., Jungk, A. and Claassen, N. Potassium fertilization on sandy soils in relation to soil test, crop yield and K-leaching. Z. Pflanzenernähr. Bodenk. 161, 591-599, 1998.

Wyn Jones, R. G. Salt tolerance. In: Physiological processes limiting plant productivity, edited by Johnson, C. B. Butterworth, London, GB, pp. 271-292, 1981.

Wyn Jones, R. G. Cytoplasmic potassium homeostasis: review of the evidence and its implications. In: Frontiers in Potassium nutrition: New Perspectives on the Effects of Potassium on Physiology of Plants, edited by D. M. Oosterhuis, & G. A. Berkowitz. Potash & Phosphate Institute, Georgia USA /Potash & Phosphate Institute of Canada, Saskatoon, Canada, 13-22, 1999.

Wyn Jones, R. G., Brady, C. J. and Speirs, J. Ionic and osmotic relations in plant cells. In: Recent Advances in the Biochemistry of Cereals, edited by D. L. Laidman and R. G. Wyn Jones. Acad. Press, London, GB, 63-103, 1979.

Wyn Jones, R. G. and Pollard, A. Proteins, enzymes and inorganic ions. In: Inorganic Plant Nutrition, Encycl. Plant Physiol. New Series, Vol. 15B, edited by Läuchli, A. and Bieleski, R. L., Springer Verlag, Berlin, Heidelberg, New York, Tokyo, p. 528-562, 1983.

Wyn Jones, R. G., Sutcliffe, M. and Marshall, C. Physiological and biochernical basis for heavy metal tolerance in clones of *Agrostis tenuis*, In: Recent Advances in Plant Nutrition, edited by R. M. Samish. Gordon and Breach, New York 1971.

Wynn Parry, D. and Smithson, F. Types of opaline silica depositions in the leaves of British grasses. Annals of Botany 28, 169-185, 1964.

Wyse, R. E., Zamski, E. and Tomos, A. D. Turgor regulation of sucrose transport in sugar beet taproot tissue. Plant Physiol. 81, 478-481, 1986.

Xu, X., Van Lammeren, A. M., Vermeer, E. and Vreugdenhil, D. The role of gibberellin, abscisic acid, and sucrose in the regulation of potato tuber formation *in vitro*. Plant Physiol. 117, 575-584, 1998.

Yamauchi, M. and Winslow, M. D. Effect of silicia and magnesium on yield of upland rice in the humid tropics. Plant and Soil 113, 265-269, 1989.

Yan, F., Feuerle, R., Shäffer, S., Fortmeier, H. and Schubert, S. Adaption of active proton pumping and plasmalemma ATPase activity of corn (*Zea mays*) roots to low root medium pH. Plant Physiol. 117, 311-319, 1998.

Yan, F., Schubert, S. and Mengel, K. Effect of low root medium pH on net proton release, root respiration, and root growth of corn (*ZeaMays* L.) and broad bean (*Vicia faba* L.). Plant Physiol. 99, 415-421, 1992.

Yan, F., Schubert, S. and Mengel, K. Soil pH changes during legume growth and application of plant matter. Biol. Fertil. Soils 23, 236-242, 1996a.

Yan, F., Schubert, S. and Mengel, K. Soil pH increase due to biological decarboxylation of organic anions. Soil Biol. Biochem. 28, 617-624, 1996b.

Yang, H. S. and Janssen, B. H. Analysis of impact of farmers practices on dynamics of soil organic matter in northern China. In: Proceedings ESA - Congress, Veldhoven 1996, edited by van Ittersum, M. and van Geijn, S. C. Amsterdam: Elsevier, p. 267 - 275, 1997.

Yang, Z.M., Sivaguru, M., Horst, W.J. and Matsumoto, H. Detoxification of aluminium achieved by specific exudation of citric acid in vegetable soybean (*Glycine max.* L.) Physiol. Plant. 110, 72-77, 2000.

Yeo, A. R. Salinity resistance: Physiology and prices. Physiol. Plant. 58, 214-222, 1983.

Yeo, A. R. Molecular biology of salt tolerance in the context of whole-plant physiology. J. Exp. Bot. 49, 915-929, 1998.

Yeo, A. R., Lee, K.-S., Izard, P., Boursier, T. J. and Flowers, T. J. Short- and long-term effects of salinity on leaf growth in rice (*Oryza sativa* L.). J. Exp. Bot. 42, 881-889, 1991.

Yoshida, S. Physiological aspects of grain yield. Annu. Rev. Plant Physiol. 23, 437-464, 1972.

Yermiyaho, U., Keren, R. and Chen, Y. Boron sorption on composted organic matter. Soil Sci. Soc. Am. J. 57, 1309-1313, 1988.

Yoshida, S. and Castaneda, L. Partial replacement of potassium by sodium in the rice plant under weakly saline conditions. Soil Sci. Plant Nutr. 15, 183-186, 1969.

Yoshida, S., Forno, D. A. and Bradrochalm, A. Zinc deficiency of the rice plant on calcareous and neutral soils in the Philippines. Soil Sci. Plant Nutr. 17, 83-87, 1971.

Young, N. D. and Galston, A. W. Putrescine in acid stress. Plant Physiol. 71, 767-771, 1983.

Zacharias, L. and Reid, M. S. Role of growth regulators in the senescence of *Arabidopsis thaliana* leaves. Physiol. Plant. 80, 549-554, 1990.

Zech, W. (G) Needle analytical investigations into the lime chlorosis of the pine (*Pinus silvestris*). Z. Pflanzenernähr. Bodenk. 125, 1-16, 1970.

Zech, W., Koch, W. and Franz, F. (G) Net assimilation and transpiration of pine twigs in dependence on potassium supply and light intensity. Kali-Briefe, Fachgeb. 6, 1. Folge, 1971.

Zech, W. and Popp, E. (G) Magnesium deficiency, one of the reasons for the spruce and fir dieback in Northeastern Bavaria. Forstw. Cbl. 102, 50-55, 1983.

Zeevaart, H. D. and Creelmann, R. A. Metabolism and physiology of abscisic acid. Annu. Rev. Plant Physiol. Plant Mol. Biol. 39, 439-473, 1988.

Zelitch, I. Photorespiration: Studies with whole tissues. In: Photosynthesis II, Encycl. Plant Physiol. New Series, Vol. 6, edited by Gibbs, M and Latzko, E., Springer-Verlag Berlin, Heidelberg, New York,

p. 351-367, 1979.

Zhang, M., Alva, A. K., Li, Y. C. and Calvert, D. V. Fractionation of iron, manganese, aluminium, and phosphorus in selected sandy soils under citrus production. Soil Sci. Soc. Am. J. 61, 794-801, 1997.

Zhang, W. H., Atwell, B. J., Patrick, J. W. and Walker, N. A. Turgor-dependent efflux of assimilates from coats of developing seeds of *Phaseolus vulgaris* L. Water relations of the cells involved in efflux. Planta 199, 25-33, 1996.

Zhang, W. H. and Rengel, Z. Cytosolic Ca^{2+} activities in intact wheat root apical cells subjected to aluminium toxicity. In: Plant Nutrition - Molecular Biology and Genetics, edited by Gissel-Nielsen, G. and Jensen, A. Dordrecht, Kluwer Academic Publishers, p. 353-358, 1999.

Zhang, Y. Q. and Moore, J. N. Reduction potential of selenate in wetland sediment. J. Environ. Qual. 26, 910-916, 1997.

Zhang, Y. S. and Scherer, H. W. Studies on the mechanism of fixation and release of ammonium in padddy soils after flooding. II Effect of transformation of nitrogen forms caused by flooding on ammonium fixation. Biol. Fert. Soils in press:-, 2000.

Zhao, F. J., Hawkesford, M. J. and Mc Grath, S. P. Sulphur assimilation and effects on yield and quality of wheat. J. Ceral Sci. 30, 1-17, 1999.

Zhao, F. J., Wu, J. and Mc Grath, S. P. Soil organic sulphur and ist turnover. In: Humic Substances in Terrestrial Ecosystems, edited by Piccolo, A. Elsevier, Amsterdam, p. 467-505. 1996.

Zhiznevskaja, G. (R) The influence of trace elements on the yield and chemical composition of maize under the conditions of the Latvian Soviet Republic. Edited by Acad. Sci. of the Latvian SSR, Riga p. 217-256, 1958.

Zhu, D., Schwab, A. P and Banks, M. K. Heavy metal leaching from mine tailings as affected by plants. J. Environ. Qual. 28, 1727-1732, 1999.

Ziegler, K., Nemeth, K. and Mengel, K. Relationship between electroultrafiltration (EUF) extractable nitrogen, grain yield, and optimum nitrogen fertilizer rates for winter wheat. Fert. Res. 32, 37-43, 1992.

Zielinski, R. E. Calmodulin and calmodulin binding proteins in plants. Annu. Rev. Plant Physiol. Plant Mal. Biol. 49, 697-725, 1998.

Zimmermann, M. Translocation of nutrients. In: Physiology of Plant Growth and Development, edited by M. B. Wilkins, p. 383-417, 1969.

Zimmermann, U. Physics of turgor- and osmoregulation. Annu. Rev. Plant Physiol. 29,121-148, 1978.

Zobrist, J., Sigg, L., Stumm, W. and Zürcher, F. (G) The fog as carrier concentrated pollutants. Gewässerschutz, Wasser, Abwasser 100, 371-393, 1987.

Zottini, M., Scoccianti, v. and Zannoni, D. Effects of 3,5 dibromo-4-hydroxybenzonitrile (Bromoxynil) on bioenergetics of higher plant mitochondria (*Pisum sativum*). Plant Physiol. 106, 1483-1488, 1994.

Zöttl, H. W. and Mies, E. (G) Supply of mineral nutrients and pollution stress of spruce ecosystems in the Black Forest being under immission influence. Mitteil. Deutsch. Bodenkundl. Ges. 38, 429-434, 1983.

Zotz, G., Tyree, M. T. and Patino, S. Hydraulic architecture and water relations of a flood-tolerant tropical tree, *Annona glabra*. Tree Physiol. 17, 359-365, 1997.

SUBJECT INDEX

A

Abscisic acid (ABA)   217, 222, 224, 273, 274, 282, 286
Abscission of leaves   641
Acetyl coenzyme A see Coenzyme A
Acetyl serine   178
Acid growth theory   219, 267
Acidification effect of legumes on soil   408
Acidification of soils   58, 62, 515
Acid mineral soils
    Aluminium toxicity   61, 660-662
    Manganese toxicity   61, 580-581
    Molybdenum availability   613-614
    Phosphate availability   455-457
    Nitrogen fixation, legumes   408
    Soil acidity   51-62
    Sulphate soils   580
    Tropical soils   534
Acid phosphatase   458
Acid rain   61-62
Actinomyces   400
Actinomyces alni   401
Actinorhizas   400
Active and passive membrane transport   120-124
Active manganese in soils   574
Activity coefficient   20, 121
Activity of water molecules   26
Activity ratio   22
Adenosine 3-phospho 5 phosphosulphate (PAPS)   177
Adenosine phosphosulphate (APS)   177
Adenosine di phosphate (ADP)   119
Adenosine tri phosphate (ATP)   119
ADP   119
ATP   119
Aerenchymatous tissues   46
Aeschynomene afraspera
    Nitrogen fixation   403
Aggregates (soil)   38
Agmatine (formula)   500
Agronomic efficiency in N fertilizer usage   428

807

B

C

H

I

J

K

L

O

Q

R

X

Y